MECHANICAL VIBRATIONS
FIFTH EDITION IN SI UNITS

Access for
Companion ___te

Thank you for purchasing a new copy of *Mechanical Vibrations, Fifth Edition in SI Units,* by Singiresu S
The one-time password below provides access to additional chapters, answers and MATLAB programs fo
the Companion Website.

For Students only:

To access the Companion Website:
1. Go to http://www.pearsoned-asia.com/rao
2. Click on the title *Mechanical Vibrations, Fifth Edition in SI Units*
3. Click on "Companion Website"
4. Register your access code given below and create an account

For instructors only:

To access the suite of Instructor's Resources available for your use, please follow steps 1 and 2 above and click on "Instructor's Resources" instead of "Companion Website". Follow the instructions on screen, and you will be prompted to enter a user ID and password. Please proceed if you are already registered and have existing instructor's access.

If you do not have instructor's access, please contact your Pearson Education representative. He/She will then register for you and provide you with a user ID and password.

Additional chapters, answers and MATLAB programs can be viewed on the Companion Website.

IMPORTANT: The password on this page can be used only once to establish a subscription to the Companion Website. Each copy of this title sold through your local bookstore is shrink-wrapped to protect the one-time password. If you have bought a copy of the book that is not shrink-wrapped, please contact your local bookstore.

Prentice Hall
is an imprint of

www.pearsonhighered.com

Mechanical Vibrations

Fifth Edition in SI Units

Singiresu S. Rao
University of Miami

SI Conversion by
Yap Fook Fah
Nanyang Technological University, Singapore

Prentice Hall

Singapore   London   New York   Toronto   Sydney   Tokyo   Madrid   Mexico City   Munich
Paris   Cape Town   Hong Kong   Montreal

Published in 2011 by
Pearson Education South Asia Pte Ltd
23/25 First Lok Yang Road, Jurong
Singapore 629733

Publishing Manager: *Gan Meng Meng*
Acquisitions Editor: *Sharie Cheung*
Project Editor: *Esther Yap*
Prepress Executive: *Florence Goh*

Pearson Asia Pacific offices: Bangkok, Beijing, Hong Kong, Jakarta, Kuala Lumpur, Manila, New Delhi, Seoul, Singapore, Taipei, Tokyo

Authorized adaptation from the United States edition, entitled *Mechanical Vibrations*, Fifth Edition, ISBN: 9780132128193 by RAO, SINGIRESU S., published by Pearson Education, Inc, publishing as Prentice Hall, Copyright © 2011 S.S. Rao.

Printed in Singapore

4   3   2   1

14   13   12   11

ISBN 13   978-981-06-8712-0
ISBN 10   981-06-8712-5

10 0623 722 2

Prentice Hall
is an imprint of

www.pearsonhighered.com

To Lord Sri Venkateswara

Contents

CHAPTER 9

Vibration Control   769

CHAPTER 10

Vibration Measurement and Applications   870

Chapters 13 and 14 are provided as downloadable
files on the Companion Website.

Preface

Changes in this Edition

This book serves as an introduction to the subject of vibration engineering at the undergraduate level. Favorable reactions by professors and students to the fourth edition have encouraged me to prepare this fifth edition of the book. I have retained the style of the prior editions, presenting the theory, computational aspects, and applications of vibration in as simple a manner as possible, and emphasizing computer techniques of analysis. Expanded explanations of the fundamentals are given, emphasizing physical significance and interpretation that build upon previous experiences in undergraduate mechanics. Numerous examples and problems are used to illustrate principles and concepts.

In this edition, some topics are modified and rewritten, many new topics are added, and several new features have been introduced. Most of the additions and modifications were suggested by users of the text and by reviewers. Important changes include the following:

1. Chapter outline and learning objectives are stated at the beginning of each chapter.
2. A chapter summary is given at the end of each chapter.
3. The presentation of some of the topics is modified for expanded coverage and better clarity. These topics include the basic components of vibration—spring elements, damping elements, mass or inertia elements, vibration isolation, and active vibration control.
4. Many new topics are presented in detail with illustrative examples. These include the response of first-order systems and time constant, graphical representation of characteristic roots and solutions, parameter variations and root locus representation, stability of systems, transfer-function approach for forced-vibration problems, Laplace transform approach for the solution of free- and forced-vibration problems, frequency transfer-function approach, Bode diagram for damped single-degree-of-freedom systems, step response and description of transient response, and inelastic and elastic impacts.
5. I have added 128 new examples, 160 new problems, 70 new review questions, and 107 new illustrations.
6. The C++ and Fortran program-based examples and problems given at the end of every chapter in the previous edition have been deleted.

Features of the Book

- Each topic in *Mechanical Vibrations* is self-contained, with all concepts fully explained and the derivations presented in complete detail.
- Computational aspects are emphasized throughout the book. MATLAB-based examples as well as several general-purpose MATLAB programs with illustrative examples are given in the last section of every chapter. Numerous problems requiring the use of MATLAB or MATLAB programs (given in the text) are included at the end of every chapter.

- Certain topics are presented in a somewhat unconventional manner—in particular, the topics of Chapters 9, 10, and 11. Most textbooks discuss isolators, absorbers, and balancing in different chapters. Since one of the main purposes of the study of vibrations is to control vibration response, all topics related to vibration control are given in Chapter 9. The vibration-measuring instruments, along with vibration exciters, experimental modal analysis procedure, and machine-condition monitoring, are presented together in Chapter 10. Similarly, all the numerical integration methods applicable to single- and multidegree-of-freedom systems, as well as continuous systems, are unified in Chapter 11.

- Specific features include the following:
  - More than 240 illustrative examples are given to accompany most topics.
  - More than 980 review questions are included to help students in reviewing and testing their understanding of the text material. The review questions are in the form of multiple-choice questions, questions with brief answers, true-false questions, questions involving matching of related descriptions, and fill-in-the-blank questions.
  - An extensive set of problems in each chapter emphasizes a variety of applications of the material covered in that chapter. In total, there are more than 1,150 problems. Solutions are provided in the instructor's manual.
  - More than 30 design-project-type problems, many with no unique solution, are given at the end of various chapters.
  - More than 25 MATLAB programs are included to aid students in the numerical implementation of the methods discussed in the text.
  - Biographical information about 20 scientists and engineers who contributed to the development of the theory of vibrations is presented on the opening pages of chapters and appendixes.
  - MATLAB programs given in the book, answers to problems, and answers to review questions can be found at the Companion Website, www.pearsoned-asia.com/rao.
  - The Instructors' Resources offer a Solutions Manual, an Image bank and PowerPoint slides, available under "Instructors' Resources" at www.pearsoned-asia.com/rao. The *Solutions Manual* with solutions to all problems and hints to design projects is available to instructors who adopt the text for their courses. The *Image Bank* contains illustrations, diagrams, and photos found in the textbook. Instructors can use these images to customize their lesson materials. Unique to the SI edition, the *PowerPoint slides* include an outline for each chapter and present concepts and theories using examples and diagrams from the book. They are fully editable and can be used for frontal teaching.

Units and Notation

The SI system of units has been used in the examples and problems. A list of symbols, along with the associated units in the SI system, appears after the Acknowledgments. A brief discussion of SI units as they apply to the field of vibrations is given in Appendix E. Arrows are used over symbols to denote column vectors, and square brackets are used to indicate matrices.

Organization of Material

Mechanical Vibrations is organized into 14 chapters and 6 appendixes. Chapters 13 and 14 are provided as downloadable files on the Companion Website. The reader is assumed to have a basic knowledge of statics, dynamics, strength of materials, and differential equations. Although some background in matrix theory and Laplace transform is desirable, an overview of these topics is given in Appendixes C and D, respectively.

Chapter 1 starts with a brief discussion of the history and importance of vibrations. The modeling of practical systems for vibration analysis along with the various steps involved in vibration analysis are discussed. A description is given of the elementary parts of a vibrating system—stiffness, damping, and mass (inertia). The basic concepts and terminology used in vibration analysis are introduced. The free-vibration analysis of single-degree-of-freedom undamped and viscously damped translational and torsional systems is given in Chapter 2. The graphical representation of characteristic roots and corresponding solutions, the parameter variations, and root locus representations are discussed. Although the root locus method is commonly used in control systems, its use in vibration is illustrated in this chapter. The response under Coulomb and hysteretic damping is also considered. The undamped and damped responses of single-degree-of-freedom systems to harmonic excitations are considered in Chapter 3. The concepts of force and displacement transmissibilities and their application in practical systems are outlined. The transfer-function approach, the Laplace transform solution of forced-vibration problems, the frequency-response, and the Bode diagram are presented.

Chapter 4 is concerned with the response of a single-degree-of-freedom system under a general forcing function. The roles of Fourier series expansion of a periodic function, convolution integral, Laplace transform, and numerical methods are outlined with illustrative examples. The specification of the response of an underdamped system in terms of peak time, rise time, and settling time is also discussed. The free and forced vibration of two-degree-of-freedom systems is considered in Chapter 5. The self-excited vibration and stability of the system are discussed. The transfer-function approach and the Laplace transform solution of undamped and dampled systems are also presented with illustrative examples. Chapter 6 presents the vibration analysis of multidegree-of-freedom systems. Matrix methods of analysis are used for presentation of the theory. The modal analysis procedure is described for the solution of forced-vibration problems in this chapter. Several methods of determining the natural frequencies and mode shapes of discrete systems are outlined in Chapter 7. The methods of Dunkerley, Rayleigh, Holzer, Jacobi, and matrix iteration are discussed with numerical examples.

While the equations of motion of discrete systems are in the form of ordinary differential equations, those of continuous or distributed systems are in the form of partial differential equations. The vibration analysis of continuous systems, including strings, bars, shafts, beams, and membranes, is given in Chapter 8. The method of separation of variables is presented for the solution of the partial differential equations associated with continuous systems. The Rayleigh and Rayleigh-Ritz methods of finding the approximate natural frequencies are also described with examples. Chapter 9 discusses the various aspects of vibration control, including the problems of elimination, isolation, and absorption. The vibration nomograph and vibration criteria which indicate the acceptable levels of vibration are also presented. The balancing of rotating and reciprocating machines and the whirling of shafts are considered. The active control techniques are also outlined for controlling the response of vibrating systems. The experimental methods used for vibration-response measurement are considered in Chapter 10. Vibration-measurement hardware and signal analysis techniques are described. Machine-condition monitoring and diagnosis techniques are also presented.

Chapter 11 presents several numerical integration techniques for finding the dynamic response of discrete and continuous systems. The central difference, Runge-Kutta, Houbolt, Wilson, and Newmark methods are discussed and illustrated. Finite element analysis, with applications involving one-dimensional elements, is discussed in Chapter 12. Bar, rod, and beam elements are used for the static and dynamic analysis of trusses, rods under torsion, and beams. The use of consistent- and lumped-mass matrices in the vibration analysis is also discussed in this chapter. Nonlinear vibration problems are governed by nonlinear differential equations and exhibit phenomena that are not predicted or even hinted at by the corresponding linearized problems. An introductory treatment of nonlinear vibration, including a discussion of subharmonic and superharmonic oscillations, limit cycles, systems with time-dependent coefficients, and chaos, is given in Chapter 13. The random vibration of linear vibration systems is considered in Chapter 14. The concepts of random process, stationary process, power spectral density, autocorrelation, and wide- and narrow-band processes are explained. The random vibration response of single- and multidegree-of-freedom systems is discussed in this chapter.

Appendixes A and B focus on mathematical relationships and deflection of beams and plates, respectively. The basics of matrix theory, Laplace transform, and SI units are presented in Appendixes C, D, and E, respectively. Finally, Appendix F provides an introduction to MATLAB programming.

Typical Syllabi

The material of the book provides flexible options for different types of vibration courses. Chapters 1 through 5, Chapter 9, and portions of Chapter 6 constitute a basic course in mechanical vibration. Different emphases/orientations can be given to the course by covering, additionally, different chapters as indicated below:

- Chapter 8 for continuous or distributed systems.
- Chapters 7 and 11 for numerical solutions.
- Chapter 10 for experimental methods and signal analysis.
- Chapter 12 for finite element analysis.
- Chapter 13 for nonlinear analysis.
- Chapter 14 for random vibration.

Alternatively, in Chapters 1 through 14, the text has sufficient material for a one-year sequence of two vibration courses at the senior or dual level.

Expected Course Outcomes

The material presented in the text helps achieve some of the program outcomes specified by ABET (Accreditation Board for Engineering and Technology):

- Ability to apply knowledge of mathematics, science, and engineering:
  The subject of vibration, as presented in the book, applies knowledge of mathematics (differential equations, matrix algebra, vector methods, and complex numbers) and science (statics and dynamics) to solve engineering vibration problems.
- Ability to identify, formulate, and solve engineering problems:
  Numerous illustrative examples, problems for practice, and design projects help the student identify various types of practical vibration problems and develop mathematical models, analyze, solve to find the response, and interpret the results.
- Ability to use the techniques, skills, and modern engineering tools necessary for engineering practice:
  - The application of the modern software, MATLAB, for the solution of vibration problems is illustrated in the last section of each chapter. The basics of MATLAB programming are summarized in Appendix F.
  - The use of the modern analysis technique, the finite element method, for the solution of vibration problems is covered in a separate chapter (Chapter 12). The finite element method is a popular technique used in industry for the modeling, analysis, and solution of complex vibrating systems.
- Ability to design and conduct experiments, as well as to analyze and interpret data:
  The experimental methods and analysis of data related to vibration are presented in Chapter 10. Discussed also are the equipment used in conducting vibration experiments, signal analysis, and identification of system parameters from the data.

Acknowledgments

I would like to express my appreciation to the many students, researchers, and faculty whose comments have helped me improve the book. I am most grateful to the following people for offering their comments, suggestions, and ideas:

Ara Arabyan, University of Arizona; Daniel Granger, Polytechnic School of Montreal, Canada; K.M. Rao, V.R.S. Engineering College Vijayawada, India; K. S. Shivakumar Aradhya, Gas Turbine Research Establishment, Bangalore, India; Donald G. Grant, University of Maine; Tom Thornton, Stress Analyst; Alejandro J. Rivas, Arizona State University; Qing Guo, University of Washington; James M. Widmann, California Polytechnic State University; G. Q. Cai, Florida Atlantic University; Richard Alexander, Texas A & M University; C. W. Bert, University of Oklahoma; Raymond M. Brach, University of Notre Dame; Alfonso Diaz-Jimenez, Universidad Distrital "Francisco Jose de Caldas," Colombia; George Doyle, University of Dayton; Hamid Hamidzadeh, South Dakota State University; H. N. Hashemi, Northeastern University; Zhikun Hou, Worchester Polytechnic Institute; J. Richard Houghton, Tennessee Technological University; Faryar Jabbari, University of California, Irvine; Robert Jeffers, University of Connecticut; Richard Keltie, North Carolina State University; J. S. Lamancusa, Pennsylvania State University; Harry Law, Clemson University; Robert Leonard, Virginia Polytechnic Institute and State University; James Li, Columbia University; Sameer Madanshetty, Boston University; Masoud Mojtahed, Purdue University, Calumet; Faissal A. Moslehy, University of Central Florida; M. G. Prasad, Stevens Institute of Technology; Mohan D. Rao, Michigan Tech; Amir G. Rezaei, California State Polytechnic University; F. P. J. Rimrott, University of Toronto; Subhash Sinha, Auburn University; Daniel Stutts, University of Missouri-Rolla; Massoud Tavakoli, Georgia Institute of Technology; Theodore Terry, Lehigh University; David F. Thompson, University of Cincinnati; Chung Tsui, University of Maryland, College Park; Alexander Vakakis, University of Illinois, Urbana-Champaign; Chuck Van Karsen, Michigan Technological University; Aleksandra Vinogradov, Montana State University; K. W. Wang, Pennsylvania State University; Gloria J. Wiens, University of Florida; and William Webster, GMI Engineering and Management Institute.

I would like to thank Purdue University for granting me permission to use the Boilermaker Special in Problem 2.104. My sincere thanks to Dr. Qing Liu for helping me write some of the MATLAB programs. Finally, I wish to thank my wife, Kamala, without whose patience, encouragement, and support this edition might never have been completed.

SINGIRESU S. RAO
srao@miami.edu

List of Symbols

Symbol	Meaning	SI Units
$a, a_0, a_1, a_2, \ldots$	constants, lengths	
a_{ij}	flexibility coefficient	m/N
$[a]$	flexibility matrix	m/N
A	area	m^2
$A, A_0, A_1, \ldots$	constants	
$b, b_1, b_2, \ldots$	constants, lengths	
$B, B_1, B_2, \ldots$	constants	
$\vec{B}$	balancing weight	N
$c, \underline{c}$	viscous damping coefficient	N$\cdot$s/m
$c, c_0, c_1, c_2, \ldots$	constants	
c	wave velocity	m/s
c_c	critical viscous damping constant	N$\cdot$s/m
c_i	damping constant of ith damper	N$\cdot$s/m
c_{ij}	damping coefficient	N$\cdot$s/m
$[c]$	damping matrix	N$\cdot$s/m
C, C_1, C_2, C_1', C_2'	constants	
d	diameter, dimension	m
D	diameter	m
$[D]$	dynamical matrix	s^2
e	base of natural logarithms	
e	eccentricity	m
$\vec{e}_x, \vec{e}_y$	unit vectors parallel to x and y directions	
E	Young's modulus	Pa
$E[x]$	expected value of x	
f	linear frequency	Hz
f	force per unit length	N/m
$f, \boldsymbol{f}$	unit impulse	N$\cdot$s
$\tilde{F}, F_d$	force	N
F_0	amplitude of force $F(t)$	N

Symbol	Meaning	SI Units
F_t, F_T	force transmitted	N
$\underset{\sim}{F}_t$	force acting on ith mass	N
$\vec{F}$	force vector	N
$\underset{\sim}{F}, \boldsymbol{F}$	impulse	N·s
g	acceleration due to gravity	m/s^2
$g(t)$	impulse response function	
G	shear modulus	N/m^2
h	hysteresis damping constant	N/m
$H(i\omega)$	frequency response function	
i	$\sqrt{-1}$	
I	area moment of inertia	m^4
$[I]$	identity matrix	
Im( )	imaginary part of ( )	
j	integer	
J	polar moment of inertia	m^4
$J, J_0, J_1, J_2, \ldots$	mass moment of inertia	kg·m^2
$k, \underset{\sim}{k}$	spring constant	N/m
k_i	spring constant of ith spring	N/m
k_t	torsional spring constant	N-m/rad
k_{ij}	stiffness coefficient	N/m
$[k]$	stiffness matrix	N/m
l, l_i	length	m
$m, \underset{\sim}{m}$	mass	kg
m_i	ith mass	kg
m_{ij}	mass coefficient	kg
$[m]$	mass matrix	kg
M	mass	kg
M	bending moment	N·m
$M_t, M_{t1}, M_{t2}, \ldots$	torque	N·m
M_{t0}	amplitude of $M_t(t)$	N·m
n	an integer	
n	number of degrees of freedom	
N	normal force	N
N	total number of time steps	
p	pressure	N/m^2
$p(x)$	probability density function of x	
$P(x)$	probability distribution function of x	
P	force, tension	N
q_j	jth generalized coordinate	
$\vec{q}$	vector of generalized displacements	
$\dot{\vec{q}}$	vector of generalized velocities	
Q_j	jth generalized force	
r	frequency ratio $= \omega/\omega_n$	
$\vec{r}$	radius vector	m

Symbol	Meaning	SI Units
Re( )	real part of ( )	
$R(\tau)$	autocorrelation function	
R	electrical resistance	ohm
R	Rayleigh's dissipation function	$N \cdot m/s$
R	Rayleigh's quotient	$1/s^2$
s	root of equation, Laplace variable	
S_a, S_d, S_v	acceleration, displacement, velocity spectrum	
$S_x(\omega)$	spectrum of x	
t	time	s
t_i	ith time station	s
T	torque	N-m
T	kinetic energy	J
T_i	kinetic energy of ith mass	J
T_d, T_f	displacement, force transmissibility	
u_{ij}	an element of matrix $[U]$	
U, U_i	axial displacement	m
U	potential energy	J
$\vec{U}$	unbalanced weight	N
$[U]$	upper triangular matrix	
v, v_0	linear velocity	m/s
V	shear force	N
V	potential energy	J
V_i	potential energy of ith spring	J
w, w_1, w_2, ω_i	transverse deflections	m
w_0	value of w at $t = 0$	m
$\dot{w}_0$	value of $\dot{w}$ at $t = 0$	m/s
w_n	nth mode of vibration	
W	weight of a mass	N
W	total energy	J
W	transverse deflection	m
W_i	value of W at $t = t_i$	m
$W(x)$	a function of x	
x, y, z	cartesian coordinates, displacements	m
$x_0, x(0)$	value of x at $t = 0$	m
$\dot{x}_0, \dot{x}(0)$	value of $\dot{x}$ at $t = 0$	m/s
x_j	displacement of jth mass	m
x_j	value of x at $t = t_j$	m
$\dot{x}_j$	value of $\dot{x}$ at $t = t_j$	m/s
x_h	homogeneous part of $x(t)$	m
x_p	particular part of $x(t)$	m
$\vec{x}$	vector of displacements	m
$\vec{x}_i$	value of $\vec{x}$ at $t = t_i$	m
$\dot{\vec{x}}_i$	value of $\dot{\vec{x}}$ at $t = t_i$	m/s
$\ddot{\vec{x}}_i$	value of $\ddot{\vec{x}}$ at $t = t_i$	m/s^2

Symbol	Meaning	SI Units
$\vec{x}^{(i)}(t)$	ith mode	
X	amplitude of $x(t)$	m
X_j	amplitude of $x_j(t)$	m
$\vec{X}^{(i)}$	ith modal vector	m
$X_i^{(j)}$	ith component of jth mode	m
$[X]$	modal matrix	m
$\vec{X}_r$	rth approximation to a mode shape	
y	base displacement	m
Y	amplitude of $y(t)$	m
z	relative displacement, $x - y$	m
Z	amplitude of $z(t)$	m
$Z(i\omega)$	mechanical impedance	N/m
α	angle, constant	
β	angle, constant	
β	hysteresis damping constant	
γ	specific weight	N/m^3
δ	logarithmic decrement	
$\delta_1, \delta_2, \ldots$	deflections	m
δ_{st}	static deflection	m
δ_{ij}	Kronecker delta	
Δ	determinant	
ΔF	increment in F	N
Δx	increment in x	m
Δt	increment in time t	s
ΔW	energy dissipated in a cycle	J
ε	a small quantity	
ε	strain	
ζ	damping ratio	
θ	constant, angular displacement	
θ_i	ith angular displacement	rad
θ_0	value of θ at $t = 0$	rad
$\dot{\theta}_0$	value of $\dot{\theta}$ at $t = 0$	rad/s
Θ	amplitude of $\theta(t)$	rad
Θ_i	amplitude of $\theta_i(t)$	rad
λ	eigenvalue $= 1/\omega^2$	s^2
$[\lambda]$	transformation matrix	
μ	viscosity of a fluid	$kg/m \cdot s$
μ	coefficient of friction	
μ_x	expected value of x	
ρ	mass density	kg/m^3
η	loss factor	
σ_x	standard deviation of x	
σ	stress	N/m^2
τ	period of oscillation, time, time constant	s

Symbol	Meaning	SI Units
τ	shear stress	N/m^2
ϕ	angle, phase angle	rad
ϕ_i	phase angle in ith mode	rad
ω	frequency of oscillation	rad/s
ω_i	ith natural frequency	rad/s
ω_n	natural frequency	rad/s
ω_d	frequency of damped vibration	rad/s

Subscripts

Symbol	Meaning
cri	critical value
eq	equivalent value
i	ith value
L	left plane
max	maximum value
n	corresponding to natural frequency
R	right plane
0	specific or reference value
t	torsional

Symbol	Meaning
$(\dot{\ })$	$\dfrac{d(\ )}{dt}$
$(\ddot{\ })$	$\dfrac{d^2(\ )}{dt^2}$
$(\vec{\ })$	column vector ( )
[ ]	matrix
$[\ ]^{-1}$	inverse of [ ]
$[\ ]^T$	transpose of [ ]
$\Delta(\ )$	increment in ( )
$\mathscr{L}(\ )$	Laplace transform of ( )
$\mathscr{L}^{-1}(\ )$	inverse Laplace transform of ( )

Galileo Galilei (1564–1642), an Italian astronomer, philosopher, and professor of mathematics at the Universities of Pisa and Padua, in 1609 became the first man to point a telescope to the sky. He wrote the first treatise on modern dynamics in 1590. His works on the oscillations of a simple pendulum and the vibration of strings are of fundamental significance in the theory of vibrations.
(Courtesy of Dirk J. Struik, *A Concise History of Mathematics* (2nd rev. ed.), Dover Publications, Inc., New York, 1948.)

C H A P T E R   1

Fundamentals of Vibration

Chapter Outline

This chapter introduces the subject of vibrations in a relatively simple manner. It begins with a brief history of the subject and continues with an examination of the importance of vibration. The basic concepts of degrees of freedom and of discrete and continuous systems are introduced, along with a description of the elementary parts of vibrating

1

systems. The various classifications of vibration—namely, free and forced vibration, undamped and damped vibration, linear and nonlinear vibration, and deterministic and random vibration—are indicated. The various steps involved in vibration analysis of an engineering system are outlined, and essential definitions and concepts of vibration are introduced.

The concept of harmonic motion and its representation using vectors and complex numbers is described. The basic definitions and terminology related to harmonic motion, such as cycle, amplitude, period, frequency, phase angle, and natural frequency, are given. Finally, the harmonic analysis, dealing with the representation of any periodic function in terms of harmonic functions, using Fourier series, is outlined. The concepts of frequency spectrum, time- and frequency-domain representations of periodic functions, half-range expansions, and numerical computation of Fourier coefficients are discussed in detail.

Learning Objectives

After completing this chapter, the reader should be able to do the following:

- Describe briefly the history of vibration
- Indicate the importance of study of vibration
- Give various classifications of vibration
- State the steps involved in vibration analysis
- Compute the values of spring constants, masses, and damping constants
- Define harmonic motion and different possible representations of harmonic motion
- Add and subtract harmonic motions
- Conduct Fourier series expansion of given periodic functions
- Determine Fourier coefficients numerically using the MATLAB program

1.1   Preliminary Remarks

The subject of vibration is introduced here in a relatively simple manner. The chapter begins with a brief history of vibration and continues with an examination of its importance. The various steps involved in vibration analysis of an engineering system are outlined, and essential definitions and concepts of vibration are introduced. We learn here that all mechanical and structural systems can be modeled as mass-spring-damper systems. In some systems, such as an automobile, the mass, spring, and damper can be identified as separate components (mass in the form of the body, spring in the form of suspension, and damper in the form of shock absorbers). In some cases, the mass, spring, and damper do not appear as separate components; they are inherent and integral to the system. For example, in an airplane wing, the mass of the wing is distributed throughout the wing. Also, due to its elasticity, the wing undergoes noticeable deformation during flight so that it can be modeled as a spring. In addition, the deflection of the wing introduces damping due to relative motion between components such as joints, connections, and support as well as internal friction due to microstructural defects in the material. The chapter describes the modeling of spring, mass and damping elements, their characteristics, and the combination of several springs, masses or damping elements appearing in a system. There follows a presentation of the concept of harmonic analysis, which can be used for the analysis of general periodic motions. No attempt at exhaustive treatment of the topics is made in Chapter 1; subsequent chapters will develop many of the ideas in more detail.

1.2 Brief History of the Study of Vibration

1.2.1 Origins of the Study of Vibration

People became interested in vibration when they created the first musical instruments, probably whistles or drums. Since then, both musicians and philosophers have sought out the rules and laws of sound production, used them in improving musical instruments, and passed them on from generation to generation. As long ago as 4000 B.C. [1.1], music had become highly developed and was much appreciated by Chinese, Hindus, Japanese, and, perhaps, Egyptians. These early peoples observed certain definite rules in connection with the art of music, although their knowledge did not reach the level of a science.

Stringed musical instruments probably originated with the hunter's bow, a weapon favored by the armies of ancient Egypt. One of the most primitive stringed instruments, the *nanga*, resembled a harp with three or four strings, each yielding only one note. An example dating back to 1500 B.C. can be seen in the British Museum. The Museum also exhibits an 11-stringed harp with a gold-decorated, bull-headed sounding box, found at Ur in a royal tomb dating from about 2600 B.C. As early as 3000 B.C., stringed instruments such as harps were depicted on walls of Egyptian tombs.

Our present system of music is based on ancient Greek civilization. The Greek philosopher and mathematician Pythagoras (582–507 B.C.) is considered to be the first person to investigate musical sounds on a scientific basis (Fig. 1.1). Among other things, Pythagoras

FIGURE 1.1 Pythagoras. (Reprinted with permission from L. E. Navia, *Pythagoras: An Annotated Bibliography*, Garland Publishing, Inc., New York, 1990.)

FIGURE 1.2   Monochord.

conducted experiments on a vibrating string by using a simple apparatus called a *mono-chord*. In the monochord shown in Fig. 1.2, the wooden bridges labeled 1 and 3 are fixed. Bridge 2 is made movable while the tension in the string is held constant by the hanging weight. Pythagoras observed that if two like strings of different lengths are subject to the same tension, the shorter one emits a higher note; in addition, if the shorter string is half the length of the longer one, the shorter one will emit a note an octave above the other. Pythagoras left no written account of his work (Fig. 1.3), but it has been described by others. Although the concept of pitch was developed by the time of Pythagoras, the relation between the pitch and the frequency was not understood until the time of Galileo in the sixteenth century.

Around 350 B.C., Aristotle wrote treatises on music and sound, making observations such as "the voice is sweeter than the sound of instruments," and "the sound of the flute is sweeter than that of the lyre." In 320 B.C., Aristoxenus, a pupil of Aristotle and a musician,

FIGURE 1.3   Pythagoras as a musician. (Reprinted with permission from D. E. Smith, *History of Mathematics*, Vol. I, Dover Publications, Inc., New York, 1958.)

wrote a three-volume work entitled *Elements of Harmony*. These books are perhaps the oldest ones available on the subject of music written by the investigators themselves. In about 300 B.C., in a treatise called *Introduction to Harmonics*, Euclid wrote briefly about music without any reference to the physical nature of sound. No further advances in scientific knowledge of sound were made by the Greeks.

It appears that the Romans derived their knowledge of music completely from the Greeks, except that Vitruvius, a famous Roman architect, wrote in about 20 B.C. on the acoustic properties of theaters. His treatise, entitled *De Architectura Libri Decem*, was lost for many years, to be rediscovered only in the fifteenth century. There appears to have been no development in the theories of sound and vibration for nearly 16 centuries after the work of Vitruvius.

China experienced many earthquakes in ancient times. Zhang Heng, who served as a historian and astronomer in the second century, perceived a need to develop an instrument to measure earthquakes precisely. In A.D. 132, he invented the world's first seismograph [1.3, 1.4]. It was made of fine cast bronze, had a diameter of eight chi (a chi is equal to 0.237 meter), and was shaped like a wine jar (Fig. 1.4). Inside the jar was a mechanism consisting of pendulums surrounded by a group of eight levers pointing in eight directions. Eight dragon figures, with a bronze ball in the mouth of each, were arranged on the outside of the seismograph. Below each dragon was a toad with mouth open upward. A strong earthquake in any direction would tilt the pendulum in that direction, triggering the lever in the dragon head. This opened the mouth of the dragon, thereby releasing its bronze ball, which fell in the mouth of the toad with a clanging sound. Thus the seismograph enabled the monitoring personnel to know both the time and direction of occurrence of the earthquake.

FIGURE 1.4   The world's first seismograph, invented in China in A.D. 132. (Reprinted with permission from R. Taton (ed.), *History of Science*, Basic Books, Inc., New York, 1957.)

1.2.2
From Galileo
to Rayleigh

Galileo Galilei (1564–1642) is considered to be the founder of modern experimental science. In fact, the seventeenth century is often considered the "century of genius" since the foundations of modern philosophy and science were laid during that period. Galileo was inspired to study the behavior of a simple pendulum by observing the pendulum movements of a lamp in a church in Pisa. One day, while feeling bored during a sermon, Galileo was staring at the ceiling of the church. A swinging lamp caught his attention. He started measuring the period of the pendulum movements of the lamp with his pulse and found to his amazement that the time period was independent of the amplitude of swings. This led him to conduct more experiments on the simple pendulum. In *Discourses Concerning Two New Sciences*, published in 1638, Galileo discussed vibrating bodies. He described the dependence of the frequency of vibration on the length of a simple pendulum, along with the phenomenon of sympathetic vibrations (resonance). Galileo's writings also indicate that he had a clear understanding of the relationship between the frequency, length, tension, and density of a vibrating stretched string [1.5]. However, the first correct published account of the vibration of strings was given by the French mathematician and theologian, Marin Mersenne (1588–1648) in his book *Harmonicorum Liber*, published in 1636. Mersenne also measured, for the first time, the frequency of vibration of a long string and from that predicted the frequency of a shorter string having the same density and tension. Mersenne is considered by many as the father of acoustics. He is often credited with the discovery of the laws of vibrating strings because he published the results in 1636, two years before Galileo. However, the credit belongs to Galileo, since the laws were written many years earlier but their publication was prohibited by the orders of the Inquisitor of Rome until 1638.

Inspired by the work of Galileo, the Academia del Cimento was founded in Florence in 1657; this was followed by the formations of the Royal Society of London in 1662 and the Paris Academie des Sciences in 1666. Later, Robert Hooke (1635–1703) also conducted experiments to find a relation between the pitch and frequency of vibration of a string. However, it was Joseph Sauveur (1653–1716) who investigated these experiments thoroughly and coined the word "acoustics" for the science of sound [1.6]. Sauveur in France and John Wallis (1616–1703) in England observed, independently, the phenomenon of mode shapes, and they found that a vibrating stretched string can have no motion at certain points and violent motion at intermediate points. Sauveur called the former points *nodes* and the latter ones *loops*. It was found that such vibrations had higher frequencies than that associated with the simple vibration of the string with no nodes. In fact, the higher frequencies were found to be integral multiples of the frequency of simple vibration, and Sauveur called the higher frequencies *harmonics* and the frequency of simple vibration the *fundamental frequency*. Sauveur also found that a string can vibrate with several of its harmonics present at the same time. In addition, he observed the phenomenon of beats when two organ pipes of slightly different pitches are sounded together. In 1700, Sauveur calculated, by a somewhat dubious method, the frequency of a stretched string from the measured sag of its middle point.

Sir Isaac Newton (1642–1727) published his monumental work, *Philosophiae Naturalis Principia Mathematica*, in 1686, describing the law of universal gravitation as well as the three laws of motion and other discoveries. Newton's second law of motion is routinely used in modern books on vibrations to derive the equations of motion of a vibrating body. The theoretical (dynamical) solution of the problem of the vibrating string was found in 1713 by the English mathematician Brook Taylor (1685–1731), who also presented the famous Taylor's theorem on infinite series. The natural frequency of vibration obtained from the equation of motion derived by Taylor agreed with the experimental values observed by Galileo and Mersenne. The procedure adopted by Taylor was perfected through the introduction of partial derivatives in the equations of motion by Daniel Bernoulli (1700–1782), Jean D'Alembert (1717–1783), and Leonard Euler (1707–1783).

The possibility of a string vibrating with several of its harmonics present at the same time (with displacement of any point at any instant being equal to the algebraic sum of displacements for each harmonic) was proved through the dynamic equations of Daniel Bernoulli in his memoir, published by the Berlin Academy in 1755 [1.7]. This characteristic was referred to as the principle of the coexistence of small oscillations, which, in present-day terminology, is the principle of superposition. This principle proved to be most valuable in the development of the theory of vibrations and led to the possibility of expressing any arbitrary function (i.e., any initial shape of the string) using an infinite series of sines and cosines. Because of this implication, D'Alembert and Euler doubted the validity of this principle. However, the validity of this type of expansion was proved by J. B. J. Fourier (1768–1830) in his *Analytical Theory of Heat* in 1822.

The analytical solution of the vibrating string was presented by Joseph Lagrange (1736–1813) in his memoir published by the Turin Academy in 1759. In his study, Lagrange assumed that the string was made up of a finite number of equally spaced identical mass particles, and he established the existence of a number of independent frequencies equal to the number of mass particles. When the number of particles was allowed to be infinite, the resulting frequencies were found to be the same as the harmonic frequencies of the stretched string. The method of setting up the differential equation of the motion of a string (called the *wave equation*), presented in most modern books on vibration theory, was first developed by D'Alembert in his memoir published by the Berlin Academy in 1750. The vibration of thin beams supported and clamped in different ways was first studied by Euler in 1744 and Daniel Bernoulli in 1751. Their approach has become known as the *Euler-Bernoulli* or *thin beam theory.*

Charles Coulomb did both theoretical and experimental studies in 1784 on the torsional oscillations of a metal cylinder suspended by a wire (Fig. 1.5). By assuming that the resisting torque of the twisted wire is proportional to the angle of twist, he derived the equation of motion for the torsional vibration of the suspended cylinder. By integrating the equation of motion, he found that the period of oscillation is independent of the angle of twist.

There is an interesting story related to the development of the theory of vibration of plates [1.8]. In 1802, the German scientist, E. F. F. Chladni (1756–1824) developed the method of placing sand on a vibrating plate to find its mode shapes and observed the beauty and intricacy of the modal patterns of the vibrating plates. In 1809, the French Academy invited Chladni to give a demonstration of his experiments. Napoléon Bonaparte, who attended the meeting, was very impressed and presented a sum of 3,000 francs to the academy, to be awarded to the first person to give a satisfactory mathematical theory of the vibration of plates. By the closing date of the competition in October

(a)

(b)

FIGURE 1.5   Coulomb's device for torsional vibration tests. (Reprinted with permission from S. P. Timoshenko, *History of Strength of Materials*, McGraw-Hill Book Company, Inc., New York, 1953.)

1811, only one candidate, Sophie Germain, had entered the contest. But Lagrange, who was one of the judges, noticed an error in the derivation of her differential equation of motion. The academy opened the competition again, with a new closing date of October 1813. Germain again entered the contest, presenting the correct form of the differential equation. However, the academy did not award the prize to her because the judges wanted physical justification of the assumptions made in her derivation. The competition was opened once more. In her third attempt, Germain was finally awarded the prize in 1815, although the judges were not completely satisfied with her theory. In fact, it was later found that her differential equation was correct but the boundary conditions were erroneous. The correct boundary conditions for the vibration of plates were given in 1850 by G. R. Kirchhoff (1824–1887).

In the meantime, the problem of vibration of a rectangular flexible membrane, which is important for the understanding of the sound emitted by drums, was solved for the first time by Simeon Poisson (1781–1840). The vibration of a circular membrane was studied by R. F. A. Clebsch (1833–1872) in 1862. After this, vibration studies were done on a number of practical mechanical and structural systems. In 1877, Lord Baron Rayleigh published his book on the theory of sound [1.9]; it is considered a classic on the subject of sound and vibration even today. Notable among the many contributions of Rayleigh is the method of finding the fundamental frequency of vibration of a conservative system by making use of the principle of conservation of energy—now known as *Rayleigh's method*.

This method proved to be a helpful technique for the solution of difficult vibration problems. An extension of the method, which can be used to find multiple natural frequencies, is known as the *Rayleigh-Ritz method.*

1.2.3 Recent Contributions

In 1902, Frahm investigated the importance of torsional vibration study in the design of the propeller shafts of steamships. The dynamic vibration absorber, which involves the addition of a secondary spring-mass system to eliminate the vibrations of a main system, was also proposed by Frahm in 1909. Among the modern contributers to the theory of vibrations, the names of Stodola, De Laval, Timoshenko, and Mindlin are notable. Aurel Stodola (1859–1943) contributed to the study of vibration of beams, plates, and membranes. He developed a method for analyzing vibrating beams that is also applicable to turbine blades. Noting that every major type of prime mover gives rise to vibration problems, C. G. P. De Laval (1845–1913) presented a practical solution to the problem of vibration of an unbalanced rotating disk. After noticing failures of steel shafts in high-speed turbines, he used a bamboo fishing rod as a shaft to mount the rotor. He observed that this system not only eliminated the vibration of the unbalanced rotor but also survived up to speeds as high as 100,000 rpm [1.10].

Stephen Timoshenko (1878–1972), by considering the effects of rotary inertia and shear deformation, presented an improved theory of vibration of beams, which has become known as the *Timoshenko* or *thick beam theory.* A similar theory was presented by R. D. Mindlin for the vibration analysis of thick plates by including the effects of rotary inertia and shear deformation.

It has long been recognized that many basic problems of mechanics, including those of vibrations, are nonlinear. Although the linear treatments commonly adopted are quite satisfactory for most purposes, they are not adequate in all cases. In nonlinear systems, phenomena may occur that are theoretically impossible in linear systems. The mathematical theory of nonlinear vibrations began to develop in the works of Poincaré and Lyapunov at the end of the nineteenth century. Poincaré developed the perturbation method in 1892 in connection with the approximate solution of nonlinear celestial mechanics problems. In 1892, Lyapunov laid the foundations of modern stability theory, which is applicable to all types of dynamical systems. After 1920, the studies undertaken by Duffing and van der Pol brought the first definite solutions into the theory of nonlinear vibrations and drew attention to its importance in engineering. In the last 40 years, authors like Minorsky and Stoker have endeavored to collect in monographs the main results concerning nonlinear vibrations. Most practical applications of nonlinear vibration involved the use of some type of a perturbation-theory approach. The modern methods of perturbation theory were surveyed by Nayfeh [1.11].

Random characteristics are present in diverse phenomena such as earthquakes, winds, transportation of goods on wheeled vehicles, and rocket and jet engine noise. It became necessary to devise concepts and methods of vibration analysis for these random effects. Although Einstein considered Brownian movement, a particular type of random vibration, as long ago as 1905, no applications were investigated until 1930. The introduction of the correlation function by Taylor in 1920 and of the spectral density by Wiener and Khinchin in the early 1930s opened new prospects for progress in the theory of random vibrations. Papers by Lin and Rice, published between 1943 and 1945, paved

FIGURE 1.6   Finite element idealization of the body of a bus [1.16]. (Reprinted with permission from Society of Automotive Engineers, Inc., 1974.)

the way for the application of random vibrations to practical engineering problems. The monographs of Crandall and Mark and of Robson systematized the existing knowledge in the theory of random vibrations [1.12, 1.13].

Until about 40 years ago, vibration studies, even those dealing with complex engineering systems, were done by using gross models, with only a few degrees of freedom. However, the advent of high-speed digital computers in the 1950s made it possible to treat moderately complex systems and to generate approximate solutions in semidefinite form, relying on classical solution methods but using numerical evaluation of certain terms that cannot be expressed in closed form. The simultaneous development of the finite element method enabled engineers to use digital computers to conduct numerically detailed vibration analysis of complex mechanical, vehicular, and structural systems displaying thousands of degrees of freedom [1.14]. Although the finite element method was not so named until recently, the concept was used centuries ago. For example, ancient mathematicians found the circumference of a circle by approximating it as a polygon, where each side of the polygon, in present-day notation, can be called a *finite element*. The finite element method as known today was presented by Turner, Clough, Martin, and Topp in connection with the analysis of aircraft structures [1.15]. Fig. 1.6 shows the finite element idealization of the body of a bus [1.16].

1.3   Importance of the Study of Vibration

Most human activities involve vibration in one form or other. For example, we hear because our eardrums vibrate and see because light waves undergo vibration. Breathing is associated with the vibration of lungs and walking involves (periodic) oscillatory motion of legs and hands. Human speech requires the oscillatory motion of larynges (and tongues) [1.17]. Early scholars in the field of vibration concentrated their efforts on understanding the natural phenomena and developing mathematical theories to describe the vibration of physical systems. In recent times, many investigations have been motivated by the engineering applications of vibration, such as the design of machines, foundations, structures, engines, turbines, and control systems.

Most prime movers have vibrational problems due to the inherent unbalance in the engines. The unbalance may be due to faulty design or poor manufacture. Imbalance in diesel engines, for example, can cause ground waves sufficiently powerful to create a nuisance in urban areas. The wheels of some locomotives can rise more than a centimeter off the track at high speeds due to imbalance. In turbines, vibrations cause spectacular mechanical failures. Engineers have not yet been able to prevent the failures that result from blade and disk vibrations in turbines. Naturally, the structures designed to support heavy centrifugal machines, like motors and turbines, or reciprocating machines, like steam and gas engines and reciprocating pumps, are also subjected to vibration. In all these situations, the structure or machine component subjected to vibration can fail because of material fatigue resulting from the cyclic variation of the induced stress. Furthermore, the vibration causes more rapid wear of machine parts such as bearings and gears and also creates excessive noise. In machines, vibration can loosen fasteners such as nuts. In metal cutting processes, vibration can cause chatter, which leads to a poor surface finish.

Whenever the natural frequency of vibration of a machine or structure coincides with the frequency of the external excitation, there occurs a phenomenon known as *resonance*, which leads to excessive deflections and failure. The literature is full of accounts of system failures brought about by resonance and excessive vibration of components and systems (see Fig. 1.7). Because of the devastating effects that vibrations can have on machines

FIGURE 1.7   Tacoma Narrows bridge during wind-induced vibration. The bridge opened on July 1, 1940, and collapsed on November 7, 1940. (Farquharson photo, Historical Photography Collection, University of Washington Libraries.)

FIGURE 1.8   Vibration testing of the space shuttle *Enterprise*. (Courtesy of NASA.)

and structures, vibration testing [1.18] has become a standard procedure in the design and development of most engineering systems (see Fig. 1.8).

In many engineering systems, a human being acts as an integral part of the system. The transmission of vibration to human beings results in discomfort and loss of efficiency. The vibration and noise generated by engines cause annoyance to people and, sometimes, damage to property. Vibration of instrument panels can cause their malfunction or difficulty in reading the meters [1.19]. Thus one of the important purposes of vibration study is to reduce vibration through proper design of machines and their mountings. In this

FIGURE 1.9   Vibratory finishing process. (Reprinted courtesy of the Society of Manufacturing Engineers, The Tool and Manufacturing Engineer, 1964.)

connection, the mechanical engineer tries to design the engine or machine so as to minimize imbalance, while the structural engineer tries to design the supporting structure so as to ensure that the effect of the imbalance will not be harmful [1.20].

In spite of its detrimental effects, vibration can be utilized profitably in several consumer and industrial applications. In fact, the applications of vibratory equipment have increased considerably in recent years [1.21]. For example, vibration is put to work in vibratory conveyors, hoppers, sieves, compactors, washing machines, electric toothbrushes, dentist's drills, clocks, and electric massaging units. Vibration is also used in pile driving, vibratory testing of materials, vibratory finishing processes, and electronic circuits to filter out the unwanted frequencies (see Fig. 1.9). Vibration has been found to improve the efficiency of certain machining, casting, forging, and welding processes. It is employed to simulate earthquakes for geological research and also to conduct studies in the design of nuclear reactors.

1.4 Basic Concepts of Vibration

1.4.1
Vibration

Any motion that repeats itself after an interval of time is called *vibration* or *oscillation*. The swinging of a pendulum and the motion of a plucked string are typical examples of vibration. The theory of vibration deals with the study of oscillatory motions of bodies and the forces associated with them.

1.4.2
Elementary Parts of Vibrating Systems

A vibratory system, in general, includes a means for storing potential energy (spring or elasticity), a means for storing kinetic energy (mass or inertia), and a means by which energy is gradually lost (damper).

The vibration of a system involves the transfer of its potential energy to kinetic energy and of kinetic energy to potential energy, alternately. If the system is damped, some energy is dissipated in each cycle of vibration and must be replaced by an external source if a state of steady vibration is to be maintained.

As an example, consider the vibration of the simple pendulum shown in Fig. 1.10. Let the bob of mass m be released after being given an angular displacement θ. At position 1, the velocity of the bob and hence its kinetic energy is zero. But it has a potential energy of magnitude $mgl(1 - \cos \theta)$ with respect to the datum position 2. Since the gravitational force mg induces a torque $mgl \sin \theta$ about the point O, the bob starts swinging to the left from position 1. This gives the bob certain angular acceleration in the clockwise direction, and by the time it reaches position 2, all of its potential energy will be converted into kinetic energy. Hence the bob will not stop in position 2 but will continue to swing to position 3. However, as it passes the mean position 2, a counterclockwise torque due to gravity starts acting on the bob and causes the bob to decelerate. The velocity of the bob reduces to zero at the left extreme position. By this time, all the kinetic energy of the bob will be converted to potential energy. Again due to the gravity torque, the bob continues to attain a counterclockwise velocity. Hence the bob starts swinging back with progressively increasing velocity and passes the mean position again. This process keeps repeating, and the pendulum will have oscillatory motion. However, in practice, the magnitude of oscillation (θ) gradually decreases and the pendulum ultimately stops due to the resistance (damping) offered by the surrounding medium (air). This means that some energy is dissipated in each cycle of vibration due to damping by the air.

FIGURE 1.10 A simple pendulum.

1.4.3
Number of
Degrees
of Freedom

The minimum number of independent coordinates required to determine completely the positions of all parts of a system at any instant of time defines the number of degrees of freedom of the system. The simple pendulum shown in Fig. 1.10, as well as each of the systems shown in Fig. 1.11, represents a single-degree-of-freedom system. For example, the motion of the simple pendulum (Fig. 1.10) can be stated either in terms of the angle θ or in terms of the Cartesian coordinates x and y. If the coordinates x and y are used to describe the motion, it must be recognized that these coordinates are not independent. They are related to each other through the relation $x^2 + y^2 = l^2$, where l is the constant length of the pendulum. Thus any one coordinate can describe the motion of the pendulum. In this example, we find that the choice of θ as the independent coordinate will be more convenient than the choice of x or y. For the slider shown in Fig. 1.11(a), either the angular coordinate θ or the coordinate x can be used to describe the motion. In Fig. 1.11(b), the linear coordinate x can

    (a) Slider-crank-   (b) Spring-mass system   (c) Torsional system
       spring mechanism

FIGURE 1.11 Single-degree-of-freedom systems.

FIGURE 1.12   Two-degree-of-freedom systems.

FIGURE 1.13   Three–degree-of-freedom systems.

be used to specify the motion. For the torsional system (long bar with a heavy disk at the end) shown in Fig. 1.11(c), the angular coordinate θ can be used to describe the motion.

Some examples of two- and three-degree-of-freedom systems are shown in Figs. 1.12 and 1.13, respectively. Figure 1.12(a) shows a two-mass, two-spring system that is described by the two linear coordinates x_1 and x_2. Figure 1.12(b) denotes a two-rotor system whose motion can be specified in terms of θ_1 and θ_2. The motion of the system shown in Fig. 1.12(c) can be described completely either by X and θ or by x, y, and X. In the latter case, x and y are constrained as $x^2 + y^2 = l^2$ where l is a constant.

For the systems shown in Figs. 1.13(a) and 1.13(c), the coordinates x_i $(i = 1, 2, 3)$ and θ_i $(i = 1, 2, 3)$ can be used, respectively, to describe the motion. In the case of the

FIGURE 1.14   A cantilever beam
(an infinite-number-of-degrees-of-freedom
system).

system shown in Fig. 1.13(b), $\theta_i\,(i = 1, 2, 3)$ specifies the positions of the masses $m_i\,(i = 1, 2, 3)$. An alternate method of describing this system is in terms of x_i and $y_i\,(i = 1, 2, 3)$; but in this case the constraints $x_i^2 + y_i^2 = l_i^2\,(i = 1, 2, 3)$ have to be considered.

The coordinates necessary to describe the motion of a system constitute a set of *generalized coordinates*. These are usually denoted as $q_1, q_2, \ldots$ and may represent Cartesian and/or non-Cartesian coordinates.

1.4.4 Discrete and Continuous Systems

A large number of practical systems can be described using a finite number of degrees of freedom, such as the simple systems shown in Figs. 1.10 to 1.13. Some systems, especially those involving continuous elastic members, have an infinite number of degrees of freedom. As a simple example, consider the cantilever beam shown in Fig. 1.14. Since the beam has an infinite number of mass points, we need an infinite number of coordinates to specify its deflected configuration. The infinite number of coordinates defines its elastic deflection curve. Thus the cantilever beam has an infinite number of degrees of freedom. Most structural and machine systems have deformable (elastic) members and therefore have an infinite number of degrees of freedom.

Systems with a finite number of degrees of freedom are called *discrete* or *lumped parameter* systems, and those with an infinite number of degrees of freedom are called *continuous* or *distributed* systems.

Most of the time, continuous systems are approximated as discrete systems, and solutions are obtained in a simpler manner. Although treatment of a system as continuous gives exact results, the analytical methods available for dealing with continuous systems are limited to a narrow selection of problems, such as uniform beams, slender rods, and thin plates. Hence most of the practical systems are studied by treating them as finite lumped masses, springs, and dampers. In general, more accurate results are obtained by increasing the number of masses, springs, and dampers—that is, by increasing the number of degrees of freedom.

1.5   Classification of Vibration

Vibration can be classified in several ways. Some of the important classifications are as follows.

1.5.1
Free and Forced
Vibration

Free Vibration.   If a system, after an initial disturbance, is left to vibrate on its own, the ensuing vibration is known as *free vibration*. No external force acts on the system. The oscillation of a simple pendulum is an example of free vibration.

Forced Vibration.   If a system is subjected to an external force (often, a repeating type of force), the resulting vibration is known as *forced vibration*. The oscillation that arises in machines such as diesel engines is an example of forced vibration.

If the frequency of the external force coincides with one of the natural frequencies of the system, a condition known as *resonance* occurs, and the system undergoes dangerously large oscillations. Failures of such structures as buildings, bridges, turbines, and airplane wings have been associated with the occurrence of resonance.

1.5.2
Undamped
and Damped
Vibration

If no energy is lost or dissipated in friction or other resistance during oscillation, the vibration is known as *undamped vibration*. If any energy is lost in this way, however, it is called *damped vibration*. In many physical systems, the amount of damping is so small that it can be disregarded for most engineering purposes. However, consideration of damping becomes extremely important in analyzing vibratory systems near resonance.

1.5.3
Linear
and Nonlinear
Vibration

If all the basic components of a vibratory system—the spring, the mass, and the damper—behave linearly, the resulting vibration is known as *linear vibration*. If, however, any of the basic components behave nonlinearly, the vibration is called *nonlinear vibration*. The differential equations that govern the behavior of linear and nonlinear vibratory systems are linear and nonlinear, respectively. If the vibration is linear, the principle of superposition holds, and the mathematical techniques of analysis are well developed. For nonlinear vibration, the superposition principle is not valid, and techniques of analysis are less well known. Since all vibratory systems tend to behave nonlinearly with increasing amplitude of oscillation, a knowledge of nonlinear vibration is desirable in dealing with practical vibratory systems.

1.5.4
Deterministic
and Random
Vibration

If the value or magnitude of the excitation (force or motion) acting on a vibratory system is known at any given time, the excitation is called *deterministic*. The resulting vibration is known as *deterministic vibration*.

In some cases, the excitation is *nondeterministic* or *random*; the value of the excitation at a given time cannot be predicted. In these cases, a large collection of records of the excitation may exhibit some statistical regularity. It is possible to estimate averages such as the mean and mean square values of the excitation. Examples of random excitations are wind velocity, road roughness, and ground motion during earthquakes. If the excitation is random, the resulting vibration is called *random vibration*. In this case, the vibratory response of the system is also random; it can be described only in terms of statistical quantities. Fig. 1.15 shows examples of deterministic and random excitations.

(a) A deterministic (periodic) excitation   (b) A random excitation

FIGURE 1.15   Deterministic and random excitations.

1.6   Vibration Analysis Procedure

A vibratory system is a dynamic one for which the variables such as the excitations (inputs) and responses (outputs) are time dependent. The response of a vibrating system generally depends on the initial conditions as well as the external excitations. Most practical vibrating systems are very complex, and it is impossible to consider all the details for a mathematical analysis. Only the most important features are considered in the analysis to predict the behavior of the system under specified input conditions. Often the overall behavior of the system can be determined by considering even a simple model of the complex physical system. Thus the analysis of a vibrating system usually involves mathematical modeling, derivation of the governing equations, solution of the equations, and interpretation of the results.

Step 1: Mathematical Modeling.   The purpose of mathematical modeling is to represent all the important features of the system for the purpose of deriving the mathematical (or analytical) equations governing the system's behavior. The mathematical model should include enough details to allow describing the system in terms of equations without making it too complex. The mathematical model may be linear or nonlinear, depending on the behavior of the system's components. Linear models permit quick solutions and are simple to handle; however, nonlinear models sometimes reveal certain characteristics of the system that cannot be predicted using linear models. Thus a great deal of engineering judgment is needed to come up with a suitable mathematical model of a vibrating system.

Sometimes the mathematical model is gradually improved to obtain more accurate results. In this approach, first a very crude or elementary model is used to get a quick insight into the overall behavior of the system. Subsequently, the model is refined by including more components and/or details so that the behavior of the system can be observed more closely. To illustrate the procedure of refinement used in mathematical modeling, consider the forging hammer shown in Fig. 1.16(a). It consists of a frame, a falling weight known as the *tup*, an anvil, and a foundation block. The anvil is a massive steel block on which material is forged into its desired shape by the repeated blows of the tup. The anvil is usually mounted on an elastic pad to reduce the transmission of vibration to the foundation block and the frame [1.22]. For a first approximation, the frame, anvil, elastic pad, foundation block, and soil are modeled as a single-degree-of-freedom system as shown in Fig. 1.16(b). For a refined approximation, the weights of the frame and anvil

FIGURE 1.16 Modeling of a forging hammer.

and the foundation block are represented separately with a two-degree-of-freedom model as shown in Fig. 1.16(c). Further refinement of the model can be made by considering eccentric impacts of the tup, which cause each of the masses shown in Fig. 1.16(c) to have both vertical and rocking (rotation) motions in the plane of the paper.

Step 2: Derivation of Governing Equations.   Once the mathematical model is available, we use the principles of dynamics and derive the equations that describe the vibration of the system. The equations of motion can be derived conveniently by drawing the free-body diagrams of all the masses involved. The free-body diagram of a mass can be obtained by isolating the mass and indicating all externally applied forces, the reactive forces, and the inertia forces. The equations of motion of a vibrating system are usually in the form of a set of ordinary differential equations for a discrete system and partial differential equations for a continuous system. The equations may be linear or nonlinear, depending on the behavior of the components of the system. Several approaches are commonly used to derive the governing equations. Among them are Newton's second law of motion, D'Alembert's principle, and the principle of conservation of energy.

Step 3: Solution of the Governing Equations.   The equations of motion must be solved to find the response of the vibrating system. Depending on the nature of the problem, we can use one of the following techniques for finding the solution: standard methods of solving differential equations, Laplace transform methods, matrix methods,[1] and numerical methods. If the governing equations are nonlinear, they can seldom be solved in closed form. Furthermore, the solution of partial differential equations is far more involved than that of ordinary differential equations. Numerical methods involving computers can be used to solve the equations. However, it will be difficult to draw general conclusions about the behavior of the system using computer results.

Step 4: Interpretation of the Results.   The solution of the governing equations gives the displacements, velocities, and accelerations of the various masses of the system. These results must be interpreted with a clear view of the purpose of the analysis and the possible design implications of the results.

Mathematical Model of a Motorcycle

EXAMPLE 1.1

Figure 1.17(a) shows a motorcycle with a rider. Develop a sequence of three mathematical models of the system for investigating vibration in the vertical direction. Consider the elasticity of the tires, elasticity and damping of the struts (in the vertical direction), masses of the wheels, and elasticity, damping, and mass of the rider.

Solution: We start with the simplest model and refine it gradually. When the equivalent values of the mass, stiffness, and damping of the system are used, we obtain a single-degree-of-freedom model

[1]The basic definitions and operations of matrix theory are given in Appendix A.

FIGURE 1.17   Motorcycle with a rider—a physical system and mathematical model.

of the motorcycle with a rider as indicated in Fig. 1.17(b). In this model, the equivalent stiffness (k_{eq}) includes the stiffnesses of the tires, struts, and rider. The equivalent damping constant (c_{eq}) includes the damping of the struts and the rider. The equivalent mass includes the masses of the wheels, vehicle body, and the rider. This model can be refined by representing the masses of wheels, elasticity of the tires, and elasticity and damping of the struts separately, as shown in Fig. 1.17(c). In this model, the mass of the vehicle body (m_v) and the mass of the rider (m_r) are shown as a single mass, $m_v + m_r$. When the elasticity (as spring constant k_r) and damping (as damping constant c_r) of the rider are considered, the refined model shown in Fig. 1.17(d) can be obtained.

Note that the models shown in Figs. 1.17(b) to (d) are not unique. For example, by combining the spring constants of both tires, the masses of both wheels, and the spring and damping constants of both struts as single quantities, the model shown in Fig. 1.17(e) can be obtained instead of Fig. 1.17(c).

■

1.7   Spring Elements

A spring is a type of mechanical link, which in most applications is assumed to have negligible mass and damping. The most common type of spring is the helical-coil spring used in retractable pens and pencils, staplers, and suspensions of freight trucks and other vehicles. Several other types of springs can be identified in engineering applications. In fact, any elastic or deformable body or member, such as a cable, bar, beam, shaft, or plate, can be considered as a spring. A spring is commonly represented as shown in Fig. 1.18(a). If the free length of the spring, with no forces acting, is denoted l, it undergoes a change in length when an axial force is applied. For example, when a tensile force F is applied at its free end 2, the spring undergoes an elongation x as shown in Fig. 1.18(b), while a compressive force F applied at the free end 2 causes a reduction in length x as shown in Fig. 1.18(c).

A spring is said to be linear if the elongation or reduction in length x is related to the applied force F as

$$F = kx \qquad (1.1)$$

where k is a constant, known as the *spring constant*, *spring stiffness*, or *spring rate*. The spring constant k is always positive and denotes the force (positive or negative) required to

(a)   (b)   (c)

FIGURE 1.18   Deformation of a spring.

cause a unit deflection (elongation or reduction in length) in the spring. When the spring is stretched (or compressed) under a tensile (or compressive) force F, according to Newton's third law of motion, a restoring force or reaction of magnitude $-F$ (or $+F$) is developed opposite to the applied force. This restoring force tries to bring the stretched (or compressed) spring back to its original unstretched or free length as shown in Fig. 1.18(b) (or 1.18(c)). If we plot a graph between F and x, the result is a straight line according to Eq. (1.1). The work done (U) in deforming a spring is stored as strain or potential energy in the spring, and it is given by

$$U = \frac{1}{2}kx^2 \tag{1.2}$$

1.7.1 Nonlinear Springs

Most springs used in practical systems exhibit a nonlinear force-deflection relation, particularly when the deflections are large. If a nonlinear spring undergoes small deflections, it can be replaced by a linear spring by using the procedure discussed in Section 1.7.2. In vibration analysis, nonlinear springs whose force-deflection relations are given by

$$F = ax + bx^3; \qquad a > 0 \tag{1.3}$$

are commonly used. In Eq. (1.3), a denotes the constant associated with the linear part and b indicates the constant associated with the (cubic) nonlinearity. The spring is said to be hard if $b > 0$, linear if $b = 0$, and soft if $b < 0$. The force-deflection relations for various values of b are shown in Fig. 1.19.

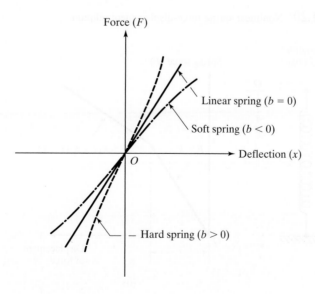

FIGURE 1.19   Nonlinear and linear springs.

Some systems, involving two or more springs, may exhibit a nonlinear force-displacement relationship although the individual springs are linear. Some examples of such systems are shown in Figs. 1.20 and 1.21. In Fig. 1.20(a), the weight (or force) W travels

FIGURE 1.20   Nonlinear spring force-displacement relation.

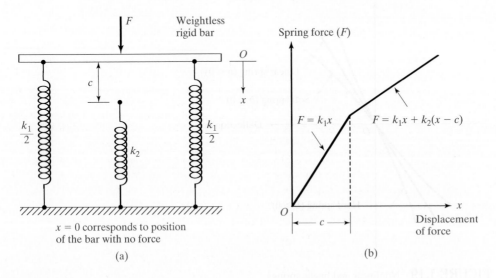

FIGURE 1.21   Nonlinear spring force-displacement relation.

freely through the clearances c_1 and c_2 present in the system. Once the weight comes into contact with a particular spring, after passing through the corresponding clearance, the spring force increases in proportion to the spring constant of the particular spring (see Fig. 1.20(b)). It can be seen that the resulting force-displacement relation, although piece-wise linear, denotes a nonlinear relationship.

In Fig. 1.21(a), the two springs, with stiffnesses k_1 and k_2, have different lengths. Note that the spring with stiffness k_1 is shown, for simplicity, in the form of two parallel springs, each with a stiffness of $k_1/2$. Spring arrangement models of this type can be used in the vibration analysis of packages and suspensions used in aircraft landing gears.

When the spring k_1 deflects by an amount $x = c$, the second spring starts providing an additional stiffness k_2 to the system. The resulting nonlinear force-displacement relationship is shown in Fig. 1.21(b).

1.7.2 Linearization of a Nonlinear Spring

Actual springs are nonlinear and follow Eq. (1.1) only up to a certain deformation. Beyond a certain value of deformation (after point A in Fig. 1.22), the stress exceeds the yield point of the material and the force-deformation relation becomes nonlinear [1.23, 1.24]. In many practical applications, we assume that the deflections are small and make use of the linear relation in Eq. (1.1). Even if the force-deflection relation of a spring is nonlinear, as shown in Fig. 1.23, we often approximate it as a linear one by using a linearization process [1.24, 1.25]. To illustrate the linearization process, let the static equilibrium load F acting on the spring cause a deflection of x^*. If an incremental force ΔF is added to F, the spring deflects by an additional quantity Δx. The new spring force $F + \Delta F$ can be expressed using Taylor's series expansion about the static equilibrium position x^* as

$$F + \Delta F = F(x^* + \Delta x)$$

$$= F(x^*) + \left.\frac{dF}{dx}\right|_{x^*} (\Delta x) + \frac{1}{2!} \left.\frac{d^2F}{dx^2}\right|_{x^*} (\Delta x)^2 + \dots \tag{1.4}$$

For small values of Δx, the higher-order derivative terms can be neglected to obtain

$$F + \Delta F = F(x^*) + \left.\frac{dF}{dx}\right|_{x^*} (\Delta x) \tag{1.5}$$

FIGURE 1.22 Nonlinearity beyond proportionality limit.

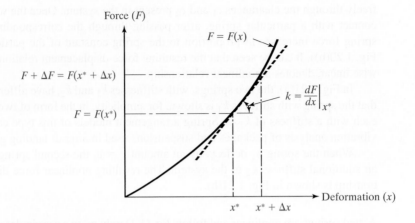

FIGURE 1.23   Linearization process.

Since $F = F(x^*)$, we can express ΔF as

$$\Delta F = k \, \Delta x \tag{1.6}$$

where k is the linearized spring constant at x^* given by

$$k = \left. \frac{dF}{dx} \right|_{x^*} \tag{1.7}$$

We may use Eq. (1.6) for simplicity, but sometimes the error involved in the approximation may be very large.

EXAMPLE 1.2

Equivalent Linearized Spring Constant

A precision milling machine, weighing 4,500 N, is supported on a rubber mount. The force-deflection relationship of the rubber mount is given by

$$F = 3{,}500x + 55x^3 \tag{E.1}$$

where the force (F) and the deflection (x) are measured in Newtons and centimeters, respectively. Determine the equivalent linearized spring constant of the rubber mount at its static equilibrium position.

Solution: The static equilibrium position of the rubber mount (x^*), under the weight of the milling machine, can be determined from Eq. (E.1):

$$4{,}500 = 3{,}500x^* + 55(x^*)^3$$

or

$$55(x^*)^3 + 3{,}500x^* - 4{,}500 = 0 \tag{E.2}$$

The roots of the cubic equation, (E.2), can be found (for example, using the function `roots` in MATLAB) as

$$x^* = 1.2547, \quad -0.6273 + 8.0509i, \quad \text{and} \quad -0.6273 - 8.0509i$$

The static equilibrium position of the rubber mount is given by the real root of Eq. (E.2): $x^* = 1.2547$ cm. The equivalent linear spring constant of the rubber mount at its static equilibrium position can be determined using Eq. (1.7):

$$k_{eq} = \left. \frac{dF}{dx} \right|_{x^*} = 3{,}500 + 165(x^*)^2 = 3{,}500 + 165(1.2547^2) = 3{,}760 \text{ N/cm}$$

Note: The equivalent linear spring constant, $k_{eq} = 3{,}760$ N/cm, predicts the static deflection of the milling machine as

$$x = \frac{F}{k_{eq}} = \frac{4{,}500}{3{,}760} = 1.1968 \text{ cm}$$

which is slightly different from the true value of 1.2547 cm. The error is due to the truncation of the higher-order derivative terms in Eq. (1.4).

■

1.7.3
Spring
Constants
of Elastic
Elements

As stated earlier, any elastic or deformable member (or element) can be considered as a spring. The equivalent spring constants of simple elastic members such as rods, beams, and hollow shafts are given on the inside front cover of the book. The procedure of finding the equivalent spring constant of elastic members is illustrated through the following examples.

EXAMPLE 1.3

Spring Constant of a Rod

Find the equivalent spring constant of a uniform rod of length l, cross-sectional area A, and Young's modulus E subjected to an axial tensile (or compressive) force F as shown in Fig. 1.24(a).

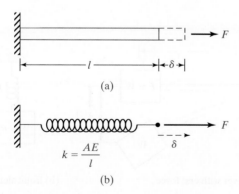

(a)

(b)

$$k = \frac{AE}{l}$$

FIGURE 1.24   Spring constant of a rod.

Solution: The elongation (or shortening) δ of the rod under the axial tensile (or compressive) force F can be expressed as

$$\delta = \frac{\delta}{l} l = \varepsilon l = \frac{\sigma}{E} l = \frac{Fl}{AE} \qquad (E.1)$$

where $\varepsilon = \dfrac{change\ in\ length}{original\ length} = \dfrac{\delta}{l}$ is the strain and $\sigma = \dfrac{force}{area} = \dfrac{F}{A}$ is the stress induced in the rod.
Using the definition of the spring constant k, we obtain from Eq. (E.1):

$$k = \frac{force\ applied}{resulting\ deflection} = \frac{F}{\delta} = \frac{AE}{l} \qquad (E.2)$$

The significance of the equivalent spring constant of the rod is shown in Fig. 1.24(b).

■

EXAMPLE 1.4

Spring Constant of a Cantilever Beam

Find the equivalent spring constant of a cantilever beam subjected to a concentrated load F at its end as shown in Fig. 1.25(a).

Solution: We assume, for simplicity, that the self weight (or mass) of the beam is negligible and the concentrated load F is due to the weight of a point mass ($W = mg$). From the strength of materials [1.26], we know that the end deflection of the beam due to a concentrated load $F = W$ is given by

$$\delta = \frac{Wl^3}{3EI} \qquad (E.1)$$

where E is Young's modulus and I is the moment of inertia of the cross section of the beam about the bending or z-axis (i.e., axis perpendicular to the page). Hence the spring constant of the beam is (Fig. 1.25(b)):

$$k = \frac{W}{\delta} = \frac{3EI}{l^3} \qquad (E.2)$$

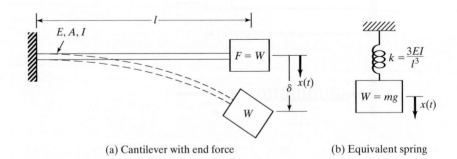

(a) Cantilever with end force   (b) Equivalent spring

FIGURE 1.25   Spring constant of a cantilever beam.

Notes:

1. It is possible for a cantilever beam to be subjected to concentrated loads in two directions at its end—one in the y direction (F_y) and the other in the z direction (F_z)—as shown in Fig. 1.26(a). When the load is applied along the y direction, the beam bends about the z-axis (Fig. 1.26(b)) and hence the equivalent spring constant will be equal to

$$k = \frac{3EI_{zz}}{l^3} \qquad (E.3)$$

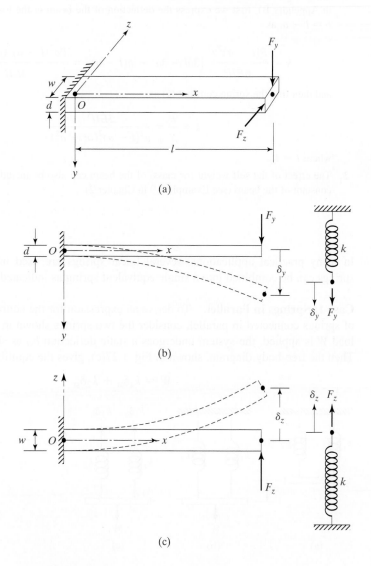

FIGURE 1.26 Spring constants of a beam in two directions.

When the load is applied along the z direction, the beam bends about the y-axis (Fig. 1.26(c)) and hence the equivalent spring constant will be equal to

$$k = \frac{3EI_{yy}}{l^3} \tag{E.4}$$

2. The spring constants of beams with different end conditions can be found in a similar manner using results from the strength of materials. The representative formulas given in Appendix B can be used to find the spring constants of the indicated beams and plates. For example, to find the spring constant of a fixed-fixed beam subjected to a concentrated force P at $x = a$ (Case 3 in Appendix B), first we express the deflection of the beam at the load point $(x = a)$, using $b = l - a$, as

$$y = \frac{P(l - a)^2 a^2}{6EIl^3}[3al - 3a^2 - a(l - a)] = \frac{Pa^2(l - a)^2(al - a^2)}{3EIl^3} \tag{E.5}$$

and then find the spring constant (k) as

$$k = \frac{P}{y} = \frac{3EIl^3}{a^2(l - a)^2(al - a^2)} \tag{E.6}$$

where $I = I_{zz}$.

3. The effect of the self weight (or mass) of the beam can also be included in finding the spring constant of the beam (see Example 2.9 in Chapter 2).

■

1.7.4 Combination of Springs

In many practical applications, several linear springs are used in combination. These springs can be combined into a single equivalent spring as indicated below.

Case 1: Springs in Parallel.   To derive an expression for the equivalent spring constant of springs connected in parallel, consider the two springs shown in Fig. 1.27(a). When a load W is applied, the system undergoes a static deflection δ_{st} as shown in Fig. 1.27(b). Then the free-body diagram, shown in Fig. 1.27(c), gives the equilibrium equation

$$W = k_1 \delta_{st} + k_2 \delta_{st} \tag{1.8}$$

FIGURE 1.27   Springs in parallel.

If k_{eq} denotes the equivalent spring constant of the combination of the two springs, then for the same static deflection δ_{st}, we have

$$W = k_{eq}\delta_{st} \qquad (1.9)$$

Equations (1.8) and (1.9) give

$$k_{eq} = k_1 + k_2 \qquad (1.10)$$

In general, if we have n springs with spring constants $k_1, k_2, \ldots, k_n$ in parallel, then the equivalent spring constant k_{eq} can be obtained:

$$k_{eq} = k_1 + k_2 + \cdots + k_n \qquad (1.11)$$

Case 2: Springs in Series.   Next we derive an expression for the equivalent spring constant of springs connected in series by considering the two springs shown in Fig. 1.28(a). Under the action of a load W, springs 1 and 2 undergo elongations δ_1 and δ_2, respectively, as shown in Fig. 1.28(b). The total elongation (or static deflection) of the system, δ_{st}, is given by

$$\delta_{st} = \delta_1 + \delta_2 \qquad (1.12)$$

Since both springs are subjected to the same force W, we have the equilibrium shown in Fig. 1.28(c):

$$W = k_1\delta_1$$
$$W = k_2\delta_2 \qquad (1.13)$$

If k_{eq} denotes the equivalent spring constant, then for the same static deflection,

$$W = k_{eq}\delta_{st} \qquad (1.14)$$

FIGURE 1.28   Springs in series.

Equations (1.13) and (1.14) give

$$k_1\delta_1 = k_2\delta_2 = k_{eq}\delta_{st}$$

or

$$\delta_1 = \frac{k_{eq}\delta_{st}}{k_1} \quad \text{and} \quad \delta_2 = \frac{k_{eq}\delta_{st}}{k_2} \tag{1.15}$$

Substituting these values of δ_1 and δ_2 into Eq. (1.12), we obtain

$$\frac{k_{eq}\delta_{st}}{k_1} + \frac{k_{eq}\delta_{st}}{k_2} = \delta_{st}$$

—that is,

$$\frac{1}{k_{eq}} = \frac{1}{k_1} + \frac{1}{k_2} \tag{1.16}$$

Equation (1.16) can be generalized to the case of n springs in series:

$$\frac{1}{k_{eq}} = \frac{1}{k_1} + \frac{1}{k_2} + \cdots + \frac{1}{k_n} \tag{1.17}$$

In certain applications, springs are connected to rigid components such as pulleys, levers, and gears. In such cases, an equivalent spring constant can be found using energy equivalence, as illustrated in Examples 1.8 and 1.9.

EXAMPLE 1.5

Equivalent k of a Suspension System

Figure 1.29 shows the suspension system of a freight truck with a parallel-spring arrangement. Find the equivalent spring constant of the suspension if each of the three helical springs is made of steel with a shear modulus $G = 80 \times 10^9 \, \text{N/m}^2$ and has five effective turns, mean coil diameter $D = 20 \, \text{cm}$, and wire diameter $d = 2 \, \text{cm}$.

Solution: The stiffness of each helical spring is given by

$$k = \frac{d^4 G}{8D^3 n} = \frac{(0.02)^4 (80 \times 10^9)}{8(0.2)^3(5)} = 40,000.0 \, \text{N/m}$$

(See inside front cover for the formula.)

Since the three springs are identical and parallel, the equivalent spring constant of the suspension system is given by

$$k_{eq} = 3k = 3(40,000.0) = 120,000.0 \, \text{N/m}$$

FIGURE 1.29   Parallel arrangement of springs in a freight truck. (Courtesy of Buckeye Steel Castings Company.)

■

Torsional Spring Constant of a Propeller Shaft

EXAMPLE 1.6

Determine the torsional spring constant of the steel propeller shaft shown in Fig. 1.30.

Solution: We need to consider the segments 12 and 23 of the shaft as springs in combination. From Fig. 1.30, the torque induced at any cross section of the shaft (such as *AA* or *BB*) can be seen to be

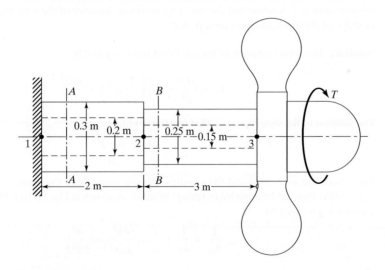

FIGURE 1.30   Propeller shaft.

equal to the torque applied at the propeller, T. Hence the elasticities (springs) corresponding to the two segments 12 and 23 are to be considered as series springs. The spring constants of segments 12 and 23 of the shaft ($k_{t_{12}}$ and $k_{t_{23}}$) are given by

$$k_{t_{12}} = \frac{GJ_{12}}{l_{12}} = \frac{G\pi(D_{12}^4 - d_{12}^4)}{32 l_{12}} = \frac{(80 \times 10^9)\pi(0.3^4 - 0.2^4)}{32(2)}$$

$$= 25.5255 \times 10^6 \text{ N-m/rad}$$

$$k_{t_{23}} = \frac{GJ_{23}}{l_{23}} = \frac{G\pi(D_{23}^4 - d_{23}^4)}{32 l_{23}} = \frac{(80 \times 10^9)\pi(0.25^4 - 0.15^4)}{32(3)}$$

$$= 8.9012 \times 10^6 \text{ N-m/rad}$$

Since the springs are in series, Eq. (1.16) gives

$$k_{t_{eq}} = \frac{k_{t_{12}} k_{t_{23}}}{k_{t_{12}} + k_{t_{23}}} = \frac{(25.5255 \times 10^6)(8.9012 \times 10^6)}{(25.5255 \times 10^6 + 8.9012 \times 10^6)} = 6.5997 \times 10^6 \text{ N-m/rad}$$

∎

EXAMPLE 1.7

Equivalent k of Hoisting Drum

A hoisting drum, carrying a steel wire rope, is mounted at the end of a cantilever beam as shown in Fig. 1.31(a). Determine the equivalent spring constant of the system when the suspended length of the wire rope is l. Assume that the net cross-sectional diameter of the wire rope is d and Young's modulus of the beam and the wire rope is E.

Solution: The spring constant of the cantilever beam is given by

$$k_b = \frac{3EI}{b^3} = \frac{3E}{b^3}\left(\frac{1}{12} at^3\right) = \frac{Eat^3}{4b^3} \tag{E.1}$$

The stiffness of the wire rope subjected to axial loading is

$$k_r = \frac{AE}{l} = \frac{\pi d^2 E}{4l} \tag{E.2}$$

Since both the wire rope and the cantilever beam experience the same load W, as shown in Fig. 1.31(b), they can be modeled as springs in series, as shown in Fig. 1.31(c). The equivalent spring constant k_{eq} is given by

$$\frac{1}{k_{eq}} = \frac{1}{k_b} + \frac{1}{k_r} = \frac{4b^3}{Eat^3} + \frac{4l}{\pi d^2 E}$$

or

$$k_{eq} = \frac{E}{4}\left(\frac{\pi a t^3 d^2}{\pi d^2 b^3 + lat^3}\right) \tag{E.3}$$

FIGURE 1.31 Hoisting drum.

EXAMPLE 1.8

Equivalent k of a Crane

The boom AB of the crane shown in Fig. 1.32(a) is a uniform steel bar of length 10 m and area of cross section 2,500 mm^2. A weight W is suspended while the crane is stationary. The cable $CDEBF$ is made of steel and has a cross-sectional area of 100 mm^2. Neglecting the effect of the cable $CDEB$, find the equivalent spring constant of the system in the vertical direction.

Solution: The equivalent spring constant can be found using the equivalence of potential energies of the two systems. Since the base of the crane is rigid, the cable and the boom can be considered to be fixed at points F and A, respectively. Also, the effect of the cable $CDEB$ is negligible; hence the weight W can be assumed to act through point B as shown in Fig. 1.32(b).

FIGURE 1.32   Crane lifting a load.

A vertical displacement x of point B will cause the spring (boom) to deform by an amount and the spring (cable) to deform by an amount. The length of the cable FB, l_1, is given by Fig. 1.32(b):

$$l_1^2 = 3^2 + 10^2 - 2(3)(10) \cos 135° = 151.426, \quad l_1 = 12.3055 \text{ m}$$

The angle θ satisfies the relation

$$l_1^2 + 3^2 - 2(l_1)(3) \cos \theta = 10^2, \quad \cos \theta = 0.8184, \quad \theta = 35.0736°$$

The total potential energy (U) stored in the springs k_1 and k_2 can be expressed, using Eq. (1.2) as

$$U = \frac{1}{2} k_1 [x \cos (90° - \theta)]^2 + \frac{1}{2} k_2 [x \cos (90° - 45°)]^2 \qquad (E.1)$$

where

$$k_1 = \frac{A_1 E_1}{l_1} = \frac{(100 \times 10^{-6})(207 \times 10^9)}{12.3055} = 1.6822 \times 10^6 \text{ N/m}$$

and

$$k_2 = \frac{A_2 E_2}{l_2} = \frac{(2{,}500 \times 10^{-6})(207 \times 10^9)}{10} = 5.1750 \times 10^7 \text{ N/m}$$

Since the equivalent spring in the vertical direction undergoes a deformation x, the potential energy of the equivalent spring (U_{eq}) is given by

$$U_{eq} = \tfrac{1}{2} k_{eq} x^2 \tag{E.2}$$

By setting $U = U_{eq}$, we obtain the equivalent spring constant of the system as

$$k_{eq} = k_1 \sin^2 \theta + k_2 \sin^2 45° = k_1 \sin^2 35.0736° + k_2 \sin^2 45° = 26.4304 \times 10^6 \text{ N/m}$$

∎

EXAMPLE 1.9

Equivalent k of a Rigid Bar Connected by Springs

A hinged rigid bar of length l is connected by two springs of stiffnesses k_1 and k_2 and is subjected to a force F as shown in Fig. 1.33(a). Assuming that the angular displacement of the bar (θ) is small, find the equivalent spring constant of the system that relates the applied force F to the resulting displacement x.

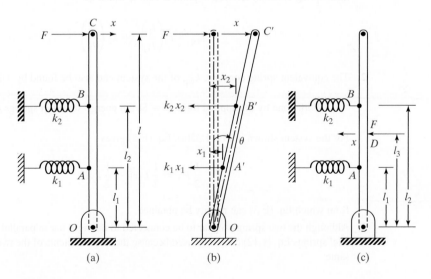

FIGURE 1.33 Rigid bar connected by springs.

Solution: For a small angular displacement of the rigid bar (θ), the points of attachment of springs k_1 and k_2 (A and B) and the point of application (C) of the force F undergo the linear or horizontal displacements $l_1 \sin \theta$, $l_2 \sin \theta$, and $l \sin \theta$, respectively. Since θ is small, the horizontal displacements of points A, B, and C can be approximated as $x_1 = l_1\theta$, $x_2 = l_2\theta$, and $x = l\theta$, respectively. The reactions of the springs, $k_1 x_1$ and $k_2 x_2$, will be as indicated in Fig. 1.33(b). The equivalent spring constant of the system (k_{eq}) referred to the point of application of the force F can be determined by considering the moment equilibrium of the forces about the hinge point O:

$$k_1 x_1 (l_1) + k_2 x_2 (l_2) = F(l)$$

or

$$F = k_1 \left(\frac{x_1 l_1}{l} \right) + k_2 \left(\frac{x_2 l_2}{l} \right) \tag{E.1}$$

By expressing F as $k_{eq}x$, Eq. (E.1) can be written as

$$F = k_{eq}x = k_1 \left(\frac{x_1 l_1}{l} \right) + k_2 \left(\frac{x_2 l_2}{l} \right) \tag{E.2}$$

Using $x_1 = l_1\theta$, $x_2 = l_2\theta$, and $x = l\theta$, Eq. (E.2) yields the desired result:

$$k_{eq} = k_1 \left(\frac{l_1}{l} \right)^2 + k_2 \left(\frac{l_2}{l} \right)^2 \tag{E.3}$$

Notes:

1. If the force F is applied at another point D of the rigid bar as shown in Fig. 1.33(c), the equivalent spring constant referred to point D can be found as

$$k_{eq} = k_1 \left(\frac{l_1}{l_3} \right)^2 + k_2 \left(\frac{l_2}{l_3} \right)^2 \tag{E.4}$$

2. The equivalent spring constant, k_{eq}, of the system can also be found by using the relation:

Work done by the applied force F = Strain energy stored in springs k_1 and k_2 (E.5)

For the system shown in Fig. 1.33(a), Eq. (E.5) gives

$$\frac{1}{2}Fx = \frac{1}{2}k_1 x_1^2 + \frac{1}{2}k_2 x_2^2 \tag{E.6}$$

from which Eq. (E.3) can readily be obtained.

3. Although the two springs appear to be connected to the rigid bar in parallel, the formula of parallel springs Eq. (1.12) cannot be used because the displacements of the two springs are not the same.

■

1.7.5
Spring Constant Associated with the Restoring Force due to Gravity

In some applications, a restoring force or moment due to gravity is developed when a mass undergoes a displacement. In such cases, an equivalent spring constant can be associated with the restoring force or moment of gravity. The following example illustrates the procedure.

EXAMPLE 1.10

Spring Constant Associated with Restoring Force due to Gravity

Figure 1.34 shows a simple pendulum of length l with a bob of mass m. Considering an angular displacement θ of the pendulum, determine the equivalent spring constant associated with the restoring force (or moment).

Solution: When the pendulum undergoes an angular displacement θ, the mass m moves by a distance $l \sin \theta$ along the horizontal (x) direction. The restoring moment or torque (T) created by the weight of the mass (mg) about the pivot point O is given by

$$T = mg(l \sin \theta) \tag{E.1}$$

For small angular displacements θ, $\sin \theta$ can be approximated as $\sin \theta \approx \theta$ (see Appendix A) and Eq. (E.1) becomes

$$T = mgl\theta \tag{E.2}$$

By expressing Eq. (E.2) as

$$T = k_t\theta \tag{E.3}$$

the desired equivalent torsional spring constant k_t can be identified as

$$k_t = mgl \tag{E.4}$$

FIGURE 1.34 Simple pendulum.

∎

1.8   Mass or Inertia Elements

The mass or inertia element is assumed to be a rigid body; it can gain or lose kinetic energy whenever the velocity of the body changes. From Newton's second law of motion, the product of the mass and its acceleration is equal to the force applied to the mass. Work is equal to the force multiplied by the displacement in the direction of the force, and the work done on a mass is stored in the form of the mass's kinetic energy.

In most cases, we must use a mathematical model to represent the actual vibrating system, and there are often several possible models. The purpose of the analysis often determines which mathematical model is appropriate. Once the model is chosen, the mass or inertia elements of the system can be easily identified. For example, consider again the cantilever beam with an end mass shown in Fig. 1.25(a). For a quick and reasonably accurate analysis, the mass and damping of the beam can be disregarded; the system can be modeled as a spring-mass system, as shown in Fig. 1.25(b). The tip mass m represents the mass element, and the elasticity of the beam denotes the stiffness of the spring. Next, consider a multistory building subjected to an earthquake. Assuming that the mass of the frame is negligible compared to the masses of the floors, the building can be modeled as a multi-degree-of-freedom system, as shown in Fig. 1.35. The masses at the various floor levels represent the mass elements, and the elasticities of the vertical members denote the spring elements.

1.8.1 Combination of Masses

In many practical applications, several masses appear in combination. For a simple analysis, we can replace these masses by a single equivalent mass, as indicated below [1.27].

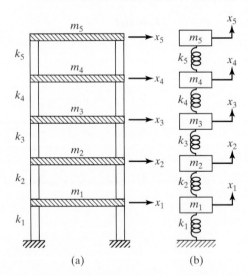

(a)   (b)

FIGURE 1.35 Idealization of a multistory building as a multi-degree-of-freedom system.

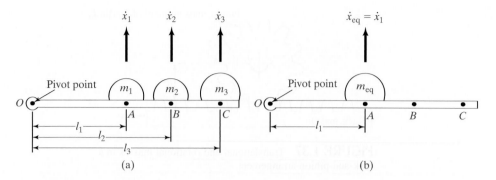

FIGURE 1.36   Translational masses connected by a rigid bar.

Case 1: Translational Masses Connected by a Rigid Bar.   Let the masses be attached to a rigid bar that is pivoted at one end, as shown in Fig. 1.36(a). The equivalent mass can be assumed to be located at any point along the bar. To be specific, we assume the location of the equivalent mass to be that of mass m_1. The velocities of masses $m_2(\dot{x}_2)$ and $m_3(\dot{x}_3)$ can be expressed in terms of the velocity of mass $m_1(\dot{x}_1)$, by assuming small angular displacements for the bar, as

$$\dot{x}_2 = \frac{l_2}{l_1}\dot{x}_1, \quad \dot{x}_3 = \frac{l_3}{l_1}\dot{x}_1 \tag{1.18}$$

and

$$\dot{x}_{eq} = \dot{x}_1 \tag{1.19}$$

By equating the kinetic energy of the three-mass system to that of the equivalent mass system, we obtain

$$\frac{1}{2}m_1\dot{x}_1^2 + \frac{1}{2}m_2\dot{x}_2^2 + \frac{1}{2}m_3\dot{x}_3^2 = \frac{1}{2}m_{eq}\dot{x}_{eq}^2 \tag{1.20}$$

This equation gives, in view of Eqs. (1.18) and (1.19):

$$m_{eq} = m_1 + \left(\frac{l_2}{l_1}\right)^2 m_2 + \left(\frac{l_3}{l_1}\right)^2 m_3 \tag{1.21}$$

It can be seen that the equivalent mass of a system composed of several masses (each moving at a different velocity) can be thought of as the imaginary mass which, while moving with a specified velocity v, will have the same kinetic energy as that of the system.

Case 2: Translational and Rotational Masses Coupled Together.   Let a mass m having a translational velocity $\dot{x}$ be coupled to another mass (of mass moment of inertia J_0) having a rotational velocity $\dot{\theta}$, as in the rack-and-pinion arrangement shown in Fig. 1.37.

FIGURE 1.37 Translational and rotational masses in a rack-and-pinion arrangement.

These two masses can be combined to obtain either (1) a single equivalent translational mass m_{eq} or (2) a single equivalent rotational mass J_{eq}, as shown below.

1. *Equivalent translational mass.* The kinetic energy of the two masses is given by

$$T = \frac{1}{2}m\dot{x}^2 + \frac{1}{2}J_0\dot{\theta}^2 \tag{1.22}$$

and the kinetic energy of the equivalent mass can be expressed as

$$T_{eq} = \frac{1}{2}m_{eq}\dot{x}_{eq}^2 \tag{1.23}$$

Since $\dot{x}_{eq} = \dot{x}$ and $\dot{\theta} = \dot{x}/R$, the equivalence of T and T_{eq} gives

$$\frac{1}{2}m_{eq}\dot{x}^2 = \frac{1}{2}m\dot{x}^2 + \frac{1}{2}J_0\left(\frac{\dot{x}}{R}\right)^2$$

—that is,

$$m_{eq} = m + \frac{J_0}{R^2} \tag{1.24}$$

2. *Equivalent rotational mass.* Here $\dot{\theta}_{eq} = \dot{\theta}$ and $\dot{x} = \dot{\theta}R$, and the equivalence of T and T_{eq} leads to

$$\frac{1}{2}J_{eq}\dot{\theta}^2 = \frac{1}{2}m(\dot{\theta}R)^2 + \frac{1}{2}J_0\dot{\theta}^2$$

or

$$J_{eq} = J_0 + mR^2 \tag{1.25}$$

Equivalent Mass of a System

EXAMPLE 1.11

Find the equivalent mass of the system shown in Fig. 1.38, where the rigid link 1 is attached to the pulley and rotates with it.

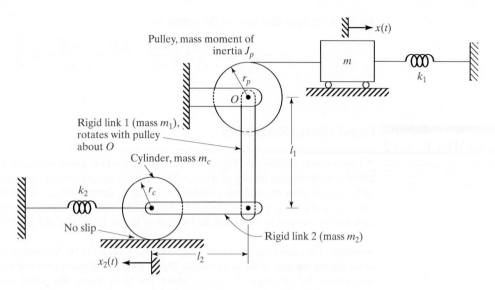

FIGURE 1.38 System considered for finding equivalent mass.

Solution: Assuming small displacements, the equivalent mass (m_{eq}) can be determined using the equivalence of the kinetic energies of the two systems. When the mass m is displaced by a distance x, the pulley and the rigid link 1 rotate by an angle $\theta_p = \theta_1 = x/r_p$. This causes the rigid link 2 and the cylinder to be displaced by a distance $x_2 = \theta_p l_1 = x l_1/r_p$. Since the cylinder rolls without slippage, it rotates by an angle $\theta_c = x_2/r_c = x l_1/r_p r_c$. The kinetic energy of the system (T) can be expressed (for small displacements) as:

$$T = \frac{1}{2}m\dot{x}^2 + \frac{1}{2}J_p\dot{\theta}_p^2 + \frac{1}{2}J_1\dot{\theta}_1^2 + \frac{1}{2}m_2\dot{x}_2^2 + \frac{1}{2}J_c\dot{\theta}_c^2 + \frac{1}{2}m_c\dot{x}_2^2 \tag{E.1}$$

where J_p, J_1, and J_c denote the mass moments of inertia of the pulley, link 1 (about O), and cylinder, respectively, $\dot{\theta}_p$, $\dot{\theta}_1$, and $\dot{\theta}_c$ indicate the angular velocities of the pulley, link 1 (about O), and cylinder, respectively, and $\dot{x}$ and $\dot{x}_2$ represent the linear velocities of the mass m and link 2, respectively. Noting that $J_c = m_c r_c^2/2$ and $J_1 = m_1 l_1^2/3$, Eq. (E.1) can be rewritten as

$$T = \frac{1}{2}m\dot{x}^2 + \frac{1}{2}J_p\left(\frac{\dot{x}}{r_p}\right)^2 + \frac{1}{2}\left(\frac{m_1 l_1^2}{3}\right)\left(\frac{\dot{x}}{r_p}\right)^2 + \frac{1}{2}m_2\left(\frac{\dot{x}l_1}{r_p}\right)^2$$

$$+ \frac{1}{2}\left(\frac{m_c r_c^2}{2}\right)\left(\frac{\dot{x}l_1}{r_p r_c}\right)^2 + \frac{1}{2}m_c\left(\frac{\dot{x}l_1}{r_p}\right)^2 \tag{E.2}$$

By equating Eq. (E.2) to the kinetic energy of the equivalent system

$$T = \frac{1}{2}m_{eq}\dot{x}^2 \tag{E.3}$$

we obtain the equivalent mass of the system as

$$m_{eq} = m + \frac{J_p}{r_p^2} + \frac{1}{3}\frac{m_1 l_1^2}{r_p^2} + \frac{m_2 l_1^2}{r_p^2} + \frac{1}{2}\frac{m_c l_1^2}{r_p^2} + m_c \frac{l_1^2}{r_p^2}$$ (E.4)

∎

Cam-Follower Mechanism

EXAMPLE 1.12

A cam-follower mechanism (Fig. 1.39) is used to convert the rotary motion of a shaft into the oscillating or reciprocating motion of a valve. The follower system consists of a pushrod of mass m_p, a rocker arm of mass m_r, and mass moment of inertia J_r about its center of gravity (C.G.), a valve of mass m_r, and a valve spring of negligible mass [1.28–1.30]. Find the equivalent mass (m_{eq}) of this cam-follower system by assuming the location of m_{eq} as (i) point A and (ii) point C.

Solution: The equivalent mass of the cam-follower system can be determined using the equivalence of the kinetic energies of the two systems. Due to a vertical displacement x of the pushrod, the rocker arm rotates by an angle $\theta_r = x/l_1$ about the pivot point, the valve moves downward by $x_v = \theta_r l_2 = xl_2/l_1$, and the C.G. of the rocker arm moves downward by $x_r = \theta_r l_3 = xl_3/l_1$. The kinetic energy of the system (T) can be expressed as[2]

$$T = \frac{1}{2}m_p \dot{x}_p^2 + \frac{1}{2}m_v \dot{x}_v^2 + \frac{1}{2}J_r \dot{\theta}_r^2 + \frac{1}{2}m_r \dot{x}_r^2$$ (E.1)

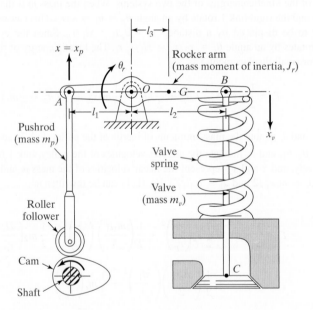

FIGURE 1.39 Cam-follower system.

---

[2]If the valve spring has a mass m_s, then its equivalent mass will be $\frac{1}{3}m_s$ (see Example 2.8). Thus the kinetic energy of the valve spring will be $\frac{1}{2}\left(\frac{1}{3}m_s\right)\dot{x}_v^2$.

where $\dot{x}_p$, $\dot{x}_r$, and $\dot{x}_v$ are the linear velocities of the pushrod, C.G. of the rocker arm, and the valve, respectively, and $\dot{\theta}_r$ is the angular velocity of the rocker arm.

(i) If m_{eq} denotes the equivalent mass placed at point A, with $\dot{x}_{eq} = \dot{x}$, the kinetic energy of the equivalent mass system T_{eq} is given by

$$T_{eq} = \frac{1}{2} m_{eq} \dot{x}_{eq}^2 \tag{E.2}$$

By equating T and T_{eq}, and noting that

$$\dot{x}_p = \dot{x}, \quad \dot{x}_v = \frac{\dot{x}l_2}{l_1}, \quad \dot{x}_r = \frac{\dot{x}l_3}{l_1}, \quad \text{and} \quad \dot{\theta}_r = \frac{\dot{x}}{l_1}$$

we obtain

$$m_{eq} = m_p + \frac{J_r}{l_1^2} + m_v \frac{l_2^2}{l_1^2} + m_r \frac{l_3^2}{l_1^2} \tag{E.3}$$

(ii) Similarly, if the equivalent mass is located at point C, $\dot{x}_{eq} = \dot{x}_v$ and

$$T_{eq} = \frac{1}{2} m_{eq} \dot{x}_{eq}^2 = \frac{1}{2} m_{eq} \dot{x}_v^2 \tag{E.4}$$

Equating (E.4) and (E.1) gives

$$m_{eq} = m_v + \frac{J_r}{l_2^2} + m_p \left(\frac{l_1}{l_2}\right)^2 + m_r \left(\frac{l_3}{l_2}\right)^2 \tag{E.5}$$

∎

1.9   Damping Elements

In many practical systems, the vibrational energy is gradually converted to heat or sound. Due to the reduction in the energy, the response, such as the displacement of the system, gradually decreases. The mechanism by which the vibrational energy is gradually converted into heat or sound is known as *damping*. Although the amount of energy converted into heat or sound is relatively small, the consideration of damping becomes important for an accurate prediction of the vibration response of a system. A damper is assumed to have neither mass nor elasticity, and damping force exists only if there is relative velocity between the two ends of the damper. It is difficult to determine the causes of damping in practical systems. Hence damping is modeled as one or more of the following types.

Viscous Damping.   Viscous damping is the most commonly used damping mechanism in vibration analysis. When mechanical systems vibrate in a fluid medium such as air, gas, water, or oil, the resistance offered by the fluid to the moving body causes energy to be dissipated. In this case, the amount of dissipated energy depends on many factors, such as the size and shape of the vibrating body, the viscosity of the fluid, the frequency of vibration, and the velocity of the vibrating body. In viscous damping, the damping force is proportional to the velocity of the vibrating body. Typical examples of viscous damping include (1) fluid film between sliding surfaces, (2) fluid flow around a piston in a cylinder, (3) fluid flow through an orifice, and (4) fluid film around a journal in a bearing.

Coulomb or Dry-Friction Damping.   Here the damping force is constant in magnitude but opposite in direction to that of the motion of the vibrating body. It is caused by friction between rubbing surfaces that either are dry or have insufficient lubrication.

Material or Solid or Hysteretic Damping.   When a material is deformed, energy is absorbed and dissipated by the material [1.31]. The effect is due to friction between the internal planes, which slip or slide as the deformations take place. When a body having material damping is subjected to vibration, the stress-strain diagram shows a hysteresis loop as indicated in Fig. 1.40(a). The area of this loop denotes the energy lost per unit volume of the body per cycle due to damping.[3]

1.9.1 Construction of Viscous Dampers

Viscous dampers can be constructed in several ways. For instance, when a plate moves relative to another parallel plate with a viscous fluid in between the plates, a viscous damper can be obtained. The following examples illustrate the various methods of constructing viscous dampers used in different applications.

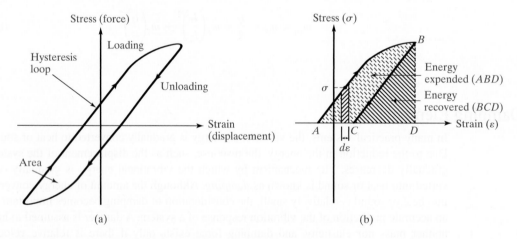

FIGURE 1.40   Hysteresis loop for elastic materials.

[3]When the load applied to an elastic body is increased, the stress (σ) and the strain (ε) in the body also increase. The area under the σ-ε curve, given by

$$u = \int \sigma \, d\varepsilon$$

denotes the energy expended (work done) per unit volume of the body. When the load on the body is decreased, energy will be recovered. When the unloading path is different from the loading path, the area ABC in Fig. 1.40(b)— the area of the hysteresis loop in Fig. 1.40(a)—denotes the energy lost per unit volume of the body.

EXAMPLE 1.13

Damping Constant of Parallel Plates Separated by Viscous Fluid

Consider two parallel plates separated by a distance h, with a fluid of viscosity μ between the plates. Derive an expression for the damping constant when one plate moves with a velocity v relative to the other as shown in Fig. 1.41.

Solution: Let one plate be fixed and let the other plate be moved with a velocity v in its own plane. The fluid layers in contact with the moving plate move with a velocity v, while those in contact with the fixed plate do not move. The velocities of intermediate fluid layers are assumed to vary linearly between 0 and v, as shown in Fig. 1.41. According to Newton's law of viscous flow, the shear stress (τ) developed in the fluid layer at a distance y from the fixed plate is given by

$$\tau = \mu \frac{du}{dy} \tag{E.1}$$

where $du/dy = v/h$ is the velocity gradient. The shear or resisting force (F) developed at the bottom surface of the moving plate is

$$F = \tau A = \frac{\mu A v}{h} \tag{E.2}$$

where A is the surface area of the moving plate. By expressing F as

$$F = cv \tag{E.3}$$

the damping constant c can be found as

$$c = \frac{\mu A}{h} \tag{E.4}$$

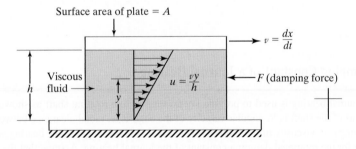

FIGURE 1.41   Parallel plates with a viscous fluid in between.

EXAMPLE 1.14

Clearance in a Bearing

A bearing, which can be approximated as two flat plates separated by a thin film of lubricant (Fig. 1.42), is found to offer a resistance of 400 N when SAE 30 oil is used as the lubricant and the relative velocity between the plates is 10 m/s. If the area of the plates (A) is 0.1 m^2, determine the clearance between the plates. Assume the absolute viscosity of SAE 30 oil as 50 μreyn or 0.3445 Pa-s.

Area (A)

v

h

FIGURE 1.42 Flat plates separated by thin
film of lubricant.

Solution: Since the resisting force (F) can be expressed as $F = cv$, where c is the damping constant
and v is the velocity, we have

$$c = \frac{F}{v} = \frac{400}{10} = 40 \text{ N-s/m} \tag{E.1}$$

By modeling the bearing as a flat-plate-type damper, the damping constant is given by Eq. (E.4) of
Example 1.13:

$$c = \frac{\mu A}{h} \tag{E.2}$$

Using the known data, Eq. (E.2) gives

$$c = 40 = \frac{(0.3445)(0.1)}{h} \quad \text{or} \quad h = 0.86125 \text{ mm} \tag{E.3}$$

∎

EXAMPLE 1.15

Damping Constant of a Journal Bearing

A journal bearing is used to provide lateral support to a rotating shaft as shown in Fig. 1.43. If the
radius of the shaft is R, angular velocity of the shaft is ω, radial clearance between the shaft and the
bearing is d, viscosity of the fluid (lubricant) is μ, and the length of the bearing is l, derive an expression for the rotational damping constant of the journal bearing. Assume that the leakage of the fluid
is negligible.

Solution: The damping constant of the journal bearing can be determined using the equation for the
shear stress in viscous fluid. The fluid in contact with the rotating shaft will have a linear velocity
(in tangential direction) of $v = R\omega$, while the fluid in contact with the stationary bearing will have
zero velocity. Assuming a linear variation for the velocity of the fluid in the radial direction, we have

$$v(r) = \frac{vr}{d} = \frac{rR\omega}{d} \tag{E.1}$$

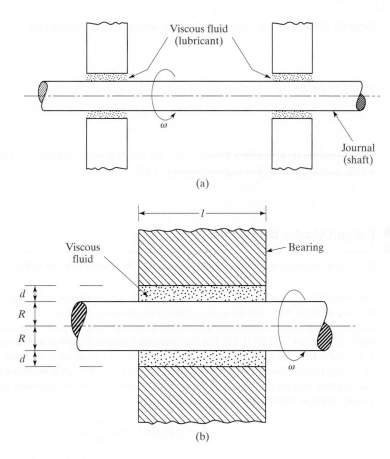

(a)

(b)

FIGURE 1.43 A journal bearing.

The shearing stress (τ) in the lubricant is given by the product of the radial velocity gradient and the viscosity of the lubricant:

$$\tau = \mu \frac{dv}{dr} = \frac{\mu R \omega}{d}$$ (E.2)

The force required to shear the fluid film is equal to the stress multiplied by the area. The torque on the shaft (T) is equal to the force times the lever arm, so that

$$T = (\tau A)R$$ (E.3)

where $A = 2\pi R l$ is the surface area of the shaft exposed to the lubricant. Thus Eq. (E.3) can be rewritten as

$$T = \left(\frac{\mu R \omega}{d}\right)(2\pi R l)R = \frac{2\pi \mu R^3 l \omega}{d}$$ (E.4)

From the definition of the rotational damping constant of the bearing (c_t):

$$c_t = \frac{T}{\omega} \qquad \text{(E.5)}$$

we obtain the desired expression for the rotational damping constant as

$$c_t = \frac{2\pi\mu R^3 l}{d} \qquad \text{(E.6)}$$

Note: Equation (E.4) is called *Petroff's law* and was published originally in 1883. This equation is widely used in the design of journal bearings [1.43].

■

EXAMPLE 1.16

Piston-Cylinder Dashpot

Develop an expression for the damping constant of the dashpot shown in Fig. 1.44(a).

Solution: The damping constant of the dashpot can be determined using the shear-stress equation for viscous fluid flow and the rate-of-fluid-flow equation. As shown in Fig. 1.44(a), the dashpot consists of a piston of diameter D and length l, moving with velocity v_0 in a cylinder filled with a liquid of viscosity μ [1.24, 1.32]. Let the clearance between the piston and the cylinder wall be d. At a distance y from the moving surface, let the velocity and shear stress be v and τ, and at a distance $(y + dy)$, let the velocity and shear stress be $(v - dv)$ and $(\tau + d\tau)$, respectively (see Fig. 1.44(b)). The negative sign for dv shows that the velocity decreases as we move toward the cylinder wall. The viscous force on this annular ring is equal to

$$F = \pi Dl \, d\tau = \pi Dl \frac{d\tau}{dy} dy \qquad \text{(E.1)}$$

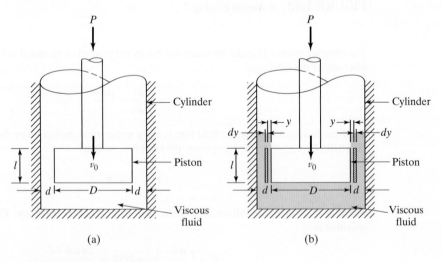

(a)   (b)

FIGURE 1.44 A dashpot.

But the shear stress is given by

$$\tau = -\mu \frac{dv}{dy} \tag{E.2}$$

where the negative sign is consistent with a decreasing velocity gradient [1.33]. Using Eq. (E.2) in Eq. (E.1), we obtain

$$F = -\pi Dl\, dy\, \mu \frac{d^2 v}{dy^2} \tag{E.3}$$

The force on the piston will cause a pressure difference on the ends of the element, given by

$$p = \frac{P}{\left(\dfrac{\pi D^2}{4}\right)} = \frac{4P}{\pi D^2} \tag{E.4}$$

Thus the pressure force on the end of the element is

$$p(\pi D\, dy) = \frac{4P}{D}\, dy \tag{E.5}$$

where $(\pi D\, dy)$ denotes the annular area between y and $(y + dy)$. If we assume uniform mean velocity in the direction of motion of the fluid, the forces given in Eqs. (E.3) and (E.5) must be equal. Thus we get

$$\frac{4P}{D}\, dy = -\pi Dl\, dy\, \mu \frac{d^2 v}{dy^2}$$

or

$$\frac{d^2 v}{dy^2} = -\frac{4P}{\pi D^2 l \mu} \tag{E.6}$$

Integrating this equation twice and using the boundary conditions $v = -v_0$ at $y = 0$ and $v = 0$ at $y = d$, we obtain

$$v = \frac{2P}{\pi D^2 l \mu}(yd - y^2) - v_0\left(1 - \frac{y}{d}\right) \tag{E.7}$$

The rate of flow through the clearance space can be obtained by integrating the rate of flow through an element between the limits $y = 0$ and $y = d$:

$$Q = \int_0^d v\pi D\, dy = \pi D\left[\frac{2Pd^3}{6\pi D^2 l \mu} - \frac{1}{2}v_0 d\right] \tag{E.8}$$

The volume of the liquid flowing through the clearance space per second must be equal to the volume per second displaced by the piston. Hence the velocity of the piston will be equal to this rate of flow divided by the piston area. This gives

$$v_0 = \frac{Q}{\left(\dfrac{\pi}{4}D^2\right)} \tag{E.9}$$

Equations (E.9) and (E.8) lead to

$$P = \left[\frac{3\pi D^3 l \left(1 + \dfrac{2d}{D} \right)}{4d^3} \right] \mu v_0 \tag{E.10}$$

By writing the force as $P = cv_0$, the damping constant c can be found as

$$c = \mu \left[\frac{3\pi D^3 l}{4d^3} \left(1 + \frac{2d}{D} \right) \right] \tag{E.11}$$

∎

1.9.2 Linearization of a Nonlinear Damper

If the force (F)-velocity (v) relationship of a damper is nonlinear:

$$F = F(v) \tag{1.26}$$

a linearization process can be used about the operating velocity (v^*), as in the case of a nonlinear spring. The linearization process gives the equivalent damping constant as

$$c = \frac{dF}{dv} \bigg|_{v^*} \tag{1.27}$$

1.9.3 Combination of Dampers

In some dynamic systems, multiple dampers are used. In such cases, all the dampers are replaced by a single equivalent damper. When dampers appear in combination, we can use procedures similar to those used in finding the equivalent spring constant of multiple springs to find a single equivalent damper. For example, when two translational dampers, with damping constants c_1 and c_2, appear in combination, the equivalent damping constant (c_{eq}) can be found as (see Problem 1.55):

Parallel dampers: $c_{eq} = c_1 + c_2$ $\tag{1.28}$

Series dampers: $\dfrac{1}{c_{eq}} = \dfrac{1}{c_1} + \dfrac{1}{c_2}$ $\tag{1.29}$

EXAMPLE 1.17

Equivalent Spring and Damping Constants of a Machine Tool Support

A precision milling machine is supported on four shock mounts, as shown in Fig. 1.45(a). The elasticity and damping of each shock mount can be modeled as a spring and a viscous damper, as shown in Fig. 1.45(b). Find the equivalent spring constant, k_{eq}, and the equivalent damping constant, c_{eq}, of the machine tool support in terms of the spring constants (k_i) and damping constants (c_i) of the mounts.

FIGURE 1.45 Horizontal milling machine.

Solution: The free-body diagrams of the four springs and four dampers are shown in Fig. 1.45(c). Assuming that the center of mass, G, is located symmetrically with respect to the four springs and dampers, we notice that all the springs will be subjected to the same displacement, x, and all the dampers will be subjected to the same relative velocity, $\dot{x}$, where x and $\dot{x}$ denote the displacement and velocity, respectively, of the center of mass, G. Hence the forces acting on the springs (F_{si}) and the dampers (F_{di}) can be expressed as

$$F_{si} = k_i x; \quad i = 1, 2, 3, 4$$
$$F_{di} = c_i \dot{x}; \quad i = 1, 2, 3, 4 \tag{E.1}$$

Let the total forces acting on all the springs and all the dampers be F_s and F_d, respectively (see Fig. 1.45(d)). The force equilibrium equations can thus be expressed as

$$F_s = F_{s1} + F_{s2} + F_{s3} + F_{s4}$$
$$F_d = F_{d1} + F_{d2} + F_{d3} + F_{d4} \tag{E.2}$$

where $F_s + F_d = W$, with W denoting the total vertical force (including the inertia force) acting on the milling machine. From Fig. 1.45(d), we have

$$F_s = k_{\text{eq}} x$$
$$F_d = c_{\text{eq}} \dot{x} \tag{E.3}$$

Equation (E.2), along with Eqs. (E.1) and (E.3), yields

$$k_{\text{eq}} = k_1 + k_2 + k_3 + k_4 = 4k$$
$$c_{\text{eq}} = c_1 + c_2 + c_3 + c_4 = 4c \tag{E.4}$$

when $k_i = k$ and $c_i = c$ for $i = 1, 2, 3, 4$.

Note: If the center of mass, G, is not located symmetrically with respect to the four springs and dampers, the ith spring experiences a displacement of x_i and the ith damper experiences a velocity of $\dot{x}_i$, where x_i and $\dot{x}_i$ can be related to the displacement x and velocity $\dot{x}$ of the center of mass of the milling machine, G. In such a case, Eqs. (E.1) and (E.4) need to be modified suitably.

■

1.10   Harmonic Motion

Oscillatory motion may repeat itself regularly, as in the case of a simple pendulum, or it may display considerable irregularity, as in the case of ground motion during an earthquake. If the motion is repeated after equal intervals of time, it is called *periodic motion*. The simplest type of periodic motion is *harmonic motion*. The motion imparted to the mass m due to the Scotch yoke mechanism shown in Fig. 1.46 is an example of simple harmonic motion [1.24, 1.34, 1.35]. In this system, a crank of radius A rotates about the point O. The other end of the crank, P, slides in a slotted rod, which reciprocates in the vertical guide R. When the crank rotates at an angular velocity ω, the end point S of the slotted link and hence the

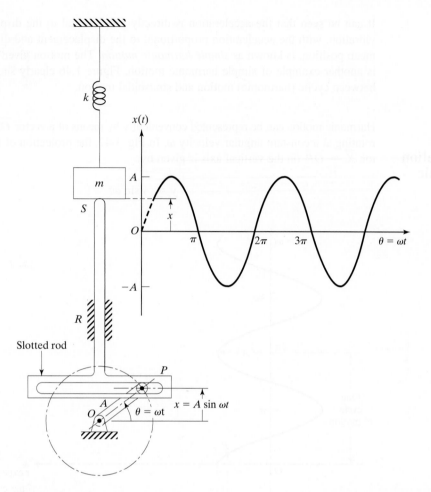

FIGURE 1.46   Scotch yoke mechanism.

mass m of the spring-mass system are displaced from their middle positions by an amount x (in time t) given by

$$x = A \sin \theta = A \sin \omega t \tag{1.30}$$

This motion is shown by the sinusoidal curve in Fig. 1.46. The velocity of the mass m at time t is given by

$$\frac{dx}{dt} = \omega A \cos \omega t \tag{1.31}$$

and the acceleration by

$$\frac{d^2 x}{dt^2} = -\omega^2 A \sin \omega t = -\omega^2 x \tag{1.32}$$

It can be seen that the acceleration is directly proportional to the displacement. Such a vibration, with the acceleration proportional to the displacement and directed toward the mean position, is known as *simple harmonic motion*. The motion given by $x = A \cos \omega t$ is another example of simple harmonic motion. Figure 1.46 clearly shows the similarity between cyclic (harmonic) motion and sinusoidal motion.

**1.10.1
Vectorial
Representation
of Harmonic
Motion**

Harmonic motion can be represented conveniently by means of a vector $\overrightarrow{OP}$ of magnitude A rotating at a constant angular velocity ω. In Fig. 1.47, the projection of the tip of the vector $\vec{X} = \overrightarrow{OP}$ on the vertical axis is given by

$$y = A \sin \omega t \tag{1.33}$$

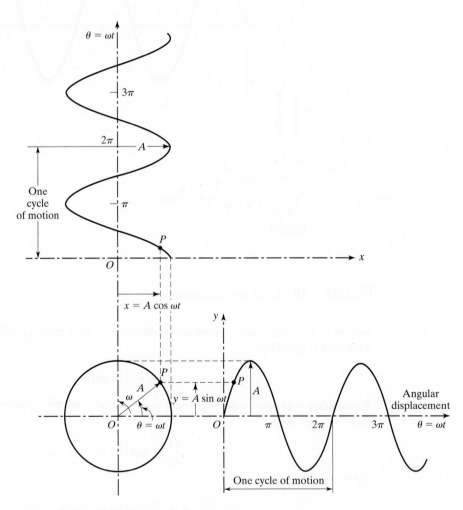

FIGURE 1.47   Harmonic motion as the projection of the end of a rotating vector.

and its projection on the horizontal axis by

$$x = A \cos \omega t \tag{1.34}$$

1.10.2 Complex-Number Representation of Harmonic Motion

As seen above, the vectorial method of representing harmonic motion requires the description of both the horizontal and vertical components. It is more convenient to represent harmonic motion using a complex-number representation. Any vector $\vec{X}$ in the xy-plane can be represented as a complex number:

$$\vec{X} = a + ib \tag{1.35}$$

where $i = \sqrt{-1}$ and a and b denote the x and y components of $\vec{X}$, respectively (see Fig. 1.48). Components a and b are also called the *real* and *imaginary* parts of the vector $\vec{X}$. If A denotes the modulus or absolute value of the vector $\vec{X}$, and θ represents the argument or the angle between the vector and the x-axis, then $\vec{X}$ can also be expressed as

$$\vec{X} = A \cos \theta + iA \sin \theta \tag{1.36}$$

with

$$A = (a^2 + b^2)^{1/2} \tag{1.37}$$

and

$$\theta = \tan^{-1} \frac{b}{a} \tag{1.38}$$

Noting that $i^2 = -1, i^3 = -i, i^4 = 1, \ldots$, $\cos \theta$ and $i \sin \theta$ can be expanded in a series as

$$\cos \theta = 1 - \frac{\theta^2}{2!} + \frac{\theta^4}{4!} - \cdots = 1 + \frac{(i\theta)^2}{2!} + \frac{(i\theta)^4}{4!} + \cdots \tag{1.39}$$

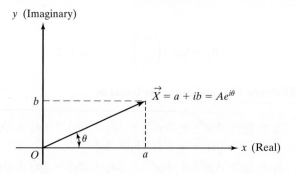

FIGURE 1.48 Representation of a complex number.

$$i \sin \theta = i \left[\theta - \frac{\theta^3}{3!} + \frac{\theta^5}{5!} - \cdots \right] = i\theta + \frac{(i\theta)^3}{3!} + \frac{(i\theta)^5}{5!} + \cdots \qquad (1.40)$$

Equations (1.39) and (1.40) yield

$$(\cos \theta + i \sin \theta) = 1 + i\theta + \frac{(i\theta)^2}{2!} + \frac{(i\theta)^3}{3!} + \cdots = e^{i\theta} \qquad (1.41)$$

and

$$(\cos \theta - i \sin \theta) = 1 - i\theta + \frac{(i\theta)^2}{2!} - \frac{(i\theta)^3}{3!} + \cdots = e^{-i\theta} \qquad (1.42)$$

Thus Eq. (1.36) can be expressed as

$$\vec{X} = A(\cos \theta + i \sin \theta) = Ae^{i\theta} \qquad (1.43)$$

1.10.3 Complex Algebra

Complex numbers are often represented without using a vector notation as

$$z = a + ib \qquad (1.44)$$

where a and b denote the real and imaginary parts of z. The addition, subtraction, multiplication, and division of complex numbers can be achieved by using the usual rules of algebra. Let

$$z_1 = a_1 + ib_1 = A_1 e^{i\theta_1} \qquad (1.45)$$
$$z_2 = a_2 + ib_2 = A_2 e^{i\theta_2} \qquad (1.46)$$

where

$$A_j = \sqrt{a_j^2 + b_j^2}; \qquad j = 1, 2 \qquad (1.47)$$

and

$$\theta_j = \tan^{-1}\left(\frac{b_j}{a_j}\right); \qquad j = 1, 2 \qquad (1.48)$$

The sum and difference of z_1 and z_2 can be found as

$$z_1 + z_2 = A_1 e^{i\theta_1} + A_2 e^{i\theta_2} = (a_1 + ib_1) + (a_2 + ib_2)$$
$$= (a_1 + a_2) + i(b_1 + b_2) \qquad (1.49)$$
$$z_1 - z_2 = A_1 e^{i\theta_1} - A_2 e^{i\theta_2} = (a_1 + ib_1) - (a_2 + ib_2)$$
$$= (a_1 - a_2) + i(b_1 - b_2) \qquad (1.50)$$

**1.10.4
Operations on
Harmonic
Functions**

Using complex-number representation, the rotating vector $\vec{X}$ of Fig. 1.47 can be written as

$$\vec{X} = Ae^{i\omega t} \tag{1.51}$$

where ω denotes the circular frequency (rad/s) of rotation of the vector $\vec{X}$ in a counter-clockwise direction. The differentiation of the harmonic motion given by Eq. (1.51) with respect to time gives

$$\frac{d\vec{X}}{dt} = \frac{d}{dt}(Ae^{i\omega t}) = i\omega Ae^{i\omega t} = i\omega\vec{X} \tag{1.52}$$

$$\frac{d^2\vec{X}}{dt^2} = \frac{d}{dt}(i\omega Ae^{i\omega t}) = -\omega^2 Ae^{i\omega t} = -\omega^2\vec{X} \tag{1.53}$$

Thus the displacement, velocity, and acceleration can be expressed as[4]

$$\text{displacement} = \text{Re}[Ae^{i\omega t}] \qquad = A\cos\omega t \tag{1.54}$$
$$\text{velocity} = \text{Re}[i\omega Ae^{i\omega t}] \qquad = -\omega A\sin\omega t$$
$$= \omega A\cos(\omega t + 90°) \tag{1.55}$$
$$\text{acceleration} = \text{Re}[-\omega^2 Ae^{i\omega t}] \qquad = -\omega^2 A\cos\omega t$$
$$= \omega^2 A\cos(\omega t + 180°) \tag{1.56}$$

where Re denotes the real part. These quantities are shown as rotating vectors in Fig. 1.49. It can be seen that the acceleration vector leads the velocity vector by 90°, and the latter leads the displacement vector by 90°.

Harmonic functions can be added vectorially, as shown in Fig. 1.50. If $\text{Re}(\vec{X_1}) = A_1\cos\omega t$ and $\text{Re}(\vec{X_2}) = A_2\cos(\omega t + \theta)$, then the magnitude of the resultant vector $\vec{X}$ is given by

$$A = \sqrt{(A_1 + A_2\cos\theta)^2 + (A_2\sin\theta)^2} \tag{1.57}$$

and the angle α by

$$\alpha = \tan^{-1}\left(\frac{A_2\sin\theta}{A_1 + A_2\cos\theta}\right) \tag{1.58}$$

---

[4]If the harmonic displacement is originally given as $x(t) = A\sin\omega t$, then we have

$$\text{displacement} = \text{Im}[Ae^{i\omega t}] = A\sin\omega t$$
$$\text{velocity} = \text{Im}[i\omega Ae^{i\omega t}] = \omega A\sin(\omega t + 90°)$$
$$\text{acceleration} = \text{Im}[-\omega^2 Ae^{i\omega t}] = \omega^2 A\sin(\omega t + 180°)$$

where Im denotes the imaginary part.

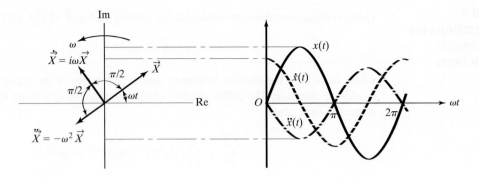

FIGURE 1.49   Displacement, velocity, and acceleration as rotating vectors.

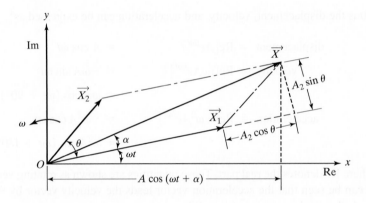

FIGURE 1.50   Vectorial addition of harmonic functions.

Since the original functions are given as real components, the sum $\vec{X}_1 + \vec{X}_2$ is given by $\text{Re}(\vec{X}) = A\cos(\omega t + \alpha)$.

Addition of Harmonic Motions

EXAMPLE 1.18

Find the sum of the two harmonic motions $x_1(t) = 10\cos\omega t$ and $x_2(t) = 15\cos(\omega t + 2)$.

Solution: *Method 1: By using trigonometric relations*: Since the circular frequency is the same for both $x_1(t)$ and $x_2(t)$, we express the sum as

$$x(t) = A\cos(\omega t + \alpha) = x_1(t) + x_2(t) \tag{E.1}$$

—That is,

$$A(\cos\omega t\cos\alpha - \sin\omega t\sin\alpha) = 10\cos\omega t + 15\cos(\omega t + 2)$$

$$= 10\cos\omega t + 15(\cos\omega t\cos 2 - \sin\omega t\sin 2) \tag{E.2}$$

—That is,

$$\cos \omega t (A \cos \alpha) - \sin \omega t (A \sin \alpha) = \cos \omega t (10 + 15 \cos 2) - \sin \omega t (15 \sin 2) \quad \text{(E.3)}$$

By equating the corresponding coefficients of $\cos \omega t$ and $\sin \omega t$ on both sides, we obtain

$$A \cos \alpha = 10 + 15 \cos 2$$

$$A \sin \alpha = 15 \sin 2$$

$$A = \sqrt{(10 + 15 \cos 2)^2 + (15 \sin 2)^2}$$

$$= 14.1477 \quad \text{(E.4)}$$

and

$$\alpha = \tan^{-1}\left(\frac{15 \sin 2}{10 + 15 \cos 2}\right) = 74.5963° \quad \text{(E.5)}$$

Method 2: By using vectors: For an arbitrary value of ωt, the harmonic motions $x_1(t)$ and $x_2(t)$ can be denoted graphically as shown in Fig. 1.51. By adding them vectorially, the resultant vector $x(t)$ can be found to be

$$x(t) = 14.1477 \cos(\omega t + 74.5963°) \quad \text{(E.6)}$$

Method 3: By using complex-number representation: The two harmonic motions can be denoted in terms of complex numbers:

$$x_1(t) = \text{Re}[A_1 e^{i\omega t}] \equiv \text{Re}[10e^{i\omega t}]$$

$$x_2(t) = \text{Re}[A_2 e^{i(\omega t + 2)}] \equiv \text{Re}[15e^{i(\omega t + 2)}] \quad \text{(E.7)}$$

The sum of $x_1(t)$ and $x_2(t)$ can be expressed as

$$x(t) = \text{Re}[A e^{i(\omega t + \alpha)}] \quad \text{(E.8)}$$

where A and α can be determined using Eqs. (1.47) and (1.48) as $A = 14.1477$ and $\alpha = 74.5963°$.

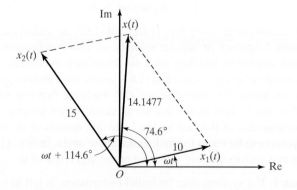

FIGURE 1.51   Addition of harmonic motions.

**1.10.5
Definitions and
Terminology**

The following definitions and terminology are useful in dealing with harmonic motion and other periodic functions.

Cycle. The movement of a vibrating body from its undisturbed or equilibrium position to its extreme position in one direction, then to the equilibrium position, then to its extreme position in the other direction, and back to equilibrium position is called a *cycle* of vibration. One revolution (i.e., angular displacement of 2π radians) of the pin P in Fig. 1.46 or one revolution of the vector $\overrightarrow{OP}$ in Fig. 1.47 constitutes a cycle.

Amplitude. The maximum displacement of a vibrating body from its equilibrium position is called the *amplitude* of vibration. In Figs. 1.46 and 1.47, the amplitude of vibration is equal to A.

Period of oscillation. The time taken to complete one cycle of motion is known as the *period of oscillation* or *time period* and is denoted by τ. It is equal to the time required for the vector $\overrightarrow{OP}$ in Fig. 1.47 to rotate through an angle of 2π and hence

$$\tau = \frac{2\pi}{\omega} \tag{1.59}$$

where ω is called the *circular frequency*.

Frequency of oscillation. The number of cycles per unit time is called the *frequency of oscillation* or simply the *frequency* and is denoted by f. Thus

$$f = \frac{1}{\tau} = \frac{\omega}{2\pi} \tag{1.60}$$

Here ω is called the *circular frequency* to distinguish it from the linear frequency $f = \omega/2\pi$. The variable ω denotes the angular velocity of the cyclic motion; f is measured in cycles per second (hertz) while ω is measured in radians per second.

Phase angle. Consider two vibratory motions denoted by

$$x_1 = A_1 \sin \omega t \tag{1.61}$$

$$x_2 = A_2 \sin(\omega t + \phi) \tag{1.62}$$

The two harmonic motions given by Eqs. (1.61) and (1.62) are called *synchronous* because they have the same frequency or angular velocity, ω. Two synchronous oscillations need not have the same amplitude, and they need not attain their maximum values at the same time. The motions given by Eqs. (1.61) and (1.62) can be represented graphically as shown in Fig. 1.52. In this figure, the second vector $\overrightarrow{OP_2}$ leads the first one $\overrightarrow{OP_1}$ by an angle ϕ, known as the *phase angle*. This means that the maximum of the second vector would occur ϕ radians earlier than that of the first vector. Note that instead of the maxima, any other corresponding points can be taken for finding the phase angle. In Eqs. (1.61) and (1.62) or in Fig. 1.52, the two vectors are said to have a *phase difference* of ϕ.

Natural frequency. If a system, after an initial disturbance, is left to vibrate on its own, the frequency with which it oscillates without external forces is known as its *natural frequency*. As will be seen later, a vibratory system having n degrees of freedom will have, in general, n distinct natural frequencies of vibration.

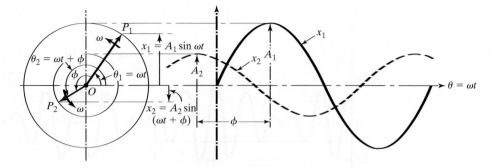

FIGURE 1.52   Phase difference between two vectors.

Beats. When two harmonic motions, with frequencies close to one another, are added, the resulting motion exhibits a phenomenon known as *beats*. For example, if

$$x_1(t) = X \cos \omega t \tag{1.63}$$

$$x_2(t) = X \cos(\omega + \delta)t \tag{1.64}$$

where δ is a small quantity, the addition of these motions yields

$$x(t) = x_1(t) + x_2(t) = X[\cos \omega t + \cos(\omega + \delta)t] \tag{1.65}$$

Using the relation

$$\cos A + \cos B = 2 \cos\left(\frac{A + B}{2}\right) \cos\left(\frac{A - B}{2}\right) \tag{1.66}$$

Eq. (1.65) can be rewritten as

$$x(t) = 2X \cos \frac{\delta t}{2} \cos\left(\omega + \frac{\delta}{2}\right)t \tag{1.67}$$

This equation is shown graphically in Fig. 1.53. It can be seen that the resulting motion, $x(t)$, represents a cosine wave with frequency $\omega + \delta/2$, which is approximately equal to ω, and with a varying amplitude of $2X \cos \delta t/2$. Whenever the amplitude reaches a maximum, it is called a *beat*. The frequency (δ) at which the amplitude builds up and dies down between 0 and $2X$ is known as *beat frequency*. The phenomenon of beats is often observed in machines, structures, and electric power houses. For example, in machines and structures, the beating phenomenon occurs when the forcing frequency is close to the natural frequency of the system (see Section 3.3.2).

Octave. When the maximum value of a range of frequency is twice its minimum value, it is known as an *octave band*. For example, each of the ranges 75–150 Hz, 150–300 Hz, and 300–600 Hz can be called an *octave band*. In each case, the maximum and minimum values of frequency, which have a ratio of 2:1, are said to differ by an *octave*.

Decibel. The various quantities encountered in the field of vibration and sound (such as displacement, velocity, acceleration, pressure, and power) are often represented

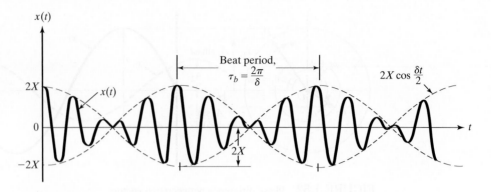

FIGURE 1.53   Phenomenon of beats.

using the notation of *decibel*. A decibel (dB) is originally defined as a ratio of electric power:

$$dB = 10 \log\left(\frac{P}{P_0}\right) \tag{1.68}$$

where P_0 is some reference value of power. Since electric power is proportional to the square of the voltage (X), the decibel can also be expressed as

$$dB = 10 \log\left(\frac{X}{X_0}\right)^2 = 20 \log\left(\frac{X}{X_0}\right) \tag{1.69}$$

where X_0 is a specified reference voltage. In practice, Eq. (1.69) is also used for express-ing the ratios of other quantities such as displacements, velocities, accelerations, and pressures. The reference values of X_0 in Eq. (1.69) are usually taken as 2×10^{-5} N/m^2 for pressure and $1 \, \mu g = 9.81 \times 10^{-6}$ m/s^2 for acceleration.

1.11   Harmonic Analysis[5]

Although harmonic motion is simplest to handle, the motion of many vibratory systems is not harmonic. However, in many cases the vibrations are periodic—for example, the type shown in Fig. 1.54(a). Fortunately, any periodic function of time can be represented by its Fourier series as an infinite sum of sine and cosine terms [1.36].

**1.11.1
Fourier Series
Expansion**

If $x(t)$ is a periodic function with period τ, its Fourier series representation is given by

$$\begin{aligned}
x(t) &= \frac{a_0}{2} + a_1 \cos \omega t + a_2 \cos 2\omega t + \cdots \\
&\quad + b_1 \sin \omega t + b_2 \sin 2\omega t + \cdots \\
&= \frac{a_0}{2} + \sum_{n=1}^{\infty} (a_n \cos n\omega t + b_n \sin n\omega t)
\end{aligned} \tag{1.70}$$

---

[5]The harmonic analysis forms a basis for Section 4.2.

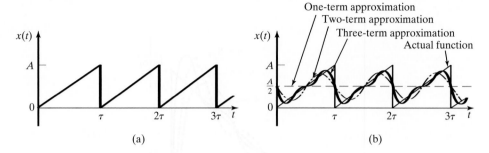

FIGURE 1.54 A periodic function.

where $\omega = 2\pi/\tau$ is the fundamental frequency and $a_0, a_1, a_2, \ldots, b_1, b_2, \ldots$ are constant coefficients. To determine the coefficients a_n and b_n, we multiply Eq. (1.70) by $\cos n\omega t$ and $\sin n\omega t$, respectively, and integrate over one period $\tau = 2\pi/\omega$—for example, from 0 to $2\pi/\omega$. Then we notice that all terms except one on the right-hand side of the equation will be zero, and we obtain

$$a_0 = \frac{\omega}{\pi} \int_0^{2\pi/\omega} x(t)\, dt = \frac{2}{\tau} \int_0^{\tau} x(t)\, dt \tag{1.71}$$

$$a_n = \frac{\omega}{\pi} \int_0^{2\pi/\omega} x(t) \cos n\omega t\, dt = \frac{2}{\tau} \int_0^{\tau} x(t) \cos n\omega t\, dt \tag{1.72}$$

$$b_n = \frac{\omega}{\pi} \int_0^{2\pi/\omega} x(t) \sin n\omega t\, dt = \frac{2}{\tau} \int_0^{\tau} x(t) \sin n\omega t\, dt \tag{1.73}$$

The physical interpretation of Eq. (1.70) is that any periodic function can be represented as a sum of harmonic functions. Although the series in Eq. (1.70) is an infinite sum, we can approximate most periodic functions with the help of only a few harmonic functions. For example, the triangular wave of Fig. 1.54(a) can be represented closely by adding only three harmonic functions, as shown in Fig. 1.54(b).

The Fourier series can also be represented by the sum of sine terms only or cosine terms only. For example, the series using cosine terms only can be expressed as

$$x(t) = d_0 + d_1 \cos(\omega t - \phi_1) + d_2 \cos(2\omega t - \phi_2) + \cdots \tag{1.74}$$

where

$$d_0 = a_0/2 \tag{1.75}$$

$$d_n = (a_n^2 + b_n^2)^{1/2} \tag{1.76}$$

and

$$\phi_n = \tan^{-1}\left(\frac{b_n}{a_n}\right) \tag{1.77}$$

FIGURE 1.55 Gibbs' phenomenon.

Gibbs' Phenomenon. When a periodic function is represented by a Fourier series, an anomalous behavior can be observed. For example, Fig. 1.55 shows a triangular wave and its Fourier series representation using a different number of terms. As the number of terms (n) increases, the approximation can be seen to improve everywhere except in the vicinity of the discontinuity (point P in Fig. 1.55). Here the deviation from the true waveform becomes narrower but not any smaller in amplitude. It has been observed that the error in amplitude remains at approximately 9 percent, even when $k \to \infty$. This behavior is known as *Gibbs' phenomenon*, after its discoverer.

1.11.2
Complex Fourier
Series

The Fourier series can also be represented in terms of complex numbers. By noting, from Eqs. (1.41) and (1.42), that

$$e^{i\omega t} = \cos \omega t + i \sin \omega t \tag{1.78}$$

and

$$e^{-i\omega t} = \cos \omega t - i \sin \omega t \tag{1.79}$$

$\cos \omega t$ and $\sin \omega t$ can be expressed as

$$\cos \omega t = \frac{e^{i\omega t} + e^{-i\omega t}}{2} \tag{1.80}$$

and

$$\sin \omega t = \frac{e^{i\omega t} - e^{-i\omega t}}{2i} \tag{1.81}$$

Thus Eq. (1.70) can be written as

$$x(t) = \frac{a_0}{2} + \sum_{n=1}^{\infty}\left\{a_n\left(\frac{e^{in\omega t} + e^{-in\omega t}}{2}\right) + b_n\left(\frac{e^{in\omega t} - e^{-in\omega t}}{2i}\right)\right\}$$

$$= e^{i(0)\omega t}\left(\frac{a_0}{2} - \frac{ib_0}{2}\right)$$

$$+ \sum_{n=1}^{\infty}\left\{e^{in\omega t}\left(\frac{a_n}{2} - \frac{ib_n}{2}\right) + e^{-in\omega t}\left(\frac{a_n}{2} + \frac{ib_n}{2}\right)\right\} \qquad (1.82)$$

where $b_0 = 0$. By defining the complex Fourier coefficients c_n and c_{-n} as

$$c_n = \frac{a_n - ib_n}{2} \qquad (1.83)$$

and

$$c_{-n} = \frac{a_n + ib_n}{2} \qquad (1.84)$$

Eq. (1.82) can be expressed as

$$x(t) = \sum_{n=-\infty}^{\infty} c_n e^{in\omega t} \qquad (1.85)$$

The Fourier coefficients c_n can be determined, using Eqs. (1.71) to (1.73), as

$$c_n = \frac{a_n - ib_n}{2} = \frac{1}{\tau}\int_0^{\tau} x(t)[\cos n\omega t - i\sin n\omega t]dt$$

$$= \frac{1}{\tau}\int_0^{\tau} x(t)e^{-in\omega t}dt \qquad (1.86)$$

1.11.3 Frequency Spectrum

The harmonic functions $a_n\cos n\omega t$ or $b_n\sin n\omega t$ in Eq. (1.70) are called the *harmonics* of order n of the periodic function $x(t)$. The harmonic of order n has a period τ/n. These harmonics can be plotted as vertical lines on a diagram of amplitude (a_n and b_n or d_n and ϕ_n) versus frequency ($n\omega$), called the *frequency spectrum* or *spectral diagram*. Fig. 1.56 shows a typical frequency spectrum.

FIGURE 1.56   Frequency spectrum of a typical periodic function of time.

1.11.4
Time- and
Frequency-
Domain
Representations

The Fourier series expansion permits the description of any periodic function using either a time-domain or a frequency-domain representation. For example, a harmonic function given by $x(t) = A \sin \omega t$ in the time domain (see Fig. 1.57(a)) can be represented by the amplitude and the frequency ω in the frequency domain (see Fig. 1.57(b)). Similarly, a periodic function, such as a triangular wave, can be represented in the time domain, as shown in Fig. 1.57(c), or in the frequency domain, as indicated in Fig. 1.57(d). Note that the amplitudes d_n and the phase angles ϕ_n corresponding to the frequencies ω_n can be used in place of the amplitudes a_n and b_n for representation in the frequency domain. Using a Fourier integral (considered in Section 14.9) permits the representation of even nonperi-

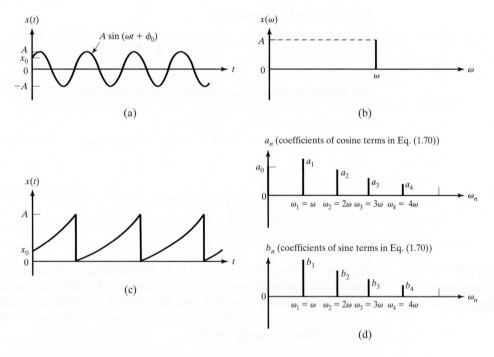

FIGURE 1.57   Representation of a function in time and frequency domains.

odic functions in either a time domain or a frequency domain. Figure 1.57 shows that the frequency-domain representation does not provide the initial conditions. However, in many practical applications the initial conditions are often considered unnecessary and only the steady-state conditions are of main interest.

1.11.5
Even and Odd Functions

An even function satisfies the relation

$$x(-t) = x(t) \tag{1.87}$$

In this case, the Fourier series expansion of $x(t)$ contains only cosine terms:

$$x(t) = \frac{a_0}{2} + \sum_{n=1}^{\infty} a_n \cos n\omega t \tag{1.88}$$

where a_0 and a_n are given by Eqs. (1.71) and (1.72), respectively. An odd function satisfies the relation

$$x(-t) = -x(t) \tag{1.89}$$

In this case, the Fourier series expansion of $x(t)$ contains only sine terms:

$$x(t) = \sum_{n=1}^{\infty} b_n \sin n\omega t \tag{1.90}$$

where b_n is given by Eq. (1.73). In some cases, a given function may be considered as even or odd depending on the location of the coordinate axes. For example, the shifting of the vertical axis from (a) to (b) or (c) in Fig. 1.58(i) will make it an odd or even function. This means that we need to compute only the coefficients b_n or a_n. Similarly, a shift in the time axis from (d) to (e) amounts to adding a constant equal to the amount of shift. In the case of Fig. 1.58(ii), when the function is considered as an odd function, the Fourier series expansion becomes (see Problem 1.107):

$$x_1(t) = \frac{4A}{\pi} \sum_{n=1}^{\infty} \frac{1}{(2n-1)} \sin \frac{2\pi(2n-1)t}{\tau} \tag{1.91}$$

On the other hand, if the function is considered an even function, as shown in Fig. 1.50(iii), its Fourier series expansion becomes (see Problem 1.107):

$$x_2(t) = \frac{4A}{\pi} \sum_{n=1}^{\infty} \frac{(-1)^{n+1}}{(2n-1)} \cos \frac{2\pi(2n-1)t}{\tau} \tag{1.92}$$

Since the functions $x_1(t)$ and $x_2(t)$ represent the same wave, except for the location of the origin, there exists a relationship between their Fourier series expansions also. Noting that

$$x_1\left(t + \frac{\tau}{4}\right) = x_2(t) \tag{1.93}$$

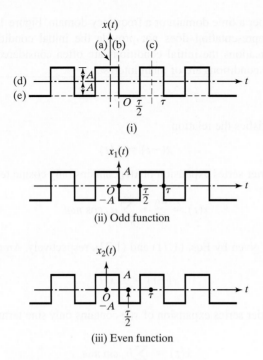

FIGURE 1.58   Even and odd functions.

we find from Eq. (1.91),

$$x_1\left(t + \frac{\tau}{4}\right) = \frac{4A}{\pi} \sum_{n=1}^{\infty} \frac{1}{(2n-1)} \sin \frac{2\pi(2n-1)}{\tau}\left(t + \frac{\tau}{4}\right)$$

$$= \frac{4A}{\pi} \sum_{n=1}^{\infty} \frac{1}{(2n-1)} \sin \left\{ \frac{2\pi(2n-1)t}{\tau} + \frac{2\pi(2n-1)}{4} \right\} \quad (1.94)$$

Using the relation $\sin(A + B) = \sin A \cos B + \cos A \sin B$, Eq. (1.94) can be expressed as

$$x_1\left(t + \frac{\tau}{4}\right) = \frac{4A}{\pi} \sum_{n=1}^{\infty} \left\{ \frac{1}{(2n-1)} \sin \frac{2\pi(2n-1)t}{\tau} \cos \frac{2\pi(2n-1)}{4} \right.$$

$$\left. + \cos \frac{2\pi(2n-1)t}{\tau} \sin \frac{2\pi(2n-1)}{4} \right\} \quad (1.95)$$

Since $\cos\left[2\pi(2n-1)/4\right] = 0$ for $n = 1, 2, 3, \ldots$, and $\sin\left[2\pi(2n-1)/4\right] = (-1)^{n+1}$ for $n = 1, 2, 3, \ldots$, Eq. (1.95) reduces to

$$x_1\left(t + \frac{\tau}{4}\right) = \frac{4A}{\pi} \sum_{n=1}^{\infty} \frac{(-1)^{n+1}}{(2n-1)} \cos \frac{2\pi(2n-1)t}{\tau} \qquad (1.96)$$

which can be identified to be the same as Eq. (1.92).

1.11.6
Half-Range
Expansions

In some practical applications, the function $x(t)$ is defined only in the interval 0 to τ as shown in Fig. 1.59(a). In such a case, there is no condition of periodicity of the function, since the function itself is not defined outside the interval 0 to τ. However, we can extend the function arbitrarily to include the interval $-\tau$ to 0 as shown in either Fig. 1.59(b) or Fig. 1.59(c). The extension of the function indicated in Fig. 1.59(b) results in an odd function, $x_1(t)$, while the extension of the function shown in Fig. 1.59(c) results in an even function, $x_2(t)$. Thus the Fourier series expansion of $x_1(t)$ yields only sine terms and that of $x_2(t)$ involves only cosine terms. These Fourier series expansions of $x_1(t)$ and $x_2(t)$ are

(a) Original function

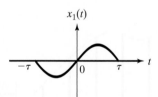

(b) Extension as an odd function

(c) Extension as an even function

FIGURE 1.59 Extension of a function for half-range expansions.

known as *half-range expansions* [1.37]. Any of these half-range expansions can be used to find $x(t)$ in the interval 0 to τ.

**1.11.7
Numerical
Computation
of Coefficients**

For very simple forms of the function $x(t)$, the integrals of Eqs. (1.71) to (1.73) can be evaluated easily. However, the integration becomes involved if $x(t)$ does not have a simple form. In some practical applications, as in the case of experimental determination of the amplitude of vibration using a vibration transducer, the function $x(t)$ is not available in the form of a mathematical expression; only the values of $x(t)$ at a number of points $t_1, t_2, \ldots, t_N$ are available, as shown in Fig. 1.60. In these cases, the coefficients a_n and b_n of Eqs. (1.71) to (1.73) can be evaluated by using a numerical integration procedure like the trapezoidal or Simpson's rule [1.38].

Let us assume that $t_1, t_2, \ldots, t_N$ are an even number of equidistant points over the period $\tau(N = \text{even})$ with the corresponding values of $x(t)$ given by $x_1 = x(t_1)$, $x_2 = x(t_2), \ldots, x_N = x(t_N)$, respectively; then the application of the trapezoidal rule gives the coefficients a_n and b_n (by setting $\tau = N \Delta t$) as:[6]

$$a_0 = \frac{2}{N} \sum_{i=1}^{N} x_i \tag{1.97}$$

$$a_n = \frac{2}{N} \sum_{i=1}^{N} x_i \cos \frac{2n\pi t_i}{\tau} \tag{1.98}$$

$$b_n = \frac{2}{N} \sum_{i=1}^{N} x_i \sin \frac{2n\pi t_i}{\tau} \tag{1.99}$$

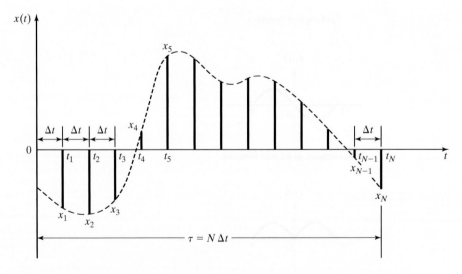

FIGURE 1.60 Values of the periodic function $x(t)$ at discrete points $t_1, t_2, \ldots, t_N$.

---

[6]N needs to be an even number for Simpson's rule but not for the trapezoidal rule. Equations (1.97) to (1.99) assume that the periodicity condition, $x_0 = x_N$, holds true.

Fourier Series Expansion

EXAMPLE 1.19

Determine the Fourier series expansion of the motion of the valve in the cam-follower system shown in Fig. 1.61.

Solution: If $y(t)$ denotes the vertical motion of the pushrod, the motion of the valve, $x(t)$, can be determined from the relation:

$$\tan \theta = \frac{y(t)}{l_1} = \frac{x(t)}{l_2}$$

or

$$x(t) = \frac{l_2}{l_1} y(t) \tag{E.1}$$

where

$$y(t) = Y\frac{t}{\tau}; \qquad 0 \le t \le \tau \tag{E.2}$$

and the period is given by $\tau = \dfrac{2\pi}{\omega}$. By defining

$$A = \frac{Yl_2}{l_1}$$

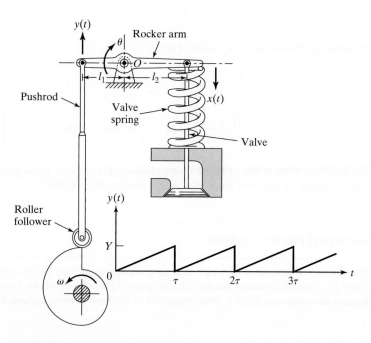

FIGURE 1.61 Cam-follower system.

$x(t)$ can be expressed as

$$x(t) = A\frac{t}{\tau}; \qquad 0 \le t \le \tau \tag{E.3}$$

Equation (E.3) is shown in Fig. 1.54(a). To compute the Fourier coefficients a_n and b_n, we use Eqs. (1.71) to (1.73):

$$a_0 = \frac{\omega}{\pi} \int_0^{2\pi/\omega} x(t)\, dt = \frac{\omega}{\pi} \int_0^{2\pi/\omega} A\frac{t}{\tau}\, dt = \frac{\omega}{\pi}\frac{A}{\tau}\left(\frac{t^2}{2}\right)_0^{2\pi/\omega} = A \tag{E.4}$$

$$a_n = \frac{\omega}{\pi} \int_0^{2\pi/\omega} x(t) \cos n\omega t \cdot dt = \frac{\omega}{\pi} \int_0^{2\pi/\omega} A\frac{t}{\tau} \cos n\omega t \cdot dt$$

$$= \frac{A\omega}{\pi\tau} \int_0^{2\pi/\omega} t \cos n\omega t \cdot dt = \frac{A}{2\pi^2}\left[\frac{\cos n\omega t}{n^2} + \frac{\omega t \sin n\omega t}{n}\right]_0^{2\pi/\omega}$$

$$= 0, \qquad n = 1, 2, \ldots \tag{E.5}$$

$$b_n = \frac{\omega}{\pi} \int_0^{2\pi/\omega} x(t) \sin n\omega t \cdot dt = \frac{\omega}{\pi} \int_0^{2\pi/\omega} A\frac{t}{\tau} \sin n\omega t \cdot dt$$

$$= \frac{A\omega}{\pi\tau} \int_0^{2\pi/\omega} t \sin n\omega t \cdot dt = \frac{A}{2\pi^2}\left[\frac{\sin n\omega t}{n^2} - \frac{\omega t \cos n\omega t}{n}\right]_0^{2\pi/\omega}$$

$$= -\frac{A}{n\pi}, \qquad n = 1, 2, \ldots \tag{E.6}$$

Therefore the Fourier series expansion of $x(t)$ is

$$x(t) = \frac{A}{2} - \frac{A}{\pi} \sin \omega t - \frac{A}{2\pi} \sin 2\omega t - \ldots$$

$$= \frac{A}{\pi}\left[\frac{\pi}{2} - \left\{\sin \omega t + \frac{1}{2} \sin 2\omega t + \frac{1}{3} \sin 3\omega t + \ldots \right\}\right] \tag{E.7}$$

The first three terms of the series are shown plotted in Fig. 1.54(b). It can be seen that the approximation reaches the sawtooth shape even with a small number of terms.

■

Numerical Fourier Analysis

EXAMPLE 1.20

The pressure fluctuations of water in a pipe, measured at 0.01-second intervals, are given in Table 1.1. These fluctuations are repetitive in nature. Make a harmonic analysis of the pressure fluctuations and determine the first three harmonics of the Fourier series expansion.

TABLE 1.1

Time Station, i	Time (s), t_i	Pressure (kN/m²), p_i
0	0	0
1	0.01	20
2	0.02	34
3	0.03	42
4	0.04	49
5	0.05	53
6	0.06	70
7	0.07	60
8	0.08	36
9	0.09	22
10	0.10	16
11	0.11	7
12	0.12	0

Solution: Since the given pressure fluctuations repeat every 0.12 s, the period is $\tau = 0.12$ s and the circular frequency of the first harmonic is 2π radians per 0.12 s or $\omega = 2\pi/0.12 = 52.36$ rad/s. As the number of observed values in each wave (N) is 12, we obtain from Eq. (1.97)

$$a_0 = \frac{2}{N}\sum_{i=1}^{N} p_i = \frac{1}{6}\sum_{i=1}^{12} p_i = 68,166.7 \tag{E.1}$$

The coefficients a_n and b_n can be determined from Eqs. (1.98) and (1.99):

$$a_n = \frac{2}{N}\sum_{i=1}^{N} p_i \cos\frac{2n\pi t_i}{\tau} = \frac{1}{6}\sum_{i=1}^{12} p_i \cos\frac{2n\pi t_i}{0.12} \tag{E.2}$$

$$b_n = \frac{2}{N}\sum_{i=1}^{N} p_i \sin\frac{2n\pi t_i}{\tau} = \frac{1}{6}\sum_{i=1}^{12} p_i \sin\frac{2n\pi t_i}{0.12} \tag{E.3}$$

The computations involved in Eqs. (E.2) and (E.3) are shown in Table 1.2. From these calculations, the Fourier series expansion of the pressure fluctuations $p(t)$ can be obtained (see Eq. (1.70)):

$$p(t) = 34,083.3 - 26,996.0 \cos 52.36t + 8,307.7 \sin 52.36t$$

$$+ 1,416.7 \cos 104.72t + 3,608.3 \sin 104.72t - 5,833.3 \cos 157.08t$$

$$- 2,333.3 \sin 157.08t + \cdots \quad \text{N/m}^2 \tag{E.4}$$

TABLE 1.2

i	t_i	p_i	$n = 1$		$n = 2$		$n = 3$	
			$p_i \cos \dfrac{2\pi t_i}{0.12}$	$p_i \sin \dfrac{2\pi t_i}{0.12}$	$p_i \cos \dfrac{4\pi t_i}{0.12}$	$p_i \sin \dfrac{4\pi t_i}{0.12}$	$p_i \cos \dfrac{6\pi t_i}{0.12}$	$p_i \sin \dfrac{6\pi t_i}{0.12}$
1	0.01	20,000	17,320	10,000	10,000	17,320	0	20,000
2	0.02	34,000	17,000	29,444	−17,000	29,444	−34,000	0
3	0.03	42,000	0	42,000	−42,000	0	0	−42,000
4	0.04	49,000	−24,500	42,434	−24,500	−42,434	49,000	0
5	0.05	53,000	−45,898	26,500	26,500	−45,898	0	53,000
6	0.06	70,000	−70,000	0	70,000	0	−70,000	0
7	0.07	60,000	−51,960	−30,000	30,000	51,960	0	−60,000
8	0.08	36,000	−18,000	−31,176	−18,000	31,176	36,000	0
9	0.09	22,000	0	−22,000	−22,000	0	0	22,000
10	0.10	16,000	8,000	−13,856	−8,000	−13,856	−16,000	0
11	0.11	7,000	6,062	−3,500	3,500	−6,062	0	−7,000
12	0.12	0	0	0	0	0	0	0
$\sum_{i=1}^{12}(\ )$		409,000	−161,976	49,846	8,500	21,650	−35,000	−14,000
$\dfrac{1}{6}\sum_{i=1}^{12}(\ )$		68,166.7	−26,996.0	8,307.7	1,416.7	3,608.3	−5,833.3	−2,333.3

∎

1.12   Examples Using MATLAB[7]

Graphical Representation of Fourier Series Using MATLAB

EXAMPLE 1.21

Plot the periodic function

$$x(t) = A\frac{t}{\tau}\text{ hr}, \qquad 0 \le t \le \tau \tag{E.1}$$

and its Fourier series representation with four terms

$$\bar{x}(t) = \frac{A}{\pi}\left\{\frac{\pi}{2} - \left(\sin \omega t + \frac{1}{2}\sin 2\omega t + \frac{1}{3}\sin 3\omega t\right)\right\} \tag{E.2}$$

for $0 \le t \le \tau$ with $A = 1$, $\omega = \pi$, and $\tau = \dfrac{2\pi}{\omega} = 2$.

---

[7]The source codes of all MATLAB programs are given at the Companion Website.

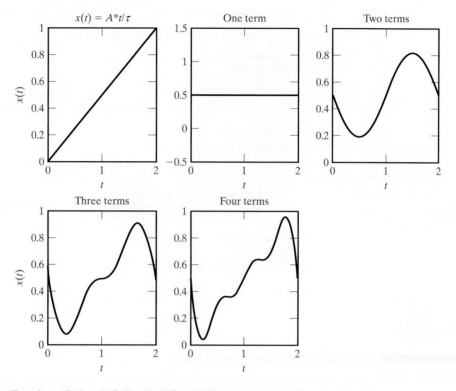

Equations (E.1) and (E.2) with different numbers of terms.

Solution: A MATLAB program is written to plot Eqs. (E.1) and (E.2) with different numbers of terms as shown below.

```
%ex1_21.m
%plot the function x(t) = A * t / tau
A = 1;
w = pi;
tau = 2;
for i = 1: 101
    t(i) = tau * (i-1)/100;
    x(i) = A * t(i) / tau;
end
subplot(231);
plot(t,x);
ylabel('x(t)');
xlabel('t');
title('x(t) = A*t/tau');
for i = 1: 101
    x1(i) = A / 2;
end
subplot(232);
plot(t,x1);
xlabel('t');
```

```
title('One term');
for i = 1: 101
    x2(i) = A/2 - A * sin(w*t(i)) / pi;
end
subplot(233);
plot(t,x2);
xlabel('t');
title('Two terms');
for i = 1: 101
    x3(i) = A/2 - A * sin(w*t(i)) / pi - A * sin(2*w*t(i)) / (2*pi);
end
subplot(234);
plot(t,x3);
ylabel('x(t)');
xlabel('t');
title('Three terms');
for i = 1: 101
    t(i) = tau * (i-1)/100;
    x4(i) = A/2 - A * sin(w*t(i)) / pi - A * sin(2*w*t(i)) / (2*pi)
    - A * sin(3*w*t(i)) / (3*pi);
end
subplot(235);
plot(t,x4);
xlabel('t');
title('Four terms');
```

■

Graphical Representation of Beats

EXAMPLE 1.22

A mass is subjected to two harmonic motions given by $x_1(t) = X \cos \omega t$ and $x_2(t) = X \cos (\omega + \delta) t$ with $X = 1$ cm, $\omega = 20$ rad/s, and $\delta = 1$ rad/s. Plot the resulting motion of the mass using MATLAB and identify the beat frequency.

Solution: The resultant motion of the mass, $x(t)$, is given by

$$x(t) = x_1(t) + x_2(t)$$
$$= X \cos \omega t + X \cos(\omega + \delta)t$$
$$= 2X \cos \frac{\delta t}{2} \cos\left(\omega + \frac{\delta}{2}\right)t \qquad (E.1)$$

The motion can be seen to exhibit the phenomenon of beats with a beat frequency $\omega_b = (\omega + \delta) - (\omega) = \delta = 1$ rad/s. Equation (E.1) is plotted using MATLAB as shown below.

```
% ex1_22.m
% Plot the Phenomenon of beats
A = 1;
w = 20;
delta = 1;
for i = 1: 1001
    t(i) = 15 * (i-1)/1000;
    x(i) = 2 * A * cos (delta*t(i)/2) * cos ((w + delta/2) *t(i));
end
plot (t,x);
xlabel ('t');
ylabel ('x(t)');
title ('Phenomenon of beats');
```

Phenomenon of beats

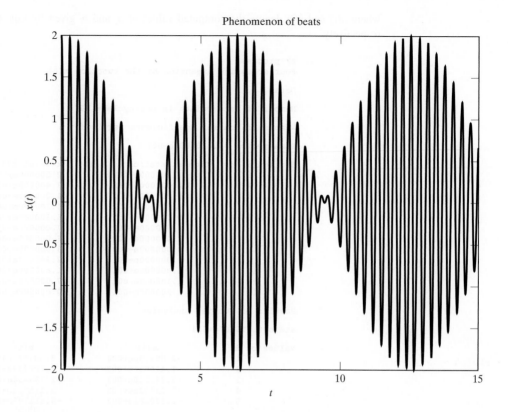

Numerical Fourier Analysis Using MATLAB

EXAMPLE 1.23

Conduct a harmonic analysis of the pressure fluctuations given in Table 1.1 on page 75 and determine the first five harmonics of the Fourier series expansion.

Solution: To find the first five harmonics of the pressure fluctuations (i.e., $a_0, a_1, \ldots, a_5, b_1, \ldots, b_5$), a general-purpose MATLAB program is developed for the harmonic analysis of a function $x(t)$ using Eqs. (1.97) to (1.99). The program, named **Program1.m**, requires the following input data:

n = number of equidistant points at which the values of $x(t)$ are known

m = number of Fourier coefficients to be computed

time = time period of the function $x(t)$

x = array of dimension n, containing the known values of $x(t)$; $x(i) = x(t_i)$

t = array of dimension n, containing the known values of t; $t(i) = t_i$

The following output is generated by the program:

azero = a_0 of Eq. (1.97)

$i, a(i), b(i); i = 1, 2, \ldots, m$

where $a(i)$ and $b(i)$ denote the computed values of a_i and b_i given by Eqs. (1.98) and (1.99), respectively.

```
>> program1
Fourier series expansion of the function x(t)

Data:

Number of data points in one cycle = 12

Number of Fourier Coefficients required = 5

Time period = 1.200000e-001

Station i        Time at station i: t(i)        x(i) at t(i)
        1            1.000000e-002               2.000000e+004
        2            2.000000e-002               3.400000e+004
        3            3.000000e-002               4.200000e+004
        4            4.000000e-002               4.900000e+004
        5            5.000000e-002               5.300000e+004
        6            6.000000e-002               7.000000e+004
        7            7.000000e-002               6.000000e+004
        8            8.000000e-002               3.600000e+004
        9            9.000000e-002               2.200000e+004
       10            1.000000e-001               1.600000e+004
       11            1.100000e-001               7.000000e+003
       12            1.200000e-001               0.000000e+000

Results of Fourier analysis:

azero=6.816667e+004

values of i                a(i)                    b(i)
        1            -2.699630e+004            8.307582e+003
        2             1.416632e+003            3.608493e+003
        3            -5.833248e+003           -2.333434e+003
        4            -5.834026e+002            2.165061e+003
        5            -2.170284e+003           -6.411708e+002
```

■

1.13   Vibration Literature

The literature on vibrations is large and diverse. Several textbooks are available [1.39], and dozens of technical periodicals regularly publish papers relating to vibrations. This is primarily because vibration affects so many disciplines, from science of materials to machinery analysis to spacecraft structures. Researchers in many fields must be attentive to vibration research.

The most widely circulated journals that publish papers relating to vibrations are *ASME Journal of Vibration and Acoustics*; *ASME Journal of Applied Mechanics*; *Journal of Sound and Vibration*; *AIAA Journal*; *ASCE Journal of Engineering Mechanics*; *Earthquake Engineering and Structural Dynamics*; *Bulletin of the Japan Society of Mechanical Engineers*; *International Journal of Solids and Structures*; *International Journal for Numerical Methods in Engineering*; *Journal of the Acoustical Society of America*; *Sound and Vibration*; *Vibrations, Mechanical Systems and Signal Processing*; *International Journal of Analytical and Experimental Modal Analysis*; *JSME International Journal Series III—Vibration Control Engineering*; and *Vehicle System Dynamics*. Many of these journals are cited in the chapter references.

In addition, *Shock and Vibration Digest, Applied Mechanics Reviews*, and *Noise and Vibration Worldwide* are monthly abstract journals containing brief discussions of nearly every published vibration paper. Formulas and solutions in vibration engineering can be readily found in [1.40–1.42].

CHAPTER SUMMARY

In this chapter, we presented the fundamental concepts of vibration along with a brief outline of the history and the importance of the study of vibration. We introduced the concepts of degree of freedom, discrete and continuous systems, and the different classes of vibration. We outlined the basic steps involved in the vibration analysis of a system. We introduced the fundamental type of vibration, namely harmonic motion, along with the associated terminology. We presented harmonic analysis and the Fourier series representation of periodic functions as well as numerical determination of Fourier coefficients with examples.

At this point, the reader should be able to achieve the objectives stated at the beginning of the chapter. To help the reader, review questions requiring brief answers, true or false statements, filling in the blanks, multiple choices, and matching of statements are given for self-testing with answers available at the Companion Website. Several problems involving different levels of difficulty in applying the basic concepts presented in the chapter are also given. The answers to selected problems can be found at the end of the book.

REFERENCES

1.1 D. C. Miller, *Anecdotal History of the Science of Sound*, Macmillan, New York, 1935.

1.2 N. F. Rieger, "The quest for $\sqrt{k/m}$: Notes on the development of vibration analysis, Part I. Genius awakening," *Vibrations*, Vol. 3, No. 3/4, December 1987, pp. 3–10.

1.3 Chinese Academy of Sciences (compiler), *Ancient China's Technology and Science*, Foreign Languages Press, Beijing, 1983.

1.4 R. Taton (ed.), *Ancient and Medieval Science: From the Beginnings to 1450*, A. J. Pomerans (trans.), Basic Books, New York, 1957.

1.5 S. P. Timoshenko, *History of Strength of Materials*, McGraw-Hill, New York, 1953.

1.6 R. B. Lindsay, "The story of acoustics," *Journal of the Acoustical Society of America*, Vol. 39, No. 4, 1966, pp. 629–644.

1.7 J. T. Cannon and S. Dostrovsky, *The Evolution of Dynamics: Vibration Theory from 1687 to 1742*, Springer-Verlag, New York, 1981.

1.8 L. L. Bucciarelli and N. Dworsky, *Sophie Germain: An Essay in the History of the Theory of Elasticity*, D. Reidel Publishing, Dordrecht, Holland, 1980.

1.9 J. W. Strutt (Baron Rayleigh), *The Theory of Sound*, Dover, New York, 1945.

1.10 R. Burton, *Vibration and Impact*, Addison-Wesley, Reading, MA, 1958.

1.11 A. H. Nayfeh, *Perturbation Methods*, Wiley, New York, 1973.

1.12 S. H. Crandall and W. D. Mark, *Random Vibration in Mechanical Systems*, Academic Press, New York, 1963.

1.13 J. D. Robson, *Random Vibration*, Edinburgh University Press, Edinburgh, 1964.

1.14 S. S. Rao, *The Finite Element Method in Engineering* (4th ed.), Elsevier Butterworth Heinemann, Burlington, MA, 2005.

1.15 M. J. Turner, R. W. Clough, H. C. Martin, and L. J. Topp, "Stiffness and deflection analysis of complex structures," *Journal of Aeronautical Sciences*, Vol. 23, 1956, pp. 805–824.

1.16 D. Radaj et al., "Finite element analysis, an automobile engineer's tool," *International Conference on Vehicle Structural Mechanics: Finite Element Application to Design*, Society of Automotive Engineers, Detroit, 1974.

1.17 R. E. D. Bishop, *Vibration* (2nd ed.), Cambridge University Press, Cambridge, 1979.

1.18 M. H. Richardson and K. A. Ramsey, "Integration of dynamic testing into the product design cycle," *Sound and Vibration*, Vol. 15, No. 11, November 1981, pp. 14–27.

1.19 M. J. Griffin and E. M. Whitham, "The discomfort produced by impulsive whole-body vibration," *Journal of the Acoustical Society of America*, Vol. 65, No. 5, 1980, pp. 1277–1284.

1.20 J. E. Ruzicka, "Fundamental concepts of vibration control," *Sound and Vibration*, Vol. 5, No. 7, July 1971, pp. 16–23.

1.21 T. W. Black, "Vibratory finishing goes automatic" (Part 1: Types of machines; Part 2: Steps to automation), *Tool and Manufacturing Engineer*, July 1964, pp. 53–56; and August 1964, pp. 72–76.

1.22 S. Prakash and V. K. Puri, *Foundations for Machines: Analysis and Design*, Wiley, New York, 1988.

1.23 L. Meirovitch, *Fundamentals of Vibrations*, McGraw-Hill, New York, 2001.

1.24 A. Dimarogonas, *Vibration for Engineers* (2nd ed.), Prentice Hall, Upper Saddle River, NJ, 1996.

1.25 E. O. Doebelin, *System Modeling and Response*, Wiley, New York, 1980.

1.26 R. W. Fitzgerald, *Mechanics of Materials* (2nd ed.), Addison-Wesley, Reading, Mass., 1982.

1.27 I. Cochin and W. Cadwallender, *Analysis and Design of Dynamic Systems* (3rd ed.), Addison-Wesley, Reading, MA, 1997.

1.28 F. Y. Chen, *Mechanics and Design of Cam Mechanisms*, Pergamon Press, New York, 1982.

1.29 W. T. Thomson and M. D. Dahleh, *Theory of Vibration with Applications* (5th ed.), Prentice Hall, Upper Saddle River, NJ, 1998.

1.30 N. O. Myklestad, *Fundamentals of Vibration Analysis*, McGraw-Hill, New York, 1956.

1.31 C. W. Bert, "Material damping: An introductory review of mathematical models, measures, and experimental techniques," *Journal of Sound and Vibration*, Vol. 29, No. 2, 1973, pp. 129–153.

1.32 J. M. Gasiorek and W. G. Carter, *Mechanics of Fluids for Mechanical Engineers*, Hart Publishing, New York, 1968.

1.33 A. Mironer, *Engineering Fluid Mechanics*, McGraw-Hill, New York, 1979.

1.34 F. P. Beer and E. R. Johnston, *Vector Mechanics for Engineers* (6th ed.), McGraw-Hill, New York, 1997.

1.35 A. Higdon and W. B. Stiles, *Engineering Mechanics* (2nd ed.), Prentice Hall, New York, 1955.

1.36 E. Kreyszig, *Advanced Engineering Mathematics* (9th ed.), Wiley, New York, 2006.

1.37 M. C. Potter, J. Goldberg, and E. F. Aboufadel, *Advanced Engineering Mathematics* (3rd ed.), Oxford University Press, New York, 2005.

1.38 S. S. Rao, *Applied Numerical Methods for Engineers and Scientists*, Prentice Hall, Upper Saddle River, NJ, 2002.

1.39 N. F. Rieger, "The literature of vibration engineering," *Shock and Vibration Digest*, Vol. 14, No. 1, January 1982, pp. 5–13.

1.40 R. D. Blevins, *Formulas for Natural Frequency and Mode Shape*, Van Nostrand Reinhold, New York, 1979.

1.41 W. D. Pilkey and P. Y. Chang, *Modern Formulas for Statics and Dynamics*, McGraw-Hill, New York, 1978.

1.42 C. M. Harris (ed.), *Shock and Vibration Handbook* (4th ed.), McGraw-Hill, New York, 1996.

1.43 R. G. Budynas and J. K. Nisbett, *Shigley's Mechanical Engineering Design* (8th ed.), McGraw-Hill, New York, 2008.

1.44 N. P. Chironis (ed.), *Machine Devices and Instrumentation*, McGraw-Hill, New York, 1966.

1.45 D. Morrey and J. E. Mottershead, "Vibratory bowl feeder design using numerical modelling techniques," in *Modern Practice in Stress and Vibration Analysis*, J. E. Mottershead (ed.), Pergamon Press, Oxford, 1989, pp. 211–217.

1.46 K. McNaughton (ed.), *Solids Handling*, McGraw-Hill, New York, 1981.

1.47 M. M. Kamal and J. A. Wolf, Jr. (eds.), *Modern Automotive Structural Analysis*, Van Nostrand Reinhold, New York, 1982.

1.48 D. J. Inman, *Engineering Vibration* (3rd ed.), Pearson Prentice Hall, Upper Saddle River, NJ, 2007.

1.49 J. H. Ginsberg, *Mechanical and Structural Vibrations: Theory and Applications*, John Wiley, New York, 2001.

1.50 S. S. Rao, *Vibration of Continuous Systems*, John Wiley, Hoboken, NJ, 2007.

1.51 S. Braun, D. J. Ewins and S. S. Rao (eds.), *Encyclopedia of Vibration*, Vols. 1–3, Academic Press, London, 2002.

1.52 B. R. Munson, D. F. Young, T. H. Okiishi and W. W. Huebsch, *Fundamentals of Fluid Mechanics* (6th ed.), Wiley, Hoboken, NJ, 2009.

1.53 C. W. de Silva (ed.), *Vibration and Shock Handbook*, Taylor & Francis, Boca Raton, FL, 2005.

REVIEW QUESTIONS

1.1 Give brief answers to the following:

1. Give two examples each of the bad and the good effects of vibration.
2. What are the three elementary parts of a vibrating system?
3. Define the number of degrees of freedom of a vibrating system.
4. What is the difference between a discrete and a continuous system? Is it possible to solve any vibration problem as a discrete one?
5. In vibration analysis, can damping always be disregarded?

6. Can a nonlinear vibration problem be identified by looking at its governing differential equation?
7. What is the difference between deterministic and random vibration? Give two practical examples of each.
8. What methods are available for solving the governing equations of a vibration problem?
9. How do you connect several springs to increase the overall stiffness?
10. Define spring stiffness and damping constant.
11. What are the common types of damping?
12. State three different ways of expressing a periodic function in terms of its harmonics.
13. Define these terms: *cycle, amplitude, phase angle, linear frequency, period,* and *natural frequency.*
14. How are τ, ω, and f related to each other?
15. How can we obtain the frequency, phase, and amplitude of a harmonic motion from the corresponding rotating vector?
16. How do you add two harmonic motions having different frequencies?
17. What are beats?
18. Define the terms *decibel* and *octave.*
19. Explain Gibbs' phenomenon.
20. What are half-range expansions?

1.2   Indicate whether each of the following statements is true or false:

1. If energy is lost in any way during vibration, the system can be considered to be damped.
2. The superposition principle is valid for both linear and nonlinear systems.
3. The frequency with which an initially disturbed system vibrates on its own is known as *natural frequency.*
4. Any periodic function can be expanded into a Fourier series.
5. A harmonic motion is a periodic motion.
6. The equivalent mass of several masses at different locations can be found using the equivalence of kinetic energy.
7. The generalized coordinates are not necessarily Cartesian coordinates.
8. Discrete systems are the same as lumped parameter systems.
9. Consider the sum of harmonic motions, $x(t) = x_1(t) + x_2(t) = A \cos(\omega t + \alpha)$, with $x_1(t) = 15 \cos \omega t$ and $x_2(t) = 20 \cos(\omega t + 1)$. The amplitude A is given by 30.8088.
10. Consider the sum of harmonic motions, $x(t) = x_1(t) + x_2(t) = A \cos(\omega t + \alpha)$, with $x_1(t) = 15 \cos \omega t$ and $x_2(t) = 20 \cos(\omega t + 1)$. The phase angle α is given by 1.57 rad.

1.3   Fill in the blank with the proper word:

1. Systems undergo dangerously large oscillations at _____.
2. Undamped vibration is characterized by no loss of _____.
3. A vibratory system consists of a spring, damper, and _____.
4. If a motion repeats after equal intervals of time, it is called a _____ motion.
5. When acceleration is proportional to the displacement and directed toward the mean position, the motion is called _____ *harmonic.*
6. The time taken to complete one cycle of motion is called the _____ of vibration.
7. The number of cycles per unit time is called the _____ of vibration.
8. Two harmonic motions having the same frequency are said to be _____.

9. The angular difference between the occurrence of similar points of two harmonic motions is called _____.

10. Continuous or distributed systems can be considered to have _____ number of degrees of freedom.

11. Systems with a finite number of degrees of freedom are called _____ *systems*.

12. The number of degrees of freedom of a system denotes the minimum number of independent _____ necessary to describe the positions of all parts of the system at any instant of time.

13. If a system vibrates due to initial disturbance only, it is called _____ *vibration*.

14. If a system vibrates due to an external excitation, it is called _____ *vibration*.

15. Resonance denotes the coincidence of the frequency of external excitation with a _____ frequency of the system.

16. A function $f(t)$ is called an *odd function* if _____.

17. The _____ range expansions can be used to represent functions defined only in the interval 0 to τ.

18. _____ analysis deals with the Fourier series representation of periodic functions.

19. The rotational speed of 1,000 rpm (revolutions per minute) is equivalent to _____ radians/sec.

20. When the speed of a turbine is 6,000 rpm, it takes _____ seconds for the turbine to complete one revolution.

1.4 Select the most appropriate answer from the multiple choices given:

1. The world's first seismograph was invented in
   (a) Japan   (b) China   (c) Egypt

2. The first experiments on simple pendulums were conducted by
   (a) Galileo   (b) Pythagoras   (c) Aristotle

3. The *Philosophiae Naturalis Principia Mathematica* was published by
   (a) Galileo   (b) Pythagoras   (c) Newton

4. Mode shapes of plates, by placing sand on vibrating plates, were first observed by
   (a) Chladni   (b) D'Alembert   (c) Galileo

5. The thick beam theory was first presented by
   (a) Mindlin   (b) Einstein   (c) Timoshenko

6. The number of degrees of freedom of a simple pendulum is
   (a) 0   (b) 1   (c) 2

7. Vibration can be classified in
   (a) one way   (b) two ways   (c) several ways

8. Gibbs' phenomenon denotes an anomalous behavior in the Fourier series representation of a
   (a) harmonic function   (b) periodic function   (c) random function

9. The graphical representation of the amplitudes and phase angles of the various frequency components of a periodic function is known as a
   (a) *spectral diagram*   (b) *frequency diagram*   (c) *harmonic diagram*

10. When a system vibrates in a fluid medium, the damping is
    (a) viscous   (b) Coulomb   (c) solid

11. When parts of a vibrating system slide on a dry surface, the damping is
    (a) viscous   (b) Coulomb   (c) solid

12. When the stress-strain curve of the material of a vibrating system exhibits a hysteresis loop, the damping is
    (a) viscous   (b) Coulomb   (c) solid

13. The equivalent spring constant of two parallel springs with stiffnesses k_1 and k_2 is

(a) $k_1 + k_2$ (b) $\dfrac{1}{\dfrac{1}{k_1} + \dfrac{1}{k_2}}$ (c) $\dfrac{1}{k_1} + \dfrac{1}{k_2}$

14. The equivalent spring constant of two series springs with stiffnesses k_1 and k_2 is

(a) $k_1 + k_2$ (b) $\dfrac{1}{\dfrac{1}{k_1} + \dfrac{1}{k_2}}$ (c) $\dfrac{1}{k_1} + \dfrac{1}{k_2}$

15. The spring constant of a cantilever beam with an end mass m is

(a) $\dfrac{3EI}{l^3}$ (b) $\dfrac{l^3}{3EI}$ (c) $\dfrac{Wl^3}{3EI}$

16. If $f(-t) = f(t)$, function $f(t)$ is said to be

(a) even (b) odd (c) continuous

1.5 Match the following:

1. Pythagoras (582–507 B.C.)	a. published a book on the theory of sound
2. Euclid (300 B.C.)	b. first person to investigate musical sounds on a scientific basis
3. Zhang Heng (A.D. 132)	c. wrote a treatise called *Introduction to Harmonics*
4. Galileo (1564–1642)	d. founder of modern experimental science
5. Rayleigh (1877)	e. invented the world's first seismograph

1.6 Match the following:

1. Imbalance in diesel engines	a. can cause failure of turbines and aircraft engines
2. Vibration in machine tools	b. cause discomfort in human activity during metal cutting
3. Blade and disk vibration	c. can cause wheels of locomotives to rise off the track
4. Wind-induced vibration	d. can cause failure of bridges
5. Transmission of vibration	e. can give rise to chatter

1.7 Consider four springs with the spring constants:

$$k_1 = 20 \text{ N/m}, k_2 = 50 \text{ N/m}, k_3 = 100 \text{ N/m}, k_4 = 200 \text{ N/m}$$

Match the equivalent spring constants:

1. $k_1, k_2, k_3,$ and k_4 are in parallel	a. 18.9189 N/m
2. $k_1, k_2, k_3,$ and k_4 are in series	b. 370.0 N/m
3. k_1 and k_2 are in parallel $(k_{eq} = k_{12})$	c. 11.7647 N/m
4. k_3 and k_4 are in parallel $(k_{eq} = k_{34})$	d. 300.0 N/m
5. $k_1, k_2,$ and k_3 are in parallel $(k_{eq} = k_{123})$	e. 70.0 N/m
6. k_{123} is in series with k_4	f. 170.0 N/m
7. $k_2, k_3,$ and k_4 are in parallel $(k_{eq} = k_{234})$	g. 350.0 N/m
8. k_1 and k_{234} are in series	h. 91.8919 N/m

PROBLEMS

Section 1.4 Basic Concepts of Vibration and
Section 1.6 Vibration Analysis Procedure

1.1* A study of the response of a human body subjected to vibration/shock is important in many applications. In a standing posture, the masses of head, upper torso, hips, and legs and the elasticity/damping of neck, spinal column, abdomen, and legs influence the response characteristics. Develop a sequence of three improved approximations for modeling the human body.

1.2* Figure 1.54 shows a human body and a restraint system at the time of an automobile collision [1.47]. Suggest a simple mathematical model by considering the elasticity, mass, and damping of the seat, human body, and restraints for a vibration analysis of the system.

FIGURE 1.62   A human body and a restraint system.

1.3* A reciprocating engine is mounted on a foundation as shown in Fig. 1.63. The unbalanced forces and moments developed in the engine are transmitted to the frame and the foundation. An elastic pad is placed between the engine and the foundation block to reduce the transmission of vibration. Develop two mathematical models of the system using a gradual refinement of the modeling process.

1.4* An automobile moving over a rough road (Fig. 1.64) can be modeled considering (a) weight of the car body, passengers, seats, front wheels, and rear wheels; (b) elasticity of

---

*The asterisk denotes a design-type problem or a problem with no unique answer.

FIGURE 1.63 A reciprocating engine on a foundation.

tires (suspension), main springs, and seats; and (c) damping of the seats, shock absorbers, and tires. Develop three mathematical models of the system using a gradual refinement in the modeling process.

FIGURE 1.64 An automobile moving on a rough road.

1.5* The consequences of a head-on collision of two automobiles can be studied by considering the impact of the automobile on a barrier, as shown in Fig. 1.65. Construct a mathematical model by considering the masses of the automobile body, engine, transmission, and suspension and the elasticity of the bumpers, radiator, sheet metal body, driveline, and engine mounts.

FIGURE 1.65 An automobile colliding with a barrier.

1.6* Develop a mathematical model for the tractor and plow shown in Fig. 1.66 by considering the mass, elasticity, and damping of the tires, shock absorbers, and plows (blades).

Shock absorber

Plow

FIGURE 1.66   A tractor and plow.

Section 1.7 Spring Elements

1.7   Determine the equivalent spring constant of the system shown in Fig. 1.67.

FIGURE 1.67   Springs in series-parallel.

1.8   Consider a system of two springs, with stiffnesses k_1 and k_2, arranged in parallel as shown in Fig. 1.68. The rigid bar to which the two springs are connected remains horizontal when the

FIGURE 1.68 Parallel springs subjected to load.

force F is zero. Determine the equivalent spring constant of the system (k_e) that relates the force applied (F) to the resulting displacement (x) as

$$F = k_e x$$

Hint: Because the spring constants of the two springs are different and the distances l_1 and l_2 are not the same, the rigid bar will not remain horizontal when the force F is applied.

1.9 In Fig. 1.69, find the equivalent spring constant of the system in the direction of θ.

FIGURE 1.69

1.10 Find the equivalent torsional spring constant of the system shown in Fig. 1.70. Assume that k_1, k_2, k_3, and k_4 are torsional and k_5 and k_6 are linear spring constants.

FIGURE 1.70

1.11 A machine of mass $m = 500$ kg is mounted on a simply supported steel beam of length $l = 2$ m having a rectangular cross section (depth $= 0.1$ m, width $= 1.2$ m) and Young's modulus $E = 2.06 \times 10^{11}$ N/m². To reduce the vertical deflection of the beam, a spring of stiffness k is attached at mid-span, as shown in Fig. 1.71. Determine the value of k needed to reduce the deflection of the beam by

a. 25 percent of its original value.
b. 50 percent of its original value.
c. 75 percent of its original value.

Assume that the mass of the beam is negligible.

FIGURE 1.71

1.12 A bar of length L and Young's modulus E is subjected to an axial force. Compare the spring constants of bars with cross sections in the form of a solid circle (of diameter d), square (of side d), and hollow circle (of mean diameter d and wall thickness $t = 0.1d$). Determine which of these cross sections leads to an economical design for a specified value of axial stiffness of the bar.

1.13 A cantilever beam of length L and Young's modulus E is subjected to a bending force at its free end. Compare the spring constants of beams with cross sections in the form of a solid circle (of diameter d), square (of side d), and hollow circle (of mean diameter d and wall thickness $t = 0.1d$). Determine which of these cross sections leads to an economical design for a specified value of bending stiffness of the beam.

1.14 An electronic instrument, weighing 900 N, is supported on a rubber mounting whose force-deflection relationship is given by $F(x) = 1{,}400\,x + 11\,x^3$, where the force ($F$) and the deflection ($x$) are in Newtons and centimeters, respectively. Determine the following:

   a. Equivalent linear spring constant of the mounting at its static equilibrium position.
   b. Deflection of the mounting corresponding to the equivalent linear spring constant.

1.15 The force-deflection relation of a steel helical spring used in an engine is found experimentally as $F(x) = 350\,x + 35\,x^2 + 3\,x^3$, where the force ($F$) and deflection ($x$) are measured in Newtons and centimeters, respectively. If the spring undergoes a steady deflection of 1.3 cm during the operation of the engine, determine the equivalent linear spring constant of the spring at its steady deflection.

1.16 Four identical rigid bars—each of length a—are connected to a spring of stiffness k to form a structure for carrying a vertical load P, as shown in Figs. 1.72(a) and (b). Find the equivalent spring constant of the system k_{eq} for each case, disregarding the masses of the bars and the friction in the joints.

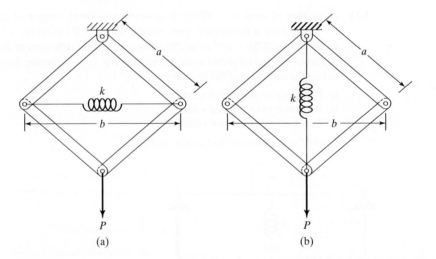

(a)   (b)

FIGURE 1.72

1.17 The tripod shown in Fig. 1.73 is used for mounting an electronic instrument that finds the distance between two points in space. The legs of the tripod are located symmetrically about the mid-vertical axis, each leg making an angle α with the vertical. If each leg has a length l and axial stiffness k, find the equivalent spring stiffness of the tripod in the vertical direction.

FIGURE 1.73

1.18 The static equilibrium position of a massless rigid bar, hinged at point O and connected with springs k_1 and k_2, is shown in Fig. 1.74. Assuming that the displacement (x) resulting from the application of a force F at point A is small, find the equivalent spring constant of the system, k_e, that relates the applied force F to the displacement x as $F = k_e x$.

FIGURE 1.74   Rigid bar connected by springs.

1.19 Figure 1.75 shows a system in which the mass m is directly connected to the springs with stiffnesses k_1 and k_2 while the spring with stiffness k_3 or k_4 comes into contact with the mass based on the value of the displacement of the mass. Determine the variation of the spring force exerted on the mass as the displacement of the mass (x) varies.

1.20 Figure 1.76 shows a uniform rigid bar of mass m that is pivoted at point O and connected by springs of stiffnesses k_1 and k_2. Considering a small angular displacement θ of the rigid bar about the point O, determine the equivalent spring constant associated with the restoring moment.

FIGURE 1.75   Mass connected by springs.

FIGURE 1.76   Rigid bar connected by springs.

1.21   Figure 1.77 shows a U-tube manometer open at both ends and containing a column of liquid mercury of length l and specific weight γ. Considering a small displacement x of the manometer meniscus from its equilibrium position (or datum), determine the equivalent spring constant associated with the restoring force.

FIGURE 1.77   U-tube manometer.

1.22 An oil drum of diameter d and mass m floats in a bath of sea water of density ρ_w as shown in Fig. 1.78. Considering a small displacement x of the oil drum from its static equilibrium position, determine the equivalent spring constant associated with the restoring force.

FIGURE 1.78   Oil drum floating in sea water.

1.23 Find the equivalent spring constant and equivalent mass of the system shown in Fig. 1.79 with references to θ. Assume that the bars AOB and CD are rigid with negligible mass.

1.24 Find the length of the equivalent uniform hollow shaft of inner diameter d and thickness t that has the same axial spring constant as that of the solid conical shaft shown in Fig. 1.80.

FIGURE 1.79

FIGURE 1.80

    1.25 Figure 1.81 shows a three-stepped bar fixed at one end and subjected to an axial force F at the other end. The length of step i is l_i and its cross-sectional area is A_i, $i = 1, 2, 3$. All the steps are made of the same material with Young's modulus $E_i = E$, $i = 1, 2, 3$.

        a. Find the spring constant (or stiffness) k_i of step i in the axial direction ($i = 1, 2, 3$).

        b. Find the equivalent spring constant (or stiffness) of the stepped bar, k_{eq}, in the axial direction so that $F = k_{eq}x$.

        c. Indicate whether the steps behave as series or parallel springs.

FIGURE 1.81 A stepped bar subjected to axial force.

    1.26 Find the equivalent spring constant of the system shown in Fig. 1.82.

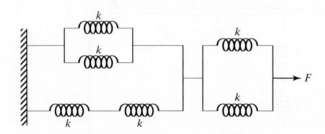

FIGURE 1.82 Springs connected in series-parallel.

    1.27 Figure 1.83 shows a three-stepped shaft fixed at one end and subjected to a torsional moment T at the other end. The length of step i is l_i and its diameter is D_i, $i = 1, 2, 3$. All the steps are made of the same material with shear modulus $G_i = G$, $i = 1, 2, 3$.

        a. Find the torsional spring constant (or stiffness) k_{t_i} of step i ($i = 1, 2, 3$).

b. Find the equivalent torsional spring constant (or stiffness) of the stepped shaft, $k_{t_{eq}}$, so that $T = k_{t_{eq}}\theta$.

c. Indicate whether the steps behave as series or parallel torsional springs.

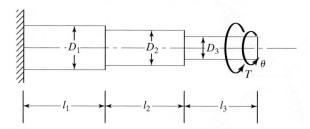

FIGURE 1.83 A stepped shaft subjected to torsional moment.

1.28 The force-deflection characteristic of a spring is described by $F = 500x + 2x^3$, where the force (F) is in Newtons and the deflection (x) is in millimeters. Find (a) the linearized spring constant at $x = 10$ mm and (b) the spring forces at $x = 9$ mm and $x = 11$ mm using the linearized spring constant. Also find the error in the spring forces found in (b).

1.29 Figure 1.84 shows an air spring. This type of spring is generally used for obtaining very low natural frequencies while maintaining zero deflection under static loads. Find the spring constant of this air spring by assuming that the pressure p and volume v change adiabatically when the mass m moves.

FIGURE 1.84

Hint: $pv^\gamma =$ constant for an adiabatic process, where γ is the ratio of specific heats. For air, $\gamma = 1.4$.

1.30 Find the equivalent spring constant of the system shown in Fig. 1.85 in the direction of the load P.

1.31 Derive the expression for the equivalent spring constant that relates the applied force F to the resulting displacement x of the system shown in Fig. 1.86. Assume the displacement of the link to be small.

1.32 The spring constant of a helical spring under axial load is given by

$$k = \frac{Gd^4}{8ND^3}$$

FIGURE 1.85

FIGURE 1.86   Rigid bar connected by springs.

where G is the shear modulus, d is the wire diameter, D is the coil diameter, and N is the number of turns. Find the spring constant and the weight of a helical spring made of steel for the following data: $D = 0.2$ m, $d = 0.005$ m, $N = 10$.

1.33   Two helical springs, one made of steel and the other made of aluminum, have identical values of d and D. (a) If the number of turns in the steel spring is 10, determine the number of turns required in the aluminum spring whose weight will be the same as that of the steel spring. (b) Find the spring constants of the two springs.

1.34 Figure 1.87 shows three parallel springs, one with stiffness $k_1 = k$ and each of the other two with stiffness $k_2 = k$. The spring with stiffness k_1 has a length l and each of the springs with stiffness k_2 has a length of $l - a$. Find the force-deflection characteristic of the system.

FIGURE 1.87 Nonlinear behavior of linear springs.

1.35* Design an air spring using a cylindrical container and a piston to achieve a spring constant of 13 kN/m. Assume that the maximum air pressure available is 1.4 MPa.

1.36 The force (F)-deflection (x) relationship of a nonlinear spring is given by

$$F = ax + bx^3$$

where a and b are constants. Find the equivalent linear spring constant when the deflection is 0.01 m with $a = 20,000$ N/m and $b = 40 \times 10^6$ N/m^3.

1.37 Two nonlinear springs, S_1 and S_2, are connected in two different ways as indicated in Fig. 1.88. The force, F_i, in spring S_i is related to its deflection (x_i) as

$$F_i = a_i x_i + b_i x_i^3, \qquad i = 1, 2$$

where a_i and b_i are constants. If an equivalent linear spring constant, k_{eq}, is defined by $W = k_{eq}x$, where x is the total deflection of the system, find an expression for k_{eq} in each case.

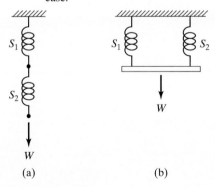

(a)   (b)

FIGURE 1.88

1.38* Design a steel helical compression spring to satisfy the following requirements:

Spring stiffness $(k) \geq 8,000$ N/mm
Fundamental natural frequency of vibration $(f_1) \geq 0.4$ Hz
Spring index $(D/d) \geq 6$
Number of active turns $(N) \geq 10$.

The stiffness and fundamental natural frequency of the spring are given by [1.43]:

$$k = \frac{Gd^4}{8D^3N} \quad \text{and} \quad f_1 = \frac{1}{2}\sqrt{\frac{kg}{W}}$$

where G = shear modulus, d = wire diameter, D = coil diameter, W = weight of the spring, and g = acceleration due to gravity.

1.39 Find the spring constant of the bimetallic bar shown in Fig. 1.89 in axial motion.

FIGURE 1.89

1.40 Consider a spring of stiffness k stretched by a distance x_0 from its free length. One end of the spring is fixed at point O and the other end is connected to a roller as shown in Fig. 1.90. The

FIGURE 1.90 One end of spring with lateral movement.

roller is constrained to move in the horizontal direction with no friction. Find the force (F)-displacement (x) relationship of the spring when the roller is moved by a horizontal distance x to position B. Discuss the resulting force-displacement relation and identify the stiffness constant $\widetilde{k}$ along the direction of x.

1.41 One end of a helical spring is fixed and the other end is subjected to five different tensile forces. The lengths of the spring measured at various values of the tensile forces are given below:

Tensile force F (N)	0	100	250	330	480	570
Total length of the spring (mm)	150	163	183	194	214	226

Determine the force-deflection relation of the helical spring.

1.42 A tapered solid steel propeller shaft is shown in Fig. 1.91. Determine the torsional spring constant of the shaft.

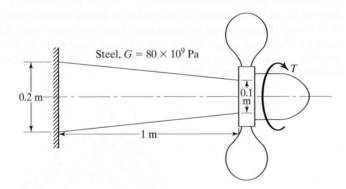

FIGURE 1.91

1.43 A composite propeller shaft, made of steel and aluminum, is shown in Fig. 1.92.
   a. Determine the torsional spring constant of the shaft.
   b. Determine the torsional spring constant of the composite shaft when the inner diameter of the aluminum tube is 5 cm instead of 10 cm.

1.44 Consider two helical springs with the following characteristics:

Spring 1: material—steel; number of turns—10; mean coil diameter—30 cm; wire diameter—5 cm; free length—38 cm; shear modulus—82 GPa.

Spring 2: material—aluminum; number of turns—10; mean coil diameter—25 cm; wire diameter—2.5 cm; free length—38 cm; shear modulus—27 GPa.

Determine the equivalent spring constant when (a) spring 2 is placed inside spring 1, and (b) spring 2 is placed on top of spring 1.

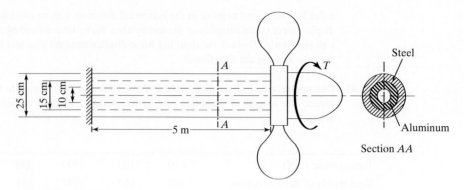

FIGURE 1.92

1.45 Solve Problem 1.44 by assuming the wire diameters of springs 1 and 2 to be 2.5 cm and 1.25 cm instead of 5 cm and 2.5 cm, respectively.

1.46 The arm *AD* of the excavator shown in Fig. 1.93 can be approximated as a steel tube of outer diameter 25 cm, inner diameter 24 cm, and length 250 cm with a viscous damping coefficient of 70 N-s/m. The arm *DE* can be approximated as a steel tube of outer diameter 18 cm, inner diameter 17 cm, and length 190 cm with a viscous damping coefficient of 52 N-s/m. Estimate the equivalent spring constant and equivalent damping coefficient of the excavator, assuming that the base *AC* is fixed.

FIGURE 1.93 An excavator.

1.47 A heat exchanger consists of six identical stainless steel tubes connected in parallel as shown in Fig. 1.94. If each tube has an outer diameter 7.5 mm, inner diameter 7.3 mm, and length 1 m, determine the axial stiffness and the torsional stiffness about the longitudinal axis of the heat exchanger.

FIGURE 1.94
A heat exchanger.

Section 1.8 Mass or Inertia Elements

1.48 Two sector gears, located at the ends of links 1 and 2, are engaged together and rotate about O_1 and O_2, as shown in Fig. 1.95. If links 1 and 2 are connected to springs k_1 to k_4 and k_{t1} and k_{t2} as shown, find the equivalent torsional spring stiffness and equivalent mass moment of inertia of the system with reference to θ_1. Assume (a) the mass moment of inertia of link 1 (including the sector gear) about O_1 is J_1 and that of link 2 (including the sector gear) about O_2 is J_2, and (b) the angles θ_1 and θ_2 are small.

FIGURE 1.95   Two sector gears.

1.49 In Fig. 1.96, find the equivalent mass of the rocker arm assembly with respect to the *x*-coordinate.

FIGURE 1.96 Rocker arm assembly.

1.50 Find the equivalent mass moment of inertia of the gear train shown in Fig. 1.97 with reference to the driving shaft. In Fig. 1.97, J_i and n_i denote the mass moment of inertia and the number of teeth, respectively, of gear i, $i = 1, 2, \ldots, 2N$.

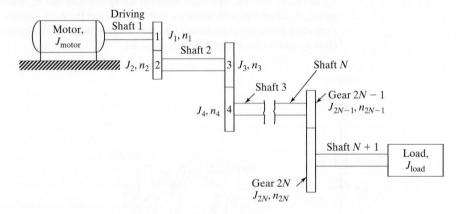

FIGURE 1.97

1.51 Two masses, having mass moments of inertia J_1 and J_2, are placed on rotating rigid shafts that are connected by gears, as shown in Fig. 1.98. If the numbers of teeth on gears 1 and 2 are n_1 and n_2, respectively, find the equivalent mass moment of inertia corresponding to θ_1.

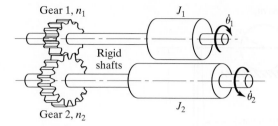

FIGURE 1.98   Rotational masses on geared shafts.

1.52   A simplified model of a petroleum pump is shown in Fig. 1.99, where the rotary motion of the crank is converted to the reciprocating motion of the piston. Find the equivalent mass, m_{eq}, of the system at location A.

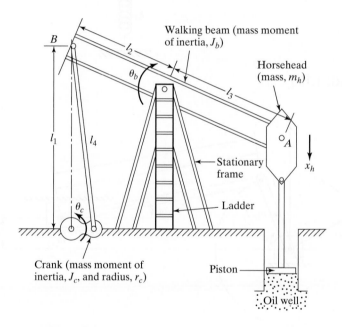

FIGURE 1.99

1.53   Find the equivalent mass of the system shown in Fig. 1.100.

1.54   Figure 1.101 shows an offset slider-crank mechanism with a crank length r, connecting rod length l, and offset δ. If the crank has a mass and mass moment of inertia of m_r and J_r, respectively, at its center of mass A, the connecting rod has a mass and mass moment of inertia of m_c and J_c, respectively, at its center of mass C, and the piston has a mass m_p, determine the equivalent rotational inertia of the system about the center of rotation of the crank, point O.

FIGURE 1.100

FIGURE 1.101 Slider-crank mechanism.

Section 1.9 Damping Elements

1.55 Find a single equivalent damping constant for the following cases:

    a. When three dampers are parallel.

    b. When three dampers are in series.

    c. When three dampers are connected to a rigid bar (Fig. 1.102) and the equivalent damper is at site c_1.

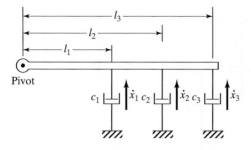

FIGURE 1.102 Dampers connected to a rigid bar.

d. When three torsional dampers are located on geared shafts (Fig. 1.103) and the equivalent damper is at location c_{t_1}.

Hint: The energy dissipated by a viscous damper in a cycle during harmonic motion is given by $\pi c \omega X^2$, where c is the damping constant, ω is the frequency, and X is the amplitude of oscillation.

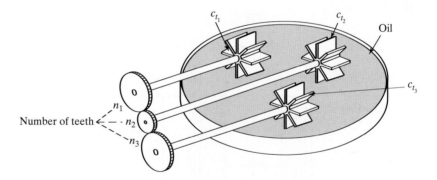

FIGURE 1.103 Dampers located on geared shafts.

1.56 Consider a system of two dampers, with damping constants c_1 and c_2, arranged in parallel as shown in Fig. 1.104. The rigid bar to which the two dampers are connected remains horizontal when the force F is zero. Determine the equivalent damping constant of the system (c_e) that relates the force applied (F) to the resulting velocity (v) as $F = c_e v$.

Hint: Because the damping constants of the two dampers are different and the distances l_1 and l_2 are not the same, the rigid bar will not remain horizontal when the force F is applied.

1.57* Design a piston-cylinder-type viscous damper to achieve a damping constant of 180 N-s/m using a fluid of viscosity 30×10^{-3} N-s/m^2.

1.58* Design a shock absorber (piston-cylinder-type dashpot) to obtain a damping constant of 20×10^6 N-s/m using SAE 30 oil at 21°C. The diameter of the piston has to be less than 65 mm.

FIGURE 1.104   Parallel dampers subjected to load.

1.59   Develop an expression for the damping constant of the rotational damper shown in Fig. 1.105 in terms of D, d, l, h, ω, and μ, where ω denotes the constant angular velocity of the inner cylinder, and d and h represent the radial and axial clearances between the inner and outer cylinders.

Fluid of viscosity μ

FIGURE 1.105

1.60   Consider two nonlinear dampers with the same force-velocity relationship given by $F = 1,000v + 400v^2 + 20v^3$ with F in Newtons and v in meters/second. Find the linearized damping constant of the dampers at an operating velocity of 10 m/s.

1.61   If the linearized dampers of Problem 1.60 are connected in parallel, determine the resulting equivalent damping constant.

1.62   If the linearized dampers of Problem 1.60 are connected in series, determine the resulting equivalent damping constant.

1.63 The force-velocity relationship of a nonlinear damper is given by $F = 500v + 100v^2 + 50v^3$, where F is in Newtons and v is in meters/second. Find the linearized damping constant of the damper at an operating velocity of 5 m/s. If the resulting linearized damping constant is used at an operating velocity of 10 m/s, determine the error involved.

1.64 The experimental determination of damping force corresponding to several values of the velocity of the damper yielded the following results:

Damping force (Newtons)	80	150	250	350	500	600
Velocity of damper (meters/second)	0.025	0.045	0.075	0.110	0.155	0.185

Determine the damping constant of the damper.

1.65 A flat plate with a surface area of 0.25 m^2 moves above a parallel flat surface with a lubricant film of thickness 1.5 mm in between the two parallel surfaces. If the viscosity of the lubricant is 0.5 Pa-s, determine the following:

a. Damping constant.
b. Damping force developed when the plate moves with a velocity of 2 m/s.

1.66 Find the torsional damping constant of a journal bearing for the following data: viscosity of the lubricant (μ): 0.35 Pa-s, diameter of the journal or shaft (2 R): 0.05 m, length of the bearing (l): 0.075 m, bearing clearance (d): 0.005 m. If the journal rotates at a speed (N) of 3,000 rpm, determine the damping torque developed.

1.67 If each of the parameters (μ, R, l, d, and N) of the journal bearing described in Problem 1.66 is subjected to a $\pm 5\%$ variation about the corresponding value given, determine the percentage fluctuation in the values of the torsional damping constant and the damping torque developed.

Note: The variations in the parameters may have several causes, such as measurement error, manufacturing tolerances on dimensions, and fluctuations in the operating temperature of the bearing.

1.68 Consider a piston with an orifice in a cylinder filled with a fluid of viscosity μ as shown in Fig. 1.106. As the piston moves in the cylinder, the fluid flows through the orifice, giving rise to a friction or damping force. Derive an expression for the force needed to move the piston with a velocity v and indicate the type of damping involved.

Hint: The mass flow rate of the fluid (q) passing through an orifice is given by $q = \alpha \sqrt{\Delta p}$, where α is a constant for a given fluid, area of cross section of the cylinder (or area of piston), and area of the orifice [1.52].

1.69 The force (F)-velocity ($\dot{x}$) relationship of a nonlinear damper is given by

$$F = a\dot{x} + b\dot{x}^2$$

where a and b are constants. Find the equivalent linear damping constant when the relative velocity is 5 m/s with $a = 5$ N-s/m and $b = 0.2$ N-s^2/m^2.

FIGURE 1.106 Piston and cylinder with orifice flow.

1.70 The damping constant (c) due to skin-friction drag of a rectangular plate moving in a fluid of viscosity μ is given by (see Fig. 1.107):

$$c = 100\mu l^2 d$$

Design a plate-type damper (shown in Fig. 1.42) that provides an identical damping constant for the same fluid.

FIGURE 1.107

1.71 The damping constant (c) of the dashpot shown in Fig. 1.108 is given by [1.27]:

$$c = \frac{6\pi\mu l}{h^3}\left[\left(a - \frac{h}{2}\right)^2 - r^2\right]\left[\frac{a^2 - r^2}{a - \frac{h}{2}} - h\right]$$

Determine the damping constant of the dashpot for the following data: $\mu = 0.3445$ Pa-s, $l = 10$ cm, $h = 0.1$ cm, $a = 2$ cm, $r = 0.5$ cm.

1.72 In Problem 1.71, using the given data as reference, find the variation of the damping constant c when

a. r is varied from 0.5 cm to 1.0 cm.
b. h is varied from 0.05 cm to 0.10 cm.
c. a is varied from 2 cm to 4 cm.

FIGURE 1.108

1.73 A massless bar of length 1 m is pivoted at one end and subjected to a force F at the other end. Two translational dampers, with damping constants $c_1 = 10$ N-s/m and $c_2 = 15$ N-s/m are connected to the bar as shown in Fig. 1.109. Determine the equivalent damping constant, c_{eq}, of the system so that the force F at point A can be expressed as $F = c_{eq}v$, where v is the linear velocity of point A.

1.74 Find an expression for the equivalent translational damping constant of the system shown in Fig. 1.110 so that the force F can be expressed as $F = c_{eq}v$, where v is the velocity of the rigid bar A.

FIGURE 1.109   Rigid bar connected by dampers.

FIGURE 1.110   Dampers connected in series-parallel.

Section 1.10 Harmonic Motion

1.75   Express the complex number $5 + 2i$ in the exponential form $Ae^{i\theta}$.

1.76   Add the two complex numbers $(1 + 2i)$ and $(3 - 4i)$ and express the result in the form $Ae^{i\theta}$.

1.77   Subtract the complex number $(1 + 2i)$ from $(3 - 4i)$ and express the result in the form $Ae^{i\theta}$.

1.78   Find the product of the complex numbers $z_1 = (1 + 2i)$ and $z_2 = (3 - 4i)$ and express the result in the form $Ae^{i\theta}$.

1.79   Find the quotient, z_1/z_2, of the complex numbers $z_1 = (1 + 2i)$ and $z_2 = (3 - 4i)$ and express the result in the form $Ae^{i\theta}$.

1.80   The foundation of a reciprocating engine is subjected to harmonic motions in x and y directions:

$$x(t) = X \cos \omega t$$
$$y(t) = Y \cos(\omega t + \phi)$$

where X and Y are the amplitudes, ω is the angular velocity, and ϕ is the phase difference.

a. Verify that the resultant of the two motions satisfies the equation of the ellipse given by (see Fig. 1.111):

$$\frac{x^2}{X^2} + \frac{y^2}{Y^2} - 2\frac{xy}{XY} \cos \phi = \sin^2 \phi \qquad (E.1)$$

b. Discuss the nature of the resultant motion given by Eq. (E.1) for the special cases of $\phi = 0, \phi = \dfrac{\pi}{2}$, and $\phi = \pi$.

Note: The elliptic figure represented by Eq. (E.1) is known as a *Lissajous figure* and is useful in interpreting certain types of experimental results (motions) displayed by oscilloscopes.

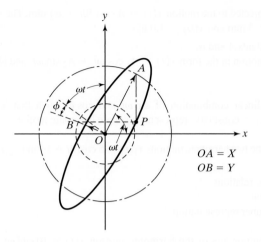

$$OA = X$$
$$OB = Y$$

FIGURE 1.111   Lissajous figure.

1.81   The foundation of an air compressor is subjected to harmonic motions (with the same frequency) in two perpendicular directions. The resultant motion, displayed on an oscilloscope, appears as shown in Fig. 1.112. Find the amplitudes of vibration in the two directions and the phase difference between them.

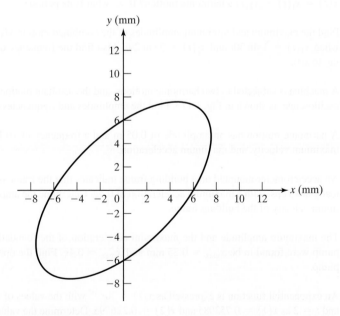

FIGURE 1.112

1.82 A machine is subjected to the motion $x(t) = A \cos(50t + \alpha)$ mm. The initial conditions are given by $x(0) = 3$ mm and $\dot{x}(0) = 1.0$ m/s.

  a. Find the constants A and α.
  b. Express the motion in the form $x(t) = A_1 \cos \omega t + A_2 \sin \omega t$, and identify the constants A_1 and A_2.

1.83 Show that any linear combination of $\sin \omega t$ and $\cos \omega t$ such that $x(t) = A_1 \cos \omega t + A_2 \sin \omega t$ ($A_1, A_2 =$ constants) represents a simple harmonic motion.

1.84 Find the sum of the two harmonic motions $x_1(t) = 5 \cos(3t + 1)$ and $x_2(t) = 10 \cos(3t + 2)$. Use:

  a. Trigonometric relations
  b. Vector addition
  c. Complex-number representation

1.85 If one of the components of the harmonic motion $x(t) = 10 \sin(\omega t + 60°)$ is $x_1(t) = 5 \sin(\omega t + 30°)$, find the other component.

1.86 Consider the two harmonic motions $x_1(t) = \frac{1}{2} \cos \frac{\pi}{2} t$ and $x_2(t) = \sin \pi t$. Is the sum $x_1(t) + x_2(t)$ a periodic motion? If so, what is its period?

1.87 Consider two harmonic motions of different frequencies: $x_1(t) = 2 \cos 2t$ and $x_2(t) = \cos 3t$. Is the sum $x_1(t) + x_2(t)$ a harmonic motion? If so, what is its period?

1.88 Consider the two harmonic motions $x_1(t) = \frac{1}{2} \cos \frac{\pi}{2} t$ and $x_2(t) = \cos \pi t$. Is the difference $x(t) = x_1(t) - x_2(t)$ a harmonic motion? If so, what is its period?

1.89 Find the maximum and minimum amplitudes of the combined motion $x(t) = x_1(t) + x_2(t)$ when $x_1(t) = 3 \sin 30t$ and $x_2(t) = 3 \sin 29t$. Also find the frequency of beats corresponding to $x(t)$.

1.90 A machine is subjected to two harmonic motions, and the resultant motion, as displayed by an oscilloscope, is shown in Fig. 1.113. Find the amplitudes and frequencies of the two motions.

1.91 A harmonic motion has an amplitude of 0.05 m and a frequency of 10 Hz. Find its period, maximum velocity, and maximum acceleration.

1.92 An accelerometer mounted on a building frame indicates that the frame is vibrating harmonically at 15 cps, with a maximum acceleration of $0.5g$. Determine the amplitude and the maximum velocity of the building frame.

1.93 The maximum amplitude and the maximum acceleration of the foundation of a centrifugal pump were found to be $x_{max} = 0.25$ mm and $\ddot{x}_{max} = 0.4g$. Find the operating speed of the pump.

1.94 An exponential function is expressed as $x(t) = Ae^{-\alpha t}$ with the values of $x(t)$ known at $t = 1$ and $t = 2$ as $x(1) = 0.752985$ and $x(2) = 0.226795$. Determine the values of A and α.

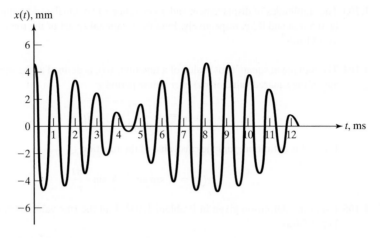

FIGURE 1.113

1.95 When the displacement of a machine is given by $x(t) = 18 \cos 8t$, where x is measured in millimeters and t in seconds, find (a) the period of the machine in s, and (b) the frequency of oscillation of the machine in rad/s as well as in Hz.

1.96 If the motion of a machine is described as $8 \sin(5t + 1) = A \sin 5t + B \cos 5t$, determine the values of A and B.

1.97 Express the vibration of a machine given by $x(t) = -3.0 \sin 5t - 2.0 \cos 5t$ in the form $x(t) = A \cos(5t + \phi)$.

1.98 If the displacement of a machine is given by $x(t) = 0.2 \sin(5t + 3)$, where x is in meters and t is in seconds, find the variations of the velocity and acceleration of the machine. Also find the amplitudes of displacement, velocity, and acceleration of the machine.

1.99 If the displacement of a machine is described as $x(t) = 0.4 \sin 4t + 5.0 \cos 4t$, where x is in centimeters and t is in seconds, find the expressions for the velocity and acceleration of the machine. Also find the amplitudes of displacement, velocity, and acceleration of the machine.

1.100 The displacement of a machine is expressed as $x(t) = 0.05 \sin(6t + \phi)$, where x is in meters and t is in seconds. If the displacement of the machine at $t = 0$ is known to be 0.04 m, determine the value of the phase angle ϕ.

1.101 The displacement of a machine is expressed as $x(t) = A \sin(6t + \phi)$, where x is in meters and t is in seconds. If the displacement and the velocity of the machine at $t = 0$ are known to be 0.05 m and 0.005 m/s, determine the values of A and ϕ.

1.102 A machine is found to vibrate with simple harmonic motion at a frequency of 20 Hz and an amplitude of acceleration of $0.5g$. Determine the displacement and velocity of the machine. Use the value of g as 9.81 m/s^2.

1.103 The amplitudes of displacement and acceleration of an unbalanced turbine rotor are found to be 0.5 mm and 0.5g, respectively. Find the rotational speed of the rotor using the value of g as 9.81 m/s^2.

1.104 The root mean square (rms) value of a function, $x(t)$, is defined as the square root of the average of the squared value of $x(t)$ over a time period τ:

$$x_{rms} = \sqrt{\frac{1}{\tau} \int_0^\tau [x(t)]^2 \, dt}$$

Using this definition, find the rms value of the function

$$x(t) = X \sin \omega t = X \sin \frac{2\pi t}{\tau}$$

1.105 Using the definition given in Problem 1.104, find the rms value of the function shown in Fig. 1.54(a).

Section 1.11 Harmonic Analysis

1.106 Prove that the sine Fourier components (b_n) are zero for even functions—that is, when $x(-t) = x(t)$. Also prove that the cosine Fourier components $(a_0$ and $a_n)$ are zero for odd functions—that is, when $x(-t) = -x(t)$.

1.107 Find the Fourier series expansions of the functions shown in Figs. 1.58(ii) and (iii). Also find their Fourier series expansions when the time axis is shifted down by a distance A.

1.108 The impact force created by a forging hammer can be modeled as shown in Fig. 1.114. Determine the Fourier series expansion of the impact force.

FIGURE 1.114

1.109 Find the Fourier series expansion of the periodic function shown in Fig. 1.115. Also plot the corresponding frequency spectrum.

1.110 Find the Fourier series expansion of the periodic function shown in Fig. 1.116. Also plot the corresponding frequency spectrum.

1.111 Find the Fourier series expansion of the periodic function shown in Fig. 1.117. Also plot the corresponding frequency spectrum.

FIGURE 1.115

FIGURE 1.116

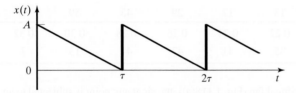

FIGURE 1.117

1.112 The Fourier series of a periodic function, $x(t)$, is an infinite series given by

$$x(t) = \frac{a_0}{2} + \sum_{n=1}^{\infty} \left(a_n \cos n\omega t + b_n \sin n\omega t \right) \tag{E.1}$$

where

$$a_0 = \frac{\omega}{\pi} \int_0^{\frac{2\pi}{\omega}} x(t)\, dt \tag{E.2}$$

$$a_n = \frac{\omega}{\pi} \int_0^{\frac{2\pi}{\omega}} x(t) \cos n\omega t\, dt \tag{E.3}$$

$$b_n = \frac{\omega}{\pi} \int_0^{\frac{2\pi}{\omega}} x(t) \sin n\omega t\, dt \tag{E.4}$$

ω is the circular frequency and $2\pi/\omega$ is the time period. Instead of including the infinite number of terms in Eq. (E.1), it is often truncated by retaining only k terms as

$$x(t) \approx \tilde{x}(t) = \frac{\tilde{a}_0}{2} + \sum_{n=1}^{k} (\tilde{a}_n \cos n\omega t + \tilde{b}_n \sin n\omega t) \qquad \text{(E.5)}$$

so that the error, $e(t)$, becomes

$$e(t) = x(t) - \tilde{x}(t) \qquad \text{(E.6)}$$

Find the coefficients $\tilde{a}_0$, $\tilde{a}_n$, and $\tilde{b}_n$ which minimize the square of the error over a time period:

$$\int_{-\frac{\pi}{\omega}}^{\frac{\pi}{\omega}} e^2(t) \, dt \qquad \text{(E.7)}$$

Compare the expressions of $\tilde{a}_0$, $\tilde{a}_n$, and $\tilde{b}_n$ with Eqs. (E.2) to (E.4) and state your observations.

1.113 Conduct a harmonic analysis, including the first three harmonics, of the function given below:

t_i	0.02	0.04	0.06	0.08	0.10	0.12	0.14	0.16	0.18
x_i	9	13	17	29	43	59	63	57	49
t_i	0.20	0.22	0.24	0.26	0.28	0.30	0.32		
x_i	35	35	41	47	41	13	7		

1.114 In a centrifugal fan (Fig. 1.118(a)), the air at any point is subjected to an impulse each time a blade passes the point, as shown in Fig. 1.118(b). The frequency of these impulses is determined by the speed of rotation of the impeller n and the number of blades, N, in the impeller. For $n = 100$ rpm and $N = 4$, determine the first three harmonics of the pressure fluctuation shown in Fig. 1.118(b).

(a) Centrifugal fan.   (b) Ideal pressure fluctuation at a point.

FIGURE 1.118

1.115 Solve Problem 1.114 by using the values of n and N as 200 rpm and 6 instead of 100 rpm and 4, respectively.

1.116 The torque (M_t) variation with time, of an internal combustion engine, is given in Table 1.3. Make a harmonic analysis of the torque. Find the amplitudes of the first three harmonics.

TABLE 1.3

t (s)	M_t (N-m)	t (s)	M_t (N-m)	t (s)	M_t (N-m)
0.00050	770	0.00450	1890	0.00850	1050
0.00100	810	0.00500	1750	0.00900	990
0.00150	850	0.00550	1630	0.00950	930
0.00200	910	0.00600	1510	0.01000	890
0.00250	1010	0.00650	1390	0.01050	850
0.00300	1170	0.00700	1290	0.01100	810
0.00350	1370	0.00750	1190	0.01150	770
0.00400	1610	0.00800	1110	0.01200	750

1.117 Make a harmonic analysis of the function shown in Fig. 1.119 including the first three harmonics.

FIGURE 1.119

Section 1.12 Examples Using MATLAB

1.118 Plot the Fourier series expansion of the function $x(t)$ given in Problem 1.113 using MATLAB.

1.119 Use MATLAB to plot the variation of the force with time using the Fourier series expansion determined in Problem 1.117.

1.120 Use MATLAB to plot the variations of the damping constant c with respect to r, h, and a as determined in Problem 1.72.

1.121 Use MATLAB to plot the variation of spring stiffness (k) with deformation (x) given by the relations:

a. $k = 1{,}000x - 100x^2; 0 \leq x \leq 4$.
b. $k = 500 + 500x^2; 0 \leq x \leq 4$.

1.122 A mass is subjected to two harmonic motions given by $x_1(t) = 3 \sin 30t$ and $x_2(t) = 3 \sin 29t$. Plot the resultant motion of the mass using MATLAB and identify the beat frequency and the beat period.

DESIGN PROJECTS

1.123* A slider-crank mechanism is shown in Fig. 1.120. Derive an expression for the motion of the piston P in terms of the crank length r, the connecting-rod length l, and the constant angular velocity of the crank ω.

a. Discuss the feasibility of using the mechanism for the generation of harmonic motion.
b. Find the value of l/r for which the amplitude of every higher harmonic is smaller than that of the first harmonic by a factor of at least 25.

FIGURE 1.120

1.124* The vibration table shown in Fig. 1.121 is used to test certain electronic components for vibration. It consists of two identical mating gears G_1 and G_2 that rotate about the axes O_1 and O_2 attached to the frame F. Two equal masses, m each, are placed symmetrically about the middle vertical axis as shown in Fig. 1.121. During rotation, an unbalanced vertical force of magnitude $P = 2m\omega^2 r \cos\theta$, where $\theta = \omega t$ and $\omega =$ angular velocity of gears, will be developed, causing the table to vibrate. Design a vibration table that can develop a force in the range 0–100 N over a frequency range 25–50 Hz.

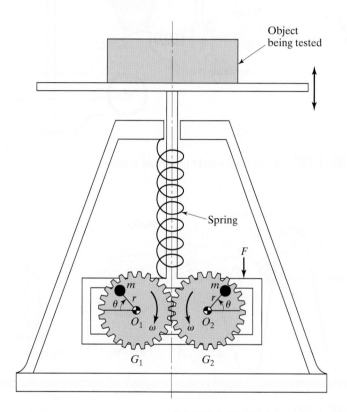

FIGURE 1.121   A vibration table.

1.125* The arrangement shown in Fig. 1.122 is used to regulate the weight of material fed from a hopper to a conveyor [1.44]. The crank imparts a reciprocating motion to the actuating rod through the wedge. The amplitude of motion imparted to the actuating rod can be varied by moving the wedge up or down. Since the conveyor is pivoted about point O, any overload on the conveyor makes the lever OA tilt downward, thereby raising the wedge. This causes a reduction in the amplitude of the actuating rod and hence the feed rate. Design such a weight-regulating system to maintain the weight at 5 ± 0.05 kg/min.

1.126* Figure 1.123 shows a vibratory compactor. It consists of a plate cam with three profiled lobes and an oscillating roller follower. As the cam rotates, the roller drops after each rise. Correspondingly, the weight attached at the end of the follower also rises and drops. The

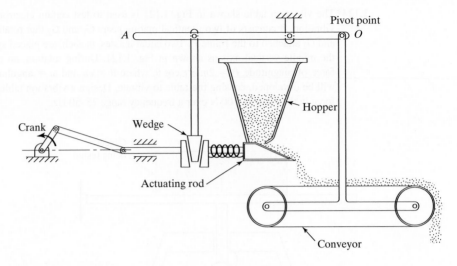

FIGURE 1.122   A vibratory weight-regulating system.

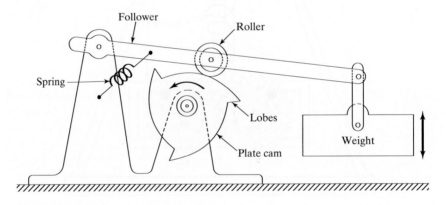

FIGURE 1.123   A vibratory compactor.

contact between the roller and the cam is maintained by the spring. Design a vibration compactor that can apply a force of 1,000 N at a frequency of 50 Hz.

1.127* Vibratory bowl feeders are widely used in automated processes where a high volume of identical parts are to be oriented and delivered at a steady rate to a workstation for further tooling [1.45, 1.46]. Basically, a vibratory bowl feeder is separated from the base by a set of inclined elastic members (springs), as shown in Fig. 1.124. An electromagnetic coil mounted between the bowl and the base provides the driving force to the bowl. The vibratory motion of the bowl causes the components to move along the spiral delivery track located inside the bowl with a hopping motion. Special tooling is fixed at suitable positions along the spiral track in order to reject the parts that are defective or out of tolerance or have

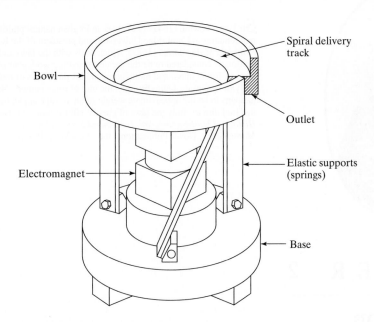

FIGURE 1.124 A vibratory bowl feeder.

incorrect orientation. What factors must be considered in the design of such vibratory bowl feeders?

1.128* The shell-and-tube exchanger shown in Fig. 1.125(a) can be modeled as shown in Fig. 1.125(b) for a simplified vibration analysis. Find the cross-sectional area of the tubes so that the total stiffness of the heat exchanger exceeds a value of 200×10^6 N/m in the axial direction and 20×10^6 N-m/rad in the tangential direction. Assume that the tubes have the same length and cross section and are spaced uniformly.

FIGURE 1.125 (Part (a) courtesy of Young Radiator Company.)

Sir Isaac Newton (1642–1727) was an English natural philosopher, a professor of mathematics at Cambridge University, and president of the Royal Society. His *Principia Mathematica* (1687), which deals with the laws and conditions of motion, is considered to be the greatest scientific work ever produced. The definitions of force, mass, and momentum and his three laws of motion crop up continually in dynamics. Quite fittingly, the unit of force named "Newton" in SI units happens to be the approximate weight of an average apple, the falling object that inspired him to study the laws of gravity. (Illustration of David Eugene Smith, *History of Mathematics, Vol. I—General Survey of the History of Elementary Mathematics*, Dover Publications, Inc., New York, 1958.)

C H A P T E R   2

Free Vibration of Single-Degree-of-Freedom Systems

Chapter Outline

This chapter starts with a consideration of the free vibration of an undamped single-degree-of-freedom (spring-mass) system. Free vibration means that the mass is set into motion due to initial disturbance with no externally applied force other than the spring force, damper force, or gravitational force. To study the free-vibration response of the mass, we need to derive the governing equation, known as the *equation of motion*. The equation of motion of the undamped translational system is derived using four methods. The natural frequency of vibration of the system is defined and the solution of the equation of motion is presented using appropriate initial conditions. The solution is shown to represent harmonic motion. The equation of motion and the solution corresponding to free vibration of an undamped torsional system are presented. The response of first-order systems and the time constant are considered. Rayleigh's method, based on the principle of conservation of energy, is presented with illustrative examples.

Next, the derivation of the equation for the free vibration of a viscously damped single-degree-of-freedom system and its solution are considered. The concepts of critical damping constant, damping ratio, and frequency of damped vibration are introduced. The distinctions between underdamped, critically damped, and overdamped systems are explained. The energy dissipated in viscous damping and the concepts of specific damping and loss coefficient are considered. Viscously damped torsional systems are also considered analogous to viscously damped translational systems with applications. The graphical representation of characteristic roots and the corresponding solutions as well as the concepts of parameter variations and root locus plots are considered. The equations of motion and their solutions of single-degree-of-freedom systems with Coulomb and hysteretic damping are presented. The concept of complex stiffness is also presented. The idea of stability and its importance are explained along with an example. The determination of the responses of single-degree-of-freedom systems using MATLAB is illustrated with examples.

Learning Objectives

After completing this chapter, you should be able to do the following:

- Derive the equation of motion of a single-degree-of-freedom system using a suitable technique such as Newton's second law of motion, D'Alembert's principle, the principle of virtual displacements, and the principle of conservation of energy.
- Linearize the nonlinear equation of motion.
- Solve a spring-mass-damper system for different types of free-vibration response depending on the amount of damping.
- Compute the natural frequency, damped frequency, logarithmic decrement, and time constant.
- Determine whether a given system is stable or not.
- Find the responses of systems with Coulomb and hysteretic damping.
- Find the free-vibration response using MATLAB.

2.1   Introduction

A system is said to undergo free vibration when it oscillates only under an initial disturbance with no external forces acting afterward. Some examples are the oscillations of the pendulum of a grandfather clock, the vertical oscillatory motion felt by a bicyclist after hitting a road bump, and the motion of a child on a swing after an initial push.

Figure 2.1(a) shows a spring-mass system that represents the simplest possible vibratory system. It is called a *single-degree-of-freedom system*, since one coordinate (x) is sufficient to specify the position of the mass at any time. There is no external force applied to the mass; hence the motion resulting from an initial disturbance will be free vibration.

(a)   (b)   (c)

FIGURE 2.1   A spring-mass system in horizontal position.

FIGURE 2.2
Equivalent spring-
mass system for
the cam-follower
system of Fig. 1.32.

Since there is no element that causes dissipation of energy during the motion of the mass, the amplitude of motion remains constant with time; it is an *undamped* system. In actual practice, except in a vacuum, the amplitude of free vibration diminishes gradually over time, due to the resistance offered by the surrounding medium (such as air). Such vibrations are said to be *damped*. The study of the free vibration of undamped and damped single-degree-of-freedom systems is fundamental to the understanding of more advanced topics in vibrations.

Several mechanical and structural systems can be idealized as single-degree-of-freedom systems. In many practical systems, the mass is distributed, but for a simple analysis, it can be approximated by a single point mass. Similarly, the elasticity of the system, which may be distributed throughout the system, can also be idealized by a single spring. For the cam-follower system shown in Fig. 1.39, for example, the various masses were replaced by an equivalent mass (m_{eq}) in Example 1.7. The elements of the follower system (pushrod, rocker arm, valve, and valve spring) are all elastic but can be reduced to a single equivalent spring of stiffness k_{eq}. For a simple analysis, the cam-follower system can thus be idealized as a single-degree-of-freedom spring-mass system, as shown in Fig. 2.2.

Similarly, the structure shown in Fig. 2.3 can be considered a cantilever beam that is fixed at the ground. For the study of transverse vibration, the top mass can be considered a

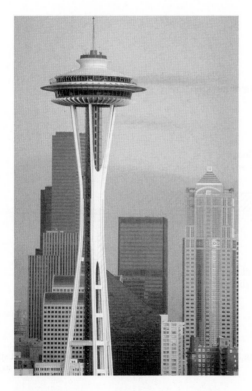

FIGURE 2.3   The space needle (structure).

point mass and the supporting structure (beam) can be approximated as a spring to obtain the single-degree-of-freedom model shown in Fig. 2.4. The building frame shown in Fig. 2.5(a) can also be idealized as a spring-mass system, as shown in Fig. 2.5(b). In this case, since the spring constant k is merely the ratio of force to deflection, it can be determined from the geometric and material properties of the columns. The mass of the idealized system is the same as that of the floor if we assume the mass of the columns to be negligible.

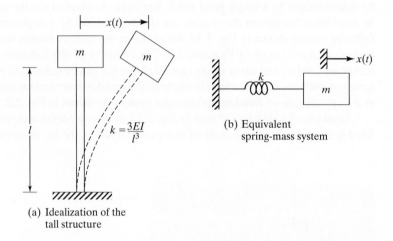

(a) Idealization of the tall structure

(b) Equivalent spring-mass system

FIGURE 2.4   Modeling of tall structure as spring-mass system.

(a) Building frame

(b) Equivalent spring-mass system

FIGURE 2.5   Idealization of a building frame.

2.2   Free Vibration of an Undamped Translational System

**2.2.1
Equation of
Motion Using
Newton's Second
Law of Motion**

Using Newton's second law of motion, in this section we will consider the derivation of the equation of motion. The procedure we will use can be summarized as follows:

1. Select a suitable coordinate to describe the position of the mass or rigid body in the system. Use a linear coordinate to describe the linear motion of a point mass or the centroid of a rigid body, and an angular coordinate to describe the angular motion of a rigid body.
2. Determine the static equilibrium configuration of the system and measure the displacement of the mass or rigid body from its static equilibrium position.
3. Draw the free-body diagram of the mass or rigid body when a positive displacement and velocity are given to it. Indicate all the active and reactive forces acting on the mass or rigid body.
4. Apply Newton's second law of motion to the mass or rigid body shown by the free-body diagram. Newton's second law of motion can be stated as follows:

The rate of change of momentum of a mass is equal to the force acting on it.

Thus, if mass m is displaced a distance $\vec{x}(t)$ when acted upon by a resultant force $\vec{F}(t)$ in the same direction, Newton's second law of motion gives

$$\vec{F}(t) = \frac{d}{dt}\left(m\frac{d\vec{x}(t)}{dt}\right)$$

If mass m is constant, this equation reduces to

$$\vec{F}(t) = m\frac{d^2\vec{x}(t)}{dt^2} = m\ddot{\vec{x}} \tag{2.1}$$

where

$$\ddot{\vec{x}} = \frac{d^2\vec{x}(t)}{dt^2}$$

is the acceleration of the mass. Equation (2.1) can be stated in words as

Resultant force on the mass = mass × acceleration

For a rigid body undergoing rotational motion, Newton's law gives

$$\vec{M}(t) = J\ddot{\vec{\theta}} \tag{2.2}$$

where $\vec{M}$ is the resultant moment acting on the body and $\vec{\theta}$ and $\ddot{\vec{\theta}} = d^2\theta(t)/dt^2$ are the resulting angular displacement and angular acceleration, respectively. Equation (2.1) or (2.2) represents the equation of motion of the vibrating system.

The procedure is now applied to the undamped single-degree-of-freedom system shown in Fig. 2.1(a). Here the mass is supported on frictionless rollers and can have translatory motion in the horizontal direction. When the mass is displaced a distance $+x$ from its static equilibrium position, the force in the spring is kx, and the free-body diagram of the mass can be represented as shown in Fig. 2.1(c). The application of Eq. (2.1) to mass m yields the equation of motion

$$F(t) = -kx = m\ddot{x}$$

or

$$m\ddot{x} + kx = 0 \qquad\qquad (2.3)$$

2.2.2
Equation of
Motion Using
Other Methods

As stated in Section 1.6, the equations of motion of a vibrating system can be derived using several methods. The applications of D'Alembert's principle, the principle of virtual displacements, and the principle of conservation of energy are considered in this section.

D'Alembert's Principle.   The equations of motion, Eqs. (2.1) and (2.2), can be rewritten as

$$\vec{F}(t) - m\ddot{\vec{x}} = 0 \qquad\qquad (2.4a)$$

$$\vec{M}(t) - J\ddot{\vec{\theta}} = 0 \qquad\qquad (2.4b)$$

These equations can be considered equilibrium equations provided that $-m\ddot{\vec{x}}$ and $-J\ddot{\vec{\theta}}$ are treated as a force and a moment. This fictitious force (or moment) is known as the *inertia force* (or *inertia moment*) and the artificial state of equilibrium implied by Eq. (2.4a) or (2.4b) is known as *dynamic equilibrium*. This principle, implied in Eq. (2.4a) or (2.4b), is called *D'Alembert's principle*. Applying it to the system shown in Fig. 2.1(c) yields the equation of motion:

$$-kx - m\ddot{x} = 0 \quad \text{or} \quad m\ddot{x} + kx = 0 \qquad\qquad (2.3)$$

Principle of Virtual Displacements.   The principle of virtual displacements states that "if a system that is in equilibrium under the action of a set of forces is subjected to a virtual displacement, then the total virtual work done by the forces will be zero." Here the virtual displacement is defined as an imaginary infinitesimal displacement given instantaneously. It must be a physically possible displacement that is compatible with the constraints of the system. The virtual work is defined as the work done by all the forces, including the inertia forces for a dynamic problem, due to a virtual displacement.

Consider a spring-mass system in a displaced position as shown in Fig. 2.6(a), where x denotes the displacement of the mass. Figure 2.6(b) shows the free-body diagram of the mass with the reactive and inertia forces indicated. When the mass is given a virtual displacement δx, as shown in Fig. 2.6(b), the virtual work done by each force can be computed as follows:

$$\text{Virtual work done by the spring force} = \delta W_s = -(kx)\,\delta x$$

$$\text{Virtual work done by the inertia force} = \delta W_i = -(m\ddot{x})\,\delta x$$

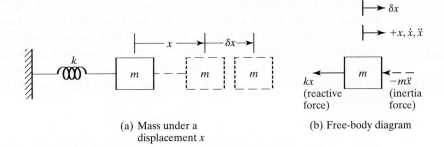

(a) Mass under a
displacement x

(b) Free-body diagram

FIGURE 2.6   Mass under virtual displacement.

When the total virtual work done by all the forces is set equal to zero, we obtain

$$-m\ddot{x}\delta x - kx\,\delta x = 0 \qquad (2.5)$$

Since the virtual displacement can have an arbitrary value, $\delta x \neq 0$, Eq. (2.5) gives the equation of motion of the spring-mass system as

$$m\ddot{x} + kx = 0 \qquad (2.3)$$

Principle of Conservation of Energy.   A system is said to be conservative if no energy is lost due to friction or energy-dissipating nonelastic members. If no work is done on a conservative system by external forces (other than gravity or other potential forces), then the total energy of the system remains constant. Since the energy of a vibrating system is partly potential and partly kinetic, the sum of these two energies remains constant. The kinetic energy T is stored in the mass by virtue of its velocity, and the potential energy U is stored in the spring by virtue of its elastic deformation. Thus the principle of conservation of energy can be expressed as:

$$T + U = \text{constant}$$

or

$$\frac{d}{dt}(T + U) = 0 \qquad (2.6)$$

The kinetic and potential energies are given by

$$T = \tfrac{1}{2}m\dot{x}^2 \qquad (2.7)$$

and

$$U = \tfrac{1}{2}kx^2 \qquad (2.8)$$

Substitution of Eqs. (2.7) and (2.8) into Eq. (2.6) yields the desired equation

$$m\ddot{x} + kx = 0 \qquad (2.3)$$

2.2.3
Equation of
Motion of a
Spring-Mass
System in
Vertical Position

Consider the configuration of the spring-mass system shown in Fig. 2.7(a). The mass hangs at the lower end of a spring, which in turn is attached to a rigid support at its upper end. At rest, the mass will hang in a position called the *static equilibrium position*, in which the upward spring force exactly balances the downward gravitational force on the mass. In this position, the length of the spring is $l_0 + \delta_{st}$, where δ_{st} is the static deflection— the elongation due to the weight W of the mass m. From Fig. 2.7(a), we find that, for static equilibrium,

$$W = mg = k\delta_{st} \tag{2.9}$$

where g is the acceleration due to gravity. Let the mass be deflected a distance $+x$ from its static equilibrium position; then the spring force is $-k(x + \delta_{st})$, as shown in Fig. 2.7(c). The application of Newton's second law of motion to mass m gives

$$m\ddot{x} = -k(x + \delta_{st}) + W$$

and since $k\delta_{st} = W$, we obtain

$$m\ddot{x} + kx = 0 \tag{2.10}$$

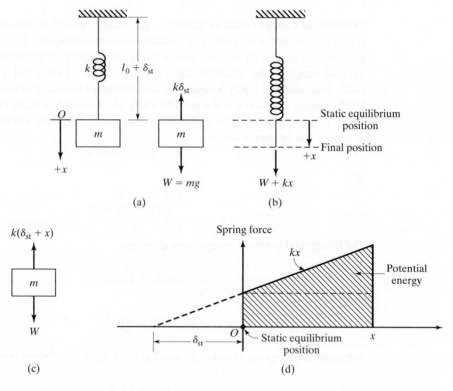

FIGURE 2.7   A spring-mass system in vertical position.

Notice that Eqs. (2.3) and (2.10) are identical. This indicates that when a mass moves in a vertical direction, we can ignore its weight, provided we measure x from its static equilibrium position.

Note: Equation (2.10), the equation of motion of the system shown in Fig. 2.7, can also be derived using D'Alembert's principle, the principle of virtual displacements, or the principle of conservation of energy. If we use the latter, for example, we note that the expression for the kinetic energy, T, remains the same as Eq. (2.7). However, the expression for the potential energy, U, is to be derived by considering the weight of the mass. For this we note that the spring force at static equilibrium position $(x = 0)$ is mg. When the spring deflects by an amount x, its potential energy is given by (see Fig. 2.7(d)):

$$mgx + \frac{1}{2}kx^2$$

Furthermore, the potential energy of the system due to the change in elevation of the mass (note that $+x$ is downward) is $-mgx$. Thus the net potential energy of the system about the static equilibrium position is given by

$$U = \text{potential energy of the spring}$$

$$+ \text{ change in potential energy due to change in elevation of the mass } m$$

$$= mgx + \frac{1}{2}kx^2 - mgx = \frac{1}{2}kx^2$$

Since the expressions of T and U remain unchanged, the application of the principle of conservation of energy gives the same equation of motion, Eq. (2.3).

2.2.4
Solution

The solution of Eq. (2.3) can be found by assuming

$$x(t) = Ce^{st} \tag{2.11}$$

where C and s are constants to be determined. Substitution of Eq. (2.11) into Eq. (2.3) gives

$$C(ms^2 + k) = 0$$

Since C cannot be zero, we have

$$ms^2 + k = 0 \tag{2.12}$$

and hence

$$s = \pm\left(-\frac{k}{m}\right)^{1/2} = \pm i\omega_n \tag{2.13}$$

where $i = (-1)^{1/2}$ and

$$\omega_n = \left(\frac{k}{m}\right)^{1/2} \tag{2.14}$$

Equation (2.12) is called the *auxiliary* or the *characteristic equation* corresponding to the differential Eq. (2.3). The two values of s given by Eq. (2.13) are the roots of the characteristic equation, also known as the *eigenvalues* or the *characteristic values* of the problem. Since both values of s satisfy Eq. (2.12), the general solution of Eq. (2.3) can be expressed as

$$x(t) = C_1 e^{i\omega_n t} + C_2 e^{-i\omega_n t} \tag{2.15}$$

where C_1 and C_2 are constants. By using the identities

$$e^{\pm i\alpha t} = \cos \alpha t \pm i \sin \alpha t$$

Eq. (2.15) can be rewritten as

$$x(t) = A_1 \cos \omega_n t + A_2 \sin \omega_n t \tag{2.16}$$

where A_1 and A_2 are new constants. The constants C_1 and C_2 or A_1 and A_2 can be determined from the initial conditions of the system. Two conditions are to be specified to evaluate these constants uniquely. Note that the number of conditions to be specified is the same as the order of the governing differential equation. In the present case, if the values of displacement $x(t)$ and velocity $\dot{x}(t) = (dx/dt)(t)$ are specified as x_0 and $\dot{x}_0$ at $t = 0$, we have, from Eq. (2.16),

$$x(t = 0) = A_1 = x_0$$
$$\dot{x}(t = 0) = \omega_n A_2 = \dot{x}_0 \tag{2.17}$$

Hence $A_1 = x_0$ and $A_2 = \dot{x}_0/\omega_n$. Thus the solution of Eq. (2.3) subject to the initial conditions of Eq. (2.17) is given by

$$x(t) = x_0 \cos \omega_n t + \frac{\dot{x}_0}{\omega_n} \sin \omega_n t \tag{2.18}$$

**2.2.5
Harmonic
Motion**

Equations (2.15), (2.16), and (2.18) are harmonic functions of time. The motion is symmetric about the equilibrium position of the mass m. The velocity is a maximum and the acceleration is zero each time the mass passes through this position. At the extreme displacements, the velocity is zero and the acceleration is a maximum. Since this represents simple harmonic motion (see Section 1.10), the spring-mass system itself is called a *harmonic oscillator.* The quantity ω_n, given by Eq. (2.14), represents the system's natural frequency of vibration.

Equation (2.16) can be expressed in a different form by introducing the notation

$$A_1 = A \cos \phi$$
$$A_2 = A \sin \phi \tag{2.19}$$

where A and ϕ are the new constants, which can be expressed in terms of A_1 and A_2 as

$$A = (A_1^2 + A_2^2)^{1/2} = \left[x_0^2 + \left(\frac{\dot{x}_0}{\omega_n} \right)^2 \right]^{1/2} = \text{amplitude}$$

$$\phi = \tan^{-1} \left(\frac{A_2}{A_1} \right) = \tan^{-1} \left(\frac{\dot{x}_0}{x_0 \omega_n} \right) = \text{phase angle} \tag{2.20}$$

Introducing Eq. (2.19) into Eq. (2.16), the solution can be written as

$$x(t) = A \cos (\omega_n t - \phi) \tag{2.21}$$

By using the relations

$$A_1 = A_0 \sin \phi_0$$

$$A_2 = A_0 \cos \phi_0 \tag{2.22}$$

Eq. (2.16) can also be expressed as

$$x(t) = A_0 \sin (\omega_n t + \phi_0) \tag{2.23}$$

where

$$A_0 = A = \left[x_0^2 + \left(\frac{\dot{x}_0}{\omega_n} \right)^2 \right]^{1/2} \tag{2.24}$$

and

$$\phi_0 = \tan^{-1} \left(\frac{x_0 \omega_n}{\dot{x}_0} \right) \tag{2.25}$$

The nature of harmonic oscillation can be represented graphically as in Fig. 2.8(a). If $\vec{A}$ denotes a vector of magnitude A, which makes an angle $\omega_n t - \phi$ with respect to the vertical (x) axis, then the solution, Eq. (2.21), can be seen to be the projection of the vector $\vec{A}$ on the x-axis. The constants A_1 and A_2 of Eq. (2.16), given by Eq. (2.19), are merely the rectangular components of $\vec{A}$ along two orthogonal axes making angles ϕ and $-(\frac{\pi}{2} - \phi)$ with respect to the vector $\vec{A}$. Since the angle $\omega_n t - \phi$ is a linear function of time, it increases linearly with time; the entire diagram thus rotates counterclockwise at an angular velocity ω_n. As the diagram (Fig. 2.8(a)) rotates, the projection of $\vec{A}$ onto the x-axis varies harmonically so that the motion repeats itself every time the vector $\vec{A}$ sweeps an angle of 2π. The projection of $\vec{A}$, namely $x(t)$, is shown plotted in Fig. 2.8(b) as a function of $\omega_n t$, and as a function of t in Fig. 2.8(c). The phase angle ϕ can also be interpreted as the angle between the origin and the first peak.

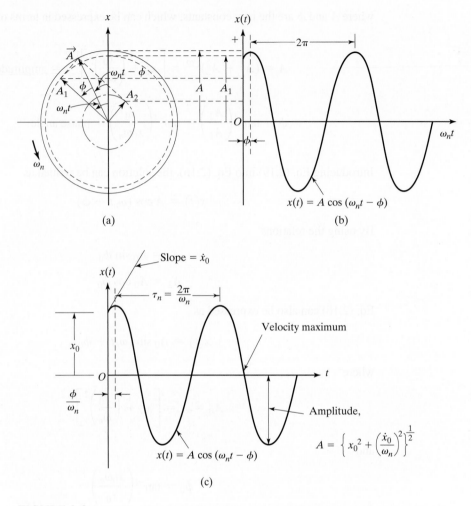

FIGURE 2.8 Graphical representation of the motion of a harmonic oscillator.

Note the following aspects of the spring-mass system:

1. If the spring-mass system is in a vertical position, as shown in Fig. 2.7(a), the circular natural frequency can be expressed as

$$\omega_n = \left(\frac{k}{m} \right)^{1/2} \tag{2.26}$$

The spring constant k can be expressed in terms of the mass m from Eq. (2.9) as

$$k = \frac{W}{\delta_{st}} = \frac{mg}{\delta_{st}} \tag{2.27}$$

Substitution of Eq. (2.27) into Eq. (2.14) yields

$$\omega_n = \left(\frac{g}{\delta_{st}}\right)^{1/2} \tag{2.28}$$

Hence the natural frequency in cycles per second and the natural period are given by

$$f_n = \frac{1}{2\pi}\left(\frac{g}{\delta_{st}}\right)^{1/2} \tag{2.29}$$

$$\tau_n = \frac{1}{f_n} = 2\pi\left(\frac{\delta_{st}}{g}\right)^{1/2} \tag{2.30}$$

Thus, when the mass vibrates in a vertical direction, we can compute the natural frequency and the period of vibration by simply measuring the static deflection δ_{st}. We do not need to know the spring stiffness k and the mass m.

2. From Eq. (2.21), the velocity $\dot{x}(t)$ and the acceleration $\ddot{x}(t)$ of the mass m at time t can be obtained as

$$\dot{x}(t) = \frac{dx}{dt}(t) = -\omega_n A \sin(\omega_n t - \phi) = \omega_n A \cos\left(\omega_n t - \phi + \frac{\pi}{2}\right)$$

$$\ddot{x}(t) = \frac{d^2 x}{dt^2}(t) = -\omega_n^2 A \cos(\omega_n t - \phi) = \omega_n^2 A \cos(\omega_n t - \phi + \pi) \tag{2.31}$$

Equation (2.31) shows that the velocity leads the displacement by $\pi/2$ and the acceleration leads the displacement by π.

3. If the initial displacement (x_0) is zero, Eq. (2.21) becomes

$$x(t) = \frac{\dot{x}_0}{\omega_n}\cos\left(\omega_n t - \frac{\pi}{2}\right) = \frac{\dot{x}_0}{\omega_n}\sin\omega_n t \tag{2.32}$$

If the initial velocity $(\dot{x}_0)$ is zero, however, the solution becomes

$$x(t) = x_0 \cos\omega_n t \tag{2.33}$$

4. The response of a single-degree-of-freedom system can be represented in the displacement (x)-velocity -plane, known as the *state space* or *phase plane*. For this, we consider the displacement given by Eq. (2.21) and the corresponding velocity:

$$x(t) = A \cos (\omega_n t - \phi)$$

or

$$\cos(\omega_n t - \phi) = \frac{x}{A}$$

$$\dot{x}(t) = -A\omega_n \sin (\omega_n t - \phi) \tag{2.34}$$

or

$$\sin(\omega_n t - \phi) = -\frac{\dot{x}}{A\omega_n} = -\frac{y}{A} \tag{2.35}$$

where $y = \dot{x}/\omega_n$. By squaring and adding Eqs. (2.34) and (2.35), we obtain

$$\cos^2(\omega_n t - \phi) + \sin^2(\omega_n t - \phi) = 1$$

or

$$\frac{x^2}{A^2} + \frac{y^2}{A^2} = 1 \tag{2.36}$$

The graph of Eq. (2.36) in the xy-plane is a circle, as shown in Fig. 2.9(a), and it constitutes the phase-plane or state-space representation of the undamped system. The radius of the circle, A, is determined by the initial conditions of motion. Note that the graph of Eq. (2.36) in the $x\dot{x}$-plane will be an ellipse, as shown in Fig. 2.9(b).

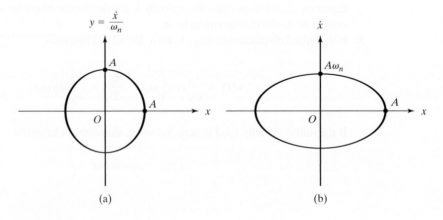

(a)   (b)

FIGURE 2.9   Phase-plane representation of an undamped system.

Harmonic Response of a Water Tank

EXAMPLE   2.1

The column of the water tank shown in Fig. 2.10(a) is 90 m high and is made of reinforced concrete with a tubular cross section of inner diameter 2.4 m and outer diameter 3 m. The tank has a mass of 3×10^5 kg when filled with water. By neglecting the mass of the column and assuming the Young's modulus of reinforced concrete as 30 GPa, determine the following:

a.   the natural frequency and the natural time period of transverse vibration of the water tank.
b.   the vibration response of the water tank due to an initial transverse displacement of 0.3 m.
c.   the maximum values of the velocity and acceleration experienced by the water tank.

Solution: Assuming that the water tank is a point mass, the column has a uniform cross section, and the mass of the column is negligible, the system can be modeled as a cantilever beam with a concentrated load (weight) at the free end as shown in Fig. 2.10(b).

a.   The transverse deflection of the beam, δ, due to a load P is given by $\frac{Pl^3}{3EI}$, where l is the length, E is the Young's modulus, and I is the area moment of inertia of the beam's cross section. The stiffness of the beam (column of the tank) is given by

$$k = \frac{P}{\delta} = \frac{3EI}{l^3}$$

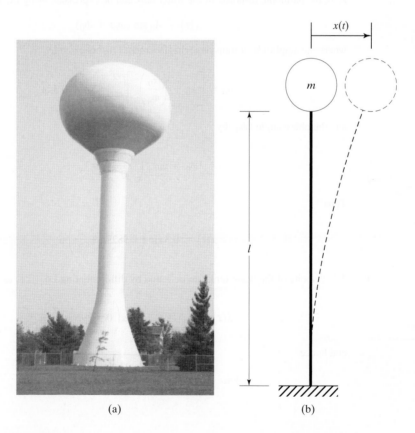

(a)   (b)

FIGURE 2.10   Elevated tank. (Photo courtesy of West Lafayette Water Company.)

In the present case, $l = 90$ m, $E = 30$ GPa,

$$I = \frac{\pi}{64}(d_0^4 - d_i^4) = \frac{\pi}{64}(3^4 - 2.4^4) = 2.3475 \text{ m}^4$$

and hence

$$k = \frac{3 \times 30 \times 10^9 \times 2.3475}{90^3} = 289,812 \text{ N/m}$$

The natural frequency of the water tank in the transverse direction is given by

$$\omega_n = \sqrt{\frac{k}{m}} = \sqrt{\frac{289,812}{3 \times 10^5}} = 0.9829 \text{ rad/s}$$

The natural time period of transverse vibration of the tank is given by

$$\tau_n = \frac{2\pi}{\omega_n} = \frac{2\pi}{0.9829} = 6.3925 \text{ s}$$

b. Using the initial displacement of $x_0 = 0.3$ m and the initial velocity of the water tank $(\dot{x}_0)$ as zero, the harmonic response of the water tank can be expressed, using Eq. (2.23), as

$$x(t) = A_0 \sin(\omega_n t + \phi_0)$$

where the amplitude of transverse displacement (A_0) is given by

$$A_0 = \left[x_0^2 + \left(\frac{\dot{x}_0}{\omega_n} \right)^2 \right]^{1/2} = x_0 = 0.3 \text{ m}$$

and the phase angle (ϕ_0) by

$$\phi_0 = \tan^{-1}\left(\frac{x_0 \omega_n}{0} \right) = \frac{\pi}{2}$$

Thus

$$x(t) = 0.3 \sin\left(0.9829t + \frac{\pi}{2} \right) \text{m} \qquad (E.1)$$

c. The velocity of the water tank can be found by differentiating Eq. (E.1) as

$$\dot{x}(t) = 0.3(0.9829)\cos\left(0.9829t + \frac{\pi}{2} \right) \qquad (E.2)$$

and hence

$$\dot{x}_{\text{max}} = A_0 \omega_n = 0.3(0.9829) = 0.2949 \text{ m/s}$$

The acceleration of the water tank can be determined by differentiating Eq. (E.2) as

$$\ddot{x}(t) = -0.3(0.9829)^2 \sin\left(0.9829t + \frac{\pi}{2}\right)$$ (E.3)

and hence the maximum value of acceleration is given by

$$\ddot{x}_{\max} = A_0(\omega_n)^2 = 0.3(0.9829)^2 = 0.2898 \text{ m/s}^2$$

■

EXAMPLE 2.2

Free-Vibration Response Due to Impact

A cantilever beam carries a mass M at the free end as shown in Fig. 2.11(a). A mass m falls from a height h onto the mass M and adheres to it without rebounding. Determine the resulting transverse vibration of the beam.

Solution: When the mass m falls through a height h, it will strike the mass M with a velocity of $v_m = \sqrt{2gh}$, where g is the acceleration due to gravity. Since the mass m adheres to M without rebounding, the velocity of the combined mass $(M + m)$ immediately after the impact $(\dot{x}_0)$ can be found using the principle of conservation of momentum:

$$mv_m = (M + m)\dot{x}_0$$

$$YY = \text{static equilibrium position of } M$$
$$ZZ = \text{static equilibrium position of } M + m$$

FIGURE 2.11 Response due to impact.

or

$$\dot{x}_0 = \left(\frac{m}{M + m} \right) v_m = \left(\frac{m}{M + m} \right) \sqrt{2gh} \qquad \text{(E.1)}$$

The static equilibrium position of the beam with the new mass $(M + m)$ is located at a distance of $\frac{mg}{k}$ below the static equilibrium position of the original mass (M) as shown in Fig. 2.11(c). Here k denotes the stiffness of the cantilever beam, given by

$$k = \frac{3EI}{l^3}$$

Since free vibration of the beam with the new mass $(M + m)$ occurs about its own static equilibrium position, the initial conditions of the problem can be stated as

$$x_0 = -\frac{mg}{k}, \qquad \dot{x}_0 = \left(\frac{m}{M + m} \right) \sqrt{2gh} \qquad \text{(E.2)}$$

Thus the resulting free transverse vibration of the beam can be expressed as (see Eq. (2.21)):

$$x(t) = A \cos (\omega_n t - \phi)$$

where

$$A = \left[x_0^2 + \left(\frac{\dot{x}_0}{\omega_n} \right)^2 \right]^{1/2}$$

$$\phi = \tan^{-1} \left(\frac{\dot{x}_0}{x_0 \omega_n} \right)$$

$$\omega_n = \sqrt{\frac{k}{M + m}} = \sqrt{\frac{3EI}{l^3(M + m)}}$$

with x_0 and $\dot{x}_0$ given by Eq. (E.2).

■

Young's Modulus from Natural Frequency Measurement

EXAMPLE 2.3

A simply supported beam of square cross section 5 mm × 5 mm and length 1 m, carrying a mass of 2.3 kg at the middle, is found to have a natural frequency of transverse vibration of 30 rad/s. Determine the Young's modulus of elasticity of the beam.

Solution: By neglecting the self weight of the beam, the natural frequency of transverse vibration of the beam can be expressed as

$$\omega_n = \sqrt{\frac{k}{m}} \qquad \text{(E.1)}$$

where

$$k = \frac{192EI}{l^3} \tag{E.2}$$

where E is the Young's modulus, l is the length, and I is the area moment of inertia of the beam:

$$I = \frac{1}{12}(5 \times 10^{-3})(5 \times 10^{-3})^3 = 0.5208 \times 10^{-10} \text{ m}^4$$

Since $m = 2.3$ kg, $l = 1.0$ m, and $\omega_n = 30.0$ rad/s, Eqs. (E.1) and (E.2) yield

$$k = \frac{192EI}{l^3} = m\omega_n^2$$

or

$$E = \frac{m\omega_n^2 l^3}{192I} = \frac{2.3(30.0)^2(1.0)^3}{192(0.5208 \times 10^{-10})} = 207.0132 \times 10^9 \text{ N/m}^2$$

This indicates that the material of the beam is probably carbon steel.

■

EXAMPLE 2.4

Natural Frequency of Cockpit of a Firetruck

The cockpit of a firetruck is located at the end of a telescoping boom, as shown in Fig. 2.12(a). The cockpit, along with the fireman, weighs 2,000 N. Find the cockpit's natural frequency of vibration in the vertical direction.

Data: Young's modulus of the material: $E = 2.1 \times 10^{11}$ N/m^2; lengths: $l_1 = l_2 = l_3 = 3$ m; cross-sectional areas: $A_1 = 20$ cm^2, $A_2 = 10$ cm^2, $A_3 = 5$ cm^2.

Solution: To determine the system's natural frequency of vibration, we find the equivalent stiffness of the boom in the vertical direction and use a single-degree-of-freedom idealization. For this, we assume that the mass of the telescoping boom is negligible and the boom can deform only in the axial direction (with no bending). Since the force induced at any cross section $O_1 O_2$ is equal to the axial load applied at the end of the boom, as shown in Fig. 2.12(b), the axial stiffness of the boom (k_b) is given by

$$\frac{1}{k_b} = \frac{1}{k_{b_1}} + \frac{1}{k_{b_2}} + \frac{1}{k_{b_3}} \tag{E.1}$$

where k_{b_i} denotes the axial stiffness of the ith segment of the boom:

$$k_{b_i} = \frac{A_i E_i}{l_i}; \qquad i = 1, 2, 3 \tag{E.2}$$

From the known data ($l_1 = l_2 = l_3 = 3$ m, $A_1 = 20$ cm^2, $A_2 = 10$ cm^2, $A_3 = 5$ cm^2, $E_1 = E_2 = E_3 = 2.1 \times 10^{11}$ N/m^2),

$$k_{b_1} = \frac{(20 \times 10^{-4})(2.1 \times 10^{11})}{3} = 14 \times 10^7 \text{ N/m}$$

$$k_{b_2} = \frac{(10 \times 10^{-4})(2.1 \times 10^{11})}{3} = 7 \times 10^7 \text{ N/m}$$

$$k_{b_3} = \frac{(5 \times 10^{-4})(2.1 \times 10^{11})}{3} = 3.5 \times 10^7 \text{ N/m}$$

Thus Eq. (E.1) gives

$$\frac{1}{k_b} = \frac{1}{14 \times 10^7} + \frac{1}{7 \times 10^7} + \frac{1}{3.5 \times 10^7} = \frac{1}{2 \times 10^7}$$

or

$$k_b = 2 \times 10^7 \text{ N/m}$$

The stiffness of the telescoping boom in the vertical direction, k, can be determined as

$$k = k_b \cos^2 45° = 10^7 \text{ N/m}$$

The natural frequency of vibration of the cockpit in the vertical direction is given by

$$\omega_n = \sqrt{\frac{k}{m}} = \sqrt{\frac{(10^7)(9.81)}{2,000}} = 221.4723 \text{ rad/s}$$

FIGURE 2.12   Telescoping boom of a fire truck.

Natural Frequency of Pulley System

EXAMPLE 2.5

Determine the natural frequency of the system shown in Fig. 2.13(a). Assume the pulleys to be frictionless and of negligible mass.

Solution: To determine the natural frequency, we find the equivalent stiffness of the system and solve it as a single-degree-of-freedom problem. Since the pulleys are frictionless and massless, the tension in the rope is constant and is equal to the weight W of the mass m. From the static equilibrium of the pulleys and the mass (see Fig. 2.13(b)), it can be seen that the upward force acting on pulley 1 is $2W$ and the downward force acting on pulley 2 is $2W$. The center of pulley 1 (point A) moves up by a distance $2W/k_1$, and the center of pulley 2 (point B) moves down by $2W/k_2$. Thus the total movement of the mass m (point O) is

$$2\left(\frac{2W}{k_1} + \frac{2W}{k_2}\right)$$

as the rope on either side of the pulley is free to move the mass downward. If k_{eq} denotes the equivalent spring constant of the system,

$$\frac{\text{Weight of the mass}}{\text{Equivalent spring constant}} = \text{Net displacement of the mass}$$

(a)   (b)

FIGURE 2.13   Pulley system.

$$\frac{W}{k_{eq}} = 4W\left(\frac{1}{k_1} + \frac{1}{k_2}\right) = \frac{4W(k_1 + k_2)}{k_1 k_2}$$

$$k_{eq} = \frac{k_1 k_2}{4(k_1 + k_2)} \tag{E.1}$$

By displacing mass m from the static equilibrium position by x, the equation of motion of the mass can be written as

$$m\ddot{x} + k_{eq}x = 0 \tag{E.2}$$

and hence the natural frequency is given by

$$\omega_n = \left(\frac{k_{eq}}{m}\right)^{1/2} = \left[\frac{k_1 k_2}{4m(k_1 + k_2)}\right]^{1/2} \text{ rad/s} \tag{E.3}$$

or

$$f_n = \frac{\omega_n}{2\pi} = \frac{1}{4\pi}\left[\frac{k_1 k_2}{m(k_1 + k_2)}\right]^{1/2} \text{ cycles/s} \tag{E.4}$$

∎

2.3   Free Vibration of an Undamped Torsional System

If a rigid body oscillates about a specific reference axis, the resulting motion is called *torsional vibration*. In this case, the displacement of the body is measured in terms of an angular coordinate. In a torsional vibration problem, the restoring moment may be due to the torsion of an elastic member or to the unbalanced moment of a force or couple.

Figure 2.14 shows a disc, which has a polar mass moment of inertia J_0, mounted at one end of a solid circular shaft, the other end of which is fixed. Let the angular rotation of the disc about the axis of the shaft be θ; θ also represents the shaft's angle of twist. From the theory of torsion of circular shafts [2.1], we have the relation

$$M_t = \frac{GI_0}{l}\theta \tag{2.37}$$

where M_t is the torque that produces the twist θ, G is the shear modulus, l is the length of the shaft, I_o is the polar moment of inertia of the cross section of the shaft, given by

$$I_0 = \frac{\pi d^4}{32} \tag{2.38}$$

and d is the diameter of the shaft. If the disc is displaced by θ from its equilibrium position, the shaft provides a restoring torque of magnitude M_t. Thus the shaft acts as a torsional spring with a torsional spring constant

$$k_t = \frac{M_t}{\theta} = \frac{GI_0}{l} = \frac{\pi G d^4}{32l} \tag{2.39}$$

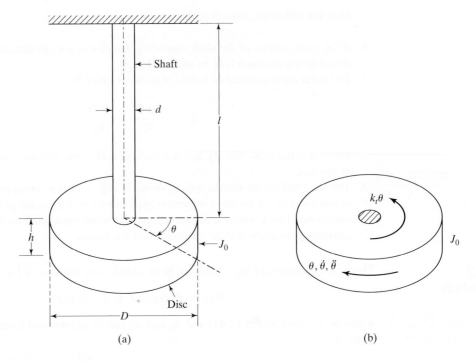

FIGURE 2.14   Torsional vibration of a disc.

2.3.1
Equation
of Motion

The equation of the angular motion of the disc about its axis can be derived by using Newton's second law or any of the methods discussed in Section 2.2.2. By considering the free-body diagram of the disc (Fig. 2.14(b)), we can derive the equation of motion by applying Newton's second law of motion:

$$J_0 \ddot{\theta} + k_t \theta = 0 \tag{2.40}$$

which can be seen to be identical to Eq. (2.3) if the polar mass moment of inertia J_0, the angular displacement θ, and the torsional spring constant k_t are replaced by the mass m, the displacement x, and the linear spring constant k, respectively. Thus the natural circular frequency of the torsional system is

$$\omega_n = \left(\frac{k_t}{J_0} \right)^{1/2} \tag{2.41}$$

and the period and frequency of vibration in cycles per second are

$$\tau_n = 2\pi \left(\frac{J_0}{k_t} \right)^{1/2} \tag{2.42}$$

$$f_n = \frac{1}{2\pi} \left(\frac{k_t}{J_0} \right)^{1/2} \tag{2.43}$$

Note the following aspects of this system:

1. If the cross section of the shaft supporting the disc is not circular, an appropriate torsional spring constant is to be used [2.4, 2.5].
2. The polar mass moment of inertia of a disc is given by

$$J_0 = \frac{\rho h \pi D^4}{32} = \frac{WD^2}{8g}$$

where ρ is the mass density, h is the thickness, D is the diameter, and W is the weight of the disc.
3. The torsional spring-inertia system shown in Fig. 2.14 is referred to as a *torsional pendulum*. One of the most important applications of a torsional pendulum is in a mechanical clock, where a ratchet and pawl convert the regular oscillation of a small torsional pendulum into the movements of the hands.

2.3.2
Solution

The general solution of Eq. (2.40) can be obtained, as in the case of Eq. (2.3):

$$\theta(t) = A_1 \cos \omega_n t + A_2 \sin \omega_n t \tag{2.44}$$

where ω_n is given by Eq. (2.41) and A_1 and A_2 can be determined from the initial conditions. If

$$\theta(t = 0) = \theta_0 \quad \text{and} \quad \dot{\theta}(t = 0) = \frac{d\theta}{dt}(t = 0) = \dot{\theta}_0 \tag{2.45}$$

the constants A_1 and A_2 can be found:

$$A_1 = \theta_0$$
$$A_2 = \dot{\theta}_0/\omega_n \tag{2.46}$$

Equation (2.44) can also be seen to represent a simple harmonic motion.

EXAMPLE 2.6

Natural Frequency of Compound Pendulum

Any rigid body pivoted at a point other than its center of mass will oscillate about the pivot point under its own gravitational force. Such a system is known as a *compound pendulum* (Fig. 2.15). Find the natural frequency of such a system.

Solution: Let O be the point of suspension and G be the center of mass of the compound pendulum, as shown in Fig. 2.15. Let the rigid body oscillate in the xy-plane so that the coordinate θ can be used to describe its motion. Let d denote the distance between O and G, and J_0 the mass moment of inertia of the body about the z-axis (perpendicular to both x and y). For a displacement θ, the restoring torque due to the weight of the body W is $Wd \sin \theta$ and the equation of motion is

$$J_0 \ddot{\theta} + Wd \sin \theta = 0 \tag{E.1}$$

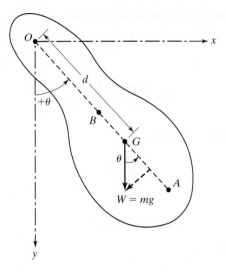

FIGURE 2.15   Compound pendulum.

Note that Eq. (E.1) is a second-order nonlinear ordinary differential equation. Although it is possible to find an exact solution of this equation (see Section 13.3), exact solutions cannot be found for most nonlinear differential equations. An approximate solution of Eq. (E.1) can be found by one of two methods. A numerical procedure can be used to integrate Eq. (E.1). Alternatively, Eq. (E.1) can be approximated by a linear equation whose exact solution can be determined readily. To use the latter approach, we assume small angular displacements so that θ is small and $\sin\theta \approx \theta$. Hence Eq. (E.1) can be approximated by the linear equation:

$$J_0 \ddot{\theta} + Wd\theta = 0 \tag{E.2}$$

This gives the natural frequency of the compound pendulum:

$$\omega_n = \left(\frac{Wd}{J_0}\right)^{1/2} = \left(\frac{mgd}{J_0}\right)^{1/2} \tag{E.3}$$

Comparing Eq. (E.3) with the natural frequency of a simple pendulum, $\omega_n = (g/l)^{1/2}$ (see Problem 2.61), we can find the length of the equivalent simple pendulum:

$$l = \frac{J_0}{md} \tag{E.4}$$

If J_0 is replaced by mk_0^2, where k_0 is the radius of gyration of the body about O, Eqs. (E.3) and (E.4) become

$$\omega_n = \left(\frac{gd}{k_0^2}\right)^{1/2} \tag{E.5}$$

$$l = \left(\frac{k_0^2}{d}\right) \tag{E.6}$$

If k_G denotes the radius of gyration of the body about G, we have

$$k_0^2 = k_G^2 + d^2 \tag{E.7}$$

and Eq. (E.6) becomes

$$l = \left(\frac{k_G^2}{d} + d \right) \tag{E.8}$$

If the line OG is extended to point A such that

$$GA = \frac{k_G^2}{d} \tag{E.9}$$

Eq. (E.8) becomes

$$l = GA + d = OA \tag{E.10}$$

Hence, from Eq. (E.5), ω_n is given by

$$\omega_n = \left\{ \frac{g}{(k_0^2/d)} \right\}^{1/2} = \left(\frac{g}{l} \right)^{1/2} = \left(\frac{g}{OA} \right)^{1/2} \tag{E.11}$$

This equation shows that, no matter whether the body is pivoted from O or A, its natural frequency is the same. The point A is called the *center of percussion*.

■

Center of Percussion. The concepts of compound pendulum and center of percussion can be used in many practical applications:

1. A hammer can be shaped to have the center of percussion at the hammer head while the center of rotation is at the handle. In this case, the impact force at the hammer head will not cause any normal reaction at the handle (Fig. 2.16(a)).
2. In a baseball bat, if on one hand the ball is made to strike at the center of percussion while the center of rotation is at the hands, no reaction perpendicular to the bat will be experienced by the batter (Fig. 2.16(b)). On the other hand, if the ball strikes the bat near the free end or near the hands, the batter will experience pain in the hands as a result of the reaction perpendicular to the bat.
3. In Izod (impact) testing of materials, the specimen is suitably notched and held in a vise fixed to the base of the machine (see Fig. 2.16(c)). A pendulum is released from a standard height, and the free end of the specimen is struck by the pendulum as it passes through its lowest position. The deformation and bending of the pendulum can be reduced if the center of percussion is located near the striking edge. In this case, the pivot will be free of any impulsive reaction.
4. In an automobile (shown in Fig. 2.16(d)), if the front wheels strike a bump, the passengers will not feel any reaction if the center of percussion of the vehicle is located near the rear axle. Similarly, if the rear wheels strike a bump at point A, no reaction

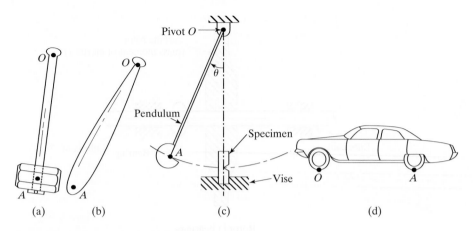

FIGURE 2.16   Applications of center of percussion.

will be felt at the front axle (point O) if the center of percussion is located near the front axle. It is desirable, therefore, to have the center of oscillation of the vehicle at one axle and the center of percussion at the other axle [2.2].

2.4   Response of First-Order Systems and Time Constant

Consider a turbine rotor mounted in bearings as shown in Fig. 2.17(a). The viscous fluid (lubricant) in the bearings offers viscous damping torque during the rotation of the turbine rotor. Assuming the mass moment of inertia of the rotor about the axis of rotation as J and the rotational damping constant of the bearings as c_t, the application of Newton's second law of motion yields the equation of motion of the rotor as

$$J\dot{\omega} + c_t\omega = 0 \tag{2.47}$$

where ω is the angular velocity of the rotor, $\dot{\omega} = \dfrac{d\omega}{dt}$ is the time rate of change of the angular velocity, and the external torque applied to the system is assumed to be zero. We assume the initial angular velocity, $\omega(t = 0) = \omega_0$, as the input and the angular velocity of the rotor as the output of the system. Note that the angular velocity, instead of the angular displacement, is considered as the output in order to obtain the equation of motion as a first-order differential equation.

The solution of the equation of motion of the rotor, Eq. (2.47), can be found by assuming the trial solution as

$$\omega(t) = Ae^{st} \tag{2.48}$$

where A and s are unknown constants. By using the initial condition, $\omega(t = 0) = \omega_0$, Eq. (2.48) can be written as

$$\omega(t) = \omega_0 e^{st} \tag{2.49}$$

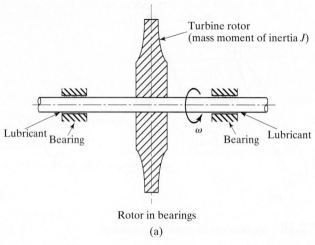

Turbine rotor
(mass moment of inertia J)

Lubricant Bearing ω Bearing Lubricant

Rotor in bearings
(a)

$\omega(t)$

ω_0

$\omega(t) = \omega_0 e^{-\frac{c_t}{J} t}$

$0.368 \omega_0$

O τ t

Variation of angular velocity
(b)

FIGURE 2.17

By substituting Eq. (2.49) into Eq. (2.47), we obtain

$$\omega_0 e^{st}(Js + c_t) = 0 \tag{2.50}$$

Since $\omega_0 = 0$ leads to "no motion" of the rotor, we assume $\omega_0 \neq 0$ and Eq. (2.50) can be satisfied only if

$$Js + c_t = 0 \tag{2.51}$$

Equation (2.51) is known as the *characteristic equation* which yields $s = -\dfrac{c_t}{J}$. Thus the solution, Eq. (2.49), becomes

$$\omega(t) = \omega_0 e^{-\frac{c_t}{J}t} \tag{2.52}$$

The variation of the angular velocity, given by Eq. (2.52), with time is shown in Fig. 2.17(b). The curve starts at ω_0, decays, and approaches zero as t increases without limit. In dealing with exponentially decaying responses, such as the one given by Eq. (2.52), it is convenient to describe the response in terms of a quantity known as the *time constant* (τ). The time constant is defined as the value of time which makes the exponent in Eq. (2.52) equal to -1. Because the exponent of Eq. (2.52) is known to be $-\dfrac{c_t}{J}t$, the time constant will be equal to

$$\tau = \frac{J}{c_t} \tag{2.53}$$

so that Eq. (2.52) gives, for $t = \tau$,

$$\omega(t) = \omega_0 e^{-\frac{c_t}{J}\tau} = \omega_0 e^{-1} = 0.368\omega_0 \tag{2.54}$$

Thus the response reduces to 0.368 times its initial value at a time equal to the time constant of the system.

2.5   Rayleigh's Energy Method

For a single-degree-of-freedom system, the equation of motion was derived using the energy method in Section 2.2.2. In this section, we shall use the energy method to find the natural frequencies of single-degree-of-freedom systems. The principle of conservation of energy, in the context of an undamped vibrating system, can be restated as

$$T_1 + U_1 = T_2 + U_2 \tag{2.55}$$

where the subscripts 1 and 2 denote two different instants of time. Specifically, we use the subscript 1 to denote the time when the mass is passing through its static equilibrium position and choose $U_1 = 0$ as reference for the potential energy. If we let the subscript 2 indicate the time corresponding to the maximum displacement of the mass, we have $T_2 = 0$. Thus Eq. (2.55) becomes

$$T_1 + 0 = 0 + U_2 \tag{2.56}$$

If the system is undergoing harmonic motion, then T_1 and U_2 denote the maximum values of T and U, respectively, and Eq. (2.56) becomes

$$T_{max} = U_{max} \tag{2.57}$$

The application of Eq. (2.57), which is also known as *Rayleigh's energy method,* gives the natural frequency of the system directly, as illustrated in the following examples.

Manometer for Diesel Engine

EXAMPLE 2.7

The exhaust from a single-cylinder four-stroke diesel engine is to be connected to a silencer, and the pressure therein is to be measured with a simple U-tube manometer (see Fig. 2.18). Calculate the minimum length of the manometer tube so that the natural frequency of oscillation of the mercury column will be 3.5 times slower than the frequency of the pressure fluctuations in the silencer at an engine speed of 600 rpm. The frequency of pressure fluctuation in the silencer is equal to

$$\frac{\text{Number of cylinders} \times \text{Speed of the engine}}{2}$$

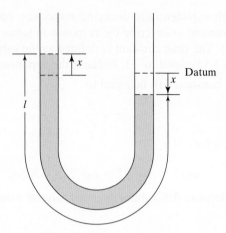

FIGURE 2.18   U-tube manometer.

Solution

1. *Natural frequency of oscillation of the liquid column:*   Let the datum in Fig. 2.18 be taken as the equilibrium position of the liquid. If the displacement of the liquid column from the equilibrium position is denoted by x, the change in potential energy is given by

U = potential energy of raised liquid column + potential energy of depressed liquid column

= (weight of mercury raised $\times$ displacement of the C.G. of the segment) + (weight of

mercury depressed $\times$ displacement of the C.G. of the segment)

$$= (Ax\gamma)\frac{x}{2} + (Ax\gamma)\frac{x}{2} = A\gamma x^2 \tag{E.1}$$

where A is the cross-sectional area of the mercury column and γ is the specific weight of mercury. The change in kinetic energy is given by

$$T = \frac{1}{2}(\text{mass of mercury})(\text{velocity})^2$$

$$= \frac{1}{2}\frac{Al\gamma}{g}\dot{x}^2 \tag{E.2}$$

where l is the length of the mercury column. By assuming harmonic motion, we can write

$$x(t) = X \cos \omega_n t \tag{E.3}$$

where X is the maximum displacement and ω_n is the natural frequency. By substituting Eq. (E.3) into Eqs. (E.1) and (E.2), we obtain

$$U = U_{max} \cos^2 \omega_n t \tag{E.4}$$

$$T = T_{max} \sin^2 \omega_n t \tag{E.5}$$

where

$$U_{max} = A\gamma X^2 \tag{E.6}$$

and

$$T_{max} = \frac{1}{2}\frac{A\gamma l \omega_n^2}{g}X^2 \tag{E.7}$$

By equating U_{max} to T_{max}, we obtain the natural frequency:

$$\omega_n = \left(\frac{2g}{l}\right)^{1/2} \tag{E.8}$$

2. *Length of the mercury column:*   The frequency of pressure fluctuations in the silencer

$$= \frac{1 \times 600}{2}$$

$$= 300 \text{ rpm}$$

$$= \frac{300 \times 2\pi}{60} = 10\pi \text{ rad/s} \tag{E.9}$$

Thus the frequency of oscillations of the liquid column in the manometer is $10\pi/3.5 = 9.0$ rad/s. By using Eq. (E.8), we obtain

$$\left(\frac{2g}{l}\right)^{1/2} = 9.0 \tag{E.10}$$

or

$$l = \frac{2.0 \times 9.81}{(9.0)^2} = 0.243 \text{ m} \tag{E.11}$$

∎

Effect of Mass on ω_n of a Spring

EXAMPLE 2.8
_____

Determine the effect of the mass of the spring on the natural frequency of the spring-mass system shown in Fig. 2.19.

Solution: To find the effect of the mass of the spring on the natural frequency of the spring-mass system, we add the kinetic energy of the system to that of the attached mass and use the energy method to determine the natural frequency. Let l be the total length of the spring. If x denotes the displacement of the lower end of the spring (or mass m), the displacement at distance y from the support is given by $y(x/l)$. Similarly, if $\dot{x}$ denotes the velocity of the mass m, the velocity of a spring element located at distance y from the support is given by $y(\dot{x}/l)$. The kinetic energy of the spring element of length dy is

$$dT_s = \frac{1}{2}\left(\frac{m_s}{l}dy\right)\left(\frac{y\dot{x}}{l}\right)^2 \tag{E.1}$$

FIGURE 2.19 Equivalent mass of a spring.

where m_s is the mass of the spring. The total kinetic energy of the system can be expressed as

$$T = \text{kinetic energy of mass } (T_m) + \text{kinetic energy of spring } (T_s)$$

$$= \frac{1}{2} m \dot{x}^2 + \int_{y=0}^{l} \frac{1}{2} \left(\frac{m_s}{l} dy \right) \left(\frac{y^2 \dot{x}^2}{l^2} \right)$$

$$= \frac{1}{2} m \dot{x}^2 + \frac{1}{2} \frac{m_s}{3} \dot{x}^2 \tag{E.2}$$

The total potential energy of the system is given by

$$U = \frac{1}{2} k x^2 \tag{E.3}$$

By assuming a harmonic motion

$$x(t) = X \cos \omega_n t \tag{E.4}$$

where X is the maximum displacement of the mass and ω_n is the natural frequency, the maximum kinetic and potential energies can be expressed as

$$T_{\text{max}} = \frac{1}{2} \left(m + \frac{m_s}{3} \right) X^2 \omega_n^2 \tag{E.5}$$

$$U_{\text{max}} = \frac{1}{2} k X^2 \tag{E.6}$$

By equating T_{max} and U_{max}, we obtain the expression for the natural frequency:

$$\omega_n = \left(\frac{k}{m + \frac{m_s}{3}} \right)^{1/2} \tag{E.7}$$

Thus the effect of the mass of the spring can be accounted for by adding one-third of its mass to the main mass [2.3].

■

Effect of Mass of Column on Natural Frequency of Water Tank

EXAMPLE 2.9

Find the natural frequency of transverse vibration of the water tank considered in Example 2.1 and Fig. 2.10 by including the mass of the column.

Solution: To include the mass of the column, we find the equivalent mass of the column at the free end using the equivalence of kinetic energy and use a single-degree-of-freedom model to find the natural frequency of vibration. The column of the tank is considered as a cantilever beam fixed at one end (ground) and carrying a mass M (water tank) at the other end. The static deflection of a cantilever beam under a concentrated end load is given by (see Fig. 2.20):

$$y(x) = \frac{Px^2}{6EI}(3l - x) = \frac{y_{max}x^2}{2l^3}(3l - x)$$

$$= \frac{y_{max}}{2l^3}(3x^2 l - x^3) \tag{E.1}$$

The maximum kinetic energy of the beam itself (T_{max}) is given by

$$T_{max} = \frac{1}{2}\int_0^l \frac{m}{l}\{\dot{y}(x)\}^2 \, dx \tag{E.2}$$

where m is the total mass and (m/l) is the mass per unit length of the beam. Equation (E.1) can be used to express the velocity variation, $\dot{y}(x)$, as

$$\dot{y}(x) = \frac{\dot{y}_{max}}{2l^3}(3x^2 l - x^3) \tag{E.3}$$

and hence Eq. (E.2) becomes

$$T_{max} = \frac{m}{2l}\left(\frac{\dot{y}_{max}}{2l^3}\right)^2 \int_0^l (3x^2 l - x^3)^2 \, dx$$

$$= \frac{1}{2}\frac{m}{l}\frac{\dot{y}_{max}^2}{4l^6}\left(\frac{33}{35}l^7\right) = \frac{1}{2}\left(\frac{33}{140}m\right)\dot{y}_{max}^2 \tag{E.4}$$

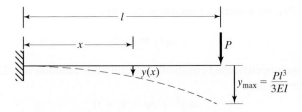

FIGURE 2.20   Equivalent mass of the column.

If m_{eq} denotes the equivalent mass of the cantilever (water tank) at the free end, its maximum kinetic energy can be expressed as

$$T_{max} = \frac{1}{2} m_{eq} \dot{y}_{max}^2 \tag{E.5}$$

By equating Eqs. (E.4) and (E.5), we obtain

$$m_{eq} = \frac{33}{140} m \tag{E.6}$$

Thus the total effective mass acting at the end of the cantilever beam is given by

$$M_{eff} = M + m_{eq} \tag{E.7}$$

where M is the mass of the water tank. The natural frequency of transverse vibration of the water tank is given by

$$\omega_n = \sqrt{\frac{k}{M_{eff}}} = \sqrt{\frac{k}{M + \dfrac{33}{140} m}} \tag{E.8}$$

∎

2.6   Free Vibration with Viscous Damping

**2.6.1
Equation
of Motion**

As stated in Section 1.9, the viscous damping force F is proportional to the velocity $\dot{x}$ or v and can be expressed as

$$F = -c\dot{x} \tag{2.58}$$

where c is the damping constant or coefficient of viscous damping and the negative sign indicates that the damping force is opposite to the direction of velocity. A single-degree-of-freedom system with a viscous damper is shown in Fig. 2.21. If x is measured from the equilibrium position of the mass m, the application of Newton's law yields the equation of motion:

$$m\ddot{x} = -c\dot{x} - kx$$

or

$$m\ddot{x} + c\dot{x} + kx = 0 \tag{2.59}$$

**2.6.2
Solution**

To solve Eq. (2.59), we assume a solution in the form

$$x(t) = Ce^{st} \tag{2.60}$$

where C and s are undetermined constants. Inserting this function into Eq. (2.59) leads to the characteristic equation

$$ms^2 + cs + k = 0 \tag{2.61}$$

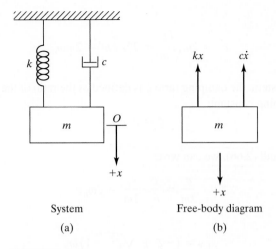

System
(a)

Free-body diagram
(b)

FIGURE 2.21   Single-degree-of-freedom system with viscous damper.

the roots of which are

$$s_{1,2} = \frac{-c \pm \sqrt{c^2 - 4mk}}{2m} = -\frac{c}{2m} \pm \sqrt{\left(\frac{c}{2m}\right)^2 - \frac{k}{m}} \qquad (2.62)$$

These roots give two solutions to Eq. (2.59):

$$x_1(t) = C_1 e^{s_1 t} \quad \text{and} \quad x_2(t) = C_2 e^{s_2 t} \qquad (2.63)$$

Thus the general solution of Eq. (2.59) is given by a combination of the two solutions $x_1(t)$ and $x_2(t)$:

$$x(t) = C_1 e^{s_1 t} + C_2 e^{s_2 t}$$

$$= C_1 e^{\left\{-\frac{c}{2m} + \sqrt{\left(\frac{c}{2m}\right)^2 - \frac{k}{m}}\right\}t} + C_2 e^{\left\{-\frac{c}{2m} - \sqrt{\left(\frac{c}{2m}\right)^2 - \frac{k}{m}}\right\}t} \qquad (2.64)$$

where C_1 and C_2 are arbitrary constants to be determined from the initial conditions of the system.

Critical Damping Constant and the Damping Ratio.   The critical damping c_c is defined as the value of the damping constant c for which the radical in Eq. (2.62) becomes zero:

$$\left(\frac{c_c}{2m}\right)^2 - \frac{k}{m} = 0$$

or

$$c_c = 2m\sqrt{\frac{k}{m}} = 2\sqrt{km} = 2m\omega_n \tag{2.65}$$

For any damped system, the damping ratio ζ is defined as the ratio of the damping constant to the critical damping constant:

$$\zeta = c/c_c \tag{2.66}$$

Using Eqs. (2.65) and (2.66), we can write

$$\frac{c}{2m} = \frac{c}{c_c} \cdot \frac{c_c}{2m} = \zeta \omega_n \tag{2.67}$$

and hence

$$s_{1,2} = \left(-\zeta \pm \sqrt{\zeta^2 - 1}\right) \omega_n \tag{2.68}$$

Thus the solution, Eq. (2.64), can be written as

$$x(t) = C_1 e^{(-\zeta + \sqrt{\zeta^2 - 1})\omega_n t} + C_2 e^{(-\zeta - \sqrt{\zeta^2 - 1})\omega_n t} \tag{2.69}$$

The nature of the roots s_1 and s_2 and hence the behavior of the solution, Eq. (2.69), depends upon the magnitude of damping. It can be seen that the case $\zeta = 0$ leads to the undamped vibrations discussed in Section 2.2. Hence we assume that $\zeta \neq 0$ and consider the following three cases.

Case 1.   *Underdamped system* ($\zeta < 1$ or $c < c_c$ or $c/2m < \sqrt{k/m}$). For this condition, ($\zeta^2 - 1$) is negative and the roots s_1 and s_2 can be expressed as

$$s_1 = \left(-\zeta + i\sqrt{1 - \zeta^2}\right)\omega_n$$

$$s_2 = \left(-\zeta - i\sqrt{1 - \zeta^2}\right)\omega_n$$

and the solution, Eq. (2.69), can be written in different forms:

$$\begin{aligned}
x(t) &= C_1 e^{(-\zeta + i\sqrt{1 - \zeta^2})\omega_n t} + C_2 e^{(-\zeta - i\sqrt{1 - \zeta^2})\omega_n t} \\
&= e^{-\zeta\omega_n t}\left\{ C_1 e^{i\sqrt{1 - \zeta^2}\omega_n t} + C_2 e^{-i\sqrt{1 - \zeta^2}\omega_n t} \right\} \\
&= e^{-\zeta\omega_n t}\left\{ (C_1 + C_2)\cos\sqrt{1 - \zeta^2}\omega_n t + i(C_1 - C_2)\sin\sqrt{1 - \zeta^2}\omega_n t \right\} \\
&= e^{-\zeta\omega_n t}\left\{ C_1' \cos\sqrt{1 - \zeta^2}\omega_n t + C_2' \sin\sqrt{1 - \zeta^2}\omega_n t \right\}
\end{aligned}$$

$$= X_0 e^{-\zeta\omega_n t} \sin\left(\sqrt{1 - \zeta^2}\,\omega_n t + \phi_0\right)$$

$$x(t) = X e^{-\zeta\omega_n t} \cos\left(\sqrt{1 - \zeta^2}\,\omega_n t - \phi\right) \tag{2.70}$$

where (C_1', C_2'), (X, ϕ), and (X_0, ϕ_0) are arbitrary constants to be determined from the initial conditions.

For the initial conditions $x(t = 0) = x_0$ and $\dot{x}(t = 0) = \dot{x}_0$, C_1' and C_2' can be found:

$$C_1' = x_0 \quad \text{and} \quad C_2' = \frac{\dot{x}_0 + \zeta\omega_n x_0}{\sqrt{1 - \zeta^2}\,\omega_n} \tag{2.71}$$

and hence the solution becomes

$$x(t) = e^{-\zeta\omega_n t}\left\{ x_0 \cos\sqrt{1 - \zeta^2}\,\omega_n t \right.$$

$$\left. + \frac{\dot{x}_0 + \zeta\omega_n x_0}{\sqrt{1 - \zeta^2}\,\omega_n} \sin\sqrt{1 - \zeta^2}\,\omega_n t \right\} \tag{2.72}$$

The constants (X, ϕ) and (X_0, ϕ_0) can be expressed as

$$X = X_0 = \sqrt{(C_1')^2 + (C_2')^2} = \frac{\sqrt{x_0^2\omega_n^2 + \dot{x}_0^2 + 2x_0\dot{x}_0\zeta\omega_n}}{\sqrt{1 - \zeta^2}\,\omega_n} \tag{2.73}$$

$$\phi_0 = \tan^{-1}\left(\frac{C_1'}{C_2'}\right) = \tan^{-1}\left(\frac{x_0\omega_n\sqrt{1 - \zeta^2}}{\dot{x}_0 + \zeta\omega_n x_0}\right) \tag{2.74}$$

$$\phi = \tan^{-1}\left(\frac{C_2'}{C_1'}\right) = \tan^{-1}\left(\frac{\dot{x}_0 + \zeta\omega_n x_0}{x_0\omega_n\sqrt{1 - \zeta^2}}\right) \tag{2.75}$$

The motion described by Eq. (2.72) is a damped harmonic motion of angular frequency $\sqrt{1 - \zeta^2}\,\omega_n$, but because of the factor $e^{-\zeta\omega_n t}$, the amplitude decreases exponentially with time, as shown in Fig. 2.22. The quantity

$$\omega_d = \sqrt{1 - \zeta^2}\,\omega_n \tag{2.76}$$

is called the *frequency of damped vibration*. It can be seen that the frequency of damped vibration ω_d is always less than the undamped natural frequency ω_n. The decrease in the frequency of damped vibration with increasing amount of damping, given by Eq. (2.76), is shown graphically in Fig. 2.23. The underdamped case is very important in the study of mechanical vibrations, as it is the only case that leads to an oscillatory motion [2.10].

FIGURE 2.22   Underdamped solution.

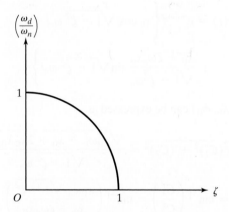

FIGURE 2.23   Variation of ω_d with damping.

Case 2.   *Critically damped system* ($\zeta = 1$ or $c = c_c$ or $c/2m = \sqrt{k/m}$). In this case, the two roots s_1 and s_2 in Eq. (2.68) are equal:

$$s_1 = s_2 = -\frac{c_c}{2m} = -\omega_n \qquad (2.77)$$

Because of the repeated roots, the solution of Eq. (2.59) is given by [2.6][1]

$$x(t) = (C_1 + C_2 t)e^{-\omega_n t} \qquad (2.78)$$

---

[1]Equation (2.78) can also be obtained by making ζ approach unity in the limit in Eq. (2.72). As $\zeta \rightarrow 1$, $\omega_n \rightarrow 0$; hence $\cos \omega_d t \rightarrow 1$ and $\sin \omega_d t \rightarrow \omega_d t$. Thus Eq. (2.72) yields

$$x(t) = e^{-\omega_n t}(C_1' + C_2' \omega_d t) = (C_1 + C_2 t)e^{-\omega_n t}$$

where $C_1 = C_1'$ and $C_2 = C_2' \omega_d$ are new constants.

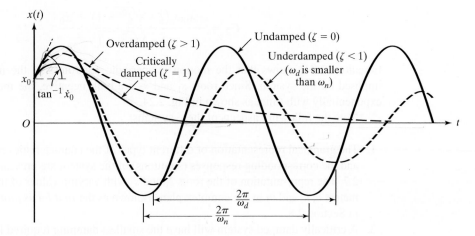

FIGURE 2.24   Comparison of motions with different types of damping.

The application of the initial conditions $x(t = 0) = x_0$ and $\dot{x}(t = 0) = \dot{x}_0$ for this case gives

$$C_1 = x_0$$
$$C_2 = \dot{x}_0 + \omega_n x_0 \tag{2.79}$$

and the solution becomes

$$x(t) = [x_0 + (\dot{x}_0 + \omega_n x_0)t]e^{-\omega_n t} \tag{2.80}$$

It can be seen that the motion represented by Eq. (2.80) is *aperiodic* (i.e., nonperiodic). Since $e^{-\omega_n t} \to 0$ as $t \to \infty$, the motion will eventually diminish to zero, as indicated in Fig. 2.24.

Case 3.   *Overdamped system* ($\zeta > 1$ or $c > c_c$ or $c/2m > \sqrt{k/m}$). As $\sqrt{\zeta^2 - 1} > 0$, Eq. (2.68) shows that the roots s_1 and s_2 are real and distinct and are given by

$$s_1 = (-\zeta + \sqrt{\zeta^2 - 1})\omega_n < 0$$
$$s_2 = (-\zeta - \sqrt{\zeta^2 - 1})\omega_n < 0$$

with $s_2 \ll s_1$. In this case, the solution, Eq. (2.69), can be expressed as

$$x(t) = C_1 e^{(-\zeta + \sqrt{\zeta^2 - 1})\omega_n t} + C_2 e^{(-\zeta - \sqrt{\zeta^2 - 1})\omega_n t} \tag{2.81}$$

For the initial conditions $x(t = 0) = x_0$ and $\dot{x}(t = 0) = \dot{x}_0$, the constants C_1 and C_2 can be obtained:

$$C_1 = \frac{x_0 \omega_n (\zeta + \sqrt{\zeta^2 - 1}) + \dot{x}_0}{2\omega_n \sqrt{\zeta^2 - 1}}$$

$$C_2 = \frac{-x_0 \omega_n (\zeta - \sqrt{\zeta^2 - 1}) - \dot{x}_0}{2\omega_n \sqrt{\zeta^2 - 1}}$$

(2.82)

Equation (2.81) shows that the motion is aperiodic regardless of the initial conditions imposed on the system. Since roots s_1 and s_2 are both negative, the motion diminishes exponentially with time, as shown in Fig. 2.24.

Note the following aspects of these systems:

1. The graphical representation of different types of the characteristics roots s_1 and s_2, and the corresponding responses (solutions) of the system are presented in Section 2.7. The representation of the roots s_1 and s_2 with varying values of the system parameters c, k, and m in the complex plane (known as the *root locus plots*) is considered in Section 2.8.

2. A critically damped system will have the smallest damping required for aperiodic motion; hence the mass returns to the position of rest in the shortest possible time without overshooting. The property of critical damping is used in many practical applications. For example, large guns have dashpots with critical damping value, so that they return to their original position after recoil in the minimum time without vibrating. If the damping provided were more than the critical value, some delay would be caused before the next firing.

3. The free damped response of a single-degree-of-freedom system can be represented in phase plane or state space as indicated in Fig. 2.25.

2.6.3
Logarithmic
Decrement

The logarithmic decrement represents the rate at which the amplitude of a free-damped vibration decreases. It is defined as the natural logarithm of the ratio of any two successive amplitudes. Let t_1 and t_2 denote the times corresponding to two consecutive amplitudes

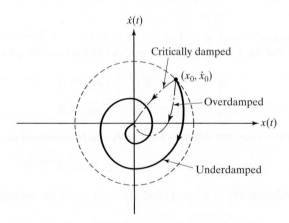

FIGURE 2.25   Phase plane of a damped system.

(displacements), measured one cycle apart for an underdamped system, as in Fig. 2.22. Using Eq. (2.70), we can form the ratio

$$\frac{x_1}{x_2} = \frac{X_0 e^{-\zeta\omega_n t_1} \cos(\omega_d t_1 - \phi_0)}{X_0 e^{-\zeta\omega_n t_2} \cos(\omega_d t_2 - \phi_0)} \tag{2.83}$$

But $t_2 = t_1 + \tau_d$, where $\tau_d = 2\pi/\omega_d$ is the period of damped vibration. Hence $\cos(\omega_d t_2 - \phi_0) = \cos(2\pi + \omega_d t_1 - \phi_0) = \cos(\omega_d t_1 - \phi_0)$, and Eq. (2.83) can be written as

$$\frac{x_1}{x_2} = \frac{e^{-\zeta\omega_n t_1}}{e^{-\zeta\omega_n(t_1+\tau_d)}} = e^{\zeta\omega_n \tau_d} \tag{2.84}$$

The logarithmic decrement δ can be obtained from Eq. (2.84):

$$\delta = \ln\frac{x_1}{x_2} = \zeta\omega_n\tau_d = \zeta\omega_n \frac{2\pi}{\sqrt{1 - \zeta^2}\omega_n} = \frac{2\pi\zeta}{\sqrt{1 - \zeta^2}} = \frac{2\pi}{\omega_d} \cdot \frac{c}{2m} \tag{2.85}$$

For small damping, Eq. (2.85) can be approximated:

$$\delta \simeq 2\pi\zeta \quad \text{if} \quad \zeta \ll 1 \tag{2.86}$$

Figure 2.26 shows the variation of the logarithmic decrement δ with ζ as given by Eqs. (2.85) and (2.86). It can be noticed that for values up to $\zeta = 0.3$, the two curves are difficult to distinguish.

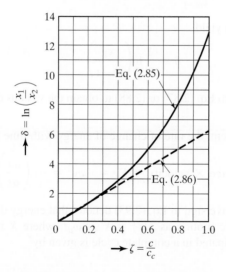

FIGURE 2.26   Variation of logarithmic decrement with damping.

The logarithmic decrement is dimensionless and is actually another form of the dimensionless damping ratio ζ. Once δ is known, ζ can be found by solving Eq. (2.85):

$$\zeta = \frac{\delta}{\sqrt{(2\pi)^2 + \delta^2}} \tag{2.87}$$

If we use Eq. (2.86) instead of Eq. (2.85), we have

$$\zeta \simeq \frac{\delta}{2\pi} \tag{2.88}$$

If the damping in the given system is not known, we can determine it experimentally by measuring any two consecutive displacements x_1 and x_2. By taking the natural logarithm of the ratio of x_1 and x_2, we obtain δ. By using Eq. (2.87), we can compute the damping ratio ζ. In fact, the damping ratio ζ can also be found by measuring two displacements separated by any number of complete cycles. If x_1 and x_{m+1} denote the amplitudes corresponding to times t_1 and $t_{m+1} = t_1 + m\tau_d$, where m is an integer, we obtain

$$\frac{x_1}{x_{m+1}} = \frac{x_1}{x_2}\frac{x_2}{x_3}\frac{x_3}{x_4}\cdots\frac{x_m}{x_{m+1}} \tag{2.89}$$

Since any two successive displacements separated by one cycle satisfy the equation

$$\frac{x_j}{x_{j+1}} = e^{\zeta\omega_n\tau_d} \tag{2.90}$$

Eq. (2.89) becomes

$$\frac{x_1}{x_{m+1}} = \left(e^{\zeta\omega_n\tau_d}\right)^m = e^{m\zeta\omega_n\tau_d} \tag{2.91}$$

Equations (2.85) and (2.91) yield

$$\delta = \frac{1}{m}\ln\left(\frac{x_1}{x_{m+1}}\right) \tag{2.92}$$

which can be substituted into Eq. (2.87) or (2.88) to obtain the viscous damping ratio ζ.

2.6.4 Energy Dissipated in Viscous Damping

In a viscously damped system, the rate of change of energy with time (dW/dt) is given by

$$\frac{dW}{dt} = \text{force} \times \text{velocity} = Fv = -cv^2 = -c\left(\frac{dx}{dt}\right)^2 \tag{2.93}$$

using Eq. (2.58). The negative sign in Eq. (2.93) denotes that energy dissipates with time. Assume a simple harmonic motion as $x(t) = X \sin \omega_d t$, where X is the amplitude of motion and the energy dissipated in a complete cycle is given by[2]

---

[2]In the case of a damped system, simple harmonic motion $x(t) = X \cos \omega_d t$ is possible only when the steady-state response is considered under a harmonic force of frequency ω_d (see Section 3.4). The loss of energy due to the damper is supplied by the excitation under steady-state forced vibration [2.7].

$$\Delta W = \int_{t=0}^{(2\pi/\omega_d)} c\left(\frac{dx}{dt}\right)^2 dt = \int_0^{2\pi} cX^2\omega_d \cos^2\omega_d t \cdot d(\omega_d t)$$

$$= \pi c\omega_d X^2 \tag{2.94}$$

This shows that the energy dissipated is proportional to the square of the amplitude of motion. Note that it is not a constant for given values of damping and amplitude, since ΔW is also a function of the frequency ω_d.

Equation (2.94) is valid even when there is a spring of stiffness k parallel to the viscous damper. To see this, consider the system shown in Fig. 2.27. The total force resisting motion can be expressed as

$$F = -kx - cv = -kx - c\dot{x} \tag{2.95}$$

If we assume simple harmonic motion

$$x(t) = X \sin \omega_d t \tag{2.96}$$

as before, Eq. (2.95) becomes

$$F = -kX \sin \omega_d t - c\omega_d X \cos \omega_d t \tag{2.97}$$

The energy dissipated in a complete cycle will be

$$\Delta W = \int_{t=0}^{2\pi/\omega_d} Fv \, dt$$

$$= \int_0^{2\pi/\omega_d} kX^2\omega_d \sin \omega_d t \cdot \cos \omega_d t \cdot d(\omega_d t)$$

$$+ \int_0^{2\pi/\omega_d} c\omega_d X^2 \cos^2 \omega_d t \cdot d(\omega_d t) = \pi c\omega_d X^2 \tag{2.98}$$

FIGURE 2.27
Spring and damper
in parallel.

which can be seen to be identical with Eq. (2.94). This result is to be expected, since the spring force will not do any net work over a complete cycle or any integral number of cycles.

We can also compute the fraction of the total energy of the vibrating system that is dissipated in each cycle of motion ($\Delta W/W$), as follows. The total energy of the system W can be expressed either as the maximum potential energy ($\frac{1}{2}kX^2$) or as the maximum kinetic energy ($\frac{1}{2}mv_{\max}^2 = \frac{1}{2}mX^2\omega_d^2$), the two being approximately equal for small values of damping. Thus

$$\frac{\Delta W}{W} = \frac{\pi c \omega_d X^2}{\frac{1}{2} m \omega_d^2 X^2} = 2\left(\frac{2\pi}{\omega_d}\right)\left(\frac{c}{2m}\right) = 2\delta \simeq 4\pi\zeta = \text{constant} \qquad (2.99)$$

using Eqs. (2.85) and (2.88). The quantity $\Delta W/W$ is called the *specific damping capacity* and is useful in comparing the damping capacity of engineering materials. Another quantity known as the *loss coefficient* is also used for comparing the damping capacity of engineering materials. The loss coefficient is defined as the ratio of the energy dissipated per radian and the total strain energy:

$$\text{loss coefficient} = \frac{(\Delta W/2\pi)}{W} = \frac{\Delta W}{2\pi W} \qquad (2.100)$$

2.6.5
Torsional
Systems with
Viscous
Damping

The methods presented in Sections 2.6.1 through 2.6.4 for linear vibrations with viscous damping can be extended directly to viscously damped torsional (angular) vibrations. For this, consider a single-degree-of-freedom torsional system with a viscous damper, as shown in Fig. 2.28(a). The viscous damping torque is given by (Fig. 2.28(b)):

$$T = -c_t \dot{\theta} \qquad (2.101)$$

where c_t is the torsional viscous damping constant, $\dot{\theta} = d\theta/dt$ is the angular velocity of the disc, and the negative sign denotes that the damping torque is opposite to the direction of angular velocity. The equation of motion can be derived as

$$J_0 \ddot{\theta} + c_t \dot{\theta} + k_t \theta = 0 \qquad (2.102)$$

where J_0 = mass moment of inertia of the disc, k_t = spring constant of the system (restoring torque per unit angular displacement), and θ = angular displacement of the disc. The solution of Eq. (2.102) can be found exactly as in the case of linear vibrations. For example, in the underdamped case, the frequency of damped vibration is given by

$$\omega_d = \sqrt{1 - \zeta^2}\, \omega_n \qquad (2.103)$$

where

$$\omega_n = \sqrt{\frac{k_t}{J_0}} \qquad (2.104)$$

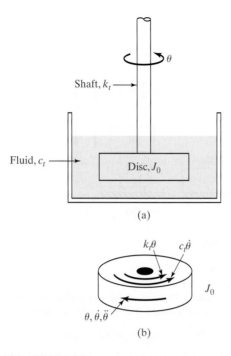

(a)

(b)

FIGURE 2.28   Torsional viscous damper.

and

$$\zeta = \frac{c_t}{c_{tc}} = \frac{c_t}{2J_0\omega_n} = \frac{c_t}{2\sqrt{k_t J_0}}$$ (2.105)

where c_{tc} is the critical torsional damping constant.

Response of Anvil of a Forging Hammer

EXAMPLE 2.10

The anvil of a forging hammer weighs 5,000 N and is mounted on a foundation that has a stiffness of 5×10^6 N/m and a viscous damping constant of 10,000 N-s/m. During a particular forging operation, the tup (i.e., the falling weight or the hammer), weighing 1,000 N, is made to fall from a height of 2 m onto the anvil (Fig. 2.29(a)). If the anvil is at rest before impact by the tup, determine the response of the anvil after the impact. Assume that the coefficient of restitution between the anvil and the tup is 0.4.

Solution: First we use the principle of conservation of momentum and the definition of the coefficient of restitution to find the initial velocity of the anvil. Let the velocities of the tup just before and just after impact with the anvil be v_{t1} and v_{t2}, respectively. Similarly, let v_{a1} and v_{a2} be the velocities of the anvil just before and just after the impact, respectively (Fig. 2.29(b)). Note that the

FIGURE 2.29   Forging hammer.

displacement of the anvil is measured from its static equilibrium position and all velocities are assumed to be positive when acting downward. The principle of conservation of momentum gives

$$M(v_{a2} - v_{a1}) = m(v_{t1} - v_{t2}) \tag{E.1}$$

where $v_{a1} = 0$ (anvil is at rest before the impact) and v_{t1} can be determined by equating its kinetic energy just before impact to its potential energy before dropping from a height of $h = 2$ m:

$$\frac{1}{2}mv_{t1}^2 = mgh \tag{E.2}$$

or

$$v_{t1} = \sqrt{2gh} = \sqrt{2 \times 9.81 \times 2} = 6.26099 \text{ m/s}$$

Thus Eq. (E.1) becomes

$$\frac{5,000}{9.81}(v_{a2} - 0) = \frac{1,000}{9.81}(6.26099 - v_{t2})$$

—that is,

$$510.204082\,v_{a2} = 638.87653 - 102.040813\,v_{t2} \tag{E.3}$$

The definition of the coefficient of restitution (r) yields:

$$r = -\left(\frac{v_{a2} - v_{t2}}{v_{a1} - v_{t1}}\right) \tag{E.4}$$

—that is,

$$0.4 = -\left(\frac{v_{a2} - v_{t2}}{0 - 6.26099}\right)$$

—that is,

$$v_{a2} = v_{t2} + 2.504396 \tag{E.5}$$

The solution of Eqs. (E.3) and (E.5) gives

$$v_{a2} = 1.460898 \text{ m/s}; \qquad v_{t2} = -1.043498 \text{ m/s}$$

Thus the initial conditions of the anvil are given by

$$x_0 = 0; \qquad \dot{x}_0 = 1.460898 \text{ m/s}$$

The damping coefficient is equal to

$$\zeta = \frac{c}{2\sqrt{kM}} = \frac{1{,}000}{2\sqrt{(5 \times 10^6)\left(\dfrac{5{,}000}{9.81}\right)}} = 0.0989949$$

The undamped and damped natural frequencies of the anvil are given by

$$\omega_n = \sqrt{\frac{k}{M}} = \sqrt{\frac{5 \times 10^6}{\left(\dfrac{5{,}000}{9.81}\right)}} = 98.994949 \text{ rad/s}$$

$$\omega_d = \omega_n\sqrt{1 - \zeta^2} = 98.994949\sqrt{1 - 0.0989949^2} = 98.024799 \text{ rad/s}$$

The displacement response of the anvil is given by Eq. (2.72):

$$x(t) = e^{-\zeta\,\omega_n\,t}\left\{\frac{\dot{x}_0}{\omega_d}\sin\omega_d t\right\}$$

$$= e^{-9.799995\,t}\{0.01490335\sin 98.024799\,t\}\text{m}$$

∎

Shock Absorber for a Motorcycle

EXAMPLE 2.11

An underdamped shock absorber is to be designed for a motorcycle of mass 200 kg (Fig. 2.30(a)). When the shock absorber is subjected to an initial vertical velocity due to a road bump, the resulting displacement-time curve is to be as indicated in Fig. 2.30(b). Find the necessary stiffness and damping constants of the shock absorber if the damped period of vibration is to be 2 s and the amplitude x_1 is to be reduced to one-fourth in one half cycle (i.e., $x_{1.5} = x_1/4$). Also find the minimum initial velocity that leads to a maximum displacement of 250 mm.

Approach: We use the equation for the logarithmic decrement in terms of the damping ratio, equation for the damped period of vibration, time corresponding to maximum displacement for an underdamped system, and envelope passing through the maximum points of an underdamped system.

Solution Since $x_{1.5} = x_1/4$, $x_2 = x_{1.5}/4 = x_1/16$. Hence the logarithmic decrement becomes

$$\delta = \ln\left(\frac{x_1}{x_2}\right) = \ln(16) = 2.7726 = \frac{2\pi\zeta}{\sqrt{1 - \zeta^2}} \tag{E.1}$$

(a)   (b)

FIGURE 2.30   Shock absorber of a motorcycle.

from which the value of ζ can be found as $\zeta = 0.4037$. The damped period of vibration is given to be 2 s. Hence

$$2 = \tau_d = \frac{2\pi}{\omega_d} = \frac{2\pi}{\omega_n \sqrt{1 - \zeta^2}}$$

$$\omega_n = \frac{2\pi}{2\sqrt{1 - (0.4037)^2}} = 3.4338 \text{ rad/s}$$

The critical damping constant can be obtained:

$$c_c = 2m\omega_n = 2(200)(3.4338) = 1{,}373.54 \text{ N-s/m}$$

Thus the damping constant is given by

$$c = \zeta c_c = (0.4037)(1{,}373.54) = 554.4981 \text{ N-s/m}$$

and the stiffness by

$$k = m\omega_n^2 = (200)(3.4338)^2 = 2{,}358.2652 \text{ N/m}$$

The displacement of the mass will attain its maximum value at time t_1, given by

$$\sin \omega_d t_1 = \sqrt{1 - \zeta^2}$$

(See Problem 2.99.) This gives

$$\sin \omega_d t_1 = \sin \pi t_1 = \sqrt{1 - (0.4037)^2} = 0.9149$$

or

$$t_1 = \frac{\sin^{-1}(0.9149)}{\pi} = 0.3678 \text{ s}$$

The envelope passing through the maximum points (see Problem 2.99) is given by

$$x = \sqrt{1 - \zeta^2} X e^{-\zeta \omega_n t} \tag{E.2}$$

Since $x = 250$ mm, Eq. (E.2) gives at t_1

$$0.25 = \sqrt{1 - (0.4037)^2}\, X e^{-(0.4037)(3.4338)(0.3678)}$$

or

$$X = 0.4550 \text{ m}$$

The velocity of the mass can be obtained by differentiating the displacement

$$x(t) = X e^{-\zeta \omega_n t} \sin \omega_d t$$

as

$$\dot{x}(t) = X e^{-\zeta \omega_n t} (-\zeta \omega_n \sin \omega_d t + \omega_d \cos \omega_d t) \tag{E.3}$$

When $t = 0$, Eq. (E.3) gives

$$\dot{x}(t = 0) = \dot{x}_0 = X \omega_d = X \omega_n \sqrt{1 - \zeta^2} = (0.4550)(3.4338) \sqrt{1 - (0.4037)^2}$$

$$= 1.4294 \text{ m/s}$$

∎

EXAMPLE 2.12

Analysis of Cannon

The schematic diagram of a large cannon is shown in Fig. 2.31 [2.8]. When the gun is fired, high-pressure gases accelerate the projectile inside the barrel to a very high velocity. The reaction force pushes the gun barrel in the direction opposite that of the projectile. Since it is desirable to bring the gun barrel to rest in the shortest time without oscillation, it is made to translate backward against a critically damped spring-damper system called the *recoil mechanism*. In a particular case, the gun barrel and the recoil mechanism have a mass of 500 kg with a recoil spring of stiffness 10,000 N/m. The gun recoils 0.4 m upon firing. Find (1) the critical damping coefficient of the damper, (2) the initial recoil velocity of the gun, and (3) the time taken by the gun to return to a position 0.1 m from its initial position.

FIGURE 2.31   Recoil of cannon.

Solution

1. The undamped natural frequency of the system is

$$\omega_n = \sqrt{\frac{k}{m}} = \sqrt{\frac{10{,}000}{500}} = 4.4721 \text{ rad/s}$$

and the critical damping coefficient (Eq. 2.65) of the damper is

$$c_c = 2m\omega_n = 2(500)(4.4721) = 4{,}472.1 \text{ N-s/m}$$

2. The response of a critically damped system is given by Eq. (2.78):

$$x(t) = (C_1 + C_2 t)e^{-\omega_n t} \tag{E.1}$$

where $C_1 = x_0$ and $C_2 = \dot{x}_0 + \omega_n x_0$. The time t_1 at which $x(t)$ reaches a maximum value can be obtained by setting $\dot{x}(t) = 0$. The differentiation of Eq. (E.1) gives

$$\dot{x}(t) = C_2 e^{-\omega_n t} - \omega_n(C_1 + C_2 t)e^{-\omega_n t}$$

Hence $\dot{x}(t) = 0$ yields

$$t_1 = \left(\frac{1}{\omega_n} - \frac{C_1}{C_2}\right) \tag{E.2}$$

In this case, $x_0 = C_1 = 0$; hence Eq. (E.2) leads to $t_1 = 1/\omega_n$. Since the maximum value of $x(t)$ or the recoil distance is given to be $x_{\max} = 0.4$ m, we have

$$x_{\max} = x(t = t_1) = C_2 t_1 e^{-\omega_n t_1} = \frac{\dot{x}_0}{\omega_n} e^{-1} = \frac{\dot{x}_0}{e\omega_n}$$

or

$$\dot{x}_0 = x_{\max}\omega_n e = (0.4)(4.4721)(2.7183) = 4.8626 \text{ m/s}$$

3. If t_2 denotes the time taken by the gun to return to a position 0.1 m from its initial position, we have

$$0.1 = C_2 t_2 e^{-\omega_n t_2} = 4.8626 t_2 e^{-4.4721 t_2} \tag{E.3}$$

The solution of Eq. (E.3) gives $t_2 = 0.8258$ s.

∎

2.7   Graphical Representation of Characteristic Roots and Corresponding Solutions[†]

**2.7.1
Roots of the
Characteristic
Equation**

The free vibration of a single-degree-of-freedom spring-mass-viscous-damper system shown in Fig. 2.21 is governed by Eq. (2.59):

$$m\ddot{x} + c\dot{x} + kx = 0 \tag{2.106}$$

whose characteristic equation can be expressed as (Eq. (2.61)):

$$ms^2 + cs + k = 0 \tag{2.107}$$

or

$$s^2 + 2\zeta\omega_n s + \omega_n^2 = 0 \tag{2.108}$$

[†]If necessary, Sections 2.7 and 2.8 can be skipped without losing continuity.

The roots of this characteristic equation, called the *characteristic roots* or, simply, *roots*, help us in understanding the behavior of the system. The roots of Eq. (2.107) or (2.108) are given by (see Eqs. (2.62) and (2.68)):

$$s_1, s_2 = \frac{-c \pm \sqrt{c^2 - 4mk}}{2m} \tag{2.109}$$

or

$$s_1, s_2 = -\zeta\omega_n \pm i\omega_n \sqrt{1 - \zeta^2} \tag{2.110}$$

**2.7.2
Graphical
Representation
of Roots and
Corresponding
Solutions**

The roots given by Eq. (2.110) can be plotted in a complex plane, also known as the *s-plane*, by denoting the real part along the horizontal axis and the imaginary part along the vertical axis. Noting that the response of the system is given by

$$x(t) = C_1 e^{s_1 t} + C_2 e^{s_2 t} \tag{2.111}$$

where C_1 and C_2 are constants, the following observations can be made by examining Eqs. (2.110) and (2.111):

1. Because the exponent of a larger real negative number (such as e^{-2t}) decays faster than the exponent of a smaller real negative number (such as e^{-t}), the roots lying further to the left in the *s*-plane indicate that the corresponding responses decay faster than those associated with roots closer to the imaginary axis.
2. If the roots have positive real values of *s*—that is, the roots lie in the right half of the *s*-plane—the corresponding response grows exponentially and hence will be unstable.
3. If the roots lie on the imaginary axis (with zero real value), the corresponding response will be naturally stable.
4. If the roots have a zero imaginary part, the corresponding response will not oscillate.
5. The response of the system will exhibit an oscillatory behavior only when the roots have nonzero imaginary parts.
6. The farther the roots lie to the left of the *s*-plane, the faster the corresponding response decreases.
7. The larger the imaginary part of the roots, the higher the frequency of oscillation of the corresponding response of the system.

Figure 2.32 shows some representative locations of the characteristic roots in the *s*-plane and the corresponding responses [2.15]. The characteristics that describe the behavior of the response of a system include oscillatory nature, frequency of oscillation, and response time. These characteristics are inherent to the system (depend on the values of *m*, *c*, and *k*) and are determined by the characteristic roots of the system but not by the initial conditions. The initial conditions determine only the amplitudes and phase angles.

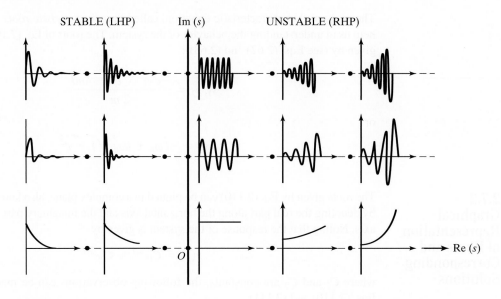

FIGURE 2.32   Locations of characteristic roots (•) and the corresponding responses of the system.

2.8   Parameter Variations and Root Locus Representations

2.8.1 Interpretations of ω_n, ω_d, ζ, and τ in the s-plane

Although the roots s_1 and s_2 appear as complex conjugates, we consider only the roots in the upper half of the s-plane. The root s_1 is plotted as point A with the real value as $\zeta\omega_n$ and the complex value as $\omega_n \sqrt{1 - \zeta^2}$, so that the length of OA is ω_n (Fig. 2.33). Thus the roots lying on the circle of radius ω_n correspond to the same natural frequency (ω_n) of the system (PAQ denotes a quarter of the circle). Thus different concentric circles represent systems with different natural frequencies as shown in Fig. 2.34. The horizontal line passing through point A corresponds to the damped natural frequency, $\omega_d = \omega_n\sqrt{1 - \zeta^2}$. Thus lines parallel to the real axis denote systems having different damped natural frequencies, as shown in Fig. 2.35.

It can be seen, from Fig. 2.33, that the angle made by the line OA with the imaginary axis is given by

$$\sin \theta = \frac{\zeta\omega_n}{\omega_n} = \zeta \tag{2.112}$$

or

$$\theta = \sin^{-1} \zeta \tag{2.113}$$

Thus, radial lines passing through the origin correspond to different damping ratios, as shown in Fig. 2.36. Therefore, when $\zeta = 0$, we have no damping ($\theta = 0$), and the damped natural frequency will reduce to the undamped natural frequency. Similarly, when $\zeta = 1$,

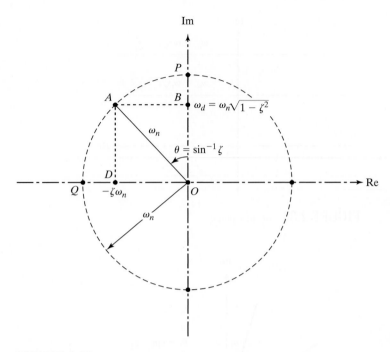

FIGURE 2.33   Interpretations of ω_n, ω_d, and ζ.

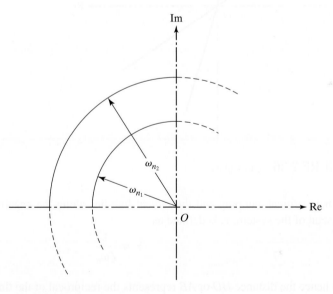

FIGURE 2.34 ω_n in s-plane.

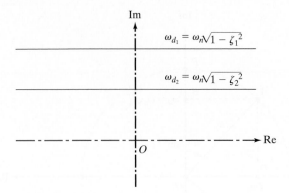

FIGURE 2.35 ω_d in s-plane.

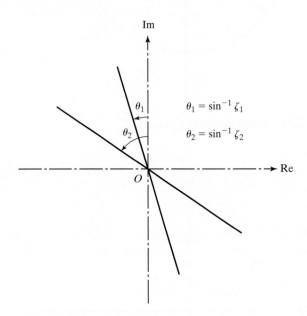

FIGURE 2.36 ζ in s-plane.

we have critical damping and the radical line lies along the negative real axis. The time constant of the system, τ, is defined as

$$\tau = \frac{1}{\zeta\omega_n} \tag{2.114}$$

and hence the distance DO or AB represents the reciprocal of the time constant, $\zeta\omega_n = \dfrac{1}{\tau}$. Hence different lines parallel to the imaginary axis denote reciprocals of different time constants (Fig. 2.37).

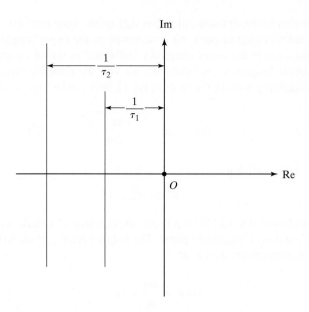

FIGURE 2.37 τ in s-plane.

2.8.2
Root Locus
and Parameter
Variations

A plot or graph that shows how changes in one of the parameters of the system will modify the roots of the characteristic equation of the system is known as the *root locus plot*. The root locus method is a powerful method of analysis and design for stability and transient response of a system. For a vibrating system, the root locus can be used to describe qualitatively the performance of the system as various parameters, such as the mass, damping constant, or spring constant, are changed. In the root locus method, the path or locus of the roots of the characteristic equation is plotted without actually finding the roots themselves. This is accomplished by using a set of rules which lead to a reasonably accurate plot in a relatively short time [2.8]. We study the behavior of the system by varying one parameter, among the damping ratio, spring constant, and mass, at a time in terms of the locations of its characteristic roots in the s-plane.

Variation of the damping ratio: We vary the damping constant from zero to infinity and study the migration of the characteristic roots in the s-plane. For this, we use Eq. (2.109). We notice that negative values of the damping constant ($c < 0$) need not be considered, because they result in roots lying in the positive real half-plane that correspond to an unstable system. Thus we start with $c = 0$ to obtain, from Eq. (2.109),

$$s_{1,2} = \pm \frac{\sqrt{-4mk}}{2m} = \pm i\sqrt{\frac{k}{m}} = \pm i\omega_n \qquad (2.115)$$

Thus the locations of the characteristic roots start on the imaginary axis. Because the roots appear in complex conjugate pairs, we concentrate on the upper imaginary half-plane and then locate the roots in the lower imaginary half-plane as mirror images. By keeping the undamped natural frequency (ω_n) constant, we vary the damping constant c. Noting that the real and imaginary parts of the roots in Eq. (2.109) can be expressed as

$$-\sigma = -\frac{c}{2m} = -\zeta\omega_n \quad \text{and} \quad \frac{\sqrt{4mk - c^2}}{2m} = \omega_n\sqrt{1 - \zeta^2} = \omega_d \quad (2.116)$$

for $0 < \zeta < 1$, we find that

$$\sigma^2 + \omega_d^2 = \omega_n^2 \quad (2.117)$$

Since ω_n is held fixed, Eq. (2.117) represents the equation of a circle with a radius $r = \omega_n$ in the σ (real) and ω_d (imaginary) plane. The radius vector $r = \omega_n$ will make an angle θ with the positive imaginary axis with

$$\sin\theta = \frac{\omega_d}{\omega_n} = \alpha \quad (2.118)$$

$$\cos\theta = \frac{\sigma}{\omega_n} = \frac{\zeta\omega_n}{\omega_n} = \zeta \quad (2.119)$$

with

$$\alpha = \sqrt{1 - \zeta^2} \quad (2.120)$$

Thus the two roots trace loci or paths in the form of circular arcs as the damping ratio is increased from zero to unity as shown in Fig. 2.38. The root with positive imaginary part moves in the counterclockwise direction while the root with negative imaginary part moves in the clockwise direction. When the damping ratio (ζ) is equal to one, the two loci meet, denoting that the two roots coincide—that is, the characteristic equation has repeated roots. As we increase the damping ratio beyond the value of unity, the system becomes overdamped and, as seen earlier in Section 2.6, both the roots will become real. From the properties of a quadratic equation, we find that the product of the two roots is equal to the coefficient of the lowest power of s (which is equal to ω_n^2 in Eq. (2.108)).

Since the value of ω_n is held constant in this study, the product of the two roots is a constant. With increasing values of the damping ratio (ζ), one root will increase and the other root will decrease, with the locus of each root remaining on the negative real axis. Thus one root will approach $-\infty$ and the other root will approach zero. The two loci will join or coincide at a point, known as the *breakaway point*, on the negative real axis. The two parts of the loci that lie on the negative real axis, one from point P to $-\infty$ and the other from point P to the origin, are known as *segments*.

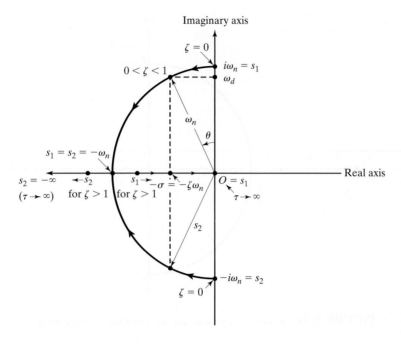

FIGURE 2.38   Root locus plot with variation of damping ratio ζ.

EXAMPLE 2.13

Study of Roots with Variation of c

Plot the root locus diagram of the system governed by the equation

$$3s^2 + cs + 27 = 0 \tag{E.1}$$

by varying the value of $c > 0$.

Solution: The roots of Eq. (E.1) are given by

$$s_{1,2} = \frac{-c \pm \sqrt{c^2 - 324}}{6} \tag{E.2}$$

We start with a value of $c = 0$. At $c = 0$, the roots are given by $s_{1,2} = \pm 3i$. These roots are shown as dots on the imaginary axis in Fig. 2.39. By using an increasing sequence of values of c, Eq. (E.2) gives the roots as indicated in Table 2.1.

It can be seen that the roots remain complex conjugates as c is increased up to a value of $c = 18$. At $c = 18$, both the roots become real and identical with a value of -3.0. As c increases beyond a value of 18, the roots remain distinct with negative real values. One root becomes more and more negative and the other root becomes less and less negative. Thus, as $c \to \infty$, one root approaches $-\infty$ while the other root approaches 0. These trends of the roots are shown in Fig 2.39.

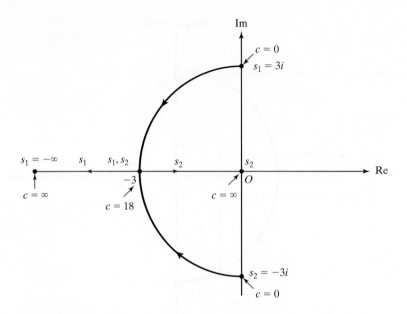

FIGURE 2.39   Root locus plot with variation of damping constant (c).

TABLE 2.1

Value of c	Value of s_1	Value of s_2
0	$+3i$	$-3i$
2	$-0.3333 + 2.9814i$	$-0.3333 - 2.9814i$
4	$-0.6667 + 2.9721i$	$-0.6667 - 2.9721i$
6	$-1.0000 + 2.8284i$	$-1.0000 - 2.8284i$
8	$-1.3333 + 2.6874i$	$-1.3333 - 2.6874i$
10	$-1.6667 + 2.4944i$	$-1.6667 - 2.4944i$
12	$-2.0000 + 2.2361i$	$-2.0000 - 2.2361i$
14	$-2.3333 + 1.8856i$	$-2.3333 - 1.8856i$
16	$-2.6667 + 1.3744i$	$-2.6667 - 1.3744i$
18	-3.0000	-3.0000
20	-1.8803	-4.7863
30	-1.0000	-9.0000
40	-0.7131	-12.6202
50	-5587	-16.1079
100	-0.2722	-33.0611
1,000	-0.0270	-333.3063

Variation of the spring constant: Since the spring constant does not appear explicitly in Eq. (2.108), we consider a specific form of the characteristic equation (2.107) as:

$$s^2 + 16s + k = 0 \qquad (2.121)$$

The roots of Eq. (2.121) are given by

$$s_{1,2} = \frac{-16 \pm \sqrt{256 - 4k}}{2} = -8 \pm \sqrt{64 - k} \qquad (2.122)$$

Since the spring stiffness cannot be negative for real vibration systems, we consider the variation of the values of k from zero to infinity. Equation (2.122) shows that for $k = 64$, both the roots are real and identical. As k is made greater than 64, the roots become complex conjugates. The roots for different values of k are shown in Table 2.2. The variations of the two roots can be plotted (as dots), as shown in Fig. 2.40.

Variation of the mass: To find the migration of the roots with a variation of the mass m, we consider a specific form of the characteristic equation, Eq. (2.107), as

$$ms^2 + 14s + 20 = 0 \qquad (2.123)$$

whose roots are given by

$$s_{1,2} = \frac{-14 \pm \sqrt{196 - 80m}}{2m} \qquad (2.124)$$

Since negative values as well as zero value of mass need not be considered for physical systems, we vary the value of m in the range $1 \le m < \infty$. Some values of m and the corresponding roots given by Eq. (2.124) are shown in Table 2.3.

It can be seen that both the roots are negative with values $(-1.6148, -12.3852)$ for $m = 1$ and $(-2, -5)$ for $m = 2$. The larger root is observed to move to the left and the

TABLE 2.2

Value of k	Value of s_1	Value of s_2
0	0	-16
16	-1.0718	-14.9282
32	-2.3431	-13.6569
48	-4	-12
64	-8	-8
80	$-8 + 4i$	$-8 - 4i$
96	$-8 + 5.6569i$	$-8 - 5.6569i$
112	$-8 + 6.9282i$	$-8 - 6.9282i$
128	$-8 + 8i$	$-8 - 8i$

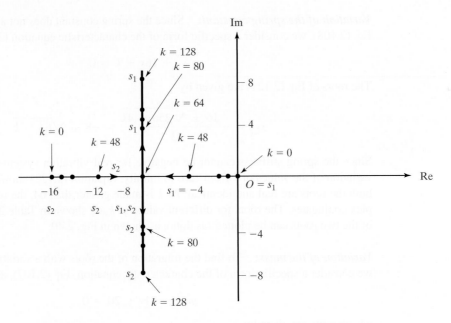

FIGURE 2.40 Root locus plot with variation of spring constant (k).

TABLE 2.3

Value of m	Value of s_1	Value of s_2
1	-1.6148	-12.3852
2	-2.0	-5.0
2.1	-2.0734	-4.5932
2.4	-2.5	-3.3333
2.45	-2.8571	-2.8571
2.5	$-2.8 + 0.4000i$	$-2.8 + 0.4000i$
3	$-2.3333 + 1.1055i$	$-2.3333 - 1.1055i$
5	$-1.4 + 1.4283i$	$-1.4 + 1.4283i$
8	$-0.8750 + 1.3169i$	$-0.8750 - 1.3169i$
10	$-0.7000 + 1.2288i$	$-0.7000 - 1.2288i$
14	$-0.5000 + 1.0856i$	$-0.5000 - 1.0856i$
20	$-0.3500 + 0.9367i$	$-0.3500 - 0.9367i$
30	$-0.2333 + 0.7824i$	$-0.2333 - 0.7824i$
40	$-0.1750 + 0.6851i$	$-0.1750 - 0.6851i$
50	$-0.1400 + 0.6167i$	$-0.1400 - 0.6167i$
100	$-0.0700 + 0.4417i$	$-0.0700 - 0.4417i$
1,000	$-0.0070 + 0.1412i$	$-0.0070 - 0.1412i$

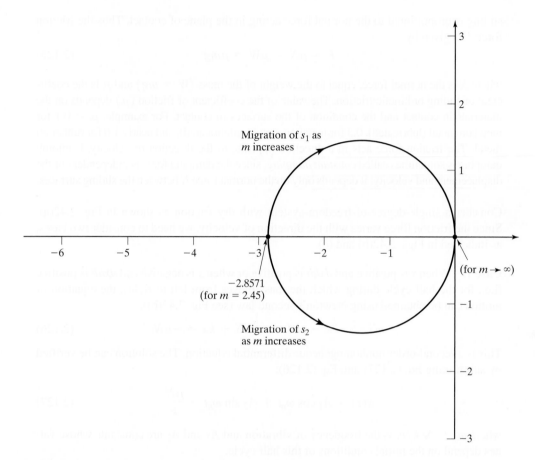

FIGURE 2.41 Root locus plot with variation of mass (*m*).

smaller root is found to move to the right, as shown in Fig. 2.41. The larger and smaller roots are found to converge to the value −2.8571 as *m* increases to a value of 2.45. Beyond this value of *m* = 2.45, the roots become complex conjugates. As the value of *m* increases from 2.45 to a large value (→ ∞), the loci of the two complex conjugates (roots) are shown by the curve (circle) in Fig. 2.41. For *m* → ∞, both the complex conjugate roots converge to zero ($s_1, s_2 \to 0$).

2.9 Free Vibration with Coulomb Damping

In many mechanical systems, *Coulomb* or *dry-friction* dampers are used because of their mechanical simplicity and convenience [2.9]. Also, in vibrating structures, whenever the components slide relative to each other, dry-friction damping appears internally. As stated in Section 1.9, Coulomb damping arises when bodies slide on dry surfaces. Coulomb's law of dry friction states that, when two bodies are in contact, the force required to produce sliding is proportional to the normal force acting in the plane of contact. Thus the friction force F is given by

$$F = \mu N = \mu W = \mu mg \tag{2.125}$$

where N is the normal force, equal to the weight of the mass ($W = mg$) and μ is the coefficient of sliding or kinetic friction. The value of the coefficient of friction (μ) depends on the materials in contact and the condition of the surfaces in contact. For example, $\mu \simeq 0.1$ for metal on metal (lubricated), 0.3 for metal on metal (unlubricated), and nearly 1.0 for rubber on metal. The friction force acts in a direction opposite to the direction of velocity. Coulomb damping is sometimes called *constant damping*, since the damping force is independent of the displacement and velocity; it depends only on the normal force N between the sliding surfaces.

**2.9.1
Equation
of Motion**

Consider a single-degree-of-freedom system with dry friction as shown in Fig. 2.42(a). Since the friction force varies with the direction of velocity, we need to consider two cases, as indicated in Figs. 2.42(b) and (c).

Case 1. When x is positive and dx/dt is positive or when x is negative and dx/dt is positive (i.e., for the half cycle during which the mass moves from left to right), the equation of motion can be obtained using Newton's second law (see Fig. 2.42(b)):

$$m\ddot{x} = -kx - \mu N \quad \text{or} \quad m\ddot{x} + kx = -\mu N \tag{2.126}$$

This is a second-order nonhomogeneous differential equation. The solution can be verified by substituting Eq. (2.127) into Eq. (2.126):

$$x(t) = A_1 \cos \omega_n t + A_2 \sin \omega_n t - \frac{\mu N}{k} \tag{2.127}$$

where $\omega_n = \sqrt{k/m}$ is the frequency of vibration and A_1 and A_2 are constants whose values depend on the initial conditions of this half cycle.

Case 2. When x is positive and dx/dt is negative or when x is negative and dx/dt is negative (i.e., for the half cycle during which the mass moves from right to left), the equation of motion can be derived from Fig. 2.42(c) as

$$-kx + \mu N = m\ddot{x} \quad \text{or} \quad m\ddot{x} + kx = \mu N \tag{2.128}$$

(a)   (b)   (c)

FIGURE 2.42   Spring-mass system with Coulomb damping.

The solution of Eq. (2.128) is given by

$$x(t) = A_3 \cos \omega_n t + A_4 \sin \omega_n t + \frac{\mu N}{k} \qquad (2.129)$$

where A_3 and A_4 are constants to be found from the initial conditions of this half cycle. The term $\mu N/k$ appearing in Eqs. (2.127) and (2.129) is a constant representing the virtual displacement of the spring under the force μN, if it were applied as a static force. Equations (2.127) and (2.129) indicate that in each half cycle the motion is harmonic, with the equilibrium position changing from $\mu N/k$ to $-(\mu N/k)$ every half cycle, as shown in Fig. 2.43.

2.9.2 Solution

Equations (2.126) and (2.128) can be expressed as a single equation (using $N = mg$):

$$m\ddot{x} + \mu m g \operatorname{sgn}(\dot{x}) + kx = 0 \qquad (2.130)$$

where $\operatorname{sgn}(y)$ is called the *signum function*, whose value is defined as 1 for $y > 0$, -1 for $y < 0$, and 0 for $y = 0$. Equation (2.130) can be seen to be a nonlinear differential equation for which a simple analytical solution does not exist. Numerical methods can be used to solve Eq. (2.130) conveniently (see Example 2.21). Equation (2.130), however, can be solved analytically if we break the time axis into segments separated by $\dot{x} = 0$ (i.e., time intervals with different directions of motion). To find the solution using this procedure, let us assume the initial conditions as

$$x(t = 0) = x_0$$

$$\dot{x}(t = 0) = 0 \qquad (2.131)$$

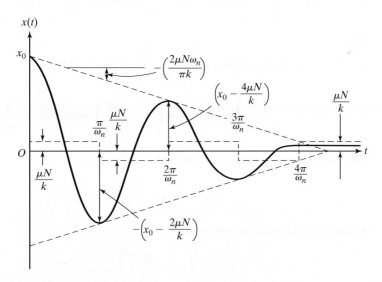

FIGURE 2.43   Motion of the mass with Coulomb damping.

That is, the system starts with zero velocity and displacement x_0 at $t = 0$. Since $x = x_0$ at $t = 0$, the motion starts from right to left. Let $x_0, x_1, x_2, \ldots$ denote the amplitudes of motion at successive half cycles. Using Eqs. (2.129) and (2.131), we can evaluate the constants A_3 and A_4:

$$A_3 = x_0 - \frac{\mu N}{k}, \qquad A_4 = 0$$

Thus Eq. (2.129) becomes

$$x(t) = \left(x_0 - \frac{\mu N}{k} \right) \cos \omega_n t + \frac{\mu N}{k} \tag{2.132}$$

This solution is valid for half the cycle only—that is, for $0 \le t \le \pi/\omega_n$. When $t = \pi/\omega_n$, the mass will be at its extreme left position and its displacement from equilibrium position can be found from Eq. (2.132):

$$-x_1 = x\left(t = \frac{\pi}{\omega_n} \right) = \left(x_0 - \frac{\mu N}{k} \right) \cos \pi + \frac{\mu N}{k} = -\left(x_0 - \frac{2\mu N}{k} \right)$$

Since the motion started with a displacement of $x = x_0$ and, in a half cycle, the value of x became $-[x_0 - (2\mu N/k)]$, the reduction in magnitude of x in time π/ω_n is $2\mu N/k$.

In the second half cycle, the mass moves from left to right, so Eq. (2.127) is to be used. The initial conditions for this half cycle are

$$x(t = 0) = \text{value of } x \text{ at } t = \frac{\pi}{\omega_n} \text{ in Eq. (2.132)} = -\left(x_0 - \frac{2\mu N}{k} \right)$$

and

$$\dot{x}(t = 0) = \text{value of } \dot{x} \text{ at } t = \frac{\pi}{\omega_n} \text{ in Eq. (2.132)}$$

$$= \left\{ \text{value of } -\omega_n \left(x_0 - \frac{\mu N}{k} \right) \sin \omega_n t \text{ at } t = \frac{\pi}{\omega_n} \right\} = 0$$

Thus the constants in Eq. (2.127) become

$$-A_1 = -x_0 + \frac{3\mu N}{k}, \qquad A_2 = 0$$

so that Eq. (2.127) can be written as

$$x(t) = \left(x_0 - \frac{3\mu N}{k} \right) \cos \omega_n t - \frac{\mu N}{k} \tag{2.133}$$

This equation is valid only for the second half cycle—that is, for $\pi/\omega_n \leq t \leq 2\pi/\omega_n$. At the end of this half cycle, the value of $x(t)$ is

$$x_2 = x\left(t = \frac{\pi}{\omega_n}\right) \text{ in Eq. (2.133)} = x_0 - \frac{4\mu N}{k}$$

and

$$\dot{x}\left(t = \frac{\pi}{\omega_n}\right) \text{ in Eq. (2.133)} = 0$$

These become the initial conditions for the third half cycle, and the procedure can be continued until the motion stops. The motion stops when $x_n \leq \mu N/k$, since the restoring force exerted by the spring (kx) will then be less than the friction force μN. Thus the number of half cycles (r) that elapse before the motion ceases is given by

$$x_0 - r\frac{2\mu N}{k} \leq \frac{\mu N}{k}$$

—that is,

$$r \geq \left\{ \frac{x_0 - \dfrac{\mu N}{k}}{\dfrac{2\mu N}{k}} \right\} \tag{2.134}$$

Note the following characteristics of a system with Coulomb damping:

1. The equation of motion is nonlinear with Coulomb damping, while it is linear with viscous damping.
2. The natural frequency of the system is unaltered with the addition of Coulomb damping, while it is reduced with the addition of viscous damping.
3. The motion is periodic with Coulomb damping, while it can be nonperiodic in a viscously damped (overdamped) system.
4. The system comes to rest after some time with Coulomb damping, whereas the motion theoretically continues forever (perhaps with an infinitesimally small amplitude) with viscous and hysteresis damping.
5. The amplitude reduces linearly with Coulomb damping, whereas it reduces exponentially with viscous damping.
6. In each successive cycle, the amplitude of motion is reduced by the amount $4\mu N/k$, so the amplitudes at the end of any two consecutive cycles are related:

$$X_m = X_{m-1} - \frac{4\mu N}{k} \tag{2.135}$$

As the amplitude is reduced by an amount $4\mu N/k$ in one cycle (i.e., in time $2\pi/\omega_n$), the slope of the enveloping straight lines (shown dotted) in Fig. 2.43 is

$$-\left(\frac{4\mu N}{k}\right)\bigg/\left(\frac{2\pi}{\omega_n}\right) = -\left(\frac{2\mu N\omega_n}{\pi k}\right)$$

The final position of the mass is usually displaced from equilibrium ($x = 0$) position and represents a permanent displacement in which the friction force is locked. Slight tapping will usually make the mass come to its equilibrium position.

2.9.3 Torsional Systems with Coulomb Damping

If a constant frictional torque acts on a torsional system, the equation governing the angular oscillations of the system can be derived, similar to Eqs. (2.126) and (2.128), as

$$J_0\ddot{\theta} + k_t\theta = -T \tag{2.136}$$

and

$$J_0\ddot{\theta} + k_t\theta = T \tag{2.137}$$

where T denotes the constant damping torque (similar to μN for linear vibrations). The solutions of Eqs. (2.136) and (2.137) are similar to those for linear vibrations. In particular, the frequency of vibration is given by

$$\omega_n = \sqrt{\frac{k_t}{J_0}} \tag{2.138}$$

and the amplitude of motion at the end of the rth half cycle (θ_r) is given by

$$\theta_r = \theta_0 - r\frac{2T}{k_t} \tag{2.139}$$

where θ_0 is the initial angular displacement at $t = 0$ (with $\dot{\theta} = 0$). The motion ceases when

$$r \geq \left\{\frac{\theta_0 - \dfrac{T}{k_t}}{\dfrac{2T}{k_t}}\right\} \tag{2.140}$$

■■■■■■ **Coefficient of Friction from Measured Positions of Mass**

EXAMPLE 2.14

A metal block, placed on a rough surface, is attached to a spring and is given an initial displacement of 10 cm from its equilibrium position. After five cycles of oscillation in 2 s, the final position of the metal block is found to be 1 cm from its equilibrium position. Find the coefficient of friction between the surface and the metal block.

Solution: Since five cycles of oscillation were observed to take place in 2 s, the period (τ_n) is $2/5 = 0.4$ s, and hence the frequency of oscillation is $\omega_n = \sqrt{\frac{k}{m}} = \frac{2\pi}{\tau_n} = \frac{2\pi}{0.4} = 15.708$ rad/s. Since the amplitude of oscillation reduces by

$$\frac{4\mu N}{k} = \frac{4\mu mg}{k}$$

in each cycle, the reduction in amplitude in five cycles is

$$5\left(\frac{4\mu mg}{k}\right) = 0.10 - 0.01 = 0.09 \text{ m}$$

or

$$\mu = \frac{0.09k}{20mg} = \frac{0.09\omega_n^2}{20g} = \frac{0.09(15.708)^2}{20(9.81)} = 0.1132$$

∎

Pulley Subjected to Coulomb Damping

EXAMPLE 2.15

A steel shaft of length 1 m and diameter 50 mm is fixed at one end and carries a pulley of mass moment of inertia 25 kg-m^2 at the other end. A band brake exerts a constant frictional torque of 400 N-m around the circumference of the pulley. If the pulley is displaced by $6°$ and released, determine (1) the number of cycles before the pulley comes to rest and (2) the final settling position of the pulley.

Solution:

1. The number of half cycles that elapse before the angular motion of the pulley ceases is given by Eq. (2.140):

$$r \geq \left\{ \frac{\theta_0 - \dfrac{T}{k_t}}{\dfrac{2T}{k_t}} \right\} \tag{E.1}$$

where θ_0 = initial angular displacement = $6°$ = 0.10472 rad, k_t = torsional spring constant of the shaft given by

$$k_t = \frac{GJ}{l} = \frac{(8 \times 10^{10})\left\{\dfrac{\pi}{32}(0.05)^4\right\}}{1} = 49,087.5 \text{ N-m/rad}$$

and T = constant friction torque applied to the pulley = 400 N-m. Equation (E.1) gives

$$r \geq \frac{0.10472 - \left(\dfrac{400}{49,087.5}\right)}{\left(\dfrac{800}{49,087.5}\right)} = 5.926$$

Thus the motion ceases after six half cycles.

2. The angular displacement after six half cycles is given by Eq. (2.120):

$$\theta = 0.10472 - 6 \times 2\left(\frac{400}{49,087.5}\right) = 0.006935 \text{ rad} = 0.39734°$$

Thus the pulley stops at 0.39734° from the equilibrium position on the same side of the initial displacement.

∎

2.10   Free Vibration with Hysteretic Damping

Consider the spring-viscous-damper arrangement shown in Fig. 2.44(a). For this system, the force F needed to cause a displacement $x(t)$ is given by

$$F = kx + c\dot{x} \tag{2.141}$$

For a harmonic motion of frequency ω and amplitude X,

$$x(t) = X \sin \omega t \tag{2.142}$$

Equations (2.141) and (2.142) yield

$$F(t) = kX \sin \omega t + cX\omega \cos \omega t$$

$$= kx \pm c\omega \sqrt{X^2 - (X \sin \omega t)^2}$$

$$= kx \pm c\omega \sqrt{X^2 - x^2} \tag{2.143}$$

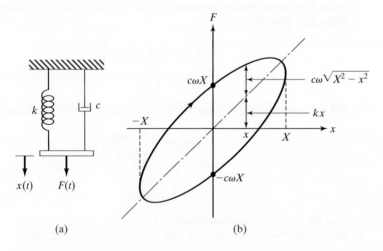

(a)   (b)

FIGURE 2.44   Spring-viscous-damper system.

When F versus x is plotted, Eq. (2.143) represents a closed loop, as shown in Fig. 2.44(b). The area of the loop denotes the energy dissipated by the damper in a cycle of motion and is given by

$$\Delta W = \oint F\, dx = \int_0^{2\pi/\omega} (kX \sin \omega t + cX\omega \cos \omega t)(\omega X \cos \omega t)\, dt$$

$$= \pi \omega c X^2 \tag{2.144}$$

Equation (2.144) has been derived in Section 2.6.4 also (see Eq. (2.98)).

As stated in Section 1.9, the damping caused by the friction between the internal planes that slip or slide as the material deforms is called *hysteresis* (or *solid* or *structural*) *damping*. This causes a hysteresis loop to be formed in the stress-strain or force-displacement curve (see Fig. 2.45(a)). The energy loss in one loading and unloading cycle is equal to the area enclosed by the hysteresis loop [2.11–2.13]. The similarity between Figs. 2.44(b) and 2.45(a) can be used to define a hysteresis damping constant. It was found experimentally that the energy loss per cycle due to internal friction is independent of the frequency but approximately proportional to the square of the amplitude. In order to achieve this observed behavior from Eq. (2.144), the damping coefficient c is assumed to be inversely proportional to the frequency as

$$c = \frac{h}{\omega} \tag{2.145}$$

where h is called the *hysteresis damping constant*. Equations (2.144) and (2.145) give

$$\Delta W = \pi h X^2 \tag{2.146}$$

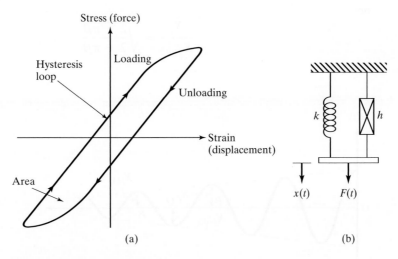

(a)   (b)

FIGURE 2.45   Hysteresis loop.

Complex Stiffness.   In Fig. 2.44(a), the spring and the damper are connected in parallel, and for a general harmonic motion, $x = Xe^{i\omega t}$, the force is given by

$$F = kXe^{i\omega t} + c\omega iXe^{i\omega t} = (k + i\omega c)x \tag{2.147}$$

Similarly, if a spring and a hysteresis damper are connected in parallel, as shown in Fig. 2.45(b), the force-displacement relation can be expressed as

$$F = (k + ih)x \tag{2.148}$$

where

$$k + ih = k\left(1 + i\frac{h}{k}\right) = k(1 + i\beta) \tag{2.149}$$

is called the *complex stiffness* of the system and $\beta = h/k$ is a constant indicating a dimensionless measure of damping.

Response of the System.   In terms of β, the energy loss per cycle can be expressed as

$$\Delta W = \pi k\beta X^2 \tag{2.150}$$

Under hysteresis damping, the motion can be considered to be nearly harmonic (since ΔW is small), and the decrease in amplitude per cycle can be determined using energy balance. For example, the energies at points P and Q (separated by half a cycle) in Fig. 2.46 are related as

$$\frac{kX_j^2}{2} - \frac{\pi k\beta X_j^2}{4} - \frac{\pi k\beta X_{j+0.5}^2}{4} = \frac{kX_{j+0.5}^2}{2}$$

or

$$\frac{X_j}{X_{j+0.5}} = \sqrt{\frac{2 + \pi\beta}{2 - \pi\beta}} \tag{2.151}$$

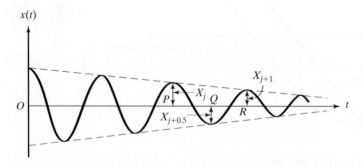

FIGURE 2.46   Response of a hysteretically damped system.

Similarly, the energies at points Q and R give

$$\frac{X_{j+0.5}}{X_{j+1}} = \sqrt{\frac{2 + \pi\beta}{2 - \pi\beta}} \tag{2.152}$$

Multiplication of Eqs. (2.151) and (2.152) gives

$$\frac{X_j}{X_{j+1}} = \frac{2 + \pi\beta}{2 - \pi\beta} = \frac{2 - \pi\beta + 2\pi\beta}{2 - \pi\beta} \simeq 1 + \pi\beta = \text{constant} \tag{2.153}$$

The hysteresis logarithmic decrement can be defined as

$$\delta = \ln\left(\frac{X_j}{X_{j+1}}\right) \simeq \ln(1 + \pi\beta) \simeq \pi\beta \tag{2.154}$$

Since the motion is assumed to be approximately harmonic, the corresponding frequency is defined by [2.10]:

$$\omega = \sqrt{\frac{k}{m}} \tag{2.155}$$

The equivalent viscous damping ratio ζ_{eq} can be found by equating the relation for the logarithmic decrement δ:

$$\delta \simeq 2\pi\zeta_{eq} \simeq \pi\beta = \frac{\pi h}{k}$$

$$\zeta_{eq} = \frac{\beta}{2} = \frac{h}{2k} \tag{2.156}$$

Thus the equivalent damping constant c_{eq} is given by

$$c_{eq} = c_c \cdot \zeta_{eq} = 2\sqrt{mk} \cdot \frac{\beta}{2} = \beta\sqrt{mk} = \frac{\beta k}{\omega} = \frac{h}{\omega} \tag{2.157}$$

Note that the method of finding an equivalent viscous damping coefficient for a structurally damped system is valid only for harmonic excitation. The above analysis assumes that the system responds approximately harmonically at the frequency ω.

Estimation of Hysteretic Damping Constant

EXAMPLE 2.16

The experimental measurements on a structure gave the force-deflection data shown in Fig. 2.47. From this data, estimate the hysteretic damping constant β and the logarithmic decrement δ.

Solution

Approach: We equate the energy dissipated in a cycle (area enclosed by the hysteresis loop) to ΔW of Eq. (2.146).

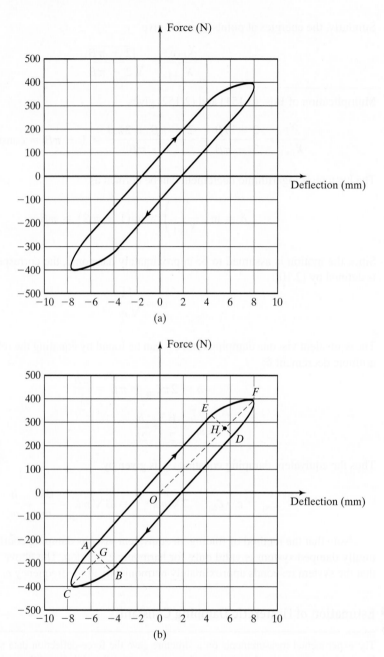

FIGURE 2.47   Force-deflection curve.

The energy dissipated in each full load cycle is given by the area enclosed by the hysteresis curve. Each square in Fig. 2.47 denotes $100 \times 2 = 200$ N-mm. The area enclosed by the loop can be found as area ACB + area $ABDE$ + area $DFE \simeq \frac{1}{2}(AB)(CG) + (AB)(AE) + \frac{1}{2}(DE)(FH) = \frac{1}{2}(1.25)(1.8) + (1.25)(8) + \frac{1}{2}(1.25)(1.8) = 12.25$ square units. This area represents an energy of $12.25 \times 200/1{,}000 = 2.5$ N-m. From Eq. (2.146), we have

$$\Delta W = \pi h X^2 = 2.5 \text{ N-m} \tag{E.1}$$

Since the maximum deflection X is 0.008 m and the slope of the force-deflection curve (given approximately by the slope of the line OF) is $k = 400/8 = 50$ N/mm $= 50{,}000$ N/m, the hysteretic damping constant h is given by

$$h = \frac{\Delta W}{\pi X^2} = \frac{2.5}{\pi (0.008)^2} = 12{,}433.95 \tag{E.2}$$

and hence

$$\beta = \frac{h}{k} = \frac{12{,}433.95}{50{,}000} = 0.248679$$

The logarithmic decrement can be found as

$$\delta \simeq \pi \beta = \pi(0.248679) = 0.78125 \tag{E.3}$$

∎

Response of a Hysteretically Damped Bridge Structure

EXAMPLE 2.17

A bridge structure is modeled as a single-degree-of-freedom system with an equivalent mass of 5×10^5 kg and an equivalent stiffness of 25×10^6 N/m. During a free-vibration test, the ratio of successive amplitudes was found to be 1.04. Estimate the structural damping constant (β) and the approximate free-vibration response of the bridge.

Solution: Using the ratio of successive amplitudes, Eq. (2.154) yields the hysteresis logarithmic decrement (δ) as

$$\delta = \ln\left(\frac{X_j}{X_{j+1}}\right) = \ln(1.04) = \ln(1 + \pi\beta)$$

or

$$1 + \pi\beta = 1.04 \quad \text{or} \quad \beta = \frac{0.04}{\pi} = 0.0127$$

The equivalent viscous damping coefficient (c_{eq}) can be determined from Eq. (2.157) as

$$c_{eq} = \frac{\beta k}{\omega} = \frac{\beta k}{\sqrt{\dfrac{k}{m}}} = \beta \sqrt{km} \tag{E.1}$$

Using the known values of the equivalent stiffness (k) and the equivalent mass (m) of the bridge, Eq. (E.1) yields

$$c_{eq} = (0.0127)\sqrt{(25 \times 10^6)(5 \times 10^5)} = 44.9013 \times 10^3 \text{ N-s/m}$$

The equivalent critical damping constant of the bridge can be computed using Eq. (2.65) as

$$c_c = 2\sqrt{km} = 2\sqrt{(25 \times 10^6)(5 \times 10^5)} = 7,071.0678 \times 10^3 \text{ N-s/m}$$

Since $c_{eq} < c_c$, the bridge is underdamped, and hence its free-vibration response is given by Eq. (2.72) as

$$x(t) = e^{-\zeta \omega_n t}\left\{ x_0 \cos \sqrt{1 - \zeta^2}\, \omega_n t + \frac{\dot{x}_0 + \zeta \omega_n x_0}{\sqrt{1 - \zeta^2}\, \omega_n} \sin \sqrt{1 - \zeta^2}\, \omega_n t \right\}$$

where

$$\zeta = \frac{c_{eq}}{c_c} = \frac{40.9013 \times 10^3}{7,071.0678 \times 10^3} = 0.0063$$

and x_0 and $\dot{x}_0$ denote the initial displacement and initial velocity given to the bridge at the start of free vibration.

■

2.11   Stability of Systems

Stability is one of the most important characteristics for any vibrating system. Although many definitions can be given for the term *stability* depending on the kind of system or the point of view, we consider our definition for linear and time-invariant systems (i.e., systems for which the parameters m, c, and k do not change with time). A system is defined to be *asymptotically stable* (called *stable* in controls literature) if its free-vibration response approaches zero as time approaches infinity. A system is considered to be *unstable* if its free-vibration response grows without bound (approaches infinity) as time approaches infinity. Finally, a system is said to be *stable* (called *marginally stable* in controls literature) if its free-vibration response neither decays nor grows, but remains constant or oscillates as time approaches infinity. It is evident that an unstable system whose free-vibration response grows without bound can cause damage to the system, adjacent property, or human life. Usually, dynamic systems are designed with limit stops to prevent their responses from growing with no limit.

As will be seen in Chapters 3 and 4, the total response of a vibrating system, subjected to external forces/excitations, is composed of two parts—one is the forced response and the other is the free-vibration response. For such systems, the definitions of asymptotically stable, unstable, and stable systems given above are still applicable. This implies that, for stable systems, only the forced response remains as the free-vibration response approaches zero as time approaches infinity.

Stability can be interpreted in terms of the roots of the characteristic roots of the system. As seen in Section 2.7, the roots lying in the left half-plane (LHP) yield either pure exponential decay or damped sinusoidal free-vibration responses. These responses decay to zero as time approaches infinity. Thus, systems whose characteristics roots lie in the left half of the s-plane (with a negative real part) will be asymptotically stable. The roots lying in the right half-plane yield either pure exponentially increasing or exponentially increasing sinusoidal free-vibration responses. These free-vibration responses approach infinity as time approaches infinity. Thus, systems whose characteristic roots lie in the right half of the s-plane (with positive real part) will be unstable. Finally, the roots lying on the imaginary axis of the s-plane yield pure sinusoidal oscillations as free-vibration responses. These responses neither increase nor decrease in amplitude as time grows. Thus, systems whose characteristic roots lie on the imaginary axis of the s-plane (with zero real parts) will be stable.[3]

Notes:

1. It is evident, from the definitions given, that the signs of the coefficients of the characteristic equation, Eq. (2.107), determine the stability behavior of a system. For example, from the theory of polynomial equations, if there is any number of negative terms or if any term in the polynomial in s is missing, then one of the roots will be positive, which results in an unstable behavior of the system. This aspect is considered further in Section 3.11 as well as in Section 5.8 in the form of the Routh-Hurwitz stability criterion.

2. In an unstable system, the free-vibration response may grow without bound with no oscillations or it may grow without bound with oscillations. The first behavior is called *divergent instability* and the second is called *flutter instability*. These cases are also known as *self-excited vibration* (see Section 3.11).

3. If a linear model of a system is asymptotically stable, then it is not possible to find a set of initial conditions for which the response approaches infinity. On the other hand, if the linear model of the system is unstable, it is possible that certain initial conditions might make the response approach zero as time increases. As an example, consider a system governed by the equation of motion $\ddot{x} - x = 0$ with characteristic roots given by $s_{1,2} = \pm 1$. Thus the response is given by $x(t) = C_1 e^{-t} + C_2 e^{t}$, where C_1 and C_2 are constants. If the initial conditions are specified as $x(0) = 1$ and $\dot{x}(0) = -1$, we find that $C_1 = 1$ and $C_2 = 0$ and hence the response becomes $x(t) = e^{-t}$, which approaches zero as time increases to infinity.

4. Typical responses corresponding to different types of stability are shown in Figs. 2.48(a)–(d).

5. Stability of a system can also be explained in terms of its energy. According to this scheme, a system is considered to be asymptotically stable, stable, or unstable if its energy decreases, remains constant, or increases, respectively, with time. This idea forms the basis for Lyapunov stability criterion [2.14, 2.15].

6. Stability of a system can also be investigated based on how sensitive the response or motion is to small perturbations (or variations) in the parameters (m, c, and k) and/or small perturbations in the initial conditions.

---

[3]Strictly speaking, the statement is true only if the roots that lie on the imaginary axis appear with multiplicity one. If such roots appear with multiplicity $n > 1$, the system will be unstable because the free-vibration response of such systems will be of the form $Ct^{n} \sin(\omega t + \phi)$.

Stable system

(a)

Asymptotically stable system

(b)

Unstable system (with divergent instability)

(c)

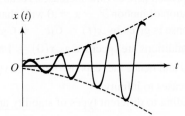

Unstable system (with flutter instability)

(d)

FIGURE 2.48   Different types of stability.

Stability of a System

EXAMPLE 2.18

Consider a uniform rigid bar, of mass m and length l, pivoted at one end and connected symmetrically by two springs at the other end, as shown in Fig. 2.49. Assuming that the springs are unstretched when the bar is vertical, derive the equation of motion of the system for small angular displacements (θ) of the bar about the pivot point, and investigate the stability behavior of the system.

FIGURE 2.49   Stability of a rigid bar.

Solution: When the bar is displaced by an angle θ, the spring force in each spring is $kl \sin \theta$; the total spring force is $2kl \sin \theta$. The gravity force $W = mg$ acts vertically downward through the center of gravity, G. The moment about the point of rotation O due to the angular acceleration $\ddot{\theta}$ is $J_0 \ddot{\theta} = (ml^2/3)\ddot{\theta}$. Thus the equation of motion of the bar, for rotation about the point O, can be written as

$$\frac{ml^2}{3}\ddot{\theta} + (2kl \sin \theta)l \cos \theta - W\frac{l}{2}\sin \theta = 0 \tag{E.1}$$

For small oscillations, Eq. (E.1) reduces to

$$\frac{ml^2}{3}\ddot{\theta} + 2kl^2\theta - \frac{Wl}{2}\theta = 0 \tag{E.2}$$

or

$$\ddot{\theta} + \alpha^2\theta = 0 \tag{E.3}$$

where

$$\alpha^2 = \left(\frac{12kl^2 - 3Wl}{2ml^2}\right) \tag{E.4}$$

The characteristic equation is given by

$$s^2 + \alpha^2 = 0 \tag{E.5}$$

and hence the solution of Eq. (E.2) depends on the sign of α^2 as indicated next.

Case 1. When $(12kl^2 - 3Wl)/2ml^2 > 0$, the solution of Eq. (E.2) represents a stable system with stable oscillations and can be expressed as

$$\theta(t) = A_1 \cos \omega_n t + A_2 \sin \omega_n t \tag{E.6}$$

where A_1 and A_2 are constants and

$$\omega_n = \left(\frac{12kl^2 - 3Wl}{2ml^2} \right)^{1/2} \tag{E.7}$$

Case 2. When $(12kl^2 - 3Wl)/2ml^2 = 0$, Eq. (E.2) reduces to $\ddot{\theta} = 0$ and the solution can be obtained directly by integrating twice as

$$\theta(t) = C_1 t + C_2 \tag{E.8}$$

For the initial conditions $\theta(t = 0) = \theta_0$ and $\dot{\theta}(t = 0) = \dot{\theta}_0$, the solution becomes

$$\theta(t) = \dot{\theta}_0 t + \theta_0 \tag{E.9}$$

Equation (E.9) shows that the system is unstable with the angular displacement increasing linearly at a constant velocity θ_0. However, if $\theta_0 = 0$, Eq. (E.9) denotes a stable or static equilibrium position with $\theta = \theta_0$—that is, the pendulum remains in its original position, defined by $\theta = \theta_0$.

Case 3. When $(12kl^2 - 3Wl)/2ml^2 < 0$, the solution of Eq. (E.2) can be expressed as

$$\theta(t) = B_1 e^{\alpha t} + B_2 e^{-\alpha t} \tag{E.10}$$

where B_1 and B_2 are constants. For the initial conditions $\theta(t = 0) = \theta_0$ and $\dot{\theta}(t = 0) = \dot{\theta}_0$, Eq. (E.10) becomes

$$\theta(t) = \frac{1}{2\alpha}[(\alpha\theta_0 + \dot{\theta}_0)e^{\alpha t} + (\alpha\theta_0 - \dot{\theta}_0)e^{-\alpha t}] \tag{E.11}$$

Equation (E.11) shows that $\theta(t)$ increases exponentially with time; hence the motion is unstable. The physical reason for this is that the restoring moment due to the spring $(2kl^2\theta)$, which tries to bring the system to the equilibrium position, is less than the nonrestoring moment due to gravity $[-W(l/2)\theta]$, which tries to move the mass away from the equilibrium position.

■

2.12   Examples Using MATLAB

Variations of Natural Frequency and Period with Static Deflection

EXAMPLE 2.19

Plot the variations of the natural frequency and the time period with static deflection of an undamped system using MATLAB.

Solution: The natural frequency (ω_n) and the time period (τ_n) are given by Eqs. (2.28) and (2.30):

$$\omega_n = \left(\frac{g}{\delta_{st}} \right)^{1/2}, \qquad \tau_n = 2\pi \left(\frac{\delta_{st}}{g} \right)^{1/2}$$

Using $g = 9.81$ m/s^2, ω_n and τ_n are plotted over the range of $\delta_{st} = 0$ to 0.5 using a MATLAB program.

```
% Ex2_19.m
g = 9.81;
for i = 1: 101
    t(i) = 0.01 + (0.5-0.01) * (i-1)/100;
    w(i) = (g/t(i))^0.5;
    tao(i) = 2 * pi * (t(i)/g)^0.5;
end
plot(t,w);
gtext('w_n');
hold on;
plot(t, tao);
gtext('T_n');
xlabel('Delta_s_t');
title('Example 2.17');
```

Example 2.19

Variations of natural frequency and time period.

EXAMPLE 2.20

Free-Vibration Response of a Spring-Mass System

A spring-mass system with a mass of 3,600 kg and stiffness 90×10^3 N/m is subject to an initial displacement of $x_0 = 0.08$ m and an initial velocity of $\dot{x}_0 = 0.1$ m/s. Plot the time variations of the mass's displacement, velocity, and acceleration using MATLAB.

Solution: The displacement of an undamped system can be expressed as (see Eq. (2.23)):

$$x(t) = A_0 \sin(\omega_n t + \phi_0) \tag{E.1}$$

where

$$\omega_n = \sqrt{\frac{k}{m}} = \sqrt{\frac{90 \times 10^3}{3{,}600}} = 5 \text{ rad/s}$$

$$A_0 = \left[x_0^2 + \left(\frac{\dot{x}_0}{\omega_n} \right)^2 \right]^{1/2} = \left[(0.08)^2 + \left(\frac{0.1}{5.0} \right)^2 \right]^{1/2} = 0.082462 \text{ m}$$

$$\phi_0 = \tan^{-1}\left(\frac{x_0 \omega_n}{\dot{x}_0} \right) = \tan^{-1}\left(\frac{(0.08)(5.0)}{0.1} \right) = 75.9638° = 1.3258 \text{ rad}$$

Thus Eq. (E.1) yields

$$x(t) = 0.082462 \sin(5t + 1.3258) \text{ m} \tag{E.2}$$

$$\dot{x}(t) = 0.41231 \cos(5t + 1.3258) \text{ m/s} \tag{E.3}$$

$$\ddot{x}(t) = -2.06155 \sin(5t + 1.3258) \text{ m/s}^2 \tag{E.4}$$

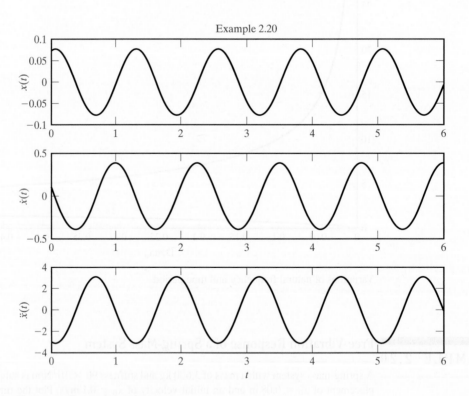

Response of an undamped system.

Equations (E.2)–(E.4) are plotted using MATLAB in the range $t = 0$ to 6 s.

```
% Ex2_20.m
for i = 1: 101
    t(i) = 6 * (i-1)/100;
    x(i) = 0.082462 * sin(5 * t(i) + 1.3258);
    x1(i) = 0.41231 * cos(5 * t(i) + 1.3258);
    x2(i) = -2.06155 * sin(5 * t(i) + 1.3258);
end
subplot (311);
plot (t,x);
ylabel ('x(t)');
title ('Example 2.20');
subplot (312);
plot (t,x1);
ylabel ('x^.(t)');
subplot (313);
plot (t,x2);
xlabel ('t');
ylabel ('x^.^.(t)');
```

■

Free-Vibration Response of a System with Coulomb Damping

EXAMPLE 2.21

Find the free-vibration response of a spring-mass system subject to Coulomb damping for the following initial conditions: $x(0) = 0.5$ m, $\dot{x}(0) = 0$.

Data: $m = 10$ kg, $k = 200$ N/m, $\mu = 0.5$

Solution: The equation of motion can be expressed as

$$m\ddot{x} + \mu mg \, \text{sgn}(\dot{x}) + kx = 0 \tag{E.1}$$

In order to solve the second-order differential equation, Eq. (E.1), using the Runge-Kutta method (see Appendix F), we rewrite Eq. (E.1) as a set of two first-order differential equations as follows:

$$x_1 = x, \qquad x_2 = \dot{x}_1 = \dot{x}$$

$$\dot{x}_1 = x_2 \equiv f_1(x_1, x_2) \tag{E.2}$$

$$\dot{x}_2 = -\mu g \, \text{sgn}(x_2) - \frac{k}{m}x_1 \equiv f_2(x_1, x_2) \tag{E.3}$$

Equations (E.2) and (E.3) can be expressed in matrix notation as

$$\dot{\vec{X}} = \vec{f}(\vec{X}) \tag{E.4}$$

where

$$\vec{X} = \begin{Bmatrix} x_1(t) \\ x_2(t) \end{Bmatrix}, \qquad \vec{f} = \begin{Bmatrix} f_1(x_1, x_2) \\ f_2(x_1, x_2) \end{Bmatrix}, \qquad \vec{X}(t = 0) = \begin{Bmatrix} x_1(0) \\ x_2(0) \end{Bmatrix}$$

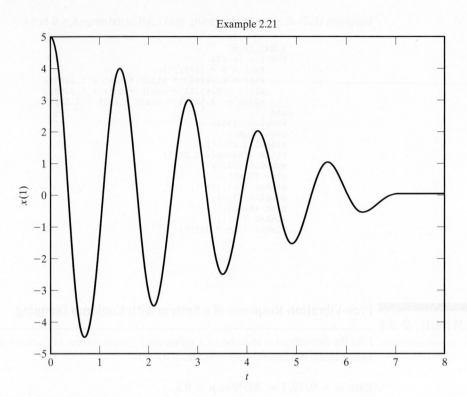

Solution of Eq. (E4):

The MATLAB program **ode23** is used to find the solution of Eq. (E.4) as shown below.

```
% Ex2_21.m
% This program will use dfunc1.m
tspan = [0: 0.05: 8];
x0 = [5.0; 0.0];
[t, x] = ode23 ('dfunc1', tspan, x0);
plot (t, x(:, 1));
xlabel ('t');
ylabel ('x(1)');
title ('Example 2.19');

% dfunc1.m
function f = dfunc1 (t, x)
f = zeros (2, 1);
f(1) = x(2);
f(2) = -0.5 * 9.81 * sign(x(2)) - 200 * x(1) / 10;
```

■

Free Vibration Response of a Viscously Damped System Using MATLAB

EXAMPLE   2.22

Develop a general-purpose MATLAB program, called **Program2.m**, to find the free-vibration response of a viscously damped system. Use the program to find the response of a system with the following data:

$$m = 450.0, \quad k = 26{,}519.2, \quad c = 1{,}000.0, \quad x_0 = 0.539657, \quad \dot{x}_0 = 1.0$$

Solution: Program2.m is developed to accept the following input data:

m = mass

k = spring stiffness

c = damping constant

x_0 = initial displacement

xd_0 = initial velocity

n = number of time steps at which values of $x(t)$ are to be found

delt = time interval between consecutive time steps (Δt)

The program gives the following output:

step number i, time (i), $x(i)$, $\dot{x}(i)$, $\ddot{x}(i)$

The program also plots the vatiations of x, $\dot{x}$, and $\ddot{x}$ with time.

```
>> program2
Free vibration analysis of a single degree of freedom analysis

Data:

m=          4.50000000e+002
k=          2.65192000e+004
c=          1.00000000e+003
x0=         5.39657000e-001
xd0=        1.00000000e+000
n=          100
delt=       2.50000000e-002

system is under damped

Results:

 i   time(i)          x(i)             xd(i)            xdd(i)

  1  2.500000e-002   5.540992e-001    1.596159e-001   -3.300863e+001
  2  5.000000e-002   5.479696e-001   -6.410545e-001   -3.086813e+001
  3  7.500000e-002   5.225989e-001   -1.375559e+000   -2.774077e+001
  4  1.000000e-001   4.799331e-001   -2.021239e+000   -2.379156e+001
  5  1.250000e-001   4.224307e-001   -2.559831e+000   -1.920599e+001
  6  1.500000e-001   3.529474e-001   -2.977885e+000   -1.418222e+001
  .
  .
  .
 96  2.400000e+000   2.203271e-002    2.313895e-001   -1.812621e+000
 97  2.425000e+000   2.722809e-002    1.834092e-001   -2.012170e+000
 98  2.450000e+000   3.117018e-002    1.314707e-001   -2.129064e+000
 99  2.475000e+000   3.378590e-002    7.764312e-002   -2.163596e+000
100  2.500000e+000   3.505350e-002    2.395118e-002   -2.118982e+000
```

■

Example 2.22

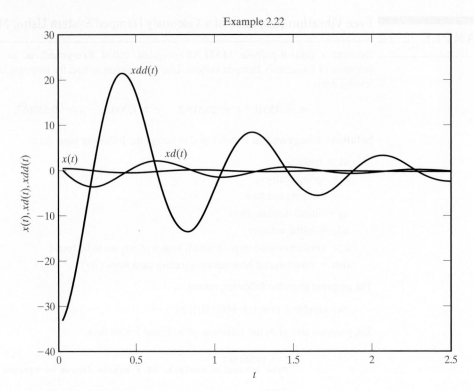

Variations of x, $\dot{x}$, and $\ddot{x}$.

CHAPTER SUMMARY

We considered the equations of motion and their solutions for the free vibration of undamped and damped single-degree-of-freedom systems. Four different methods—namely, Newton's second law of motion, D'Alembert's principle, the principle of virtual displacements, and the principle of conservation of energy—were presented for deriving the equation of motion of undamped systems. Both translational and torsional systems were considered. The free-vibration solutions have been presented for undamped systems. The equation of motion, in the form of a first-order differential equation, was considered for a mass-damper system (with no spring), and the idea of a time constant was introduced.

The free-vibration solution of viscously damped systems was presented along with the concepts of underdamped, overdamped, and critically damped systems. The free-vibration solutions of systems with Coulomb and hysteretic damping were also considered. The graphical representation of characteristic roots in the complex plane and the corresponding solutions were explained. The effects of variation of the parameters m, c, and k on the characteristic roots and their representations using root locus plots were also considered. The identification of the stability status of a system was also explained.

Now that you have finished this chapter, you should be able to answer the review questions and solve the problems given next.

REFERENCES

2.1 R. W. Fitzgerald, *Mechanics of Materials* (2nd ed.), Addison-Wesley, Reading, MA, 1982.

2.2 R. F. Steidel, Jr., *An Introduction to Mechanical Vibrations* (4th ed.), Wiley, New York, 1989.

2.3 W. Zambrano, "A brief note on the determination of the natural frequencies of a spring-mass system," *International Journal of Mechanical Engineering Education*, Vol. 9, October 1981, pp. 331–334; Vol. 10, July 1982, p. 216.

2.4 R. D. Blevins, *Formulas for Natural Frequency and Mode Shape*, Van Nostrand Reinhold, New York, 1979.

2.5 A. D. Dimarogonas, *Vibration Engineering*, West Publishing, Saint Paul, MN, 1976.

2.6 E. Kreyszig, *Advanced Engineering Mathematics* (9th ed.), Wiley, New York, 2006.

2.7 S. H. Crandall, "The role of damping in vibration theory," *Journal of Sound and Vibration*, Vol. 11, 1970, pp. 3–18.

2.8 I. Cochin, *Analysis and Design of Dynamic Systems* (3rd ed.), Addison-Wesley, Reading, MA, 1997.

2.9 D. Sinclair, "Frictional vibrations," *Journal of Applied Mechanics*, Vol. 22, 1955, pp. 207–214.

2.10 T. K. Caughey and M. E. J. O'Kelly, "Effect of damping on the natural frequencies of linear dynamic systems," *Journal of the Acoustical Society of America*, Vol. 33, 1961, pp. 1458–1461.

2.11 E. E. Ungar, "The status of engineering knowledge concerning the damping of built-up structures," *Journal of Sound and Vibration*, Vol. 26, 1973, pp. 141–154.

2.12 W. Pinsker, "Structural damping," *Journal of the Aeronautical Sciences*, Vol. 16, 1949, p. 699.

2.13 R. H. Scanlan and A. Mendelson, "Structural damping," *AIAA Journal*, Vol. 1, 1963, pp. 938–939.

2.14 D. J. Inman, *Vibration with Control, Measurement, and Stability*, Prentice Hall, Englewood Cliffs, NJ, 1989.

2.15 G. F. Franklin, J. D. Powell, and A. Emami-Naeini, *Feedback Control of Dynamic Systems* (5th ed.), Pearson Prentice Hall, Upper Saddle River, NJ, 2006.

2.16 N. S. Nise, *Control Systems Engineering* (3rd ed.), Wiley, New York, 2000.

2.17 K. Ogata, *System Dynamics* (4th ed.), Pearson Prentice Hall, Upper Saddle River, NJ, 2004.

REVIEW QUESTIONS

2.1 Give brief answers to the following:

1. Suggest a method for determining the damping constant of a highly damped vibrating system that uses viscous damping.

2. Can you apply the results of Section 2.2 to systems where the restoring force is not proportional to the displacement—that is, where k is not a constant?

3. State the parameters corresponding to m, c, k, and x for a torsional system.
4. What effect does a decrease in mass have on the frequency of a system?
5. What effect does a decrease in the stiffness of the system have on the natural period?
6. Why does the amplitude of free vibration gradually diminish in practical systems?
7. Why is it important to find the natural frequency of a vibrating system?
8. How many arbitrary constants must a general solution to a second-order differential equation have? How are these constants determined?
9. Can the energy method be used to find the differential equation of motion of all single-degree-of-freedom systems?
10. What assumptions are made in finding the natural frequency of a single-degree-of-freedom system using the energy method?
11. Is the frequency of a damped free vibration smaller or greater than the natural frequency of the system?
12. What is the use of the logarithmic decrement?
13. Is hysteresis damping a function of the maximum stress?
14. What is critical damping, and what is its importance?
15. What happens to the energy dissipated by damping?
16. What is equivalent viscous damping? Is the equivalent viscous-damping factor a constant?
17. What is the reason for studying the vibration of a single-degree-of-freedom system?
18. How can you find the natural frequency of a system by measuring its static deflection?
19. Give two practical applications of a torsional pendulum.
20. Define these terms: *damping ratio*, *logarithmic decrement*, *loss coefficient*, and *specific damping capacity*.
21. In what ways is the response of a system with Coulomb damping different from that of systems with other types of damping?
22. What is complex stiffness?
23. Define the hysteresis damping constant.
24. Give three practical applications of the concept of center of percussion.
25. What is the order of the equation of motion given by $m\dot{v} + cv = 0$?
26. Define the time constant.
27. What is a root locus plot?
28. What is the significance of $c < 0$?
29. What is a time-invariant system?

2.2   Indicate whether each of the following statements is true or false:

1. The amplitude of an undamped system will not change with time.
2. A system vibrating in air can be considered a damped system.
3. The equation of motion of a single-degree-of-freedom system will be the same whether the mass moves in a horizontal plane or an inclined plane.
4. When a mass vibrates in a vertical direction, its weight can always be ignored in deriving the equation of motion.
5. The principle of conservation of energy can be used to derive the equation of motion of both damped and undamped systems.
6. The damped frequency can in some cases be larger than the undamped natural frequency of the system.
7. The damped frequency can be zero in some cases.

8. The natural frequency of vibration of a torsional system is given by $\sqrt{\dfrac{k}{m}}$, where k and m denote the torsional spring constant and the polar mass moment of inertia, respectively.

9. Rayleigh's method is based on the principle of conservation of energy.

10. The final position of the mass is always the equilibrium position in the case of Coulomb damping.

11. The undamped natural frequency of a system is given by $\sqrt{g/\delta_{st}}$, where δ_{st} is the static deflection of the mass.

12. For an undamped system, the velocity leads the displacement by $\pi/2$.

13. For an undamped system, the velocity leads the acceleration by $\pi/2$.

14. Coulomb damping can be called *constant damping*.

15. The loss coefficient denotes the energy dissipated per radian per unit strain energy.

16. The motion diminishes to zero in both underdamped and overdamped cases.

17. The logarithmic decrement can be used to find the damping ratio.

18. The hysteresis loop of the stress-strain curve of a material causes damping.

19. The complex stiffness can be used to find the damping force in a system with hysteresis damping.

20. Motion in the case of hysteresis damping can be considered harmonic.

21. In the s-plane, the locus corresponding to constant natural frequency will be a circle.

22. The characteristic equation of a single-degree-of-freedom system can have one real root and one complex root.

2.3   Fill in the blanks with proper words:

1. The free vibration of an undamped system represents interchange of _____ and _____ energies.

2. A system undergoing simple harmonic motion is called a _____ *oscillator*.

3. The mechanical clock represents a _____ pendulum.

4. The center of _____ can be used advantageously in a baseball bat.

5. With viscous and hysteresis damping, the motion _____ forever, theoretically.

6. The damping force in Coulomb damping is given by _____.

7. The _____ coefficient can be used to compare the damping capacity of different engineering materials.

8. Torsional vibration occurs when a _____ body oscillates about an axis.

9. The property of _____ damping is used in many practical applications, such as large guns.

10. The logarithmic decrement denotes the rate at which the _____ of a free damped vibration decreases.

11. Rayleigh's method can be used to find the _____ frequency of a system directly.

12. Any two successive displacements of the system, separated by a cycle, can be used to find the _____ decrement.

13. The damped natural frequency (ω_d) can be expressed in terms of the undamped natural frequency (ω_n) as _____.

14. The time constant denotes the time at which the initial response reduces by _____ percent.

15. The term e^{-2t} decays _____ than the term e^{-t} as time t increases.

16. In the s-plane, lines parallel to the real axis denote systems having different _____ frequencies.

2.4   Select the most appropriate answer out of the multiple choices given:

1. The natural frequency of a system with mass m and stiffness k is given by:

   a. $\dfrac{k}{m}$ **b.** $\sqrt{\dfrac{k}{m}}$ **c.** $\sqrt{\dfrac{m}{k}}$

2. In Coulomb damping, the amplitude of motion is reduced in each cycle by:

   a. $\dfrac{\mu N}{k}$ **b.** $\dfrac{2\mu N}{k}$ **c.** $\dfrac{4\mu N}{k}$

3. The amplitude of an undamped system subject to an initial displacement 0 and initial velocity $\dot{x}_0$ is given by:

   a. $\dot{x}_0$ **b.** $\dot{x}_0 \omega_n$ **c.** $\dfrac{\dot{x}_0}{\omega_n}$

4. The effect of the mass of the spring can be accounted for by adding the following fraction of its mass to the vibrating mass:

   a. $\dfrac{1}{2}$ **b.** $\dfrac{1}{3}$ **c.** $\dfrac{4}{3}$

5. For a viscous damper with damping constant c, the damping force is:

   a. $c\dot{x}$ **b.** cx **c.** $c\ddot{x}$

6. The relative sliding of components in a mechanical system causes:

   a. dry-friction damping   **b.** viscous damping   **c.** hysteresis damping

7. In torsional vibration, the displacement is measured in terms of a:

   a. linear coordinate   **b.** angular coordinate   **c.** force coordinate

8. The damping ratio, in terms of the damping constant c and critical damping constant c_c, is given by:

   a. $\dfrac{c_c}{c}$ **b.** $\dfrac{c}{c_c}$ **c.** $\sqrt{\dfrac{c}{c_c}}$

9. The amplitude of an underdamped system subject to an initial displacement x_0 and initial velocity 0 is given by:

   a. x_0 **b.** $2x_0$ **c.** $x_0 \omega_n$

10. The phase angle of an undamped system subject to an initial displacement x_0 and initial velocity 0 is given by:

    a. x_0 **b.** $2x_0$ **c.** 0

11. The energy dissipated due to viscous damping is proportional to the following power of the amplitude of motion:

    a. 1   **b.** 2   **c.** 3

12. For a critically damping system, the motion will be:

    a. periodic   **b.** aperiodic   **c.** harmonic

13. The energy dissipated per cycle in viscous damping with damping constant c during the simple harmonic motion $x(t) = X \sin \omega_d t$, is given by:

    a. $\pi c \omega_d X^2$ **b.** $\pi \omega_d X^2$ **c.** $\pi c \omega_d X$

14. For a vibrating system with a total energy W and a dissipated energy ΔW per cycle, the specific damping capacity is given by:

    a. $\dfrac{W}{\Delta W}$ **b.** $\dfrac{\Delta W}{W}$ **c.** ΔW

15. If the characteristic roots have positive real values, the system response will be:

    a. stable   **b.** unstable   **c.** asymptotically stable

16. The frequency of oscillation of the response of a system will be higher if the imaginary part of the roots is:

    a. smaller   **b.** zero   **c.** larger

17. If the characteristic roots have a zero imaginary part, the response of the system will be:
   a. oscillatory   **b.** nonoscillatory   **c.** steady
18. The shape of the root locus of a single-degree-of-freedom system for $0 \leq \zeta \leq 1$ is:
   a. circular   **b.** horizontal line   **c.** radial line
19. The shape of the root locus of a single-degree-of-freedom system as k is varied is:
   a. vertical and horizontal lines   **b.** circular arc   **c.** radial lines

2.5   Match the following for a single-degree-of-freedom system with $m = 1$, $k = 2$, and $c = 0.5$:

1. Natural frequency, ω_n	**a.** 1.3919
2. Linear frequency, f_n	**b.** 2.8284
3. Natural time period, τ_n	**c.** 2.2571
4. Damped frequency, ω_d	**d.** 0.2251
5. Critical damping constant, c_c	**e.** 0.1768
6. Damping ratio, ζ	**f.** 4.4429
7. Logarithmic decrement, δ	**g.** 1.4142

2.6   Match the following for a mass $m = 5$ kg moving with velocity $v = 10$ m/s:

1. 20 N	**a.** Coulomb damping with a coefficient of friction of 0.3
2. 1.5 N	**b.** Viscous damping with a damping coefficient of 1 N-s/m
3. 30 N	**c.** Viscous damping with a damping coefficient of 2 N-s/m
4. 25 N	**d.** Hysteretic damping with a hysteretic damping coefficient of 12 N/m at a frequency of 4 rad/s
5. 10 N	**e.** Quadratic damping (force $= av^2$) with damping constant $a = 0.25$ N-s^2/m^2

2.7   Match the following characteristics of the s-plane:

1. Concentric circles	**a.** Different values of damped natural frequency
2. Lines parallel to real axis	**b.** Different values of reciprocals of time constant
3. Lines parallel to imaginary axis	**c.** Different values of damping ratio
4. Radial lines through origin	**d.** Different values of natural frequency

2.8   Match the following terms related to stability of systems:

1. Asymptotically stable	**a.** Neither decays nor grows
2. Unstable	**b.** Grows with oscillations
3. Stable	**c.** Grows without oscillations
4. Divergent instability	**d.** Approaches zero
5. Flutter instability	**e.** Grows without bound

PROBLEMS

Section 2.2 Free Vibration of an Undamped Translational System

2.1   An industrial press is mounted on a rubber pad to isolate it from its foundation. If the rubber pad is compressed 5 mm by the self weight of the press, find the natural frequency of the system.

2.2   A spring-mass system has a natural period of 0.21 s. What will be the new period if the spring constant is (a) increased by 50 percent and (b) decreased by 50 percent?

2.3   A spring-mass system has a natural frequency of 10 Hz. When the spring constant is reduced by 800 N/m, the frequency is altered by 45 percent. Find the mass and spring constant of the original system.

2.4   A helical spring, when fixed at one end and loaded at the other, requires a force of 100 N to produce an elongation of 10 mm. The ends of the spring are now rigidly fixed, one end vertically above the other, and a mass of 10 kg is attached at the middle point of its length. Determine the time taken to complete one vibration cycle when the mass is set vibrating in the vertical direction.

2.5   An air-conditioning chiller unit weighing 9,000 N is to be supported by four air springs (Fig. 2.50). Design the air springs such that the natural frequency of vibration of the unit lies between 5 rad/s and 10 rad/s.

FIGURE 2.50   (Courtesy of *Sound and Vibration*.)

2.6   The maximum velocity attained by the mass of a simple harmonic oscillator is 10 cm/s, and the period of oscillation is 2 s. If the mass is released with an initial displacement of 2 cm, find (a) the amplitude, (b) the initial velocity, (c) the maximum acceleration, and (d) the phase angle.

2.7   Three springs and a mass are attached to a rigid, weightless bar PQ as shown in Fig. 2.51. Find the natural frequency of vibration of the system.

FIGURE 2.51

2.8   An automobile having a mass of 2,000 kg deflects its suspension springs 0.02 m under static conditions. Determine the natural frequency of the automobile in the vertical direction by assuming damping to be negligible.

2.9   Find the natural frequency of vibration of a spring-mass system arranged on an inclined plane, as shown in Fig. 2.52.

FIGURE 2.52

2.10   A loaded mine cart, weighing 20 kN, is being lifted by a frictionless pulley and a wire rope, as shown in Fig. 2.53. Find the natural frequency of vibration of the cart in the given position.

2.11   An electronic chassis weighing 500 N is isolated by supporting it on four helical springs, as shown in Fig. 2.54. Design the springs so that the unit can be used in an environment in which the vibratory frequency ranges from 0 to 5 Hz.

Pulley

9 m

Steel wire rope,
1 mm diameter

7.5 m

Loaded
mine cart

50°

FIGURE 2.53

FIGURE 2.54 An electronic chassis mounted on
vibration isolators. (Courtesy of Titan SESCO.)

2.12 Find the natural frequency of the system shown in Fig. 2.55 with and without the springs k_1
and k_2 in the middle of the elastic beam.

2.13 Find the natural frequency of the pulley system shown in Fig. 2.56 by neglecting the friction
and the masses of the pulleys.

2.14 A weight W is supported by three frictionless and massless pulleys and a spring of stiffness k,
as shown in Fig. 2.57. Find the natural frequency of vibration of weight W for small oscillations.

FIGURE 2.55

FIGURE 2.56

2.15   A rigid block of mass M is mounted on four elastic supports, as shown in Fig. 2.58. A mass m drops from a height l and adheres to the rigid block without rebounding. If the spring constant of each elastic support is k, find the natural frequency of vibration of the system (a) without the mass m and (b) with the mass m. Also find the resulting motion of the system in case (b).

2.16   A sledgehammer strikes an anvil with a velocity of 20 m/s (Fig. 2.59). The hammer and the anvil weigh 50 N and 500 N, respectively. The anvil is supported on four springs, each of stiffness $k = 20$ kN/m. Find the resulting motion of the anvil (a) if the hammer remains in contact with the anvil and (b) if the hammer does not remain in contact with the anvil after the initial impact.

2.17   Derive the expression for the natural frequency of the system shown in Fig. 2.60. Note that the load W is applied at the tip of beam 1 and midpoint of beam 2.

FIGURE 2.57

FIGURE 2.58

2.18   A heavy machine weighing 9,810 N is being lowered by a winch at a uniform velocity of 2 m/s. The steel cable supporting the machine has a diameter of 0.01 m. The winch is suddenly stopped when the steel cable's length is 20 m. Find the period and amplitude of the ensuing vibration of the machine.

FIGURE 2.59

FIGURE 2.60

2.19 The natural frequency of a spring-mass system is found to be 2 Hz. When an additional mass of 1 kg is added to the original mass m, the natural frequency is reduced to 1 Hz. Find the spring constant k and the mass m.

2.20 An electrical switch gear is supported by a crane through a steel cable of length 4 m and diameter 0.01 m (Fig. 2.61). If the natural time period of axial vibration of the switch gear is found to be 0.1 s, find the mass of the switch gear.

2.21 Four weightless rigid links and a spring are arranged to support a weight W in two different ways, as shown in Fig. 2.62. Determine the natural frequencies of vibration of the two arrangements.

FIGURE 2.61   (Photo courtesy of the
Institution of Electrical Engineers.)

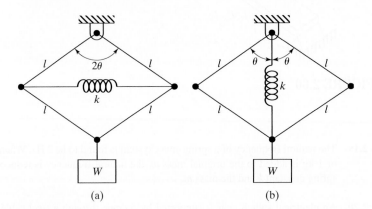

(a)   (b)

FIGURE 2.62

2.22 A scissors jack is used to lift a load W. The links of the jack are rigid and the collars can slide
freely on the shaft against the springs of stiffnesses k_1 and k_2 (see Fig. 2.63). Find the natural
frequency of vibration of the weight in the vertical direction.

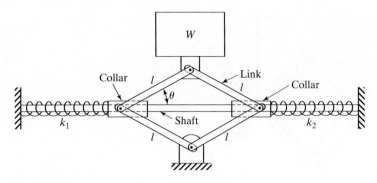

FIGURE 2.63

2.23 A weight is suspended using six rigid links and two springs in two different ways, as shown in Fig. 2.64. Find the natural frequencies of vibration of the two arrangements.

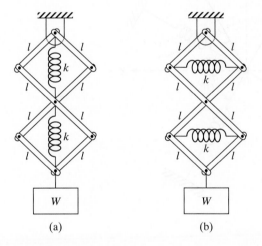

(a)   (b)

FIGURE 2.64

2.24 Figure 2.65 shows a small mass m restrained by four linearly elastic springs, each of which has an unstretched length l, and an angle of orientation of $45°$ with respect to the x-axis. Determine the equation of motion for small displacements of the mass in the x direction.

2.25 A mass m is supported by two sets of springs oriented at $30°$ and $120°$ with respect to the X-axis, as shown in Fig. 2.66. A third pair of springs, each with a stiffness of k_3, is to be designed so as to make the system have a constant natural frequency while vibrating in any direction x. Determine the necessary spring stiffness k_3 and the orientation of the springs with respect to the X-axis.

FIGURE 2.65

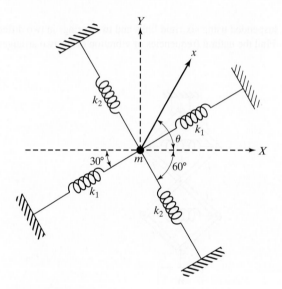

FIGURE 2.66

2.26 A mass m is attached to a cord that is under a tension T, as shown in Fig. 2.67. Assuming that T remains unchanged when the mass is displaced normal to the cord, (a) write the differential equation of motion for small transverse vibrations and (b) find the natural frequency of vibration.

2.27 A bungee jumper weighing 720 N ties one end of an elastic rope of length 60 m and stiffness 2 kN/m to a bridge and the other end to himself and jumps from the bridge (Fig. 2.68). Assuming the bridge to be rigid, determine the vibratory motion of the jumper about his static equilibrium position.

FIGURE 2.67

Unstretched length, 60 m

FIGURE 2.68

2.28 An acrobat weighing 540 N walks on a tightrope, as shown in Fig. 2.69. If the natural frequency of vibration in the given position, in the vertical direction, is 10 rad/s, find the tension in the rope.

2.29 The schematic diagram of a centrifugal governor is shown in Fig. 2.70. The length of each rod is l, the mass of each ball is m, and the free length of the spring is h. If the shaft speed is ω, determine the equilibrium position and the frequency for small oscillations about this position.

2.30 In the Hartnell governor shown in Fig. 2.71, the stiffness of the spring is 10^4 N/m and the weight of each ball is 25 N. The length of the ball arm is 20 cm, and that of the sleeve arm is 12 cm. The distance between the axis of rotation and the pivot of the bell crank lever is 16 cm. The spring is compressed by 1 cm when the ball arm is vertical. Find (a) the speed of the governor at which the ball arm remains vertical and (b) the natural frequency of vibration for small displacements about the vertical position of the ball arms.

FIGURE 2.69

FIGURE 2.70

2.31 A square platform *PQRS* and a car that it is supporting have a combined mass of *M*. The platform is suspended by four elastic wires from a fixed point *O*, as indicated in Fig. 2.72. The vertical distance between the point of suspension *O* and the horizontal equilibrium position of the platform is *h*. If the side of the platform is *a* and the stiffness of each wire is *k*, determine the period of vertical vibration of the platform.

2.32 The inclined manometer, shown in Fig. 2.73, is used to measure pressure. If the total length of mercury in the tube is *L*, find an expression for the natural frequency of oscillation of the mercury.

FIGURE 2.71   Hartnell governor.

FIGURE 2.72

2.33   The crate, of mass 250 kg, hanging from a helicopter (shown in Fig. 2.74(a)) can be modeled as shown in Fig. 2.74(b). The rotor blades of the helicopter rotate at 300 rpm. Find the diameter of the steel cables so that the natural frequency of vibration of the crate is at least twice the frequency of the rotor blades.

FIGURE 2.73

(a)   (b)

FIGURE 2.74

2.34   A pressure-vessel head is supported by a set of steel cables of length 2 m as shown in Fig. 2.75. The time period of axial vibration (in vertical direction) is found to vary from 5 s to 4.0825 s when an additional mass of 5,000 kg is added to the pressure-vessel head. Determine the equivalent cross-sectional area of the cables and the mass of the pressure-vessel head.

2.35   A flywheel is mounted on a vertical shaft, as shown in Fig. 2.76. The shaft has a diameter d and length l and is fixed at both ends. The flywheel has a weight of W and a radius of gyration of r. Find the natural frequency of the longitudinal, the transverse, and the torsional vibration of the system.

FIGURE 2.75 (Photo courtesy of CBI Industries Inc.)

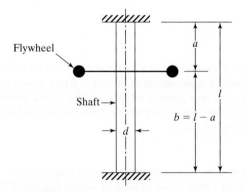

FIGURE 2.76

2.36 A TV antenna tower is braced by four cables, as shown in Fig. 2.77. Each cable is under tension and is made of steel with a cross-sectional area of 3×10^{-4} m^2. The antenna tower can be modeled as a steel beam of square section of side 25 mm for estimating its mass and stiffness. Find the tower's natural frequency of bending vibration about the y-axis.

FIGURE 2.77

2.37 Figure 2.78(a) shows a steel traffic sign, of thickness 0.003 m, fixed to a steel post. The post is 1.8 m high with a cross section of 50 mm $\times$ 6 mm, and it can undergo torsional vibration (about the z-axis) or bending vibration (either in the zx-plane or the yz-plane). Determine the mode of vibration of the post in a storm during which the wind velocity has a frequency component of 1.1 Hz.

Hints:

1. Neglect the weight of the post in finding the natural frequencies of vibration.
2. Torsional stiffness of a shaft with a rectangular section (see Fig. 2.78(b)) is given by

$$k_t = 5.33 \frac{ab^3 G}{l}\left[1 - 0.63\frac{b}{a}\left(1 - \frac{b^4}{12a^4}\right)\right]$$

where G is the shear modulus.

FIGURE 2.78

3. Mass moment of inertia of a rectangular block about axis OO (see Fig. 2.78(c)) is given by

$$I_{OO} = \frac{\rho l}{3}(b^3 h + h^3 b)$$

where ρ is the density of the block.

2.38 A building frame is modeled by four identical steel columns, each of weight w, and a rigid floor of weight W, as shown in Fig. 2.79. The columns are fixed at the ground and have a bending rigidity of EI each. Determine the natural frequency of horizontal vibration of the building frame by assuming the connection between the floor and the columns to be (a) pivoted as shown in Fig. 2.79(a) and (b) fixed against rotation as shown in Fig. 2.79(b). Include the effect of self weights of the columns.

FIGURE 2.79

2.39 A pick-and-place robot arm, shown in Fig. 2.80, carries an object weighing 50 N. Find the natural frequency of the robot arm in the axial direction for the following data: $l_1 = 0.30$ m, $l_2 = 0.25$ m, $l_3 = 0.20$ m; $E_1 = E_2 = E_3 = 70$ GPa; $D_1 = 0.05$ m, $D_2 = 0.04$ m, $D_3 = 0.03$ m; $d_1 = 0.045$ m, $d_2 = 0.035$ m, $d_3 = 0.025$ m.

Section AA Section BB Section CC

FIGURE 2.80

2.40 A helical spring of stiffness k is cut into two halves and a mass m is connected to the two halves as shown in Fig. 2.81(a). The natural time period of this system is found to be 0.5 s. If an identical spring is cut so that one part is one-fourth and the other part three-fourths of the original length, and the mass m is connected to the two parts as shown in Fig. 2.81(b), what would be the natural period of the system?

(a)   (b)

FIGURE 2.81

2.41* Figure 2.82 shows a metal block supported on two identical cylindrical rollers rotating in opposite directions at the same angular speed. When the center of gravity of the block is initially displaced by a distance x, the block will be set into simple harmonic motion. If the frequency of motion of the block is found to be ω, determine the coefficient of friction between the block and the rollers.

2.42* If two identical springs of stiffness k each are attached to the metal block of Problem 2.41 as shown in Fig. 2.83, determine the coefficient of friction between the block and the rollers.

*The asterisk denotes a design problem or a problem with no unique answer.

FIGURE 2.82

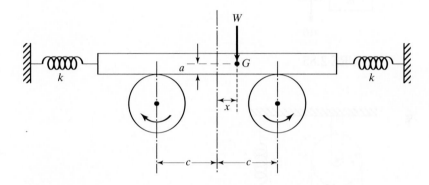

FIGURE 2.83

2.43 An electromagnet having a mass of 1,400 kg is at rest while holding an automobile of mass 900 kg in a junkyard. The electric current is turned off, and the automobile is dropped. Assuming that the crane and the supporting cable have an equivalent spring constant of 1,800 kN/m, find the following: (a) the natural frequency of vibration of the electromagnet, (b) the resulting motion of the electromagnet, and (c) the maximum tension developed in the cable during the motion.

2.44 Derive the equation of motion of the system shown in Fig. 2.84, using the following methods: (a) Newton's second law of motion, (b) D'Alembert's principle, (c) principle of virtual work, and (d) principle of conservation of energy.

FIGURE 2.84

2.45–2.46   Draw the free-body diagram and derive the equation of motion using Newton's second law of motion for each of the systems shown in Figs. 2.85 and 2.86.

FIGURE 2.85

FIGURE 2.86

2.47–2.48   Derive the equation of motion using the principle of conservation of energy for each of the systems shown in Figs. 2.85 and 2.86.

2.49   A steel beam of length 1 m carries a mass of 50 kg at its free end, as shown in Fig. 2.87. Find the natural frequency of transverse vibration of the mass by modeling it as a single-degree-of-freedom system.

FIGURE 2.87

2.50 A steel beam of length 1 m carries a mass of 50 kg at its free end, as shown in Fig. 2.88. Find the natural frequency of transverse vibration of the system by modeling it as a single-degree-of-freedom system.

FIGURE 2.88

2.51 Determine the displacement, velocity, and acceleration of the mass of a spring-mass system with $k = 500$ N/m, $m = 2$ kg, $x_0 = 0.1$ m, and $\dot{x}_0 = 5$ m/s.

2.52 Determine the displacement (x), velocity $(\dot{x})$, and acceleration $(\ddot{x})$ of a spring-mass system with $\omega_n = 10$ rad/s for the initial conditions $x_0 = 0.05$ m and $\dot{x}_0 = 1$ m/s. Plot $x(t)$, $\dot{x}(t)$, and $\ddot{x}(t)$ from $t = 0$ to 5 s.

2.53 The free-vibration response of a spring-mass system is observed to have a frequency of 2 rad/s, an amplitude of 10 mm, and a phase shift of 1 rad from $t = 0$. Determine the initial conditions that caused the free vibration. Assume the damping ratio of the system as 0.1.

2.54 An automobile is found to have a natural frequency of 20 rad/s without passengers and 17.32 rad/s with passengers of mass 500 kg. Find the mass and stiffness of the automobile by treating it as a single-degree-of-freedom system.

2.55 A spring-mass system with mass 2 kg and stiffness 3,200 N/m has an initial displacement of $x_0 = 0$. What is the maximum initial velocity that can be given to the mass without the amplitude of free vibration exceeding a value of 0.1 m?

2.56 A helical spring, made of music wire of diameter d, has a mean coil diameter (D) of 1.5 cm and N active coils (turns). It is found to have a frequency of vibration (f) of 193 Hz and a spring rate k of 4.62 kN/m. Determine the wire diameter d and the number of coils N, assuming the shear modulus G is 79.3 GPa and weight density ρ is 7,800 kg/m³. The spring rate (k) and frequency (f) are given by

$$k = \frac{d^4 G}{8 D^3 N}, \qquad f = \frac{1}{2}\sqrt{\frac{kg}{W}}$$

where W is the weight of the helical spring and g is the acceleration due to gravity.

2.57 Solve Problem 2.56 if the material of the helical spring is changed from music wire to aluminum with $G = 30$ GPa and $\rho = 3,000$ kg/m³.

2.58 A steel cantilever beam is used to carry a machine at its free end. To save weight, it is proposed to replace the steel beam by an aluminum beam of identical dimensions. Find the expected change in the natural frequency of the beam-machine system.

2.59 An oil drum of diameter 1 m and a mass of 500 kg floats in a bath of salt water of density $\rho_w = 1,050$ kg/m³. Considering small displacements of the drum in the vertical direction (x), determine the natural frequency of vibration of the system.

2.60 The equation of motion of a spring-mass system is given by (units: SI system)

$$500\ddot{x} + 1,000\left(\frac{x}{0.025}\right)^3 = 0$$

  a. Determine the static equilibrium position of the system.
  b. Derive the linearized equation of motion for small displacements (x) about the static equilibrium position.
  c. Find the natural frequency of vibration of the system for small displacements.
  d. Find the natural frequency of vibration of the system for small displacements when the mass is 600 (instead of 500).

2.61 A deceleration of 10 m/s² is caused when brakes are applied to a vehicle traveling at a speed of 100 km/h. Determine the time taken and the distance traveled before the vehicle comes to a complete stop.

2.62 A steel hollow cylindrical post is welded to a steel rectangular traffic sign as shown in Fig. 2.89 with the following data:
Dimensions: $l = 2$ m, $r_0 = 0.050$ m, $r_i = 0.045$ m, $b = 0.75$ m, $d = 0.40$ m, $t = 0.005$ m; material properties: ρ (specific weight) $= 76.50$ kN/m³, $E = 207$ GPa, $G = 79.3$ GPa.

Find the natural frequencies of the system in transverse vibration in the yz- and xz-planes by considering the masses of both the post and the sign.

Hint: Consider the post as a cantilever beam in transverse vibration in the appropriate plane.

2.63 Solve Problem 2.62 by changing the material from steel to bronze for both the post and the sign. Material properties of bronze: ρ (specific weight) $= 80.1$ kN/m³, $E = 111.0$ GPa, $G = 41.4$ GPa.

Section 2.3 Free Vibration of an Undamped Torsional System

2.64 A simple pendulum is set into oscillation from its rest position by giving it an angular velocity of 1 rad/s. It is found to oscillate with an amplitude of 0.5 rad. Find the natural frequency and length of the pendulum.

Traffic sign

Hollow cylindrical post

Section *A-A*

FIGURE 2.89

2.65 A pulley 250 mm in diameter drives a second pulley 1,000 mm in diameter by means of a belt (see Fig. 2.90). The moment of inertia of the driven pulley is 0.2 kg-m². The belt connecting these pulleys is represented by two springs, each of stiffness k. For what value of k will the natural frequency be 6 Hz?

2.66 Derive an expression for the natural frequency of the simple pendulum shown in Fig. 1.10. Determine the period of oscillation of a simple pendulum having a mass $m = 5$ kg and a length $l = 0.5$ m.

2.67 A mass m is attached at the end of a bar of negligible mass and is made to vibrate in three different configurations, as indicated in Fig. 2.91. Find the configuration corresponding to the highest natural frequency.

2.68 Figure 2.92 shows a spacecraft with four solar panels. Each panel has the dimensions 150 cm × 90 cm × 2.5 cm with a weight density of 3,000 kg/m³ and is connected to the body of the spacecraft by aluminum rods of length 30 cm and diameter 2.5 cm. Assuming that the body of the spacecraft is very large (rigid), determine the natural frequency of vibration of each panel about the axis of the connecting aluminum rod.

2.69 One of the blades of an electric fan is removed (as shown by dotted lines in Fig. 2.93). The steel shaft AB, on which the blades are mounted, is equivalent to a uniform shaft of diameter 2.5 cm and length 15 cm. Each blade can be modeled as a uniform slender rod of mass 1 kg and length 30 cm. Determine the natural frequency of vibration of the remaining three blades about the y-axis.

(a)

(b)

FIGURE 2.90 (Photo courtesy of Reliance Electric Company.)

FIGURE 2.91

FIGURE 2.92

Aluminum rod

FIGURE 2.93

2.70 A heavy ring of mass moment of inertia 1.0 kg-m^2 is attached at the end of a two-layered hollow shaft of length 2 m (Fig. 2.94). If the two layers of the shaft are made of steel and brass, determine the natural time period of torsional vibration of the heavy ring.

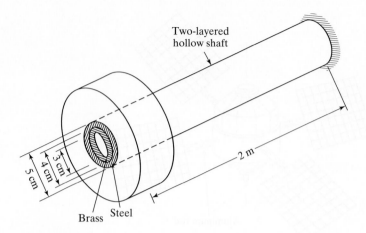

FIGURE 2.94

2.71 Find the natural frequency of the pendulum shown in Fig. 2.95 when the mass of the connecting bar is not negligible compared to the mass of the pendulum bob.

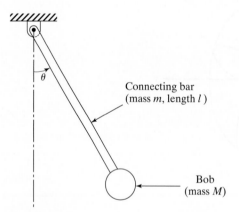

FIGURE 2.95

2.72 A steel shaft of 0.05 m diameter and 2 m length is fixed at one end and carries at the other end a steel disc of 1 m diameter and 0.1 m thickness, as shown in Fig. 2.14. Find the system's natural frequency of torsional vibration.

2.73 A uniform slender rod of mass m and length l is hinged at point A and is attached to four linear springs and one torsional spring, as shown in Fig. 2.96. Find the natural frequency of the system if $k = 2,000$ N/m, $k_t = 1,000$ N-m/rad, $m = 10$ kg, and $l = 5$ m.

FIGURE 2.96

2.74 A cylinder of mass m and mass moment of inertia J_0 is free to roll without slipping but is restrained by two springs of stiffnesses k_1 and k_2, as shown in Fig. 2.97. Find its natural frequency of vibration. Also find the value of a that maximizes the natural frequency of vibration.

FIGURE 2.97

2.75 If the pendulum of Problem 2.66 is placed in a rocket moving vertically with an acceleration of 5 m/s², what will be its period of oscillation?

2.76 Find the equation of motion of the uniform rigid bar OA of length l and mass m shown in Fig. 2.98. Also find its natural frequency.

FIGURE 2.98

2.77 A uniform circular disc is pivoted at point O, as shown in Fig. 2.99. Find the natural frequency of the system. Also find the maximum frequency of the system by varying the value of b.

FIGURE 2.99

2.78 Derive the equation of motion of the system shown in Fig. 2.100, using the following methods: (a) Newton's second law of motion, (b) D'Alembert's principle, and (c) principle of virtual work.

FIGURE 2.100

2.79 Find the natural frequency of the traffic sign system described in Problem 2.62 in torsional vibration about the z-axis by considering the masses of both the post and the sign.

Hint: The spring stiffness of the post in torsional vibration about the z-axis is given by $k_t = \dfrac{\pi G}{2l}(r_0^4 - r_i^4)$. The mass moment of inertia of the sign about the z-axis is given by $I_0 = \dfrac{1}{12}m_0(d^2 + b^2)$, where m_0 is the mass of the sign.

2.80 Solve Problem 2.79 by changing the material from steel to bronze for both the post and the sign. Material properties of bronze: ρ (specific weight) = 80.1 kN/m³, E = 111.0 GPa, G = 41.4 GPa.

2.81 A mass m_1 is attached at one end of a uniform bar of mass m_2 whose other end is pivoted at point O as shown in Fig. 2.101. Determine the natural frequency of vibration of the resulting pendulum for small angular displacements.

FIGURE 2.101

2.82 The angular motion of the forearm of a human hand carrying a mass m_0 is shown in Fig. 2.102. During motion, the forearm can be considered to rotate about the joint (pivot point) O with muscle forces modeled in the form of a force by triceps $(c_1 \dot{x})$ and a force in biceps $(-c_2 \theta)$, where c_1 and c_2 are constants and $\dot{x}$ is the velocity with which triceps are stretched (or contracted). Approximating the forearm as a uniform bar of mass m and length l, derive the equation of motion of the forearm for small angular displacements θ. Also find the natural frequency of the forearm.

Section 2.4 Response of First-Order Systems and Time Constant

2.83 Find the free-vibration response and the time constant, where applicable, of systems governed by the following equations of motion:

a. $100\dot{v} + 20v = 0, \quad v(0) = v(t = 0) = 10$
b. $100\dot{v} + 20v = 10, \quad v(0) = v(t = 0) = 10$
c. $100\dot{v} - 20v = 0, \quad v(0) = v(t = 0) = 10$
d. $500\dot{\omega} + 50\omega = 0, \quad \omega(0) = \omega(t = 0) = 0.5$

Hint: The time constant can also be defined as the value of time at which the step response of a system rises to 63.2% $(100.0\% - 36.8\%)$ of its final value.

2.84 A viscous damper, with damping constant c, and a spring, with spring stiffness k, are connected to a massless bar AB as shown in Fig. 2.103. The bar AB is displaced by a distance of $x = 0.1$ m when a constant force $F = 500$ N is applied. The applied force F is then abruptly released from its displaced position. If the displacement of the bar AB is reduced from its initial value of 0.1 m at $t = 0$ to 0.01 m at $t = 10$, find the values of c and k.

2.85 The equation of motion of a rocket, of mass m, traveling vertically under a thrust F and air resistance or drag D is given by

$$m\dot{v} = F - D - mg$$

FIGURE 2.102 Motion of arm.

FIGURE 2.103

If $m = 1,000$ kg, $F = 50,000$ N, $D = 2,000$ v, and $g = 9.81$ m/s^2, find the time variation of the velocity of the rocket, $v(t) = \dfrac{dx(t)}{dt}$, using the initial conditions $x(0) = 0$ and $v(0) = 0$, where $x(t)$ is the distance traveled by the rocket in time t.

Section 2.5 Rayleigh's Energy Method

2.86 Determine the effect of self weight on the natural frequency of vibration of the pinned-pinned beam shown in Fig. 2.104.

Uniform beam,
flexural stiffness = EI,
total weight = mg

M

$\dfrac{l}{2}$

l

FIGURE 2.104

2.87 Use Rayleigh's method to solve Problem 2.7.

2.88 Use Rayleigh's method to solve Problem 2.13.

2.89 Find the natural frequency of the system shown in Fig. 2.54.

2.90 Use Rayleigh's method to solve Problem 2.26.

2.91 Use Rayleigh's method to solve Problem 2.73.

2.92 Use Rayleigh's method to solve Problem 2.76.

2.93 A wooden rectangular prism of density ρ_w, height h, and cross section $a \times b$ is initially depressed in an oil tub and made to vibrate freely in the vertical direction (see Fig. 2.105).

b a

h

x

FIGURE 2.105

Use Rayleigh's method to find the natural frequency of vibration of the prism. Assume the density of oil is ρ_0. If the rectangular prism is replaced by a uniform circular cylinder of radius r, height h, and density ρ_w, will there be any change in the natural frequency?

2.94 Use the energy method to find the natural frequency of the system shown in Fig. 2.97.

2.95 Use the energy method to find the natural frequency of vibration of the system shown in Fig. 2.85.

2.96 A cylinder of mass m and mass moment of inertia J is connected to a spring of stiffness k and rolls on a rough surface as shown in Fig. 2.106. If the translational and angular displacements of the cylinder are x and θ from its equilibrium position, determine the following:

a. Equation of motion of the system for small displacements in terms of x using the energy method.

b. Equation of motion of the system for small displacements in terms of θ using the energy method.

c. Find the natural frequencies of the system using the equation of motion derived in parts (a) and (b). Are the resulting natural frequencies the same?

FIGURE 2.106

Section 2.6 Free Vibration with Viscous Damping

2.97 A simple pendulum is found to vibrate at a frequency of 0.5 Hz in a vacuum and 0.45 Hz in a viscous fluid medium. Find the damping constant, assuming the mass of the bob of the pendulum is 1 kg.

2.98 The ratio of successive amplitudes of a viscously damped single-degree-of-freedom system is found to be 18:1. Determine the ratio of successive amplitudes if the amount of damping is (a) doubled and (b) halved.

2.99 Assuming that the phase angle is zero, show that the response $x(t)$ of an underdamped single-degree-of-freedom system reaches a maximum value when

$$\sin \omega_d t = \sqrt{1 - \zeta^2}$$

and a minimum value when

$$\sin \omega_d t = -\sqrt{1 - \zeta^2}$$

Also show that the equations of the curves passing through the maximum and minimum values of $x(t)$ are given, respectively, by

$$x = \sqrt{1 - \zeta^2} X e^{-\zeta \omega_n t}$$

and

$$x = -\sqrt{1 - \zeta^2} X e^{-\zeta \omega_n t}$$

2.100 Derive an expression for the time at which the response of a critically damped system will attain its maximum value. Also find the expression for the maximum response.

2.101 A shock absorber is to be designed to limit its overshoot to 15 percent of its initial displacement when released. Find the damping ratio ζ_0 required. What will be the overshoot if ζ is made equal to (a) $\frac{3}{4}\zeta_0$ and (b) $\frac{5}{4}\zeta_0$?

2.102 The free-vibration responses of an electric motor of weight 500 N mounted on different types of foundations are shown in Figs. 2.107(a) and (b). Identify the following in each case: (i) the nature of damping provided by the foundation, (ii) the spring constant and damping coefficient of the foundation, and (iii) the undamped and damped natural frequencies of the electric motor.

(a)

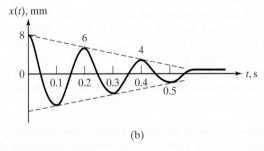

(b)

FIGURE 2.107

2.103 For a spring-mass-damper system, $m = 50$ kg and $k = 5,000$ N/m. Find the following: (a) critical damping constant c_c, (b) damped natural frequency when $c = c_c/2$, and (c) logarithmic decrement.

2.104 A railroad car of mass 2,000 kg traveling at a velocity $v = 10$ m/s is stopped at the end of the tracks by a spring-damper system, as shown in Fig. 2.108. If the stiffness of the spring is $k = 80$ N/mm and the damping constant is $c = 20$ N-s/mm, determine (a) the maximum displacement of the car after engaging the springs and damper and (b) the time taken to reach the maximum displacement.

FIGURE 2.108

2.105 A torsional pendulum has a natural frequency of 200 cycles/min when vibrating in a vacuum. The mass moment of inertia of the disc is 0.2 kg-m^2. It is then immersed in oil and its natural frequency is found to be 180 cycles/min. Determine the damping constant. If the disc, when placed in oil, is given an initial displacement of 2°, find its displacement at the end of the first cycle.

2.106 A boy riding a bicycle can be modeled as a spring-mass-damper system with an equivalent weight, stiffness, and damping constant of 800 N, 50,000 N/m, and 1,000 N-s/m, respectively. The differential setting of the concrete blocks on the road caused the level surface to decrease suddenly, as indicated in Fig. 2.109. If the speed of the bicycle is 5 m/s (18 km/h), determine the displacement of the boy in the vertical direction. Assume that the bicycle is free of vertical vibration before encountering the step change in the vertical displacement.

2.107 A wooden rectangular prism of mass 9 kg, height 90 cm, and cross section 30 cm × 60 cm floats and remains vertical in a tub of oil. The frictional resistance of the oil can be assumed to be equivalent to a viscous damping coefficient ζ. When the prism is depressed by a distance of 15 cm from its equilibrium and released, it is found to reach a depth of 14 cm at the end of its first cycle of oscillation. Determine the value of the damping coefficient of the oil.

FIGURE 2.109

2.108 A body vibrating with viscous damping makes five complete oscillations per second, and in 50 cycles its amplitude diminishes to 10 percent. Determine the logarithmic decrement and the damping ratio. In what proportion will the period of vibration be decreased if damping is removed?

2.109 The maximum permissible recoil distance of a gun is specified as 0.5 m. If the initial recoil velocity is to be between 8 m/s and 10 m/s, find the mass of the gun and the spring stiffness of the recoil mechanism. Assume that a critically damped dashpot is used in the recoil mechanism and the mass of the gun has to be at least 500 kg.

2.110 A viscously damped system has a stiffness of 5,000 N/m, critical damping constant of 0.2 N-s/mm, and a logarithmic decrement of 2.0. If the system is given an initial velocity of 1 m/s, determine the maximum displacement of the system.

2.111 Explain why an overdamped system never passes through the static equilibrium position when it is given (a) an initial displacement only and (b) an initial velocity only.

2.112–2.114 Derive the equation of motion and find the natural frequency of vibration of each of the systems shown in Figs. 2.110 to 2.112.

2.115–2.117 Using the principle of virtual work, derive the equation of motion for each of the systems shown in Figs. 2.110 to 2.112.

FIGURE 2.110

FIGURE 2.111

FIGURE 2.112

2.118 A wooden rectangular prism of cross section 40 cm $\times$ 60 cm, height 120 cm, and mass 40 kg floats in a fluid as shown in Fig. 2.105. When disturbed, it is observed to vibrate freely with a natural period of 0.5 s. Determine the density of the fluid.

2.119 The system shown in Fig. 2.113 has a natural frequency of 5 Hz for the following data: $m = 10$ kg, $J_0 = 5$ kg-m^2, $r_1 = 10$ cm, $r_2 = 25$ cm. When the system is disturbed by giving it an initial displacement, the amplitude of free vibration is reduced by 80 percent in 10 cycles. Determine the values of k and c.

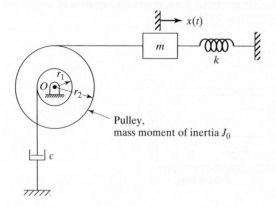

FIGURE 2.113

2.120 The rotor of a dial indicator is connected to a torsional spring and a torsional viscous damper to form a single-degree-of-freedom torsional system. The scale is graduated in equal divisions, and the equilibrium position of the rotor corresponds to zero on the scale. When a torque of 2×10^{-3} N-m is applied, the angular displacement of the rotor is found to be $50°$ with the pointer showing 80 divisions on the scale. When the rotor is released from this position, the pointer swings first to -20 divisions in one second and then to 5 divisions in another second. Find (a) the mass moment of inertia of the rotor, (b) the undamped natural time period of the rotor, (c) the torsional damping constant, and (d) the torsional spring stiffness.

2.121 Determine the values of ζ and ω_d for the following viscously damped systems:

  a. $m = 10$ kg, $c = 150$ N-s/m, $k = 1,000$ N/m

  b. $m = 10$ kg, $c = 200$ N-s/m, $k = 1,000$ N/m

  c. $m = 10$ kg, $c = 250$ N-s/m, $k = 1,000$ N/m

2.122 Determine the free-vibration response of the viscously damped systems described in Problem 2.121 when $x_0 = 0.1$ m and $\dot{x}_0 = 10$ m/s.

2.123 Find the energy dissipated during a cycle of simple harmonic motion given by $x(t) = 0.2 \sin \omega_d t$ m by a viscously damped single-degree-of-freedom system with the following parameters:

  a. $m = 10$ kg, $c = 50$ N-s/m, $k = 1,000$ N/m

  b. $m = 10$ kg, $c = 150$ N-s/m, $k = 1,000$ N/m

2.124 The equation of motion of a spring-mass-damper system, with a hardening-type spring, is given by (in SI units)

$$100\ddot{x} + 500\dot{x} + 10,000x + 400x^3 = 0$$

  a. Determine the static equilibrium position of the system.
  b. Derive the linearized equation of motion for small displacements (x) about the static equilibrium position.
  c. Find the natural frequency of vibration of the system for small displacements.

2.125 The equation of motion of a spring-mass-damper system, with a softening-type spring, is given by (in SI units)

$$100\ddot{x} + 500\dot{x} + 10,000x - 400x^3 = 0$$

  a. Determine the static equilibrium position of the system.
  b. Derive the linearized equation of motion for small displacements (x) about the static equilibrium position.
  c. Find the natural frequency of vibration of the system for small displacements.

2.126 The needle indicator of an electronic instrument is connected to a torsional viscous damper and a torsional spring. If the rotary inertia of the needle indicator about its pivot point is 25 kg-m^2 and the spring constant of the torsional spring is 100 N-m/rad, determine the damping constant of the torsional damper if the instrument is to be critically damped.

2.127 Find the responses of systems governed by the following equations of motion for the initial conditions $x(0) = 0$, $\dot{x}(0) = 1$:

  a. $2\ddot{x} + 8\dot{x} + 16x = 0$
  b. $3\ddot{x} + 12\dot{x} + 9x = 0$
  c. $2\ddot{x} + 8\dot{x} + 8x = 0$

2.128 Find the responses of systems governed by the following equations of motion for the initial conditions $x(0) = 1$, $\dot{x}(0) = 0$:

  a. $2\ddot{x} + 8\dot{x} + 16x = 0$
  b. $3\ddot{x} + 12\dot{x} + 9x = 0$
  c. $2\ddot{x} + 8\dot{x} + 8x = 0$

2.129 Find the responses of systems governed by the following equations of motion for the initial conditions $x(0) = 1$, $\dot{x}(0) = -1$:

  a. $2\ddot{x} + 8\dot{x} + 16x = 0$
  b. $3\ddot{x} + 12\dot{x} + 9x = 0$
  c. $2\ddot{x} + 8\dot{x} + 8x = 0$

2.130 A spring-mass system is found to vibrate with a frequency of 120 cycles per minute in air and 100 cycles per minute in a liquid. Find the spring constant k, the damping constant c, and the damping ratio ζ when vibrating in the liquid. Assume $m = 10$ kg.

2.131 Find the frequency of oscillation and time constant for the systems governed by the following equations:

  a. $\ddot{x} + 2\dot{x} + 9x = 0$
  b. $\ddot{x} + 8\dot{x} + 9x = 0$
  c. $\ddot{x} + 6\dot{x} + 9x = 0$

2.132 The mass moment of inertia of a nonhomogeneous and/or complex-shaped body of revolution about the axis of rotation can be determined by first finding its natural frequency of torsional vibration about its axis of rotation. In the torsional system shown in Fig. 2.114, the body of revolution (or rotor), of rotary inertia J, is supported on two frictionless bearings and

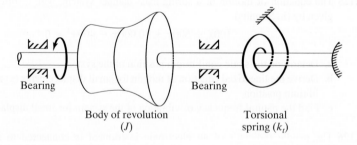

Bearing   Bearing

Body of revolution
(J)

Torsional
spring (k_t)

FIGURE 2.114

connected to a torsional spring of stiffness k_t. By giving an initial twist (angular displacement) of θ_0 and releasing the rotor, the period of the resulting vibration is measured as τ.

a. Find an expression for the mass moment of inertia of the rotor (J) in terms of τ and k_t.

b. Determine the value of J if $\tau = 0.5$ s and $k_t = 5,000$ N-m/rad.

Section 2.7 Graphical Representation of Characteristic Roots and Corresponding Solutions

2.133 The characteristic roots of a single-degree-of-freedom system are given below. Find all the applicable features of the system among the characteristic equation, time constant, undamped natural frequency, damped frequency, and damping ratio.

a. $s_{1,2} = -4 \pm 5i$

b. $s_{1,2} = 4 \pm 5i$

c. $s_{1,2} = -4, -5$

d. $s_{1,2} = -4, -4$

2.134 Show the characteristic roots indicated in Problem 2.133 (a)–(d) in the s-plane and describe the nature of the response of the system in each case.

2.135 The characteristic equation of a single-degree-of-freedom system, given by Eq. (2.107), can be rewritten as

$$s^2 + as + b = 0 \tag{E.1}$$

where $a = c/m$ and $b = k/m$ can be considered as the parameters of the system. Identify regions that represent a stable, unstable, and marginally stable system in the parameter plane—i.e., the plane in which a and b are denoted along the vertical and horizontal axes, respectively.

Section 2.8 Parameter Variations and Root Locus Representations

2.136 Consider the characteristic equation: $2s^2 + cs + 18 = 0$. Draw the root locus of the system for $c \geq 0$.

2.137 Consider the characteristic equation: $2s^2 + 12s + k = 0$. Draw the root locus of the system for $k \geq 0$.

2.138 Consider the characteristic equation: $ms^2 + 12s + 4 = 0$. Draw the root locus of the system for $m \geq 0$.

Section 2.9 Free Vibration with Coulomb Damping

2.139 A single-degree-of-freedom system consists of a mass of 20 kg and a spring of stiffness 4,000 N/m. The amplitudes of successive cycles are found to be 50, 45, 40, 35, ... mm. Determine the nature and magnitude of the damping force and the frequency of the damped vibration.

2.140 A mass of 20 kg slides back and forth on a dry surface due to the action of a spring having a stiffness of 10 N/mm. After four complete cycles, the amplitude has been found to be 100 mm. What is the average coefficient of friction between the two surfaces if the original amplitude was 150 mm? How much time has elapsed during the four cycles?

2.141 A 10-kg mass is connected to a spring of stiffness 3,000 N/m and is released after giving an initial displacement of 100 mm. Assuming that the mass moves on a horizontal surface, as shown in Fig. 2.42(a), determine the position at which the mass comes to rest. Assume the coefficient of friction between the mass and the surface to be 0.12.

2.142 A weight of 25 N is suspended from a spring that has a stiffness of 1,000 N/m. The weight vibrates in the vertical direction under a constant damping force. When the weight is initially pulled downward a distance of 10 cm from its static equilibrium position and released, it comes to rest after exactly two complete cycles. Find the magnitude of the damping force.

2.143 A mass of 20 kg is suspended from a spring of stiffness 10,000 N/m. The vertical motion of the mass is subject to Coulomb friction of magnitude 50 N. If the spring is initially displaced downward by 5 cm from its static equilibrium position, determine (a) the number of half cycles elapsed before the mass comes to rest, (b) the time elapsed before the mass comes to rest, and (c) the final extension of the spring.

2.144 The Charpy impact test is a dynamic test in which a specimen is struck and broken by a pendulum (or hammer) and the energy absorbed in breaking the specimen is measured. The energy values serve as a useful guide for comparing the impact strengths of different materials. As shown in Fig. 2.115, the pendulum is suspended from a shaft, is released from a particular position, and is allowed to fall and break the specimen. If the pendulum is made to oscillate freely (with no specimen), find (a) an expression for the decrease in the angle of swing for each cycle caused by friction, (b) the solution for $\theta(t)$ if the pendulum is released from an angle θ_0, and (c) the number of cycles after which the motion ceases. Assume the mass of the pendulum is m and the coefficient of friction between the shaft and the bearing of the pendulum is μ.

2.145 Find the equivalent viscous-damping constant for Coulomb damping for sinusoidal vibration.

2.146 A single-degree-of-freedom system consists of a mass, a spring, and a damper in which both dry friction and viscous damping act simultaneously. The free-vibration amplitude is found to decrease by 1 percent per cycle when the amplitude is 20 mm and by 2 percent per cycle when the amplitude is 10 mm. Find the value of $(\mu N/k)$ for the dry-friction component of the damping.

2.147 A metal block, placed on a rough surface, is attached to a spring and is given an initial displacement of 10 cm from its equilibrium position. It is found that the natural time period of motion is 1 s and that the amplitude reduces by 0.5 cm in each cycle. Find (a) the kinetic coefficient of friction between the metal block and the surface and (b) the number of cycles of motion executed by the block before it stops.

2.148 The mass of a spring-mass system with $k = 10,000$ N/m and $m = 5$ kg is made to vibrate on a rough surface. If the friction force is $F = 20$ N and the amplitude of the mass is observed to decrease by 50 mm in 10 cycles, determine the time taken to complete the 10 cycles.

2.149 The mass of a spring-mass system vibrates on a dry surface inclined at 30° to the horizontal as shown in Fig. 2.116.

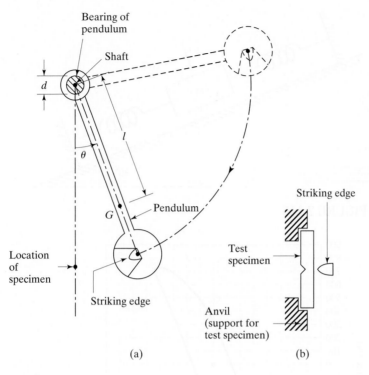

Bearing of
pendulum

Shaft

d

l

θ

Location
of
specimen

G

Pendulum

Striking edge

Striking edge

Test
specimen

Anvil
(support for
test specimen)

(a)

(b)

FIGURE 2.115

a. Derive the equation of motion.
b. Find the response of the system for the following data:

$$m = 20 \text{ kg}, \quad k = 1{,}000 \text{ N/m}, \quad \mu = 0.1, \quad x_0 = 0.1 \text{ m}, \quad \dot{x}_0 = 5 \text{ m/s}.$$

2.150 The mass of a spring-mass system is initially displaced by 10 cm from its unstressed position by applying a force of 25 N, which is equal to five times the weight of the mass. If the mass is released from this position, how long will the mass vibrate and at what distance will it stop from the unstressed position? Assume a coefficient of friction of 0.2.

Section 2.10 Free Vibration with Hysteretic Damping

2.151 The experimentally observed force-deflection curve for a composite structure is shown in Fig. 2.117. Find the hysteresis damping constant, the logarithmic decrement, and the equivalent viscous-damping ratio corresponding to this curve.

2.152 A panel made of fiber-reinforced composite material is observed to behave as a single-degree-of-freedom system of mass 1 kg and stiffness 2 N/m. The ratio of successive amplitudes is found to be 1.1. Determine the value of the hysteresis-damping constant β, the equivalent viscous-damping constant c_{eq}, and the energy loss per cycle for an amplitude of 10 mm.

FIGURE 2.116

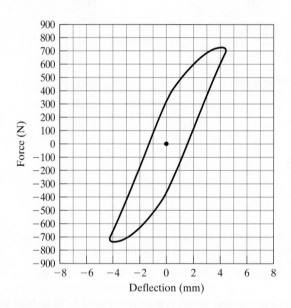

FIGURE 2.117

2.153 A built-up cantilever beam having a bending stiffness of 200 N/m supports a mass of 2 kg at its free end. The mass is displaced initially by 30 mm and released. If the amplitude is found to be 20 mm after 100 cycles of motion, estimate the hysteresis-damping constant β of the beam.

2.154 A mass of 5 kg is attached to the top of a helical spring, and the system is made to vibrate by giving to the mass an initial deflection of 25 mm. The amplitude of the mass is found to reduce to 10 mm after 100 cycles of vibration. Assuming a spring rate of 200 N/m for the helical spring, find the value of the hysteresis-damping coefficient (h) of the spring.

Section 2.11 Stability of Systems

2.155 Consider the equation of motion of a simple pendulum:

$$\ddot{\theta} + \frac{g}{l}\sin\theta = 0 \qquad\qquad (E.1)$$

a. Linearize Eq. (E.1) about an arbitrary angular displacement θ_0 of the pendulum.
b. Investigate the stability of the pendulum about $\theta_0 = 0$ and $\theta_0 = \pi$ using the linearized equation of motion.

2.156 Figure 2.118 shows a uniform rigid bar of mass m and length l, pivoted at one end (point O) and carrying a circular disk of mass M and mass moment of inertia J (about its rotational axis) at the other end (point P). The circular disk is connected to a spring of stiffness k and a viscous damper of damping constant c as indicated.

a. Derive the equation of motion of the system for small angular displacements of the rigid bar about the pivot point O and express it in the form:

$$m_0\,\dddot{\theta} + c_0\ddot{\theta} + k_0\theta = 0$$

b. Derive conditions corresponding to the stable, unstable, and marginally stable behavior of the system.

FIGURE 2.118

Section 2.12 Examples Using MATLAB

2.157 Find the free-vibration response of a spring-mass system subject to Coulomb damping using MATLAB for the following data:

$$m = 5 \text{ kg}, \quad k = 100 \text{ N/m}, \quad \mu = 0.5, \quad x_0 = 0.4 \text{ m}, \quad \dot{x}_0 = 0.$$

2.158 Plot the response of a critically damped system (Eq. (2.80)) for the following data using MATLAB:

  a. $x_0 = 10$ mm, 50 mm, 100 mm; $\dot{x}_0 = 0$, $\omega_n = 10$ rad/s.
  b. $x_0 = 0$, $\dot{x}_0 = 10$ mm/s, 50 mm/s, 100 mm/s; $\omega_n = 10$ rad/s.

2.159 Plot Eq. (2.81) as well as each of the two terms of Eq. (2.81) as functions of t using MATLAB for the following data:

$$\omega_n = 10 \text{ rad/s}, \quad \zeta = 2.0, \quad x_0 = 20 \text{ mm}, \quad \dot{x}_0 = 50 \text{ mm/s}$$

2.160–2.163 Using the MATLAB Program2.m, plot the free-vibration response of a viscously damped system with $m = 4$ kg, $k = 2,500$ N/m, $x_0 = 100$ mm, $\dot{x}_0 = -10$ m/s, $\Delta t = 0.01$ s, $n = 50$ for the following values of the damping constant:

  a. $c = 0$
  b. $c = 100$ N-s/m
  c. $c = 200$ N-s/m
  d. $c = 400$ N-s/m

2.164 Find the response of the system described in Problem 2.149 using MATLAB.

DESIGN PROJECTS

2.165* A water turbine of mass 1,000 kg and mass moment of inertia 500 kg-m^2 is mounted on a steel shaft, as shown in Fig. 2.119. The operational speed of the turbine is 2,400 rpm. Assuming the ends of the shaft to be fixed, find the values of l, a, and d, such that the natural frequency of vibration of the turbine in each of the axial, transverse, and circumferential directions is greater than the operational speed of the turbine.

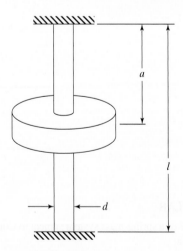

FIGURE 2.119

2.166* Design the columns for each of the building frames shown in Figs. 2.79(a) and (b) for minimum weight such that the natural frequency of vibration is greater than 50 Hz. The mass of the floor (W) is 1,800 kg and the length of the columns (l) is 240 cm. Assume that the columns are made of steel and have a tubular cross section with outer diameter d and wall thickness t.

2.167* One end of a uniform rigid bar of mass m is connected to a wall by a hinge joint O, and the other end carries a concentrated mass M, as shown in Fig. 2.120. The bar rotates about the hinge point O against a torsional spring and a torsional damper. It is proposed to use this mechanism, in conjunction with a mechanical counter, to control entry to an amusement park. Find the masses m and M, the stiffness of the torsional spring (k_t), and the damping force (F_d) necessary to satisfy the following specifications: (1) A viscous damper or a Coulomb damper can be used. (2) The bar has to return to within 5° of closing in less than 2 s when released from an initial position of $\theta = 75°$.

FIGURE 2.120

2.168* The lunar excursion module has been modeled as a mass supported by four symmetrically located legs, each of which can be approximated as a spring-damper system with negligible mass (see Fig. 2.121). Design the springs and dampers of the system in order to have the damped period of vibration between 1 s and 2 s.

FIGURE 2.121

2.169* Consider the telescoping boom and cockpit of the firetruck shown in Fig. 2.12(a). Assume that the telescoping boom $PQRS$ is supported by a strut QT, as shown in Fig. 2.122. Determine the cross section of the strut QT so that the natural time period of vibration of the cockpit with the fireperson is equal to 1 s for the following data. Assume that each segment of the telescoping boom and the strut is hollow circular in cross section. In addition, assume that the strut acts as a spring that deforms only in the axial direction.

Data:

Lengths of segments: $PQ = 3.6$ m, $QR = 3$ m, $RS = 2.4$ m, $TP = 0.9$ m

Young's modulus of the telescoping arm and strut $= 200$ GPa

Outer diameters of sections: $PQ = 5$ cm, $QR = 3.75$ cm, $RS = 2.5$ cm

Inner diameters of sections: $PQ = 4.5$ cm, $QR = 3.25$ cm, $RS = 2$ cm

Mass of the cockpit $= 50$ kg

Mass of fireperson $= 100$ kg

FIGURE 2.122

Charles Augustin de Coulomb (1736–1806) was a French military engineer and physicist. His early work on statics and mechanics was presented in 1779 in his great memoir *The Theory of Simple Machines*, which describes the effect of resistance and the so-called "Coulomb's law of proportionality" between friction and normal pressure. In 1784, he obtained the correct solution to the problem of the small oscillations of a body subjected to torsion. He is well known for his laws of force for electrostatic and magnetic charges. His name has been given to the unit of electric charge. (Courtesy of *Applied Mechanics Reviews*.)

C H A P T E R   3

Harmonically Excited Vibration

Chapter Outline

This chapter deals with the response of single-degree-of-freedom systems subjected to harmonic excitations. First, it presents the derivation of the equation of motion and its solution when a single degree of freedom system is subjected to harmonic excitation. Both undamped and damped systems are considered. The magnification or amplification factor, and the phenomena of resonance and beating are introduced in the context of an undamped spring-mass system. The total solution of the governing nonhomogeneous second-order differential equation is presented as a sum of the homogeneous equation (free-vibration solution) and the particular integral (forced-vibration solution). The known initial conditions of the system are used to evaluate the constants in the total solution. The important characteristics of the magnification factor and the phase angle for a viscously damped system are presented in detail. Quality factor, bandwidth, and half-power point are defined and the use of quality factor in estimating the viscous damping factor in a mechanical system is outlined. The response of the spring-mass-damper system with the harmonic forcing function in complex form is presented and the concept of complex frequency response is introduced. The response of a damped system under the harmonic motion of the base and the ideas of displacement transmissibility and force transmissibility are introduced. The applications of this problem include vibration of airplanes caused by runway roughness during taxiing and landing, vibration of ground vehicles due to unevenness and potholes in roads, and vibration of buildings caused by ground motion during earthquakes. The response of a damped system under rotating unbalance is also presented. The applications of this problem include a variety of rotating machines with unbalance in the rotors. The forced vibration of a spring-mass system under Coulomb, hysteresis, and other types of damping is also presented. Self-excitation and dynamic stability analysis of a single-degree-of-freedom system along with applications are presented. The general transfer-function approach, the Laplace transform approach, and the harmonic transfer-function approach for the solution of harmonically excited systems are outlined. Finally, the solution of different types of harmonically excited undamped and damped vibration problems using MATLAB is presented.

Learning Objectives

After completing this chapter, you should be able to do the following:

- Find the responses of undamped and viscously damped single-degree-of-freedom systems subjected to different types of harmonic force, including base excitation and rotating unbalance.
- Distinguish between transient, steady-state, and total solutions.
- Understand the variations of magnification factor and phase angles with the frequency of excitation and the phenomena of resonance and beats.
- Find the response of systems involving Coulomb, hysteresis, and other types of damping.
- Identify self-excited problems and investigate their stability aspects.
- Derive transfer functions of systems governed by linear differential equations with constant coefficients.
- Solve harmonically excited single-degree-of-freedom vibration problems using Laplace transforms.

- Derive frequency transfer function from the general transfer function and represent frequency-response characteristics using Bode diagrams.
- Solve harmonically excited vibration response using MATLAB.

3.1 Introduction

A mechanical or structural system is said to undergo forced vibration whenever external energy is supplied to the system during vibration. External energy can be supplied through either an applied force or an imposed displacement excitation. The applied force or displacement excitation may be harmonic, nonharmonic but periodic, nonperiodic, or random in nature. The response of a system to a harmonic excitation is called *harmonic response*. The nonperiodic excitation may have a long or short duration. The response of a dynamic system to suddenly applied nonperiodic excitations is called *transient response*.

In this chapter, we shall consider the dynamic response of a single-degree-of-freedom system under harmonic excitations of the form $F(t) = F_0 e^{i(\omega t + \phi)}$ or $F(t) = F_0 \cos(\omega t + \phi)$ or $F(t) = F_0 \sin(\omega t + \phi)$, where F_0 is the amplitude, ω is the frequency, and ϕ is the phase angle of the harmonic excitation. The value of ϕ depends on the value of $F(t)$ at $t = 0$ and is usually taken to be zero. Under a harmonic excitation, the response of the system will also be harmonic. If the frequency of excitation coincides with the natural frequency of the system, the response will be very large. This condition, known as *resonance*, is to be avoided to prevent failure of the system. The vibration produced by an unbalanced rotating machine, the oscillations of a tall chimney due to vortex shedding in a steady wind, and the vertical motion of an automobile on a sinusoidal road surface are examples of harmonically excited vibration.

The applications of transfer-function, Laplace transform, and frequency-function approaches in the solution of harmonically excited systems are also discussed in this chapter.

3.2 Equation of Motion

If a force $F(t)$ acts on a viscously damped spring-mass system as shown in Fig. 3.1, the equation of motion can be obtained using Newton's second law:

$$m\ddot{x} + c\dot{x} + kx = F(t) \tag{3.1}$$

Since this equation is nonhomogeneous, its general solution $x(t)$ is given by the sum of the homogeneous solution, $x_h(t)$, and the particular solution, $x_p(t)$. The homogeneous solution, which is the solution of the homogeneous equation

$$m\ddot{x} + c\dot{x} + kx = 0 \tag{3.2}$$

represents the free vibration of the system and was discussed in Chapter 2. As seen in Section 2.6.2, this free vibration dies out with time under each of the three possible conditions of damping (underdamping, critical damping, and overdamping) and under all possi-

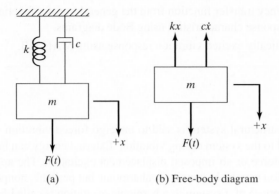

FIGURE 3.1   A spring-mass-damper system.

ble initial conditions. Thus the general solution of Eq. (3.1) eventually reduces to the particular solution $x_p(t)$, which represents the steady-state vibration. The steady-state motion is present as long as the forcing function is present. The variations of homogeneous, particular, and general solutions with time for a typical case are shown in Fig. 3.2. It can be seen that $x_h(t)$ dies out and $x(t)$ becomes $x_p(t)$ after some time (τ in Fig. 3.2). The part of the motion that dies out due to damping (the free-vibration part) is called *transient*. The rate at which the transient motion decays depends on the values of the system parameters k, c, and m. In this chapter, except in Section 3.3, we ignore the transient motion and derive only the particular solution of Eq. (3.1), which represents the steady-state response, under harmonic forcing functions.

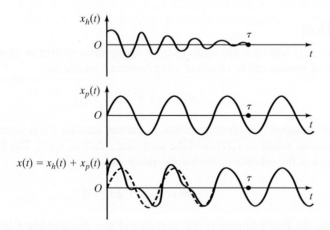

FIGURE 3.2   Homogenous, particular, and general solutions of Eq. (3.1) for an underdamped case.

3.3   Response of an Undamped System Under Harmonic Force

Before studying the response of a damped system, we consider an undamped system subjected to a harmonic force, for the sake of simplicity. If a force $F(t) = F_0 \cos \omega t$ acts on the mass m of an undamped system, the equation of motion, Eq. (3.1), reduces to

$$m\ddot{x} + kx = F_0 \cos \omega t \qquad (3.3)$$

The homogeneous solution of this equation is given by

$$x_h(t) = C_1 \cos \omega_n t + C_2 \sin \omega_n t \qquad (3.4)$$

where $\omega_n = (k/m)^{1/2}$ is the natural frequency of the system. Because the exciting force $F(t)$ is harmonic, the particular solution $x_p(t)$ is also harmonic and has the same frequency ω. Thus we assume a solution in the form

$$x_p(t) = X \cos \omega t \qquad (3.5)$$

where X is a constant that denotes the maximum amplitude of $x_p(t)$. By substituting Eq. (3.5) into Eq. (3.3) and solving for X, we obtain

$$X = \frac{F_0}{k - m\omega^2} = \frac{\delta_{st}}{1 - \left(\dfrac{\omega}{\omega_n}\right)^2} \qquad (3.6)$$

where $\delta_{st} = F_0/k$ denotes the deflection of the mass under a force F_0 and is sometimes called *static deflection* because F_0 is a constant (static) force. Thus the total solution of Eq. (3.3) becomes

$$x(t) = C_1 \cos \omega_n t + C_2 \sin \omega_n t + \frac{F_0}{k - m\omega^2} \cos \omega t \qquad (3.7)$$

Using the initial conditions $x(t = 0) = x_0$ and $\dot{x}(t = 0) = \dot{x}_0$, we find that

$$C_1 = x_0 - \frac{F_0}{k - m\omega^2}, \qquad C_2 = \frac{\dot{x}_0}{\omega_n} \qquad (3.8)$$

and hence

$$x(t) = \left(x_0 - \frac{F_0}{k - m\omega^2}\right) \cos \omega_n t + \left(\frac{\dot{x}_0}{\omega_n}\right) \sin \omega_n t$$

$$+ \left(\frac{F_0}{k - m\omega^2}\right) \cos \omega t \qquad (3.9)$$

The maximum amplitude X in Eq. (3.6) can be expressed as

$$\frac{X}{\delta_{st}} = \frac{1}{1 - \left(\dfrac{\omega}{\omega_n}\right)^2} \tag{3.10}$$

The quantity X/δ_{st} represents the ratio of the dynamic to the static amplitude of motion and is called the *magnification factor*, *amplification factor*, or *amplitude ratio*. The variation of the amplitude ratio, X/δ_{st}, with the frequency ratio $r = \omega/\omega_n$ Eq. (3.10) is shown in Fig. 3.3. From this figure, the response of the system can be identified to be of three types.

Case 1. When $0 < \omega/\omega_n < 1$, the denominator in Eq. (3.10) is positive and the response is given by Eq. (3.5) without change. The harmonic response of the system $x_p(t)$ is said to be in phase with the external force as shown in Fig. 3.4.

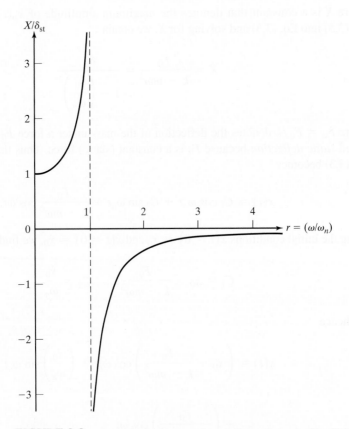

FIGURE 3.3   Magnification factor of an undamped system, Eq. (3.10).

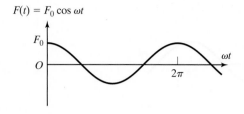

$F(t) = F_0 \cos \omega t$

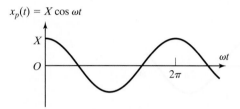

$x_p(t) = X \cos \omega t$

FIGURE 3.4 Harmonic response when $0 < \omega/\omega_n < 1$.

Case 2. When $\omega/\omega_n > 1$, the denominator in Eq. (3.10) is negative, and the steady-state solution can be expressed as

$$x_p(t) = -X \cos \omega t \qquad (3.11)$$

where the amplitude of motion X is redefined to be a positive quantity as

$$X = \frac{\delta_{\text{st}}}{\left(\dfrac{\omega}{\omega_n}\right)^2 - 1} \qquad (3.12)$$

The variations of $F(t)$ and $x_p(t)$ with time are shown in Fig. 3.5. Since $x_p(t)$ and $F(t)$ have opposite signs, the response is said to be 180° out of phase with the external force. Further, as $\omega/\omega_n \to \infty$, $X \to 0$. Thus the response of the system to a harmonic force of very high frequency is close to zero.

Case 3. When $\omega/\omega_n = 1$, the amplitude X given by Eq. (3.10) or (3.12) becomes infinite. This condition, for which the forcing frequency ω is equal to the natural frequency of the system ω_n, is called *resonance*. To find the response for this condition, we rewrite Eq. (3.9) as

$$x(t) = x_0 \cos \omega_n t + \frac{\dot{x}_0}{\omega_n} \sin \omega_n t + \delta_{\text{st}} \left[\frac{\cos \omega t - \cos \omega_n t}{1 - \left(\dfrac{\omega}{\omega_n}\right)^2} \right] \qquad (3.13)$$

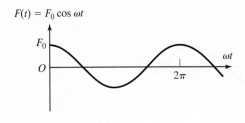

$F(t) = F_0 \cos \omega t$

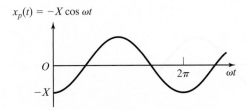

$x_p(t) = -X \cos \omega t$

FIGURE 3.5   Harmonic response when
$\omega/\omega_n > 1$.

Since the last term of this equation takes an indefinite form for $\omega = \omega_n$, we apply L'Hospital's rule [3.1] to evaluate the limit of this term:

$$\lim_{\omega \to \omega_n} \left[\frac{\cos \omega t - \cos \omega_n t}{1 - \left(\dfrac{\omega}{\omega_n}\right)^2} \right] = \lim_{\omega \to \omega_n} \left[\frac{\dfrac{d}{d\omega}(\cos \omega t - \cos \omega_n t)}{\dfrac{d}{d\omega}\left(1 - \dfrac{\omega^2}{\omega_n^2}\right)} \right]$$

$$= \lim_{\omega \to \omega_n} \left[\frac{t \sin \omega t}{2\dfrac{\omega}{\omega_n^2}} \right] = \frac{\omega_n t}{2} \sin \omega_n t \qquad (3.14)$$

Thus the response of the system at resonance becomes

$$x(t) = x_0 \cos \omega_n t + \frac{\dot{x}_0}{\omega_n} \sin \omega_n t + \frac{\delta_{st} \omega_n t}{2} \sin \omega_n t \qquad (3.15)$$

It can be seen from Eq. (3.15) that at resonance, $x(t)$ increases indefinitely. The last term of Eq. (3.15) is shown in Fig. 3.6, from which the amplitude of the response can be seen to increase linearly with time.

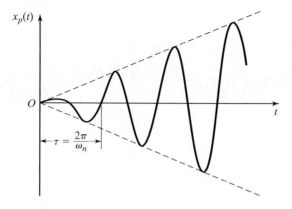

FIGURE 3.6   Response when $\omega/\omega_n = 1$.

3.3.1
Total Response

The total response of the system, Eq. (3.7) or (3.9), can also be expressed as

$$x(t) = A\cos(\omega_n t - \phi) + \frac{\delta_{st}}{1 - \left(\dfrac{\omega}{\omega_n}\right)^2}\cos\omega t; \quad \text{for } \frac{\omega}{\omega_n} < 1 \tag{3.16}$$

$$x(t) = A\cos(\omega_n t - \phi) - \frac{\delta_{st}}{-1 + \left(\dfrac{\omega}{\omega_n}\right)^2}\cos\omega t; \quad \text{for } \frac{\omega}{\omega_n} > 1 \tag{3.17}$$

where A and ϕ can be determined as in the case of Eq. (2.21). Thus the complete motion can be expressed as the sum of two cosine curves of different frequencies. In Eq. (3.16), the forcing frequency ω is smaller than the natural frequency, and the total response is shown in Fig. 3.7(a). In Eq. (3.17), the forcing frequency is greater than the natural frequency, and the total response appears as shown in Fig. 3.7(b).

3.3.2
Beating
Phenomenon

If the forcing frequency is close to, but not exactly equal to, the natural frequency of the system, a phenomenon known as *beating* may occur. In this kind of vibration, the amplitude builds up and then diminishes in a regular pattern (see Section 1.10.5). The phenomenon of beating can be explained by considering the solution given by Eq. (3.9). If the initial conditions are taken as $x_0 = \dot{x}_0 = 0$, Eq. (3.9) reduces to

$$x(t) = \frac{(F_0/m)}{\omega_n^2 - \omega^2}(\cos\omega t - \cos\omega_n t)$$

$$= \frac{(F_0/m)}{\omega_n^2 - \omega^2}\left[2\sin\frac{\omega + \omega_n}{2}t \cdot \sin\frac{\omega_n - \omega}{2}t\right] \tag{3.18}$$

(a) $\dfrac{\omega}{\omega_n} < 1$

(b) $\dfrac{\omega}{\omega_n} > 1$

FIGURE 3.7 Total response.

Let the forcing frequency ω be slightly less than the natural frequency:

$$\omega_n - \omega = 2\varepsilon \tag{3.19}$$

where ε is a small positive quantity. Then $\omega_n \approx \omega$ and

$$\omega + \omega_n \simeq 2\omega \tag{3.20}$$

Multiplication of Eqs. (3.19) and (3.20) gives

$$\omega_n^2 - \omega^2 = 4\varepsilon\omega \tag{3.21}$$

The use of Eqs. (3.19) to (3.21) in Eq. (3.18) gives

$$x(t) = \left(\frac{F_0/m}{2\varepsilon\omega} \sin \varepsilon t \right) \sin \omega t \tag{3.22}$$

Since ε is small, the function $\sin \varepsilon t$ varies slowly; its period, equal to $2\pi/\varepsilon$, is large. Thus Eq. (3.22) may be seen as representing vibration with period $2\pi/\omega$ and of variable amplitude equal to

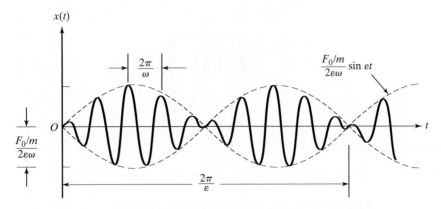

FIGURE 3.8   Phenomenon of beats.

$$\left(\frac{F_0/m}{2\varepsilon\omega}\right)\sin \varepsilon t$$

It can also be observed that the sin ωt curve will go through several cycles, while the sin εt wave goes through a single cycle, as shown in Fig. 3.8. Thus the amplitude builds up and dies down continuously. The time between the points of zero amplitude or the points of maximum amplitude is called the *period of beating* (τ_b) and is given by

$$\tau_b = \frac{2\pi}{2\varepsilon} = \frac{2\pi}{\omega_n - \omega} \tag{3.23}$$

with the frequency of beating defined as

$$\omega_b = 2\varepsilon = \omega_n - \omega$$

EXAMPLE 3.1

Plate Supporting a Pump

A reciprocating pump, having a mass of 68 kg, is mounted at the middle of a steel plate of thickness 1 cm, width 50 cm, and length 250 cm, clamped along two edges as shown in Fig. 3.9. During operation of the pump, the plate is subjected to a harmonic force, $F(t) = 220 \cos 62.832t$ N. Find the amplitude of vibration of the plate.

Solution: The plate can be modeled as a fixed-fixed beam having Young's modulus $(E) = 200$ GPa, length $(l) = 250$ cm, and area moment of inertia $(I) = \frac{1}{12}(50 \times 10^{-2})(10^{-2})^3 = 41.667 \times 10^{-9}$m^4. The bending stiffness of the beam is given by

$$k = \frac{192EI}{l^3} = \frac{192(200 \times 10^9)(41.667 \times 10^{-9})}{(250 \times 10^{-2})^3} = 102,400.82 \text{ N/m} \tag{E.1}$$

FIGURE 3.9   Plate supporting an unbalanced pump.

The amplitude of harmonic response is given by Eq. (3.6) with $F_0 = 220$ N, $m = 68$ kg (neglecting the mass of the steel plate), $k = 102,400.82$ N/m, and $\omega = 62.832$ rad/s. Thus Eq. (3.6) gives

$$X = \frac{F_0}{k - \left(\dfrac{m}{g}\right)\omega^2} = \frac{220}{102,400.82 - 68(62.832)^2} = -0.00132487 \text{ m} = -1.32487 \text{ mm} \qquad \text{(E.2)}$$

The negative sign indicates that the response $x(t)$ of the plate is out of phase with the excitation $F(t)$.

∎

EXAMPLE 3.2

Determination of Mass from Known Harmonic Response

A spring-mass system, with a spring stiffness of 5,000 N/m, is subjected to a harmonic force of magnitude 30 N and frequency 20 Hz. The mass is found to vibrate with an amplitude of 0.2 m. Assuming that vibration starts from rest $(x_0 = \dot{x}_0 = 0)$, determine the mass of the system.

Solution: The vibration response of the system can be found from Eq. (3.9) with $x_0 = \dot{x}_0 = 0$:

$$x(t) = \frac{F_0}{k - m\omega^2}(\cos \omega t - \cos \omega_n t) \qquad \text{(E.1)}$$

which can be rewritten as

$$x(t) = \frac{2F_0}{k - m\omega^2} \sin \frac{\omega_n + \omega}{2}t \sin \frac{\omega_n - \omega}{2}t \qquad \text{(E.2)}$$

Since the amplitude of vibration is known to be 0.2 m, Eq. (E.2) gives

$$\frac{2F_0}{k - m\omega^2} = 0.2 \qquad \text{(E.3)}$$

Using the known values of $F_0 = 30$ N, $\omega = 20$ Hz $= 125.665$ rad/s, and $k = 5,000$ N/m, Eq. (E.3) yields

$$\frac{2(30)}{5,000 - m(125.664)^2} = 0.2 \tag{E.4}$$

Equation (E.4) can be solved to find $m = 0.2976$ kg.

■

3.4   Response of a Damped System Under Harmonic Force

If the forcing function is given by $F(t) = F_0 \cos \omega t$, the equation of motion becomes

$$m\ddot{x} + c\dot{x} + kx = F_0 \cos \omega t \tag{3.24}$$

The particular solution of Eq. (3.24) is also expected to be harmonic; we assume it in the form[1]

$$x_p(t) = X \cos (\omega t - \phi) \tag{3.25}$$

where X and ϕ are constants to be determined. X and ϕ denote the amplitude and phase angle of the response, respectively. By substituting Eq. (3.25) into Eq. (3.24), we arrive at

$$X[(k - m\omega^2) \cos (\omega t - \phi) - c\omega \sin (\omega t - \phi)] = F_0 \cos \omega t \tag{3.26}$$

Using the trigonometric relations

$$\cos (\omega t - \phi) = \cos \omega t \cos \phi + \sin \omega t \sin \phi$$
$$\sin (\omega t - \phi) = \sin \omega t \cos \phi - \cos \omega t \sin \phi$$

in Eq. (3.26) and equating the coefficients of $\cos \omega t$ and $\sin \omega t$ on both sides of the resulting equation, we obtain

$$X[(k - m\omega^2) \cos \phi + c\omega \sin \phi] = F_0$$
$$X[(k - m\omega^2) \sin \phi - c\omega \cos \phi] = 0 \tag{3.27}$$

Solution of Eq. (3.27) gives

$$X = \frac{F_0}{[(k - m\omega^2)^2 + c^2\omega^2]^{1/2}} \tag{3.28}$$

---

[1]Alternatively, we can assume $x_p(t)$ to be of the form $x_p(t) = C_1 \cos \omega t + C_2 \sin \omega t$, which also involves two constants C_1 and C_2. But the final result will be the same in both cases.

and

$$\phi = \tan^{-1}\left(\frac{c\omega}{k - m\omega^2}\right) \tag{3.29}$$

By inserting the expressions of X and ϕ from Eqs. (3.28) and (3.29) into Eq. (3.25), we obtain the particular solution of Eq. (3.24). Figure 3.10(a) shows typical plots of the forcing function and steady-state response. The various terms of Eq. (3.26) are shown vectorially in Fig. 3.10(b). Dividing both the numerator and denominator of Eq. (3.28) by k and making the following substitutions

$$\omega_n = \sqrt{\frac{k}{m}} = \text{undamped natural frequency,}$$

$$\zeta = \frac{c}{c_c} = \frac{c}{2m\omega_n} = \frac{c}{2\sqrt{mk}}; \qquad \frac{c}{m} = 2\zeta\omega_n,$$

$$\delta_{st} = \frac{F_0}{k} = \text{deflection under the static force } F_0, \text{ and}$$

$$r = \frac{\omega}{\omega_n} = \text{frequency ratio}$$

we obtain

$$\frac{X}{\delta_{st}} = \frac{1}{\left\{\left[1 - \left(\frac{\omega}{\omega_n}\right)^2\right]^2 + \left[2\zeta\frac{\omega}{\omega_n}\right]^2\right\}^{1/2}} = \frac{1}{\sqrt{(1 - r^2)^2 + (2\zeta r)^2}} \tag{3.30}$$

(a) Graphical representation   (b) Vectorial representation

FIGURE 3.10   Representation of forcing function and response.

and

$$\phi = \tan^{-1}\left\{\frac{2\zeta\dfrac{\omega}{\omega_n}}{1 - \left(\dfrac{\omega}{\omega_n}\right)^2}\right\} = \tan^{-1}\left(\frac{2\zeta r}{1 - r^2}\right) \tag{3.31}$$

As stated in Section 3.3, the quantity $M = X/\delta_{st}$ is called the *magnification factor, amplification factor*, or *amplitude ratio*. The variations of X/δ_{st} and ϕ with the frequency ratio r and the damping ratio ζ are shown in Fig. 3.11.

The following characteristics of the magnification factor (M) can be noted from Eq. (3.30) and Fig. 3.11(a):

1. For an undamped system ($\zeta = 0$), Eq. (3.30) reduces to Eq. (3.10), and $M \rightarrow \infty$ as $r \rightarrow 1$.
2. Any amount of damping ($\zeta > 0$) reduces the magnification factor (M) for all values of the forcing frequency.
3. For any specified value of r, a higher value of damping reduces the value of M.
4. In the degenerate case of a constant force (when $r = 0$), the value of $M = 1$.
5. The reduction in M in the presence of damping is very significant at or near resonance.
6. The amplitude of forced vibration becomes smaller with increasing values of the forcing frequency (i.e., $M \rightarrow 0$ as $r \rightarrow \infty$).

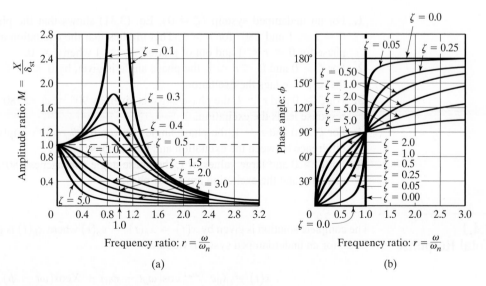

FIGURE 3.11   Variation of X and ϕ with frequency ratio r.

7. For $0 < \zeta < \frac{1}{\sqrt{2}}$, the maximum value of M occurs when (see Problem 3.32)

$$r = \sqrt{1 - 2\zeta^2} \quad \text{or} \quad \omega = \omega_n\sqrt{1 - 2\zeta^2} \tag{3.32}$$

which can be seen to be lower than the undamped natural frequency ω_n and the damped natural frequency $\omega_d = \omega_n\sqrt{1 - \zeta^2}$.

8. The maximum value of X (when $r = \sqrt{1 - 2\zeta^2}$) is given by

$$\left(\frac{X}{\delta_{st}}\right)_{max} = \frac{1}{2\zeta\sqrt{1 - \zeta^2}} \tag{3.33}$$

and the value of X at $\omega = \omega_n$ by

$$\left(\frac{X}{\delta_{st}}\right)_{\omega=\omega_n} = \frac{1}{2\zeta} \tag{3.34}$$

Equation (3.33) can be used for the experimental determination of the measure of damping present in the system. In a vibration test, if the maximum amplitude of the response $(X)_{max}$ is measured, the damping ratio of the system can be found using Eq. (3.33). Conversely, if the amount of damping is known, one can make an estimate of the maximum amplitude of vibration.

9. For $\zeta = \frac{1}{\sqrt{2}}$, $\frac{dM}{dr} = 0$ when $r = 0$. For $\zeta > \frac{1}{\sqrt{2}}$, the graph of M monotonically decreases with increasing values of r.

The following characteristics of the phase angle can be observed from Eq. (3.31) and Fig. 3.11(b):

1. For an undamped system ($\zeta = 0$), Eq. (3.31) shows that the phase angle is 0 for $0 < r < 1$ and $180°$ for $r > 1$. This implies that the excitation and response are in phase for $0 < r < 1$ and out of phase for $r > 1$ when $\zeta = 0$.
2. For $\zeta > 0$ and $0 < r < 1$, the phase angle is given by $0 < \phi < 90°$, implying that the response lags the excitation.
3. For $\zeta > 0$ and $r > 1$, the phase angle is given by $90° < \phi < 180°$, implying that the response leads the excitation.
4. For $\zeta > 0$ and $r = 1$, the phase angle is given by $\phi = 90°$, implying that the phase difference between the excitation and the response is $90°$.
5. For $\zeta > 0$ and large values of r, the phase angle approaches $180°$, implying that the response and the excitation are out of phase.

3.4.1
Total Response

The complete solution is given by $x(t) = x_h(t) + x_p(t)$ where $x_h(t)$ is given by Eq. (2.70). Thus, for an underdamped system, we have

$$x(t) = X_0 e^{-\zeta\omega_n t}\cos(\omega_d t - \phi_0) + X\cos(\omega t - \phi) \tag{3.35}$$

$$\omega_d = \sqrt{1 - \zeta^2}\,\omega_n$$

X and ϕ are given by Eqs. (3.30) and (3.31), respectively, and X_0 and ϕ_0 [different from those of Eq. (2.70)] can be determined from the initial conditions. For the initial conditions, $x(t = 0) = x_0$ and $\dot{x}(t = 0) = \dot{x}_0$, Eq. (3.35) yields

$$x_0 = X_0 \cos \phi_0 + X \cos \phi$$

$$\dot{x}_0 = -\zeta\omega_n X_0 \cos \phi_0 + \omega_d X_0 \sin \phi_0 + \omega X \sin \phi \qquad (3.36)$$

The solution of Eq. (3.36) gives X_0 and ϕ_0 as

$$\left.\begin{array}{l} X_0 = \left[(x_0 - X \cos \phi)^2 + \dfrac{1}{\omega_d^2}(\zeta\omega_n x_0 + \dot{x}_0 - \zeta\omega_n X \cos \phi - \omega X \sin \phi)^2 \right]^{\frac{1}{2}} \\[4mm] \tan \phi_0 = \dfrac{\zeta\omega_n x_0 + \dot{x}_0 - \zeta\omega_n X \cos \phi - \omega X \sin \phi}{\omega_d(x_0 - X \cos \phi)} \end{array}\right\} \qquad (3.37)$$

EXAMPLE 3.3

Total Response of a System

Find the total response of a single-degree-of-freedom system with $m = 10$ kg, $c = 20$ N-s/m, $k = 4,000$ N/m, $x_0 = 0.01$ m, and $\dot{x}_0 = 0$ under the following conditions:

a. An external force $F(t) = F_0 \cos \omega t$ acts on the system with $F_0 = 100$ N and $\omega = 10$ rad/s.
b. Free vibration with $F(t) = 0$.

Solution:

a. From the given data, we obtain

$$\omega_n = \sqrt{\frac{k}{m}} = \sqrt{\frac{4,000}{10}} = 20 \text{ rad/s}$$

$$\delta_{st} = \frac{F_0}{k} = \frac{100}{4,000} = 0.025 \text{ m}$$

$$\zeta = \frac{c}{c_c} = \frac{c}{2\sqrt{km}} = \frac{20}{2\sqrt{(4,000)(10)}} = 0.05$$

$$\omega_d = \sqrt{1 - \zeta^2}\omega_n = \sqrt{1 - (0.05)^2}(20) = 19.974984 \text{ rad/s}$$

$$r = \frac{\omega}{\omega_n} = \frac{10}{20} = 0.5$$

$$X = \frac{\delta_{st}}{\sqrt{(1 - r^2)^2 + (2\zeta r)^2}} = \frac{0.025}{[(1 - 0.05^2)^2 + (2\cdot0.5\cdot0.5)^2]^{1/2}} = 0.03326 \text{ m} \quad \text{(E.1)}$$

$$\phi = \tan^{-1}\left(\frac{2\zeta r}{1 - r^2}\right) = \tan^{-1}\left(\frac{2\cdot0.05\cdot0.5}{1 - 0.5^2}\right) = 3.814075° \quad \text{(E.2)}$$

Using the initial conditions, $x_0 = 0.01$ and $\dot{x}_0 = 0$, Eq. (3.36) yields:

$$0.01 = X_0 \cos \phi_0 + (0.03326)(0.997785)$$

or

$$X_0 \cos \phi_0 = -0.023186 \tag{E.3}$$

$$0 = -(0.05)(20) X_0 \cos \phi_0 + X_0 (19.974984) \sin \phi_0 + (0.03326)(10) \sin (3.814075°) \tag{E.4}$$

Substituting the value of $X_0 \cos \phi_0$ from Eq. (E.3) into Eq. (E.4), we obtain

$$X_0 \sin \phi_0 = -0.002268 \tag{E.5}$$

Solution of Eqs. (E.3) and (E.5) yields

$$X_0 = \left[(X_0 \cos \phi_0)^2 + (X_0 \sin \phi_0)^2 \right]^{1/2} = 0.023297 \tag{E.6}$$

and

$$\tan \phi_0 = \frac{X_0 \sin \phi_0}{X_0 \cos \phi_0} = 0.0978176$$

or

$$\phi_0 = 5.586765° \tag{E.7}$$

b. For free vibration, the total response is given by

$$x(t) = X_0 e^{-\zeta \omega_n t} \cos (\omega_d t - \phi_0) \tag{E.8}$$

Using the initial conditions $x(0) = x_0 = 0.01$ and $\dot{x}(0) = \dot{x}_0 = 0$, X_0 and ϕ_0 of Eq. (E.8) can be determined as (see Eqs. (2.73) and (2.75)):

$$X_0 = \left[x_0^2 + \left(\frac{\zeta \omega_n x_0}{\omega_d} \right)^2 \right]^{1/2} = \left[0.01^2 + \left(\frac{0.05 \cdot 20 \cdot 0.01}{19.974984} \right)^2 \right]^{1/2} = 0.010012 \tag{E.9}$$

$$\phi_0 = \tan^{-1} \left(-\frac{\dot{x}_0 + \zeta \omega_n x_0}{\omega_d x_0} \right) = \tan^{-1} \left(-\frac{0.05 \cdot 20}{19.974984} \right) = -2.865984° \tag{E.10}$$

Note that the constants X_0 and ϕ_0 in cases (a) and (b) are very different.

■

3.4.2 Quality Factor and Bandwidth

For small values of damping ($\zeta < 0.05$), we can take

$$\left(\frac{X}{\delta_{st}} \right)_{max} \simeq \left(\frac{X}{\delta_{st}} \right)_{\omega = \omega_n} = \frac{1}{2\zeta} = Q \tag{3.38}$$

The value of the amplitude ratio at resonance is also called the *Q factor* or *quality factor* of the system, in analogy with some electrical-engineering applications, such as the tuning circuit of a radio, where the interest lies in an amplitude at resonance that is as large as possible [3.2]. The points R_1 and R_2, where the amplification factor falls to $Q/\sqrt{2}$, are called *half-power points* because the power absorbed (ΔW) by the damper (or by the resistor in an electrical circuit), responding harmonically at a given frequency, is proportional to the square of the amplitude (see Eq. (2.94)):

$$\Delta W = \pi c \omega X^2 \tag{3.39}$$

FIGURE 3.12   Harmonic-response curve showing half-power points and bandwidth.

The difference between the frequencies associated with the half-power points R_1 and R_2 is called the *bandwidth* of the system (see Fig. 3.12). To find the values of R_1 and R_2, we set $X/\delta_{st} = Q/\sqrt{2}$ in Eq. (3.30) so that

$$\frac{1}{\sqrt{(1 - r^2)^2 + (2\zeta r)^2}} = \frac{Q}{\sqrt{2}} = \frac{1}{2\sqrt{2}\zeta}$$

or

$$r^4 - r^2(2 - 4\zeta^2) + (1 - 8\zeta^2) = 0 \tag{3.40}$$

The solution of Eq. (3.40) gives

$$r_1^2 = 1 - 2\zeta^2 - 2\zeta\sqrt{1 + \zeta^2}, \qquad r_2^2 = 1 - 2\zeta^2 + 2\zeta\sqrt{1 + \zeta^2} \tag{3.41}$$

For small values of ζ, Eq. (3.41) can be approximated as

$$r_1^2 = R_1^2 = \left(\frac{\omega_1}{\omega_n}\right)^2 \simeq 1 - 2\zeta, \qquad r_2^2 = R_2^2 = \left(\frac{\omega_2}{\omega_n}\right)^2 \simeq 1 + 2\zeta \tag{3.42}$$

where $\omega_1 = \omega|_{R_1}$ and $\omega_2 = \omega|_{R_2}$. From Eq. (3.42),

$$\omega_2^2 - \omega_1^2 = (\omega_2 + \omega_1)(\omega_2 - \omega_1) = (R_2^2 - R_1^2)\omega_n^2 \simeq 4\zeta\omega_n^2 \tag{3.43}$$

Using the relation

$$\omega_2 + \omega_1 = 2\omega_n \tag{3.44}$$

in Eq. (3.43), we find that the bandwidth $\Delta\omega$ is given by

$$\Delta\omega = \omega_2 - \omega_1 \simeq 2\zeta\omega_n \tag{3.45}$$

Combining Eqs. (3.38) and (3.45), we obtain

$$Q \simeq \frac{1}{2\zeta} \simeq \frac{\omega_n}{\omega_2 - \omega_1} \tag{3.46}$$

It can be seen that the quality factor Q can be used for estimating the equivalent viscous damping in a mechanical system.[2]

3.5   Response of a Damped System Under $F(t) = F_0 e^{i\omega t}$

Let the harmonic forcing function be represented in complex form as $F(t) = F_0 e^{i\omega t}$ so that the equation of motion becomes

$$m\ddot{x} + c\dot{x} + kx = F_0 e^{i\omega t} \tag{3.47}$$

Since the actual excitation is given only by the real part of $F(t)$, the response will also be given only by the real part of $x(t)$, where $x(t)$ is a complex quantity satisfying the differential equation (3.47). F_0 in Eq. (3.47) is, in general, a complex number. By assuming the particular solution $x_p(t)$

$$x_p(t) = X e^{i\omega t} \tag{3.48}$$

we obtain, by substituting Eq. (3.48) into Eq. (3.47),[3]

$$X = \frac{F_0}{(k - m\omega^2) + ic\omega} \tag{3.49}$$

Multiplying the numerator and denominator on the right side of Eq. (3.49) by $[(k - m\omega^2) - ic\omega]$ and separating the real and imaginary parts, we obtain

$$X = F_0 \left[\frac{k - m\omega^2}{(k - m\omega^2)^2 + c^2\omega^2} - i\frac{c\omega}{(k - m\omega^2)^2 + c^2\omega^2} \right] \tag{3.50}$$

Using the relation $x + iy = Ae^{i\phi}$, where $A = \sqrt{x^2 + y^2}$ and $\tan\phi = y/x$, Eq. (3.50) can be expressed as

$$X = \frac{F_0}{[(k - m\omega^2)^2 + c^2\omega^2]^{1/2}} e^{-i\phi} \tag{3.51}$$

---

[2]The determination of the system parameters (m, c, and k) based on half-power points and other response characteristics of the system is considered in Section 10.8.

[3]Equation (3.49) can be written as $Z(i\omega)X = F_0$, where $Z(i\omega) = -m\omega^2 + i\omega c + k$ is called the *mechanical impedance* of the system [3.8].

where

$$\phi = \tan^{-1}\left(\frac{c\omega}{k - m\omega^2}\right) \tag{3.52}$$

Thus the steady-state solution, Eq. (3.48), becomes

$$x_p(t) = \frac{F_0}{[(k - m\omega^2)^2 + (c\omega)^2]^{1/2}} e^{i(\omega t - \phi)} \tag{3.53}$$

Frequency Response.   Equation (3.49) can be rewritten in the form

$$\frac{kX}{F_0} = \frac{1}{1 - r^2 + i2\zeta r} \equiv H(i\omega) \tag{3.54}$$

where $H(i\omega)$ is known as the *complex frequency response* of the system. The absolute value of $H(i\omega)$ given by

$$|H(i\omega)| = \left|\frac{kX}{F_0}\right| = \frac{1}{[(1 - r^2)^2 + (2\zeta r)^2]^{1/2}} \tag{3.55}$$

denotes the magnification factor defined in Eq. (3.30). Recalling that $e^{i\phi} = \cos\phi + i\sin\phi$, we can show that Eqs. (3.54) and (3.55) are related:

$$H(i\omega) = |H(i\omega)|e^{-i\phi} \tag{3.56}$$

where ϕ is given by Eq. (3.52), which can also be expressed as

$$\phi = \tan^{-1}\left(\frac{2\zeta r}{1 - r^2}\right) \tag{3.57}$$

Thus Eq. (3.53) can be expressed as

$$x_p(t) = \frac{F_0}{k}|H(i\omega)|e^{i(\omega t - \phi)} \tag{3.58}$$

It can be seen that the complex frequency-response function, $H(i\omega)$, contains both the magnitude and phase of the steady-state response. The use of this function in the experimental determination of the system parameters (m, c, and k) is discussed in Section 10.8. If $F(t) = F_0 \cos\omega t$, the corresponding steady-state solution is given by the real part of Eq. (3.53):

$$x_p(t) = \frac{F_0}{[(k - m\omega^2)^2 + (c\omega)^2]^{1/2}} \cos(\omega t - \phi)$$

$$= \text{Re}\left[\frac{F_0}{k}H(i\omega)e^{i\omega t}\right] = \text{Re}\left[\frac{F_0}{k}|H(i\omega)|e^{i(\omega t - \phi)}\right] \tag{3.59}$$

which can be seen to be the same as Eq. (3.25). Similarly, if $F(t) = F_0 \sin \omega t$, the corresponding steady-state solution is given by the imaginary part of Eq. (3.53):

$$x_p(t) = \frac{F_0}{\left[(k - m\omega^2)^2 + (c\omega)^2\right]^{1/2}} \sin(\omega t - \phi)$$

$$= \text{Im}\left[\frac{F_0}{k}|H(i\omega)|e^{i(\omega t - \phi)}\right] \tag{3.60}$$

Complex Vector Representation of Harmonic Motion.   The harmonic excitation and the response of the damped system to that excitation can be represented graphically in the complex plane, and an interesting interpretation can be given to the resulting diagram. We first differentiate Eq. (3.58) with respect to time and obtain

$$\text{Velocity} = \dot{x}_p(t) = i\omega \frac{F_0}{k}|H(i\omega)|e^{i(\omega t - \phi)} = i\omega x_p(t)$$

$$\text{Acceleration} = \ddot{x}_p(t) = (i\omega)^2 \frac{F_0}{k}|H(i\omega)|e^{i(\omega t - \phi)} = -\omega^2 x_p(t) \tag{3.61}$$

Because i can be expressed as

$$i = \cos \frac{\pi}{2} + i \sin \frac{\pi}{2} = e^{i\frac{\pi}{2}} \tag{3.62}$$

we can conclude that the velocity leads the displacement by the phase angle $\pi/2$ and that it is multiplied by ω. Similarly, -1 can be written as

$$-1 = \cos \pi + i \sin \pi = e^{i\pi} \tag{3.63}$$

Hence the acceleration leads the displacement by the phase angle π, and it is multiplied by ω^2.

Thus the various terms of the equation of motion (3.47) can be represented in the complex plane, as shown in Fig. 3.13. The interpretation of this figure is that the sum of the complex vectors $m\ddot{x}(t)$, $c\dot{x}(t)$, and $kx(t)$ balances $F(t)$, which is precisely what is required to satisfy Eq. (3.47). It is to also be noted that the entire diagram rotates with angular velocity ω in the complex plane. If only the real part of the response is to be considered, then the entire diagram must be projected onto the real axis. Similarly, if only the imaginary part of the response is to be considered, then the diagram must be projected onto the imaginary axis. In Fig. 3.13, notice that the force $F(t) = F_0 e^{i\omega t}$ is represented as a vector located at an angle ωt to the real axis. This implies that F_0 is real. If F_0 is also complex, then the force vector $F(t)$ will be located at an angle of $(\omega + \psi)$, where ψ is some phase angle introduced by F_0. In such a case, all the other vectors—namely, $m\ddot{x}$, $c\dot{x}$, and kx—will be shifted by the same angle ψ. This is equivalent to multiplying both sides of Eq. (3.47) by $e^{i\psi}$.

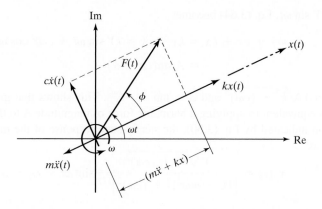

FIGURE 3.13   Representation of Eq. (3.47) in a complex plane.

3.6   Response of a Damped System Under the Harmonic Motion of the Base

Sometimes the base or support of a spring-mass-damper system undergoes harmonic motion, as shown in Fig. 3.14(a). Let $y(t)$ denote the displacement of the base and $x(t)$ the displacement of the mass from its static equilibrium position at time t. Then the net elongation of the spring is $x - y$ and the relative velocity between the two ends of the damper is $\dot{x} - \dot{y}$. From the free-body diagram shown in Fig. 3.14(b), we obtain the equation of motion:

$$m\ddot{x} + c(\dot{x} - \dot{y}) + k(x - y) = 0 \tag{3.64}$$

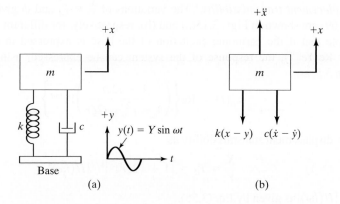

(a)   (b)

FIGURE 3.14   Base excitation.

If $y(t) = Y \sin \omega t$, Eq. (3.64) becomes

$$m\ddot{x} + c\dot{x} + kx = ky + c\dot{y} = kY \sin \omega t + c\omega Y \cos \omega t$$

$$= A \sin(\omega t - \alpha) \tag{3.65}$$

where $A = Y\sqrt{k^2 + (c\omega)^2}$ and $\alpha = \tan^{-1}\left[-\frac{c\omega}{k}\right]$. This shows that giving excitation to the base is equivalent to applying a harmonic force of magnitude A to the mass. By using the solution indicated by Eq. (3.60), the steady-state response of the mass, $x_p(t)$, can be expressed as

$$x_p(t) = \frac{Y\sqrt{k^2 + (c\omega)^2}}{\left[(k - m\omega^2)^2 + (c\omega)^2\right]^{1/2}} \sin(\omega t - \phi_1 - \alpha) \tag{3.66}$$

where

$$\phi_1 = \tan^{-1}\left(\frac{c\omega}{k - m\omega^2}\right)$$

Using trigonometric identities, Eq. (3.66) can be rewritten in a more convenient form as

$$x_p(t) = X \sin(\omega t - \phi) \tag{3.67}$$

where X and ϕ are given by

$$\frac{X}{Y} = \left[\frac{k^2 + (c\omega)^2}{(k - m\omega^2)^2 + (c\omega)^2}\right]^{1/2} = \left[\frac{1 + (2\zeta r)^2}{(1 - r^2)^2 + (2\zeta r)^2}\right]^{1/2} \tag{3.68}$$

and

$$\phi = \tan^{-1}\left[\frac{mc\omega^3}{k(k - m\omega^2) + (\omega c)^2}\right] = \tan^{-1}\left[\frac{2\zeta r^3}{1 + (4\zeta^2 - 1)r^2}\right] \tag{3.69}$$

The ratio of the amplitude of the response $x_p(t)$ to that of the base motion $y(t)$, $\frac{X}{Y}$, is called the *displacement transmissibility*.[4] The variations of $\frac{X}{Y} \equiv T_d$ and ϕ given by Eqs. (3.68) and (3.69) are shown in Figs. 3.15(a) and (b), respectively, for different values of r and ζ.

Note that if the harmonic excitation of the base is expressed in complex form as $y(t) = \text{Re}(Ye^{i\omega t})$, the response of the system can be expressed, using the analysis of Section 3.5, as

$$x_p(t) = \text{Re}\left\{\left(\frac{1 + i2\zeta r}{1 - r^2 + i2\zeta r}\right)Ye^{i\omega t}\right\} \tag{3.70}$$

and the displacement transmissibility as

$$\frac{X}{Y} = T_d = \left[1 + (2\zeta r)^2\right]^{1/2}|H(i\omega)| \tag{3.71}$$

where $|H(i\omega)|$ is given by Eq. (3.55).

---

[4]The expression for the displacement transmissibility can also be derived using the transfer-function approach described in Section 3.14.

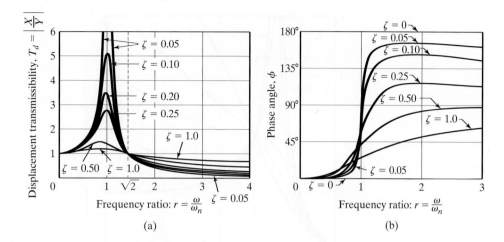

FIGURE 3.15   Variations of T_d and ϕ with r.

The following aspects of displacement transmissibility, $T_d = \frac{X}{Y}$, can be noted from Fig. 3.15(a):

1. The value of T_d is unity at $r = 0$ and close to unity for small values of r.
2. For an undamped system ($\zeta = 0$), $T_d \to \infty$ at resonance ($r = 1$).
3. The value of T_d is less than unity ($T_d < 1$) for values of $r > \sqrt{2}$ (for any amount of damping ζ).
4. The value of T_d is unity for all values of ζ at $r = \sqrt{2}$.
5. For $r < \sqrt{2}$, smaller damping ratios lead to larger values of T_d. On the other hand, for $r > \sqrt{2}$, smaller values of damping ratio lead to smaller values of T_d.
6. The displacement transmissibility, T_d, attains a maximum for $0 < \zeta < 1$ at the frequency ratio $r = r_m < 1$ given by (see Problem 3.60):

$$r_m = \frac{1}{2\zeta} \left[\sqrt{1 + 8\zeta^2} - 1 \right]^{1/2}$$

3.6.1
Force
Transmitted

In Fig. 3.14, a force, F, is transmitted to the base or support due to the reactions from the spring and the dashpot. This force can be determined as

$$F = k(x - y) + c(\dot{x} - \dot{y}) = -m\ddot{x} \tag{3.72}$$

From Eq. (3.67), Eq. (3.72) can be written as

$$F = m\omega^2 X \sin(\omega t - \phi) = F_T \sin(\omega t - \phi) \tag{3.73}$$

where F_T is the amplitude or maximum value of the force transmitted to the base given by

$$\frac{F_T}{kY} = r^2 \left[\frac{1 + (2\zeta r)^2}{(1 - r^2)^2 + (2\zeta r)^2} \right]^{1/2} \tag{3.74}$$

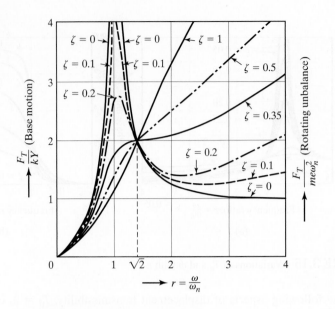

FIGURE 3.16   Force transmissibility.

The ratio (F_T/kY) is known as the *force transmissibility*.[5] Note that the transmitted force is in phase with the motion of the mass $x(t)$. The variation of the force transmitted to the base with the frequency ratio r is shown in Fig. 3.16 for different values of ζ.

3.6.2
Relative Motion

If $z = x - y$ denotes the motion of the mass relative to the base, the equation of motion, Eq. (3.64), can be rewritten as

$$m\ddot{z} + c\dot{z} + kz = -m\ddot{y} = m\omega^2 Y \sin \omega t \tag{3.75}$$

The steady-state solution of Eq. (3.75) is given by

$$z(t) = \frac{m\omega^2 Y \sin(\omega t - \phi_1)}{[(k - m\omega^2)^2 + (c\omega)^2]^{1/2}} = Z \sin(\omega t - \phi_1) \tag{3.76}$$

where Z, the amplitude of $z(t)$, can be expressed as

$$Z = \frac{m\omega^2 Y}{\sqrt{(k - m\omega^2)^2 + (c\omega)^2}} = Y\frac{r^2}{\sqrt{(1 - r^2)^2 + (2\zeta r)^2}} \tag{3.77}$$

---

[5]The use of the concept of transmissibility in the design of vibration isolation systems is discussed in Chapter 9. The expression for the force transmissibility can also be derived using the transfer-function approach described in Section 3.14.

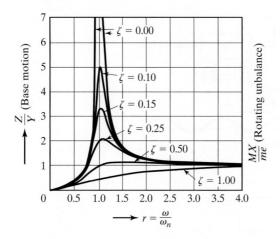

FIGURE 3.17   Variation of (Z/Y) or (MX/me) with frequency ratio $r = (\omega/\omega_n)$.

and ϕ_1 by

$$\phi_1 = \tan^{-1}\left(\frac{c\omega}{k - m\omega^2}\right) = \tan^{-1}\left(\frac{2\zeta r}{1 - r^2}\right)$$

The ratio Z/Y is shown graphically in Fig. 3.17. The variation of ϕ_1 is same as that of ϕ shown in Fig. 3.11(b).

EXAMPLE 3.4

Vehicle Moving on a Rough Road

Figure 3.18 shows a simple model of a motor vehicle that can vibrate in the vertical direction while traveling over a rough road. The vehicle has a mass of 1,200 kg. The suspension system has a spring constant of 400 kN/m and a damping ratio of $\zeta = 0.5$. If the vehicle speed is 20 km/h, determine the displacement amplitude of the vehicle. The road surface varies sinusoidally with an amplitude of $Y = 0.05$ m and a wavelength of 6 m.

Solution: The frequency ω of the base excitation can be found by dividing the vehicle speed v km/h by the length of one cycle of road roughness:

$$\omega = 2\pi f = 2\pi \left(\frac{v \times 1{,}000}{3{,}600}\right)\frac{1}{6} = 0.290889v \text{ rad/s}$$

For $v = 20$ km/h, $\omega = 5.81778$ rad/s. The natural frequency of the vehicle is given by

$$\omega_n = \sqrt{\frac{k}{m}} = \left(\frac{400 \times 10^3}{1{,}200}\right)^{1/2} = 18.2574 \text{ rad/s}$$

FIGURE 3.18   Vehicle moving over a rough road.

and hence the frequency ratio r is

$$r = \frac{\omega}{\omega_n} = \frac{5.81778}{18.2574} = 0.318653$$

The amplitude ratio can be found from Eq. (3.68):

$$\frac{X}{Y} = \left\{ \frac{1 + (2\zeta r)^2}{(1 - r^2)^2 + (2\zeta r)^2} \right\}^{1/2} = \left\{ \frac{1 + (2 \times 0.5 \times 0.318653)^2}{(1 - 0.318653)^2 + (2 \times 0.5 \times 0.318653)^2} \right\}^{1/2}$$

$$= 1.100964$$

Thus the displacement amplitude of the vehicle is given by

$$X = 1.100964Y = 1.100964(0.05) = 0.055048 \text{ m}$$

This indicates that a 5-cm bump in the road is transmitted as a 5.5-cm bump to the chassis and the passengers of the car. Thus in the present case the passengers feel an amplified motion (see Problem 3.107 for other situations).

■

EXAMPLE 3.5

Machine on Resilient Foundation

A heavy machine, weighing 3,000 N, is supported on a resilient foundation. The static deflection of the foundation due to the weight of the machine is found to be 7.5 cm. It is observed that the machine vibrates with an amplitude of 1 cm when the base of the foundation is subjected to harmonic oscillation at the undamped natural frequency of the system with an amplitude of 0.25 cm. Find

a. the damping constant of the foundation,
b. the dynamic force amplitude on the base, and
c. the amplitude of the displacement of the machine relative to the base.

Solution:

a. The stiffness of the foundation can be found from its static deflection: k = weight of machine/δ_{st} = 3,000/0.075 = 40,000 N/m.

At resonance ($\omega = \omega_n$ or $r = 1$), Eq. (3.68) gives

$$\frac{X}{Y} = \frac{0.010}{0.0025} = 4 = \left[\frac{1 + (2\zeta)^2}{(2\zeta)^2}\right]^{1/2} \tag{E.1}$$

The solution of Eq. (E.1) gives $\zeta = 0.1291$. The damping constant is given by

$$c = \zeta \cdot c_c = \zeta 2\sqrt{km} = 0.1291 \times 2 \times \sqrt{40,000 \times (3,000/9.81)}$$
$$= 903.0512 \text{ N-s/m} \tag{E.2}$$

b. The dynamic force amplitude on the base at $r = 1$ can be found from Eq. (3.74):

$$F_T = Yk\left[\frac{1 + 4\zeta^2}{4\zeta^2}\right]^{1/2} = kX = 40,000 \times 0.01 = 400 \text{ N} \tag{E.3}$$

c. The amplitude of the relative displacement of the machine at $r = 1$ can be obtained from Eq. (3.77):

$$Z = \frac{Y}{2\zeta} = \frac{0.0025}{2 \times 0.1291} = 0.00968 \text{ m} \tag{E.4}$$

It can be noticed that $X = 0.01$ m, $Y = 0.0025$ m, and $Z = 0.00968$ m; therefore, $Z \neq X - Y$. This is due to the phase differences between x, y, and z.

∎

3.7 Response of a Damped System Under Rotating Unbalance

Unbalance in rotating machinery is one of the main causes of vibration. A simplified model of such a machine is shown in Fig. 3.19. The total mass of the machine is M, and there are two eccentric masses $m/2$ rotating in opposite directions with a constant angular velocity ω. The centrifugal force $(me\omega^2)/2$ due to each mass will cause excitation of the mass M. We consider two equal masses $m/2$ rotating in opposite directions in order to have the horizontal components of excitation of the two masses cancel each other. However, the vertical components of excitation add together and act along the axis of symmetry $A - A$ in

FIGURE 3.19   Rotating unbalanced masses.

Fig. 3.19. If the angular position of the masses is measured from a horizontal position, the total vertical component of the excitation is always given by $F(t) = me\omega^2 \sin \omega t$. The equation of motion can be derived by the usual procedure:

$$M\ddot{x} + c\dot{x} + kx = me\omega^2 \sin \omega t \tag{3.78}$$

The solution of this equation will be identical to Eq. (3.60) if we replace m and F_0 by M and $me\omega^2$, respectively. This solution can also be expressed as

$$x_p(t) = X \sin(\omega t - \phi) = \text{Im}\left[\frac{me}{M}\left(\frac{\omega}{\omega_n}\right)^2 |H(i\omega)| e^{i(\omega t - \phi)} \right] \tag{3.79}$$

where $\omega_n = \sqrt{k/M}$ and X and ϕ denote the amplitude and the phase angle of vibration given by

$$X = \frac{me\omega^2}{\left[(k - M\omega^2)^2 + (c\omega)^2\right]^{1/2}} = \frac{me}{M}\left(\frac{\omega}{\omega_n}\right)^2 |H(i\omega)|$$

$$\phi = \tan^{-1}\left(\frac{c\omega}{k - M\omega^2}\right) \tag{3.80}$$

By defining $\zeta = c/c_c$ and $c_c = 2M\omega_n$, Eq. (3.80) can be rewritten as

$$\frac{MX}{me} = \frac{r^2}{\left[(1 - r^2)^2 + (2\zeta r)^2\right]^{1/2}} = r^2|H(i\omega)|$$

$$\phi = \tan^{-1}\left(\frac{2\zeta r}{1 - r^2}\right) \tag{3.81}$$

The variation of MX/me with r for different values of ζ is shown in Fig. 3.17. On the other hand, the graph of ϕ versus r remains as in Fig. 3.11(b). The following observations can be made from Eq. (3.81) and Fig. 3.17:

1. All the curves begin at zero amplitude. The amplitude near resonance ($\omega = \omega_n$) is markedly affected by damping. Thus if the machine is to be run near resonance, damping should be introduced purposefully to avoid dangerous amplitudes.
2. At very high speeds (ω large), MX/me is almost unity, and the effect of damping is negligible.
3. For $0 < \zeta < \dfrac{1}{\sqrt{2}}$, the maximum of $\dfrac{MX}{me}$ occurs when

$$\frac{d}{dr}\left(\frac{MX}{me}\right) = 0 \tag{3.82}$$

The solution of Eq. (3.82) gives

$$r = \frac{1}{\sqrt{1 - 2\zeta^2}} > 1$$

with the corresponding maximum value of $\dfrac{MX}{me}$ given by

$$\left(\frac{MX}{me}\right)_{\text{max}} = \frac{1}{2\zeta\sqrt{1 - \zeta^2}} \tag{3.83}$$

Thus the peaks occur to the right of the resonance value of $r = 1$.

4. For $\zeta > \dfrac{1}{\sqrt{2}}$, $\left[\dfrac{MX}{me}\right]$ does not attain a maximum. Its value grows from 0 at $r = 0$ to 1 at $r \to \infty$.

5. The force transmitted to the foundation due to the rotating unbalanced force (F) can be found as $F(t) = kx(t) + c\dot{x}(t)$. The magnitude (or maximum value) of F can be derived as (see Problem 3.73):

$$|F| = me\omega^2 \left[\frac{1 + 4\zeta^2 r^2}{(1 - r^2)^2 + 4\zeta^2 r^2} \right]^{\frac{1}{2}} \tag{3.84}$$

EXAMPLE 3.6

Deflection of an Electric Motor due to Rotating Unbalance

An electric motor of mass M, mounted on an elastic foundation, is found to vibrate with a deflection of 0.15 m at resonance (Fig. 3.20). It is known that the unbalanced mass of the motor is 8% of the mass of the rotor due to manufacturing tolerances used, and the damping ratio of the foundation is $\zeta = 0.025$. Determine the following:

a. the eccentricity or radial location of the unbalanced mass (e),
b. the peak deflection of the motor when the frequency ratio varies from resonance, and
c. the additional mass to be added uniformly to the motor if the deflection of the motor at resonance is to be reduced to 0.1 m.

Assume that the eccentric mass remains unaltered when the additional mass is added to the motor.

a. From Eq. (3.81), the deflection at resonance ($r = 1$) is given by

$$\frac{MX}{me} = \frac{1}{2\zeta} = \frac{1}{2(0.025)} = 20$$

FIGURE 3.20

from which the eccentricity can be found as

$$e = \frac{MX}{20m} = \frac{M(0.15)}{20(0.08M)} = 0.09375 \text{ m}$$

b. The peak deflection of the motor is given by Eq. (3.83):

$$\left(\frac{MX}{me}\right)_{\text{max}} = \frac{1}{2\zeta\sqrt{1-\zeta^2}} = \frac{1}{2(0.025)\sqrt{1-0.025^2}} = 20.0063$$

from which the peak deflection can be determined as

$$X_{\text{max}} = \frac{20.0063me}{M} = \frac{20.0063(0.08M)(0.09375)}{M} = 0.150047 \text{ m}$$

c. If the additional mass added to the motor is denoted as M_a, the corresponding deflection is given by Eq. (3.81):

$$\frac{(M + M_a)(0.1)}{(0.08M)(0.09375)} = 20$$

which yields $M_a = 0.5M$. Thus the mass of the motor is to be increased by 50% in order to reduce the deflection at resonance from 0.15 m to 0.10 m.

∎

EXAMPLE 3.7

Francis Water Turbine

Figure 3.21 is a schematic diagram of a Francis water turbine, in which water flows from A into the blades B and down into the tail race C. The rotor has a mass of 250 kg and an unbalance (me) of 5 kg-mm. The radial clearance between the rotor and the stator is 5 mm. The turbine operates in the speed range 600 to 6,000 rpm. The steel shaft carrying the rotor can be assumed to be clamped at the bearings. Determine the diameter of the shaft so that the rotor is always clear of the stator at all the operating speeds of the turbine. Assume damping to be negligible.

Solution: The maximum amplitude of the shaft (rotor) due to rotating unbalance can be obtained from Eq. (3.80) by setting $c = 0$ as

$$X = \frac{me\omega^2}{(k - M\omega^2)} = \frac{me\omega^2}{k(1 - r^2)} \tag{E.1}$$

where $me = 5$ kg-mm, $M = 250$ kg, and the limiting value of $X = 5$ mm. The value of ω ranges from

$$600 \text{ rpm} = 600 \times \frac{2\pi}{60} = 20\pi \text{ rad/s}$$

Bearing
Shaft
$l = 2$ m
Rotor
5 mm
5 mm
A
A
B
B
Stator
C
Tail race

FIGURE 3.21   Francis water turbine.

to

$$6{,}000 \text{ rpm} = 6{,}000 \times \frac{2\pi}{60} = 200\pi \text{ rad/s}$$

while the natural frequency of the system is given by

$$\omega_n = \sqrt{\frac{k}{M}} = \sqrt{\frac{k}{250}} = 0.063245\sqrt{k} \text{ rad/s} \tag{E.2}$$

if k is in N/m. For $\omega = 20\pi$ rad/s, Eq. (E.1) gives

$$0.005 = \frac{(5.0 \times 10^{-3}) \times (20\pi)^2}{k\left[1 - \frac{(20\pi)^2}{0.004\,k}\right]} = \frac{2\pi^2}{k - 10^5\pi^2}$$

$$k = 10.04 \times 10^4 \pi^2 \text{ N/m} \tag{E.3}$$

For $\omega = 200\pi$ rad/s, Eq. (E.1) gives

$$0.005 = \frac{(5.0 \times 10^{-3}) \times (200\pi)^2}{k\left[1 - \frac{(200\pi)^2}{0.004k}\right]} = \frac{200\pi^2}{k - 10^7\pi^2}$$

$$k = 10.04 \times 10^6 \pi^2 \text{ N/m} \tag{E.4}$$

From Fig. 3.17, we find that the amplitude of vibration of the rotating shaft can be minimized by making $r = \omega/\omega_n$ very large. This means that ω_n must be made small compared to ω—that is, k must be made small. This can be achieved by selecting the value of k as $10.04 \times 10^4 \pi^2$ N/m. Since the stiffness of a cantilever beam (shaft) supporting a load (rotor) at the end is given by

$$k = \frac{3EI}{l^3} = \frac{3E}{l^3}\left(\frac{\pi d^4}{64}\right) \tag{E.5}$$

the diameter of the beam (shaft) can be found:

$$d^4 = \frac{64kl^3}{3\pi E} = \frac{(64)(10.04 \times 10^4 \pi^2)(2^3)}{3\pi(2.07 \times 10^{11})} = 2.6005 \times 10^{-4} \text{ m}^4$$

or

$$d = 0.1270 \text{ m} = 127 \text{ mm} \tag{E.6}$$

■

3.8   Forced Vibration with Coulomb Damping

For a single-degree-of-freedom system with Coulomb or dry-friction damping, subjected to a harmonic force $F(t) = F_0 \sin \omega t$ as in Fig. 3.22, the equation of motion is given by

$$m\ddot{x} + kx \pm \mu N = F(t) = F_0 \sin \omega t \tag{3.85}$$

where the sign of the friction force ($\mu N = \mu mg$) is positive (negative) when the mass moves from left to right (right to left). The exact solution of Eq. (3.85) is quite involved. However, we can expect that if the dry-friction damping force is large, the motion of the mass will be discontinuous. On the other hand, if the dry-friction force is small compared to the amplitude of the applied force F_0, the steady-state solution is expected to be nearly harmonic. In this case, we can find an approximate solution of Eq. (3.85) by finding an equivalent viscous-damping ratio. To find such a ratio, we equate the energy dissipated due to dry friction to the energy dissipated by an equivalent viscous damper during a full cycle of motion. If the amplitude of motion is denoted as X, the energy dissipated by the friction force μN in a quarter cycle is μNX. Hence in a full cycle, the energy dissipated by dry-friction damping is given by

$$\Delta W = 4\mu NX \tag{3.86}$$

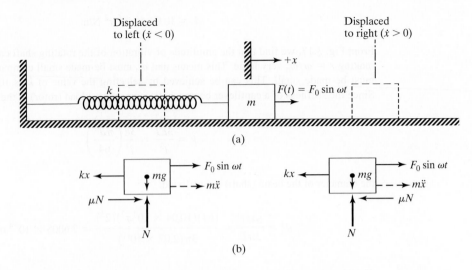

FIGURE 3.22   Single-degree-of-freedom system with Coulomb damping.

If the equivalent viscous-damping constant is denoted as c_{eq}, the energy dissipated during a full cycle (see Eq. (2.94)) will be

$$\Delta W = \pi c_{eq} \omega X^2 \tag{3.87}$$

By equating Eqs. (3.86) and (3.87), we obtain

$$c_{eq} = \frac{4\mu N}{\pi \omega X} \tag{3.88}$$

Thus the steady-state response is given by

$$x_p(t) = X \sin(\omega t - \phi) \tag{3.89}$$

where the amplitude X can be found from Eq. (3.60):

$$X = \frac{F_0}{\left[(k - m\omega^2)^2 + (c_{eq}\omega)^2 \right]^{1/2}} = \frac{(F_0/k)}{\left[\left(1 - \frac{\omega^2}{\omega_n^2} \right)^2 + \left(2\zeta_{eq}\frac{\omega}{\omega_n} \right)^2 \right]^{1/2}} \tag{3.90}$$

with

$$\zeta_{eq} = \frac{c_{eq}}{c_c} = \frac{c_{eq}}{2m\omega_n} = \frac{4\mu N}{2m\omega_n \pi \omega X} = \frac{2\mu N}{\pi m\omega\omega_n X} \tag{3.91}$$

Substitution of Eq. (3.91) into Eq. (3.90) gives

$$X = \frac{(F_0/k)}{\left[\left(1 - \frac{\omega^2}{\omega_n^2}\right)^2 + \left(\frac{4\mu N}{\pi k X}\right)^2\right]^{1/2}} \tag{3.92}$$

The solution of this equation gives the amplitude X as

$$X = \frac{F_0}{k}\left[\frac{1 - \left(\frac{4\mu N}{\pi F_0}\right)^2}{\left(1 - \frac{\omega^2}{\omega_n^2}\right)^2}\right]^{1/2} \tag{3.93}$$

As stated earlier, Eq. (3.93) can be used only if the friction force is small compared to F_0. In fact, the limiting value of the friction force μN can be found from Eq. (3.93). To avoid imaginary values of X, we need to have

$$1 - \left(\frac{4\mu N}{\pi F_0}\right)^2 > 0 \quad \text{or} \quad \frac{F_0}{\mu N} > \frac{4}{\pi}$$

If this condition is not satisfied, the exact analysis, given in [3.3], is to be used. The phase angle ϕ appearing in Eq. (3.89) can be found using Eq. (3.52):

$$\phi = \tan^{-1}\left(\frac{c_{eq}\omega}{k - m\omega^2}\right) = \tan^{-1}\left[\frac{2\zeta_{eq}\frac{\omega}{\omega_n}}{1 - \frac{\omega^2}{\omega_n^2}}\right] = \tan^{-1}\left\{\frac{\frac{4\mu N}{\pi k X}}{1 - \frac{\omega^2}{\omega_n^2}}\right\} \tag{3.94}$$

Substituting Eq. (3.93) into Eq. (3.94) for X, we obtain

$$\phi = \tan^{-1}\left[\frac{\frac{4\mu N}{\pi F_0}}{\left\{1 - \left(\frac{4\mu N}{\pi F_0}\right)^2\right\}^{1/2}}\right] \tag{3.95}$$

Equation (3.94) shows that $\tan \phi$ is a constant for a given value of $F_0/\mu N$. ϕ is discontinuous at $\omega/\omega_n = 1$ (resonance), since it takes a positive value for $\omega/\omega_n < 1$ and a negative value for $\omega/\omega_n > 1$. Thus Eq. (3.95) can also be expressed as

$$\phi = \tan^{-1}\left[\frac{\pm\dfrac{4\mu N}{\pi F_0}}{\left\{1 - \left(\dfrac{4\mu N}{\pi F_0}\right)^2\right\}^{1/2}}\right] \tag{3.96}$$

Equation (3.93) shows that friction serves to limit the amplitude of forced vibration for $\omega/\omega_n \neq 1$. However, at resonance ($\omega/\omega_n = 1$), the amplitude becomes infinite. This can be explained as follows. The energy directed into the system over one cycle when it is excited harmonically at resonance is

$$\Delta W' = \int_{\text{cycle}} F \cdot dx = \int_0^\tau F\frac{dx}{dt}dt$$

$$= \int_0^{\tau = 2\pi/\omega} F_0 \sin \omega t \cdot [\omega X \cos (\omega t - \phi)]\, dt \tag{3.97}$$

Since Eq. (3.94) gives $\phi = 90°$ at resonance, Eq. (3.97) becomes

$$\Delta W' = F_0 X \omega \int_0^{2\pi/\omega} \sin^2 \omega t\, dt = \pi F_0 X \tag{3.98}$$

The energy dissipated from the system is given by Eq. (3.86). Since $\pi F_0 X > 4\mu N X$ for X to be real-valued, $\Delta W' > \Delta W$ at resonance (see Fig. 3.23). Thus more energy is directed into the system per cycle than is dissipated per cycle. This extra energy is used to build up the amplitude of vibration. For the nonresonant condition ($\omega/\omega_n \neq 1$), the energy input can be found from Eq. (3.97):

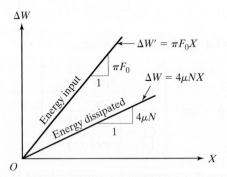

FIGURE 3.23   Energy input and energy dissipated with Coulomb damping.

$$\Delta W' = \omega F_0 X \int_0^{2\pi/\omega} \sin \omega t \cos(\omega t - \phi) \, dt = \pi F_0 X \sin \phi \qquad (3.99)$$

Due to the presence of $\sin \phi$ in Eq. (3.99), the input energy curve in Fig. 3.23 is made to coincide with the dissipated energy curve, so the amplitude is limited. Thus the phase of the motion ϕ can be seen to limit the amplitude of the motion.

The periodic response of a spring-mass system with Coulomb damping subjected to base excitation is given in [3.10, 3.11].

EXAMPLE 3.8

Spring-Mass System with Coulomb Damping

A spring-mass system, having a mass of 10 kg and a spring of stiffness 4,000 N/m, vibrates on a horizontal surface. The coefficient of friction is 0.12. When subjected to a harmonic force of frequency 2 Hz, the mass is found to vibrate with an amplitude of 40 mm. Find the amplitude of the harmonic force applied to the mass.

Solution: The vertical force (weight) of the mass is $N = mg = 10 \times 9.81 = 98.1$ N. The natural frequency is

$$\omega_n = \sqrt{\frac{k}{m}} = \sqrt{\frac{4,000}{10}} = 20 \text{ rad/s}$$

and the frequency ratio is

$$\frac{\omega}{\omega_n} = \frac{2 \times 2\pi}{20} = 0.6283$$

The amplitude of vibration X is given by Eq. (3.93):

$$X = \frac{F_0}{k} \left[\frac{1 - \left(\dfrac{4\mu N}{\pi F_0}\right)^2}{\left\{1 - \left(\dfrac{\omega}{\omega_n}\right)^2\right\}^2} \right]^{1/2}$$

$$0.04 = \frac{F_0}{4,000} \left[\frac{1 - \left\{\dfrac{4(0.12)(98.1)}{\pi F_0}\right\}^2}{(1 - 0.6283^2)^2} \right]^{1/2}$$

The solution of this equation gives $F_0 = 97.9874$ N.

■

3.9   Forced Vibration with Hysteresis Damping

Consider a single-degree-of-freedom system with hysteresis damping and subjected to a harmonic force $F(t) = F_0 \sin \omega t$, as indicated in Fig. 3.24. The equation of motion of the mass can be derived, using Eq. (2.157), as

$$m\ddot{x} + \frac{\beta k}{\omega}\dot{x} + kx = F_0 \sin \omega t \tag{3.100}$$

where $(\beta k/\omega)\dot{x} = (h/\omega)\dot{x}$ denotes the damping force.[6] Although the solution of Eq. (3.100) is quite involved for a general forcing function $F(t)$, our interest is to find the response under a harmonic force.

The steady-state solution of Eq. (3.100) can be assumed:

$$x_p(t) = X \sin(\omega t - \phi) \tag{3.101}$$

By substituting Eq. (3.101) into Eq. (3.100), we obtain

$$X = \frac{F_0}{k\left[\left(1 - \dfrac{\omega^2}{\omega_n^2}\right)^2 + \beta^2\right]^{1/2}} \tag{3.102}$$

and

$$\phi = \tan^{-1}\left[\frac{\beta}{\left(1 - \dfrac{\omega^2}{\omega_n^2}\right)}\right] \tag{3.103}$$

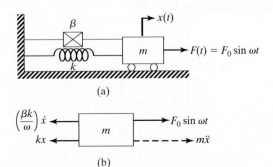

(a)

(b)

FIGURE 3.24   System with hysteresis damping.

---

[6]In contrast to viscous damping, the damping force here can be seen to be a function of the forcing frequency ω (see Section 2.10).

Equations (3.102) and (3.103) are shown plotted in Fig. 3.25 for several values of β. A comparison of Fig. 3.25 with Fig. 3.11 for viscous damping reveals the following:

1. The amplitude ratio

$$\frac{X}{(F_0/k)}$$

attains its maximum value of $F_0/k\beta$ at the resonant frequency ($\omega = \omega_n$) in the case of hysteresis damping, while it occurs at a frequency below resonance ($\omega < \omega_n$) in the case of viscous damping.

2. The phase angle ϕ has a value of $\tan^{-1}(\beta)$ at $\omega = 0$ in the case of hysteresis damping, while it has a value of zero at $\omega = 0$ in the case of viscous damping. This indicates that the response can never be in phase with the forcing function in the case of hysteresis damping.

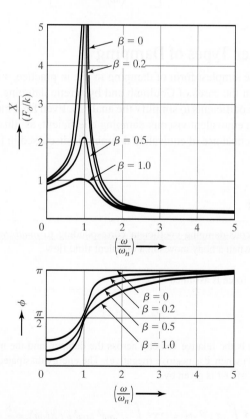

FIGURE 3.25   Steady-state response.

Note that if the harmonic excitation is assumed to be $F(t) = F_0 e^{i\omega t}$ in Fig. 3.24, the equation of motion becomes

$$m\ddot{x} + \frac{\beta k}{\omega}\dot{x} + kx = F_0 e^{i\omega t} \tag{3.104}$$

In this case, the response $x(t)$ is also a harmonic function involving the factor $e^{i\omega t}$. Hence $\dot{x}(t)$ is given by $i\omega x(t)$, and Eq. (3.104) becomes

$$m\ddot{x} + k(1 + i\beta)x = F_0 e^{i\omega t} \tag{3.105}$$

where the quantity $k(1 + i\beta)$ is called the *complex stiffness* or *complex damping* [3.7]. The steady-state solution of Eq. (3.105) is given by the real part of

$$x(t) = \frac{F_0 e^{i\omega t}}{k\left[1 - \left(\dfrac{\omega}{\omega_n}\right)^2 + i\beta\right]} \tag{3.106}$$

3.10   Forced Motion with Other Types of Damping

Viscous damping is the simplest form of damping to use in practice, since it leads to linear equations of motion. In the cases of Coulomb and hysteretic damping, we defined equivalent viscous-damping coefficients to simplify the analysis. Even for a more complex form of damping, we define an equivalent viscous-damping coefficient, as illustrated in the following examples. The practical use of equivalent damping is discussed in [3.12].

EXAMPLE 3.9

Quadratic Damping

Find the equivalent viscous-damping coefficient corresponding to *quadratic* or *velocity-squared damping* that is present when a body moves in a turbulent fluid flow.

Solution: The damping force is assumed to be

$$F_d = \pm a(\dot{x})^2 \tag{E.1}$$

where a is a constant, $\dot{x}$ is the relative velocity across the damper, and the negative (positive) sign must be used in Eq. (E.1) when $\dot{x}$ is positive (negative). The energy dissipated per cycle during harmonic motion $x(t) = X \sin \omega t$ is given by

$$\Delta W = 2\int_{-x}^{x} a(\dot{x})^2 \, dx = 2X^3 \int_{-\pi/2}^{\pi/2} a\omega^2 \cos^3 \omega t \, d(\omega t) = \frac{8}{3}\omega^2 a X^3 \tag{E.2}$$

By equating this energy to the energy dissipated in an equivalent viscous damper (see Eq. (2.94))

$$\Delta W = \pi c_{eq} \omega X^2 \tag{E.3}$$

we obtain the equivalent viscous-damping coefficient (c_{eq})

$$c_{eq} = \frac{8}{3\pi} a\omega X \tag{E.4}$$

It can be noted that c_{eq} is not a constant but varies with ω and X. The amplitude of the steady-state response can be found from Eq. (3.30):

$$\frac{X}{\delta_{st}} = \frac{1}{\sqrt{(1 - r^2)^2 + (2\zeta_{eq}r)^2}} \tag{E.5}$$

where $r = \omega/\omega_n$ and

$$\zeta_{eq} = \frac{c_{eq}}{c_c} = \frac{c_{eq}}{2m\omega_n} \tag{E.6}$$

Using Eqs. (E.4) and (E.6), Eq. (E.5) can be solved to obtain

$$X = \frac{3\pi m}{8ar^2} \left[-\frac{(1 - r^2)^2}{2} + \sqrt{\frac{(1 - r^2)^4}{4} + \left(\frac{8ar^2\delta_{st}}{3\pi m} \right)^2} \right]^{1/2} \tag{E.7}$$

∎

3.11   Self-Excitation and Stability Analysis

The force acting on a vibrating system is usually external to the system and independent of the motion. However, there are systems for which the exciting force is a function of the motion parameters of the system, such as displacement, velocity, or acceleration. Such systems are called self-excited vibrating systems, since the motion itself produces the exciting force (see Problem 3.92). The instability of rotating shafts, the flutter of turbine blades, the flow-induced vibration of pipes, and the automobile wheel shimmy, and the aerodynamically-induced motion of bridges are typical examples of self-excited vibrations.

**3.11.1
Dynamic
Stability
Analysis**

A system is dynamically stable if the motion (or displacement) converges or remains steady with time. On the other hand, if the amplitude of displacement increases continuously (diverges) with time, it is said to be dynamically unstable. The motion diverges and the system becomes unstable if energy is fed into the system through self-excitation. To see the circumstances that lead to instability, we consider the equation of motion of a single-degree-of-freedom system:

$$m\ddot{x} + c\dot{x} + kx = 0 \tag{3.107}$$

If a solution of the form $x(t) = Ce^{st}$, where C is a constant, is assumed, Eq. (3.107) leads to the characteristic equation

$$s^2 + \frac{c}{m}s + \frac{k}{m} = 0 \qquad (3.108)$$

The roots of this equation are

$$s_{1,2} = -\frac{c}{2m} \pm \frac{1}{2}\left[\left(\frac{c}{m}\right)^2 - 4\left(\frac{k}{m}\right)\right]^{1/2} \qquad (3.109)$$

Since the solution is assumed to be $x(t) = Ce^{st}$, the motion will be diverging and aperiodic if the roots s_1 and s_2 are real and positive. This situation can be avoided if c/m and k/m are positive. The motion will also diverge if the roots s_1 and s_2 are complex conjugates with positive real parts. To analyze the situation, let the roots s_1 and s_2 of Eq. (3.108) be expressed as

$$s_1 = p + iq, \qquad s_2 = p - iq \qquad (3.110)$$

where p and q are real numbers, so that

$$(s - s_1)(s - s_2) = s^2 - (s_1 + s_2)s + s_1 s_2 = s^2 + \frac{c}{m}s + \frac{k}{m} = 0 \quad (3.111)$$

Equations (3.110) and (3.111) give

$$\frac{c}{m} = -(s_1 + s_2) = -2p, \qquad \frac{k}{m} = s_1 s_2 = p^2 + q^2 \qquad (3.112)$$

Equations (3.112) show that for negative p, c/m must be positive and for positive $p^2 + q^2$, k/m must be positive. Thus the system will be dynamically stable if c and k are positive (assuming that m is positive).

Instability of Spring-Supported Mass on Moving Belt

EXAMPLE 3.10

Consider a spring-supported mass on a moving belt, as shown in Fig. 3.26(a). The kinetic coefficient of friction between the mass and the belt varies with the relative (rubbing) velocity, as shown in Fig. 3.26(b). As rubbing velocity increases, the coefficient of friction first decreases from its static value linearly and then starts to increase. Assuming that the rubbing velocity, v, is less than the transition value, v_Q, the coefficient of friction can be expressed as

$$\mu = \mu_0 - \frac{a}{W}v$$

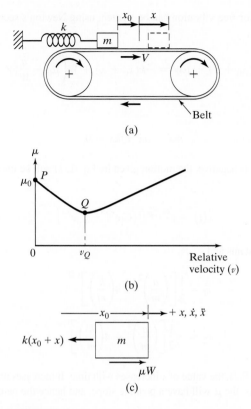

FIGURE 3.26   Motion of a spring-supported mass due to belt friction.

where a is a constant and $W = mg$ is the weight of the mass. Determine the nature of free vibration about the equilibrium position of the mass.

Solution: Let the equilibrium position of mass m correspond to an extension of x_0 of the spring. Then

$$\mu W = k x_0$$

or

$$x_0 = \frac{\mu W}{k} = \frac{\mu_0 W}{k} - \frac{aV}{k}$$

where V is the velocity of the belt. If the mass is displaced by a distance x from its equilibrium position (x_0), the rubbing velocity v is given by

$$v = V - \dot{x}$$

The equation of motion for free vibration can be written, using Newton's second law of motion, as (see Fig. 3.26(c)):

$$m\ddot{x} = -k(x_0 + x) + \mu W = -k(x_0 + x) + W\left(\mu_0 - \frac{a}{W}(V - \dot{x})\right)$$

i.e.,

$$m\ddot{x} - a\dot{x} + kx = 0 \tag{E.1}$$

Since the coefficient of $\dot{x}$ is negative, the motion given by Eq. (E.1) will be unstable. The solution of Eq. (E.1) is given by

$$x(t) = e^{(a/2m)t}\{C_1 e^{r_1 t} + C_2 e^{r_2 t}\} \tag{E.2}$$

where C_1 and C_2 are constants and

$$r_1 = \frac{1}{2}\left[\left(\frac{a}{m}\right)^2 - 4\left(\frac{k}{m}\right)\right]^{1/2}$$

$$r_2 = -\frac{1}{2}\left[\left(\frac{a}{m}\right)^2 - 4\left(\frac{k}{m}\right)\right]^{1/2}$$

As can be seen from Eq. (E.2), the value of x increases with time. It increases until either $V - \dot{x} = 0$ or $V + \dot{x} = v_Q$. After this, the μ will have a positive slope, and hence the nature of the motion will be different [3.13].

Note: A similar motion can be observed in belt and pulley-type absorption brakes and in machine tool slides [3.14]. In machine tools, for example, a work table is mounted on suitable guideways and a feed screw is used to impart motion to the work table, as shown in Fig. 3.27. In some cases, the work table may slide in a jerky fashion even when the feed screw has a uniform and smooth motion. Such a motion is known as *stick-slip motion.* A simplified analysis of the stick-slip motion can be conducted by modeling the work table as a mass (m) and the connection between the work table and the feed screw (which is never perfectly rigid) as a spring (k) and viscous damper (c). The coefficient of friction between the mass and the sliding surface varies as a function of the sliding speed,

FIGURE 3.27   Motion of work table on feed screw in a machine tool.

as indicated in Fig. 3.26(b). The equation of motion of the mass (work table) can be derived as in the case of Eq. (E.1) of Example 3.8 as

$$m\ddot{x} + c\dot{x} + kx = \mu W = W \left[\mu_0 - \frac{a}{W} (V - \dot{x}) \right]$$

—i.e.,

$$m\ddot{x} + (c - a)\dot{x} + kx = 0$$

It can be seen that dynamic instability occurs if $c < a$.

■

3.11.2 Dynamic Instability Caused by Fluid Flow

The vibration caused by a fluid flowing around a body is known as *flow-induced vibration* [3.4]. For example, tall chimneys, submarine periscopes, electric transmission lines, and nuclear fuel rods are found to vibrate violently under certain conditions of fluid flow around them. Similarly, water and oil pipelines and tubes in air compressors undergo severe vibration under certain conditions of fluid flow through them. In all these examples, the vibration of the system continuously extracts energy from the source, leading to larger and larger amplitudes of vibration.

The flow-induced vibration may be caused by various phenomena. For example, in ice-covered electric transmission lines, low-frequency vibration (1 to 2 Hz) known as *galloping*, occurs as a result of the lift and drag forces developed by air flowing around the lines. The unstable vibration, known as *flutter*, of airfoil sections is also due to the lift and drag forces developed by the air flowing around the airfoil. In addition, a high-frequency vibration known as *singing of transmission lines* occurs as a result of the phenomenon of vortex shedding.

To see the phenomenon of galloping of wires, consider a cylindrical section with wind blowing against it at a velocity U, as shown in Fig. 3.28(a) [3.3]. Due to symmetry of the section, the direction of force due to wind will be the same as that of the wind. If a small downward velocity u is given to the cylinder, the wind will have an upward component of velocity u (relative to the cylinder) along with the horizontal component U. Thus the direction of the resultant force due to wind on the cylinder will be upward, as shown in Fig. 3.28(b). Since this force (upward) is opposite to the direction of motion of the cylinder (downward), the motion of the cylinder will be damped. In contrast, if a noncircular section such as an ice-covered cylindrical wire is considered, the resultant wind force may not always oppose the motion of the wire, as shown in Fig. 3.28(c). In such a case, the motion of the wire is aided by the wind forces, implying a negative damping in the system.

To visualize the phenomenon of singing of wires, consider a fluid flowing past a smooth cylinder. Under certain conditions, alternating vortices in a regular pattern are formed downstream, as shown in Fig. 3.29. These are called *Karman vortices*, in honor of the prominent fluid mechanician, Theodor von Karman, who was first to predict the stable spacing of the vortices on theoretical grounds in 1911. The Karman vortices are alternately clockwise and counterclockwise and thus cause harmonically varying lift forces on the

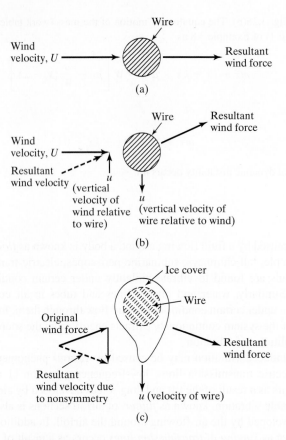

FIGURE 3.28   Galloping of a wire.

cylinder perpendicular to the velocity of the fluid. Experimental data show that regular vortex shedding occurs strongly in the range of Reynolds number (Re) from about 60 to 5,000. In this case

$$\text{Re} = \frac{\rho V d}{\mu} \tag{3.113}$$

FIGURE 3.29   Fluid flow past a cylinder.

where d is the diameter of the cylinder, ρ is the density, V is the velocity, and μ is the absolute viscosity of the fluid. For Re $> 1,000$, the dimensionless frequency of vortex shedding, expressed as a Strouhal number (St), is approximately equal to 0.21 [3.15]

$$\text{St} \equiv \frac{fd}{V} = 0.21 \tag{3.114}$$

where f is the frequency of vortex shedding. The harmonically varying lift force (F) is given by

$$F(t) = \frac{1}{2} c\rho V^2 A \sin \omega t \tag{3.115}$$

where c is a constant ($c \approx 1$ for a cylinder), A is the projected area of the cylinder perpendicular to the direction of V, ω is the circular frequency ($\omega = 2\pi f$), and t is the time. The mechanism of vortex shedding from a cylinder can be called a *self-excited* one, since the fluid flow (V) has no alternating component. From a design point of view, we have to ensure the following:

1. The magnitude of the force exerted on the cylinder, given by Eq. (3.115), is less than the static-failure load.
2. Even if the magnitude of force F is small, the frequency of oscillation (f) should not cause fatigue failure during the expected lifetime of the structure (or cylinder).
3. The frequency of vortex shedding (f) does not coincide with the natural frequency of the structure or cylinder to avoid resonance.

Reduction of Flow-Induced Vibration.   Several methods can be used to reduce failures caused by flow-induced vibration.

1. To reduce the singing vibration of transmission lines due to vortex shedding, a damped vibration absorber, known as a *Stockbridge damper*, can be used. A typical Stockbridge damper consists of a short steel cable with two masses attached at the ends. This damper is clamped to the transmission line, as shown in Fig. 3.30(a). The device thus acts as a spring-mass system and can be tuned to the frequency of flow-induced vibration by adjusting its length (the length of the cable) or the value of the masses. The Stockbridge damper is clamped to the transmission line at a point where the amplitude of vibration is expected to be large.
2. For tall steel chimneys, the effect of flow-induced vibration can be minimized by attaching vibration dampers through guy cables between the top of the chimney and the ground, as shown in Fig. 3.30(b).
3. For tall chimneys, helical spoilers or strakes can be provided around the chimney, as shown in Fig. 3.31. The helical spoilers break the vortex pattern so that no well-defined excitation is applied to the chimney wall.
4. In high-speed (racing) cars, the flow-induced lift forces can unload the tires, thereby causing problems with steering control and stability of the vehicle. Although lift

FIGURE 3.30 Stockbridge damper.

FIGURE 3.31 Helical spoilers. (Photo courtesy of Bethlehem Steel Corporation.)

FIGURE 3.32   Contemporary sports-racing car with aerodynamic features for low drag and high stability. (Photo courtesy of Goodyear Tire & Rubber Co., Inc.)

forces can be countered partly by adding spoilers, the drag force will increase. In recent years, movable inverted airfoils are being used to develop a downward aerodynamic force with improved stability characteristics (see Fig. 3.32).

EXAMPLE 3.11

Dynamic Instability of an Airfoil

Find the value of free-stream velocity u at which the airfoil section (single-degree-of-freedom system) shown in Fig. 3.33 becomes unstable.

Solution

Approach: Find the vertical force acting on the airfoil (or mass m) and obtain the condition that leads to zero damping.

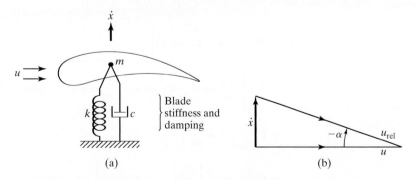

FIGURE 3.33   Modeling of airfoil as a single-degree-of-freedom system.

The vertical force acting on the airfoil (or mass m) due to fluid flow can be expressed as [3.4]

$$F = \frac{1}{2}\rho u^2 D C_x \qquad \text{(E.1)}$$

where ρ = density of the fluid, u = free-stream velocity, D = width of the cross section normal to the fluid flow direction, and C_x = vertical force coefficient, which can be expressed as

$$C_x = \frac{u_{rel}^2}{u^2}(C_L \cos\alpha + C_D \sin\alpha) \qquad \text{(E.2)}$$

where u_{rel} is the relative velocity of the fluid, C_L is the lift coefficient, C_D is the drag coefficient, and α is the angle of attack (see Fig. 3.33):

$$\alpha = -\tan^{-1}\left(\frac{\dot{x}}{u}\right) \qquad \text{(E.3)}$$

For small angles of attack,

$$\alpha = -\frac{\dot{x}}{u} \qquad \text{(E.4)}$$

and C_x can be approximated, using Taylor's series expansion, about $\alpha = 0$, as

$$C_x \simeq C_x\Big|_{\alpha=0} + \frac{\partial C_x}{\partial \alpha}\Big|_{\alpha=0}\cdot\alpha \qquad \text{(E.5)}$$

where, for small values of α, $u_{rel} \simeq u$ and Eq. (E.2) become

$$C_x = C_L \cos\alpha + C_D \sin\alpha \qquad \text{(E.6)}$$

Equation (E.5) can be rewritten, using Eqs. (E.4) and (E.6), as

$$C_x = (C_L \cos\alpha + C_D \sin\alpha)\Big|_{\alpha=0}$$

$$+ \alpha\left[\frac{\partial C_L}{\partial \alpha}\cos\alpha - C_L \sin\alpha + \frac{\partial C_D}{\partial \alpha}\sin\alpha + C_D \cos\alpha\right]\Big|_{\alpha=0}$$

$$= C_L\Big|_{\alpha=0} + \alpha\frac{\partial C_x}{\partial \alpha}\Big|_{\alpha=0}$$

$$= C_L\Big|_{\alpha=0} - \frac{\dot{x}}{u}\left\{\frac{\partial C_L}{\partial \alpha}\Big|_{\alpha=0} + C_D\Big|_{\alpha=0}\right\} \qquad \text{(E.7)}$$

Substitution of Eq. (E.7) into Eq. (E.1) gives

$$F = \frac{1}{2}\rho u^2 D C_L \bigg|_{\alpha=0} - \frac{1}{2}\rho u D \frac{\partial C_x}{\partial \alpha}\bigg|_{\alpha=0} \dot{x} \tag{E.8}$$

The equation of motion of the airfoil (or mass m) is

$$m\ddot{x} + c\dot{x} + kx = F = \frac{1}{2}\rho u^2 D C_L \bigg|_{\alpha=0} - \frac{1}{2}\rho u D \frac{\partial C_x}{\partial \alpha}\bigg|_{\alpha=0} \dot{x} \tag{E.9}$$

The first term on the right-hand side of Eq. (E.9) produces a static displacement, and hence only the second term can cause instability of the system. The equation of motion, considering only the second term on the right-hand side, is

$$m\ddot{x} + c\dot{x} + kx \equiv m\ddot{x} + \left[c + \frac{1}{2}\rho u D \frac{\partial C_x}{\partial \alpha}\bigg|_{\alpha=0} \right]\dot{x} + kx = 0 \tag{E.10}$$

Note that m includes the mass of the entrained fluid. We can see from Eq. (E.10) that the displacement of the airfoil (or mass m) will grow without bound (i.e., the system becomes unstable) if c is negative. Hence the minimum velocity of the fluid for the onset of unstable oscillations is given by $c = 0$, or,

$$u = -\left\{ \frac{2c}{\rho D \frac{\partial C_x}{\partial \alpha}\bigg|_{\alpha=0}} \right\} \tag{E.11}$$

The value of $\frac{\partial C_x}{\partial \alpha}\bigg|_{\alpha=0} = -2.7$ for a square section in a steady flow [3.4].

∎

Note: An analysis similar to that of Example 3.11 is applicable to other vibrating structures such as water tanks (Fig. 3.34(a)) and galloping ice-coated power lines (Fig. 3.34(b)) under wind loading.

Flow-Induced Vibration of a Chimney

EXAMPLE 3.12

A steel chimney has a height 2 m, an inner diameter 0.75 m, and an outer diameter 0.80 m. Find the velocity of the wind flowing around the chimney which will induce transverse vibration of the chimney in the direction of airflow.

Solution

Approach: Model the chimney as a cantilever beam and equate the natural frequency of the transverse vibration of the chimney to the frequency of vortex shedding.

FIGURE 3.34 Instability of typical vibrating structures.

To find the natural frequency of the transverse vibration of the chimney, the Rayleigh's energy method can be used by assuming a suitable transverse deflection of the cantilever beam (see Section 8.7). However, in this case, we use the natural frequencies of the cantilever beam given in Fig. 8.15. Figure 8.15 gives the fundamental natural frequency of transverse vibration (ω_1) of a cantilever (fixed-free) beam as

$$\omega_1 = (\beta_1 l)^2 \sqrt{\frac{EI}{\rho A l^4}} \tag{E.1}$$

where

$$\beta_1 l = 1.875104 \tag{E.2}$$

For the chimney, $E = 207 \times 10^9$ Pa, $\rho g = $ unit weight $= 76.5 \times 10^3$ N/m³, $l = 20$ m, $d = 0.75$ m, $D = 0.80$ m,

$$A = \frac{\pi}{4}(D^2 - d^2) = \frac{\pi}{4}(0.80^2 - 0.75^2) = 0.0608685 \text{ m}^2$$

and

$$I = \frac{\pi}{64}(D^4 - d^4) = \frac{\pi}{64}(0.80^4 - 0.75^4) = 0.004574648 \text{ m}^4$$

Thus

$$\omega_1 = (1.875104)^2 \left\{ \frac{(207 \times 10^9)(0.004574648)}{\left(\dfrac{76.5 \times 10^3}{9.81}\right)(0.0608685)(20)^4} \right\}^{1/2}$$

$$= 12.415417 \text{ rad/s} = 1.975970 \text{ Hz}$$

The frequency of vortex shedding (f) is given by the Strouhal number:

$$\text{St} = \frac{fd}{V} = 0.21$$

Using $d = 0.80$ m and $f = f_1 = 1.975970$ Hz, the velocity of wind (V) which causes resonance can be determined as

$$V = \frac{f_1 d}{0.21} = \frac{1.975970(0.80)}{0.21} = 7.527505 \text{ m/s}$$

■

3.12   Transfer-Function Approach

The transfer-function approach, based on Laplace transforms, is commonly used for the formulation and solution of dynamic problems in the controls literature. It can also be conveniently used for solving forced-vibration problems. Transfer function relates a system's output to its input. This function permits separation of the input, system, and output into three separate and distinct parts (unlike the differential equation, in which the three aspects cannot be separated easily).

Definition: The transfer function of a linear, time-invariant differential equation is defined as the ratio of the Laplace transform of the output or response function to the Laplace transform of the input or forcing function, assuming zero initial conditions.

The general procedure used to find the transfer function of a linear differential equation involves taking the Laplace transforms of both sides, assuming zero initial conditions, and solving for the ratio of the output Laplace transform and the input Laplace transform. Since the linear differential equation consists of the variable and its derivatives, the Laplace transform converts the differential equation into a polynomial equation in the Laplace variable s. The expressions given in Appendix D for the Laplace transforms of derivatives can be used in deriving the transfer function.

EXAMPLE 3.13

Transfer Function Corresponding to a Differential Equation

Consider the following nth-order linear, time-invariant differential equation that governs the behavior of a dynamic system:

$$a_n \frac{d^n x(t)}{dt^n} + a_{n-1} \frac{d^{n-1} x(t)}{dt^{n-1}} + \ldots + a_0 x(t)$$
$$= b_m \frac{d^m f(t)}{dt^m} + b_{m-1} \frac{d^{m-1} f(t)}{dt^{m-1}} + \ldots + b_0 f(t) \qquad \text{(E.1)}$$

where $x(t)$ is the output, $f(t)$ is the input, t is the time, and a_i's and b_i's are constants. Find the transfer function of the system and show the input, system, and output in a block diagram.

Solution: By taking Laplace transforms of both sides of Eq. (E.1), we obtain

$$a_n s^n X(s) + a_{n-1} s^{n-1} X(s) + \ldots + a_0 X(s) + \textit{initial conditions involving } x(t)$$
$$= b_m s^m F(s) + b_{m-1} s^{m-1} F(s) + \ldots + b_0 F(s)$$
$$+ \textit{initial conditions involving } f(t) \tag{E.2}$$

It can be seen that Eq. (E.2) is a purely algebraic expression. If all initial conditions are assumed to be zero, Eq. (E.2) reduces to the following form:

$$\{a_n s^n + a_{n-1} s^{n-1} + \ldots + a_0\} X(s) = \{b_m s^m + b_{m-1} s^{m-1} + \ldots + b_0\} F(s) \tag{E.3}$$

By solving Eq. (E.3), the transfer function of the system evaluated at zero initial conditions, $T(s)$, can be found as the ratio of the output transform, X(s), and the input transform, F(s):

$$T(s) = \frac{X(s)}{F(s)} = \frac{(b_m s^m + b_{m-1} s^{m-1} + \ldots + b_0)}{(a_n s^n + a_{n-1} s^{n-1} + \ldots + a_0)} \tag{E.4}$$

It can be seen that the transfer function identifies the input, $F(s)$, the output, $X(s)$, and the system (defined by the right-hand side expression in Eq. (E.4)) as separate entities. From Eq. (E.4), the output of the system can be determined as

$$X(s) = T(s)F(s) \tag{E.5}$$

By taking the inverse Laplace transform of Eq. (E.5), we can find the output of the system in the time domain for any known input.

The transfer function can be represented as a block diagram as shown in Fig. 3.35, where the input and output are shown on the left and the right sides, respectively, of the block with the transfer function shown inside the block. Note that the denominator of the transfer function is identical to the characteristic polynomial of the differential equation.

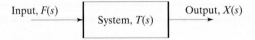

FIGURE 3.35 Block-diagram representation of input, system, and output.

EXAMPLE 3.14

Transfer Function of a Damped Single-Degree-of-Freedom System

Derive the transfer function of a viscously damped single-degree-of-freedom system subjected to external force $f(t)$ as shown in Fig. 3.1.

Solution: The equation of motion of the system is given by

$$m\ddot{x} + c\dot{x} + kx = f(t) \tag{E.1}$$

By taking the Laplace transforms of both sides of Eq. (E.1), we obtain

$$m\mathscr{L}[\ddot{x}(t)] + c\mathscr{L}[\dot{x}(t)] + k\mathscr{L}[x(t)] = \mathscr{L}[f(t)] \tag{E.2}$$

or

$$m\left\{s^2 X(s) - sx(0) - \dot{x}(0)\right\} + \left\{sX(s) - x(0)\right\} + kX(s) = F(s) \tag{E.3}$$

Equation (E.2) can be rewritten as

$$(ms^2 + cs + k)\, X(s) - \left\{msx(0) + m\dot{x}(0) + sx(0)\right\} = F(s) \tag{E.4}$$

where $X(s) = \mathscr{L}[x(t)]$ and $F(s) = \mathscr{L}[f(t)]$. The transfer function of the system can be obtained from Eq. (E.4), by setting $x(0) = \dot{x}(0) = 0$, as

$$T(s) = \frac{\mathscr{L}[\text{output}]}{\mathscr{L}[\text{input}]}\bigg|_{\text{zero initial conditions}} = \frac{X(s)}{F(s)} = \frac{1}{ms^2 + cs + k} \tag{E.5}$$

Notes:

1. The transfer function is a property of the system and is unrelated to the input or forcing function.
2. The transfer function does not provide any information on the physical structure of the system. In fact, the transfer functions of many physically different systems can be identical.
3. Representation of a dynamic system using the transfer function is very useful in control theory as well as in vibration testing for measuring the dynamic response and for system identification. For example, for a system whose parameters such as mass (m), damping constant (c), and spring stiffness (k) are unknown, the transfer function can be determined experimentally by measuring the response or output due to a known input. Once the transfer function is determined, it provides a complete description of the dynamic characteristics of the system.

    In vibration testing, the measured vibration response (due to a known input or forcing function) could be the displacement, velocity or, more commonly, the acceleration. The transfer function corresponding to the acceleration response can be defined as the ratio, $\dfrac{s^2 X(s)}{F(s)}$, where

$F(s)$ is the Laplace transform of the input and $s^2 X(s)$ is the Laplace transform of the acceleration.

4. If the transfer function of a system is known, the output or response of the system can be found for any type of input.

5. The variable s in the Laplace transform is a complex number and hence the transfer function will be a complex quantity. The variable s is similar to the characteristic root s used in representing the solution of a differential equation [see Eq. (2.61)]. By denoting the variable s in the Laplace transform in complex form as

$$s = \sigma + i\omega_d \tag{3.116}$$

where σ and ω_d denote the real and imaginary parts, respectively, of s, the complete analysis considered in Section 2.8 can be seen to be valid for the variable s of the Laplace transform as well.

6. As can be seen from Eq. (3.1), the equation of motion of a vibrating system is in the time domain. Although the output or response of the system can be found in the time domain directly as indicated in Section 3.4, sometimes, it is easier to find the response using the Laplace transform approach. The Laplace transform converts a linear differential equation into an algebraic expression, which is easier to manipulate. It transforms functions defined in terms of the independent variable (such as time) into functions in terms of the complex quantity s as the independent variable. In order to use the Laplace transform, we need first to determine the transfer function of the system.

7. Although the transfer function is derived formally through application of the Laplace transform, it can be derived informally in a simple manner. For this, consider the equation

$$m\ddot{x} + c\dot{x}(t) + kx(t) = f(t) \tag{3.117}$$

The transfer function associated with this equation can be derived by replacing $x(t)$ by $X(s)e^{st}$ and $f(t)$ by $F(s)e^{st}$. The time derivatives of $x(t)$ can be obtained by differentiating $X(s)e^{st}$ with respect to time as $\dot{x}(t) = X(s)se^{st}$ and $\ddot{x}(t) = X(s)s^2 e^{st}$. Thus Eq. (3.117) can be rewritten as

$$ms^2 X(s)e^{st} + csX(s)e^{st} + kX(s)e^{st} = F(s)e^{st} \tag{3.118}$$

Equation (3.118) can be solved for the ratio $X(s)/F(s)$ to obtain the transfer function, $T(s)$, as

$$T(s) = \frac{X(s)}{F(s)} = \frac{1}{ms^2 + cs + k} \tag{3.119}$$

This equation can be seen to be identical to Eq. (E.5) of Example 3.14.

∎

3.13 Solutions Using Laplace Transforms

The computation of responses of single-degree-of-freedom systems using Laplace transforms is illustrated through the following examples.

EXAMPLE 3.15

Response of a Damped System Using Laplace Transforms

Derive an expression for the complete response of a damped single-degree-of-freedom system subjected to a general force, $f(t)$, as shown in Fig. 3.1 using Laplace transforms.

Solution: The Laplace transform of Eq. (3.1) leads to the relation (see Eq. (E.4) of Example 3.14)

$$X(s) = \frac{F(s)}{m(s^2 + 2\zeta\omega_n s + \omega_n^2)} + \frac{s + 2\zeta\omega_n}{s^2 + 2\zeta\omega_n s + \omega_n^2}x(0) + \frac{1}{s^2 + 2\zeta\omega_n s + \omega_n^2}\dot{x}(0) \quad \text{(E.1)}$$

The complete response of the system can be found by taking inverse Laplace transforms of each term on the right-hand side of Eq. (E.1). For convenience, we define the following functions with the subscripts i and s denoting the input and system, respectively:

$$F_i(s) = F(s) \quad \text{(E.2)}$$

$$F_s(s) = \frac{1}{m(s^2 + 2\zeta\omega_n s + \omega_n^2)} \quad \text{(E.3)}$$

We note that the inverse Laplace transform of $F_i(s)$ will be equal to the known forcing function

$$f_i(t) = F_0 \cos \omega t \quad \text{(E.4)}$$

and the inverse Laplace transform of $F_s(s)$ is given by (see Appendix D)

$$f_s(t) = \frac{1}{m\omega_d}e^{-\zeta\omega_n t} \sin \omega_d t \quad \text{(E.5)}$$

where

$$\omega_d = \sqrt{1 - \zeta^2}\,\omega_n \quad \text{(E.6)}$$

The inverse Laplace transform of the first term on the right-hand side of Eq. (E.1) can be expressed as (see Appendix D):

$$\mathcal{L}^{-1}F_i(s)F_s(s) = \int_{\tau=0}^{t} f_i(\tau)f_s(t - \tau)\,d\tau = \frac{1}{m\omega_d}\int_{\tau=0}^{t} f(\tau)e^{-\zeta\omega_n (t-\tau)} \sin \omega_d(t - \tau)\,d\tau \quad \text{(E.7)}$$

The inverse Laplace transform of the coefficient of $x(0)$ in Eq. (E.1) yields

$$\mathcal{L}^{-1} \frac{s + 2\,\zeta\omega_n}{s^2 + 2\,\zeta\omega_n s + \omega_n^2} = \frac{\omega_n}{\omega_d} e^{-\zeta\omega_n t} \cos(\omega_d t - \phi_1) \qquad (E.8)$$

where

$$\phi_1 = \tan^{-1} \frac{\zeta\omega_n}{\omega_d} = \tan^{-1} \frac{\zeta}{\sqrt{1 - \zeta^2}} \qquad (E.9)$$

The inverse Laplace transform of the coefficient of $\dot{x}(0)$ can be obtained by multiplying $f_s(t)$ by m so that

$$\mathcal{L}^{-1} \frac{1}{(s^2 + 2\,\zeta\omega_n s + \omega_n^2)} = \frac{1}{\omega_d} e^{-\zeta\omega_n t} \sin \omega_d t \qquad (E.10)$$

Thus the complete response of the system, using the responses given on the right-hand sides of Eqs. (E.7), (E.8), and (E.10), can be expressed as

$$x(t) = \frac{1}{m\omega_d} \int_{\tau=0}^{t} f(\tau) e^{-\zeta\omega_n(t-\tau)} \sin \omega_d(t - \tau)\, d\tau$$

$$+ \frac{\omega_n}{\omega_d} e^{-\zeta\omega_n t} \cos(\omega_d t - \phi_1) + \frac{1}{\omega_d} e^{-\zeta\omega_n t} \sin \omega_d t \qquad (E.11)$$

Noting that the inverse Laplace transform of the product function in Eq. (E.7) can also be expressed as

$$\mathcal{L}^{-1} F_i(s) F_s(s) = \int_{\tau=0}^{t} f_i(t - \tau) f_s(\tau)\, d\tau = \frac{1}{m\omega_d} \int_{\tau=0}^{t} f(t - \tau) e^{-\zeta\omega_n \tau} \sin \omega_d \tau\, d\tau \quad (E.12)$$

the complete response of the system can also be expressed as

$$x(t) = \frac{1}{m\omega_d} \int_{\tau=0}^{t} f(t - \tau)\, e^{-\zeta\omega_n \tau} \sin \omega_d \tau\, d\tau$$

$$+ \frac{\omega_n}{\omega_d} e^{-\zeta\omega_n t} \cos(\omega_d t - \phi_1) + \frac{1}{\omega_d} e^{-\zeta\omega_n t} \sin \omega_d t \qquad (E.13)$$

∎

EXAMPLE 3.16

Steady-State Response Using Laplace Transform

Find the steady-state response of a damped single-degree-of-freedom system subjected to a harmonic force $f(t) = F_0 \cos \omega t$ using the Laplace transform.

Solution: The Laplace transform of Eq. (3.1) leads to the relation (with zero initial conditions for steady-state response in Eq. (E.1) of Example 3.15)

$$X(s) = \frac{F(s)}{m(s^2 + 2\zeta\omega_n s + \omega_n^2)} \tag{E.1}$$

The Laplace transform of the input $f(t) = F_0 \cos \omega t$ is given by $F(s) = F_0 \dfrac{s}{s^2 + \omega^2}$. Thus Eq. (E.1) becomes

$$X(s) = \frac{F_0}{m} \frac{s}{s^2 + \omega^2} \frac{1}{s^2 + 2\zeta\omega_n s + \omega_n^2} \tag{E.2}$$

where the relations $\omega_n = \sqrt{\dfrac{k}{m}}$ and $\zeta = \dfrac{c}{2\sqrt{mk}}$ have been used in expressing Eq. (E.2). By expressing the right-hand side of Eq. (E.2) as

$$F(s) = \frac{F_0}{m} \left(\frac{a_1 s + a_2}{s^2 + \omega^2} + \frac{a_3 s + a_4}{s^2 + 2\zeta\omega_n s + \omega_n^2} \right) \tag{E.3}$$

the constants a_1, a_2, a_3, and a_4 can be identified as (see Problem 3.99)

$$a_1 = \frac{\omega_n^2 - \omega^2}{(2\zeta\omega_n)^2 \omega^2 + (\omega_n^2 - \omega^2)^2} \tag{E.4}$$

$$a_2 = \frac{2\zeta\omega_n\omega^2}{(2\zeta\omega_n)^2 \omega^2 + (\omega_n^2 - \omega^2)^2} \tag{E.5}$$

$$a_3 = -\frac{\omega_n^2 - \omega^2}{(2\zeta\omega_n)^2 \omega^2 + (\omega_n^2 - \omega^2)^2} \tag{E.6}$$

$$a_4 = -\frac{\omega_n^2 - \omega^2}{(2\zeta\omega_n)^2 \omega^2 + (\omega_n^2 - \omega^2)^2} \tag{E.7}$$

Thus, $X(s)$ can be expressed as

$$X(s) = \frac{F_0}{m} \frac{1}{(2\zeta\omega_n)^2 \omega^2 + (\omega_n^2 - \omega^2)^2} \left\{ (\omega_n^2 - \omega^2) \left(\frac{s}{s^2 + \omega^2} \right) + (2\zeta\omega_n\omega) \left(\frac{\omega}{s^2 + \omega^2} \right) \right.$$

$$\left. - (\omega_n^2 - \omega^2) \left(\frac{s}{s^2 + 2\zeta\omega_n s + \omega_n^2} \right) - (2\zeta\omega_n) \left(\frac{\omega_n^2}{s^2 + 2\zeta\omega_n s + \omega_n^2} \right) \right. \tag{E.8}$$

Using the relations 14, 15, 27, and 28 of Appendix D, the response of the system can be expressed as

$$x(t) = \frac{F_0}{m} \frac{1}{(2\zeta\omega_n)^2 + (\omega_n^2 - \omega^2)^2} [(\omega_n^2 - \omega^2)\cos \omega t + 2\zeta\omega_n\omega \sin \omega t$$

$$+ \frac{(\omega_n^2 - \omega^2)}{\sqrt{1 - \zeta^2}} e^{-\zeta\omega_n t} \sin(\omega_n\sqrt{1 - \zeta^2}\, t - \phi)$$

$$- \frac{(2\zeta\omega_n^2)}{\sqrt{1 - \zeta^2}} e^{-\zeta\omega_n t} \sin(\omega_n\sqrt{1 - \zeta^2}\, t)] \tag{E.9}$$

where

$$\phi = \tan^{-1}\left(\frac{1 - \zeta^2}{\zeta}\right) \tag{E.10}$$

It can be observed that as $t \to \infty$, the terms involving $e^{-\zeta\omega_n t}$ in Eq. (E.9) approach zero. Thus the steady-state response of the system can be expressed as

$$x(t) = \frac{F_0}{m} \frac{1}{(2\zeta\omega_n)^2 + (\omega_n^2 - \omega^2)^2} [(\omega_n^2 - \omega^2)\cos \omega t + 2\zeta\omega_n\omega \sin \omega t] \tag{E.11}$$

which can be simplified as

$$x(t) = \frac{F_0}{\sqrt{c^2\omega^2 + (k - m\omega^2)^2}} \cos(\omega t - \phi) \tag{E.12}$$

This solution can be seen to be the one found in Section 3.4 [Eqs. (3.25), (3.28), and (3.29)].

∎

3.14   Frequency Transfer Functions

As seen earlier, for example, in Section 3.4, the steady-state response of a linear system subjected to a sinusoidal (or harmonic) input will also be sinusoidal (or harmonic) of the same frequency. Although the response is at the same frequency as the input, it differs in amplitude and phase angle from the input. These differences are functions of the frequency (see Fig. 3.11). Also, as indicated in Section 1.10.2, any sinusoid can be represented as a complex number (called *phasor*). The magnitude of the complex number is the amplitude of the sinusoid, and the angle of the complex number is the phase angle of the sinusoid. Thus, the input phasor, $M_i \sin(\omega t + \phi_i)$, can be denoted in polar form as $M_i e^{i\phi_i}$, where the frequency, ω, is considered to be implicit.

Because a system causes both the amplitude and phase angle of the input to be changed (see, for example, Section 3.4), we can think of representing the system itself by a complex number or function defined such that the output phasor is given by the product of the system function and the input phasor. For example, for the spring-mass-damper system shown in Fig. 3.36(a), the input-output relation can be shown in the form of a block diagram as in Fig. 3.36(b). Thus the steady-state output or response sinusoid of the system can be denoted as

$$M_o(\omega)e^{i\phi_o(\omega)} = M_s(\omega)e^{i\phi_s(\omega)}M_i(\omega)e^{i\phi_i(\omega)} = M_s(\omega)M_i(\omega)e^{i\{\phi_s(\omega)+\phi_i(\omega)\}} \tag{3.120}$$

$$x(t) = M_o \cos(\omega t + \phi_o) = M_o(\omega)e^{i\phi_o(\omega)}$$

$$f(t) = M_i \cos(\omega t + \phi_i) = M_i(\omega)e^{i\phi_i(\omega)}$$

(a) Physical system

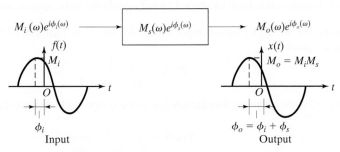

(b) Block diagram

FIGURE 3.36

where M's denote the amplitudes and ϕ's indicate the phase angles, respectively, of the sinusoids. It can be seen from Eq. (3.120) that the system function, $M_s(\omega)e^{i\phi_s(\omega)}$, is defined by its magnitude

$$M_s(\omega) = \frac{M_o(\omega)}{M_i(\omega)} \tag{3.121}$$

and the phase by

$$\phi_s(\omega) = \phi_o(\omega) - \phi_i(\omega) \tag{3.122}$$

The system function, $M_s(\omega)e^{i\phi_s(\omega)}$, is called the *frequency-response function*, with $M_s(\omega)$ being called the *amplitude* or *magnitude frequency response* and $\phi_s(\omega)$ the *phase-response function* of the system.

The frequency-response magnitude is given by the ratio of the output sinusoid's magnitude to the input sinusoid's magnitude. The phase response is given by the difference in phase angles between the output and the input sinusoids. Each of these responses is a function of frequency and will apply only to the steady-state sinusoidal responses of the system. For convenience of notation, the frequency-response function is sometimes called the *frequency transfer function*, denoted $T(i\omega)$, so that

$$T(i\omega) = M_s(\omega)e^{i\phi_s(\omega)} \tag{3.123}$$

**3.14.1
Relation Between
the General
Transfer Function
$T(s)$ and the
Frequency
Transfer
Function $T(i\omega)$**

The frequency transfer function, $T(i\omega)$, can be obtained by substituting $s = i\omega$ in the general transfer function $T(s)$. The following example illustrates not only the generation of the frequency transfer function from the general transfer function but also the identification of the input, system, and output sinusoids.

EXAMPLE 3.17

Generation of Frequency Transfer Function from General Transfer Function

Generate the frequency transfer function from the general transfer function for a spring-mass-damper system considered in Example 3.14 and identify the input, system, and output sinusoids.

Solution: For the spring-mass-damper system considered in Example 3.14, the general transfer function is given by

$$T(s) = \frac{1}{ms^2 + cs + k} \tag{E.1}$$

Using $s = i\omega$, the frequency transfer function of the system can be generated as

$$T(i\omega) = \frac{1}{k - m\omega^2 + i\omega c} \tag{E.2}$$

This frequency transfer function can be rewritten as

$$T(i\omega) = M_s(\omega)e^{i\phi_s(\omega)} = \frac{M_o(\omega)e^{i\phi_o(\omega)}}{M_i(\omega)e^{i\phi_i(\omega)}} \tag{E.3}$$

where

$$M_o(\omega) = 1, \quad \phi_o(\omega) = 0 \tag{E.4}$$

$$M_i(\omega) = \frac{1}{\sqrt{(k - m\omega^2)^2 + (\omega c)^2}}, \quad \phi_i(\omega) = \tan^{-1}\left(\frac{\omega c}{k - m\omega^2}\right) \tag{E.5}$$

It can be seen that the amplitude or magnitude of $T(i\omega)$ is given by

$$M_s(s) = |T(i\omega)| = \frac{1}{[(k - m\omega^2)^2 + (\omega c)^2]^{\frac{1}{2}}} \tag{E.6}$$

and the phase angle by

$$\phi_s = \tan^{-1}\left(\frac{\omega c}{m\omega^2 - k}\right) \tag{E.7}$$

It can be observed that Eq. (E.5) is identical to Eqs. (3.30) and (3.31). Thus the frequency transfer function of the system, $T(i\omega)$, can be found from the general transfer function, $T(s)$, by substituting

$i\omega$ for s. Although this observation is made only for a damped single-degree-of-freedom (second-order differential equation) system, it can be proved for any nth-order linear, time-invariant differential equation.

∎

3.14.2 Representation of Frequency-Response Characteristics

The frequency response of a second-order system, such as the spring-mass-damper system, denotes the steady-state response of the system to a sinusoidal input for different possible frequencies of the sinusoidal input. It can be shown graphically in several ways. In Section 3.4, the variations of the magnitude or amplitude ratio (M) and the phase angle (ϕ) with frequency (ω) have been plotted as two separate graphs. For some systems, the frequency ω will vary over a considerably large range. In such cases, it is desirable to use logarithmic scales to accommodate the complete range of ω in graphs plotted on standard-size paper.

Bode Diagrams.   A Bode diagram consists of two graphs—a graph of the logarithm of the magnitude of the frequency transfer function (M) versus the logarithm of the frequency (ω) and a graph of the phase angle (ϕ) versus the logarithm of the frequency (ω). Bode diagrams are also known as *logarithmic plots of the frequency response*.

As the standard representation of the logarithmic magnitude of $T(i\omega)$, a logarithmic unit known as *decibel*, abbreviated dB, is used. The magnitude ratio in decibels, m, is defined as

$$m = 10 \log_{10} (M^2) = 20 \log_{10} M \ \text{dB} \tag{3.124}$$

Number-Decibel Conversion Line.   It can be seen from Eq. (3.124) that for any number N, its decibel value is given by $20 \log_{10} N$. For some representative values of N, the decibel equivalencies are shown below:

Value of N	0.001	0.01	0.1	0.5	$\dfrac{1}{\sqrt{2}}$	1	$\sqrt{2}$	2	10	100	1,000
dB Value	-60	-40	-20	-6	-2	0	3	6	20	40	60

The main advantages of representing the frequency-response characteristics in the form of a Bode diagram are the following:

1. The transfer function of a system can be identified (i.e., experimentally determined) from the Bode diagram.
2. The frequency-response curves can be plotted over a wide range of the frequency, ω.
3. In some applications, we need to multiply the magnitudes of the frequency response. In such cases, the result can be obtained by a simple addition in Bode diagrams.

███████████████   **Bode Diagrams of a Damped Single-Degree-of-Freedom System**

EXAMPLE 3.18
_____

Plot the Bode diagrams corresponding to a damped second-order (spring-mass-damper) system in standard form with the transfer function

$$T(s) = \frac{\omega_n^2}{s^2 + 2\zeta\omega_n s + \omega_n^2} \tag{E.1}$$

Solution: The frequency transfer function, $T(i\omega)$, can be obtained by substituting $i\omega$ for s as

$$T(i\omega) = \frac{\omega_n^2}{(i\omega)^2 + 2\zeta\omega_n(i\omega) + \omega_n^2} \tag{E.2}$$

or

$$T(i\omega) = \frac{1}{1 - r^2 + i2\zeta r} \tag{E.3}$$

where $r = \omega/\omega_n$. The magnitude, M, of $T(i\omega)$ is given by

$$M = |T(i\omega)| = \left| \frac{1}{1 - r^2 + i2\zeta r} \right| = \frac{1}{\sqrt{(1 - r^2)^2 + (2\zeta r)^2}} \tag{E.4}$$

so that

$$20 \log_{10} M = -20 \log_{10} \sqrt{(1 - r^2)^2 + (2\zeta r)^2} \tag{E.5}$$

Note that for low frequencies with $\omega \ll \omega_n$ or $r \ll 1$, Eq. (E.5) reduces to

$$-20 \log_{10} 1 = 0 \text{ dB}$$

For very high frequencies with $\omega \gg \omega_n$ or $r \gg 1$, Eq. (E.5) becomes

$$-20 \log_{10} r^2 = -40 \log_{10} r \text{ dB}$$

The phase angle given by Eq. (E.3) is

$$\phi = \frac{1}{1 - r^2 + i2\zeta r} = -\tan^{-1}\frac{2\zeta r}{1 - r^2} \tag{E.6}$$

Equation (E.6) shows that ϕ is a function of ω and ζ. At $\omega = 0$, $\phi = 0$. When $\omega = \omega_n$, $\phi = -90°$ regardless of the value of ζ, since

$$\phi = -\tan^{-1}\left(\frac{2\zeta}{0}\right) = -\tan^{-1} \infty = -90°$$

At $\omega = \infty$, the phase angle becomes $-180°$. The phase angle will be skew symmetric about the inflection point, where $\phi = -90°$.

The Bode diagrams of Eqs. (E.5) and (E.6) are shown in Figs. 3.37(a) and (b), respectively.

FIGURE 3.37 Bode diagrams.

3.15   Examples Using MATLAB

Total Response of an Undamped System

EXAMPLE 3.19

Using MATLAB, plot the response of a spring-mass system under a harmonic force for the following data:

$$m = 5 \text{ kg}, \quad k = 2{,}000 \text{ N/m}, \quad F(t) = 100 \cos 30t \ N, \quad x_0 = 0.1 \text{ m}, \quad \dot{x}_0 = 0.1 \text{ m/s}$$

Solution: The response of the system is given by Eq. (3.9), which can be rewritten as

$$x(t) = \frac{\dot{x}_0}{\omega_n} \sin \omega_n t + \left(x_0 - \frac{f_0}{\omega_n^2 - \omega^2} \right) \cos \omega_n t + \frac{f_0}{\omega_n^2 - \omega^2} \cos \omega t \qquad (E.1)$$

where $f_0 = \dfrac{F_0}{m} = \dfrac{100}{5} = 20$, $\omega_n = \sqrt{\dfrac{k}{m}} = 20$ rad/s, and $\omega = 30$ rad/s.

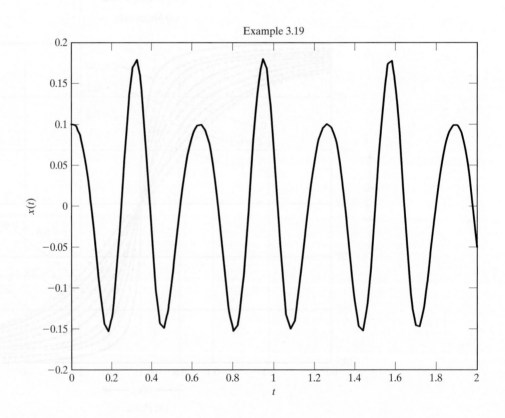

Example 3.19

Equation (E.1) is plotted using the following MATLAB program:

```
% Ex3_19.m
F0 = 100;
wn = 20;
m = 5;
w = 30;
x0 = 0.1;
x0_dot = 0.1;
f_0 = F0/m;
for i = 1: 101
    t(i) = 2 * (i-1)/100;
    x(i) = x0_dot*sin(wn*t(i))/wn + (x0 - f_0/(wn&and;2-w&and;2))*cos
        (wn*t(i)).. + f_0/ (wn^2-w^2)*cos(w*t(i));
end
plot (t, x);
xlabel ('t');
ylabel ('x(t)');
title ('Ex3.11')
```

■

EXAMPLE 3.20

Forced Response of a System with Coulomb Damping

Using MATLAB, plot the forced response of a spring-mass system with Coulomb damping for the following data: $m = 5$ kg, $k = 2,000$ N/m, $\mu = 0.5$, $F(t) = 100 \sin 30t$ N, $x_0 = 0.1$ m, $\dot{x}_0 = 0.1$ m/s.

Solution: The equation of motion of the system can be expressed as

$$m\ddot{x} + kx + \mu mg \, \text{sgn}(\dot{x}) = F_0 \sin \omega t \tag{E.1}$$

which can be rewritten as a system of two first-order differential equations (using $x_1 = x$ and $x_2 = \dot{x}$) as

$$\dot{x}_1 = x_2$$
$$\dot{x}_2 = \frac{F_0}{m} \sin \omega t - \frac{k}{m} x_1 - \mu g \, \text{sgn}(x_2) \tag{E.2}$$

with the initial conditions $x_1(0) = 0.1$ and $x_2(0) = 0.1$. The MATLAB solution of Eq. (E.2), using `ode23`, is given next.

```
% Ex3_20.m
% This program will use the function dfunc3_20.m, they should
% be in the same folder
tspan  =  [0: 0.01: 4];
x0  =  [0.1; 0.1];
[t, x]  =  ode23 ('dfunc3_12', tspan, x0);
disp ('      t            x(t)        xd(t)');
disp ([t x]);
plot (t, x(:, 1));
xlabel ('t');
gtext ('x(t)');
title ('Ex3.12');

% dfunc3_12.m
function f = dfunc3_12 (t, x)
f = zeros (2, 1);
f(1) = x(2);
f(2) = 100*sin(30*t)/5 - 9.81*0.5*sign(x(2)) - (2000/5)*x(1);
```

```
>> Ex3_12
    t           x(t)        xd(t)
    0         0.1000       0.1000
0.0100         0.0991      -0.2427
0.0200         0.0954      -0.4968
0.0300         0.0894      -0.6818
0.0400         0.0819      -0.8028
0.0500         0.0735      -0.8704
    .
    .
    .
3.9500         0.0196      -0.9302
3.9600         0.0095      -1.0726
3.9700        -0.0016      -1.1226
3.9800        -0.0126      -1.0709
3.9900        -0.0226      -0.9171
4.0000        -0.0307      -0.6704
```

Example 3.20

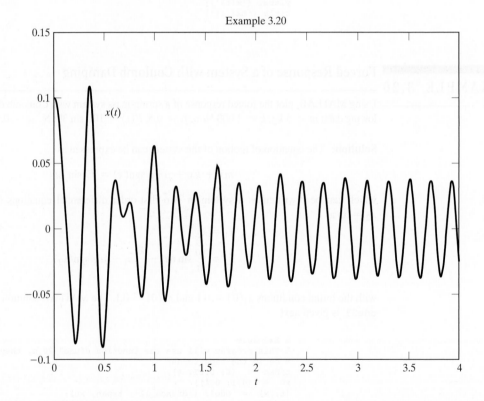

$x(t)$

Response of a System Under Base Excitation

EXAMPLE 3.21

Using MATLAB, find and plot the response of a viscously damped spring-mass system under the base excitation $y(t) = Y \sin \omega t$ for the following data: $m = 1{,}200$ kg, $k = 4 \times 10^5$ N/m, $\zeta = 0.5$, $Y = 0.05$ m, $\omega = 29.0887$ rad/s, $x_0 = 0$, $\dot{x}_0 = 0.1$ m/s.

Example 3.21

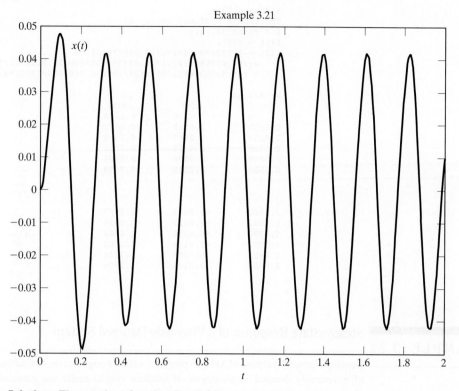

Solution: The equation of motion, Eq. (3.64),

$$m\ddot{x} + c\dot{x} + kx = ky + c\dot{y} \tag{E.1}$$

can be expressed as a system of two first-order ordinary differential equations (using $x_1 = x$ and $x_2 = \dot{x}$) as

$$\dot{x}_1 = x_2$$

$$\dot{x}_2 = -\frac{c}{m}x_2 - \frac{k}{m}x_1 + \frac{k}{m}y + \frac{c}{m}\dot{y} \tag{E.2}$$

with $c = \zeta c_c = 2\zeta\sqrt{km} = 2(0.5)\sqrt{(4 \times 10^5)(1{,}200)}$, $y = 0.5 \sin 29.0887t$, and $\dot{y} = (29.0887)$ $(0.05) \cos 29.0887t$. The MATLAB solution of Eq. (E.2), using **ode23**, is given below.

```
% Ex3_21.m
% This program will use the function dfunc3_21.m, they should
% be in the same folder
tspan = [0: 0.01: 2];
x0 = [0; 0.1];
[t, x] = ode23 ('dfunc3_13', tspan, x0);
disp ('        t            x(t)        xd(t)');
disp ([t x]);
plot (t, x (:, 1));
xlabel ('t');
gtext ('x(t)');
title ('Ex3.13');
```

```
% dfunc3_20.m
function f = dfunc3_20 (t, x)
f = zeros (2, 1);
f(1) = x(2);
f(2) = 400000*0.05*sin(29.0887*t)/1200 + ...
    sqrt (400000*1200)*29.0887*0.05*cos(29.0887*t)/1200 ...
    - sqrt(400000*1200)*x(2)/1200 - (400000/1200)*x(1);

>> Ex3_13
    t           x(t)          xd(t)
    0             0          0.1000
0.0100          0.0022      0.3422
0.0200          0.0067      0.5553
0.0300          0.0131      0.7138
0.0400          0.0208      0.7984
0.0500          0.0288      0.7976
    .
    .
    .
1.9500         -0.0388      0.4997
1.9600         -0.0322      0.8026
1.9700         -0.0230      1.0380
1.9800         -0.0118      1.1862
1.9900          0.0004      1.2348
2.0000          0.0126      1.1796
```

∎

Steady-State Response of a Viscously Damped System

EXAMPLE 3.22

Develop a general-purpose MATLAB program, called `Program3.m`, to find the steady-state response of a viscously damped single-degree-of-freedom system under the harmonic force $F_0 \cos \omega t$ or $F_0 \sin \omega t$. Use the program to find and plot the response of a system with the following data:

$$m = 5 \text{ kg}, \quad c = 20 \text{ N-s/m}, \quad k = 500 \text{ N/m}, \quad F_0 = 250 \text{ N}, \quad \omega = 40 \text{ rad/s}, \quad n = 40, \quad ic = 0.$$

Solution: `Program3.m` is developed to accept the following input data:

xm = mass

xc = damping constant

xk = spring constant

f_0 = amplitude of the forcing function

om = forcing frequency

n = number of time steps in a cycle at which the response is to be computed

ic = 1 for cosine-type forcing function; 0 for sine-type forcing function

The program gives the following output:

$$\text{step number } i, x(i), \dot{x}(i), \ddot{x}(i)$$

The program also plots the variations of x, $\dot{x}$, and $\ddot{x}$ with time.

```
>> program3
Steady state response of an undamped
Single degree of freedom system under harmonic force

Given data
xm  =   5.00000000e+000
xc  =   2.00000000e+001
xk  =   5.00000000e+002
```

```
f0  =   2.50000000e+002
om  =   4.00000000e+001
ic  =   0
n   =   20
```

Response:

i	x(i)	xd(i)	xdd(i)
1	1.35282024e-002	1.21035472e+000	-2.16451238e+001
2	2.22166075e-002	9.83897315e-001	-3.55465721e+001
3	2.87302863e-002	6.61128738e-001	-4.59684581e+001
4	3.24316314e-002	2.73643972e-001	-5.18906102e+001
5	3.29583277e-002	-1.40627096e-001	-5.27333244e+001
6	3.02588184e-002	-5.41132540e-001	-4.84141094e+001
7	2.45973513e-002	-8.88667916e-001	-3.93557620e+001
8	1.65281129e-002	-1.14921388e+000	-2.64449806e+001
9	6.84098018e-003	-1.29726626e+000	-1.09455683e+001
10	-3.51579846e-003	-1.31833259e+000	5.62527754e+000
11	-1.35284247e-002	-1.21035075e+000	2.16454794e+001
12	-2.22167882e-002	-9.83890787e-001	3.55468612e+001
13	-2.87304077e-002	-6.61120295e-001	4.59686523e+001
14	-3.24316817e-002	-2.73634442e-001	5.18906907e+001
15	-3.29583019e-002	1.40636781e-001	5.27332831e+001
16	-3.02587190e-002	5.41141432e-001	4.84139504e+001
17	-2.45971881e-002	8.88675144e-001	3.93555009e+001
18	-1.65279018e-002	1.14921874e+000	2.64446429e+001
19	-6.84074192e-003	1.29726827e+000	1.09451871e+001
20	3.51604059e-003	1.31833156e+000	-5.62566494e+000

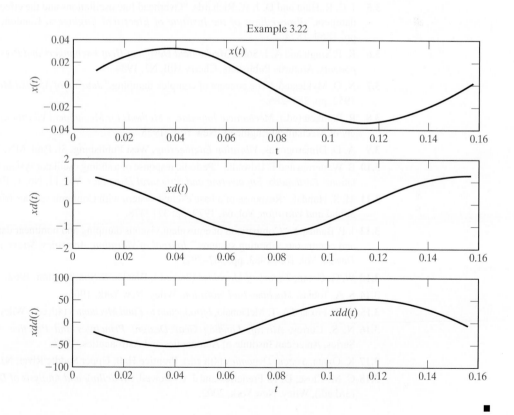

Example 3.22

■

CHAPTER SUMMARY

We considered the forced-vibration responses of undamped and viscously damped systems subjected to harmonic excitations. The harmonic excitations are in the form of external force applied to the mass, base motion, and force exerted on the mass of the system by a rotating unbalanced mass. We also discussed the aspects of resonance, beats, magnification or amplitude ratio, phase angle, transient vibration, and steady-state vibration. Finally, we studied application of the transfer-function approach, Laplace transforms, and the frequency transfer function in finding the response of harmonically excited systems.

Now that you have finished this chapter, you should be able to answer the review questions and solve the problems given below.

REFERENCES

3.1 G. B. Thomas and R. L. Finney, *Calculus and Analytic Geometry* (6th ed.), Addison-Wesley, Reading, MA, 1984.

3.2 J. W. Nilsson, *Electric Circuits*, Addison-Wesley, Reading, MA, 1983.

3.3 J. P. Den Hartog, "Forced vibrations with combined Coulomb and viscous friction," *Journal of Applied Mechanics (Transactions of ASME)*, Vol. 53, 1931, pp. 107–115.

3.4 R. D. Blevins, *Flow-Induced Vibration* (2nd ed.), Van Nostrand Reinhold, New York, 1990.

3.5 J. C. R. Hunt and D. J. W. Richards, "Overhead line oscillations and the effect of aerodynamic dampers," *Proceedings of the Institute of Electrical Engineers*, London, Vol. 116, 1969, pp. 1869–1874.

3.6 K. P. Singh and A. I. Soler, *Mechanical Design of Heat Exchangers and Pressure Vessel Components*, Arcturus Publishers, Cherry Hill, NJ, 1984.

3.7 N. O. Myklestad, "The concept of complex damping," *Journal of Applied Mechanics*, Vol. 19, 1952, pp. 284–286.

3.8 R. Plunkett (ed.), *Mechanical Impedance Methods for Mechanical Vibrations*, American Society of Mechanical Engineers, New York, 1958.

3.9 A. D. Dimarogonas, *Vibration Engineering*, West Publishing, St. Paul, MN, 1976.

3.10 B. Westermo and F. Udwadia, "Periodic response of a sliding oscillator system to harmonic excitation," *Earthquake Engineering and Structural Dynamics*, Vol. 11, No. 1, 1983, pp. 135–146.

3.11 M. S. Hundal, "Response of a base excited system with Coulomb viscous friction," *Journal of Sound and Vibration*, Vol. 64, 1979, pp. 371–378.

3.12 J. P. Bandstra, "Comparison of equivalent viscous damping and nonlinear damping in discrete and continuous vibrating systems," *Journal of Vibration, Acoustics, Stress, and Reliability in Design*, Vol. 105, 1983, pp. 382–392.

3.13 W. G. Green, *Theory of Machines* (2nd ed.), Blackie & Son, London, 1962.

3.14 S. A. Tobias, *Machine-Tool Vibration*, Wiley, New York, 1965.

3.15 R. W. Fox and A. T. McDonald, *Introduction to Fluid Mechanics* (4th ed.), Wiley, New York, 1992.

3.16 N. S. Currey, *Aircraft Landing Gear Design: Principles and Practice*, AIAA Education Series, American Institute of Aeronautics and Astronautics, 1988.

3.17 K. Ogata, *System Dynamics* (4th ed.), Prentice Hall, Upper Saddle River, NJ, 2004.

3.18 C. M. Close, D. K. Frederick and J. C. Newell, *Modelling and Analysis of Dynamics Systems* (3rd ed.), Wiley, New York, 2002.

REVIEW QUESTIONS

3.1   Give brief answers to the following:

1. How are the amplitude, frequency, and phase of a steady-state vibration related to those of the applied harmonic force for an undamped system?
2. Explain why a constant force on the vibrating mass has no effect on the steady-state vibration.
3. Define the term *magnification factor.* How is the magnification factor related to the frequency ratio?
4. What will be the frequency of the applied force with respect to the natural frequency of the system if the magnification factor is less than unity?
5. What are the amplitude and the phase angle of the response of a viscously damped system in the neighborhood of resonance?
6. Is the phase angle corresponding to the peak amplitude of a viscously damped system ever larger than 90°?
7. Why is damping considered only in the neighborhood of resonance in most cases?
8. Show the various terms in the forced equation of motion of a viscously damped system in a vector diagram.
9. What happens to the response of an undamped system at resonance?
10. Define the following terms: *beating, quality factor, transmissibility, complex stiffness, quadratic damping.*
11. Give a physical explanation of why the magnification factor is nearly equal to 1 for small values of r and is small for large values of r.
12. Will the force transmitted to the base of a spring-mounted machine decrease with the addition of damping?
13. How does the force transmitted to the base change as the speed of the machine increases?
14. If a vehicle vibrates badly while moving on a uniformly bumpy road, will a change in the speed improve the condition?
15. Is it possible to find the maximum amplitude of a damped forced vibration for any value of r by equating the energy dissipated by damping to the work done by the external force?
16. What assumptions are made about the motion of a forced vibration with nonviscous damping in finding the amplitude?
17. Is it possible to find the approximate value of the amplitude of a damped forced vibration without considering damping at all? If so, under what circumstances?
18. Is dry friction effective in limiting the reasonant amplitude?
19. How do you find the response of a viscously damped system under rotating unbalance?
20. What is the frequency of the response of a viscously damped system when the external force is $F_0 \sin \omega t$? Is this response harmonic?
21. What is the difference between the peak amplitude and the resonant amplitude?
22. Why is viscous damping used in most cases rather than other types of damping?
23. What is self-excited vibration?
24. How is transfer function defined?
25. How can we generate the frequency transfer function from the general transfer function?
26. What is a Bode diagram?
27. How is decibel defined?

3.2   Indicate whether each of the following statements is true or false:

1. The magnification factor is the ratio of maximum amplitude and static deflection.
2. The response will be harmonic if excitation is harmonic.
3. The phase angle of the response depends on the system parameters m, c, k, and ω.
4. The phase angle of the response depends on the amplitude of the forcing function.
5. During beating, the amplitude of the response builds up and then diminishes in a regular pattern.
6. The Q-factor can be used to estimate the damping in a system.
7. The half-power points denote the values of frequency ratio where the amplification factor falls to $Q/\sqrt{2}$, where Q is the Q-factor.
8. The amplitude ratio attains its maximum value at resonance in the case of viscous damping.
9. The response is always in phase with the harmonic forcing function in the case of hysteresis damping.
10. Damping reduces the amplitude ratio for all values of the forcing frequency.
11. The unbalance in a rotating machine causes vibration.
12. The steady-state solution can be assumed to be harmonic for small values of dry-friction force.
13. In a system with rotational unbalance, the effect of damping becomes negligibly small at higher speeds.
14. The transfer function is a property of the system and is not related to the input.
15. The transfer functions of several different systems can be the same.
16. If the transfer function of a system is known, the system response can be found for all types of input.

3.3   Fill in each of the following blanks with the appropriate word:

1. The excitation can be _____, periodic, nonperiodic, or random in nature.
2. The response of a system to a harmonic excitation is called _____ response.
3. The response of a system to suddenly applied nonperiodic excitation is called _____ *response*.
4. When the frequency of excitation coincides with the natural frequency of the system, the condition is known as _____.
5. The magnification factor is also known as _____ *factor*.
6. The phenomenon of _____ can occur when the forcing frequency is close to the natural frequency of the system.
7. When the base of system is subject to harmonic motion with amplitude Y resulting in a response amplitude X, the ratio $\frac{X}{Y}$ is called the *displacement* _____.
8. $Z(i\omega) = -m\omega^2 + i\omega c + k$ is called the *mechanical* _____ of the system.
9. The difference between the frequencies associated with half-power points is called the _____ of the system.
10. The value of the amplitude ratio at resonance is called _____ *factor*.
11. The dry-friction damping is also known as _____ *damping*.
12. For _____ values of dry-friction damping, the motion of the mass will be discontinuous.
13. The quantity $k(1 + i\beta)$ in hysteresis damping is called _____ *stiffness*.
14. Quadratic or velocity-squared damping is present whenever a body moves in a _____ fluid flow.
15. In self-excited systems, the _____ itself produces the exciting force.

16. The flutter of turbine blades is an example of _____ vibration.

17. The motion _____ and the system becomes unstable during self-excitation.

18. The transfer function approach is based on _____ transform.

19. _____ identifies the input, system, and output clearly.

20. The Laplace transform of $f(t)$ is denoted as _____.

21. The Laplace transform converts a linear differential equation into an _____ expression.

3.4   Select the most appropriate answer out of the choices given:

1. The response of an undamped system under resonance will be
   a. very large   **b.** infinity   **c.** zero

2. The reduction of the amplitude ratio in the presence of damping is very significant
   a. near $\omega = \omega_n$ **b.** near $\omega = 0$ **c.** near $\omega = \infty$

3. The frequency of beating is
   a. $\omega_n - \omega$ **b.** ω_n **c.** ω

4. The energy dissipated in a cycle by dry-friction damping is given by
   a. $4\mu N X$ **b.** $4\mu N$ **c.** $4\mu N X^2$

5. The complex frequency response, $H(i\omega)$, is defined by

   a. $\dfrac{kX}{F_0}$ **b.** $\dfrac{X}{F_0}$ **c.** $\left|\dfrac{kX}{F_0}\right|$

6. The energy dissipated over the following duration is considered in finding the equivalent viscous-damping constant of a system with Coulomb damping:
   a. half cycle   **b.** full cycle   **c.** one second

7. The damping force depends on the frequency of the applied force in the case of
   a. viscous damping   **b.** Coulomb damping   **c.** hysteresis damping

8. The system governed by the equation $m\ddot{x} + c\dot{x} + kx = 0$ is dynamically stable if
   a. k is positive   **b.** c and k are positive **c.** c is positive

9. Complex stiffness or complex damping is defined in the case of
   a. hysteresis damping   **b.** Coulomb damping   **c.** viscous damping

10. The equation of motion of a machine (rotating at frequency ω) of mass M, with an unbalanced mass m, at radius e, is given by

   a. $m\ddot{x} + c\dot{x} + kx = me\omega^2 \sin\omega t$ **b.** $M\ddot{x} + c\dot{x} + kx = me\omega^2 \sin\omega t$
   c. $M\ddot{x} + c\dot{x} + kx = Me\omega^2 \sin\omega t$

11. The force transmissibility of a system, subjected to base excitation (with amplitude Y) resulting in a transmitted force F_T, is defined as

   a. $\dfrac{F_T}{kY}$ **b.** $\dfrac{X}{kY}$ **c.** $\dfrac{F_T}{k}$

3.5   Using the notation:

   $r = \text{frequency ratio} = \dfrac{\omega}{\omega_n}$

   $\omega = \text{forcing frequency}$

   $\omega_n = \text{natural frequency}$

   $\zeta = \text{damping ratio}$

   $\omega_1, \omega_2 = \text{frequencies corresponding to half-power points}$

match the items in the two columns below:

1. Magnification factor of an undamped system   a. $\dfrac{2\pi}{\omega_n - \omega}$

2. Period of beating   b. $\left[\dfrac{1 + (2\zeta r)^2}{(1 - r^2)^2 + (2\zeta r)^2}\right]^{1/2}$

3. Magnification factor of a damped system   c. $\dfrac{\omega_n}{\omega_2 - \omega_1}$

4. Damped frequency   d. $\dfrac{1}{1 - r^2}$

5. Quality factor   e. $\omega_n\sqrt{1 - \zeta^2}$

6. Displacement transmissibility   f. $\left[\dfrac{1}{(1 - r^2)^2 + (2\zeta r)^2}\right]^{1/2}$

3.6   Match the following equations of motion:

1. $m\ddot{z} + c\dot{z} + kz = -m\ddot{y}$ a. System with Coulomb damping
2. $M\ddot{x} + c\dot{x} + kx = me\omega^2 \sin \omega t$ b. System with viscous damping
3. $m\ddot{x} + kx \pm \mu N = F(t)$ c. System subject to base excitation
4. $m\ddot{x} + k(1 + i\beta)x = F_0 \sin \omega t$ d. System with hysteresis damping
5. $m\ddot{x} + c\dot{x} + kx = F_0 \sin \omega t$ e. System with rotating unbalance

PROBLEMS

Section 3.3 Response of an Undamped System Under Harmonic Force

3.1   A weight of 50 N is suspended from a spring of stiffness 4,000 N/m and is subjected to a harmonic force of amplitude 60 N and frequency 6 Hz. Find (a) the extension of the spring due to the suspended weight, (b) the static displacement of the spring due to the maximum applied force, and (c) the amplitude of forced motion of the weight.

3.2   A spring-mass system is subjected to a harmonic force whose frequency is close to the natural frequency of the system. If the forcing frequency is 39.8 Hz and the natural frequency is 40.0 Hz, determine the period of beating.

3.3   Consider a spring-mass system, with $k = 4,000$ N/m and $m = 10$ kg, subject to a harmonic force $F(t) = 400 \cos 10t$ N. Find and plot the total response of the system under the following initial conditions:

a. $x_0 = 0.1$ m, $\dot{x}_0 = 0$
b. $x_0 = 0, \dot{x}_0 = 10$ m/s
c. $x_0 = 0.1$ m, $\dot{x}_0 = 10$ m/s

3.4   Consider a spring-mass system, with $k = 4{,}000$ N/m and $m = 10$ kg, subject to a harmonic force $F(t) = 400 \cos 20t$ N. Find and plot the total response of the system under the following initial conditions:

  a. $x_0 = 0.1$ m, $\dot{x}_0 = 0$
  b. $x_0 = 0$, $\dot{x}_0 = 10$ m/s
  c. $x_0 = 0.1$ m, $\dot{x}_0 = 10$ m/s

3.5   Consider a spring-mass system, with $k = 4{,}000$ N/m and $m = 10$ kg, subject to a harmonic force $F(t) = 400 \cos 20.1t$ N. Find and plot the total response of the system under the following initial conditions:

  a. $x_0 = 0.1$ m, $\dot{x}_0 = 0$
  b. $x_0 = 0$, $\dot{x}_0 = 10$ m/s
  c. $x_0 = 0.1$ m, $\dot{x}_0 = 10$ m/s

3.6   Consider a spring-mass system, with $k = 4{,}000$ N/m and $m = 10$ kg, subject to a harmonic force $F(t) = 400 \cos 30t$ N. Find and plot the total response of the system under the following initial conditions:

  a. $x_0 = 0.1$ m, $\dot{x}_0 = 0$
  b. $x_0 = 0$, $\dot{x}_0 = 10$ m/s
  c. $x_0 = 0.1$ m, $\dot{x}_0 = 10$ m/s

3.7   A spring-mass system consists of a mass weighing 100 N and a spring with a stiffness of 2,000 N/m. The mass is subjected to resonance by a harmonic force $F(t) = 25 \cos \omega t$ N. Find the amplitude of the forced motion at the end of (a) $\frac{1}{4}$ cycle, (b) $2\frac{1}{2}$ cycles, and (c) $5\frac{3}{4}$ cycles.

3.8   A mass m is suspended from a spring of stiffness 4,000 N/m and is subjected to a harmonic force having an amplitude of 100 N and a frequency of 5 Hz. The amplitude of the forced motion of the mass is observed to be 20 mm. Find the value of m.

3.9   A spring-mass system with $m = 10$ kg and $k = 5{,}000$ N/m is subjected to a harmonic force of amplitude 250 N and frequency ω. If the maximum amplitude of the mass is observed to be 100 mm, find the value of ω.

3.10   In Fig. 3.1(a), a periodic force $F(t) = F_0 \cos \omega t$ is applied at a point on the spring that is located at a distance of 25 percent of its length from the fixed support. Assuming that $c = 0$, find the steady-state response of the mass m.

3.11   A spring-mass system, resting on an inclined plane, is subjected to a harmonic force as shown in Fig. 3.38. Find the response of the system by assuming zero initial conditions.

3.12   The natural frequency of vibration of a person is found to be 5.2 Hz while standing on a horizontal floor. Assuming damping to be negligible, determine the following:

  a. If the weight of the person is 70 kg_f, determine the equivalent stiffness of his body in the vertical direction.
  b. If the floor is subjected to a vertical harmonic vibration of frequency 5.3 Hz and amplitude 0.1 m due to an unbalanced rotating machine operating on the floor, determine the vertical displacement of the person.

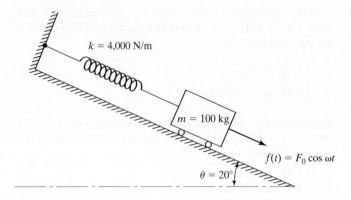

$k = 4{,}000$ N/m

$m = 100$ kg

$f(t) = F_0 \cos \omega t$

$\theta = 20°$

FIGURE 3.38

3.13 Plot the forced-vibration response of a spring-mass system given by Eq. (3.13) for the following sets of data:

a. Set 1: $\delta_{st} = 0.1$, $\omega = 5$, $\omega_n = 6$, $x_0 = 0.1$, $\dot{x}_0 = 0.5$
b. Set 2: $\delta_{st} = 0.1$, $\omega = 6.1$, $\omega_n = 6$, $x_0 = 0.1$, $\dot{x}_0 = 0.5$
c. Set 3: $\delta_{st} = 0.1$, $\omega = 5.9$, $\omega_n = 6$, $x_0 = 0.1$, $\dot{x}_0 = 0.5$

3.14 A spring-mass system is set to vibrate from zero initial conditions under a harmonic force. The response is found to exhibit the phenomenon of beats with the period of beating equal to 0.5 s and the period of oscillation equal to 0.05 s. Find the natural frequency of the system and the frequency of the harmonic force.

3.15 A spring-mass system, with $m = 100$ kg and $k = 400$ N/m, is subjected to a harmonic force $f(t) = F_0 \cos \omega t$ with $F_0 = 10$ N. Find the response of the system when ω is equal to (a) 2 rad/s, (b) 0.2 rad/s, and (c) 20 rad/s. Discuss the results.

3.16 An aircraft engine has a rotating unbalanced mass m at radius r. If the wing can be modeled as a cantilever beam of uniform cross section $a \times b$, as shown in Fig. 3.39(b), determine the maximum deflection of the engine at a speed of N rpm. Assume damping and effect of the wing between the engine and the free end to be negligible.

3.17 A three-bladed wind turbine (Fig. 3.40(a)) has a small unbalanced mass m located at a radius r in the plane of the blades. The blades are located from the central vertical (y) axis at a distance R and rotate at an angular velocity of ω. If the supporting truss can be modeled as a hollow steel shaft of outer diameter 0.1 m and inner diameter 0.08 m, determine the maximum stresses developed at the base of the support (point A). The mass moment of inertia of the turbine system about the vertical (y) axis is J_0. Assume $R = 0.5$ m, $m = 0.1$ kg, $r = 0.1$ m, $J_0 = 100$ kg-m^2, $h = 8$ m, and $\omega = 31.416$ rad/s.

(a)

(b)

FIGURE 3.39

(a)

(b)

FIGURE 3.40   Three-bladed wind turbine. (Photo courtesy of *Power Transmission Design*.)

3.18   An electromagnetic fatigue-testing machine is shown in Fig. 3.41 in which an alternating force is applied to the specimen by passing an alternating current of frequency f through the armature. If the mass of the armature is 18 kg, the stiffness of the spring (k_1) is 1,790 kN/m and the stiffness of the steel specimen is 130,000 kN/m, determine the frequency of the alternating current that induces a stress in the specimen that is twice the amount generated by the magnets.

FIGURE 3.41   Electromagnetic fatigue-testing machine.

3.19   The spring actuator shown in Fig. 3.42 operates by using the air pressure from a pneumatic controller (p) as input and providing an output displacement to a valve (x) proportional to the input air pressure. The diaphragm, made of a fabric-base rubber, has an area A and deflects under the input air pressure against a spring of stiffness k. Find the response of the valve under a harmonically fluctuating input air pressure $p(t) = p_0 \sin \omega t$ for the following data: $p_0 = 70$ kPa, $\omega = 8$ rad/s, $A = 650$ cm^2, $k = 70$ kN/m, mass of spring = 7.5 kg, and mass of valve and valve rod = 10 kg.

3.20   In the cam-follower system shown in Fig. 3.43, the rotation of the cam imparts a vertical motion to the follower. The pushrod, which acts as a spring, has been compressed by an amount x_0 before assembly. Determine the following: (a) equation of motion of the follower, including the gravitational force; (b) force exerted on the follower by the cam; and (c) conditions under which the follower loses contact with the cam.

FIGURE 3.42   A spring actuator.

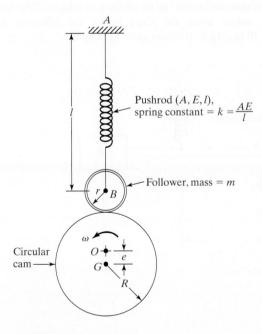

FIGURE 3.43

3.21* Design a solid steel shaft supported in bearings which carries the rotor of a turbine at the middle. The rotor has a mass of 225 kg and delivers a power of 150 kW at 3,000 rpm. In order to keep the stress due to the unbalance in the rotor small, the critical speed of the shaft is to be made one-fifth of the operating speed of the rotor. The length of the shaft is to be made equal to at least 30 times its diameter.

3.22 A hollow steel shaft, of length 250 cm, outer diameter 10 cm, and inner diameter 9 cm, carries the rotor of a turbine, having a mass of 200 kg, at the middle and is supported at the ends in bearings. The clearance between the rotor and the stator is 1 cm. The rotor has an eccentricity equivalent to a mass of 0.2 kg at a radius of 5 cm. A limit switch is installed to stop the rotor whenever the rotor touches the stator. If the rotor operates at resonance, how long will it take to activate the limit switch? Assume the initial displacement and velocity of the rotor perpendicular to the shaft to be zero.

3.23 A steel cantilever beam, carrying a mass of 0.05 kg at the free end, is used as a frequency meter.[7] The beam has a length of 25 cm, width of 0.5 cm, and thickness of 0.125 cm. The internal friction is equivalent to a damping ratio of 0.01. When the fixed end of the beam is subjected to a harmonic displacement $y(t) = 0.125 \cos \omega t$, the maximum tip displacement has been observed to be 6.25 cm. Find the forcing frequency.

3.24 Derive the equation of motion and find the steady-state response of the system shown in Fig. 3.44 for rotational motion about the hinge O for the following data: $k_1 = k_2 = 5,000$ N/m, $a = 0.25$ m, $b = 0.5$ m, $l = 1$ m, $M = 50$ kg, $m = 10$ kg, $F_0 = 500$ N, $\omega = 1,000$ rpm.

3.25 Derive the equation of motion and find the steady-state solution of the system shown in Fig. 3.45 for rotational motion about the hinge O for the following data: $k = 5,000$ N/m, $l = 1$ m, $m = 10$ kg, $M_0 = 100$ N-m, $\omega = 1,000$ rpm.

FIGURE 3.44

FIGURE 3.45

Section 3.4 Response of a Damped System under Harmonic Force

3.26 Consider a spring-mass-damper system with $k = 4,000$ N/m, $m = 10$ kg, and $c = 40$ N-s/m. Find the steady-state and total responses of the system under the harmonic force $F(t) = 200 \cos 10t$ N and the initial conditions $x_0 = 0.1$ m and $\dot{x}_0 = 0$.

3.27 Consider a spring-mass-damper system with $k = 4,000$ N/m, $m = 10$ kg, and $c = 40$ N-s/m. Find the steady-state and total responses of the system under the harmonic force $F(t) = 200 \cos 10t$ N and the initial conditions $x_0 = 0$ and $\dot{x}_0 = 10$ m/s.

3.28 Consider a spring-mass-damper system with $k = 4,000$ N/m, $m = 10$ kg, and $c = 40$ N-s/m. Find the steady-state and total responses of the system under the harmonic force $F(t) = 200 \cos 20t$ N and the initial conditions $x_0 = 0.1$ m and $\dot{x}_0 = 0$.

3.29 Consider a spring-mass-damper system with $k = 4,000$ N/m, $m = 10$ kg, and $c = 40$ N-s/m. Find the steady-state and total responses of the system under the harmonic force $F(t) = 200 \cos 20t$ N and the initial conditions $x_0 = 0$ and $\dot{x}_0 = 10$ m/s.

3.30 A four-cylinder automobile engine is to be supported on three shock mounts, as indicated in Fig. 3.46. The engine-block assembly has a mass of 225 kg. If the unbalanced force generated by the engine is given by $900 \sin 100\,\pi t$ N, design the three shock mounts (each of stiffness k and viscous-damping constant c) such that the amplitude of vibration is less than 0.25 cm.

3.31 The propeller of a ship, of weight 10^5 N and polar mass moment of inertia 10,000 kg-m^2, is connected to the engine through a hollow stepped steel propeller shaft, as shown in Fig. 3.47. Assuming that water provides a viscous damping ratio of 0.1, determine the torsional vibratory response of the propeller when the engine induces a harmonic angular displacement of $0.05 \sin 314.16t$ rad at the base (point A) of the propeller shaft.

3.32 Find the frequency ratio $r = \omega/\omega_n$ at which the amplitude of a single-degree-of-freedom damped system attains the maximum value. Also find the value of the maximum amplitude.

3.33 Figure 3.48 shows a permanent-magnet moving-coil ammeter. When current (I) flows through the coil wound on the core, the core rotates by an angle proportional to the magnitude of the current that is indicated by the pointer on a scale. The core, with the coil, has a mass moment of inertia J_0, the torsional spring constant is k_t, and the torsional damper has a

FIGURE 3.46   Four-cylinder automobile engine.

damping constant of c_t. The scale of the ammeter is calibrated such that when a direct current of magnitude 1 ampere is passed through the coil, the pointer indicates a current of 1 ampere. The meter has to be recalibrated for measuring the magnitude of alternating current. Determine the steady-state value of the current indicated by the pointer when an alternating current of magnitude 5 amperes and frequency 50 Hz is passed through the coil. Assume $J_0 = 0.001$ N-m^2, $k_t = 62.5$ N-m/rad, and $c_t = 0.5$ N-m-s/rad.

3.34   A spring-mass-damper system is subjected to a harmonic force. The amplitude is found to be 20 mm at resonance and 10 mm at a frequency 0.75 times the resonant frequency. Find the damping ratio of the system.

3.35   For the system shown in Fig. 3.49, x and y denote, respectively, the absolute displacements of the mass m and the end Q of the dashpot c_1. (a) Derive the equation of motion of the mass m, (b) find the steady-state displacement of the mass m, and (c) find the force transmitted to the support at P, when the end Q is subjected to the harmonic motion $y(t) = Y \cos \omega t$.

FIGURE 3.47   Propeller of a ship.

FIGURE 3.48   Permanent-magnet moving-coil ammeter.

FIGURE 3.49

3.36 The equation of motion of a spring-mass-damper system subjected to a harmonic force can be expressed as

$$\ddot{x} + 2\zeta\omega_n\dot{x} + \omega_n^2 = f_0 \cos \omega t \qquad (E.1)$$

where $f_0 = \dfrac{F_0}{m}$, $\omega_n = \sqrt{\dfrac{k}{m}}$, and $\zeta = c/(2m\omega_n)$.

  i.   Find the steady-state response of the system in the form $x_s(t) = C_1 \cos \omega t + C_2 \sin \omega t$
  ii. Find the total response of the system in the form

$$x(t) = x_h(t) + x_p(t) = A \cos \omega_d t + B \sin \omega_d t + C_1 \cos \omega t + C_2 \sin \omega t \qquad (E.2)$$

Assume the initial conditions of the system as $x\,(t = 0) = x_0$ and $\dot{x}(t = 0) = \dot{x}_0$.

3.37 A video camera, of mass 2.0 kg, is mounted on the top of a bank building for surveillance. The video camera is fixed at one end of a tubular aluminum rod whose other end is fixed to the building as shown in Fig. 3.50. The wind-induced force acting on the video camera, $f(t)$, is found to be harmonic with $f(t) = 25 \cos 75.3984t$ N. Determine the cross-sectional dimensions of the aluminum tube if the maximum amplitude of vibration of the video camera is to be limited to 0.005 m.

FIGURE 3.50

3.38   A turbine rotor is mounted on a stepped shaft that is fixed at both ends as shown in Fig. 3.51. The torsional stiffnesses of the two segments of the shaft are given by $k_{t1} = 3,000$ N-m/rad and $k_{t2} = 4,000$ N-m/rad. The turbine generates a harmonic torque given by $M(t) = M_0 \cos \omega t$ about the shaft axis with $M_0 = 200$ N-m and $\omega = 500$ rad/s. The mass moment of inertia of the rotor about the shaft axis is $J_0 = 0.05$ kg-m^2. Assuming the equivalent torsional damping constant of the system as $c_t = 2.5$ N-m-s/rad, determine the steady-state response of the rotor, $\theta(t)$.

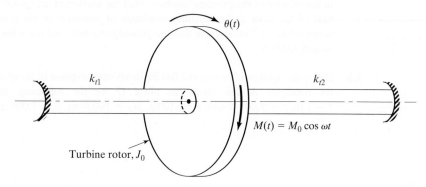

FIGURE 3.51

3.39   It is required to design an electromechanical system to achieve a natural frequency of 1,000 Hz and a Q-factor of 1,200. Determine the damping factor and the bandwidth of the system.

3.40   Show that, for small values of damping, the damping ratio ζ can be expressed as

$$\zeta = \frac{\omega_2 - \omega_1}{\omega_2 + \omega_1}$$

where ω_1 and ω_2 are the frequencies corresponding to the half-power points.

3.41   A torsional system consists of a disc of mass moment of inertia $J_0 = 10$ kg-m^2, a torsional damper of damping constant $c_t = 300$ N-m-s/rad, and a steel shaft of diameter 4 cm and length 1 m (fixed at one end and attached to the disc at the other end). A steady angular oscillation of amplitude 2° is observed when a harmonic torque of magnitude 1,000 N-m is applied to the disc. (a) Find the frequency of the applied torque, and (b) find the maximum torque transmitted to the support.

3.42   For a vibrating system, $m = 10$ kg, $k = 2,500$ N/m, and $c = 45$ N-s/m. A harmonic force of amplitude 180 N and frequency 3.5 Hz acts on the mass. If the initial displacement and velocity of the mass are 15 mm and 5 m/s, find the complete solution representing the motion of the mass.

3.43   The peak amplitude of a single-degree-of-freedom system, under a harmonic excitation, is observed to be 5.08 mm. If the undamped natural frequency of the system is 5 Hz, and the static deflection of the mass under the maximum force is 2.54 mm, (a) estimate the damping ratio of the system, and (b) find the frequencies corresponding to the amplitudes at half power.

3.44   The landing gear of an airplane can be idealized as the spring-mass-damper system shown in Fig. 3.52. If the runway surface is described as $y(t) = y_0 \cos \omega t$, determine the values of k and c that limit the amplitude of vibration of the airplane (x) to 0.1 m. Assume $m = 2{,}000$ kg, $y_0 = 0.2$ m, and $\omega = 157.08$ rad/s.

3.45   A precision grinding machine (Fig. 3.53) is supported on an isolator that has a stiffness of 1 MN/m and a viscous-damping constant of 1 kN-s/m. The floor on which the machine is mounted is subjected to a harmonic disturbance due to the operation of an unbalanced engine in the vicinity of the grinding machine. Find the maximum acceptable displacement amplitude of the floor if the resulting amplitude of vibration of the grinding wheel is to be restricted to 10^{-6} m. Assume that the grinding machine and the wheel are a rigid body of weight 5,000 N.

3.46   Derive the equation of motion and find the steady-state response of the system shown in Fig. 3.54 for rotational motion about the hinge O for the following data: $k = 5{,}000$ N/m, $l = 1$ m, $c = 1{,}000$ N-s/m, $m = 10$ kg, $M_0 = 100$ N-m, $\omega = 1{,}000$ rpm.

(a)

(b)

FIGURE 3.52   Modeling of landing gear.

FIGURE 3.53

FIGURE 3.54

3.47 An air compressor of mass 100 kg is mounted on an elastic foundation. It has been observed that, when a harmonic force of amplitude 100 N is applied to the compressor, the maximum steady-state displacement of 5 mm occurred at a frequency of 300 rpm. Determine the equivalent stiffness and damping constant of the foundation.

3.48 Find the steady-state response of the system shown in Fig. 3.55 for the following data: $k_1 = 1,000$ N/m, $k_2 = 500$ N/m, $c = 500$ N-s/m, $m = 10$ kg, $r = 5$ cm, $J_0 = 1$ kg-m^2, $F_0 = 50$ N, $\omega = 20$ rad/s.

3.49 A uniform slender bar of mass m may be supported in one of two ways as shown in Fig. 3.56. Determine the arrangement that results in a reduced steady-state response of the bar under a harmonic force, $F_0 \sin \omega t$, applied at the middle of the bar, as shown in the figure.

FIGURE 3.55

FIGURE 3.56

Section 3.5 Response of a Damped System Under $F(t) = F_0 e$

3.50   Derive the expression for the complex frequency response of an undamped torsional system.

3.51   A damped single-degree-of-freedom system, with parameters $m = 150$ kg, k $= 25$ kN/m, and $c = 2,000$ N-s/m, is subjected to the harmonic force $f(t) = 100 \cos 20t$ N. Find the amplitude and phase angle of the steady-state response of the system using a graphical method.

Section 3.6 Response of a System Under the Harmonic Motion of the Base

3.52   A single-story building frame is subjected to a harmonic ground acceleration, as shown in Fig. 3.57. Find the steady-state motion of the floor (mass m).

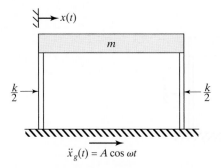

FIGURE 3.57

3.53 Find the horizontal displacement of the floor (mass m) of the building frame shown in Fig. 3.57 when the ground acceleration is given by $\ddot{x}_g = 100 \sin \omega t$ mm/s. Assume $m = 2{,}000$ kg, $k = 0.1$ MN/m, $\omega = 25$ rad/s, and $x_g(t = 0) = \dot{x}_g (t = 0) = x(t = 0) = \dot{x}(t = 0) = 0$.

3.54 If the ground in Fig. 3.57 is subjected to a horizontal harmonic displacement with frequency $\omega = 200$ rad/s and amplitude $X_g = 15$ mm, find the amplitude of vibration of the floor (mass m). Assume the mass of the floor as 2,000 kg and the stiffness of the columns as 0.5 MN/m.

3.55 An automobile is modeled as a single-degree-of-freedom system vibrating in the vertical direction. It is driven along a road whose elevation varies sinusoidally. The distance from peak to trough is 0.2 m and the distance along the road between the peaks is 35 m. If the natural frequency of the automobile is 2 Hz and the damping ratio of the shock absorbers is 0.15, determine the amplitude of vibration of the automobile at a speed of 60 km/h. If the speed of the automobile is varied, find the most unfavorable speed for the passengers.

3.56 Derive Eq. (3.74).

3.57 A single-story building frame is modeled by a rigid floor of mass m and columns of stiffness k, as shown in Fig. 3.58. It is proposed that a damper shown in the figure is attached to absorb vibrations due to a horizontal ground motion $y(t) = Y \cos \omega t$. Derive an expression for the damping constant of the damper that absorbs maximum power.

3.58 A uniform bar of mass m is pivoted at point O and supported at the ends by two springs, as shown in Fig. 3.59. End P of spring PQ is subjected to a sinusoidal displacement, $x(t) = x_0 \sin \omega t$. Find the steady-state angular displacement of the bar when $l = 1$ m, $k = 1{,}000$ N/m, $m = 10$ kg, $x_0 = 1$ cm, and $\omega = 10$ rad/s.

3.59 A uniform bar of mass m is pivoted at point O and supported at the ends by two springs, as shown in Fig. 3.60. End P of spring PQ is subjected to a sinusoidal displacement, $x(t) = x_0 \sin \omega t$. Find the steady-state angular displacement of the bar when $l = 1$ m, $k = 1{,}000$ N/m, $c = 500$ N-s/m, $m = 10$ kg, $x_0 = 1$ cm, and $\omega = 10$ rad/s.

FIGURE 3.58

FIGURE 3.59

FIGURE 3.60

3.60 Find the frequency ratio, $r = r_m$, at which the displacement transmissibility given by Eq. (3.68) attains a maximum value.

3.61 An automobile, having a mass of 500 kg empty and 1,500 kg fully loaded, vibrates in a vertical direction while traveling at 90 km/h on a rough road having a sinusoidal waveform with an amplitude Y m and a period 3.7 m. Assuming that the automobile can be modeled as a single-degree-of-freedom system with stiffness 450 kN/m and damping ratio $\zeta = 0.2$, determine the amplitude of vibration of the automobile when it is (a) empty and (b) fully loaded.

3.62 The base of a damped spring-mass system, with $m = 25$ kg and $k = 2,500$ N/m, is subjected to a harmonic excitation $y(t) = Y_0 \cos \omega t$. The amplitude of the mass is found to be 0.05 m when the base is excited at the natural frequency of the system with $Y_0 = 0.01$ m. Determine the damping constant of the system.

Section 3.7 Response of a Damped System Under Rotating Unbalance

3.63 A single-cylinder air compressor of mass 100 kg is mounted on rubber mounts, as shown in Fig. 3.61. The stiffness and damping constants of the rubber mounts are given by 10^6 N/m and 2,000 N-s/m, respectively. If the unbalance of the compressor is equivalent to a mass 0.1 kg located at the end of the crank (point A), determine the response of the compressor at a crank speed of 3,000 rpm. Assume $r = 10$ cm and $l = 40$ cm.

FIGURE 3.61

3.64   One of the tail rotor blades of a helicopter has an unbalanced mass of $m = 0.5$ kg at a distance of $e = 0.15$ m from the axis of rotation, as shown in Fig. 3.62. The tail section has a length of 4 m, a mass of 240 kg, a flexural stiffness (EI) of 2.5 MN-m^2, and a damping ratio of 0.15. The mass of the tail rotor blades, including their drive system, is 20 kg. Determine the forced response of the tail section when the blades rotate at 1,500 rpm.

FIGURE 3.62

3.65   When an exhaust fan of mass 380 kg is supported on springs with negligible damping, the resulting static deflection is found to be 45 mm. If the fan has a rotating unbalance of 0.15 kg-m, find (a) the amplitude of vibration at 1,750 rpm, and (b) the force transmitted to the ground at this speed.

3.66   A fixed-fixed steel beam, of length 5 m, width 0.5 m, and thickness 0.1 m, carries an electric motor of mass 75 kg and speed 1,200 rpm at its mid-span, as shown in Fig. 3.63. A rotating force of magnitude $F_0 = 5,000$ N is developed due to the unbalance in the rotor of the motor. Find the amplitude of steady-state vibrations by disregarding the mass of the beam. What will be the amplitude if the mass of the beam is considered?

FIGURE 3.63

3.67 If the electric motor of Problem 3.66 is to be mounted at the free end of a steel cantilever beam of length 5 m (Fig. 3.64), and the amplitude of vibration is to be limited to 0.5 cm, find the necessary cross-sectional dimensions of the beam. Include the weight of the beam in the computations.

FIGURE 3.64

3.68 A centrifugal pump, weighing 600 N and operating at 1,000 rpm, is mounted on six springs of stiffness 6,000 N/m each. Find the maximum permissible unbalance in order to limit the steady-state deflection to 5 mm peak-to-peak.

3.69* An air compressor, weighing 5,000 N and operating at 1,500 rpm, is to be mounted on a suitable isolator. A helical spring with a stiffness of 8×10^6 N/m, another helical spring with a stiffness of 2.5×10^6 N/m, and a shock absorber with a damping ratio of 0.15 are available for use. Select the best possible isolation system for the compressor.

3.70 A variable-speed electric motor, having an unbalance, is mounted on an isolator. As the speed of the motor is increased from zero, the amplitudes of vibration of the motor are observed to be 0.015 m at resonance and 0.004 m beyond resonance. Find the damping ratio of the isolator.

3.71 An electric motor weighing 3,500 N and running at 1,800 rpm is supported on four steel helical springs, each having eight active coils with a wire diameter of 0.0065 m and a coil diameter of 0.08 m. The rotor has a weight of 500 N with its center of mass located at a distance of 0.2 mm from the axis of rotation. Find the amplitude of vibration of the motor and the force transmitted through the springs to the base.

3.72 A small exhaust fan, rotating at 1,500 rpm, is mounted on a 5 mm steel shaft. The rotor of the fan weighs 100 N and has an eccentricity of 0.2 mm from the axis of rotation. Find (a) the maximum force transmitted to the bearings, and (b) the power needed to drive the shaft.

3.73 Derive Eq. (3.84) for the force transmitted to the foundation due to rotating unbalance.

3.74 A rigid plate, weighing 500 N, is hinged along an edge (*P*) and is supported on a dashpot with $c = 200$ N-s/m at the opposite edge (*Q*), as shown in Fig. 3.65. A small fan weighing 200 N and rotating at 750 rpm is mounted on the plate through a spring with $k = 40$ kN/m. If the center of gravity of the fan is located at 3 mm from its axis of rotation, find the steady-state motion of the edge *Q* and the force transmitted to the point *S*.

FIGURE 3.65

3.75 An electric motor is mounted at the end of a cantilever beam. The beam is observed to deflect by 0.02 m when the motor runs at a speed of 1,500 rpm. By neglecting the mass and damping of the beam, determine the speed of the motor so that the dynamic amplification is less than 10% about the static equilibrium value. Assume $m = 5$ kg and $\zeta = 30,000$ N/m.

3.76 An air compressor of mass 50 kg is mounted on an elastic support and operates at a speed of 1,000 rpm. It has an unbalanced mass of 2 kg at a radial distance (eccentricity) of 0.1 m from the axis of rotation. If the damping factor of the elastic support is $\zeta = 0.1$, determine the following: (a) the spring constant of the elastic support which transmits no more than 25% of the unbalanced force to the foundation, and (b) the magnitude of the force transmitted to the foundation.

3.77 A turbine rotor of mass 200 kg has an unbalanced mass of 15 kg. It is supported on a foundation which has an equivalent stiffness of 5,000 N/m and a damping ratio of $\zeta = 0.05$. If the rotor is found to vibrate with a deflection of 0.1 m at resonance, determine: (a) the radial location (eccentricity) of the unbalanced mass, (b) the additional mass to be added (uniformly) to the rotor if the deflection of the rotor at resonance is to be reduced to 0.05 m, and (c) the peak deflection of the turbine when the frequency ratio (r) varies.

Section 3.8 Forced Vibration with Coulomb Damping

3.78 Derive Eq. (3.99).

3.79 Derive the equation of motion of the mass m shown in Fig. 3.66 when the pressure in the cylinder fluctuates sinusoidally. The two springs with stiffnesses k_1 are initially under a tension of T_0, and the coefficient of friction between the mass and the contacting surfaces is μ.

3.80 The mass of a spring-mass system, with $m = 15$ kg and $k = 25$ kN/m, vibrates on a horizontal surface under a harmonic force of magnitude 200 N and frequency 20 Hz. Find the resulting amplitude of steady-state vibration. Assume the coefficient of friction between the mass and the horizontal surface as 0.25.

FIGURE 3.66

3.81   A spring-mass system with $m = 25$ kg and $k = 10$ kN/m vibrates on a horizontal surface with coefficient of friction $\mu = 0.3$. Under a harmonic force of frequency 8 Hz, the steady-state vibration of the mass is found to be 0.2 m. Determine the equivalent viscous-damping constant of the system.

3.82   A spring-mass system is subjected to Coulomb damping. When a harmonic force of amplitude 120 N and frequency 2.5173268 Hz is applied, the system is found to oscillate with an amplitude of 75 mm. Determine the coefficient of dry friction if $m = 2$ kg and $k = 2,100$ N/m.

Section 3.9 Forced Vibration with Hysteresis Damping

3.83   A load of 5,000 N resulted in a static displacement of 0.05 m in a composite structure. A harmonic force of amplitude 1,000 N is found to cause a resonant amplitude of 0.1 m. Find (a) the hysteresis-damping constant of the structure, (b) the energy dissipated per cycle at resonance, (c) the steady-state amplitude at one-quarter of the resonant frequency, and (d) the steady-state amplitude at thrice the resonant frequency.

3.84   The energy dissipated in hysteresis damping per cycle under harmonic excitation can be expressed in the general form

$$\Delta W = \pi \beta k X^{\gamma} \tag{E.1}$$

where γ is an exponent ($\gamma = 2$ was considered in Eq. (2.150)), and β is a coefficient of dimension $(\text{meter})^{2-\gamma}$. A spring-mass system having $k = 60$ kN/m vibrates under hysteresis damping. When excited harmonically at resonance, the steady-state amplitude is found to be 40 mm for an energy input of 3.8 N-m. When the resonant energy input is increased to 9.5 N-m, the amplitude is found to be 60 mm. Determine the values of β and γ in Eq. (E.1).

Section 3.10 Forced Motion with Other Types of Damping

3.85   When a spring-mass-damper system is subjected to a harmonic force $F(t) = 5 \cos 3\pi t$ 16, the resulting displacement is given by $x(t) = 0.5 \cos (3\pi t - \pi/3)$ in. Find the work done (a) during the first second, and (b) during the first 4 seconds.

3.86 Find the equivalent viscous-damping coefficient of a damper that offers a damping force of $F_d = c(\dot{x})^n$, where c and n are constants and $\dot{x}$ is the relative velocity across the damper. Also find the amplitude of vibration.

3.87 Show that for a system with both viscous and Coulomb damping the approximate value of the steady-state amplitude is given by

$$X^2\left[k^2(1 - r^2)^2 + c^2\omega^2\right] + X\frac{8\mu Nc\omega}{\pi} + \left(\frac{16\mu^2 N^2}{\pi^2} - F_0^2\right) = 0$$

3.88 The equation of motion of a spring-mass-damper system is given by

$$m\ddot{x} \pm \mu N + c\dot{x}^3 + kx = F_0 \cos \omega t$$

Derive expressions for (a) the equivalent viscous-damping constant, (b) the steady-state amplitude, and (c) the amplitude ratio at resonance.

Section 3.11 Self-Excitation and Stability Analysis

3.89 A fluid, with density ρ, flows through a cantilevered steel pipe of length l and cross-sectional area A (Fig. 3.67). Determine the velocity (v) of the fluid at which instability occurs. Assume that the total mass and the bending stiffness of the pipe are m and EI, respectively.

FIGURE 3.67

3.90 The first two natural frequencies of the telescoping car antenna shown in Fig. 3.68 are given by 3.0 Hz and 7.0 Hz. Determine whether the vortex shedding around the antenna causes instability over the speed range 80–120 km/h of the automobile.

3.91 The signpost of a fast food restaurant consists of a hollow steel cylinder of height h, inside diameter d, and outside diameter D, fixed to the ground and carries a concentrated mass M at the top. It can be modeled as a single-degree-of-freedom spring-mass-damper system with an equivalent viscous-damping ratio of 0.1 for analyzing its transverse vibration characteristics under wind excitation. Determine the following: (a) the natural frequency of transverse vibration of the signpost; (b) the wind velocity at which the signpost undergoes maximum steady-state displacement; and (c) the maximum wind-induced steady-state displacement of the signpost. Data: $h = 10$ m, $D = 25$ cm, $d = 20$ cm, $M = 200$ kg.

2 mm → ← 0.2 m

5 mm → ← 0.2 m

8 mm → ← 0.2 m

FIGURE 3.68

3.92 Consider the equation of motion of a single-degree-of-freedom system:

$$m\ddot{x} + c\dot{x} + kx = F$$

Derive the condition that leads to divergent oscillations in each of the following cases: (a) when the forcing function is proportional to the displacement, $F(t) = F_0 x(t)$; (b) when the forcing function is proportional to the velocity, $F(t) = F_0 \dot{x}(t)$; and (c) when the forcing function is proportional to the acceleration, $F(t) = F_0 \ddot{x}(t)$.

Section 3.12 Transfer-Function Approach

3.93 Derive the transfer function of a viscously damped system subject to a harmonic base motion, with the equation of motion:

$$m\ddot{x} + c(\dot{x} - \dot{y}) + k(x - y) = 0$$

where $y(t) = Y \sin \omega t$.

3.94 Derive the transfer function of a viscously damped system under rotating unbalance, with the equation of motion:

$$M\ddot{x} + c\dot{x} + kx = me\omega^2 \sin \omega t$$

Section 3.13 Solutions Using Laplace Transforms

3.95 Find the steady-state response of a damped single-degree-of-freedom system subjected to a harmonic base motion, considered in Section 3.6, using the Laplace transform.

3.96 Find the steady-state response of a damped single-degree-of-freedom system under rotating unbalance, considered in Section 3.7, using the Laplace transform.

3.97 Find the steady-state response of an undamped single-degree-of-freedom system subjected to a harmonic force, considered in Section 3.3, using the Laplace transform.

3.98 A spring and a viscous damper, connected to a massless rigid bar, are subjected to a harmonic force $f(t)$ as shown in Fig. 3.69. Find the steady-state response of the system using the Laplace transform.

$F = 200$ N $x(t)$

k c

FIGURE 3.69

3.99 Derive Eqs. (E.4)–(E.7) in Example 3.16.

3.100 An experiment is conducted to find the dynamic response characteristics of an automobile wheel assembly system. For this, the wheel is connected to a shaft through a tie rod and is subjected to a harmonic force $f(t)$ as shown in Fig. 3.70. The shaft offers a torsional stiffness of k_t while the wheel undergoes torsional vibration about the axis of the shaft. Find the response of the system, $\theta(t)$, using the Laplace transform. Assume the initial conditions to be zero.

Section 3.14 Frequency Transfer Functions

3.101 Generate the frequency transfer function from the general transfer function derived for a viscously damped system subject to a harmonic base motion considered in Problem 3.93 and identify the input, system, and output sinusoids.

3.102 Generate the frequency transfer function from the general transfer function derived for a viscously damped system under rotating unbalance considered in Problem 3.94 and identify the input, system, and output sinusoids.

Section 3.15 Examples Using MATLAB

3.103 Plot the forced response of an undamped spring-mass system under the following conditions using MATLAB: $m = 10$ kg, $k = 4{,}000$ N/m, $F(t) = 200 \cos 10t$ N, $x_0 = 0.1$ m, $\dot{x}_0 = 10$ m/s.

3.104 Plot the forced response of a spring-mass system subject to Coulomb damping using MATLAB. Assume the following data: $m = 10$ kg, $k = 4{,}000$ N/m, $F(t) = 200 \sin 10t$ N, $\mu = 0.3$, $x_0 = 0.1$ m, $\dot{x}_0 = 10$ m/s.

3.105 Plot the response of a viscously damped system under harmonic base excitation, $y(t) = Y \sin \omega t$ m, using MATLAB for the following data: $m = 100$ kg, $k = 4 \times 10^4$ N/m, $\zeta = 0.25$, $Y = 0.05$ m, $\omega = 10$ rad/s, $x_0 = 1$ m, $\dot{x}_0 = 0$.

3.106 Plot the steady-state response of a viscously damped system under the harmonic force $F(t) = F_0 \cos \omega t$ using MATLAB. Assume the following data: $m = 10$ kg, $k = 1{,}000$ N/m, $\zeta = 0.1$, $F_0 = 100$ N, $\omega = 20$ rad/s.

Shaft, diameter: 0.05 m

1 m

$f(t) = 500 \sin 30t$ N

Wheel

$\theta(t)$

Tie rod

0.25 m

FIGURE 3.70

3.107 Consider an automobile traveling over a rough road at a speed of v km/h. The suspension system has a spring constant of 40 kN/m and a damping ratio of $\zeta = 0.1$. The road surface varies sinusoidally with an amplitude of $Y = 0.05$ m and a wavelength of 6 m. Write a MATLAB program to find the displacement amplitude of the automobile for the following conditions: (a) mass of the automobile = 600 kg (empty), 1,000 kg (loaded), (b) velocity of the automobile (v) = 10 km/h, 50 km/h, 100 km/h.

3.108 Write a computer program for finding the total response of a spring-mass-viscous-damper system subjected to base excitation. Use this program to find the solution of a problem with $m = 2$ kg, $c = 10$ N-s/m, $k = 100$ N/m, $y(t) = 0.1 \sin 25t$ m, $x_0 = 10$ mm, and $\dot{x}_0 = 5$ m/s.

3.109 Plot the graphs of $\dfrac{MX}{me}$ versus r and ϕ versus r for a damped system under rotating unbalance (Eq. (3.81)) for the damping ratios $\zeta = 0, 0.2, 0.4, 0.6, 0.8,$ and 1 using MATLAB.

3.110 Plot the graphs of $\dfrac{X}{Y}$ versus r and ϕ versus r for a damped system subjected to base excitation (Eqs. (3.68) and (3.69)) for the damping ratios $\zeta = 0, 0.2, 0.4, 0.6, 0.8,$ and 1 using MATLAB.

DESIGN PROJECTS

3.111 The arrangement shown in Fig. 3.71 consists of two eccentric masses rotating in opposite directions at the same speed ω. It is to be used as a mechanical shaker over the frequency range 20 to 30 Hz. Find the values of ω, e, M, m, k, and c to satisfy the following requirements: (a) The mean power output of the shaker should be at least 700 W over the specified frequency range. (b) The amplitude of vibration of the masses should be between 2 and 5 mm. (c) The mass of the shaker (M) should be at least 50 times that of the eccentric mass (m).

3.112 Design a minimum-weight, hollow circular steel column for the water tank shown in Fig. 3.72. The weight of the tank (W) is 500 kN and the height is 20 m. The stress induced in the column should not exceed the yield strength of the material, which is 200 GPa, when subjected to a harmonic ground acceleration (due to an earthquake) of amplitude $0.5g$ and frequency 15 Hz. In addition, the natural frequency of the water tank should be greater than 15 Hz. Assume a damping ratio of 0.15 for the column.

FIGURE 3.71

FIGURE 3.72

Jean Baptiste Joseph Fourier (1768–1830) was a French mathematician and a professor at the Ecole Polytechnique in Paris. His works on heat flow, published in 1822, and on trigonometric series are well known. The expansion of a periodic function in terms of harmonic functions has been named after him as the "Fourier series."
(Reproduced with permission from *Applied Mechanics Reviews.*)

C H A P T E R   4

Vibration Under General Forcing Conditions

Chapter Outline

This chapter is devoted to the vibration response of a single-degree-of-freedom system under arbitrary forcing conditions. The response of the system under a general periodic force is presented by first expanding the periodic force into a series of harmonic forces using the Fourier series and then superposing the responses due to the individual harmonic forces. The response of the system under a nonperiodic force is presented using two

363

methods—those of the convolution integral and of the Laplace transform. The method of convolution or Duhamel integral makes use of the impulse response function of the system. The method is also used to find the response to base excitation. Several examples are presented to illustrate its use. The concept of response spectra corresponding to specific forcing functions and their use in finding the maximum response of the system is also outlined. The response spectrum corresponding to the base excitation, such as the one caused by an earthquake, is also considered. Typical earthquake response spectra and their use in finding the responses of building frames are illustrated. The concept of pseudo velocity and the associated pseudo spectrum are also defined. The design of mechanical systems under a shock environment is presented with an illustrative example. The Laplace transform method and its use in finding the response of both first- and second-order systems are presented. The responses under impulse, step, and ramp forcing functions are considered. Inelastic and elastic collision problems are considered as applications of impulse response computations. The analysis of the step response and the description of the transient response in terms of peak time, rise time, maximum overshoot, settling time, and delay time are presented. The response of systems under irregular forcing conditions using numerical methods, including the fourth-order Runge-Kutta method, is presented with illustrative examples. Finally, the use of MATLAB programs in finding the response of a system under arbitrary forcing functions is illustrated with examples.

Learning Objectives

After completing this chapter, you should be able to do the following:

- Find the responses of single-degree-of-freedom systems subjected to general periodic forces using the Fourier series.
- Use the method of convolution or Duhamel integral to solve vibration problems of systems subjected to arbitrary forces.
- Find the response of systems subjected to earthquakes using response spectra.
- Solve undamped and damped systems subjected to arbitrary forces, including impulse, step, and ramp forces, using the Laplace transform.
- Understand the characteristics of transient response, such as peak time, overshoot, settling time, rise time, and decay time, and procedures for their estimation.
- Apply numerical methods to solve vibration problems of systems subjected to forces that are described numerically.
- Solve forced-vibration problems using MATLAB.

4.1   Introduction

In Chapter 3, we considered the response of single-degree-of-freedom systems subjected to harmonic excitation. However, many practical systems are subjected to several types of forcing functions that are not harmonic. The general forcing functions may be periodic (nonharmonic) or nonperiodic. The nonperiodic forces include forces such as a suddenly applied constant force (called a *step force*), a linearly increasing force (called a *ramp force*), and an exponentially varying force. A nonperiodic forcing function may be acting for a short, long, or infinite duration. A forcing function or excitation of short duration compared to the natural time period of the system is called a *shock*. Examples of general forcing functions are the motion imparted by a cam to the follower, the vibration felt by an instrument when its package is dropped from a height, the force applied to the foundation of a forging press, the motion of an automobile when it hits a pothole, and the ground vibration of a building frame during an earthquake.

If the forcing function is periodic but not harmonic, it can be replaced by a sum of harmonic functions using the harmonic analysis procedure discussed in Section 1.11. Using the principle of superposition, the response of the system can then be determined by superposing the responses due to the individual harmonic forcing functions.

The response of a system subjected to any type of nonperiodic force is commonly found using the following methods:

1. Convolution integral.
2. Laplace transform.
3. Numerical methods.

The first two methods are analytical ones, in which the response or solution is expressed in a way that helps in studying the behavior of the system under the applied force with respect to various parameters and in designing the system. The third method, on the other hand, can be used to find the response of a system under any arbitrary force for which an analytical solution is difficult or impossible to find. However, the solution found is applicable only for the particular set of parameter values used in finding the solution. This makes it difficult to study the behavior of the system when the parameters are varied. This chapter presents all three methods of solution.

4.2   Response Under a General Periodic Force

When the external force $F(t)$ is periodic with period $\tau = 2\pi/\omega$, it can be expanded in a Fourier series (see Section 1.11):

$$F(t) = \frac{a_0}{2} + \sum_{j=1}^{\infty} a_j \cos j\omega t + \sum_{j=1}^{\infty} b_j \sin j\omega t \tag{4.1}$$

where

$$a_j = \frac{2}{\tau} \int_0^{\tau} F(t) \cos j\omega t \, dt, \qquad j = 0, 1, 2, \ldots \tag{4.2}$$

and

$$b_j = \frac{2}{\tau} \int_0^{\tau} F(t) \sin j\omega t \, dt, \qquad j = 1, 2, \ldots \tag{4.3}$$

The response of systems under general periodic forces is considered in this section for both first- and second-order systems. First-order systems are those for which the equation of

FIGURE 4.1   Examples of first-order systems.

motion is a first-order differential equation. Similarly, second-order systems are those for which the equation of motion is a second-order differential equation. Typical examples of first- and second-order systems are shown in Figs. 4.1 and 4.2, respectively.

4.2.1 First-Order Systems

Consider a spring-damper system subjected to a periodic excitation as shown in Fig. 4.1(a). The equation of motion of the system is given by

$$c\dot{x} + k(x - y) = 0 \tag{4.4}$$

where $y(t)$ is the periodic motion (or excitation) imparted to the system at point A (for example, by a cam). If the periodic displacement of point A, $y(t)$, is expressed in Fourier series as indicated by the right-hand side of Eq. (4.1), the equation of motion of the system can be expressed as

$$\dot{x} + ax = ay = A_0 + \sum_{j=1}^{\infty} A_j \sin \omega_j t + \sum_{j=1}^{\infty} B_j \cos \omega_j t \tag{4.5}$$

FIGURE 4.2   Examples of second-order systems.

where

$$a = \frac{k}{c}, \quad A_0 = \frac{aa_0}{2}, \quad A_j = aa_j, \quad B_j = ab_j, \quad \omega_j = j\omega, \quad j = 1, 2, 3, \ldots \text{(4.6)}$$

The solution of Eq. (4.5) is presented in Example 4.1.

Response of a First-Order System under Periodic Force

EXAMPLE 4.1

Find the response of the spring-damper system shown in Fig. 4.1(a) subjected to a periodic force with the equation of motion given by Eq. (4.5).

Solution: It can be seen that the right-hand side of the equation of motion, Eq. (4.5), is a constant plus a linear sum of harmonic (sine and cosine) functions. Using the principle of superposition, the steady-state solution of Eq. (4.5) can be found by summing the steady-state solutions corresponding to the individual forcing terms on the right-hand side of Eq. (4.5).

The equation of motion corresponding to the constant force A_0 can be expressed, using x_0 for x, as

$$\dot{x}_0 + ax_0 = A_0 \tag{E.1}$$

The solution of Eq. (E.1) is given by (can be verified by substituting in Eq. (E.1)):

$$x_0(t) = \frac{A_0}{a} \tag{E.2}$$

The equation of motion under the force $A_j \sin \omega_j t$ can be expressed as

$$\dot{x}_j + ax_j = A_j \sin \omega_j t \tag{E.3}$$

in which the steady-state solution of Eq. (E.3) can be assumed in the form

$$x_j(t) = X_j \sin(\omega_j t - \phi_j) \tag{E.4}$$

where the magnitude X_j and the phase angle ϕ_j denote the unknown constants to be determined. The solution in Eq. (E.4) can be expressed as the imaginary part of the following solution in complex form:

$$x_j(t) = \text{Im} \lfloor X_j e^{i(\omega_j t - \phi_j)} \rfloor = X_j e^{i\omega_j t} e^{-i\phi_j} = U_j e^{i\omega_j t} \tag{E.5}$$

where U_j denotes the complex number:

$$U_j = X_j e^{-i\phi_j} \tag{E.6}$$

Noting that the time derivative of $x_j(t)$ is given by

$$\dot{x}_j(t) = i\omega_j U_j e^{i\omega_j t} \tag{E.7}$$

Eq. (E.3) can be expressed with the forcing term in complex form (with the understanding that we are interested only in the imaginary part of the solution):

$$\dot{x}_j + ax_j = A_j e^{i\omega_j t} = A_j(\cos \omega_j t + i \sin \omega_j t) \tag{E.8}$$

By inserting Eqs. (E.5) and (E.7) into Eq. (E.8), we obtain

$$i\omega_j U_j e^{i\omega_j t} + aU_j e^{i\omega_j t} = A_j e^{i\omega_j t} \tag{E.9}$$

Since $e^{i\omega_j t} \neq 0$, Eq. (E.9) can be reduced to

$$i\omega_j U_j + aU_j = A_j \tag{E.10}$$

or

$$U_j = \frac{A_j}{a + i\omega_j} \tag{E.11}$$

Equations (E.6) and (E.11) yield

$$U_j = X_j e^{-i\phi_j} = \frac{A_j}{a + i\omega_j} \tag{E.12}$$

By expressing $\dfrac{1}{a + i\omega_j}$ as

$$\frac{1}{a + i\omega_j} = \frac{a - i\omega_j}{(a + i\omega_j)(a - i\omega_j)} = \frac{1}{\sqrt{a^2 + \omega_j^2}}\left[\frac{a}{\sqrt{a^2 + \omega_j^2}} - i\frac{\omega_j}{\sqrt{a^2 + \omega_j^2}}\right] \tag{E.13}$$

Eq. (E.13) can be rewritten as

$$\frac{1}{a + i\omega_j} = \frac{1}{\sqrt{a^2 + \omega_j^2}}[\cos\phi_j - i\sin\phi_j] = \frac{1}{\sqrt{a^2 + \omega_j^2}}e^{-i\phi_j} \tag{E.14}$$

where

$$\phi_j = \tan^{-1}\left(\frac{\omega_j}{a}\right) \tag{E.15}$$

By using Eq. (E.14) in Eq. (E.12), we find that

$$X_j = \frac{A_j}{\sqrt{a^2 + \omega_j^2}}, \qquad \phi_j = \tan^{-1}\left(\frac{\omega_j}{a}\right) \tag{E.16}$$

The solution of Eq. (E.3) is thus given by Eq. (E.4) with X_j and ϕ_j given by Eq. (E.16). The equation of motion under the force $B_j \cos\omega_j t$ can be expressed as

$$\dot{x}_j + ax_j = B_j \cos\omega_j t \tag{E.17}$$

By assuming the steady-state solution of Eq. (E.17) in the form

$$x_j(t) = Y_j \cos(\omega_j t - \phi_j) \tag{E.18}$$

the constants Y_j and ϕ_j can be determined, by proceeding as in the case of the solution of Eq. (E.3), as

$$Y_j = \frac{B_j}{\sqrt{a^2 + \omega_j^2}}, \qquad \phi_j = \tan^{-1}\left(\frac{\omega_j}{a}\right) \tag{E.19}$$

The complete steady-state (or particular) solution of Eq. (4.5) can be expressed as

$$x_p(t) = \frac{A_0}{a} + \sum_{j=1}^{\infty} \frac{A_j}{\sqrt{a^2 + \omega_j^2}} \sin\left\{\omega_j t - \tan^{-1}\left(\frac{\omega_j}{a}\right)\right\}$$

$$+ \sum_{j=1}^{\infty} \frac{B_j}{\sqrt{a^2 + \omega_j^2}} \cos\left\{\omega_j t - \tan^{-1}\left(\frac{\omega_j}{a}\right)\right\} \tag{E.20}$$

where a, A_0, A_j, B_j, and ω_j are given by Eq. (4.6).

Note: The total solution of Eq. (4.5) is given by the sum of the homogeneous and particular (or steady-state) solutions:

$$x(t) = x_h(t) + x_p(t) \tag{E.21}$$

where the particular solution is given by Eq. (E.20) and the homogeneous solution of Eq. (4.5) can be expressed as

$$x_h(t) = Ce^{-at} \tag{E.22}$$

where C is an unknown constant to be determined using the initial condition of the system. The total solution can be expressed as

$$x(t) = Ce^{-at} + \frac{A_0}{a} + \sum_{j=1}^{\infty} X_j \sin(\omega_j t - \phi_j) + \sum_{j=1}^{\infty} Y_j \cos(\omega_j t - \phi_j) \tag{E.23}$$

When the initial condition $x(t = 0) = x_0$ is used in Eq. (E.23), we obtain

$$x_0 = C + \frac{A_0}{a} - \sum_{j=1}^{\infty} X_j \sin \phi_j + \sum_{j=1}^{\infty} Y_i \cos \phi_j \tag{E.24}$$

which yields

$$C = x_0 - \frac{A_0}{a} + \sum_{j=1}^{\infty} X_j \sin \phi_j - \sum_{j=1}^{\infty} Y_j \cos \phi_j \tag{E.25}$$

Thus the total solution of Eq. (4.5) becomes

$$x(t) = \left[x_0 - \frac{A_0}{a} + \sum_{j=1}^{\infty} X_j \sin \phi_j - \sum_{j=1}^{\infty} Y_j \cos \phi_j \right] e^{-at}$$

$$+ \frac{A_0}{a} + \sum_{j=1}^{\infty} X_j \sin(\omega_d t - \phi_j) + \sum_{j=1}^{\infty} Y_j \cos(\omega_d t - \phi_j) \tag{E.26}$$

The features of the response of the system can be studied by considering a simpler type of forcing function through the following example.

∎

Response of a First-Order System

EXAMPLE 4.2

Determine the response of a spring-damper system, similar to the one shown in Fig. 4.1(a), with the equation of motion:

$$\dot{x} + 1.5x = 7.5 + 4.5 \cos t + 3 \sin 5t$$

Assume the initial condition as $x(t = 0) = 0$.

Solution: The equation of motion of the system is given by

$$\dot{x} + 1.5x = 7.5 + 4.5 \cos t + 3 \sin 5t \tag{E.1}$$

We first find the solution of the differential equation by considering one forcing term at a time given on the right-hand side of Eq. (E.1) and then adding the solutions to find the total solution of Eq. (E.1). For the constant term, the equation to be solved is

$$\dot{x} + 1.5x = 7.5 \tag{E.2}$$

The solution of Eq. (E.2) is $x(t) = 7.5/1.5 = 5$. For the cosine term, the equation to be solved is given by

$$\dot{x} + 1.5x = 4.5 \cos t \tag{E.3}$$

Using the steady-state solution indicated in Eq. (E.21) of Example 4.1, we can express the solution of Eq. (E.3) as

$$x(t) = Y \cos(t - \phi) \tag{E.4}$$

where

$$Y = \frac{4.5}{\sqrt{(1.5)^2 + (1)^2}} = \frac{4.5}{\sqrt{3.25}} = 2.4961 \tag{E.5}$$

and

$$\phi = \tan^{-1}\left(\frac{1}{1.5}\right) = 0.5880 \text{ rad} \tag{E.6}$$

Similarly, for the sine term, the equation to be solved is

$$\dot{x} + 1.5x = 3 \sin 5t \tag{E.7}$$

Using the steady-state solution indicated in Eq. (E.4) of Example 4.1, we can express the solution of Eq. (E.7) as

$$x(t) = X \sin(5t - \phi) \tag{E.8}$$

where

$$X = \frac{3}{\sqrt{(1.5)^2 + (5)^2}} = \frac{3}{\sqrt{27.25}} = 0.5747 \tag{E.9}$$

and

$$\phi = \tan^{-1}\left(\frac{5}{1.5}\right) = 1.2793 \text{ rad} \tag{E.10}$$

Thus the total particular solution of Eq. (E.1) is given by the sum of the solutions of Eqs. (E.2), (E.3), and (E.7):

$$x(t) = 5 + 2.4961 \cos(t - 0.5880) + 0.5747 \sin(5t - 1.2793) \tag{E.11}$$

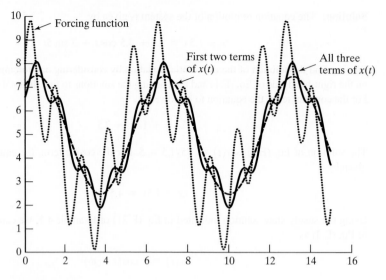

FIGURE 4.3

The forcing function given by the right-hand-side expression in Eq. (E.1) and the steady-state response of the system given by Eq. (E.11) are shown graphically in Fig. 4.3. The first two terms of the response (given by the first two terms on the right-hand side of Eq. (E.11)) are also shown in Fig. 4.3. It can be seen that the system does not filter the constant term. However, it filters the lower-frequency (cosine) term to some extent and the higher-frequency (sine) term to a larger extent.

■

4.2.2 Second-Order Systems

Let a spring-mass-damper system, Fig. 4.2(a), be subjected to a periodic force. This is a second-order system because the governing equation is a second-order differential equation:

$$m\ddot{x} + c\dot{x} + kx = f(t) \tag{4.7}$$

If the forcing function $f(t)$ is periodic, it can be expressed in a Fourier series so that the equation of motion becomes

$$m\ddot{x} + c\dot{x} + kx = F(t) = \frac{a_0}{2} + \sum_{i=1}^{\infty} a_j \cos j\omega t + \sum_{j=1}^{\infty} b_j \sin j\omega t \tag{4.8}$$

The determination of the solution of Eq. (4.8) is illustrated in Example 4.3.

EXAMPLE   4.3

Response of a Second-Order System Under Periodic Force

Determine the response of a spring-mass-damper system subjected to a periodic force with the equation of motion given by Eq. (4.8). Assume the initial conditions as zero.

Solution: The right-hand side of Eq. (4.8) is a constant plus a sum of harmonic functions. Using the principle of superposition, the steady-state solution of Eq. (4.4) is the sum of the steady-state solutions of the following equations:

$$m\ddot{x} + c\dot{x} + kx = \frac{a_0}{2} \tag{E.1}$$

$$m\ddot{x} + c\dot{x} + kx = a_j \cos j\omega t \tag{E.2}$$

$$m\ddot{x} + c\dot{x} + kx = b_j \sin j\omega t \tag{E.3}$$

Noting that the solution of Eq. (E.1) is given by

$$x_p(t) = \frac{a_0}{2k} \tag{E.4}$$

and, using the results of Section 3.4, we can express the solutions of Eqs. (E.2) and (E.3), respectively, as

$$x_p(t) = \frac{(a_j/k)}{\sqrt{(1 - j^2 r^2)^2 + (2\zeta j r)^2}} \cos(j\omega t - \phi_j) \tag{E.5}$$

$$x_p(t) = \frac{(b_j/k)}{\sqrt{(1 - j^2 r^2)^2 + (2\zeta j r)^2}} \sin(j\omega t - \phi_j) \tag{E.6}$$

where

$$\phi_j = \tan^{-1}\left(\frac{2\zeta j r}{1 - j^2 r^2}\right) \tag{E.7}$$

and

$$r = \frac{\omega}{\omega_n} \tag{E.8}$$

Thus the complete steady-state solution of Eq. (4.8) is given by

$$x_p(t) = \frac{a_0}{2k} + \sum_{j=1}^{\infty} \frac{(a_j/k)}{\sqrt{(1 - j^2 r^2)^2 + (2\zeta j r)^2}} \cos(j\omega t - \phi_j)$$

$$+ \sum_{j=1}^{\infty} \frac{(b_j/k)}{\sqrt{(1 - j^2 r^2)^2 + (2\zeta j r)^2}} \sin(j\omega t - \phi_j) \tag{E.9}$$

It can be seen from the solution, Eq. (E.9), that the amplitude and phase shift corresponding to the jth term depend on j. If $j\omega = \omega_n$, for any j, the amplitude of the corresponding harmonic will be comparatively large. This will be particularly true for small values of j and ζ. Further, as j becomes larger, the amplitude becomes smaller and the corresponding terms tend to zero. Thus the first few terms are usually sufficient to obtain the response with reasonable accuracy.

The solution given by Eq. (E.9) denotes the steady-state response of the system. The transient part of the solution arising from the initial conditions can also be included to find the complete solution. To find the complete solution, we need to evaluate the arbitrary constants by setting the value of the complete solution and its derivative to the specified values of initial displacement $x(0)$ and the initial velocity $\dot{x}(0)$. This results in a complicated expression for the transient part of the total solution.

∎

EXAMPLE 4.4

Periodic Vibration of a Hydraulic Valve

In the study of vibrations of valves used in hydraulic control systems, the valve and its elastic stem are modeled as a damped spring-mass system, as shown in Fig. 4.4(a). In addition to the spring force and damping force, there is a fluid-pressure force on the valve that changes with the amount of opening or closing of the valve. Find the steady-state response of the valve when the pressure in the chamber varies as indicated in Fig. 4.4(b). Assume $k = 2{,}500$ N/m, $c = 10$ N-s/m, and $m = 0.25$ kg.

Solution: The valve can be considered as a mass connected to a spring and a damper on one side and subjected to a forcing function $F(t)$ on the other side. The forcing function can be expressed as

$$F(t) = Ap(t) \tag{E.1}$$

where A is the cross-sectional area of the chamber, given by

$$A = \frac{\pi(50)^2}{4} = 625\pi \text{ mm}^2 = 0.000625\pi \text{ m}^2 \tag{E.2}$$

and $p(t)$ is the pressure acting on the valve at any instant t. Since $p(t)$ is periodic with period $\tau = 2$ s and A is a constant, $F(t)$ is also a periodic function of period $\tau = 2$ s. The

FIGURE 4.4   Periodic vibration of a hydraulic valve.

frequency of the forcing function is $\omega = (2\pi/\tau) = \pi$ rad/s. $F(t)$ can be expressed in a Fourier series as

$$F(t) = \frac{a_0}{2} + a_1 \cos \omega t + a_2 \cos 2\omega t + \cdots$$
$$+ b_1 \sin \omega t + b_2 \sin 2\omega t + \cdots \tag{E.3}$$

where a_j and b_j are given by Eqs. (4.2) and (4.3). Since the function $F(t)$ is given by

$$F(t) = \begin{cases} 50{,}000At & \text{for } 0 \leq t \leq \dfrac{\tau}{2} \\[2mm] 50{,}000A(2 - t) & \text{for } \dfrac{\tau}{2} \leq t \leq \tau \end{cases} \tag{E.4}$$

the Fourier coefficients a_j and b_j can be computed with the help of Eqs. (4.2) and (4.3):

$$a_0 = \frac{2}{2}\left[\int_0^1 50{,}000At\, dt + \int_1^2 50{,}000A(2 - t)\, dt \right] = 50{,}000A \tag{E.5}$$

$$a_1 = \frac{2}{2}\left[\int_0^1 50{,}000At \cos \pi t\, dt + \int_1^2 50{,}000A(2 - t) \cos \pi t\, dt \right]$$
$$= -\frac{2 \times 10^5 A}{\pi^2} \tag{E.6}$$

$$b_1 = \frac{2}{2}\left[\int_0^1 50{,}000At \sin \pi t\, dt + \int_1^2 50{,}000A(2 - t) \sin \pi t\, dt \right] = 0 \tag{E.7}$$

$$a_2 = \frac{2}{2}\left[\int_0^1 50{,}000At \cos 2\pi t\, dt + \int_1^2 50{,}000A(2 - t) \cos 2\pi t\, dt \right] = 0 \tag{E.8}$$

$$b_2 = \frac{2}{2}\left[\int_0^1 50{,}000At \sin 2\pi t\, dt + \int_1^2 50{,}000A(2 - t) \sin 2\pi t\, dt \right] = 0 \tag{E.9}$$

$$a_3 = \frac{2}{2}\left[\int_0^1 50{,}000At \cos 3\pi t\, dt + \int_1^2 50{,}000A(2 - t) \cos 3\pi t\, dt \right]$$
$$= -\frac{2 \times 10^5 A}{9\pi^2} \tag{E.10}$$

$$b_3 = \frac{2}{2}\left[\int_0^1 50{,}000At \sin 3\pi t\, dt + \int_1^2 50{,}000A(2 - t) \sin 3\pi t\, dt \right] = 0 \tag{E.11}$$

Likewise, we can obtain $a_4 = a_6 = \cdots = b_4 = b_5 = b_6 = \cdots = 0$. By considering only the first three harmonics, the forcing function can be approximated:

$$F(t) \simeq 25{,}000A - \frac{2 \times 10^5 A}{\pi^2} \cos \omega t - \frac{2 \times 10^5 A}{9\pi^2} \cos 3\omega t \tag{E.12}$$

The steady-state response of the valve to the forcing function of Eq. (E.12) can be expressed as

$$x_p(t) = \frac{25{,}000A}{k} - \frac{(2 \times 10^5 A/(k\pi^2))}{\sqrt{(1-r^2)^2 + (2\zeta r)^2}} \cos(\omega t - \phi_1)$$

$$- \frac{(2 \times 10^5 A/(9k\pi^2))}{\sqrt{(1 - 9r^2)^2 + (6\zeta r)^2}} \cos(3\omega t - \phi_3) \qquad \text{(E.13)}$$

The natural frequency of the valve is given by

$$\omega_n = \sqrt{\frac{k}{m}} = \sqrt{\frac{2{,}500}{0.25}} = 100 \text{ rad/s} \qquad \text{(E.14)}$$

and the forcing frequency ω by

$$\omega = \frac{2\pi}{\tau} = \frac{2\pi}{2} = \pi \text{ rad/s} \qquad \text{(E.15)}$$

Thus the frequency ratio can be obtained:

$$r = \frac{\omega}{\omega_n} = \frac{\pi}{100} = 0.031416 \qquad \text{(E.16)}$$

and the damping ratio:

$$\zeta = \frac{c}{c_c} = \frac{c}{2m\omega_n} = \frac{10.0}{2(0.25)(100)} = 0.2 \qquad \text{(E.17)}$$

The phase angles ϕ_1 and ϕ_3 can be computed as follows:

$$\phi_1 = \tan^{-1}\left(\frac{2\zeta r}{1 - r^2}\right)$$

$$= \tan^{-1}\left(\frac{2 \times 0.2 \times 0.031416}{1 - 0.031416^2}\right) = 0.0125664 \text{ rad} \qquad \text{(E.18)}$$

and

$$\phi_3 = \tan^{-1}\left(\frac{6\zeta r}{1 - 9r^2}\right)$$

$$= \tan^{-1}\left(\frac{6 \times 0.2 \times 0.031416}{1 - 9(0.031416)^2}\right) = 0.0380483 \text{ rad} \qquad \text{(E.19)}$$

In view of Eqs. (E.2) and (E.14) to (E.19), the solution can be written as

$$x_p(t) = 0.019635 - 0.015930 \cos(\pi t - 0.0125664)$$

$$- 0.0017828 \cos(3\pi t - 0.0380483) \text{ m} \qquad \text{(E.20)}$$

■

EXAMPLE 4.5

Total Response Under Harmonic Base Excitation

Find the total response of a viscously damped single-degree-of-freedom system subjected to a harmonic base excitation for the following data: $m = 10$ kg, $c = 20$ N-s/m, $k = 4{,}000$ N/m, $y(t) = 0.05 \sin 5t$ m, $x_0 = 0.02$ m, $\dot{x}_0 = 10$ m/s.

Solution: The equation of motion of the system is given by (see Eq. (3.65)):

$$m\ddot{x} + c\dot{x} + kx = ky + c\dot{y} = kY \sin \omega t + c\omega Y \cos \omega t \tag{E.1}$$

Noting that Eq. (E.1) is similar to Eq. (4.8) with $a_0 = 0$, $a_1 = c\omega Y$, $b_1 = kY$, and $a_i = b_i = 0; i = 2, 3, \ldots$, the steady-state response of the system can be expressed, using Eq. (E.9) of Example 4.3, as

$$x_p(t) = \frac{1}{\sqrt{(1 - r^2)^2 + (2\zeta r)^2}} \left[\frac{a_1}{k} \cos(\omega t - \phi_1) + \frac{b_1}{k} \sin(\omega t - \phi_1) \right] \tag{E.2}$$

For the given data, we find

$$Y = 0.05 \text{ m}, \qquad \omega = 5 \text{ rad/s}, \qquad \omega_n = \sqrt{\frac{k}{m}} = \sqrt{\frac{4{,}000}{10}} = 20 \text{ rad/s}$$

$$r = \frac{\omega}{\omega_n} = \frac{5}{20} = 0.25, \qquad \zeta = \frac{c}{c_c} = \frac{c}{2\sqrt{km}} = \frac{20}{2\sqrt{(4{,}000)(10)}} = 0.05$$

$$\omega_d = \sqrt{1 - \zeta^2}\,\omega_n = 19.975 \text{ rad/s}$$

$$a_1 = c\omega Y = (20)(5)(0.05) = 5, \qquad b_1 = kY = (4{,}000)(0.05) = 200$$

$$\phi_1 = \tan^{-1}\left(\frac{2\zeta r}{1 - r^2}\right) = \tan^{-1}\left(\frac{2(0.05)(0.25)}{1 - (0.25)^2}\right) = 0.02666 \text{ rad}$$

$$\sqrt{(1 - r^2)^2 + (2\zeta r)^2} = \sqrt{(1 - 0.25^2)^2 + (2(0.05)(0.25))^2} = 0.937833.$$

The solution of the homogeneous equation is given by (see Eq. (2.70)):

$$x_h(t) = X_0 e^{-\zeta \omega_n t} \cos(\omega_d t - \phi_0) = X_0 e^{-t} \cos(19.975t - \phi_0) \tag{E.3}$$

where X_0 and ϕ_0 are unknown constants. The total solution can be expressed as the superposition of $x_h(t)$ and $x_p(t)$ as

$$x(t) = X_0 e^{-t} \cos(19.975t - \phi_0) + \frac{1}{0.937833}\left[\frac{5}{4{,}000}\cos(5t - \phi_1) + \frac{200}{4{,}000}\sin(5t - \phi_1) \right]$$

$$= X_0 e^{-t} \cos(19.975t - \phi_0) + 0.001333 \cos(5t - 0.02666)$$
$$+ 0.053314 \sin(5t - 0.02666) \tag{E.4}$$

where the unknowns X_0 and ϕ_0 are to be found from the initial conditions. The velocity of the mass can be expressed from Eq. (E.4) as

$$\dot{x}(t) = \frac{dx}{dt}(t) = -X_0 e^{-t} \cos(19.975t - \phi_0) - 19.975 X_0 e^{-t} \sin(19.975t - \phi_0)$$

$$- 0.006665 \sin(5t - 0.02666) + 0.266572 \cos(5t - 0.02666) \tag{E.5}$$

Using Eqs. (E.4) and (E.5), we find

$$x_0 = x(t = 0) = 0.02 = X_0 \cos \phi_0 + 0.001333 \cos(0.02666) - 0.053314 \sin(0.02666)$$

or

$$X_0 \cos \phi_0 = 0.020088 \qquad\qquad (E.6)$$

and

$$\dot{x}_0 = \dot{x}(t = 0) = 10 = -X_0 \cos \phi_0 + 19.975 X_0 \sin \phi_0$$
$$+ 0.006665 \sin(0.02666) + 0.266572 \cos(0.02666)$$

or

$$- X_0 \cos \phi_0 + 19.975 \sin \phi_0 = 9.733345 \qquad\qquad (E.7)$$

The solution of Eqs. (E.6) and (E.7) yields $X_0 = 0.488695$ and $\phi_0 = 1.529683$ rad. Thus the total response of the mass under base excitation, in meters, is given by

$$x(t) = 0.488695 e^{-t} \cos(19.975t - 1.529683)$$
$$+ 0.001333 \cos(5t - 0.02666) + 0.053314 \sin(5t - 0.02666) \qquad (E.8)$$

Note: Equation (E.8) is plotted in Example 4.32.

∎

4.3 Response Under a Periodic Force of Irregular Form

In some cases, the force acting on a system may be quite irregular and may be determined only experimentally. Examples of such forces include wind and earthquake-induced forces. In such cases, the forces will be available in graphical form and no analytical expression can be found to describe $F(t)$. Sometimes, the value of $F(t)$ may be available only at a number of discrete points $t_1, t_2, \ldots, t_N$. In all these cases, it is possible to find the Fourier coefficients by using a numerical integration procedure, as described in Section 1.11. If $F_1, F_2, \ldots, F_N$ denote the values of $F(t)$ at $t_1, t_2, \ldots, t_N$, respectively, where N denotes an even number of equidistant points in one time period $\tau (\tau = N\Delta t)$, as shown in Fig. 4.5, the application of trapezoidal rule [4.1] gives

$$a_0 = \frac{2}{N} \sum_{i=1}^{N} F_i \qquad\qquad (4.9)$$

$$a_j = \frac{2}{N} \sum_{i=1}^{N} F_i \cos \frac{2j\pi t_i}{\tau}, \qquad j = 1, 2, \ldots \qquad (4.10)$$

$$b_j = \frac{2}{N} \sum_{i=1}^{N} F_i \sin \frac{2j\pi t_i}{\tau}, \qquad j = 1, 2, \ldots \qquad (4.11)$$

FIGURE 4.5   An irregular forcing function.

Once the Fourier coefficients a_0, a_j, and b_j are known, the steady-state response of the system can be found using Eq. (4.9) with

$$r = \left(\frac{2\pi}{\tau \omega_n}\right) \tag{4.12}$$

EXAMPLE   4.6

Steady-State Vibration of a Hydraulic Valve

Find the steady-state response of the valve in Example 4.4 if the pressure fluctuations in the chamber are found to be periodic. The values of pressure measured at 0.01-s intervals in one cycle are given below.

Time, t_i (s)	0	0.01	0.02	0.03	0.04	0.05	0.06	0.07	0.08	0.09	0.10	0.11	0.12
$p_i = p(t_i)$ (kN/m^2)	0	20	34	42	49	53	70	60	36	22	16	7	0

Solution: Since the pressure fluctuations on the valve are periodic, the Fourier analysis of the given data of pressures in a cycle gives

$$p(t) = 34,083.3 - 26,996.0 \cos 52.36t + 8,307.7 \sin 52.36t$$
$$+ 1,416.7 \cos 104.72t + 3,608.3 \sin 104.72t$$
$$- 5,833.3 \cos 157.08t - 2,333.3 \sin 157.08t + \dots \text{ N/m}^2 \tag{E.1}$$

(See Example 1.20.) Other quantities needed for the computation are

$$\omega = \frac{2\pi}{\tau} = \frac{2\pi}{0.12} = 52.36 \text{ rad/s}$$

$$\omega_n = 100 \text{ rad/s}$$

$$r = \frac{\omega}{\omega_n} = 0.5236$$

$$\zeta = 0.2$$

$$A = 0.000625\pi \text{ m}^2$$

$$\phi_1 = \tan^{-1}\left(\frac{2\zeta r}{1 - r^2}\right) = \tan^{-1}\left(\frac{2 \times 0.2 \times 0.5236}{1 - 0.5236^2}\right) = 16.1°$$

$$\phi_2 = \tan^{-1}\left(\frac{4\zeta r}{1 - 4r^2}\right) = \tan^{-1}\left(\frac{4 \times 0.2 \times 0.5236}{1 - 4 \times 0.5236^2}\right) = -77.01°$$

$$\phi_3 = \tan^{-1}\left(\frac{6\zeta r}{1 - 9r^2}\right) = \tan^{-1}\left(\frac{6 \times 0.2 \times 0.5236}{1 - 9 \times 0.5236^2}\right) = -23.18°$$

The steady-state response of the valve can be expressed, using Eq. (E.9) of Example 4.3, as

$$
\begin{aligned}
x_p(t) = {} & \frac{34{,}083.3A}{k} - \frac{(26{,}996.0A/k)}{\sqrt{(1 - r^2)^2 + (2\zeta r)^2}}\cos(52.36t - \phi_1) \\
& + \frac{(8{,}309.7A/k)}{\sqrt{(1 - r^2)^2 + (2\zeta r)^2}}\sin(52.36t - \phi_1) \\
& + \frac{(1{,}416.7A/k)}{\sqrt{(1 - 4r^2)^2 + (4\zeta r)^2}}\cos(104.72t - \phi_2) \\
& + \frac{(3{,}608.3A/k)}{\sqrt{(1 - 4r^2)^2 + (4\zeta r)^2}}\sin(104.72t - \phi_2) \\
& - \frac{(5{,}833.3A/k)}{\sqrt{(1 - 9r^2)^2 + (6\zeta r)^2}}\cos(157.08t - \phi_3) \\
& - \frac{(2{,}333.3A/k)}{\sqrt{(1 - 9r^2)^2 + (6\zeta r)^2}}\sin(157.08t - \phi_3)
\end{aligned}
$$

∎

4.4   Response Under a Nonperiodic Force

We have seen that periodic forces of any general waveform can be represented by a Fourier series as a superposition of harmonic components of various frequencies. The response of a linear system is then found by superposing the harmonic response to each of the exciting forces. When the exciting force $F(t)$ is nonperiodic, such as that due to the blast from an explosion, a different method of calculating the response is required. Various methods can be used to find the response of the system to an arbitrary excitation. Some of these methods are as follows:

1. Representing the excitation by a Fourier integral.
2. Using the method of convolution integral.

3. Using the method of Laplace transforms.
4. Numerically integrating the equations of motion (numerical solution of differential equations).

We shall discuss methods 2, 3, and 4 in the following sections. The numerical methods are also considered in Chapter 11.

4.5   Convolution Integral

A nonperiodic exciting force usually has a magnitude that varies with time; it acts for a specified period and then stops. The simplest form is the impulsive force—a force that has a large magnitude F and acts for a very short time Δt. From dynamics, we know that impulse can be measured by finding the change it causes in momentum of the system [4.2]. If $\dot{x}_1$ and $\dot{x}_2$ denote the velocities of the mass m before and after the application of the impulse, we have

$$\text{Impulse} = F\,\Delta t = m\dot{x}_2 - m\dot{x}_1 \qquad (4.12)$$

By designating the magnitude of the impulse $F\,\Delta t$ by F, we can write, in general,

$$F = \int_t^{t+\Delta t} F\,dt \qquad (4.13)$$

A unit impulse acting at $t = 0$ (f) is defined as

$$f = \lim_{\Delta t \to 0} \int_t^{t+\Delta t} F\,dt = F\,dt = 1 \qquad (4.14)$$

It can be seen that in order for $F\,dt$ to have a finite value, F tends to infinity (since dt tends to zero).

The unit impulse, $f = 1$, acting at $t = 0$, is also denoted by the Dirac delta function as

$$f = f\,\delta(t) = \delta(t) \qquad (4.15)$$

and the impulse of magnitude F, acting at $t = 0$, is denoted as[1]

$$F = F\,\delta(t) \qquad (4.16)$$

---

[1]The unit impulse, f, acting at $t = 0$, is also denoted by the Dirac delta function, $\delta(t)$. The Dirac delta function at time $t = \tau$, denoted as $\delta(t - \tau)$, has the properties

$$\delta(t - \tau) = 0 \quad \text{for } t \neq \tau;$$

$$\int_0^\infty \delta(t - \tau)\,dt = 1, \qquad \int_0^\infty \delta(t - \tau)F(t)\,dt = F(\tau)$$

where $0 < \tau < \infty$. Thus an impulse of magnitude F, acting at $t = \tau$, can be denoted as $F(t) = F\,\delta(t - \tau)$.

4.5.1
Response to
an Impulse

We first consider the response of a single-degree-of-freedom system to an impulse excitation; this case is important in studying the response under more general excitations. Consider a viscously damped spring-mass system subjected to a unit impulse at $t = 0$, as shown in Figs. 4.6(a) and (b). For an underdamped system, the solution of the equation of motion

$$m\ddot{x} + c\dot{x} + kx = 0 \tag{4.17}$$

is given by Eq. (2.72) as

$$x(t) = e^{-\zeta\omega_n t}\left\{ x_0 \cos \omega_d t + \frac{\dot{x}_0 + \zeta\omega_n x_0}{\omega_d} \sin \omega_d t \right\} \tag{4.18}$$

where

$$\zeta = \frac{c}{2m\omega_n} \tag{4.19}$$

$$\omega_d = \omega_n\sqrt{1 - \zeta^2} = \sqrt{\frac{k}{m} - \left(\frac{c}{2m}\right)^2} \tag{4.20}$$

$$\omega_n = \sqrt{\frac{k}{m}} \tag{4.21}$$

If the mass is at rest before the unit impulse is applied ($x = \dot{x} = 0$ for $t < 0$ or at $t = 0^-$), we obtain, from the impulse-momentum relation,

$$\text{Impulse} = f = 1 = m\dot{x}(t = 0) - m\dot{x}(t = 0^-) = m\dot{x}_0 \tag{4.22}$$

Thus the initial conditions are given by

$$x(t = 0) = x_0 = 0 \tag{4.23}$$

$$\dot{x}(t = 0) = \dot{x}_0 = \frac{1}{m} \tag{4.24}$$

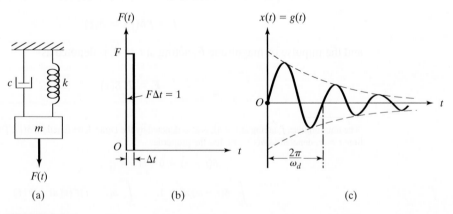

FIGURE 4.6   A single-degree-of-freedom system subjected to an impulse.

In view of Eqs. (4.23) and (4.24), Eq. (4.18) reduces to

$$x(t) = g(t) = \frac{e^{-\zeta\omega_n t}}{m\omega_d} \sin \omega_d t \tag{4.25}$$

Equation (4.25) gives the response of a single-degree-of-freedom system to a unit impulse, which is also known as the *impulse response function*, denoted by $g(t)$. The function $g(t)$, Eq. (4.25), is shown in Fig. 4.6(c).

If the magnitude of the impulse is F instead of unity, the initial velocity $\dot{x}_0$ is F/m and the response of the system becomes

$$x(t) = \frac{Fe^{-\zeta\omega_n t}}{m\omega_d} \sin \omega_d t = Fg(t) \tag{4.26}$$

If the impulse F is applied at an arbitrary time $t = \tau$, as shown in Fig. 4.7(a), it will change the velocity at $t = \tau$ by an amount F/m. Assuming that $x = 0$ until the impulse is applied, the displacement x at any subsequent time t, caused by a change in the velocity at time τ, is given by Eq. (4.26) with t replaced by the time elapsed after the application of the impulse—that is, $t - \tau$. Thus we obtain

$$x(t) = Fg(t - \tau) \tag{4.27}$$

This is shown in Fig. 4.7(b).

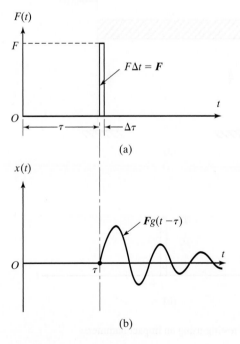

(a)

(b)

FIGURE 4.7 Impulse response.

EXAMPLE   4.7

Response of a Structure Under Impact

In the vibration testing of a structure, an impact hammer with a load cell to measure the impact force is used to cause excitation, as shown in Fig. 4.8(a). Assuming $m = 5 \, \text{kg}$, $k = 2,000 \, \text{N/m}$, $c = 10 \, \text{N-s/m}$, and $F = 20 \, \text{N-s}$, find the response of the system.

Solution: From the known data, we can compute

$$\omega_n = \sqrt{\frac{k}{m}} = \sqrt{\frac{2,000}{5}} = 20 \, \text{rad/s}, \qquad \zeta = \frac{c}{c_c} = \frac{c}{2\sqrt{km}} = \frac{10}{2\sqrt{2,000(5)}} = 0.05$$

$$\omega_d = \sqrt{1 - \zeta^2}\,\omega_n = 19.975 \, \text{rad/s}$$

Assuming that the impact is given at $t = 0$, we find (from Eq. (4.26)) the response of the system as

$$
\begin{aligned}
x_1(t) &= F\,\frac{e^{-\zeta\omega_n t}}{m\omega_d}\sin\omega_d t \\[2mm]
&= \frac{20}{(5)(19.975)}\,e^{-0.05(20)t}\sin 19.975t = 0.20025 e^{-t}\sin 19.975t \ \text{m}
\end{aligned}
\qquad\text{(E.1)}
$$

Note: The graph of Eq. (E.1) is shown in Example 4.33.

FIGURE 4.8   Structural testing using an impact hammer.

EXAMPLE 4.8

Response of a Structure Under Double Impact

In many cases, providing only one impact to the structure using an impact hammer is difficult. Sometimes a second impact takes place after the first, as shown in Fig. 4.8(b), and the applied force, $F(t)$, can be expressed as

$$F(t) = F_1\, \delta(t) + F_2\, \delta(t - \tau)$$

where $\delta(t)$ is the Dirac delta function and τ indicates the time between the two impacts of magnitudes F_1 and F_2. For a structure with $m = 5$ kg, $k = 2{,}000$ N/m, $c = 10$ N-s/m, and $F(t) = 20\,\delta(t) + 10\,\delta(t - 0.2)$ N, find the response of the structure.

Solution: From the known data, we find $\omega_n = 20$ rad/s (see the solution for Example 4.7), $\zeta = 0.05$, and $\omega_d = 19.975$ rad/s. The response due to the impulse $F_1\,\delta(t)$ is given by Eq. (E.1) of Example 4.7, while the response due to the impulse $F_2\,\delta(t - 0.2)$ can be determined from Eqs. (4.26) and (4.27) as

$$x_2(t) = F_2 \frac{e^{-\zeta\omega_n(t-\tau)}}{m\omega_d} \sin \omega_d (t - \tau) \qquad (E.1)$$

For $\tau = 0.2$, Eq. (E.1) becomes

$$x_2(t) = \frac{10}{(5)(19.975)}\, e^{-0.05(20)(t-0.2)} \sin 19.975(t - 0.2)$$

$$= 0.100125 e^{-(t-0.2)} \sin 19.975(t - 0.2); \qquad t > 0.2 \qquad (E.2)$$

Using the superposition of the two responses $x_1(t)$ and $x_2(t)$, the response due to two impacts, in meters, can be expressed as

$$x(t) = \begin{cases} 0.20025\, e^{-t} \sin 19.975t;\ 0 \le t \le 0.2 \\ 0.20025\, e^{-t} \sin 19.975t + 0.100125\, e^{-(t-0.2)} \sin 19.975(t - 0.2);\ t > 0.2 \end{cases} \qquad (E.3)$$

Note: The graph of Eq. (E.3) is shown in Example 4.33.

■

4.5.2
Response to a General Forcing Condition

Now we consider the response of the system under an arbitrary external force $F(t)$, shown in Fig. 4.9. This force may be assumed to be made up of a series of impulses of varying magnitude. Assuming that at time τ, the force $F(\tau)$ acts on the system for a short period of time $\Delta\tau$, the impulse acting at $t = \tau$ is given by $F(\tau)\,\Delta\tau$. At any time t, the elapsed time since the impulse is $t - \tau$, so the response of the system at t due to this impulse alone is given by Eq. (4.27) with $F = F(\tau)\,\Delta\tau$:

$$\Delta x(t) = F(\tau)\,\Delta\tau\, g(t - \tau) \qquad (4.28)$$

The total response at time t can be found by summing all the responses due to the elementary impulses acting at all times τ:

$$x(t) \simeq \sum F(\tau) g(t - \tau)\,\Delta\tau \qquad (4.29)$$

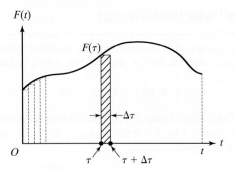

FIGURE 4.9   An arbitrary (nonperiodic) forcing function.

Letting $\Delta\tau \to 0$ and replacing the summation by integration, we obtain

$$x(t) = \int_0^t F(\tau)g(t - \tau)\, d\tau \qquad (4.30)$$

By substituting Eq. (4.25) into Eq. (4.30), we obtain

$$x(t) = \frac{1}{m\omega_d}\int_0^t F(\tau)e^{-\zeta\omega_n(t-\tau)}\sin \omega_d(t - \tau)\, d\tau \qquad (4.31)$$

which represents the response of an underdamped single-degree-of-freedom system to the arbitrary excitation $F(t)$. Note that Eq. (4.31) does not consider the effect of initial conditions of the system, because the mass is assumed to be at rest before the application of the impulse, as implied by Eqs. (4.25) and (4.28). The integral in Eq. (4.30) or (4.31) is called the *convolution* or *Duhamel integral*. In many cases, the function $F(t)$ has a form that permits an explicit integration of Eq. (4.31). If such integration is not possible, we can evaluate numerically without much difficulty, as illustrated in Section 4.9 and in Chapter 11. An elementary discussion of the Duhamel integral in vibration analysis is given in [4.6].

4.5.3
Response to
Base Excitation

If a spring-mass-damper system is subjected to an arbitrary base excitation described by its displacement, velocity, or acceleration, the equation of motion can be expressed in terms of the relative displacement of the mass $z = x - y$ as follows (see Section 3.6.2):

$$m\ddot{z} + c\dot{z} + kz = -m\ddot{y} \qquad (4.32)$$

This is similar to the equation

$$m\ddot{x} + c\dot{x} + kx = F \qquad (4.33)$$

with the variable z replacing x and the term $-m\ddot{y}$ replacing the forcing function F. Hence all of the results derived for the force-excited system are applicable to the base-excited system also for z when the term F is replaced by $-m\ddot{y}$. For an underdamped system subjected to base excitation, the relative displacement can be found from Eq. (4.31):

$$z(t) = -\frac{1}{\omega_d} \int_0^t \ddot{y}(\tau) e^{-\zeta\omega_n(t-\tau)} \sin \omega_d(t - \tau)\, d\tau \qquad (4.34)$$

EXAMPLE 4.9

Step Force on a Compacting Machine

A compacting machine, modeled as a single-degree-of-freedom system, is shown in Fig. 4.10(a). The force acting on the mass m (m includes the masses of the piston, the platform, and the material being compacted) due to a sudden application of the pressure can be idealized as a step force, as shown in Fig. 4.10(b). Determine the response of the system.

FIGURE 4.10 Step force applied to a compacting machine.

Solution: Since the compacting machine is modeled as a mass-spring-damper system, the problem is to find the response of a damped single-degree-of-freedom system subjected to a step force. By noting that $F(t) = F_0$, we can write Eq. (4.31) as

$$x(t) = \frac{F_0}{m\omega_d} \int_0^t e^{-\zeta\omega_n(t-\tau)} \sin \omega_d(t - \tau) \, d\tau$$

$$= \frac{F_0}{m\omega_d} \left[e^{-\zeta\omega_n(t-\tau)} \left\{ \frac{\zeta\omega_n \sin \omega_d(t - \tau) + \omega_d \cos \omega_d(t - \tau)}{(\zeta\omega_n)^2 + (\omega_d)^2} \right\} \right]_{\tau=0}^t$$

$$= \frac{F_0}{k} \left[1 - \frac{1}{\sqrt{1 - \zeta^2}} \cdot e^{-\zeta\omega_n t} \cos(\omega_d t - \phi) \right] \qquad (E.1)$$

where

$$\phi = \tan^{-1} \left(\frac{\zeta}{\sqrt{1 - \zeta^2}} \right) \qquad (E.2)$$

This response is shown in Fig. 4.10(c). If the system is undamped ($\zeta = 0$ and $\omega_d = \omega_n$), Eq. (E.1) reduces to

$$x(t) = \frac{F_0}{k} [1 - \cos \omega_n t] \qquad (E.3)$$

Equation (E.3) is shown graphically in Fig. 4.10(d). It can be seen that if the load is instantaneously applied to an undamped system, a maximum displacement of twice the static displacement will be attained—that is, $x_{\max} = 2F_0/k$.

∎

EXAMPLE 4.10

Time-Delayed Step Force

Find the response of the compacting machine shown in Fig. 4.10(a) when it is subjected to the force shown in Fig. 4.11.

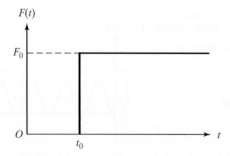

FIGURE 4.11   Step force applied with a time delay.

Solution: Since the forcing function starts at $t = t_0$ instead of at $t = 0$, the response can be obtained from Eq. (E.1) of Example 4.9 by replacing t with $t - t_0$. This gives

$$x(t) = \frac{F_0}{k\sqrt{1 - \zeta^2}}\left[\sqrt{1 - \zeta^2} - e^{-\zeta\omega_n(t-t_0)}\cos\{\omega_d(t - t_0) - \phi\}\right] \tag{E.1}$$

If the system is undamped, Eq. (E.1) reduces to

$$x(t) = \frac{F_0}{k}[1 - \cos\omega_n(t - t_0)] \tag{E.2}$$

■

EXAMPLE 4.11

Rectangular Pulse Load

If the compacting machine shown in Fig. 4.10(a) is subjected to a constant force only during the time $0 \le t \le t_0$ (Fig. 4.12(a)), determine the response of the machine.

Solution: The given forcing function, $F(t)$, can be considered as the sum of a step function $F_1(t)$ of magnitude $+F_0$ beginning at $t = 0$ and a second step function $F_2(t)$ of magnitude $-F_0$ starting at time $t = t_0$, as shown in Fig. 4.12(b).

Thus the response of the system can be obtained by subtracting Eq. (E.1) of Example 4.10 from Eq. (E.1) of Example 4.9. This gives

$$x(t) = \frac{F_0 e^{-\zeta\omega_n t}}{k\sqrt{1 - \zeta^2}}\left[-\cos(\omega_d t - \phi) + e^{\zeta\omega_n t_0}\cos\{\omega_d(t - t_0) - \phi\}\right] \tag{E.1}$$

with

$$\phi = \tan^{-1}\left(\frac{\zeta}{\sqrt{1 - \zeta^2}}\right) \tag{E.2}$$

To see the vibration response graphically, we consider the system as undamped, so that Eq. (E.1) reduces to

$$x(t) = \frac{F_0}{k}\left[\cos\omega_n(t - t_0) - \cos\omega_n t\right] \tag{E.3}$$

The response is shown in Fig. 4.12(c) for two different pulse widths of t_0 for the following data (Problem 4.90): $m = 100$ kg, $c = 50$ N-s/m, $k = 1{,}200$ N/m, and $F_0 = 100$ N. The responses will be different for the two cases $t_0 < \tau_n/2$ and $t_0 > \tau_n/2$, where τ_n is the undamped natural time period of the system. If $t_0 > \tau_n/2$, the peak will be larger and occur during the forced-vibration era (that is, during 0 to t_0) while the peak will be smaller and occur in the residual-vibration era (that is, after t_0) if $t_0 < \tau_n/2$. In Fig. 4.12(c), $\tau_n = 1.8138$ s and the peak corresponding to $t_0 = 1.5$ s is about six times larger than the one with $t_0 = 0.1$ s.

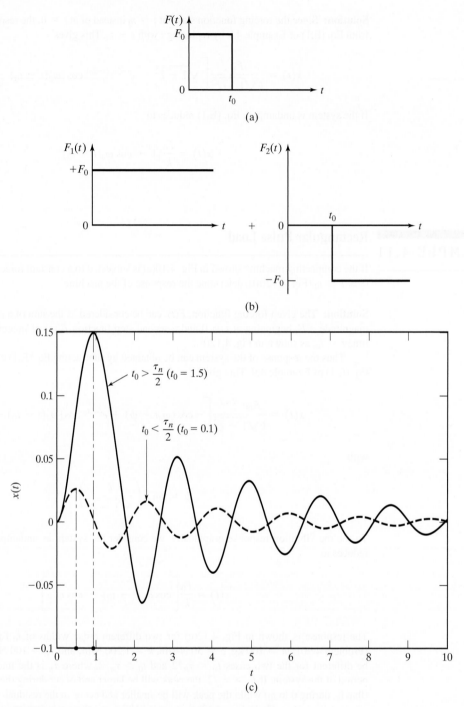

FIGURE 4.12   Response due to a pulse load.

EXAMPLE 4.12

Compacting Machine Under Linear Force

Determine the response of the compacting machine shown in Fig. 4.13(a) when a linearly varying force (shown in Fig. 4.13(b)) is applied due to the motion of the cam.

Solution: The linearly varying force shown in Fig. 4.13(b) is known as the *ramp function*. This forcing function can be represented as $F(\tau) = \delta F \cdot \tau$, where δF denotes the rate of increase of the force F per unit time. By substituting this into Eq. (4.31), we obtain

$$x(t) = \frac{\delta F}{m\omega_d} \int_0^t \tau e^{-\zeta\omega_n(t-\tau)} \sin \omega_d(t - \tau)\, d\tau$$

(a)

(b)   (c)

FIGURE 4.13 Compacting machine subjected to a linear force.

$$= \frac{\delta F}{m \omega_d} \int_0^t (t - \tau) e^{-\zeta \omega_n (t - \tau)} \sin \omega_d (t - \tau) (-d\tau)$$

$$- \frac{\delta F \cdot t}{m \omega_d} \int_0^t e^{-\zeta \omega_n (t - \tau)} \sin \omega_d (t - \tau) (-d\tau)$$

These integrals can be evaluated and the response expressed as follows:

$$x(t) = \frac{\delta F}{k} \left[t - \frac{2\zeta}{\omega_n} + e^{-\zeta \omega_n t} \left(\frac{2\zeta}{\omega_n} \cos \omega_d t - \left\{ \frac{\omega_d^2 - \zeta^2 \omega_n^2}{\omega_n^2 \omega_d} \right\} \sin \omega_d t \right) \right] \tag{E.1}$$

(See Problem 4.28.) For an undamped system, Eq. (E.1) reduces to

$$x(t) = \frac{\delta F}{\omega_n k} [\omega_n t - \sin \omega_n t] \tag{E.2}$$

Figure 4.13(c) shows the response given by Eq. (E.2).

■

EXAMPLE 4.13

Blast Load on a Building Frame

A building frame is modeled as an undamped single-degree-of-freedom system (Fig. 4.14(a)). Find the response of the frame if it is subjected to a blast loading represented by the triangular pulse shown in Fig. 4.14(b).

Solution: The forcing function is given by

$$F(\tau) = F_0 \left(1 - \frac{\tau}{t_0} \right) \qquad \text{for } 0 \leq \tau \leq t_0 \tag{E.1}$$

$$F(\tau) = 0 \qquad \text{for } \tau > t_0 \tag{E.2}$$

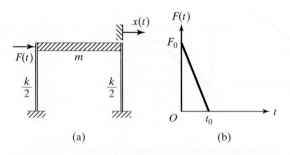

(a) (b)

FIGURE 4.14   Building frame subjected to a blast load.

Equation (4.31) gives, for an undamped system,

$$x(t) = \frac{1}{m\omega_n} \int_0^t F(\tau) \sin \omega_n(t - \tau)\,d\tau \qquad \text{(E.3)}$$

Response during $0 \leq t \leq t_0$: Using Eq. (E.1) for $F(\tau)$ in Eq. (E.3) gives

$$x(t) = \frac{F_0}{m\omega_n^2} \int_0^t \left(1 - \frac{\tau}{t_0}\right) \left[\sin \omega_n t \cos \omega_n \tau - \cos \omega_n t \sin \omega_n \tau\right] d(\omega_n \tau)$$

$$= \frac{F_0}{k} \sin \omega_n t \int_0^t \left(1 - \frac{\tau}{t_0}\right) \cos \omega_n \tau \cdot d(\omega_n \tau)$$

$$- \frac{F_0}{k} \cos \omega_n t \int_0^t \left(1 - \frac{\tau}{t_0}\right) \sin \omega_n \tau \cdot d(\omega_n \tau) \qquad \text{(E.4)}$$

By noting that integration by parts gives

$$\int \tau \cos \omega_n \tau \cdot d(\omega_n \tau) = \tau \sin \omega_n \tau + \frac{1}{\omega_n} \cos \omega_n \tau \qquad \text{(E.5)}$$

and

$$\int \tau \sin \omega_n \tau \cdot d(\omega_n \tau) = -\tau \cos \omega_n \tau + \frac{1}{\omega_n} \sin \omega_n \tau \qquad \text{(E.6)}$$

Eq. (E.4) can be written as

$$x(t) = \frac{F_0}{k} \left\{ \sin \omega_n t \left[\sin \omega_n t - \frac{t}{t_0} \sin \omega_n t - \frac{1}{\omega_n t_0} \cos \omega_n t + \frac{1}{\omega_n t_0} \right] \right.$$

$$\left. - \cos \omega_n t \left[-\cos \omega_n t + 1 + \frac{t}{t_0} \cos \omega_n t - \frac{1}{\omega_n t_0} \sin \omega_n t \right] \right\} \qquad \text{(E.7)}$$

Simplifying this expression, we obtain

$$x(t) = \frac{F_0}{k} \left[1 - \frac{t}{t_0} - \cos \omega_n t + \frac{1}{\omega_n t_0} \sin \omega_n t \right] \qquad \text{(E.8)}$$

Response during $t > t_0$: Here also we use Eq. (E.1) for $F(\tau)$, but the upper limit of integration in Eq. (E.3) will be t_0, since $F(\tau) = 0$ for $\tau > t_0$. Thus the response can be found from Eq. (E.7) by setting $t = t_0$ within the square brackets. This results in

$$x(t) = \frac{F_0}{k\omega_n t_0} \left[(1 - \cos \omega_n t_0) \sin \omega_n t - (\omega_n t_0 - \sin \omega_n t_0) \cos \omega_n t \right] \qquad \text{(E.9)}$$

∎

4.6   Response Spectrum

The graph showing the variation of the maximum response (maximum displacement, velocity, acceleration, or any other quantity) with the natural frequency (or natural period) of a single-degree-of-freedom system to a specified forcing function is known as the *response spectrum*. Since the maximum response is plotted against the natural frequency (or natural period), the response spectrum gives the maximum response of all possible single-degree-of-freedom systems. The response spectrum is widely used in earthquake engineering design [4.2, 4.5]. A review of recent literature on shock and seismic response spectra in engineering design is given in [4.7].

Once the response spectrum corresponding to a specified forcing function is available, we need to know just the natural frequency of the system to find its maximum response. Example 4.14 illustrates the construction of a response spectrum.

EXAMPLE 4.14

████████████   Response Spectrum of Sinusoidal Pulse

Find the undamped response spectrum for the sinusoidal pulse force shown in Fig. 4.15(a) using the initial conditions $x(0) = \dot{x}(0) = 0$.

Solution

Approach: Find the response and express its maximum value in terms of its natural time period.

The equation of motion of an undamped system can be expressed as

$$m\ddot{x} + kx = F(t) = \begin{cases} F_0 \sin \omega t, & 0 \le t \le t_0 \\ 0, & t > t_0 \end{cases} \qquad \text{(E.1)}$$

where

$$\omega = \frac{\pi}{t_0} \qquad \text{(E.2)}$$

FIGURE 4.15   Response spectrum due to a sinusoidal pulse.

The solution of Eq. (E.1) can be obtained by superposing the homogeneous solution $x_c(t)$ and the particular solution $x_p(t)$ as

$$x(t) = x_c(t) + x_p(t) \tag{E.3}$$

That is,

$$x(t) = A \cos \omega_n t + B \sin \omega_n t + \left(\frac{F_0}{k - m\omega^2} \right) \sin \omega t \tag{E.4}$$

where A and B are constants and ω_n is the natural frequency of the system:

$$\omega_n = \frac{2\pi}{\tau_n} = \sqrt{\frac{k}{m}} \tag{E.5}$$

Using the initial conditions $x(0) = \dot{x}(0) = 0$ in Eq. (E.4), we can find the constants A and B as

$$A = 0, \qquad B = -\frac{F_0 \omega}{\omega_n(k - m\omega^2)} \tag{E.6}$$

Thus the solution becomes

$$x(t) = \frac{F_0/k}{1 - (\omega/\omega_n)^2} \left\{ \sin \omega t - \frac{\omega}{\omega_n} \sin \omega_n t \right\}, \qquad 0 \le t \le t_0 \tag{E.7}$$

which can be rewritten as

$$\frac{x(t)}{\delta_{st}} = \frac{1}{1 - \left(\dfrac{\tau_n}{2t_0} \right)^2} \left\{ \sin \frac{\pi t}{t_0} - \frac{\tau_n}{2t_0} \sin \frac{2\pi t}{\tau_n} \right\}, \qquad 0 \le t \le t_0 \tag{E.8}$$

where

$$\delta_{st} = \frac{F_0}{k} \tag{E.9}$$

The solution given by Eq. (E.8) is valid only during the period of force application, $0 \le t \le t_0$. Since there is no force applied for $t > t_0$, the solution can be expressed as a free-vibration solution:

$$x(t) = A' \cos \omega_n t + B' \sin \omega_n t, \qquad t > t_0 \tag{E.10}$$

where the constants A' and B' can be found by using the values of $x(t = t_0)$ and $\dot{x}(t = t_0)$, given by Eq. (E.8), as initial conditions for the duration $t > t_0$. This gives

$$x(t = t_0) = \alpha \left[-\frac{\tau_n}{2t_0} \sin \frac{2\pi t_0}{\tau_n} \right] = A' \cos \omega_n t_0 + B' \sin \omega_n t_0 \tag{E.11}$$

$$\dot{x}(t = t_0) = \alpha \left\{ \frac{\pi}{t_0} - \frac{\pi}{t_0} \cos \frac{2\pi t_0}{\tau_n} \right\}$$

$$= -\omega_n A' \sin \omega_n t + \omega_n B' \cos \omega_n t \tag{E.12}$$

where

$$\alpha = \frac{\delta_{st}}{1 - \left(\dfrac{\tau_n}{2t_0}\right)^2}$$

(E.13)

Equations (E.11) and (E.12) can be solved to find A' and B' as

$$A' = \frac{\alpha\pi}{\omega_n t_0}\sin\omega_n t_0, \qquad B' = -\frac{\alpha\pi}{\omega_n t_0}[1 + \cos\omega_n t_0]$$

(E.14)

Equations (E.14) can be substituted into Eq. (E.10) to obtain

$$\frac{x(t)}{\delta_{st}} = \frac{(\tau_n/t_0)}{2\{1 - (\tau_n/2t_0)^2\}}\left[\sin 2\pi\left(\frac{t_0}{\tau_n} - \frac{t}{\tau_n}\right) - \sin 2\pi\frac{t}{\tau_n}\right], \qquad t \geq t_0$$

(E.15)

Equations (E.8) and (E.15) give the response of the system in nondimensional form—that is, x/δ_{st} is expressed in terms of t/τ_n. Thus for any specified value of t_0/τ_n, the maximum value of x/δ_{st} can be found. This maximum value of x/δ_{st}, when plotted against t_0/τ_n, gives the response spectrum shown in Fig. 4.15(b). It can be observed that the maximum value of $(x/\delta_{st})_{max} \simeq 1.75$ occurs at a value of $t_0/\tau_n \simeq 0.75$.

■

In Example 4.14, the input force is simple and hence a closed-form solution has been obtained for the response spectrum. However, if the input force is arbitrary, we can find the response spectrum only numerically. In such a case, Eq. (4.31) can be used to express the peak response of an undamped single-degree-of-freedom system due to an arbitrary input force $F(t)$ as

$$x(t)\bigg|_{max} = \frac{1}{m\omega_n}\int_0^t F(\tau)\sin\omega_n(t - \tau)\,d\tau\bigg|_{max}$$

(4.35)

4.6.1
Response Spectrum for Base Excitation

In the design of machinery and structures subjected to a ground shock, such as that caused by an earthquake, the response spectrum corresponding to the base excitation is useful. If the base of a damped single-degree-of-freedom system is subjected to an acceleration $\ddot{y}(t)$, the equation of motion, in terms of the relative displacement $z = x - y$, is given by Eq. (4.32) and the response $z(t)$ by Eq. (4.34). In the case of a ground shock, the velocity response spectrum is generally used. The displacement and acceleration spectra are then expressed in terms of the velocity spectrum. For a harmonic oscillator (an undamped system under free vibration), we notice that

$$\ddot{x}\big|_{max} = -\omega_n^2 x\big|_{max}$$

(4.36)

and

$$\dot{x}\big|_{max} = \omega_m x\big|_{max}$$

(4.37)

Thus the acceleration and displacement spectra S_a and S_d can be obtained in terms of the velocity spectrum (S_v):

$$S_d = \frac{S_v}{\omega_n}, \qquad S_a = \omega_n S_v \tag{4.38}$$

To consider damping in the system, if we assume that the maximum relative displacement occurs after the shock pulse has passed, the subsequent motion must be harmonic. In such a case, we can use Eq. (4.38). The fictitious velocity associated with this apparent harmonic motion is called the *pseudo velocity* and its response spectrum, S_v, is called the *pseudo spectrum*. The velocity spectra of damped systems are used extensively in earthquake analysis.

To find the relative velocity spectrum, we differentiate Eq. (4.34) and obtain[2]

$$\dot{z}(t) = -\frac{1}{\omega_d} \int_0^t \ddot{y}(\tau) e^{-\zeta \omega_n (t-\tau)} \left[-\zeta \omega_n \sin \omega_d (t-\tau) \right.$$
$$\left. + \omega_d \cos \omega_d (t-\tau) \right] d\tau \tag{4.39}$$

Equation (4.39) can be rewritten as

$$\dot{z}(t) = \frac{e^{-\zeta \omega_n t}}{\sqrt{1-\zeta^2}} \sqrt{P^2 + Q^2} \sin(\omega_d t - \phi) \tag{4.40}$$

where

$$P = \int_0^t \ddot{y}(\tau) e^{\zeta \omega_n t} \cos \omega_d \tau \, d\tau \tag{4.41}$$

$$Q = \int_0^t \ddot{y}(\tau) e^{\zeta \omega_n t} \sin \omega_d \tau \, d\tau \tag{4.42}$$

and

$$\phi = \tan^{-1} \left\{ \frac{-(P\sqrt{1-\zeta^2} + Q\zeta)}{(P\zeta - Q\sqrt{1-\zeta^2})} \right\} \tag{4.43}$$

The velocity response spectrum, S_v, can be obtained from Eq. (4.40):

$$S_v = |\dot{z}(t)|_{max} = \left| \frac{e^{-\zeta \omega_n t}}{\sqrt{1-\zeta^2}} \sqrt{P^2 + Q^2} \right|_{max} \tag{4.44}$$

Thus the pseudo response spectra are given by

$$S_d = |z|_{max} = \frac{S_v}{\omega_n}, \qquad S_v = |\dot{z}|_{max}, \qquad S_a = |\ddot{z}|_{max} = \omega_n S_v \tag{4.45}$$

---

[2]The following relation is used in deriving Eq. (4.39) from Eq. (4.34):

$$\frac{d}{dt} \int_0^t f(t, \tau) \, d\tau = \int_0^t \frac{\partial f}{\partial t}(t, \tau) d\tau + f(t, \tau)|_{\tau=t}$$

EXAMPLE 4.15

Water Tank Subjected to Base Acceleration

The water tank, shown in Fig. 4.16(a), is subjected to a linearly varying ground acceleration as shown in Fig. 4.16(b) due to an earthquake. The mass of the tank is m, the stiffness of the column is k, and damping is negligible. Find the response spectrum for the relative displacement, $z = x - y$, of the water tank.

Solution

Approach: Model the water tank as an undamped single-degree-of-freedom system. Find the maximum relative displacement of the tank and express it as a function of ω_n.

The base acceleration can be expressed as

$$\ddot{y}(t) = \ddot{y}_{max}\left(1 - \frac{t}{t_0}\right) \qquad \text{for} \qquad 0 \le t \le 2t_0 \tag{E.1}$$

$$\ddot{y}(t) = 0 \qquad \text{for} \qquad t > 2t_0 \tag{E.2}$$

Response during $0 \le t \le 2t_0$: By substituting Eq. (E.1) into Eq. (4.34), the response can be expressed, for an undamped system, as

$$z(t) = -\frac{1}{\omega_n}\ddot{y}_{max}\left[\int_0^t \left(1 - \frac{\tau}{t_0}\right)\right.$$

$$\left. (\sin \omega_n t \cos \omega_n \tau - \cos \omega_n t \sin \omega_n \tau)\, d\tau \right] \tag{E.3}$$

This equation is the same as Eq. (E.4) of Example 4.13 except that $(-\ddot{y}_{max})$ appears in place of F_0/m. Hence $z(t)$ can be written, using Eq. (E.8) of Example 4.13, as

$$z(t) = -\frac{\ddot{y}_{max}}{\omega_n^2}\left[1 - \frac{t}{t_0} - \cos \omega_n t + \frac{1}{\omega_n t_0}\sin \omega_n t\right] \tag{E.4}$$

To find the maximum response z_{max}, we set

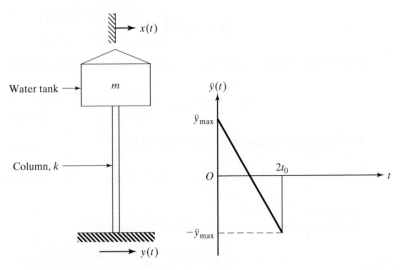

Water tank $\rightarrow$

m

Column, k $\rightarrow$

$x(t)$

$y(t)$

$\ddot{y}(t)$

$\ddot{y}_{max}$

O

$2t_0$

t

$-\ddot{y}_{max}$

FIGURE 4.16   Water tank subjected to base motion.

$$\dot{z}(t) = -\frac{\ddot{y}_{max}}{t_0 \omega_n^2}\left[-1 + \omega_n t_0 \sin \omega_n t + \cos \omega_n t\right] = 0 \tag{E.5}$$

This equation gives the time t_m at which z_{max} occurs:

$$t_m = \frac{2}{\omega_n}\tan^{-1}(\omega_n t_0) \tag{E.6}$$

By substituting Eq. (E.6) into Eq. (E.4), the maximum response of the tank can be found:

$$z_{max} = -\frac{\ddot{y}_{max}}{\omega_n^2}\left[1 - \frac{t_m}{t_0} - \cos \omega_n t_m + \frac{1}{\omega_n t_0}\sin \omega_n t_m\right] \tag{E.7}$$

Response during $t > 2t_0$: Since there is no excitation during this time, we can use the solution of the free-vibration problem Eq. (2.18)

$$z(t) = z_0 \cos \omega_n t + \left(\frac{\dot{z}_0}{\omega_n}\right)\sin \omega_n t \tag{E.8}$$

provided that we take the initial displacement and initial velocity as

$$z_0 = z(t = 2t_0) \quad \text{and} \quad \dot{z}_0 = \dot{z}(t = 2t_0) \tag{E.9}$$

using Eq. (E.7). The maximum of $z(t)$ given by Eq. (E.8) can be identified as

$$z_{max} = \left[z_0^2 + \left(\frac{\dot{z}_0}{\omega_n}\right)^2\right]^{1/2} \tag{E.10}$$

where z_0 and $\dot{z}_0$ are computed as indicated in Eq. (E.9).

■

4.6.2 Earthquake Response Spectra

The most direct description of an earthquake motion in time domain is provided by accelerograms that are recorded by instruments called *strong motion accelerographs*. They record three orthogonal components of ground acceleration at a certain location. A typical accelerogram is shown in Fig. 4.17. Accelerograms are generally recorded on photographic paper or film and are digitized for engineering applications. The peak ground acceleration, duration, and frequency content of the earthquake can be obtained from an accelerogram. An accelerogram can be integrated to obtain the time variations of the ground velocity and ground displacement.

A response spectrum is used to provide the most descriptive representation of the influence of a given earthquake on a structure or machine. It is possible to plot the maximum response of a single-degree-of-freedom system in terms of the acceleration, relative pseudo velocity, and relative displacement using logarithmic scales. A typical response spectrum, plotted on a four-way logarithmic paper, is shown in Fig. 4.18. In this figure, the

FIGURE 4.17   A typical accelerogram.

FIGURE 4.18   Response spectrum of a typical earthquake [4.12]. (Imperial Valley Earthquake, May 18, 1940; ζ = 0, 0.02, 0.05, 0.10, and 0.20.) (Reprinted with permission from *The Shock and Vibration Digest*.)

vertical axis denotes the spectral velocity, the horizontal axis represents the natural time period, the 45°-inclined axis indicates the spectral displacement, and the 135° inclined axis shows the spectral acceleration.

As can be seen from Fig. 4.18, the response spectrum of a particular accelerogram (earthquake) exhibits considerable irregularities in the frequency domain. However, spectra corresponding to an ensemble of accelerograms produced by ground shakings of sites with similar geological and seismological features are smooth functions of time and provide statistical trends that characterize them collectively. This idea has led to the development of the concept of a design spectrum, a typical one is shown in Fig. 4.19, for use in earthquake-resistant design of structures and machines. The following examples illustrate the use of the response and design spectra of earthquakes.

FIGURE 4.19 Design spectrum [4.12]. (Reprinted with permission from *The Shock and Vibration Digest.*)

EXAMPLE 4.16

Response of a Building Frame to an Earthquake

A building frame has a mass of 6,800 kg and two columns of total stiffness k, as indicated in Fig. 4.20. It has a damping ratio of 0.05 and a natural time period of 1.0 s. For the earthquake characterized in Fig. 4.18, determine the following:

a. Maximum relative displacement of the mass, x_{max}
b. Maximum shear force in the columns
c. Maximum bending stress in the columns

Solution

Approach: Find the spectral displacement, spectral velocity, and spectral acceleration corresponding to the given natural time period.

For $\tau_n = 1.0$ s and $\zeta = 0.05$, Fig. 4.18 gives $S_v = 0.635$ m/s, $S_d = 0.1016$ m, and $S_a = 0.42g$ = 4.116 m/s².

a. Maximum relative displacement of the mass, $x_{max} = S_d = 0.1016$ m
b. Maximum shear force in both columns:

$$|kx_{max}| = m\ddot{x}_{max} = m\,S_a = 6,800 \times 4.116 = 27.9888 \times 10^3\text{ N}$$

Thus the maximum shear force in each column is given by

$$F_{max} = \frac{27.9888 \times 10^3}{2} = 13.9944 \times 10^3\text{ N}$$

c. Maximum bending moment in each column = $M_{max} = F_{max}l$. Thus the maximum bending stress is given by the beam formula

$$\sigma_{max} = \frac{M_{max}c}{I}$$

where I is the area moment of inertia and c is the distance of the outer fiber from the neutral axis of the column section.

FIGURE 4.20   Building frame subjected to base motion.

Derailment of Trolley of a Crane During Earthquake

EXAMPLE 4.17

The trolley of an electric overhead traveling (EOT) crane travels horizontally on the girder as indicated in Fig. 4.21. Assuming the trolley as a point mass, the crane can be modeled as a single-degree-of-freedom system with a period 2 s and a damping ratio 2%. Determine whether the trolley derails under a vertical earthquake excitation whose design spectrum is given by Fig. 4.19.

Solution

Approach: Determine whether the spectral acceleration of the trolley (mass) exceeds a value of $1g$.

For $\tau_n = 2$ s and $\zeta = 0.02$, Fig. 4.19 gives the spectral acceleration as $S_a = 0.25g$ and hence the trolley will not derail.

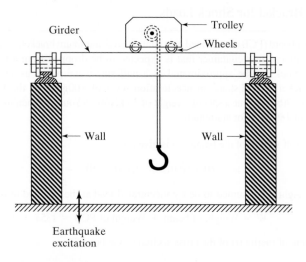

FIGURE 4.21   A crane subjected to an earthquake excitation.

■

4.6.3
Design Under
a Shock
Environment

When a force is applied for a short duration, usually for a period of less than one natural time period, it is called a *shock load*. A shock causes a significant increase in the displacement, velocity, acceleration, or stress in a mechanical system. Although fatigue is a major cause of failure under harmonic forces, usually it is not very important under shock loads. A shock may be described by a pulse shock, velocity shock, or a shock response spectrum. Pulse shocks are introduced by suddenly applied forces or displacements in the form of a square, half-sine, triangular, or similar shape (see Fig. 4.22). A velocity shock is caused by sudden changes in the velocity such as those caused when packages are dropped from a height. The shock response spectrum describes the way in which a machine or structure responds to a specific shock instead of describing the shock itself. Different types of shock pulses are used in qualifying most commercial, industrial, and military products. Many military specifications such as MIL-E-5400 and MIL-STD-810 define different types of shock pulses and detailed methods for testing with these pulses. The following example illustrates the method of limiting dynamic stresses in mechanical systems under a shock environment.

(a) Half-sine pulse   (b) Triangular pulse   (c) Rectangular pulse

FIGURE 4.22   Typical shock pulses.

Design of a Bracket for Shock Loads

EXAMPLE 4.18

A printed circuit board (PCB) is mounted on a cantilevered aluminum bracket, as shown in Fig. 4.23(a). The bracket is placed in a container that is expected to be dropped from a low-flying helicopter. The resulting shock can be approximated as a half-sine-wave pulse, as shown in Fig. 4.23(b). Design the bracket to withstand an acceleration level of 100g under the half-sine-wave pulse shown in Fig. 4.23(b). Assume a specific weight of 30 kN/m^3, a Young's modulus of 70 GPa, and a permissible stress of 180 MPa for aluminum.

Solution: The self weight of the beam (w) is given by

$$w = (0.3)(0.015 \times d)(30 \times 10^3) = 135d$$

and the total weight, W, assumed to be a concentrated load at the free end of the beam, is given by

$$W = \text{Weight of beam} + \text{Weight of PCB} = 135d + 2$$

The area moment of inertia (I) of the cross section of the beam is

$$I = \frac{1}{12} \times 0.015 \times d^3 = 0.00125d^3$$

The static deflection of the beam under the end load W, δ_{st}, can be computed as

$$\delta_{st} = \frac{Wl^3}{3EI} = \frac{(135d + 2)(0.3^3)}{3 \times (70 \times 10^9)(0.00125d^3)} = (1.0286 \times 10^{-10})\frac{135d + 2}{d^3}$$

PCB, weight 2 N

A

d

15 mm

Section A–A

A

0.3 m

(a)

Acceleration

100g

$t_0 = 0.1$ s

t

(b)

FIGURE 4.23   A cantilever subjected to an acceleration pulse.

Since the shock amplification factor (the ordinate in Fig. 4.15(b)) cannot be determined unless the value of t_0/τ_n is known, we adopt a trial-and-error procedure to determine the value of τ_n and hence that of t_0/τ_n. If d is assumed as 10 mm,

$$\delta_{st} = (1.0286 \times 10^{-10}) \frac{135 \times 0.01 + 2}{0.01^3} = 3.446 \times 10^{-4} \text{ m}$$

Equation (2.30) gives

$$\tau_n = 2\pi \sqrt{\frac{\delta_{st}}{g}} = 2\pi \sqrt{\left(\frac{3.446 \times 10^{-4}}{9.8}\right)} = 0.03726 \text{ s}$$

Hence

$$\frac{t_0}{\tau_n} = \frac{0.1}{0.03726} = 2.6841$$

The shock amplification factor (A_a) can be found from Fig. 4.15(b) as 1.1. The dynamic load (P_d) acting on the cantilever is given by

$$P_d = A_a M a_s = (1.1)\left(\frac{3.35}{g}\right)(100g) = 368.5 \text{ N}$$

where a_s is the acceleration corresponding to the shock, M is the mass at the end of the beam, and Ma_s is the inertia force on the beam. Noting that $I = 0.00125d^3 = 1.25 \times 10^{-9} \text{ (m}^4)$, the maximum bending stress at the root of the cantilever bracket can be computed as

$$\sigma_{max} = \frac{M_b c}{I} = \frac{(368.5 \times 0.3)\left(\frac{0.01}{2}\right)}{1.25 \times 10^{-9}} = 442.2 \text{ MPa}$$

Since this stress exceeds the permissible value, we assume the next trial value of d as 20 mm. This yields

$$\delta_{st} = (1.0286 \times 10^{-10}) \frac{135 \times 0.02 + 2}{0.02^3} = 6.0430 \times 10^{-5} \text{ m}$$

$$\tau_n = 2\pi \sqrt{\frac{\delta_{st}}{g}} = 2\pi \sqrt{\frac{6.0430 \times 10^{-5}}{9.8}} = 0.01560 \text{ s}$$

$$\frac{t_0}{\tau_n} = \frac{0.1}{0.01560} = 6.4$$

From Fig. 4.15(b), the shock amplification factor is found as $A_a \approx 1.1$, and hence the dynamic load acting on the beam can be determined as

$$P_d = (1.1)\left(\frac{4.7}{g}\right)(100g) = 517 \text{ N}$$

Since $d = 20$ mm gives $I = 0.00125d^3 = 10^{-8}$ m^4, the maximum bending stress at the root of the bracket will be

$$\sigma_{max} = \frac{M_b c}{I} = \frac{(517 \times 0.3)\left(\dfrac{0.02}{2}\right)}{10^{-8}} = 155.1 \text{ MPa}$$

Since this stress is within the permissible limit, the thickness of the bracket can be taken as $d = 20$ mm.

■

4.7   Laplace Transform

As stated earlier, the Laplace transform method can be used to find the response of a system under any type of excitation, including the harmonic and periodic type. A major advantage of the method is that it takes into account the initial conditions automatically. An introduction to the Laplace transform is given in Appendix D along with a table of Laplace transform pairs. The application of the Laplace transform method for finding the response of a system basically involves the following steps:

1. Write the equation of motion of the system.
2. Transform each term of the equation, using known initial conditions.
3. Solve for the transformed response of the system.
4. Obtain the desired solution (response) by using inverse Laplace transformation.

4.7.1
Transient and
Steady-State
Responses

The transient response denotes the portion of the solution caused by the initial conditions and dies down with time. The steady-state response represents the portion of the solution caused by the applied force or excitation and approaches the condition in which equilibrium prevails.

Initial value of the response: If the response or solution of a system is known in the time domain, the initial value of the response, $x\,(t = 0)$, can be determined by setting $t = 0$. If the response of the system is given in the Laplace domain, the initial value can be found as follows:

$$x(t = 0) \lim_{s \to \infty} \left[sX(s) \right] \tag{4.46}$$

Equation (4.46) is known as the *initial value theorem*.

Steady-state value of the response: If the response of a system is known in the time domain, the steady-state value of the response, x_{ss}, can be determined by taking the limit as time approaches infinity. If the response of the system is given in the Laplace domain, the steady-state value can be found by taking the limit, as s approaches zero, of s times the response in the Laplace domain:

$$x_{ss} = \lim_{s \to 0} \left[sX(s) \right] \tag{4.47}$$

Equation (4.47) is called the *final value theorem*.

The application of the Laplace transform for the computation of the response of first- and second-order systems under different forcing functions is considered below.

4.7.2 Response of First-Order Systems

Consider a spring-damper system subjected to a forcing function $\overline{F}(t)$ with the equation of motion (Fig. 4.1(b)):

$$c\dot{x} + kx = \overline{F}(t) \tag{4.48}$$

Equation (4.48) can be rewritten as

$$\dot{x} + ax = F(t) \tag{4.49}$$

where

$$a = \frac{k}{c}, \qquad F(t) = F\overline{F}(t), \qquad F = \frac{1}{c} \tag{4.50}$$

The solution of Eq. (4.49) under different forcing functions $\overline{F}(t)$ is illustrated in the following examples.

EXAMPLE 4.19

Unit Impulse Response of a First-Order System

Find the solution of Eq. (4.49) when the forcing function is a unit impulse at $t = 0$ and determine the initial and steady-state values of the response.

Solution: The equation of motion, Eq. (4.49), in this case becomes

$$\dot{x} + ax = F\,\delta(t) \tag{E.1}$$

where $F = 1/c$. By taking the Laplace transform of Eq. (E.1), we obtain

$$sX(s) - x(0) + aX(s) = F \tag{E.2}$$

Assuming the initial condition to be zero, $x(0) = 0$, Eq. (E.2) can be expressed as

$$X(s) = \frac{F}{s+a} = F\left(\frac{1}{s+a}\right) \tag{E.3}$$

The inverse Laplace transform of Eq. (E.3) gives the steady-state response of the system as

$$x(t) = Fe^{-at} \tag{E.4}$$

The initial value of the response can be found from the time response, Eq. (E.4), by setting $t = 0$. This gives

$$x(t = 0+) = F \tag{E.5}$$

From the solution in the Laplace domain, the initial value of the response is given by the initial value theorem:

$$x(t = 0+) = \lim_{s\to\infty}[sX(s)] = \lim_{s\to\infty} F\left(\frac{s}{s+a}\right) = \lim_{s\to\infty} F\left(\frac{1}{1+(a/s)}\right) = F \tag{E.6}$$

Similarly, from the response in the time domain, Eq. (E.4), the steady-state value can be found by taking the limit as $t \rightarrow \infty$. Thus Eq. (E.4) yields

$$x_{ss} = \lim_{t \to \infty} Fe^{-at} = 0 \tag{E.7}$$

The steady-state value of the response can be determined from Eq. (E.3) using the final value theorem as

$$x_{ss} = \lim_{s \to 0}[sX(s)] = \lim_{s \to 0}\left(\frac{Fs}{s+a}\right) = 0 \tag{E.8}$$

∎

EXAMPLE 4.20

Response of a First-Order System due to Ramp Function

Find the solution of Eq. (4.49) when the applied force is a ramp function.

Solution: The equation of motion, Eq. (4.49), in this case can be written as

$$\dot{x} + ax = Fbt = dt \tag{E.1}$$

where $d = Fb$, $F = 1/c$, and b denotes the slope of the ramp (Fig. 4.24). By taking the Laplace transform of Eq. (E.1), we obtain

$$sX(s) - x(0) + aX(s) = \frac{d}{x^2} \tag{E.2}$$

Assuming the initial condition to be zero, $x(0) = 0$, Eq. (E.2) can be expressed as

$$X(s) = d\left(\frac{1}{s^2(s+a)}\right) = \frac{d}{a^2}\frac{a^2}{s^2(s+a)} \tag{E.3}$$

The inverse Laplace transform of Eq. (E.3) gives the steady-state response of the system as

$$x(t) = \frac{d}{a^2}[at - (1 - e^{-at})] \tag{E.4}$$

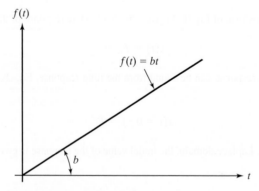

FIGURE 4.24   Ramp function.

∎

**4.7.3
Response of
Second-Order
Systems**

Consider a spring-mass-damper system subjected to a forcing function $\overline{F}(t)$ with the equation of motion (Fig. 4.2(a)):

$$m\ddot{x} + c\dot{x} + kx = \overline{F}(t) \tag{4.51}$$

The solution of Eq. (4.51) under different forcing functions $\overline{F}(t)$ is illustrated in the following examples.

EXAMPLE 4.21

Unit Impulse Response of a Second-Order System

Find the response of an underdamped single-degree-of-freedom system to a unit impulse.

Solution: The equation of motion is given by

$$m\ddot{x} + c\dot{x} + kx = \delta(t) \tag{E.1}$$

By taking the Laplace transform of both sides of Eq. (E.1), we obtain

$$[m(s^2 - sx_0 - \dot{x}_0) + c(s - x_0) + k]X(s) = 1$$

or

$$(ms^2 + cs + k)X(s) = m\dot{x}_0 + (ms + c)x_0 + 1 \tag{E.2}$$

Assuming zero initial conditions, $x_0 = \dot{x}_0 = 0$, Eq. (E.2) can be expressed as

$$(ms^2 + cs + k)X(s) = 1$$

or

$$X(s) = \frac{1}{m(s^2 + 2\zeta\omega_n s + \omega_n^2)} \tag{E.3}$$

We can express the right-hand side of Eq. (E.3) in partial fractions as

$$X(s) = \frac{C_1}{s - s_1} + \frac{C_2}{s - s_2} \tag{E.4}$$

where s_1 and s_2 are the roots of the polynomial equation:

$$s^2 + 2\zeta\omega_n s + \omega_n^2 = 0 \tag{E.5}$$

which are given by

$$s_1 = -\zeta\omega_n + i\omega_d, \qquad s_2 = -\zeta\omega_n - i\omega_d \tag{E.6}$$

where

$$\omega_d = \omega_n\sqrt{1 - \zeta^2} \tag{E.7}$$

is the damped frequency of the system. Substitution of Eq. (E.6) into Eq. (E.4) yields

$$C_1(s - s_2) + C_2(s - s_1) = \frac{1}{m}$$

or

$$(C_1 + C_2)s - \{C_1(-\zeta\omega_n - i\omega_d) + C_2(-\zeta\omega_n + i\omega_d)\} = (0)\,s + \frac{1}{m} \tag{E.8}$$

By equating the coefficients on both sides of Eq. (E.8), we obtain

$$C_1 + C_2 = 0 \quad \text{or} \quad C_1 = -C_2$$

$$C_1(-\zeta\omega_n - i\omega_d) + C_2(-\zeta\omega_n + i\omega_d) = -\frac{1}{m} \tag{E.9}$$

or

$$C_2(\zeta\omega_n + i\omega_d - \zeta\omega_n + i\omega_d) = -\frac{1}{m} \tag{E.10}$$

Equations (E.9) and (E.10) give

$$C_2 = -\frac{1}{2im\omega_d} = -C_1 \tag{E.11}$$

Using Eq. (E.11) in Eq. (E.4), $X(s)$ can be expressed as

$$X(s) = \frac{1}{2im\omega_d}\left(\frac{1}{s - s_1} - \frac{1}{s - s_2}\right) \tag{E.12}$$

By taking the inverse Laplace transform of Eq. (E.12), we obtain

$$x(t) = \frac{1}{2im\omega_d}(e^{s_1 t} - e^{s_2 t}) = \frac{1}{2im\omega_d}[e^{(-\zeta\omega_n + i\omega_d)} - e^{(-\zeta\omega_n + i\omega_d)}]$$

$$= \frac{1}{2im\omega_d}e^{-\zeta\omega_n t}(e^{i\omega_d t} - e^{-i\omega_d t})$$

$$= \frac{1}{m\omega_d}e^{-\zeta\omega_n t}\sin\omega_d t; \qquad t \geq 0 \tag{E.13}$$

Notes:
1. The response $x(t) = 0$ for $t < 0$ (because the unit impulse is applied at $t = 0$).
2. Equation (E.13) can be seen to be the same as the unit impulse response function derived using the traditional method, Eq. (4.25).

The following two examples illustrate the application of impulse response computations in the context of inelastic and elastic impacts.

∎

Response to Inelastic Impact

EXAMPLE 4.22

A mass m, moving with velocity v_1, strikes the mass M of a damped single-degree-of-freedom system as shown in Fig. 4.25(a) and gets attached to the mass M after the impact as shown in Fig. 4.25(b). Find the resulting displacement response of the system.

Approach: Use the relation:

Change in momentum = impulse

—i.e.,

$$mv_2 - mv_1 = \int_0^1 f(\tau)\, d\tau \tag{E.1}$$

where m is the striking mass, v_2 is the final velocity (after impact), v_1 is the initial velocity (before impact), $f(t)$ is the force applied during the short duration 0 to t, and the integral denotes the impulse (same as the area under the force-time curve).

Solution: Since the mass m gets attached to the mass M after the impact, the impact can be considered to be perfectly plastic or inelastic. The combined system (with the two masses together as shown in Fig. 4.25(b)) can be considered to be subjected to an impulse with changes in the velocities of the masses. The impact force, $f(t)$, is internal to the system and can be assumed to be zero. Thus Eq. (E.1) can be rewritten as

$$(m + M)V_s - \{mv_1 + M(0)\} = 0 \tag{E.2}$$

where V_s is the velocity of the combined system $(m + M)$ after impact. Equation (E.2) gives the velocity of the system immediately after the impact as

$$V_s = \frac{mv_1}{m + M} \tag{E.3}$$

The equation of motion for the combined system is given by

$$(m + M)\ddot{x} + c\dot{x} + kx = 0 \tag{E.4}$$

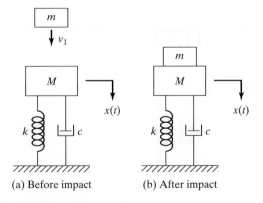

(a) Before impact   (b) After impact

FIGURE 4.25 Inelastic impact.

Since the impact causes a change in the velocity, but not the displacement, of the system just imme-
diately after impact, the initial conditions can be taken as $x(t = 0) = 0$ and $\dot{x}(t = 0) = V_s = \dfrac{mv_1}{m + M}$. The free-vibration response of the system (solution of Eq. (E.4)) can be obtained from
Eq. (4.18) as

$$x(t) = e^{-\zeta \omega_n t} \frac{\dot{x}_0}{\omega_d} \sin \omega_d t = \frac{mv_1}{(m + M)\omega_d} e^{-\zeta \omega_n t} \sin \omega_d t \tag{E.5}$$

∎

Response to Perfectly Elastic Impact

EXAMPLE 4.23

A mass m, moving with velocity v_1, strikes the mass M of a damped single-degree-of-freedom system as shown in Fig. 4.26(a). The impact is perfectly elastic, so that after impact the mass m
rebounds with velocity v_2. Find the resulting displacement response of the mass M.

Approach: When two masses m and M moving initially with velocities v_1 and V_1 strike each other,
and attain the velocities v_2 and V_2 immediately after impact, respectively (Fig. 4.26(b)), the con-
servation-of-momentum principle yields

$$mv_1 + MV_1 = mv_2 + MV_2$$

or

$$m(v_1 - v_2) = -M(V_1 - V_2) \tag{E.1}$$

Because the impact is perfectly elastic, the principle of conservation of kinetic energy is applicable,
so that

$$\frac{1}{2}mv_1^2 + \frac{1}{2}MV_1^2 = \frac{1}{2}mv_2^2 + \frac{1}{2}MV_2^2$$

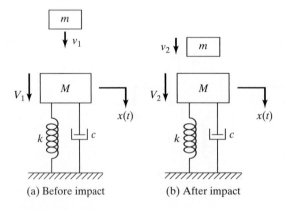

(a) Before impact   (b) After impact

FIGURE 4.26   Elastic impact.

or

$$\frac{1}{2}m(v_1^2 - v_2^2) = -\frac{1}{2}M(V_1^2 - V_2^2)$$

which can be rewritten in the form

$$\frac{1}{2}m(v_1 + v_2)(v_1 - v_2) = -\frac{1}{2}M(V_1 + V_2)(V_1 - V_2) \qquad (E.2)$$

Using Eq. (E.1) in Eq. (E.2), we find that

$$v_1 + v_2 = V_1 + V_2$$

or

$$(v_1 - V_1) = -(v_2 - V_2) \qquad (E.3)$$

Equation (E.3) indicates that the magnitude of the relative velocity of the masses remains constant and only the sign changes during a perfectly elastic impact.

Solution: Because the velocities of the masses m and M are known to be v_1 and $V_1 = 0$ before impact, their velocities immediately after impact can be determined from Eqs. (E.1) and (E.3):

$$m(v_1 - v_2) = -M(0 - V_2) = MV_2$$

or

$$V_2 = \frac{mv_1}{M} - \frac{m}{M}v_2 \qquad (E.4)$$

and

$$(v_1 - 0) = v_1 = -(v_2 - V_2) = V_2 - v_2 \qquad (E.5)$$

The solution of Eqs. (E.4) and (E.5) yields

$$v_2 = \frac{m - M}{m + M}v_1, \qquad V_2 = \frac{2m}{m + M}v_1 \qquad (E.6)$$

The change in the momentum of mass m is given by

$$m(v_2 - v_1) = m\left(\frac{m - M}{m + M} - 1\right)v_1 = -\left(\frac{2mM}{m + M}\right)v_1 \qquad (E.7)$$

Thus the impulse applied to mass m during the impact is given by

$$\int_0^t f(\tau)\, d\tau = -\left(\frac{2mM}{m + M}\right)v_1 \qquad (E.8)$$

According to Newton's third law of motion, the impulse applied to mass M during impact will be the same as, but opposite in sign to, the impulse applied to mass m. Due to this impulse applied, the equation of motion of mass M can be expressed as

$$M\dot{x} + c\dot{x} + kx = \int_0^t F(\tau)\, d\tau = F = \frac{2mM}{m + M}v_1\delta(t) \qquad (E.9)$$

Using the initial conditions of M as $x(t = 0) = x_0 = 0$ and $\dot{x}(t = 0) = \dot{x}_0 = 0$, the solution of Eq. (E.8) can be expressed, using Eq. (4.26), as

$$x(t) = \frac{Fe^{-\zeta\omega_n t}}{M\omega_d}\sin\omega_d t = \left(\frac{2mM}{m+M}\right)\frac{v_1}{M\omega_d}e^{-\zeta\omega_n t}\sin\omega_d t \qquad (E.10)$$

∎

4.7.4 Response to Step Force

EXAMPLE 4.24

Step Response of an Underdamped System

Find the response of an underdamped single-degree-of-freedom system to a unit step function.

Solution: The equation of motion is given by

$$m\ddot{x} + c\dot{x} + kx = f(t) = 1 \qquad (E.1)$$

By taking the Laplace transform of both sides of Eq. (E.1) and assuming zero initial conditions $(x_0 = \dot{x}_0 = 0)$, we obtain

$$(ms^2 + cs + k)X(s) = \mathcal{L}[1] = \frac{1}{s} \qquad (E.2)$$

which can be rewritten as

$$X(s) = \frac{1}{ms(s^2 + 2\zeta\omega_n s + \omega_n^2)} \qquad (E.3)$$

We can express the right-hand side of Eq. (E.3) in partial fractions as

$$X(s) = \frac{1}{ms(s^2 + 2\zeta\omega_n s + \omega_n^2)} = \frac{C_1}{s - s_1} + \frac{C_2}{s - s_2} + \frac{C_3}{s - s_3} \qquad (E.4)$$

where s_1, s_2, and s_3 are the roots of the polynomial equation

$$s(s^2 + 2\zeta\omega_n s + \omega_n^2) = 0 \qquad (E.5)$$

which are given by

$$s_1 = 0, \qquad s_2 = -\zeta\omega_n + i\omega_d, \qquad s_3 = -\zeta\omega_n - i\omega_d \qquad (E.6)$$

The constants C_1, C_2, and C_3 in Eq. (E.4) can be determined as follows. Substituting the values of s_1, s_2, and s_3 given by Eq. (E.6) into Eq. (E.4) and rearranging the terms leads to

$$\frac{1}{m} = C_1(s^2 + 2\zeta\omega_n s + \omega_n^2) + C_2[s^2 + s(\zeta\omega_n + i\omega_d)] + C_3[s^2 + s(\zeta\omega_n - i\omega_d)] \qquad (E.7)$$

Equation (E.7) can be rewritten as

$$s^2(C_1 + C_2 + C_3) + s[(2\zeta\omega_n)C_1 + (\zeta\omega_n + i\omega_d)C_2 + (\zeta\omega_n - i\omega_d)C_3] + \omega_n^2 C_1$$

$$= (0)s^2 + (0)s + \frac{1}{m} \qquad (E.8)$$

By equating the coefficients of corresponding terms on both sides of Eq. (E.8), we obtain

$$C_1 + C_2 + C_3 = 0 \tag{E.9}$$

$$C_1(2\zeta\omega_n) + C_2(-\zeta\omega_n + i\omega_d) + C_3(\zeta\omega_n - i\omega_d) = 0 \tag{E.10}$$

$$C_1\omega_n^2 = \frac{1}{m} \tag{E.11}$$

The solution of Eqs. (E.9)–(E.11) gives

$$C_1 = \frac{1}{m\omega_n^2} \tag{E.12}$$

$$C_2 = \frac{1}{2im\omega_d(-\zeta\omega_n + i\omega_d)} \tag{E.13}$$

$$C_3 = \frac{1}{2im\omega_d(\zeta\omega_n + i\omega_d)} \tag{E.14}$$

Using Eqs. (E.12)–(E.14) in Eq. (E.3), $X(s)$ can be expressed as

$$X(s) = \frac{1}{m\omega_n^2}\frac{1}{s}$$
$$+ \frac{1}{2im\omega_d}\left[\frac{1}{-\zeta\omega_n + i\omega_d}\frac{1}{s - (-\zeta\omega_n + i\omega_d)} - \frac{1}{-\zeta\omega_n - i\omega_d}\frac{1}{s - (-\zeta\omega_n - i\omega_d)}\right] \tag{E.15}$$

By taking the inverse Laplace transform of Eq. (E.15) and using the results given in Appendix D, we obtain

$$x(t) = \frac{1}{m\omega_n^2} + \frac{e^{-\zeta\omega_n t}}{2im\omega_d}\left(\frac{e^{i\omega_d t}}{-\zeta\omega_n + i\omega_d} - \frac{e^{-i\omega_d t}}{-\zeta\omega_n - i\omega_d}\right)$$

$$= \frac{1}{m\omega_n^2}\left\{1 + \frac{e^{-\zeta\omega_d t}}{2i\omega_d}[(-\zeta\omega_n - i\omega_d)e^{i\omega_d t} - (-\zeta\omega_n + i\omega_d)e^{-i\omega_d t}]\right\}$$

$$= \frac{1}{k}\left\{1 - \frac{e^{-\zeta\omega_n t}}{\omega_d}[\zeta\omega_n \sin\omega_d t + \omega_d \cos\omega_d t]\right\}$$

$$= \frac{1}{k}\left[1 - \frac{e^{-\zeta\omega_n t}}{\sqrt{1-\zeta^2}}\cos(\omega_d t - \phi)\right] \tag{E.16}$$

where

$$\phi = \tan^{-1}\left(\frac{\zeta}{\sqrt{1-\zeta^2}}\right) \tag{E.17}$$

It can be seen that Eq. (E.16) is the same as the unit step response (with $F_0 = 0$) derived using the traditional method, Eq. (E.1) of Example 4.9. The response given by Eq. (E.16) is shown in Fig. 4.27.

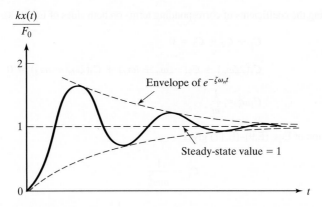

FIGURE 4.27   Response of an underdamped system subjected to a step force.

EXAMPLE 4.25

Initial and Final Values of Step Response of Underdamped System

Find the initial and steady-state values of the unit step response of an underdamped system from the responses indicated by Eqs. (E.3) and (E.16) of Example 4.24.

Solution: The response of the system in the time domain, Eq. (E.16) of Example 4.24, can be written as

$$x(t) = \frac{1}{k}\left\{1 - \frac{e^{-\zeta\omega_n t}}{\omega_d}\left[\zeta\omega_n \sin \omega_d t + \omega_d \cos \omega_d t\right]\right\} \tag{E.1}$$

By setting $t = 0$ in Eq. (E.1), we find the initial value as 0. By taking the limit as $t \to \infty$, the term $e^{-\zeta\omega_n t} \to 0$ and hence the steady-state value of $x(t)$ is given by $1/k$. The response of the system in the Laplace domain is given by Eq. (E.3) of Example 4.24. Using the initial value theorem, we find the initial value as

$$x\,(t = 0+) = \lim_{s\to\infty}\left[sX(s)\right] = \lim_{s\to\infty}\left[\frac{1}{m(s^2 + 2\zeta\omega_n s + \omega_n^2)}\right] = 0$$

$$x_{ss} = \lim_{s\to 0}\left[sX(s)\right] = \lim_{s\to 0}\left[\frac{1}{m(s^2 + 2\zeta\omega_n s + \omega_n^2)}\right] = \frac{1}{m\omega_n^2} = \frac{1}{k}$$

∎

EXAMPLE 4.26

Response of a Compacting Machine

Find the response of the compacting machine of Example 4.9 assuming the system to be under-damped (i.e., $\zeta < 1$).

Approach: Use a spring-mass-damper model of the compacting machine and use the Laplace transform technique.

Solution: The forcing function is given by

$$F(t) = \begin{cases} F_0 & \text{for } 0 \le t \le t_0 \\ 0 & \text{for } t > t_0 \end{cases} \tag{E.1}$$

By taking the Laplace transform of the governing differential equation, Eq. (4.51), and using Appendix D, we obtain the following equation:

$$X(S) = \frac{F(s)}{m(s^2 + 2\zeta\omega_n s + \omega_n^2)} + \frac{s + 2\zeta\omega_n}{s^2 + 2\zeta\omega_n s + \omega_n^2}x_0$$

$$+ \frac{1}{s^2 + 2\zeta\omega_n s + \omega_n^2}\dot{x}_0 \tag{E.2}$$

where

$$F(s) = \mathscr{L}F(t) = \frac{F_0(1 - e^{-t_0 s})}{s} \tag{E.3}$$

Thus Eq. (E.2) can be written as

$$X(s) = \frac{F_0(1 - e^{-t_0 s})}{ms(s^2 + 2\zeta\omega_n s + \omega_n^2)} + \frac{s + 2\zeta\omega_n}{s^2 + 2\zeta\omega_n s + \omega_n^2}x_0$$

$$+ \frac{1}{s^2 + 2\zeta\omega_n + \omega_n^2}\dot{x}_0$$

$$= \frac{F_0}{m\omega_n^2}\frac{1}{s\left(\frac{s^2}{\omega_n^2} + \frac{2\zeta s}{\omega_n} + 1\right)} - \frac{F_0}{m\omega_n^2}\frac{e^{-t_0 s}}{s\left(\frac{s^2}{\omega_n^2} + \frac{2\zeta s}{\omega_n} + 1\right)}$$

$$+ \frac{x_0}{\omega_n^2}\frac{s}{\left(\frac{s^2}{\omega_n^2} + \frac{2\zeta s}{\omega_n} + 1\right)} + \left(\frac{2\zeta x_0}{\omega_n} + \frac{\dot{x}_0}{\omega_n^2}\right)\frac{1}{\left(\frac{s^2}{\omega_n^2} + \frac{2\zeta s}{\omega_n} + 1\right)} \tag{E.4}$$

The inverse transform of Eq. (E.4) can be expressed by using the results in Appendix D as

$$x(t) = \frac{F_0}{m\omega_n^2}\left[1 - \frac{e^{-\zeta\omega_n t}}{\sqrt{1 - \zeta^2}}\sin\{\omega_n\sqrt{1 - \zeta^2}t + \phi_1\}\right]$$

$$- \frac{F_0}{m\omega_n^2}\left[1 - \frac{e^{-\zeta\omega_n(t - t_0)}}{\sqrt{1 - \zeta^2}}\sin\{\omega_n\sqrt{1 - \zeta^2}(t - t_0) + \phi_1\}\right]$$

$$- \frac{x_0}{\omega_n^2}\left[\frac{\omega_n^2 e^{-\zeta\omega_n t}}{\sqrt{1 - \zeta^2}}\sin\{\omega_n\sqrt{1 - \zeta^2}t - \phi_1\}\right]$$

$$+ \left(\frac{2\zeta x_0}{\omega_n} + \frac{\dot{x}_0}{\omega_n^2}\right)\left[\frac{\omega_n}{\sqrt{1 - \zeta^2}}e^{-\zeta\omega_n t}\sin(\omega_n\sqrt{1 - \zeta^2}t)\right] \tag{E.5}$$

where

$$\phi_1 = \cos^{-1}(\zeta) \tag{E.6}$$

Thus the response of the compacting machine can be expressed as

$$x(t) = \frac{F_0}{m\omega_n^2\sqrt{1 - \zeta^2}}[-e^{-\zeta\omega_n t}\sin(\omega_n\sqrt{1 - \zeta^2}\,t + \phi_1)$$

$$+ e^{-\zeta\omega_n(t-t_0)}\sin\{\omega_n\sqrt{1 - \zeta^2}\,(t - t_0) + \phi_1\}]$$

$$- \frac{x_0}{\sqrt{1 - \zeta^2}}e^{-\zeta\omega_n t}\sin(\omega_n\sqrt{1 - \zeta^2}\,t - \phi_1)$$

$$+ \frac{(2\zeta\omega_n x_0 + \dot{x}_0)}{\omega_n\sqrt{1 - \zeta^2}}e^{-\zeta\omega_n t}\sin(\omega_n\sqrt{1 - \zeta^2}\,t) \tag{E.7}$$

Although the first part of Eq. (E.7) is expected to be the same as Eq. (E.1) of Example 4.11, it is difficult to see the equivalence in the present form of Eq. (E.7). However, for the undamped system, Eq. (E.7) reduces to

$$x(t) = \frac{F_0}{m\omega_n^2}\left[-\sin\left(\omega_n t + \frac{\pi}{2}\right) + \sin\left\{\omega_n(t - t_0) + \frac{\pi}{2}\right\}\right]$$

$$- x_0\sin\left(\omega_n t - \frac{\pi}{2}\right) + \frac{\dot{x}_0}{\omega_n}\sin\omega_n t$$

$$= \frac{F_0}{k}[\cos\omega_n(t - t_0) - \cos\omega_n t] + x_0\cos\omega_n t + \frac{\dot{x}_0}{\omega_n}\sin\omega_n t \tag{E.8}$$

The first or steady-state part of Eq. (E.8) can be seen to be identical to Eq. (E.3) of Example 4.11.

∎

Overdamped System Subjected to Step Force

EXAMPLE 4.27

Find the response of an overdamped single-degree-of-freedom system subjected to a step force with the equation of motion

$$2\ddot{x} + 8\dot{x} + 6x = 5u_s(t) \tag{E.1}$$

Assume the initial conditions as $x_0 = 1$ and $\dot{x}_0 = 2$.

Solution: By taking the Laplace transform of both sides of Eq. (E.1), we obtain

$$[2\{s^2 X(s) - s x_0 - \dot{x}_0\} + 8\{s X(s) - x_0\} + 6X(s)] = \frac{5}{s}$$

or

$$s(2s^2 + 8s + 6)X(s) = 5 + 2s(s x_0 + \dot{x}_0) + 8s x_0 \tag{E.2}$$

Using the initial values, $x_0 = 1$ and $\dot{x}_0 = 2$, Eq. (E.2) can be expressed as

$$s(2s^2 + 8s + 6)X(s) = 2s^2 + 12s + 5$$

or

$$X(s) = \frac{2s^2 + 12s + 5}{2s(s^2 + 4s + 3)} = \frac{s^2 + 6s + 2.5}{s(s + 1)(s + 3)} \tag{E.3}$$

Noting that the roots of the polynomial in the denominator on the right-hand side of Eq. (E.3) are $s_1 = 0$, $s_2 = -1$, and $s_3 = -3$, $X(s)$ can be expressed, using partial fractions, as

$$X(s) = \frac{C_1}{s - s_1} + \frac{C_2}{s - s_2} + \frac{C_3}{s - s_3} \tag{E.4}$$

where the constants can be found, using Eq. (E.1), as

$$C_k = \frac{A(s)}{B'(s)}\Big|_{s=s_k} ; \quad k = 1, 2, 3 \tag{E.5}$$

where $A(s)$ is the numerator and $B(s)$ is the denominator of the middle expression in Eq. (E.3) and a prime denotes the derivative with respect to s. The middle expression in Eq. (E.3) yields

$$\frac{A(s)}{B'(s)} = \frac{s^2 + 6s + 2.5}{3s^2 + 8s + 3} \tag{E.6}$$

Equations (E.5) and (E.6) give

$$C_1 = \frac{A(s)}{B'(s)}\Big|_{s=s_1=0} = \frac{2.5}{3} = \frac{5}{6}$$

$$C_2 = \frac{A(s)}{B'(s)}\Big|_{s=s_2=-1} = \frac{-2.5}{-2} = \frac{5}{4}$$

$$C_3 = \frac{A(s)}{B'(s)}\Big|_{s=s_3=-3} = \frac{-6.5}{6} = -\frac{13}{12} \tag{E.7}$$

In view of Eqs. (E.7), Eq. (E.4) becomes

$$X(s) = \frac{5}{6}\frac{1}{s} + \frac{5}{4}\frac{1}{s + 1} - \frac{13}{12}\frac{1}{s + 3} \tag{E.8}$$

By taking the inverse Laplace transform of Eq. (E.8), we obtain the response of the system as

$$x(t) = \frac{5}{6} + \frac{5}{4}e^{-t} - \frac{13}{12}e^{-3t} \tag{E.9}$$

The response given by Eq. (E.9) is shown graphically in Fig. 4.28.

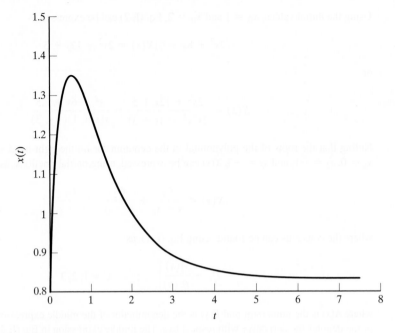

FIGURE 4.28

■

4.7.5
Analysis of the
Step Response

The response of a damped single-degree-of-freedom system under a step force, given by Eqs. (E.1) and (E.2) of Example 4.9 and Eqs. (E.16) and (E.17) of Example 4.24, can be expressed in the form

$$\frac{kx(t)}{F_0} = 1 - \frac{1}{\sqrt{1 - \zeta^2}} e^{-\zeta \omega_n t} \cos(\omega_d t - \phi) \tag{4.52}$$

where

$$\phi = \tan^{-1}\left(\frac{\zeta}{\sqrt{1 - \zeta^2}}\right) \tag{4.53}$$

The variations of the nondimensional response, $kx(t)/F_0$, with the nondimensional time, $\omega_n t$, are shown graphically in Fig. 4.29 for several values of the damping ratio ζ. It can be seen that for an undamped system ($\zeta = 0$), the response exhibits oscillations which never die out. For an underdamped system ($\zeta < 1$), the response overshoots and oscillates about the final or steady-state value. Furthermore, the smaller the value of the damping ratio, the larger will be the overshoot, so that the oscillations take longer to die out. For a critically damped system ($\zeta = 1$), the response reaches the final or steady-state value quickest with no oscillation. For an overdamped system ($\zeta > 1$), the response reaches the steady-state value slowly with no overshoot.

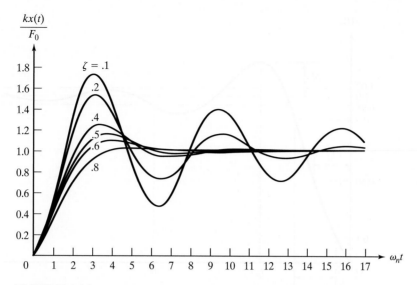

FIGURE 4.29   Response of an underdamped system due to a unit step force.

**4.7.6
Description
of Transient
Response**

The performance and behavior of a vibrating system for the transient response is described in terms of parameters such as maximum overshoot, peak time, rise time, delay time, and settling time. These parameters are shown in Fig. 4.30, which denotes a typical step response of an underdamped system. They are discussed below.

1. **Peak time (t_p):** The peak time is the time required for the response to attain the first peak of the overshoot.

    The maximum amount the response overshoots, M_p, occurs when the derivative of $x(t)$ is zero. The time variation of the unit step response of an underdamped system is given by Eq. (E.16) of Example 4.24:

$$kx(t) = 1 - e^{-\zeta\omega_n t}\left(\frac{\zeta\omega_n}{\omega_d}\sin\omega_d t + \cos\omega_d t\right) \qquad (4.54)$$

where $\omega_d = \omega_n\sqrt{1 - \zeta^2}$. Equation (4.54) can also be expressed in compact form as

$$kx(t) = 1 - e^{-\zeta\omega_n t}\sqrt{1 + \left(\frac{\zeta\omega_n}{\omega_d}\right)^2}\cos(\omega_d t - \alpha) \qquad (4.55)$$

where

$$\alpha = \tan^{-1}\left(\frac{\zeta\omega_n}{\omega_d}\right) = \tan^{-1}\left(\frac{\zeta}{\sqrt{1 - \zeta^2}}\right) \qquad (4.56)$$

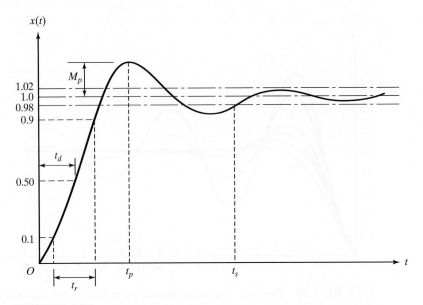

FIGURE 4.30   Specifications of the response of an underdamped system.

The derivative of $x(t)$ will be zero when $x(t)$ reaches its maximum, so that

$$k\dot{x}(t) = \zeta\omega_n e^{-\zeta\omega_n t}\left\{\frac{\zeta\omega_n}{\omega_d}\sin\omega_d t + \cos\omega_d t\right\}$$

$$- e^{-\zeta\omega_n t}\{\zeta\omega_n\cos\omega_d t - \omega_d\sin\omega_d t\} = 0$$

or

$$e^{-\zeta\omega_n t}\left\{\frac{(\zeta\omega_n)^2}{\omega_d}\sin\omega_d t + \omega_d\sin\omega_d t\right\} = 0 \tag{4.57}$$

Equation (4.57) is satisfied when $\sin\omega_d t = 0$ so that

$$\omega_d t_p = 0 \tag{4.58}$$

Thus the peak time is given by

$$t_p = \frac{\pi}{\omega_d} \tag{4.59}$$

2. **Rise time (t_r):** The rise time is the time needed for the response to rise from 10% to 90% of the final or steady-state value for overdamped systems. For underdamped systems, usually, the rise time is taken as the time required for the response to rise from 0% to 100% of the final or steady-state value.

By assuming the rise time to be equal to the time required for the response to rise from 0% to 100%, we can find the rise time by setting the value of $x(t)$, given by Eq. (4.54) at t_r, equal to one:

$$x(t_r) = 1 = 1 - e^{-\zeta\omega_n t_r}\left(\frac{\zeta\omega_n}{\omega_d}\sin\omega_d t_r + \cos\omega_d t_r\right) \qquad (4.60)$$

Noting that $e^{-\zeta\omega_n t_r} \neq 0$, Eq. (4.60) yields

$$\left(\frac{\zeta\omega_n}{\omega_d}\sin\omega_d t_r + \cos\omega_d t_r\right) = 0$$

or

$$\tan\omega_d t = -\frac{\sqrt{1-\zeta^2}}{\zeta} \qquad (4.61)$$

This gives the rise time t_r as

$$t_r = \frac{1}{\omega_d}\tan^{-1}\left(-\frac{\sqrt{1-\zeta^2}}{\zeta}\right) = \frac{\pi - \alpha}{\omega_d} \qquad (4.62)$$

where α is given by Eq. (4.56). Equation (4.56) indicates that the rise time t_r can be reduced by increasing the value of ω_d or ζ.

3. **Maximum overshoot (M_p):** The maximum overshoot is the maximum peak value of the response compared to the final or steady-state value ($x(\infty)$ or x_{ss}), expressed as a percentage of the steady-state value. It can be computed as

$$\text{Overshoot} = \frac{x(t_p) - x(\infty)}{x(\infty)} \qquad (4.63)$$

Substituting Eq. (4.59) into the expression for $x(t)$, Eq. (4.54), we obtain

$$x(t_p) \equiv 1 + M_p = 1 - e^{-\frac{\zeta\omega_n\pi}{\omega_d}}\left(\frac{\zeta\omega_n}{\omega_d}\sin\pi + \cos\pi\right) = 1 + e^{-\frac{\zeta\omega_n\pi}{\omega_d}} \qquad (4.64)$$

Thus the overshoot is given by

$$M_p = e^{-\frac{\zeta\omega_n\pi}{\omega_d}} = e^{-\frac{\zeta\pi}{\sqrt{1-\zeta^2}}} \qquad (4.65)$$

The percent overshoot can be obtained as

$$\% M_p = 100e^{-\frac{\zeta\pi}{\sqrt{1-\zeta^2}}} \qquad (4.66)$$

By inverting the relation in Eq. (4.66), we can find the damping ratio (ζ) for a given percent overshoot as

$$\zeta = -\frac{\ln\left(\% \, M_p/100\right)}{\sqrt{\pi^2 + \ln^2\left(\% \, M_p/100\right)}}$$ (4.67)

The overshoot, given by Eq. (4.65), is shown graphically in Fig. 4.31.

4. *Settling time:* The settling time, by definition, is the time during which $x(t)$ in Eq. (4.55) reaches and stays within $\pm 2\%$ of the steady-state value, x_{final}. By assuming the cosine term in Eq. (4.55) to be approximately equal to one, the settling time is given by the time it takes for the multiplication factor of the cosine term to reach a value of 0.02:

$$e^{-\zeta\omega_n t_s}\sqrt{1 + \left(\frac{\zeta\omega_n}{\omega_d}\right)^2} = e^{-\zeta\omega_n t_s}\frac{1}{\sqrt{1-\zeta^2}} = 0.02$$

which gives

$$t_s = \frac{-\ln\left(0.02\sqrt{1-\zeta^2}\right)}{\zeta\omega_n}$$ (4.68)

FIGURE 4.31   Variation of percent overshoot with damping ratio.

As ζ varies from 0 to 0.9, the numerator in Eq. (4.68) can be seen to vary from 3.01 to 4.74. Thus the settling time, valid approximately for all values of ζ, can be taken as

$$t_s \cong \frac{4}{\zeta \omega_n} \tag{4.69}$$

5. **Delay time (t_d):** The delay time is the time required for the response to reach 50% of the final or steady-state value for the first time.

Response Characteristics from Transfer Function

EXAMPLE 4.28

Find the peak time (t_p), percent overshoot (% M_p), settling time (t_s), and rise time (t_r) from the transfer function of the system given by

$$T(s) = \frac{X(s)}{F(s)} = \frac{225}{s^2 + 15s + 225} \tag{E.1}$$

Solution: The natural frequency of the system can be found from the last term in the denominator of Eq. (E.1):

$$\omega_n = \sqrt{225} = 15 \text{ rad/s} \tag{E.2}$$

The damping ratio can be determined from the middle term in the denominator of Eq. (E.1) as

$$2\zeta \omega_n = 15 \text{ or } \zeta = \frac{15}{2\omega_n} = \frac{15}{2(15)} = 0.5 \tag{E.3}$$

Substitution of the values of ω_n and ζ into Eqs. (4.59), (4.62), (4.66), and (4.69) yields

$$\text{Peak time} = t_p = \frac{\pi}{\omega_d} = \frac{\pi}{\omega_n \sqrt{1 - \zeta^2}} = \frac{\pi}{15\sqrt{1 - 0.5^2}} = 0.2418 \text{ s} \tag{E.4}$$

$$\text{Percent overshoot} = \% M_p = 100 e^{-\frac{\pi \zeta}{\sqrt{1-\zeta^2}}} = 100 e^{-\left(\frac{\pi(0.5)}{\sqrt{1-0.5^2}}\right)} = 100(0.1231) = 12.31 \tag{E.5}$$

$$\text{Settling time} = t_s = \frac{4}{\zeta \omega_n} = \frac{4}{0.5(15)} = 0.5333 \text{ s} \tag{E.6}$$

$$\text{Rise time} = t_r = \frac{\pi - \alpha}{\omega_d} = \frac{\pi - \tan^{-1}\left(\frac{\zeta}{\sqrt{1 - \zeta^2}}\right)}{\omega_d}$$

$$= \frac{\pi - \tan^{-1}\left(\frac{0.5}{\sqrt{1 - 0.5^2}}\right)}{15\sqrt{1 - 0.5^2}} = 0.2015 \text{ s} \tag{E.7}$$

This example demonstrates that the response characteristics—peak time, percent overshoot, settling time, and rise time—can be determined without the tedious task of finding the time response through an inverse Laplace transform, plotting the time response, and taking measurements from the time response plot.

∎

System Parameters from Known Response Characteristics

EXAMPLE 4.29

Determine the values of the mass moment of inertia and torsional damping constant of a torsional system, shown in Fig. 4.32, to achieve a 25% overshoot and a settling time of 2.5 s for a step input torque of $T_0(t)$. The torsional stiffness of the system is 10 N-m/rad.

Solution: The transfer function of the system can be expressed as

$$T(s) = \frac{\Theta(s)}{T_0(s)} = \frac{(1/J)}{s^2 + \dfrac{c_t}{J}s + \dfrac{k_t}{J}}$$

(E.1)

From the last term in the denominator of Eq. (E.1), we obtain

$$\omega_n = \sqrt{\frac{k_t}{J}}$$

(E.2)

FIGURE 4.32

The middle term in the denominator of Eq. (E.1) gives

$$2\zeta\omega_n = \frac{c_t}{J} \qquad \text{(E.3)}$$

Since the settling time is given as 2.5 s, we have (from Eq. (4.69)):

$$t_s = \frac{4}{\zeta\omega_n} = 2.5 \qquad \text{or} \qquad \zeta\omega_n = 1.6 \qquad \text{(E.4)}$$

Equations (E.3) and (E.4) give

$$2\zeta\omega_n = 3.2 = \frac{c_t}{J} \qquad \text{(E.5)}$$

Equations (E.2) and (E.4) yield

$$\zeta = \frac{1.6}{\omega_n} = 1.6\sqrt{\frac{J}{k_t}} \qquad \text{(E.6)}$$

Using the known percent overshoot, the damping ratio can be determined from Eq. (4.67) as

$$\zeta = -\frac{\ln(\% M_p/100)}{\sqrt{\pi^2 + \ln^2(\% M_p/100)}} = -\frac{\ln(25/100)}{\sqrt{\pi^2 + \ln^2(25/100)}} = 0.4037 \qquad \text{(E.7)}$$

Equation (E.4) gives

$$\omega_n = \frac{1.6}{\zeta} = \frac{1.6}{0.4037} = 3.9633 \text{ rad/s} \qquad \text{(E.8)}$$

Equation (E.2) yields

$$J = \frac{k_t}{\omega_n^2} = \frac{10}{3.9633^2} = 0.6366 \text{ kg-m}^2 \qquad \text{(E.9)}$$

The torsional damping constant c_t can be found from Eq. (E.5) as

$$c_t = 3.2J = 3.2(0.6366) = 2.0372 \text{ N-m-s/rad} \qquad \text{(E.10)}$$

∎

4.8   Numerical Methods

The determination of the response of a system subjected to arbitrary forcing functions using numerical methods is called *numerical simulation*. The analytical methods discussed so far become tedious and sometimes even impossible to use in finding the response of a system if the forcing function or excitation cannot be described in simple analytical form, or if experimentally determined force data (such as the ground acceleration history measured during an earthquake) is to be used. Numerical simulations can be used to check the accuracy of analytical solutions, especially if the system is complex. In a similar manner, the numerical solutions are to be checked using analytical methods wherever possible. The numerical methods of solving single-degree-of-freedom systems subjected to arbitrary forcing functions are considered in this section.

Analytical solutions are extremely useful in understanding the system's behavior with respect to changes in its parameters. The analytical solutions directly help in designing systems to meet any specified response characteristics by selecting the parameter values appropriately. If analytical solution becomes difficult, the response of the system can be found using a suitable numerical integration procedure. Several methods are available for numerically integrating ordinary differential equations. The Runge-Kutta methods are quite popular for the numerical solution of differential equations.

Consider the equation of motion of a damped single-degree-of-freedom system subject to an arbitrary force $f(t)$:

$$m\ddot{x}(t) + c\dot{x}(t) + kx(t) = f(t) \qquad (4.70)$$

with the initial conditions $x\,(t = 0) = x_0$ and $\dot{x}\,(t = 0) = \dot{x}_0$. Most numerical methods assume that the differential equation is in the form of a first-order differential equation (or a set of simultaneous first-order differential equations). As such, we need to convert the second-order differential equation, Eq. (4.70), into an equivalent set of two first-order differential equations. For this, we introduce the unknown functions

$$x_1(t) = x(t), \qquad x_2(t) = \dot{x}(t) = \frac{dx(t)}{dt} \equiv \dot{x}_1(t) \qquad (4.71)$$

and rewrite Eq. (4.70) as

$$m\ddot{x}(t) = -c\dot{x}(t) - kx(t) + f(t) \qquad (4.72)$$

or, in view of the functions $x_1(t)$ and $x_2(t)$ introduced in Eq. (4.71),

$$m\dot{x}_2 = -cx_2(t) - kx_1(t) + f(t) \qquad (4.73)$$

Equation (4.73) along with the second relation given in Eq. (4.71) can be expressed as

$$\dot{x}_1(t) = x_2(t) \qquad (4.74)$$

$$\dot{x}_2(t) = \frac{c}{m}x_2(t) - \frac{k}{m}x_1(t) + \frac{1}{m}f(t) \tag{4.75}$$

Equations (4.74) and (4.75) represent two first-order differential equations which together denote Eq. (4.70). Equations (4.74) and (4.75) can be expressed in vector form as

$$\vec{\dot{X}}(t) = \vec{F}(\vec{X}, t) \tag{4.76}$$

where

$$\vec{X}(t) = \begin{Bmatrix} x_1(t) \\ x_2(t) \end{Bmatrix}, \qquad \vec{\dot{X}}(t) = \begin{Bmatrix} \dot{x}_1(t) \\ \dot{x}_2(t) \end{Bmatrix},$$

$$\vec{F}(\vec{X}, t) = \begin{Bmatrix} F_1(t) \\ F_2(t) \end{Bmatrix} \equiv \begin{Bmatrix} x_2(t) \\ -\frac{k}{m}x_1(t) - \frac{c}{m}x_2(t) + \frac{1}{m}f(t) \end{Bmatrix} \tag{4.77}$$

4.8.1 Runge-Kutta Methods

In most of the numerical methods, improved solutions are found from the present solution (starting with a known initial value at time zero) according to the formula

$$x_{i+1} = x_i + \Delta x_i \tag{4.78}$$

where x_{i+1} is the value of x at $t = t_{i+1}$, x_i is the value of x at $t = t_i$, and Δx is the incremental improvement added to x_i. If the solution, $x(t)$, is to be found over the time interval $0 \le t \le T$, the total time T is divided into n equal parts with $\Delta t = T/n$, so that $t_0 = 0, t_1 = \Delta t, t_2 = 2\,\Delta t, \ldots, t_i = i\,\Delta t, \ldots, t_n = n\,\Delta t = T$.

In the Runge-Kutta methods, the approximate formula used for obtaining the solution x_{i+1} from x_i is made to coincide with the Taylor's series expansion of x at x_{i+1} up to terms of order $(\Delta t)^k$, where k denotes the order of the Runge-Kutta method. The Taylor's series expansion of $x(t)$ at $t + \Delta t$ is given by

$$x(t + \Delta t) = x(t) + \dot{x}\,\Delta t + \ddot{x}\frac{(\Delta t)^2}{2!} + \dddot{x}\frac{(\Delta t)^3}{3!} + \cdots \tag{4.79}$$

In contrast to Eq. (4.79), which requires higher-order derivatives, the Runge-Kutta methods do not require explicitly derivatives beyond the first.

In the fourth-order Runge-Kutta method, which is most commonly used, the following recurrence formula is used to find the values of $\vec{X}(t)$ at different time stations t_i starting from the known initial vector, $\vec{X}_0 = \begin{Bmatrix} x(t = 0) \\ \dot{x}(t = 0) \end{Bmatrix} = \begin{Bmatrix} x_0 \\ \dot{x}_0 \end{Bmatrix}$:

$$\vec{X}_{i+1} = \vec{X}_i + \frac{1}{6}[\vec{K}_1 + 2\vec{K}_2 + 2\vec{K}_3 + \vec{K}_4] \tag{4.80}$$

where

$$\vec{K}_1 = h\vec{F}(\vec{X}_i, t_i) \tag{4.81}$$

$$\vec{K}_2 = h\vec{F}\left(\vec{X}_i + \frac{1}{2}\vec{K}_1, t_i + \frac{1}{2}h\right) \tag{4.82}$$

$$\vec{K}_3 = h\vec{F}\left(\vec{X}_i + \frac{1}{2}\vec{K}_2, t_i + \frac{1}{2}h\right) \tag{4.83}$$

$$\vec{K}_4 = h\vec{F}(\vec{X}_i + \vec{K}_3, t_{i+1}) \tag{4.84}$$

The method is stable and self-starting—that is, only the value of the vector function $\vec{F}$ at a single previous time station is required to find the function value at the current time station. The following example illustrates the procedure.

◼◼◼◼◼◼ **Response Using Runge-Kutta Method**

EXAMPLE 4.30

Find the response of a single-degree-of-freedom system subjected to a force with the equation of motion

$$500\ddot{x} + 200\dot{x} + 750x = F(t) = 2,000 \tag{E.1}$$

so that $m = 500, c = 200, k = 750$, and $F(t) = F_0 = 2,000$. Use the fourth-order Runge-Kutta method. Assume the initial conditions as $x(t = 0) = x_0 = 0$ and $\dot{x}(t = 0) = \dot{x}_0 = 0$.

Solution: The equation of motion given by (E.1) can be expressed as a system of two first-order differential equations as shown in Eq. (4.76) with

$$\vec{f} = \left\{\begin{matrix} f_1(t) \\ f_2(t) \end{matrix}\right\} = \left\{\begin{matrix} x_2(t) \\ \dfrac{1}{500}(2,000 - 200x_2 - 750x_1) \end{matrix}\right\}$$

and

$$\vec{X}_0 = \left\{\begin{matrix} x_1(0) \\ x_2(0) \end{matrix}\right\} = \left\{\begin{matrix} 0 \\ 0 \end{matrix}\right\}$$

The response is computed over the time $(0, T)$. The time duration of $T = 20$ s is divided into 400 equal time steps so that

$$\Delta t = h = \frac{T}{400} = \frac{20}{400} = 0.05 \text{ s}$$

TABLE 4.1

i	$x_1(i) = x(t_i)$	$x_2(i) = x(t_i)$
1	0.000000e+000	0.000000e+000
2	4.965271e−003	1.978895e−001
3	1.971136e−002	3.911261e−001
4	4.398987e−002	5.790846e−001
5	7.752192e−002	7.611720e−001
6	1.199998e−001	9.368286e−001
7	1.710888e−001	1.105530e+000
8	2.304287e−001	1.266787e+000
9	2.976359e−001	1.420150e+000
10	3.723052e−001	1.565205e+000
⋮		
391	2.675602e+000	−6.700943e−002
392	2.672270e+000	−6.622167e−002
393	2.668983e+000	−6.520372e−002
394	2.665753e+000	−6.396391e−002
395	2.662590e+000	−6.251125e−002
396	2.659505e+000	−6.085533e−002
397	2.656508e+000	−5.900634e−002
398	2.653608e+000	−5.697495e−002
399	2.650814e+000	−5.477231e−002
400	2.648133e+000	−5.241000e−002

Thus $t_0 = 0$, $t_1 = 0.05$, $t_2 = 0.10$, $t_3 = 0.15$, ..., $t_{400} = 20.0$. The Runge-Kutta method is applied to find the response of the system ($x_1(t)$). Typical solution vectors generated $\vec{X}_i$ for $i = 1, 2, 3, \ldots,$ 400 are shown in Table 4.1. The response of the system is shown plotted in Fig. 4.32.

■

4.9   Response to Irregular Forcing Conditions Using Numerical Methods

In the method of direct numerical integration of the equation of motion (numerical solution of differential equations) presented in Section 4.8, it was assumed that the forcing functions $F(t)$ are available as functions of time in an explicit manner. In many practical problems, however, the forcing functions $F(t)$ are not available in the form of analytical expressions. When a forcing function is determined experimentally, $F(t)$ may be known as an *irregular curve*. Sometimes, only the values of $F(t) = F_i$ at a series of points $t = t_i$

may be available, in the form of a diagram or a table. In such cases, we can fit polynomials or some such curves to the data and use them in the Duhamel integral, Eq. (4.31), to find the response of the system. Another more common method of finding the response involves dividing the time axis into a number of discrete points and using a simple variation of $F(t)$ during each time step. We shall present this numerical approach in this section, using a linear interpolation function for $F(t)$ [4.8].

Let the function $F(t)$ vary with time in an arbitrary manner, as indicated in Fig. 4.33. This forcing function is approximated by a piecewise linear function. In the piecewise linear interpolation, the variation of $F(t)$ in any time interval is assumed to be linear, as shown in Fig. 4.34. In this case, the response of the system in the time interval $t_{j-1} \leq t \leq t_j$ can be found by adding the response due to the linear (ramp) function applied during the current interval to the response existing at $t = t_{j-1}$ (initial condition). This gives

$$
\begin{aligned}
x(t) = {} & \frac{\Delta F_j}{k \, \Delta t_j} \left[t - t_{j-1} - \frac{2\zeta}{\omega_n} + e^{-\zeta \omega_n (t - t_{j-1})} \times \left\{ \frac{2\zeta}{\omega_n} \cos \omega_d (t - t_{j-1}) - \frac{\omega_d^2 - \zeta^2 \omega_n^2}{\omega_n^2 \omega_d} \sin \omega_d (t - t_{j-1}) \right\} \right] \\
& + \frac{F_{j-1}}{k} \left[1 - e^{-\zeta \omega_n (t - t_{j-1})} \left\{ \cos \omega_d (t - t_{j-1}) + \frac{\zeta \omega_n}{\omega_d} \sin \omega_d (t - t_{j-1}) \right\} \right] \\
& + e^{-\zeta \omega_n (t - t_{j-1})} \left[x_{j-1} \cos \omega_d (t - t_{j-1}) + \frac{\dot{x}_{j-1} + \zeta \omega_n x_{j-1}}{\omega_d} \sin \omega_d (t - t_{j-1}) \right] \qquad (4.85)
\end{aligned}
$$

FIGURE 4.33   Arbitrary forcing function.

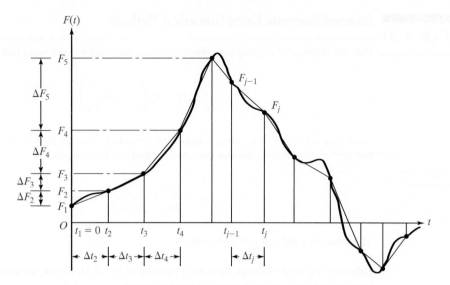

FIGURE 4.34   Approximation of forcing function as a piecewise linear function.

where $\Delta F_j = F_j - F_{j-1}$. By setting $t = t_j$ in Eq. (4.85), we obtain the response at the end of the interval Δt_j:

$$
\begin{aligned}
x_j &= \frac{\Delta F_j}{k\,\Delta t_j}\left[\Delta t_j - \frac{2\zeta}{\omega_n} + e^{-\zeta\omega_n\Delta t_j}\left\{\frac{2\zeta}{\omega_n}\cos\omega_d\,\Delta t_j - \frac{\omega_d^2 - \zeta^2\omega_n^2}{\omega_n^2\omega_d}\sin\omega_d\,\Delta t_j\right\}\right] \\
&\quad + \frac{F_{j-1}}{k}\left[1 - e^{-\zeta\omega_n\Delta t_j}\left\{\cos\omega_d\,\Delta t_j + \frac{\zeta\omega_n}{\omega_d}\sin\omega_d\,\Delta t_j\right\}\right] \\
&\quad + e^{-\zeta\omega_n\Delta t_j}\left[x_{j-1}\cos\omega_d\,\Delta t_j + \frac{\dot{x}_{j-1} + \zeta\omega_n x_{j-1}}{\omega_d}\sin\omega_d\,\Delta t_j\right]
\end{aligned}
\tag{4.86}
$$

By differentiating Eq. (4.85) with respect to t and substituting $t = t_j$, we obtain the velocity at the end of the interval:

$$
\begin{aligned}
\dot{x}_j &= \frac{\Delta F_j}{k\,\Delta t_j}\left[1 - e^{-\zeta\omega_n\Delta t_j}\left\{\cos\omega_d\,\Delta t_j + \frac{\zeta\omega_n}{\omega_d}\sin\omega_d\,\Delta t_j\right\}\right] \\
&\quad + \frac{F_{j-1}}{k}e^{-\zeta\omega_n\Delta t_j}\frac{\omega_n^2}{\omega_d}\sin\omega_d\,\Delta t_j + e^{-\zeta\omega_n\Delta t_j} \\
&\quad \times \left[\dot{x}_{j-1}\cos\omega_d\,\Delta t_j - \frac{\zeta\omega_n}{\omega_d}\left(\dot{x}_{j-1} + \frac{\omega_n}{\zeta}x_{j-1}\right)\sin\omega_d\,\Delta t_j\right]
\end{aligned}
\tag{4.87}
$$

Equations (4.86) and (4.87) are the recurrence relations for finding the response of the system at the end of jth time step.

EXAMPLE 4.31

Damped Response Using Numerical Methods

Find the response of a spring-mass-damper system subjected to the forcing function

$$F(t) = F_0\left(1 - \sin\frac{\pi t}{2t_0}\right) \tag{E.1}$$

in the interval $0 \le t \le t_0$, using a numerical procedure. Assume $F_0 = 1$, $k = 1$, $m = 1$, $\zeta = 0.1$, and $t_0 = \tau_n/2$, where τ_n denotes the natural period of vibration given by

$$\tau_n = \frac{2\pi}{\omega_n} = \frac{2\pi}{(k/m)^{1/2}} = 2\pi \tag{E.2}$$

The values of x and $\dot{x}$ at $t = 0$ are zero.

Solution: Figure 4.35 shows the forcing function of Eq. (E.1). For the numerical computations, the time interval 0 to t_0 is divided into 10 equal steps with

$$\Delta t_i = \frac{t_0}{10} = \frac{\pi}{10}; \qquad i = 2, 3, \ldots, 11 \tag{E.3}$$

FIGURE 4.35   Forcing function.

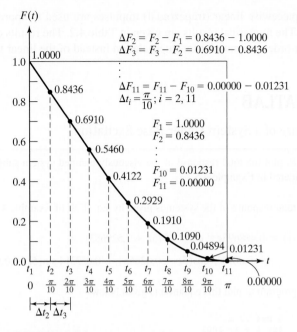

FIGURE 4.36   Piecewise linear approximation.

TABLE 4.2   Response of the system

i	t_i	$x(t_i)$ Obtained According to Fig. 4.36 (Idealization 4)
1	0	0.00000
2	0.1π	0.04541
3	0.2π	0.16377
4	0.3π	0.32499
5	0.4π	0.49746
6	0.5π	0.65151
7	0.6π	0.76238
8	0.7π	0.81255
9	0.8π	0.79323
10	0.9π	0.70482
11	π	0.55647

In Fig. 4.36, piecewise linear (trapezoidal) impulses are used to approximate the forcing function $F(t)$. The numerical results are given in Table 4.2. The results can be improved by using a higher-order polynomial for interpolation instead of the linear function.

4.10   Examples Using MATLAB

Total Response of a System Under Base Excitation

EXAMPLE 4.32

Using MATLAB, plot the total response of the viscously damped system subject to harmonic base excitation considered in Example 4.5.

Solution: The total response of the system is given by Eq. (E.8) of Example 4.5:

$$x(t) = 0.488695e^{-t}\cos(19.975t - 1.529683)$$
$$+ 0.001333\cos(5t - 0.02666) + 0.053314\sin(5t - 0.02666)$$

The MATLAB program to plot this equation is given below.

```
% Ex4_32.m
for i = 1: 1001
    t(i) = (i - 1)*10/1000;
```

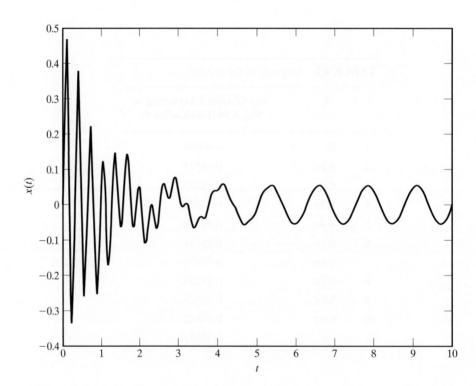

```
        x(i) = 0.488695 * exp(-t(i)) * cos(19.975*t(i)-1.529683) + ...
               0.001333*cos(5*t(i)-0.02666) + 0.053314 * sin(5*t(i)
               - 0.02666);
    end
    plot(t,x);
    xlabel('t');
    ylabel('x(t)');
```

■

EXAMPLE 4.33

Impulse Response of a Structure

Using MATLAB, plot the impulse response of the single-degree-of-freedom structure due to (a) a single impact and (b) a double impact considered in Examples 4.7 and 4.8.

Solution: The impulse responses of the structure due to single and double impacts are given by Eqs. (E.1) and (E.3) of Examples 4.7 and 4.8, respectively:

$$x(t) = 0.20025e^{-t} \sin 19.975t \tag{E.1}$$

$$x(t) = \left\{ \begin{array}{l} 0.20025e^{-t} \sin 19.975t;\ 0 \le t \le 0.2 \\ 0.20025e^{-t} \sin 19.975t + 0.100125e^{-(t-0.2)} \sin 19.975(t - 0.2);\ t \ge 0.2 \end{array} \right\} \tag{E.2}$$

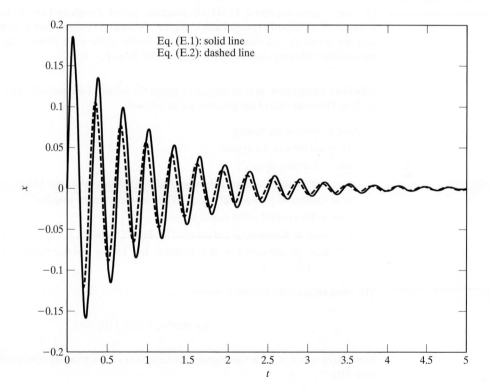

The MATLAB program to plot Eqs. (E.1) and (E.2) is given below.

```
% Ex4_33.m
for i = 1: 1001
    t (i) = (i-1)*5/1000;
    x1(i) = 0.20025 * exp(-t(i)) * sin(19.975*t(i));
    if t(i) > 0.2
        a = 0.100125;
    else
        a = 0.0;
    end
    x2(i) = 0.20025 * exp(-t(i)) * sin(19.975*t(i)) + ...
        a * exp(-(t(i)-0.2)) * sin(19.975*(t(i)-0.2));
end
plot(t,x1);
gtext('Eq. (E.1): solid line');
hold on;
plot(t,x2,'-');
gtext('Eq. (E.2): dash line');
xlabel('t');
```

∎

EXAMPLE 4.34

Response Under a Periodic Force

Develop a general-purpose MATLAB program, called **Program4.m**, to find the steady-state response of a viscously damped single-degree-of-freedom system under a periodic force. Use the program to find the response of a system that is subject to the force shown in the accompanying figure with the following data: $m = 100$ kg, $k = 10^5$ N/m, $\zeta = 0.1$.

Solution: Program4.m is developed to accept the values of the periodic force at n discrete values of time. The input data of the program are as follows:

xm = mass of the system

xk = stiffness of the system

xai = damping ratio (ζ)

n = number of equidistant points at which the values of the force $F(t)$ are known

m = number of Fourier coefficients to be considered in the solution

time = time period of the function $F(t)$

f = array of dimension n that contains the known values of $F(t)$; $f(i) = F(t_i), i = 1, 2, \ldots, n$

t = array of dimension n that contains the known discrete values of time t; $t(i) = t_i$, $i = 1, 2, \ldots, n$

The program gives the following output:

$$\text{step number } i, t(i), f(i), x(i)$$

where $x(i) = x(t = t_i)$ is the response at time step i. The program also plots the variation of x with time.

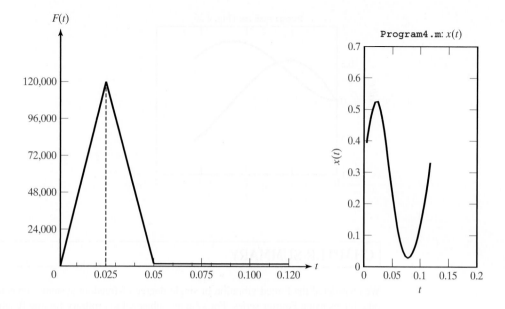

Response Under Arbitrary Forcing Function

EXAMPLE 4.35

Develop a general-purpose MATLAB program, called **Program5.m**, to find the response of a viscously damped spring-mass system under an arbitrary forcing function using the methods of Section 4.9. Use the program to find the solution of Example 4.31.

Solution: **Program5.m** is developed to accept the values of the applied force at n discrete values of time. The program requires the following input data:

n = number of time stations at which the values of the forcing function are known

t = array of size n containing the values of time at which the forcing function is known

f = array of size n containing the values of the forcing function at various time stations according to the idealization of Fig. 4.34 (Fig. 4.36 for Example 4.31)

ff = array of size n containing the values of the forcing function at various time stations according to the idealization of Fig. 4.34 (Fig. 4.36 for Example 4.31)

xai = damping factor (ζ)

omn = undamped natural frequency of the system

$delt$ = incremental time between consecutive time stations

xk = spring stiffness

The program gives the values of $x(i)$ obtained by the numerical method at the various time stations i. The program also plots the variation of x with time.

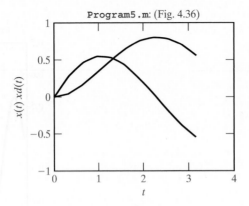

Program5.m: (Fig. 4.36)

$x(t)$ $xd(t)$

■

CHAPTER SUMMARY

We considered the forced vibration of single-degree-of-freedom systems subjected to general periodic forces using Fourier series. For systems subjected to arbitrary forcing functions, we discussed the methods of convolution integral and Laplace transform for finding the response of undamped and damped systems. We studied the concept of response spectra and its use in finding the response of systems subjected to earthquake excitations. Finally, we considered numerical methods, including the fourth-order Runge-Kutta method, for finding the response of systems subjected to arbitrary forces, including those described numerically.

Now that you have finished this chapter, you should be able to answer the review questions and solve the problems given below.

REFERENCES

4.1   S. S. Rao, *Applied Numerical Methods for Engineers and Scientists*, Prentice Hall, Upper Saddle River, NJ, 2002.

4.2   M. Paz, *Structural Dynamics: Theory and Computation* (2nd ed.), Van Nostrand Reinhold, New York, 1985.

4.3   E. Kreyszig, *Advanced Engineering Mathematics* (7th ed.), Wiley, New York, 1993.

4.4   F. Oberhettinger and L. Badii, *Tables of Laplace Transforms*, Springer Verlag, New York, 1973.

4.5   G. M. Hieber et al., "Understanding and measuring the shock response spectrum. Part I," *Sound and Vibration*, Vol. 8, March 1974, pp. 42–49.

4.6   R. E. D. Bishop, A. G. Parkinson, and J. W. Pendered, "Linear analysis of transient vibration," *Journal of Sound and Vibration*, Vol. 9, 1969, pp. 313–337.

4.7   Y. Matsuzaki and S. Kibe, "Shock and seismic response spectra in design problems," *Shock and Vibration Digest*, Vol. 15, October 1983, pp. 3–10.

4.8   S. Timoshenko, D. H. Young, and W. Weaver, Jr., *Vibration Problems in Engineering* (4th ed.), Wiley, New York, 1974.

4.9 R. A. Spinelli, "Numerical inversion of a Laplace transform," *SIAM Journal of Numerical Analysis*, Vol. 3, 1966, pp. 636–649.

4.10 R. Bellman, R. E. Kalaba, and J. A. Lockett, *Numerical Inversion of the Laplace Transform*, American Elsevier, New York, 1966.

4.11 S. G. Kelly, *Fundamentals of Mechanical Vibrations*, McGraw-Hill, New York, 1993.

4.12 P.-T. D. Spanos, "Digital synthesis of response-design spectrum compatible earthquake records for dynamic analyses," *Shock and Vibration Digest*, Vol. 15, No. 3, March 1983, pp. 21–30.

4.13 C. M. Close, D. K. Frederick, and J. C. Newell, *Modeling and Analysis of Dynamic Systems* (3rd ed.), Wiley, New York, 2002.

4.14 G. F. Franklin, J. D. Powell, and A. Emami-Naeini, *Feedback Control of Dynamic Systems* (5th ed.), Pearson Prentice Hall, Upper Saddle River, NJ, 2006.

4.15 N. S. Nise, *Control Systems Engineering* (3rd ed.), Wiley, New York, 2000.

4.16 K. Ogata, *System Dynamics* (4th ed.), Pearson Prentice Hall, Upper Saddle River, NJ, 2004.

REVIEW QUESTIONS

4.1 Give brief answers to the following:

1. What is the basis for expressing the response of a system under periodic excitation as a summation of several harmonic responses?
2. Indicate some methods for finding the response of a system under nonperiodic forces.
3. What is the Duhamel integral? What is its use?
4. How are the initial conditions determined for a single-degree-of-freedom system subjected to an impulse at $t = 0$?
5. Derive the equation of motion of a system subjected to base excitation.
6. What is a response spectrum?
7. What are the advantages of the Laplace transform method?
8. What is the use of a pseudo spectrum?
9. How is the Laplace transform of a function $x(t)$ defined?
10. Define the terms *generalized impedance* and *admittance of a system*.
11. State the interpolation models that can be used for approximating an arbitrary forcing function.
12. How many resonant conditions are there when the external force is not harmonic?
13. How do you compute the frequency of the first harmonic of a periodic force?
14. What is the relation between the frequencies of higher harmonics and frequency of the first harmonic for a periodic excitation?
15. What is the difference between transient and steady-state responses?
16. What is a first-order system?
17. What is an impulse?
18. What are the properties of the Dirac delta function $\delta(t)$?

4.2 Indicate whether each of the following statements is true or false:

1. The change in momentum is called *impulse*.
2. The response of a system under arbitrary force can be found by summing the responses due to several elementary impulses.

3. The response spectrum corresponding to base excitation is useful in the design of machinery subject to earthquakes.
4. Some periodic functions cannot be replaced by a sum of harmonic functions.
5. The amplitudes of higher harmonics will be smaller in the response of a system.
6. The Laplace transform method takes the initial conditions into account automatically.
7. The equation of motion can be integrated numerically even when the exciting force is nonperiodic.
8. The response spectrum gives the maximum response of all possible single-degree-of-freedom systems.
9. For a harmonic oscillator, the acceleration and displacement spectra can be obtained from the velocity spectrum.
10. If two masses m_1 and m_2 get attached to one another after a collision, it is called an *elastic collision*.
11. The transient response characteristics can be found from the transfer function.
12. The Runge-Kutta method can be used to numerically solve differential equations of any order.
13. The Laplace transform of 1 is $\frac{1}{s}$.

4.3 Fill in each of the following blanks with the appropriate word:

1. The response of a linear system under any periodic force can be found by _____ appropriate harmonic responses.
2. Any nonperiodic function can be represented by a _____ integral.
3. An impulse force has a large magnitude and acts for a very _____ period of time.
4. The response of a single-degree-of-freedom system to a unit _____ is known as the *impulse response function*.
5. The Duhamel integral is also known as the _____ *integral*.
6. The variation of the maximum response with the natural frequency of a single-degree-of-freedom system is known as _____ *spectrum*.
7. The transient response of a system can be found using the _____ integral.
8. The complete solution of a vibration problem is composed of the _____ state and transient solutions.
9. The Laplace transform method converts a differential equation into an _____ equation.
10. The transfer function is the _____ of the generalized impedance.
11. An impulse can be measured by finding the change in _____ of the system.
12. The Duhamel integral is based on the _____ response function of the system.
13. The Duhamel integral can be used to find the response of _____ single-degree-of-freedom systems under arbitrary excitations.
14. The velocity response spectrum, determined from the acceleration spectrum, is known as the _____ *spectrum*.
15. Any periodic forcing function can be expanded in _____ series.
16. In the Laplace domain, $\lim_{s \to 0}[sX(s)]$ gives _____ value of the response.
17. A change in momentum of a system gives the _____.
18. Total response of a system is composed of transient and _____ values.
19. The Laplace transform of $x(t)$ is denoted as _____.
20. $f(t)$ denotes the inverse Laplace transform of _____.
21. The equation of motion $m\ddot{x} + c\dot{x} + kx = f(t)$ corresponds to _____ order system.
22. The Laplace transform of $\delta(t)$ is _____.

4.4   Select the most appropriate answer out of the choices given:

1. The transient part of the solution arises from
   a. forcing function   **b.** initial conditions   **c.** bounding conditions
2. If a system is subjected to a suddenly applied nonperiodic force, the response will be
   a. periodic   **b.** transient   **c.** steady
3. The initial conditions are to be applied to a
   a. steady-state solution   **b.** transient solution   **c.** total solution
4. The acceleration spectrum (S_a) can be expressed in terms of the displacement spectrum (S_d) as
   a. $S_a = -\omega_n^2 S_d$ **b.** $S_a = \omega_n S_d$ **c.** $S_a = \omega_n^2 S_d$
5. The pseudo spectrum is associated with
   a. pseudo acceleration   **b.** pseudo velocity   **c.** pseudo displacement
6. The Fourier coefficients are to be found numerically when the values of the function $f(t)$ are available
   a. in analytical form
   b. at discrete values of t
   c. in the form of a complex equation
7. The response of a single-degree-of-freedom system under base excitation, $y(t)$, can be determined by using the external force as
   a. $-m\ddot{y}$ **b.** $m\ddot{y}$ **c.** $m\ddot{y} + c\dot{y} + ky$
8. The response spectrum is widely used in
   a. building design under large live loads
   b. earthquake design
   c. design of machinery under fatigue
9. The equation of motion of a system subjected to base excitation, $y(t)$, is given by
   a. $m\ddot{x} + c\dot{x} + kx = -m\ddot{y}$
   b. $m\ddot{z} + c\dot{z} + kz = -m\ddot{y}; z = x - y$
   c. $m\ddot{x} + c\dot{x} + kx = -m\ddot{z}; z = x - y$
10. The function e^{-st} used in the Laplace transform is known as
    a. *kernel*   **b.** *integrand*   **c.** *subsidiary term*
11. The Laplace transform of $x(t)$ is defined by
    a. $\bar{x}(s) = \displaystyle\int_0^{\infty} e^{-st} x(t)\, dt$

    b. $\bar{x}(s) = \displaystyle\int_{-\infty}^{\infty} e^{-st} x(t)\, dt$

    c. $\bar{x}(s) = \displaystyle\int_0^{\infty} e^{st} x(t)\, dt$

12. In the Laplace domain, $\lim_{s \to 0}[sX(s)]$ gives the
    a. initial value   **b.** transient value   **c.** steady-state value
13. $F(t) = \alpha t$ corresponds to
    a. an impulse   **b.** step force   **c.** ramp force
14. $f(t) = \delta(t - \tau)$ corresponds to a force applied at
    a. $t - \tau = 0$ **b.** $t - \tau < 0$ **c.** $t - \tau > 0$
15. In a perfect elastic collision of two masses m_1 and m_2, the quantity conserved is
    a. energy   **b.** momentum   **c.** velocity

16. The step response of an overdamped system exhibits
   a. no oscillations   **b.** oscillations   **c.** overshoot

17. The method used to express $\dfrac{3s + 4}{(s + 1)(s + 2)}$ as $\dfrac{C_1}{s + 1} + \dfrac{C_2}{s + 2}$ is called

   a. separation   **b.** partial fractions   **c.** decomposition

18. Most numerical methods of solving differential equations assume that the order of the equation is
   a. one   **b.** two   **c.** arbitrary

4.5 Match the items in the two columns below:

1. $x(t) = \dfrac{1}{m\omega_d} e^{-\zeta\omega_n t} \sin \omega_d t$ **a.** Inverse Laplace transform of $\bar{x}(s)$

2. $x(t) = \displaystyle\int_0^t F(\tau)g(t - \tau)\, d\tau$ **b.** Generalized impedance function

3. $x(t) = \mathcal{L}^{-1}\bar{Y}(s)\bar{F}(s)$ **c.** Unit impulse response function

4. $\bar{Y}(s) = \dfrac{1}{ms^2 + cs + k}$ **d.** Laplace transform

5. $\bar{z}(s) = ms^2 + cs + k$ **e.** Convolution integral

6. $\bar{x}(s) = \displaystyle\int_0^\infty e^{-st} x(t)\, dt$ **f.** Admittance function

4.6 Match the following characteristics of transient response:

a. Peak time   **1.** Maximum peak value
b. Rise time   **2.** Time to attain maximum value
c. Maximum overshoot   **3.** Time to reach within $\pm 2\%$ of steady-state value
d. Settling time   **4.** Time to reach 50% of steady-state value
e. Decay time   **5.** Time to increase from 10% to 90% of steady-state value

---

PROBLEMS

Section 4.2 Response Under a General Periodic Force

**4.1
–4.4**   Find the steady-state response of the hydraulic control valve shown in Fig. 4.4(a) to the forcing functions obtained by replacing $x(t)$ with $F(t)$ and A with F_0 in Figs. 1.114–1.117.

4.5   Find the steady-state response of a viscously damped system to the forcing function obtained by replacing $x(t)$ and A with $F(t)$ and F_0, respectively, in Fig. 1.54(a).

4.6   The torsional vibrations of a driven gear mounted on a shaft (see Fig. 4.37) under steady conditions are governed by the equation

$$J_0\ddot{\theta} + k_t\theta = M_t$$

where k_t is the torsional stiffness of the driven shaft, M_t is the torque transmitted, J_0 is the mass moment of inertia, and θ is the angular deflection of the driven gear. If one of the 16 teeth on the driving gear breaks, determine the resulting torsional vibration of the driven gear for the following data.

Driven gear: $J_0 = 0.1$ N-m-s^2, speed $= 1{,}000$ rpm, driven shaft: material—steel, solid circular section with diameter 5 cm and length 1 m, $M_{t0} = 1{,}000$ N-m.

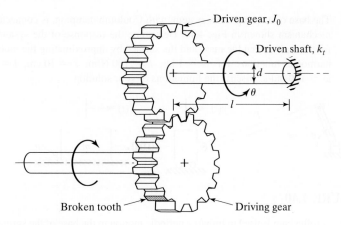

FIGURE 4.37

4.7   A slider-crank mechanism is used to impart motion to the base of a spring-mass-damper system, as shown in Fig. 4.38. Approximating the base motion $y(t)$ as a series of harmonic functions, find the response of the mass for $m = 1$ kg, $c = 10$ N-s/m, $k = 100$ N/m, $r = 10$ cm, $l = 1$ m, and $\omega = 100$ rad/s.

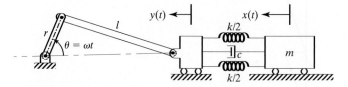

FIGURE 4.38

4.8   The base of a spring-mass-damper system is subjected to the periodic displacement shown in Fig. 4.39. Determine the response of the mass using the principle of superposition.

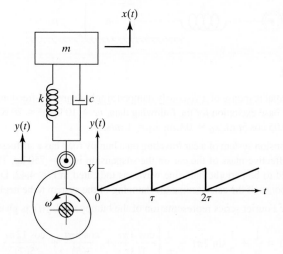

FIGURE 4.39

4.9 The base of a spring-mass system, with Coulomb damping, is connected to the slider-crank mechanism shown in Fig. 4.40. Determine the response of the system for a coefficient of friction μ between the mass and the surface by approximating the motion $y(t)$ as a series of harmonic functions for $m = 1$ kg, $k = 100$ N/m, $r = 10$ cm, $l = 1$ m, $\mu = 0.1$, and $\omega = 100$ rad/s. Discuss the limitations of your solution.

FIGURE 4.40

4.10 A roller cam is used to impart a periodic motion to the base of the spring-mass system shown in Fig. 4.41. If the coefficient of friction between the mass and the surface is μ, find the response of the system using the principle of superposition. Discuss the validity of the result.

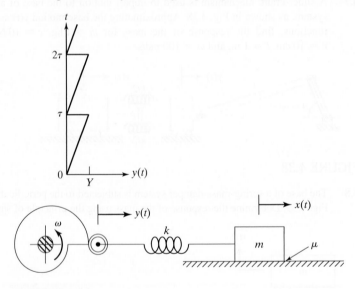

FIGURE 4.41

4.11 Find the total response of a viscously damped single-degree-of-freedom system subjected to a harmonic base excitation for the following data: $m = 10$ kg, $c = 20$ N-s/m, $k = 4{,}000$ N/m, $y(t) = 0.05 \cos 5t$ m, $x_0 = 0.1$ m, $\dot{x}_0 = 1$ m/s.

4.12 The suspension system of a car traveling on a bumpy road has a stiffness of $k = 5 \times 10^6$ N/m and the effective mass of the car on the suspension is $m = 750$ kg. The road bumps can be considered to be periodic half-sine waves as indicated in Fig. 4.42. Determine the displacement response of the car. Assume the damping of the system to be negligible.

Hint: The Fourier series representation of the bumpy road, $y(t)$, is given by

$$y(t) = \frac{1}{\pi} + \frac{1}{2} \sin 2\pi t - \frac{2}{\pi}\left\{ \frac{\cos 4\pi t}{1(3)} + \frac{\cos 8\pi t}{3(5)} + \frac{\cos 12\pi t}{5(7)} + \cdots \right\}$$

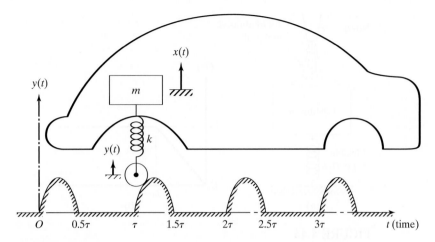

FIGURE 4.42

Section 4.3 Response Under a Periodic Force of Irregular Form

4.13 Find the response of a damped system with $m = 1$ kg, $k = 15$ kN/m, and $\zeta = 0.1$ under the action of a periodic forcing function, as shown in Fig. 1.119.

4.14 Find the response of a viscously damped system under the periodic force whose values are given in Problem 1.116. Assume that M_t denotes the value of the force in Newtons at time t_i. Use $m = 0.5$ kg, $k = 8,000$ N/m, and $\zeta = 0.06$.

4.15 Find the displacement of the water tank shown in Fig. 4.43(a) under the periodic force shown in Fig. 4.43(b) by treating it as an undamped single-degree-of-freedom system. Use the numerical procedure described in Section 4.3.

FIGURE 4.43

Section 4.5 Convolution Integral

4.16 Sandblasting is a process in which an abrasive material, entrained in a jet, is directed onto the surface of a casting to clean its surface. In a particular setup for sandblasting, the casting of mass m is placed on a flexible support of stiffness k as shown in Fig. 4.44(a). If the force exerted on the casting due to the sandblasting operation varies as shown in Fig. 4.44(b), find the response of the casting.

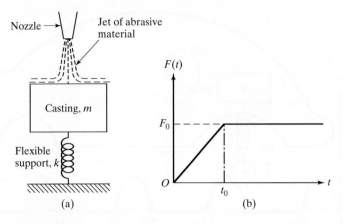

FIGURE 4.44

4.17 Find the displacement of a damped single-degree-of-freedom system under the forcing function $F(t) = F_0 e^{-\alpha t}$, where α is a constant.

4.18 A compressed air cylinder is connected to the spring-mass system shown in Fig. 4.45(a). Due to a small leak in the valve, the pressure on the piston, $p(t)$, builds up as indicated in Fig. 4.45(b). Find the response of the piston for the following data: $m = 10$ kg, $k = 1,000$ N/m, and $d = 0.1$ m.

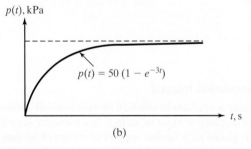

FIGURE 4.45

4.19   Find the transient response of an undamped spring-mass system for $t > \pi/\omega$ when the mass is subjected to a force

$$F(t) = \begin{cases} \dfrac{F_0}{2}(1 - \cos \omega t) & \text{for } 0 \le t \le \dfrac{\pi}{\omega} \\[3mm] F_0 & \text{for } t > \dfrac{\pi}{\omega} \end{cases}$$

Assume that the displacement and velocity of the mass are zero at $t = 0$.

4.20–4.22   Use the Duhamel integral method to derive expressions for the response of an undamped system subjected to the forcing functions shown in Figs. 4.46(a) to (c).

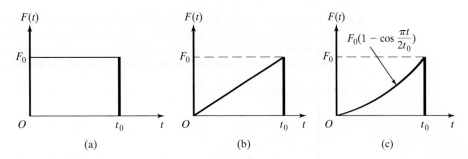

(a)   (b)   (c)

FIGURE 4.46

4.23   Figure 4.47 shows a single-degree-of-freedom model of a motor vehicle traveling in the horizontal direction. Find the relative displacement of the vehicle as it travels over a road bump of the form $y(s) = Y \sin \pi s/\delta$.

FIGURE 4.47

4.24   A vehicle traveling at a constant speed v in the horizontal direction encounters a triangular road bump, as shown in Fig. 4.48. Treating the vehicle as an undamped spring-mass system, determine the response of the vehicle in the vertical direction.

FIGURE 4.48

4.25 An automobile, having a mass of 1,000 kg, runs over a road bump of the shape shown in Fig. 4.49. The speed of the automobile is 50 km/h. If the undamped natural period of vibration in the vertical direction is 1.0 s, find the response of the car by assuming it as a single-degree-of-freedom undamped system vibrating in the vertical direction.

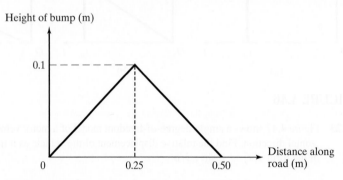

FIGURE 4.49

4.26 A camcorder of mass m is packed in a container using a flexible packing material. The stiffness and damping constant of the packing material are given by k and c, respectively, and the mass of the container is negligible. If the container is dropped accidentally from a height of h onto a rigid floor (see Fig. 4.50), find the motion of the camcorder.

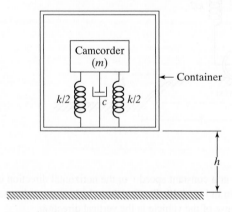

FIGURE 4.50

4.27 An airplane, taxiing on a runway, encounters a bump. As a result, the root of the wing is subjected to a displacement that can be expressed as

$$y(t) = \begin{cases} Y(t^2/t_0^2), & 0 \le t \le t_0 \\ 0, & t > t_0 \end{cases}$$

Find the response of the mass located at the tip of the wing if the stiffness of the wing is k (see Fig. 4.51).

FIGURE 4.51

4.28 Derive Eq. (E.1) of Example 4.12.

4.29 In a static firing test, a rocket is anchored to a rigid wall by a spring-damper system, as shown in Fig. 4.52(a). The thrust acting on the rocket reaches its maximum value F in a negligibly short time and remains constant until the burnout time t_0, as indicated in Fig. 4.52(b). The thrust acting on the rocket is given by $F = m_0 v$, where m_0 is the constant rate at which fuel is burnt and v is the velocity of the jet stream. The initial mass of the rocket is M, so that its mass at any time t is given by $m = M - m_0 t, 0 \le t \le t_0$. If the data are $k = 7.5 \times 10^6$ N/m, $c = 0.1 \times 10^6$ N-s/m, $m_0 = 10$ kg/s, $v = 2,000$ m/s, $M = 2,000$ kg, and $t_0 = 100$ s, (1) derive the equation of motion of the rocket, and (2) find the maximum steady-state displacement of the rocket by assuming an average (constant) mass of $(M - \frac{1}{2}m_0 t_0)$.

FIGURE 4.52

4.30 Show that the response to a unit step function $h(t)$ ($F_0 = 1$ in Fig. 4.10(b)) is related to the impulse response function $g(t)$, Eq. (4.25), as follows:

$$g(t) = \frac{dh(t)}{dt}$$

4.31   Show that the convolution integral, Eq. (4.31), can also be expressed in terms of the response to a unit step function $h(t)$ as

$$x(t) = F(0)h(t) + \int_0^t \frac{dF(\tau)}{d\tau} h(t - \tau)\, d\tau$$

4.32   Find the response of the rigid bar shown in Fig. 4.53 using the convolution integral for the following data: $k_1 = k_2 = 5{,}000$ N/m, $a = 0.25$ m, $b = 0.5$ m, $l = 1.0$ m, $M = 50$ kg, $m = 10$ kg, $F_0 = 500$ N.

FIGURE 4.53

4.33   Find the response of the rigid bar shown in Fig. 4.54 using the convolution integral for the following data: $k = 5{,}000$ N/m, $l = 1$ m, $m = 10$ kg, $M_0 = 100$ N-m.

FIGURE 4.54

4.34   Find the response of the rigid bar shown in Fig. 4.55 using the convolution integral when the end P of the spring PQ is subjected to the displacement, $x(t) = x_0 e^{-t}$. Data: $k = 5{,}000$ N/m, $l = 1$ m, $m = 10$ kg, $x_0 = 1$ cm.

FIGURE 4.55

4.35 Find the response of the mass shown in Fig. 4.56 under the force $F(t) = F_0 e^{-t}$ using the convolution integral. Data: $k_1 = 1,000$ N/m, $k_2 = 500$ N/m, $r = 5$ cm, $m = 10$ kg, $J_0 = 1$ kg-m^2, $F_0 = 50$ N.

Pulley, mass moment of inertia J_0

k_2

r

$2r$ $\quad F_0 e^{-t}$

m

$x(t)$

k_1

FIGURE 4.56

4.36 Find the impulse response functions of a viscously damped spring-mass system for the following cases:

a. Undamped $(c = 0)$ **c.** Critically damped $(c = c_c)$

b. Underdamped $(c < c_c)$ **d.** Overdamped $(c > c_c)$

4.37 Find the response of a single-degree-of-freedom system under an impulse F for the following data: $m = 2$ kg, $c = 4$ N-s/m, $k = 32$ N/m, $F = 4\,\delta(t)$, $x_0 = 0.01$ m, $\dot{x}_0 = 1$ m/s.

4.38 The wing of a fighter aircraft, carrying a missile at its tip, as shown in Fig. 4.57, can be approximated as an equivalent cantilever beam with $EI = 15 \times 10^9$ N-m^2 about the vertical axis and length $l = 10$ m. If the equivalent mass of the wing, including the mass of the missile and its carriage system, at the tip of the wing is $m = 2,500$ kg, determine the vibration response of the wing (of m) due to the release of the missile. Assume that the force on m due to the release of the missile can be approximated as an impulse function of magnitude $F = 50$ N-s.

(a) Real system

FIGURE 4.57

(b) Beam model

(c) Vibration model

FIGURE 4.57   (Continued)

4.39   The frame, anvil, and base of the forging hammer shown in Fig. 4.58(a) have a total mass of m. The support elastic pad has a stiffness of k. If the force applied by the hammer is given by Fig. 4.58(b), find the response of the anvil.

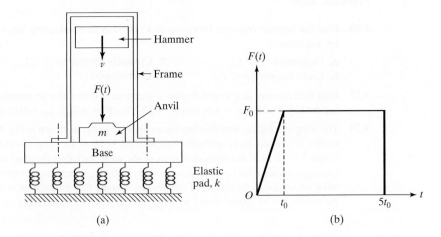

(a)

(b)

FIGURE 4.58

4.40   The input to the valve of an internal combustion engine is a force of $F = 15{,}000$ N applied over a period of 0.001 s by a cam as shown in Fig. 4.59 (see Fig. 1.39 for the arrangement of the valve). The valve has a mass of 15 kg, stiffness of 10,000 N/m, and damping constant of 20 N-s/m. The cam applies the force F once every 0.5 s. (a) Find the displacement response of the valve from its rest position when the cam applies the force F for the first time. (b) Determine the displacement of the valve from its rest position when the cam applies the force F for the second time.

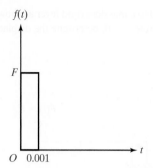

FIGURE 4.59

4.41 A bird strike on the engine of an airplane can be considered as an impulse (Fig. 4.60(a)). If the stiffness and damping coefficient of the engine mount are given by $k = 50,000$ N/m and $c = 1,000$ N-s/m, and the engine mass is $m = 500$ kg, find the response of the engine. Assume the mass of the bird as 4 kg and the velocity of the airplane as 250 km/h.

(a) Physical system   (b) Model

FIGURE 4.60

4.42 The rail car, shown in Fig. 4.61, is initially at rest and is set into motion by an impulse $5\,\delta(t)$. (a) Determine the motion of the car, $x(t)$. (b) If it is desired to stop the car by applying another impulse, determine the impulse that needs to be applied to the car.

FIGURE 4.61

4.43   A spring-damper system is connected to a massless rigid lever as shown in Fig. 4.62. If a step force of magnitude F_0 is applied at time $t = 0$, determine the displacement, $x(t)$, of point A of the lever.

FIGURE 4.62

4.44   A space experimental package of mass m is supported on an elastic suspension of stiffness k in the space shuttle. During launching, the space shuttle (base of the elastically supported package) experiences an acceleration of $\ddot{y}(t) = \alpha t$, where α is a constant. Find the time variation of the displacement, $x(t)$, and the relative displacement, $x(t) - y(t)$, of the package. Assume zero initial conditions.

4.45   A person, carrying a precision instrument of mass m, rides in the elevator of a building in a standing position (Fig. 4.63). The elevator, while moving with velocity v_0 at time $t = 0$, decelerates to zero velocity (stops) in time τ, so that the variation of its velocity can be expressed as

$$v(t) = \begin{cases} v_0\left(1 - \dfrac{t}{\tau}\right); & 0 \le t \le \tau \\ 0; & t > \tau \end{cases}$$

Assuming that the stiffness of the person in standing position is k, determine the displacement variation of the precision instrument, $x(t)$.

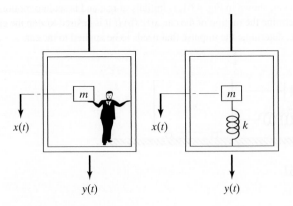

FIGURE 4.63

4.46 The water tank shown in Fig. 4.43(a) is subjected to a sudden hurricane force which varies with time as shown in Fig. 4.64. Assuming zero initial conditions, determine the displacement response, $x(t)$, of the water tank.

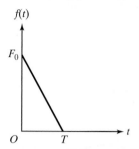

FIGURE 4.64

Section 4.6 Response Spectrum

4.47 Derive the response spectrum of an undamped system for the rectangular pulse shown in Fig. 4.46(a). Plot $(x/\delta_{st})_{max}$ with respect to (t_0/τ_n).

4.48 Find the displacement response spectrum of an undamped system for the pulse shown in Fig. 4.46(c).

4.49 The base of an undamped spring-mass system is subjected to an acceleration excitation given by $a_0[1 - \sin(\pi t/2t_0)]$. Find the relative displacement of the mass z.

4.50 Find the response spectrum of the system considered in Example 4.13. Plot $(kx/F_0)_{max}$ versus $\omega_n t_0$ in the range $0 \leq \omega_n t_0 \leq 15$.

4.51* A building frame is subjected to a blast load, and the idealization of the frame and the load are shown in Fig. 4.14. If $m = 5{,}000$ kg, $F_0 = 4$ MN, and $t_0 = 0.4$ s, find the minimum stiffness required if the displacement is to be limited to 10 mm.

4.52 Consider the printed circuit board (PCB) mounted on a cantilevered aluminum bracket shown in Fig. 4.23(a). Design the bracket to withstand an acceleration level of $100g$ under the rectangular pulse shown in Fig. 4.65. Assume the specific weight, Young's modulus, and permissible stress of aluminum as 30 kN/m^3, 70 GPa, and 180 MPa, respectively.

FIGURE 4.65

---

*The asterisk denotes a problem with no unique answer.

4.53 Consider the printed circuit board (PCB) mounted on a cantilevered aluminum bracket shown in Fig. 4.23(a). Design the bracket to withstand an acceleration level of $100g$ under the triangular pulse shown in Fig. 4.66. Assume the material properties as given in Problem 4.52.

FIGURE 4.66

4.54 An electronic box, having a mass of 0.5 kg, is to be shock-tested using a $100g$ half-sine pulse with a 0.1-s time base for a qualification test. The box is mounted at the middle of a fixed-fixed beam as shown in Fig. 4.67. The beam, along with the box, is placed in a container and subjected to the shock test. Design the beam to withstand the stated shock pulse. Assume the material properties as given in Problem 4.52.

FIGURE 4.67

4.55* The water tank shown in Fig. 4.68 is subjected to an earthquake whose response spectrum is indicated in Fig. 4.18. The mass of the tank with water is 50,000 kg. Design a uniform steel hollow circular column of height 15 m so that the maximum bending stress does not exceed

FIGURE 4.68

the yield stress of the material $\delta y = 230$ MPa. Assume a damping ratio of 0.05 and a factor of safety of 2.

4.56 Consider the overhead traveling crane shown in Fig. 4.21. Assuming the mass of the trolley as 2,300 kg and the overall damping ratio as 2%, determine the overall stiffness of the system necessary in order to avoid derailment of the trolley under a vertical earthquake excitation whose design spectrum is given by Fig. 4.19.

4.57 An electric pole of circular cross section, with a bending stiffness $k = 5,000$ N/m and a damping ratio $\zeta = 0.05$, carries a transformer of mass $m = 250$ kg as shown in Fig. 4.69. It is subjected to an earthquake that is characterized by a response spectrum given in Fig. 4.18. Find (a) maximum relative displacement of the transformer, (b) maximum shear force in the pole, and (c) maximum bending moment in the pole. Assume the length of pole (height of transformer) as $l = 20$ m.

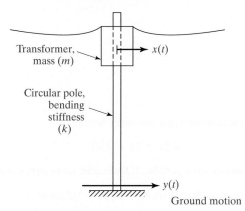

FIGURE 4.69

Section 4.7 Laplace Transform

4.58 Find the steady-state response of an undamped single-degree-of-freedom system subjected to the force $F(t) = F_0 e^{i\omega t}$ by using the method of Laplace transformation.

4.59 Find the response of a damped spring-mass system subjected to a step function of magnitude F_0 by using the method of Laplace transformation.

4.60 Find the response of an undamped system subjected to a square pulse $F(t) = F_0$ for $0 \le t \le t_0$ and 0 for $t > t_0$ by using the method of Laplace transformation. Assume the initial conditions as zero.

4.61 Derive the expression for the Laplace transform of the response of a damped single-degree-of-freedom system subjected to the following types of forcing functions:

    a. $f(t) = A \sin \omega t$ **c.** $f(t) = Ae^{-\omega t}$
    b. $f(t) = A \cos \omega t$ **d.** $f(t) = A\delta(t - t_0)$

4.62 Derive an expression for the impulse response function of a critically damped single-degree-of-freedom system.

4.63 Find the response of a system with the following equation of motion:

$$2\ddot{x} + 8\dot{x} + 16x = 5\,\delta(t)$$

using the initial conditions $x(t = 0) = x_0 = 0.05$ m and $\dot{x}(t = 0) = \dot{x}_0 = 0$. Plot the response of the system.

4.64 A bronze ball of mass m_0 is dropped on the mass of a single-degree-of-freedom system from a height h as shown in Fig. 4.70. If the ball is caught after its first bounce, determine the resulting displacement response of the mass M. Assume that the collision is perfectly elastic and the system is at rest initially. Data: $m = 2$ kg, $m_0 = 0.1$ kg, $k = 100$ N/m, $c = 5$ N-s/m, and $h = 2$ m.

FIGURE 4.70

4.65 Consider the equation of motion of a first-order system:

$$0.5\dot{x} + 4x = f(t)$$

where the forcing function $f(t)$ is periodic. If the Fourier series representation of $f(t)$ is given by

$$f(t) = 4 \sin 2t + 2 \sin 4t + \sin 6t + 0.5 \sin 8t + \dots$$

a. what is the bandwidth of the system?
b. find the steady-state response of the system considering only those components of $f(t)$ that lie within the bandwidth of the system.

4.66 Find the step response of a system with the stated equation of motion:
a. $2\ddot{x} + 10\dot{x} + 12.5 = 10u_s(t)$
b. $2\ddot{x} + 10\dot{x} + 8 = 10u_s(t)$
c. $2\ddot{x} + 10\dot{x} + 18 = 10u_s(t)$

4.67 Derive the Laplace transform of the ramp function $F(t) = bt$, $t \geq 0$, starting from the definition of a Laplace transform.

4.68 Find the inverse Laplace transform of

$$F(s) = \frac{-s + 3}{(s + 1)(s + 2)}$$

4.69 Find the inverse Laplace transform of

$$F(s) = \frac{3s + 8}{(s + 2)^2(s + 5)}$$

4.70 Find the response of a spring-damper (first-order) system shown in Fig. 4.1(a) with the equation of motion

$$c\dot{x} + kx = \overline{F}(t)$$

where the forcing function $\overline{F}(t)$ is a unit step function. Also determine the initial and steady-state values of the response from the time and Laplace domain solutions.

4.71 Determine the initial and steady-state values of the ramp response of a first-order system considered in Example 4.20 from the time and Laplace domain solutions.

4.72 Find the initial and steady-state values of the impulse response of the underdamped system considered in Example 4.19 using both the time and Laplace domain solutions.

4.73 Find the response of a critically damped single-degree-of-freedom system subjected to a step force with the equation of motion

$$2\ddot{x} + 8\dot{x} + 8x = 5$$

Assume the initial conditions as $x_0 = 1$ and $\dot{x}_0 = 2$.

4.74 Find the steady-state response of an underdamped single-degree-of-freedom system subjected to a ramp input $\overline{F}(t) = bt$ where b is the slope of the ramp.

4.75 Derive the expression for the total response of an underdamped single-degree-of-freedom system subjected to a forcing function $F(t)$. Assume the initial conditions as $x\,(t = 0) = x_0$ and $\dot{x}\,(t = 0) = \dot{x}_0$.

4.76 For the damped second-order system with the transfer function given below, find the values of $\zeta, \omega_n, t_s, t_r, t_p$, and percent overshoot:

$$T(s) = \frac{X(s)}{F(s)} = \frac{121}{s^2 + 17.6s + 121}$$

4.77 For the damped second-order system with the transfer function given below, find the values of $\zeta, \omega_n, t_s, t_r, t_p$, and percent overshoot:

$$T(s) = \frac{X(s)}{F(s)} = \frac{3.24 \times 10^6}{s^2 + 2{,}700s + 3.24 \times 10^6}$$

4.78 For the translational second-order system shown in Fig. 4.2(a) with $m = 6$ kg, $c = 30$ N-s/m, and $k = 45$ N/m, find the values of $\zeta, \omega_n, t_s, t_r, t_p$, and percent overshoot for $x(t)$.

4.79 For the torsional second-order system shown in Fig. 4.2(c) with $J = 2$ kg-m^2, $c_t = 2$ N-m-s/rad, and $k_t = 2$ N-m/rad, find the values of $\zeta, \omega_n, t_s, t_r, t_p$, and percent overshoot for $\theta(t)$.

4.80 For the translational system shown in Fig. 4.2(a) with $k = 1$ and $f(t) = $ unit step function, determine the values of m and c to achieve a 40% overshoot and a settling time of 5 s.

Section 4.8 Numerical Methods

4.81 Find the response of a damped single-degree-of-freedom system with the equation of motion

$$m\ddot{x} + c\dot{x} + kx = F(t)$$

using the Runge-Kutta method. Assume that $m = 5$ kg, $c = 200$ N-s/m, $k = 750$ N/m, and

$$F(t) = \begin{cases} \dfrac{F_0 t}{t_1}; & 0 \le t \le t_1 \\ F_0; & t \ge t_1 \end{cases}$$

with $F_0 = 2{,}000$ N and $t_1 = 6$ s.

4.82   Solve Problem 4.81 (using the Runge-Kutta method) for the forcing function

$$F(t) = \begin{cases} F_0 \sin\dfrac{\pi t}{t_1}; & 0 \le t \le t_1 \\ 0; & t \ge t_1 \end{cases}$$

with $F_0 = 2,000$ N and $t_1 = 6$ s.

4.83   Solve Problem 4.81 (using the Runge-Kutta method) for the forcing function

$$F(t) = \begin{cases} \dfrac{F_0 t}{t_1}; & 0 \le t \le t_1 \\ F_0\left(\dfrac{t_2 - t}{t_2 - t_1}\right); & t_1 \le t \le t_2 \\ 0; & t \ge t_2 \end{cases}$$

with $F_0 = 2,000$ N, $t_1 = 3$ s, and $t_2 = 6$ s.

Section 4.9 Response to Irregular Forcing Conditions Using Numerical Methods

4.84   Derive the expressions for x_j and $\dot{x}_j$ according to the linear interpolation function, considered in Section 4.9 for the undamped case. Using these expressions, find the solution of Example 4.31 by assuming the damping to be zero.

4.85   Find the response of a damped single-degree-of-freedom system with the equation of motion

$$m\ddot{x} + c\dot{x} + kx = F(t)$$

using the numerical method of Section 4.9. Assume that $m = 500$ kg, $c = 200$ N-s/m, $k = 750$ N/m, and the values of the forcing function $F(t)$ at discrete times are as indicated below:

t	0	1	2	3	4	5	6	7	8	9	10
$F(t)$	0	400	800	1,200	1,600	2,000	2,000	2,000	2,000	2,000	2,000

4.86   Find the response of a damped single-degree-of-freedom system with the equation of motion

$$m\ddot{x} + c\dot{x} + kx = F(t)$$

using the numerical method of Section 4.9. Assume that $m = 500$ kg, $c = 200$ N-s/m, $k = 750$ N/m, and the values of the forcing function $F(t)$ at discrete times are as indicated below:

t	0	1	2	3	4	5	6	7	8	9	10
$F(t)$	0	1,000	1,732	2,000	1,732	1,000	0	0	0	0	0

4.87 Find the response of a damped single-degree-of-freedom system with the equation of motion

$$m\ddot{x} + c\dot{x} + kx = F(t)$$

using the numerical method of Section 4.9. Assume that $m = 500$ kg, $c = 200$ N-s/m, $k = 750$ N/m, and the values of the forcing function $F(t)$ at discrete times are as indicated below:

t	0	1	2	3	4	5	6	7	8	9	10
$F(t)$	0	666.7	1,333.3	2,000	1,333.3	666.7	0	0	0	0	0

Section 4.10 Examples Using MATLAB

4.88 A machine is given an impact force by an impact hammer. If the machine can be modeled as a single-degree-of-freedom system with $m = 10$ kg, $k = 4,000$ N/m, and $c = 40$ N-s/m, and the magnitude of the impact is $F = 100$ N-s, determine the response of the machine. Also plot the response using MATLAB.

4.89 If the machine described in Problem 4.88 is given a double impact by the impact hammer, find the response of the machine. Assume the impact force, $F(t)$, as $F(t) = 100\,\delta(t) + 50\,\delta(t - 0.5)$ N, where $\delta(t)$ is the Dirac delta function. Also plot the response of the machine using MATLAB.

4.90 Using MATLAB, plot the response of a viscously damped spring-mass system subject to the rectangular pulse shown in Fig. 4.12(a) with (a) $t_0 = 0.1$ s and (b) $t_0 = 1.5$ s. Assume the following data: $m = 100$ kg, $k = 1,200$ N/m, $c = 50$ N-s/m, $F_0 = 100$ N.

4.91 Using `Program4.m`, find the steady-state response of a viscously damped system with $m = 1$ kg, $k = 400$ N/m, and $c = 5$ N-s/m subject to the periodic force shown in Fig. 4.71.

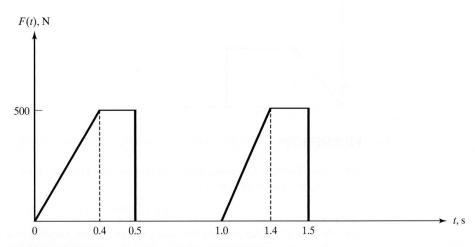

FIGURE 4.71

4.92 Using `Program5.m`, find the response of a viscously damped system with $m = 100$ kg, $k = 10^5$ N, and $\zeta = 0.1$ subject to the force $F(t) = 1,000\,(1 - \cos \pi t)$ N.

4.93 A damped single-degree-of-freedom system has a mass $m = 2$, a spring of stiffness $k = 50$, and a damper with $c = 2$. A forcing function $F(t)$, whose magnitude is indicated in the table below, acts on the mass for 1 s. Find the response of the system by using the piecewise linear interpolation method described in Section 4.9 using `Program5.m`.

Time (t_i)	$F(t_i)$
0.0	−8.0
0.1	−12.0
0.2	−15.0
0.3	−13.0
0.4	−11.0
0.5	−7.0
0.6	−4.0
0.7	3.0
0.8	10.0
0.9	15.0
1.0	18.0

4.94 The equation of motion of an undamped system is given by $2\ddot{x} + 1{,}500x = F(t)$, where the forcing function is defined by the curve shown in Fig. 4.72. Find the response of the system numerically for $0 \leq t \leq 0.5$. Assume the initial conditions as $x_0 = \dot{x}_0 = 0$ and the step size as $\Delta t = 0.01$. Use the MATLAB program `ode23`.

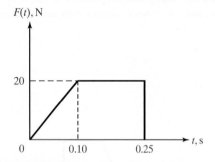

FIGURE 4.72

4.95 Solve Problem 4.94 using MATLAB program `ode23` if the system is viscously damped so that the equation of motion is

$$2\ddot{x} + 10\dot{x} + 1{,}500x = F(t)$$

4.96 Write a MATLAB program for finding the steady-state response of a single-degree-of-freedom system subjected to an arbitrary force, by numerically evaluating the Duhamel integral. Using this program, solve Example 4.31.

4.97 Find the relative displacement of the water tank shown in Fig. 4.43(a) when its base is subjected to the earthquake acceleration record shown in Fig. 1.115 by assuming the ordinate represents acceleration in g's. Use the program of Problem 4.96.

4.98 The differential equation of motion of an undamped system is given by $2\ddot{x} + 150x = F(t)$ with the initial conditions $x_0 = \dot{x}_0 = 0$. If $F(t)$ is as shown in Fig. 4.73, find the response of the problem using the program of Problem 4.96.

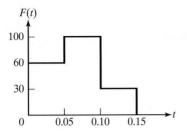

FIGURE 4.73

DESIGN PROJECTS

4.99 Design a seismometer of the type shown in Fig. 4.74(a) (by specifying the values of a, m, and k) to measure earthquakes. The seismometer should have a natural frequency of 10 Hz, and the maximum relative displacement of the mass should be at least 2 cm when its base is subjected to the displacement shown in Fig. 4.74(b).

FIGURE 4.74

4.100 The cutting forces developed during two different machining operations are shown in Figs. 4.75(a) and (b). The inaccuracies (in the vertical direction) in the surface finish in the two cases were observed to be 0.1 mm and 0.05 mm, respectively. Find the equivalent mass and stiffness of the cutting head (Fig. 4.76), assuming it to be an undamped single-degree-of-freedom system.

FIGURE 4.75

FIGURE 4.76

4.101 A milling cutter, mounted at the middle of an arbor, is used to remove metal from a workpiece (Fig. 4.77). A torque of 500 N-m is developed in the cutter under steady-state cutting conditions. One of the 16 teeth on the cutter breaks during the cutting operation. Determine the cross section of the arbor to limit the amplitude of angular displacement of the cutter to 1°. Assume that the arbor can be modeled as a hollow steel shaft fixed at both ends.

Data: Length of arbor = 0.5 m, mass moment of inertia of the cutter = 0.1 N-m^2, speed of cutter = 1,000 rpm.

FIGURE 4.77

Daniel Bernoulli (1700–1782) was a Swiss who became a professor of mathematics at Saint Petersburg in 1725 after receiving his doctorate in medicine for his thesis on the action of lungs. He later became professor of anatomy and botany at Basel. He developed the theory of hydrostatics and hydrodynamics, and "Bernoulli's theorem" is well known to engineers. He derived the equation of motion for the vibration of beams (the Euler-Bernoulli theory) and studied the problem of vibrating strings. Bernoulli was the first person to propose the principle of superposition of harmonics in free vibration.
(A photo of a portrait courtesy of David Eugene Smith, *History of Mathematics*, Volume 1—*General Survey of the History of Elementary Mathematics*, Dover Publications, New York, 1958.)

C H A P T E R   5

Two-Degree-of-Freedom Systems

Chapter Outline

This chapter deals with two-degree-of-freedom systems, which require two independent coordinates to describe their motion. The coupled equations of motion of the system are derived using Newton's second law of motion. By expressing these equations in matrix form, the mass, damping, and stiffness matrices of the system are identified. By assuming harmonic motion of the two masses, the eigenvalues or natural frequencies of vibration, the modal vectors, and the free-vibration solution of the undamped system are found. The method of incorporating the initial conditions is also outlined. The two-degree-of-freedom torsional systems are considered in an analogous manner. The concepts of coordinate coupling, generalized coordinates, and principal coordinates are introduced with examples. The forced-vibration analysis of the system under the complex form of harmonic force is presented and the impedance matrix is identified. The semidefinite, unrestricted, or degenerate systems are introduced along with a method of finding their natural frequencies of vibration. The self-excitation and stability analysis of two-degree-of-freedom systems are considered along with a derivation of the conditions of stability. The Routh-Hurwitz criterion, which can be used for deriving the conditions of stability of any n-degree-of-freedom system, is also introduced. The transfer-function approach, the computation of the response of two-degree-of-freedom systems using the Laplace transform, and solutions using frequency transfer functions are also presented. Finally, the free- and forced-vibration solutions of two-degree-of-freedom systems using MATLAB are illustrated with examples.

Learning Objectives

After completing this chapter, you should be able to do the following:

- Formulate the equations of motion of two-degree-of-freedom systems.
- Identify the mass, damping, and stiffness matrices from the equations of motion.
- Compute the eigenvalues or natural frequencies of vibration and the modal vectors.
- Determine the free-vibration solution using the known initial conditions.
- Understand the concepts of coordinate coupling and principal coordinates.
- Determine the forced-vibration solutions under harmonic forces.
- Understand the concepts of self-excitation and stability of the system.
- Use the Laplace transform approach for solution of two-degree-of-freedom systems.
- Solve two-degree-of-freedom free- and forced-vibration problems using MATLAB.

5.1   Introduction

Systems that require two independent coordinates to describe their motion are called *two-degree-of-freedom systems*. Some examples of systems having two degrees of freedom were shown in Fig. 1.12. We shall consider only two-degree-of-freedom systems in this chapter, so as to provide a simple introduction to the behavior of systems with an arbitrarily large number of degrees of freedom, which is the subject of Chapter 6.

Consider a simplified model of a lathe shown in Fig. 5.1(a), in which the lathe bed, represented as an elastic beam, is supported on short elastic columns with the headstock and tailstock denoted as lumped masses attached to the beam [5.1–5.3]. For a simplified vibration analysis, the lathe can be treated as a rigid body of total mass m and mass moment of inertia J_0 about its center of gravity (C.G.), resting on springs of stiffness k_1 and k_2, as shown in Fig. 5.1(b). The displacement of the system at any time can be specified by a linear coordinate $x(t)$, indicating the vertical displacement of the C.G. of the mass,

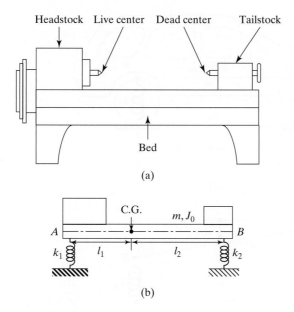

Headstock   Live center   Dead center   Tailstock

Bed

(a)

C.G.

m, J_0

A

B

k_1 l_1 l_2 k_2

(b)

FIGURE 5.1   Lathe.

and an angular coordinate $\theta(t)$, denoting the rotation of the mass m about its C.G. Instead of $x(t)$ and $\theta(t)$, we can also use $x_1(t)$ and $x_2(t)$, the displacements of points A and B, as independent coordinates to specify the motion of the system. Thus the system has two degrees of freedom. It is important to note that in this case the mass m is treated not as a point mass but as a rigid body having two possible types of motion. (If it is a particle, there is no need to specify the rotation of the mass about its C.G.)

Similarly, consider the automobile shown in Fig. 5.2(a). For the vibration of the automobile in the vertical plane, a two-degree-of-freedom model shown in Fig. 5.2(b) can be used. Here the body is idealized as a bar of mass m and mass moment of inertia J_0, supported on the rear and front wheels (suspensions) of stiffness k_1 and k_2. The displacement of the automobile at any time can be specified by the linear coordinate $x(t)$ denoting the vertical displacement of the C.G. of the body and the angular coordinate $\theta(t)$ indicating the rotation (pitching) of the body about its C.G. Alternatively, the motion of the automobile can be specified using the independent coordinates, $x_1(t)$ and $x_2(t)$, of points A and B.

Next, consider the motion of a multistory building under an earthquake. For simplicity, a two-degree-of-freedom model can be used as shown in Fig. 5.3. Here the building is modeled as a rigid bar having a mass m and mass moment of inertia J_0. The resistance offered to the motion of the building by the foundation and surrounding soil is approximated by a linear spring of stiffness k and a torsional spring of stiffness k_t. The displacement of the building at any time can be specified by the horizontal motion of the base $x(t)$ and the angular motion $\theta(t)$ about the point O. Finally, consider the system shown in Fig. 5.4(a), which illustrates the packaging of an instrument of mass m. Assuming that the motion of the instrument is confined to the xy-plane, the system can be modeled as a mass

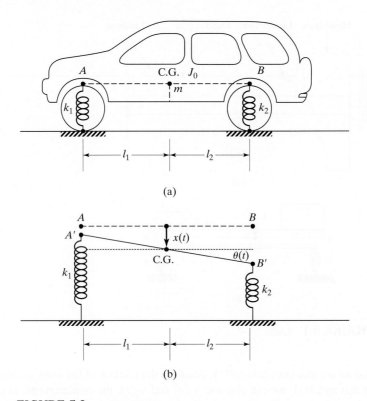

FIGURE 5.2   Automobile.

m supported by springs in the x and y directions, as indicated in Fig. 5.4(b). Thus the system has one point mass m and two degrees of freedom, because the mass has two possible types of motion (translations along the x and y directions). The general rule for the computation of the number of degrees of freedom can be stated as follows:

$$
\begin{matrix}
\text{Number of} & & \text{Number of masses in the system} \\
\text{degrees of freedom} = & \times & \text{number of possible types} \\
\text{of the system} & & \text{of motion of each mass}
\end{matrix}
$$

There are two equations of motion for a two-degree-of-freedom system, one for each mass (more precisely, for each degree of freedom). They are generally in the form of *coupled differential equations*—that is, each equation involves all the coordinates. If a harmonic solution is assumed for each coordinate, the equations of motion lead to a frequency equation that gives two natural frequencies for the system. If we give suitable initial excitation, the system vibrates at one of these natural frequencies. During free vibration at one of the natural frequencies, the amplitudes of the two degrees of freedom (coordinates) are related in a specific manner and the configuration is called a *normal mode*, *principal mode*, or

FIGURE 5.3   Multistory building subjected to an earthquake.

natural mode of vibration. Thus a two-degree-of-freedom system has two normal modes of vibration corresponding to the two natural frequencies.

If we give an arbitrary initial excitation to the system, the resulting free vibration will be a superposition of the two normal modes of vibration. However, if the system vibrates under the action of an external harmonic force, the resulting forced harmonic vibration takes place at the frequency of the applied force. Under harmonic excitation, resonance

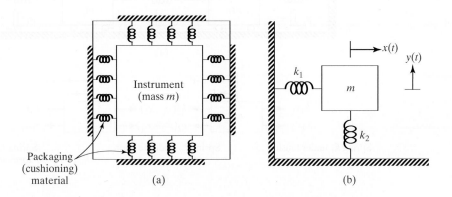

FIGURE 5.4   Packaging of an instrument.

occurs (i.e., the amplitudes of the two coordinates will be maximum) when the forcing frequency is equal to one of the natural frequencies of the system.

As is evident from the systems shown in Figs. 5.1–5.4, the configuration of a system can be specified by a set of independent coordinates such as length, angle, or some other physical parameters. Any such set of coordinates is called *generalized coordinates*. Although the equations of motion of a two-degree-of-freedom system are generally coupled so that each equation involves all the coordinates, it is always possible to find a particular set of coordinates such that each equation of motion contains only one coordinate. The equations of motion are then *uncoupled* and can be solved independently of each other. Such a set of coordinates, which leads to an uncoupled system of equations, is called *principal coordinates*.

5.2   Equations of Motion for Forced Vibration

Consider a viscously damped two-degree-of-freedom spring-mass system, shown in Fig. 5.5(a). The motion of the system is completely described by the coordinates $x_1(t)$ and $x_2(t)$, which define the positions of the masses m_1 and m_2 at any time t from the respective equilibrium positions. The external forces $F_1(t)$ and $F_2(t)$ act on the masses m_1 and m_2, respectively. The free-body diagrams of the masses m_1 and m_2 are shown in Fig. 5.5(b). The application of Newton's second law of motion to each of the masses gives the equations of motion:

$$m_1\ddot{x}_1 + (c_1 + c_2)\dot{x}_1 - c_2\dot{x}_2 + (k_1 + k_2)x_1 - k_2x_2 = f_1 \tag{5.1}$$

$$m_2\ddot{x}_2 - c_2\dot{x}_1 + (c_2 + c_3)\dot{x}_2 - k_2x_1 + (k_2 + k_3)x_2 = f_2 \tag{5.2}$$

It can be seen that Eq. (5.1) contains terms involving x_2 (namely, $-c_2\dot{x}_2$ and $-k_2x_2$), whereas Eq. (5.2) contains terms involving x_1 (namely, $-c_2\dot{x}_1$ and $-k_2x_1$). Hence they

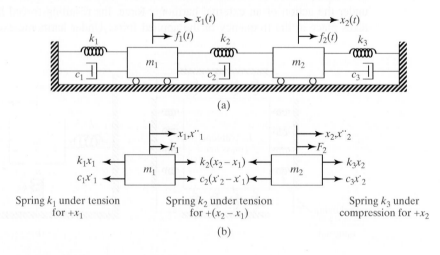

(a)

(b)

FIGURE 5.5   A two-degree-of-freedom spring-mass-damper system.

represent a system of two coupled second-order differential equations. We can therefore expect that the motion of the mass m_1 will influence the motion of the mass m_2, and vice versa. Equations (5.1) and (5.2) can be written in matrix form as

$$[m]\,\ddot{\vec{x}}(t) + [c]\,\dot{\vec{x}}(t) + [k]\,\vec{x}(t) = \vec{f}(t) \tag{5.3}$$

where $[m]$, $[c]$, and $[k]$ are called the *mass, damping,* and *stiffness matrices*, respectively, and are given by

$$[m] = \begin{bmatrix} m_1 & 0 \\ 0 & m_2 \end{bmatrix}$$

$$[c] = \begin{bmatrix} c_1 + c_2 & -c_2 \\ -c_2 & c_2 + c_3 \end{bmatrix}$$

$$[k] = \begin{bmatrix} k_1 + k_2 & -k_2 \\ -k_2 & k_2 + k_3 \end{bmatrix}$$

and $\vec{x}(t)$ and $\vec{f}(t)$ are called the *displacement* and *force vectors*, respectively, and are given by

$$\vec{x}(t) = \begin{Bmatrix} x_1(t) \\ x_2(t) \end{Bmatrix}$$

and

$$\vec{f}(t) = \begin{Bmatrix} f_1(t) \\ f_2(t) \end{Bmatrix}$$

It can be seen that $[m]$, $[c]$, and $[k]$ are all 2×2 matrices whose elements are the known masses, damping coefficients, and stiffnesses of the system, respectively. Further, these matrices can be seen to be symmetric, so that

$$[m]^T = [m], \qquad [c]^T = [c], \qquad [k]^T = [k]$$

where the superscript T denotes the transpose of the matrix.

Notice that the equations of motion (5.1) and (5.2) become uncoupled (independent of one another) only when $c_2 = k_2 = 0$, which implies that the two masses m_1 and m_2 are not physically connected. In such a case, the matrices $[m]$, $[c]$, and $[k]$ become diagonal. The solution of the equations of motion (5.1) and (5.2) for any arbitrary forces $f_1(t)$ and $f_2(t)$ is difficult to obtain, mainly due to the coupling of the variables $x_1(t)$ and $x_2(t)$. The solution of Eqs. (5.1) and (5.2) involves four constants of integration (two for each equation). Usually the initial displacements and velocities of the two masses are specified as $x_1(t = 0) = x_1(0)$, $\dot{x}_1(t = 0) = \dot{x}_1(0)$, $x_2(t = 0) = x_2(0)$, and $\dot{x}_2(t = 0) = \dot{x}_2(0)$. We shall first consider the free-vibration solution of Eqs. (5.1) and (5.2).

5.3   Free-Vibration Analysis of an Undamped System

For the free-vibration analysis of the system shown in Fig. 5.5(a), we set $f_1(t) = f_2(t) = 0$. Further, if damping is disregarded, $c_1 = c_2 = c_3 = 0$, and the equations of motion (5.1) and (5.2) reduce to

$$m_1\ddot{x}_1(t) + (k_1 + k_2)x_1(t) - k_2x_2(t) = 0 \qquad (5.4)$$

$$m_2\ddot{x}_2(t) - k_2x_1(t) + (k_2 + k_3)x_2(t) = 0 \qquad (5.5)$$

We are interested in knowing whether m_1 and m_2 can oscillate harmonically with the same frequency and phase angle but with different amplitudes. Assuming that it is possible to have harmonic motion of m_1 and m_2 at the same frequency ω and the same phase angle ϕ, we take the solutions of Eqs. (5.4) and (5.5) as

$$x_1(t) = X_1 \cos(\omega t + \phi)$$

$$x_2(t) = X_2 \cos(\omega t + \phi) \qquad (5.6)$$

where X_1 and X_2 are constants that denote the maximum amplitudes of $x_1(t)$ and $x_2(t)$, and ϕ is the phase angle. Substituting Eq. (5.6) into Eqs. (5.4) and (5.5), we obtain

$$\left[\{-m_1\omega^2 + (k_1 + k_2)\}X_1 - k_2X_2\right]\cos(\omega t + \phi) = 0$$

$$\left[-k_2X_1 + \{-m_2\omega^2 + (k_2 + k_3)\}X_2\right]\cos(\omega t + \phi) = 0 \qquad (5.7)$$

Since Eq. (5.7) must be satisfied for all values of the time t, the terms between brackets must be zero. This yields

$$\{-m_1\omega^2 + (k_1 + k_2)\}X_1 - k_2X_2 = 0$$

$$-k_2X_1 + \{-m_2\omega^2 + (k_2 + k_3)\}X_2 = 0 \qquad (5.8)$$

which represent two simultaneous homogenous algebraic equations in the unknowns X_1 and X_2. It can be seen that Eq. (5.8) is satisfied by the trivial solution $X_1 = X_2 = 0$, which implies that there is no vibration. For a nontrivial solution of X_1 and X_2, the determinant of the coefficients of X_1 and X_2 must be zero:

$$\det \begin{bmatrix} \{-m_1\omega^2 + (k_1 + k_2)\} & -k_2 \\ -k_2 & \{-m_2\omega^2 + (k_2 + k_3)\} \end{bmatrix} = 0$$

or

$$(m_1m_2)\omega^4 - \{(k_1 + k_2)m_2 + (k_2 + k_3)m_1\}\omega^2$$

$$+ \{(k_1 + k_2)(k_2 + k_3) - k_2^2\} = 0 \qquad (5.9)$$

Equation (5.9) is called the *frequency* or *characteristic equation* because its solution yields the frequencies or the characteristic values of the system. The roots of Eq. (5.9) are given by

$$\omega_1^2, \omega_2^2 = \frac{1}{2} \left\{ \frac{(k_1 + k_2)m_2 + (k_2 + k_3)m_1}{m_1 m_2} \right\}$$

$$\pm \frac{1}{2} \left[\left\{ \frac{(k_1 + k_2)m_2 + (k_2 + k_3)m_1}{m_1 m_2} \right\}^2 \right.$$

$$\left. - 4 \left\{ \frac{(k_1 + k_2)(k_2 + k_3) - k_2^2}{m_1 m_2} \right\} \right]^{1/2} \tag{5.10}$$

This shows that it is possible for the system to have a nontrivial harmonic solution of the form of Eq. (5.6) when ω is equal to ω_1 and ω_2 given by Eq. (5.10). We call ω_1 and ω_2 the *natural frequencies* of the system.

The values of X_1 and X_2 remain to be determined. These values depend on the natural frequencies ω_1 and ω_2. We shall denote the values of X_1 and X_2 corresponding to ω_1 as $X_1^{(1)}$ and $X_2^{(1)}$ and those corresponding to ω_2 as $X_1^{(2)}$ and $X_2^{(2)}$. Further, since Eq. (5.8) is homogenous, only the ratios $r_1 = \{X_2^{(1)}/X_1^{(1)}\}$ and $r_2 = \{X_2^{(2)}/X_1^{(2)}\}$ can be found. For $\omega^2 = \omega_1^2$ and $\omega^2 = \omega_2^2$, Eq. (5.8) gives

$$r_1 = \frac{X_2^{(1)}}{X_1^{(1)}} = \frac{-m_1\omega_1^2 + (k_1 + k_2)}{k_2} = \frac{k_2}{-m_2\omega_1^2 + (k_2 + k_3)}$$

$$r_2 = \frac{X_2^{(2)}}{X_1^{(2)}} = \frac{-m_1\omega_2^2 + (k_1 + k_2)}{k_2} = \frac{k_2}{-m_2\omega_2^2 + (k_2 + k_3)} \tag{5.11}$$

Notice that the two ratios given for each r_i ($i = 1, 2$) in Eq. (5.11) are identical. The normal modes of vibration corresponding to ω_1^2 and ω_2^2 can be expressed, respectively, as

$$\vec{X}^{(1)} = \left\{ \begin{matrix} X_1^{(1)} \\ X_2^{(1)} \end{matrix} \right\} = \left\{ \begin{matrix} X_1^{(1)} \\ r_1 X_1^{(1)} \end{matrix} \right\}$$

and

$$\vec{X}^{(2)} = \left\{ \begin{matrix} X_1^{(2)} \\ X_2^{(2)} \end{matrix} \right\} = \left\{ \begin{matrix} X_1^{(2)} \\ r_2 X_1^{(2)} \end{matrix} \right\} \tag{5.12}$$

The vectors $\vec{X}^{(1)}$ and $\vec{X}^{(2)}$, which denote the normal modes of vibration, are known as the *modal vectors* of the system. The free-vibration solution or the motion in time can be expressed, using Eq. (5.6), as

$$\vec{x}^{(1)}(t) = \begin{Bmatrix} x_1^{(1)}(t) \\ x_2^{(1)}(t) \end{Bmatrix} = \begin{Bmatrix} X_1^{(1)} \cos(\omega_1 t + \phi_1) \\ r_1 X_1^{(1)} \cos(\omega_1 t + \phi_1) \end{Bmatrix} = \text{first mode}$$

$$\vec{x}^{(2)}(t) = \begin{Bmatrix} x_1^{(2)}(t) \\ x_2^{(2)}(t) \end{Bmatrix} = \begin{Bmatrix} X_1^{(2)} \cos(\omega_2 t + \phi_2) \\ r_2 X_1^{(2)} \cos(\omega_2 t + \phi_2) \end{Bmatrix} = \text{second mode} \qquad (5.13)$$

where the constants $X_1^{(1)}$, $X_1^{(2)}$, ϕ_1, and ϕ_2 are determined by the initial conditions.

Initial Conditions. As stated earlier, each of the two equations of motion, Eqs. (5.1) and (5.2), involves second-order time derivatives; hence we need to specify two initial conditions for each mass. As stated in Section 5.1, the system can be made to vibrate in its ith normal mode $(i = 1, 2)$ by subjecting it to the specific initial conditions

$$x_1(t = 0) = X_1^{(i)} = \text{some constant}, \qquad \dot{x}_1(t = 0) = 0,$$

$$x_2(t = 0) = r_i X_1^{(i)}, \qquad \dot{x}_2(t = 0) = 0$$

However, for any other general initial conditions, both modes will be excited. The resulting motion, which is given by the general solution of Eqs. (5.4) and (5.5), can be obtained by a linear superposition of the two normal modes, Eq. (5.13):

$$\vec{x}(t) = c_1 \vec{x}^{(1)}(t) + c_2 \vec{x}^{(2)}(t) \qquad (5.14)$$

where c_1 and c_2 are constants. Since $\vec{x}^{(1)}(t)$ and $\vec{x}^{(2)}(t)$ already involve the unknown constants $X_1^{(1)}$ and $X_1^{(2)}$ (see Eq. (5.13)), we can choose $c_1 = c_2 = 1$ with no loss of generality. Thus the components of the vector $\vec{x}(t)$ can be expressed, using Eq. (5.14) with $c_1 = c_2 = 1$ and Eq. (5.13), as

$$x_1(t) = x_1^{(1)}(t) + x_1^{(2)}(t) = X_1^{(1)} \cos(\omega_1 t + \phi_1) + X_1^{(2)} \cos(\omega_2 t + \phi_2)$$

$$x_2(t) = x_2^{(1)}(t) + x_2^{(2)}(t)$$

$$= r_1 X_1^{(1)} \cos(\omega_1 t + \phi_1) + r_2 X_1^{(2)} \cos(\omega_2 t + \phi_2) \qquad (5.15)$$

where the unknown constants $X_1^{(1)}$, $X_1^{(2)}$, ϕ_1, and ϕ_2 can be determined from the initial conditions:

$$x_1(t = 0) = x_1(0), \qquad \dot{x}_1(t = 0) = \dot{x}_1(0),$$

$$x_2(t = 0) = x_2(0), \qquad \dot{x}_2(t = 0) = \dot{x}_2(0) \qquad (5.16)$$

Substitution of Eq. (5.16) into Eq. (5.15) leads to

$$x_1(0) = X_1^{(1)} \cos \phi_1 + X_1^{(2)} \cos \phi_2$$

$$\dot{x}_1(0) = -\omega_1 X_1^{(1)} \sin \phi_1 - \omega_2 X_1^{(2)} \sin \phi_2$$

$$x_2(0) = r_1 X_1^{(1)} \cos \phi_1 + r_2 X_1^{(2)} \cos \phi_2$$
$$\dot{x}_2(0) = -\omega_1 r_1 X_1^{(1)} \sin \phi_1 - \omega_2 r_2 X_1^{(2)} \sin \phi_2 \tag{5.17}$$

Equation (5.17) can be regarded as four algebraic equations in the unknowns $X_1^{(1)} \cos \phi_1$, $X_1^{(2)} \cos \phi_2$, $X_1^{(1)} \sin \phi_1$, and $X_1^{(2)} \sin \phi_2$. The solution of Eq. (5.17) can be expressed as

$$X_1^{(1)} \cos \phi_1 = \left\{ \frac{r_2 x_1(0) - x_2(0)}{r_2 - r_1} \right\}, \qquad X_1^{(2)} \cos \phi_2 = \left\{ \frac{-r_1 x_1(0) + x_2(0)}{r_2 - r_1} \right\}$$

$$X_1^{(1)} \sin \phi_1 = \left\{ \frac{-r_2 \dot{x}_1(0) + \dot{x}_2(0)}{\omega_1(r_2 - r_1)} \right\}, \qquad X_1^{(2)} \sin \phi_2 = \left\{ \frac{r_1 \dot{x}_1(0) - \dot{x}_2(0)}{\omega_2(r_2 - r_1)} \right\}$$

from which we obtain the desired solution:

$$X_1^{(1)} = [\{X_1^{(1)} \cos \phi_1\}^2 + \{X_1^{(1)} \sin \phi_1\}^2]^{1/2}$$

$$= \frac{1}{(r_2 - r_1)} \left[\{r_2 x_1(0) - x_2(0)\}^2 + \frac{\{-r_2 \dot{x}_1(0) + \dot{x}_2(0)\}^2}{\omega_1^2} \right]^{1/2}$$

$$X_1^{(2)} = [\{X_1^{(2)} \cos \phi_2\}^2 + \{X_1^{(2)} \sin \phi_2\}^2]^{1/2}$$

$$= \frac{1}{(r_2 - r_1)} \left[\{-r_1 x_1(0) + x_2(0)\}^2 + \frac{\{r_1 \dot{x}_1(0) - \dot{x}_2(0)\}^2}{\omega_2^2} \right]^{1/2}$$

$$\phi_1 = \tan^{-1} \left\{ \frac{X_1^{(1)} \sin \phi_1}{X_1^{(1)} \cos \phi_1} \right\} = \tan^{-1} \left\{ \frac{-r_2 \dot{x}_1(0) + \dot{x}_2(0)}{\omega_1[r_2 x_1(0) - x_2(0)]} \right\}$$

$$\phi_2 = \tan^{-1} \left\{ \frac{X_1^{(2)} \sin \phi_2}{X_1^{(2)} \cos \phi_2} \right\} = \tan^{-1} \left\{ \frac{r_1 \dot{x}_1(0) - \dot{x}_2(0)}{\omega_2[-r_1 x_1(0) + x_2(0)]} \right\} \tag{5.18}$$

EXAMPLE 5.1

Frequencies of Spring-Mass System

Find the natural frequencies and mode shapes of a spring-mass system, shown in Fig. 5.6, which is constrained to move in the vertical direction only. Take $n = 1$.

Solution: If we measure x_1 and x_2 from the static equilibrium positions of the masses m_1 and m_2, respectively, the equations of motion and the solution obtained for the system of Fig. 5.5(a) are also applicable to this case if we substitute $m_1 = m_2 = m$ and $k_1 = k_2 = k_3 = k$. Thus the equations of motion, Eqs. (5.4) and (5.5), are given by

$$m\ddot{x}_1 + 2kx_1 - kx_2 = 0$$
$$m\ddot{x}_2 - kx_1 + 2kx_2 = 0 \tag{E.1}$$

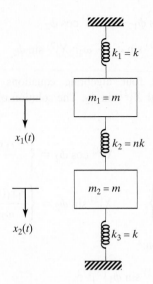

FIGURE 5.6   Two-degree-of-freedom system.

By assuming harmonic solution as

$$x_i(t) = X_i \cos(\omega t + \phi); i = 1, 2 \tag{E.2}$$

the frequency equation can be obtained by substituting Eq. (E.2) into Eq. (E.1):

$$\begin{vmatrix} (-m\omega^2 + 2k) & (-k) \\ (-k) & (-m\omega^2 + 2k) \end{vmatrix} = 0$$

or

$$m^2\omega^4 - 4km\omega^2 + 3k^2 = 0 \tag{E.3}$$

The solution of Eq. (E.3) gives the natural frequencies

$$\omega_1 = \left\{ \frac{4km - [16k^2m^2 - 12m^2k^2]^{1/2}}{2m^2} \right\}^{1/2} = \sqrt{\frac{k}{m}} \tag{E.4}$$

$$\omega_2 = \left\{ \frac{4km + [16k^2m^2 - 12m^2k^2]^{1/2}}{2m^2} \right\}^{1/2} = \sqrt{\frac{3k}{m}} \tag{E.5}$$

From Eq. (5.11), the amplitude ratios are given by

$$r_1 = \frac{X_2^{(1)}}{X_1^{(1)}} = \frac{-m\omega_1^2 + 2k}{k} = \frac{k}{-m\omega_1^2 + 2k} = 1 \tag{E.6}$$

$$r_2 = \frac{X_2^{(2)}}{X_1^{(2)}} = \frac{-m\omega_2^2 + 2k}{k} = \frac{k}{-m\omega_2^2 + 2k} = -1 \tag{E.7}$$

The natural modes are given by Eq. (5.13):

$$\text{First mode} = \vec{x}^{(1)}(t) = \begin{Bmatrix} X_1^{(1)} \cos\left(\sqrt{\dfrac{k}{m}}\,t + \phi_1\right) \\[2ex] X_1^{(1)} \cos\left(\sqrt{\dfrac{k}{m}}\,t + \phi_1\right) \end{Bmatrix} \tag{E.8}$$

$$\text{Second mode} = \vec{x}^{(2)}(t) = \begin{Bmatrix} X_1^{(2)} \cos\left(\sqrt{\dfrac{3k}{m}}\,t + \phi_2\right) \\[2ex] -X_1^{(2)} \cos\left(\sqrt{\dfrac{3k}{m}}\,t + \phi_2\right) \end{Bmatrix} \tag{E.9}$$

It can be seen from Eq. (E.8) that when the system vibrates in its first mode, the amplitudes of the two masses remain the same. This implies that the length of the middle spring remains constant. Thus the motions of m_1 and m_2 are in phase (see Fig. 5.7(a)). When the system vibrates in its second mode, Eq. (E.9) shows that the displacements of the two masses have the same magnitude with opposite signs. Thus the motions of m_1 and m_2 are 180° out of phase (see Fig. 5.7(b)). In this case, the midpoint of the middle spring remains stationary for all time t. Such a point is called a *node*. Using Eq. (5.15), the motion (general solution) of the system can be expressed as

$$x_1(t) = X_1^{(1)} \cos\left(\sqrt{\frac{k}{m}}\,t + \phi_1\right) + X_1^{(2)} \cos\left(\sqrt{\frac{3k}{m}}\,t + \phi_2\right)$$

$$x_2(t) = X_1^{(1)} \cos\left(\sqrt{\frac{k}{m}}\,t + \phi_1\right) - X_1^{(2)} \cos\left(\sqrt{\frac{3k}{m}}\,t + \phi_2\right) \tag{E.10}$$

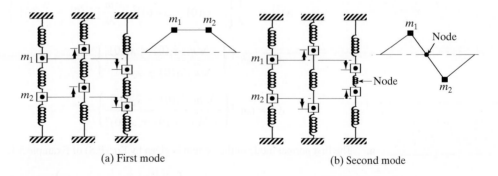

(a) First mode   (b) Second mode

FIGURE 5.7   Modes of vibration.

Note: It can be seen that the computation of the natural frequencies and mode shapes is lengthy and tedious. Computer programs can be used conveniently for the numerical computation of the natural frequencies and mode shapes of multidegree-of-freedom systems (see Section 5.12).

EXAMPLE 5.2

Initial Conditions to Excite Specific Mode

Find the initial conditions that need to be applied to the system shown in Fig. 5.6 so as to make it vibrate in (a) the first mode, and (b) the second mode.

Solution:

Approach: Specify the solution to be obtained for the first or second mode from the general solution for arbitrary initial conditions and solve the resulting equations.

For arbitrary initial conditions, the motion of the masses is described by Eq. (5.15). In the present case, $r_1 = 1$ and $r_2 = -1$, so Eq. (5.15) reduces to Eq. (E.10) of Example 5.1:

$$x_1(t) = X_1^{(1)} \cos\left(\sqrt{\frac{k}{m}}\, t + \phi_1\right) + X_1^{(2)} \cos\left(\sqrt{\frac{3k}{m}}\, t + \phi_2\right)$$

$$x_2(t) = X_1^{(1)} \cos\left(\sqrt{\frac{k}{m}}\, t + \phi_1\right) - X_1^{(2)} \cos\left(\sqrt{\frac{3k}{m}}\, t + \phi_2\right) \tag{E.1}$$

Assuming the initial conditions as in Eq. (5.16), the constants $X_1^{(1)}$, $X_1^{(2)}$, ϕ_1, and ϕ_2 can be obtained from Eq. (5.18), using $r_1 = 1$ and $r_2 = -1$:

$$X_1^{(1)} = -\frac{1}{2}\left\{[x_1(0) + x_2(0)]^2 + \frac{m}{k}[\dot{x}_1(0) + \dot{x}_2(0)]^2\right\}^{1/2} \tag{E.2}$$

$$X_1^{(2)} = -\frac{1}{2}\left\{[-x_1(0) + x_2(0)]^2 + \frac{m}{3k}[\dot{x}_1(0) - \dot{x}_2(0)]^2\right\}^{1/2} \tag{E.3}$$

$$\phi_1 = \tan^{-1}\left\{\frac{-\sqrt{m}\,[\dot{x}_1(0) + \dot{x}_2(0)]}{\sqrt{k}\,[x_1(0) + x_2(0)]}\right\} \tag{E.4}$$

$$\phi_2 = \tan^{-1}\left\{\frac{\sqrt{m}\,[\dot{x}_1(0) - \dot{x}_2(0)]}{\sqrt{3k}\,[-x_1(0) + x_2(0)]}\right\} \tag{E.5}$$

a.   The first normal mode of the system is given by Eq. (E.8) of Example 5.1:

$$\vec{x}^{(1)}(t) = \left\{ \begin{array}{c} X_1^{(1)} \cos\left(\sqrt{\dfrac{k}{m}}\, t + \phi_1\right) \\[4mm] X_1^{(1)} \cos\left(\sqrt{\dfrac{k}{m}}\, t + \phi_1\right) \end{array} \right\} \tag{E.6}$$

Comparison of Eqs. (E.1) and (E.6) shows that the motion of the system is identical with the first normal mode only if $X_1^{(2)} = 0$. This requires that (from Eq. (E.3))

$$x_1(0) = x_2(0) \qquad \text{and} \qquad \dot{x}_1(0) = \dot{x}_2(0) \tag{E.7}$$

b. The second normal mode of the system is given by Eq. (E.9) of Example 5.1:

$$\vec{x}^{(2)}(t) = \left\{ \begin{array}{c} X_1^{(2)} \cos\left(\sqrt{\dfrac{3k}{m}}\, t + \phi_2\right) \\[2ex] -X_1^{(2)} \cos\left(\sqrt{\dfrac{3k}{m}}\, t + \phi_2\right) \end{array} \right\} \tag{E.8}$$

Comparison of Eqs. (E.1) and (E.8) shows that the motion of the system coincides with the second normal mode only if $X_1^{(1)} = 0$. This implies that (from Eq. (E.2))

$$x_1(0) = -x_2(0) \qquad \text{and} \qquad \dot{x}_1(0) = -\dot{x}_2(0) \tag{E.9}$$

∎

EXAMPLE 5.3

Free-Vibration Response of a Two-Degree-of-Freedom System

Find the free-vibration response of the system shown in Fig. 5.5(a) with $k_1 = 30$, $k_2 = 5$, $k_3 = 0$, $m_1 = 10$, $m_2 = 1$, and $c_1 = c_2 = c_3 = 0$ for the initial conditions $x_1(0) = 1$, $\dot{x}_1(0) = x_2(0) = \dot{x}_2(0) = 0$.

Solution: For the given data, the eigenvalue problem, Eq. (5.8), becomes

$$\begin{bmatrix} -m_1\omega^2 + k_1 + k_2 & -k_2 \\ -k_2 & -m_2\omega^2 + k_2 + k_3 \end{bmatrix} \begin{Bmatrix} X_1 \\ X_2 \end{Bmatrix} = \begin{Bmatrix} 0 \\ 0 \end{Bmatrix}$$

or

$$\begin{bmatrix} -10\omega^2 + 35 & -5 \\ -5 & -\omega^2 + 5 \end{bmatrix} \begin{Bmatrix} X_1 \\ X_2 \end{Bmatrix} = \begin{Bmatrix} 0 \\ 0 \end{Bmatrix} \tag{E.1}$$

By setting the determinant of the coefficient matrix in Eq. (E.1) to zero, we obtain the frequency equation (see Eq. (5.9)):

$$10\omega^4 - 85\omega^2 + 150 = 0 \tag{E.2}$$

from which the natural frequencies can be found as

$$\omega_1^2 = 2.5, \qquad \omega_2^2 = 6.0$$

or

$$\omega_1 = 1.5811, \qquad \omega_2 = 2.4495 \tag{E.3}$$

The substitution of $\omega^2 = \omega_1^2 = 2.5$ in Eq. (E.1) leads to $X_2^{(1)} = 2X_1^{(1)}$, while $\omega^2 = \omega_2^2 = 6.0$ in Eq. (E.1) yields $X_2^{(2)} = -5X_1^{(2)}$. Thus the normal modes (or eigenvectors) are given by

$$\vec{X}^{(1)} = \begin{Bmatrix} X_1^{(1)} \\ X_2^{(1)} \end{Bmatrix} = \begin{Bmatrix} 1 \\ 2 \end{Bmatrix} X_1^{(1)} \tag{E.4}$$

$$\vec{X}^{(2)} = \begin{Bmatrix} X_1^{(2)} \\ X_2^{(2)} \end{Bmatrix} = \begin{Bmatrix} 1 \\ -5 \end{Bmatrix} X_1^{(2)} \tag{E.5}$$

The free-vibration responses of the masses m_1 and m_2 are given by (see Eq. (5.15)):

$$x_1(t) = X_1^{(1)} \cos(1.5811t + \phi_1) + X_1^{(2)} \cos(2.4495t + \phi_2) \tag{E.6}$$

$$x_2(t) = 2X_1^{(1)} \cos(1.5811t + \phi_1) - 5X_1^{(2)} \cos(2.4495t + \phi_2) \tag{E.7}$$

where $X_1^{(1)}$, $X_1^{(2)}$, ϕ_1, and ϕ_2 are constants to be determined from the initial conditions. By using the given initial conditions in Eqs. (E.6) and (E.7), we obtain

$$x_1(t = 0) = 1 = X_1^{(1)} \cos \phi_1 + X_1^{(2)} \cos \phi_2 \tag{E.8}$$

$$x_2(t = 0) = 0 = 2X_1^{(1)} \cos \phi_1 - 5X_1^{(2)} \cos \phi_2 \tag{E.9}$$

$$\dot{x}_1(t = 0) = 0 = -1.5811X_1^{(1)} \sin \phi_1 - 2.4495X_1^{(2)} \sin \phi_2 \tag{E.10}$$

$$\dot{x}_2(t = 0) = -3.1622X_1^{(1)} + 12.2475X_1^{(2)} \sin \phi_2 \tag{E.11}$$

The solution of Eqs. (E.8) and (E.9) yields

$$X_1^{(1)} \cos \phi_1 = \frac{5}{7}, \qquad X_1^{(2)} \cos \phi_2 = \frac{2}{7} \tag{E.12}$$

while the solution of Eqs. (E.10) and (E.11) leads to

$$X_1^{(1)} \sin \phi_1 = 0, \qquad X_1^{(2)} \sin \phi_2 = 0 \tag{E.13}$$

Equations (E.12) and (E.13) give

$$X_1^{(1)} = \frac{5}{7}, \qquad X_1^{(2)} = \frac{2}{7}, \qquad \phi_1 = 0, \qquad \phi_2 = 0 \tag{E.14}$$

Thus the free-vibration responses of m_1 and m_2 are given by

$$x_1(t) = \frac{5}{7} \cos 1.5811t + \frac{2}{7} \cos 2.4495t \tag{E.15}$$

$$x_2(t) = \frac{10}{7} \cos 1.5811t - \frac{10}{7} \cos 2.4495t \tag{E.16}$$

The graphical representation of Eqs. (E.15) and (E.16) is considered in Example 5.17.

■

5.4   Torsional System

Consider a torsional system consisting of two discs mounted on a shaft, as shown in Fig. 5.8. The three segments of the shaft have rotational spring constants k_{t1}, k_{t2}, and k_{t3}, as indicated in the figure. Also shown are the discs of mass moments of inertia J_1 and J_2, the applied torques M_{t1} and M_{t2}, and the rotational degrees of freedom θ_1 and θ_2. The differential equations of rotational motion for the discs J_1 and J_2 can be derived as

$$J_1\ddot{\theta}_1 = -k_{t1}\theta_1 + k_{t2}(\theta_2 - \theta_1) + M_{t1}$$

$$J_2\ddot{\theta}_2 = -k_{t2}(\theta_2 - \theta_1) - k_{t3}\theta_2 + M_{t2}$$

which upon rearrangement become

$$J_1\ddot{\theta}_1 + (k_{t1} + k_{t2})\theta_1 - k_{t2}\theta_2 = M_{t1}$$

$$J_2\ddot{\theta}_2 - k_{t2}\theta_1 + (k_{t2} + k_{t3})\theta_2 = M_{t2} \tag{5.19}$$

For the free-vibration analysis of the system, Eq. (5.19) reduces to

$$J_1\ddot{\theta}_1 + (k_{t1} + k_{t2})\theta_1 - k_{t2}\theta_2 = 0$$

$$J_2\ddot{\theta}_2 - k_{t2}\theta_1 + (k_{t2} + k_{t3})\theta_2 = 0 \tag{5.20}$$

Note that Eq. (5.20) is similar to Eqs. (5.4) and (5.5). In fact, Eq. (5.20) can be obtained by substituting θ_1, θ_2, J_1, J_2, k_{t1}, k_{t2}, and k_{t3} for x_1, x_2, m_1, m_2, k_1, k_2, and k_3, respectively. Thus the analysis presented in Section 5.3 is also applicable to torsional systems with proper substitutions. The following two examples illustrate the procedure.

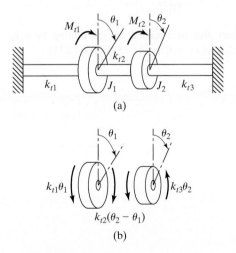

(a)

(b)

FIGURE 5.8   Torsional system with discs mounted on a shaft.

EXAMPLE 5.4

Natural Frequencies of a Torsional System

Find the natural frequencies and mode shapes for the torsional system shown in Fig. 5.9 for $J_1 = J_0$, $J_2 = 2J_0$, and $k_{t1} = k_{t2} = k_t$.

Solution: The differential equations of motion, Eq. (5.20), reduce to (with $k_{t3} = 0$, $k_{t1} = k_{t2} = k_t$, $J_1 = J_0$, and $J_2 = 2J_0$):

$$J_0\ddot{\theta}_1 + 2k_t\theta_1 - k_t\theta_2 = 0$$

$$2J_0\ddot{\theta}_2 - k_t\theta_1 + k_t\theta_2 = 0 \tag{E.1}$$

Rearranging and substituting the harmonic solution

$$\theta_i(t) = \Theta_i \cos(\omega t + \phi); \qquad i = 1, 2 \tag{E.2}$$

gives the frequency equation:

$$2\omega^4 J_0^2 - 5\omega^2 J_0 k_t + k_t^2 = 0 \tag{E.3}$$

The solution of Eq. (E.3) gives the natural frequencies

$$\omega_1 = \sqrt{\frac{k_t}{4J_0}(5 - \sqrt{17})} \qquad \text{and} \qquad \omega_2 = \sqrt{\frac{k_t}{4J_0}(5 + \sqrt{17})} \tag{E.4}$$

The amplitude ratios are given by

$$r_1 = \frac{\Theta_2^{(1)}}{\Theta_1^{(1)}} = 2 - \frac{(5 - \sqrt{17})}{4}$$

$$r_2 = \frac{\Theta_2^{(2)}}{\Theta_1^{(2)}} = 2 - \frac{(5 + \sqrt{17})}{4} \tag{E.5}$$

Equations (E.4) and (E.5) can also be obtained by substituting $k_1 = k_{t1} = k_t$, $k_2 = k_{t2} = k_t$, $m_1 = J_1 = J_0$, $m_2 = J_2 = 2J_0$, and $k_3 = 0$ in Eqs. (5.10) and (5.11).

FIGURE 5.9
Torsional system.

Note: For a two-degree-of-freedom system, the two natural frequencies ω_1 and ω_2 are not equal to either of the natural frequencies of the two single-degree-of-freedom systems constructed from the same components. In Example 5.4, the single-degree-of-freedom systems k_{t1} and J_1

$$\left(\text{with } \overline{\omega}_1 = \sqrt{\frac{k_{t1}}{J_1}} = \sqrt{\frac{k_t}{J_0}} \right)$$

and k_{t2} and J_2

$$\left(\text{with } \overline{\omega}_2 = \sqrt{\frac{k_{t2}}{J_2}} = \frac{1}{\sqrt{2}}\sqrt{\frac{k_t}{J_0}} \right)$$

are combined to obtain the system shown in Fig. 5.9. It can be seen that ω_1 and ω_2 are different from $\overline{\omega}_1$ and $\overline{\omega}_2$.

EXAMPLE 5.5

Natural Frequencies of a Marine Engine Propeller

The schematic diagram of a marine engine connected to a propeller through gears is shown in Fig. 5.10(a). The mass moments of inertia of the flywheel, engine, gear 1, gear 2, and the propeller (in kg-m^2) are 9,000, 1,000, 250, 150, and 2,000, respectively. Find the natural frequencies and mode shapes of the system in torsional vibration.

Solution

Approach: Find the equivalent mass moments of inertia of all rotors with respect to one rotor and use a two-degree-of-freedom model.

Assumptions:

1. The flywheel can be considered to be stationary (fixed), since its mass moment of inertia is very large compared to that of other rotors.
2. The engine and gears can be replaced by a single equivalent rotor.

Since gears 1 and 2 have 40 and 20 teeth, respectively, shaft 2 rotates at twice the speed of shaft 1. Thus the mass moments of inertia of gear 2 and the propeller, referred to the engine, are given by

$$(J_{G2})_{\text{eq}} = (2)^2(150) = 600 \text{ kg-m}^2$$

$$(J_P)_{\text{eq}} = (2)^2(2{,}000) = 8{,}000 \text{ kg-m}^2$$

Since the distance between the engine and the gear unit is small, the engine and the two gears can be replaced by a single rotor with a mass moment of inertia of

$$J_1 = J_E + J_{G1} + (J_{G2})_{\text{eq}} = 1{,}000 + 250 + 600 = 1{,}850 \text{ kg-m}^2$$

Assuming a shear modulus of 80×10^9 N/m^2 for steel, the torsional stiffnesses of shafts 1 and 2 can be determined as

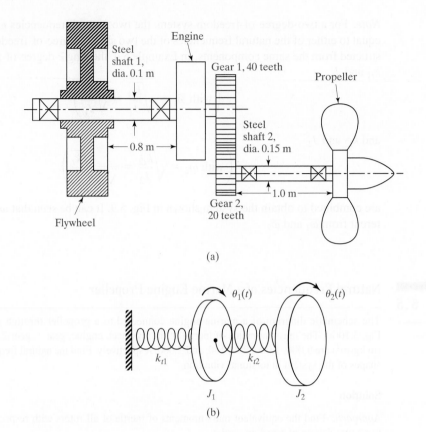

(a)

(b)

FIGURE 5.10   Marine engine propeller system.

$$k_{t1} = \frac{GI_{01}}{l_1} = \frac{G}{l_1}\left(\frac{\pi d_1^4}{32}\right) = \frac{(80 \times 10^9)(\pi)(0.10)^4}{(0.8)(32)} = 981{,}750.0 \text{ N-m/rad}$$

$$k_{t2} = \frac{GI_{02}}{l_2} = \frac{G}{l_2}\left(\frac{\pi d_2^4}{32}\right) = \frac{(80 \times 10^9)(\pi)(0.15)^4}{(1.0)(32)} = 3{,}976{,}087.5 \text{ N-m/rad}$$

Since the length of shaft 2 is not negligible, the propeller is assumed to be a rotor connected at the end of shaft 2. Thus the system can be represented as a two-degree-of-freedom torsional system, as indicated in Fig. 5.10(b). By setting $k_3 = 0$, $k_1 = k_{t1}$, $k_2 = k_{t2}$, $m_1 = J_1$, and $m_2 = J_2$ in Eq. (5.10), the natural frequencies of the system can be found as

$$\omega_1^2, \omega_2^2 = \frac{1}{2}\left\{\frac{(k_{t1} + k_{t2})J_2 + k_{t2}J_1}{J_1 J_2}\right\}$$

$$\pm \left[\left\{\frac{(k_{t1} + k_{t2})J_2 + k_{t2}J_1}{J_1 J_2}\right\}^2 - 4\left\{\frac{(k_{t1} + k_{t2})k_{t2} - k_{t2}^2}{J_1 J_2}\right\}\right]^{1/2}$$

$$= \left\{ \frac{(k_{t1} + k_{t2})}{2J_1} + \frac{k_{t2}}{2J_2} \right\}$$

$$\pm \left[\left\{ \frac{(k_{t1} + k_{t2})}{2J_1} + \frac{k_{t2}}{2J_2} \right\}^2 - \frac{k_{t1}k_{t2}}{J_1 J_2} \right]^{1/2} \tag{E.1}$$

Since

$$\frac{(k_{t1} + k_{t2})}{2J_1} + \frac{k_{t2}}{2J_2} = \frac{(98.1750 + 397.6087) \times 10^4}{2 \times 1,850} + \frac{397.6087 \times 10^4}{2 \times 8,000}$$

$$= 1,588.46$$

and

$$\frac{k_{t1}k_{t2}}{J_1 J_2} = \frac{(98.1750 \times 10^4)(397.6087 \times 10^4)}{(1,850)(8,000)} = 26.3750 \times 10^4$$

Eq. (E.1) gives

$$\omega_1^2, \omega_2^2 = 1,588.46 \pm [(1,588.46)^2 - 26.3750 \times 10^4]^{1/2}$$

$$= 1,588.46 \pm 1,503.1483$$

Thus

$$\omega_1^2 = 85.3117 \qquad \text{or} \qquad \omega_1 = 9.2364 \text{ rad/s}$$

$$\omega_2^2 = 3,091.6083 \qquad \text{or} \qquad \omega_2 = 55.6022 \text{ rad/s}$$

For the mode shapes, we set $k_1 = k_{t1}$, $k_2 = k_{t2}$, $k_3 = 0$, $m_1 = J_1$, and $m_2 = J_2$ in Eq. (5.11) to obtain

$$r_1 = \frac{-J_1\omega_1^2 + (k_{t1} + k_{t2})}{k_{t2}}$$

$$= \frac{-(1,850)(85.3117) + (495.7837 \times 10^4)}{397.6087 \times 10^4} = 1.2072$$

and

$$r_2 = \frac{-J_1\omega_2^2 + (k_{t1} + k_{t2})}{k_{t2}}$$

$$= \frac{-(1,850)(3,091.6083) + (495.7837 \times 10^4)}{397.6087 \times 10^4} = -0.1916$$

Thus the mode shapes can be determined from an equation similar to Eq. (5.12) as

$$\left\{ \frac{\Theta_1}{\Theta_2} \right\}^{(1)} = \left\{ \frac{1}{r_1} \right\} = \frac{1}{1.2072}$$

and

$$\left\{\frac{\Theta_1}{\Theta_2}\right\}^{(2)} = \left\{\frac{1}{r_2}\right\} = \frac{1}{-0.1916}$$

∎

5.5   Coordinate Coupling and Principal Coordinates

As stated earlier, an *n*-degree-of-freedom system requires *n* independent coordinates to describe its configuration. Usually, these coordinates are independent geometrical quantities measured from the equilibrium position of the vibrating body. However, it is possible to select some other set of *n* coordinates to describe the configuration of the system. The latter set may be, for example, different from the first set in that the coordinates may have their origin away from the equilibrium position of the body. There could be still other sets of coordinates to describe the configuration of the system. Each of these sets of *n* coordinates is called the *generalized coordinates*.

As an example, consider the lathe shown in Fig. 5.11(a). For simplicity, the lathe bed can be replaced by an elastic beam supported on short elastic columns and the headstock and tailstock can be replaced by two lumped masses as shown in Fig. 5.11(b). The modeling of the lathe as a two-degree-of-freedom system has been indicated in Section 5.1. As shown in Figs. 5.12(a) and (b), any of the following sets of coordinates can be used to describe the motion of this two-degree-of-freedom system:

1. Deflections $x_1(t)$ and $x_2(t)$ of the two ends of the lathe *AB*.
2. Deflection $x(t)$ of the C.G. and rotation $\theta(t)$.
3. Deflection $x_1(t)$ of the end *A* and rotation $\theta(t)$.
4. Deflection $y(t)$ of point *P* located at a distance *e* to the left of the center of gravity (C.G.) and rotation $\theta(t)$.

(a)

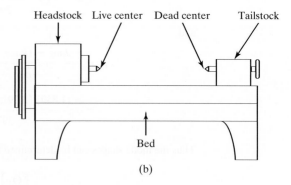

(b)

FIGURE 5.11   Lathe. (Photo courtesy of South Bend Lathe Corp.)

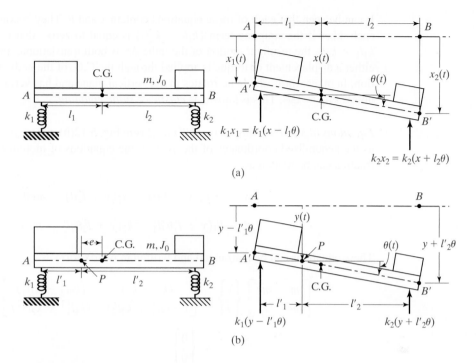

FIGURE 5.12   Modeling of a lathe.

Thus any set of these coordinates—(x_1, x_2), (x, θ), (x_1, θ), and (y, θ)—represents the generalized coordinates of the system. Now we shall derive the equations of motion of the lathe using two different sets of coordinates to illustrate the concept of coordinate coupling.

Equations of Motion Using x(t) and θ(t).   From the free-body diagram shown in Fig. 5.12(a), with the positive values of the motion variables as indicated, the force equilibrium equation in the vertical direction can be written as

$$m\ddot{x} = -k_1(x - l_1\theta) - k_2(x + l_2\theta) \tag{5.21}$$

and the moment equation about the C.G. can be expressed as

$$J_0\ddot{\theta} = k_1(x - l_1\theta)l_1 - k_2(x + l_2\theta)l_2 \tag{5.22}$$

Equations (5.21) and (5.22) can be rearranged and written in matrix form as

$$\begin{bmatrix} m & 0 \\ 0 & J_0 \end{bmatrix} \begin{Bmatrix} \ddot{x} \\ \ddot{\theta} \end{Bmatrix} + \begin{bmatrix} (k_1 + k_2) & -(k_1l_1 - k_2l_2) \\ -(k_1l_1 - k_2l_2) & (k_1l_1^2 + k_2l_2^2) \end{bmatrix} \begin{Bmatrix} x \\ \theta \end{Bmatrix}$$

$$= \begin{Bmatrix} 0 \\ 0 \end{Bmatrix} \tag{5.23}$$

It can be seen that each of these equations contain x and θ. They become independent of each other if the coupling term $(k_1l_1 - k_2l_2)$ is equal to zero—that is, if $k_1l_1 = k_2l_2$. If $k_1l_1 \neq k_2l_2$, the resultant motion of the lathe AB is both translational and rotational when either a displacement or torque is applied through the C.G. of the body as an initial condition. In other words, the lathe rotates in the vertical plane and has vertical motion as well, unless $k_1l_1 = k_2l_2$. This is known as *elastic* or *static coupling*.

Equations of Motion Using y(t) and θ(t).   From Fig. 5.12(b), where $y(t)$ and $\theta(t)$ are used as the generalized coordinates of the system, the equations of motion for translation and rotation can be written as

$$m\ddot{y} = -k_1(y - l_1'\theta) - k_2(y + l_2'\theta) - me\ddot{\theta}$$

$$J_p\ddot{\theta} = k_1(y - l_1'\theta)l_1' - k_2(y + l_2'\theta)l_2' - me\ddot{y} \tag{5.24}$$

These equations can be rearranged and written in matrix form as

$$\begin{bmatrix} m & me \\ me & J_p \end{bmatrix} \begin{Bmatrix} \ddot{y} \\ \ddot{\theta} \end{Bmatrix} + \begin{bmatrix} (k_1 + k_2) & (k_2l_2' - k_1l_1') \\ (-k_1l_1' + k_2l_2') & (k_1l_1'^2 + k_2l_2'^2) \end{bmatrix} \begin{Bmatrix} y \\ \theta \end{Bmatrix}$$

$$= \begin{Bmatrix} 0 \\ 0 \end{Bmatrix} \tag{5.25}$$

Both the equations of motion represented by Eq. (5.25) contain y and θ, so they are coupled equations. They contain static (or elastic) as well as dynamic (or mass) coupling terms. If $k_1l_1' = k_2l_2'$, the system will have *dynamic* or *inertia coupling* only. In this case, if the lathe moves up and down in the y direction, the inertia force $m\ddot{y}$, which acts through the center of gravity of the body, induces a motion in the θ direction, by virtue of the moment $m\ddot{y}e$. Similarly, a motion in the θ direction induces a motion of the lathe in the y direction due to the force $me\ddot{\theta}$.

Note the following characteristics of these systems:

1. In the most general case, a viscously damped two-degree-of-freedom system has equations of motion in the following form:

$$\begin{bmatrix} m_{11} & m_{12} \\ m_{12} & m_{22} \end{bmatrix} \begin{Bmatrix} \ddot{x}_1 \\ \ddot{x}_2 \end{Bmatrix} + \begin{Bmatrix} c_{11} & c_{12} \\ c_{12} & c_{22} \end{Bmatrix} \begin{Bmatrix} \dot{x}_1 \\ \dot{x}_2 \end{Bmatrix} + \begin{bmatrix} k_{11} & k_{12} \\ k_{12} & k_{22} \end{bmatrix} \begin{Bmatrix} x_1 \\ x_2 \end{Bmatrix} = \begin{Bmatrix} 0 \\ 0 \end{Bmatrix} \tag{5.26}$$

This equation reveals the type of coupling present. If the stiffness matrix is not diagonal, the system has elastic or static coupling. If the damping matrix is not diagonal, the system has damping or velocity coupling. Finally, if the mass matrix is not diagonal, the system has mass or inertia coupling. Both velocity and mass coupling come under the heading of dynamic coupling.

2. The system vibrates in its own natural way regardless of the coordinates used. The choice of the coordinates is a mere convenience.

3. From Eqs. (5.23) and (5.25), it is clear that the nature of the coupling depends on the coordinates used and is not an inherent property of the system. It is possible to choose a system of coordinates $q_1(t)$ and $q_2(t)$ which give equations of motion that are uncoupled both statically and dynamically. Such coordinates are called *principal* or *natural coordinates*. The main advantage of using principal coordinates is that the resulting uncoupled equations of motion can be solved independently of one another.

The following example illustrates the method of finding the principal coordinates in terms of the geometrical coordinates.

EXAMPLE 5.6

Principal Coordinates of Spring-Mass System

Determine the principal coordinates for the spring-mass system shown in Fig. 5.6.

Solution

Approach: Define two independent solutions as principal coordinates and express them in terms of the solutions $x_1(t)$ and $x_2(t)$.

The general motion of the system shown in Fig. 5.6 is given by Eq. (E.10) of Example 5.1:

$$x_1(t) = B_1 \cos\left(\sqrt{\frac{k}{m}}t + \phi_1\right) + B_2 \cos\left(\sqrt{\frac{3k}{m}}t + \phi_2\right)$$

$$x_2(t) = B_1 \cos\left(\sqrt{\frac{k}{m}}t + \phi_1\right) - B_2 \cos\left(\sqrt{\frac{3k}{m}}t + \phi_2\right) \tag{E.1}$$

where $B_1 = X_1^{(1)}$, $B_2 = X_1^{(2)}$, ϕ_1, and ϕ_2 are constants. We define a new set of coordinates $q_1(t)$ and $q_2(t)$ such that

$$q_1(t) = B_1 \cos\left(\sqrt{\frac{k}{m}}t + \phi_1\right)$$

$$q_2(t) = B_2 \cos\left(\sqrt{\frac{3k}{m}}t + \phi_2\right) \tag{E.2}$$

Since $q_1(t)$ and $q_2(t)$ are harmonic functions, their corresponding equations of motion can be written as[1]

$$\ddot{q}_1 + \left(\frac{k}{m}\right)q_1 = 0$$

$$\ddot{q}_2 + \left(\frac{3k}{m}\right)q_2 = 0 \tag{E.3}$$

---

[1] Note that the equation of motion corresponding to the solution $q = B\cos(\omega t + \phi)$ is given by $\ddot{q} + \omega^2 q = 0$.

These equations represent a two-degree-of-freedom system whose natural frequencies are $\omega_1 = \sqrt{k/m}$ and $\omega_2 = \sqrt{3k/m}$. Because there is neither static nor dynamic coupling in the equations of motion (E.3), $q_1(t)$ and $q_2(t)$ are principal coordinates. From Eqs. (E.1) and (E.2), we can write

$$x_1(t) = q_1(t) + q_2(t)$$
$$x_2(t) = q_1(t) - q_2(t) \tag{E.4}$$

The solution of Eqs. (E.4) gives the principal coordinates:

$$q_1(t) = \frac{1}{2}[x_1(t) + x_2(t)]$$

$$q_2(t) = \frac{1}{2}[x_1(t) - x_2(t)] \tag{E.5}$$

∎

EXAMPLE 5.7

Frequencies and Modes of an Automobile

Determine the pitch (angular motion) and bounce (up-and-down linear motion) frequencies and the location of oscillation centers (nodes) of an automobile with the following data (see Fig. 5.13):

Mass $(m) = 1{,}000$ kg
Radius of gyration $(r) = 0.9$ m
Distance between front axle and C.G. $(l_1) = 1.0$ m
Distance between rear axle and C.G. $(l_2) = 1.5$ m
Front spring stiffness $(k_f) = 18$ kN/m
Rear spring stiffness $(k_r) = 22$ kN/m

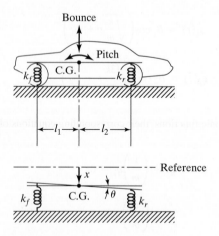

FIGURE 5.13   Pitch and bounce motions of an automobile.

Solution: If x and θ are used as independent coordinates, the equations of motion are given by Eq. (5.23) with $k_1 = k_f$, $k_2 = k_r$, and $J_0 = mr^2$. For free vibration, we assume a harmonic solution:

$$x(t) = X \cos(\omega t + \phi), \qquad \theta(t) = \Theta \cos(\omega t + \phi) \tag{E.1}$$

Using Eqs. (E.1) and (5.23), we obtain

$$\begin{bmatrix} (-m\omega^2 + k_f + k_r) & (-k_f l_1 + k_r l_2) \\ (-k_f l_1 + k_r l_2) & (-J_0\omega^2 + k_f l_1^2 + k_r l_2^2) \end{bmatrix} \begin{Bmatrix} X \\ \Theta \end{Bmatrix} = \begin{Bmatrix} 0 \\ 0 \end{Bmatrix} \tag{E.2}$$

For the known data, Eq. (E.2) becomes

$$\begin{bmatrix} (-1{,}000\omega^2 + 40{,}000) & 15{,}000 \\ 15{,}000 & (-810\omega^2 + 67{,}500) \end{bmatrix} \begin{Bmatrix} X \\ \Theta \end{Bmatrix} = \begin{Bmatrix} 0 \\ 0 \end{Bmatrix} \tag{E.3}$$

from which the frequency equation can be derived:

$$8.1\omega^4 - 999\omega^2 + 24{,}750 = 0 \tag{E.4}$$

The natural frequencies can be found from Eq. (E.4):

$$\omega_1 = 5.8593 \text{ rad/s}, \qquad \omega_2 = 9.4341 \text{ rad/s} \tag{E.5}$$

With these values, the ratio of amplitudes can be found from Eq. (E.3):

$$\frac{X^{(1)}}{\Theta^{(1)}} = -2.6461, \qquad \frac{X^{(2)}}{\Theta^{(2)}} = 0.3061 \tag{E.6}$$

The node locations can be obtained by noting that the tangent of a small angle is approximately equal to the angle itself. Thus, from Fig. 5.14, we find that the distance between the C.G. and the node is -2.6461 m for ω_1 and 0.3061 m for ω_2. The mode shapes are shown by dashed lines in Fig. 5.14.

FIGURE 5.14   Mode shapes of an automobile.

∎

5.6   Forced-Vibration Analysis

The equations of motion of a general two-degree-of-freedom system under external forces can be written as

$$
\begin{bmatrix} m_{11} & m_{12} \\ m_{12} & m_{22} \end{bmatrix} \begin{Bmatrix} \ddot{x}_1 \\ \ddot{x}_2 \end{Bmatrix} + \begin{bmatrix} c_{11} & c_{12} \\ c_{12} & c_{22} \end{bmatrix} \begin{Bmatrix} \dot{x}_1 \\ \dot{x}_2 \end{Bmatrix}
$$

$$
+ \begin{bmatrix} k_{11} & k_{12} \\ k_{12} & k_{22} \end{bmatrix} \begin{Bmatrix} x_1 \\ x_2 \end{Bmatrix} = \begin{Bmatrix} F_1 \\ F_2 \end{Bmatrix} \tag{5.27}
$$

Equations (5.1) and (5.2) can be seen to be special cases of Eq. (5.27), with $m_{11} = m_1$, $m_{22} = m_2$, and $m_{12} = 0$. We shall consider the external forces to be harmonic:

$$
F_j(t) = F_{j0} e^{i\omega t}, \qquad j = 1, 2 \tag{5.28}
$$

where ω is the forcing frequency. We can write the steady-state solutions as

$$
x_j(t) = X_j e^{i\omega t}, \qquad j = 1, 2 \tag{5.29}
$$

where X_1 and X_2 are, in general, complex quantities that depend on ω and the system parameters. Substitution of Eqs. (5.28) and (5.29) into Eq. (5.27) leads to

$$
\begin{bmatrix} (-\omega^2 m_{11} + i\omega c_{11} + k_{11}) & (-\omega^2 m_{12} + i\omega c_{12} + k_{12}) \\ (-\omega^2 m_{12} + i\omega c_{12} + k_{12}) & (-\omega_2 m_{22} + i\omega c_{22} + k_{22}) \end{bmatrix} \begin{Bmatrix} X_1 \\ X_2 \end{Bmatrix}
$$

$$
= \begin{Bmatrix} F_{10} \\ F_{20} \end{Bmatrix} \tag{5.30}
$$

As in Section 3.5, we define the mechanical impedance $Z_{rs}(i\omega)$ as

$$
Z_{rs}(i\omega) = -\omega^2 m_{rs} + i\omega c_{rs} + k_{rs}, \qquad r, s = 1, 2 \tag{5.31}
$$

and write Eq. (5.30) as

$$
[Z(i\omega)]\vec{X} = \vec{F}_0 \tag{5.32}
$$

where

$$
[Z(i\omega)] = \begin{bmatrix} Z_{11}(i\omega) & Z_{12}(i\omega) \\ Z_{12}(i\omega) & Z_{22}(i\omega) \end{bmatrix} = \text{Impedance matrix}
$$

$$
\vec{X} = \begin{Bmatrix} X_1 \\ X_2 \end{Bmatrix}
$$

and

$$
\vec{F}_0 = \begin{Bmatrix} F_{10} \\ F_{20} \end{Bmatrix}
$$

Equation (5.32) can be solved to obtain

$$\vec{X} = [Z(i\omega)]^{-1}\vec{F}_0 \tag{5.33}$$

where the inverse of the impedance matrix is given by

$$[Z(i\omega)]^{-1} = \frac{1}{Z_{11}(i\omega)Z_{22}(i\omega) - Z_{12}^2(i\omega)}\begin{bmatrix} Z_{22}(i\omega) & -Z_{12}(i\omega) \\ -Z_{12}(i\omega) & Z_{11}(i\omega) \end{bmatrix} \tag{5.34}$$

Equations (5.33) and (5.34) lead to the solution

$$X_1(i\omega) = \frac{Z_{22}(i\omega)F_{10} - Z_{12}(i\omega)F_{20}}{Z_{11}(i\omega)Z_{22}(i\omega) - Z_{12}^2(i\omega)}$$

$$X_2(i\omega) = \frac{-Z_{12}(i\omega)F_{10} + Z_{11}(i\omega)F_{20}}{Z_{11}(i\omega)Z_{22}(i\omega) - Z_{12}^2(i\omega)} \tag{5.35}$$

By substituting Eq. (5.35) into Eq. (5.29), we can find the complete solution, $x_1(t)$ and $x_2(t)$.
    The analysis of a two-degree-of-freedom system used as a vibration absorber is given in Section 9.11. [5.4] deals with the impact response of a two-degree-of-freedom system, while [5.5] considers the steady-state response under harmonic excitation.

Steady-State Response of Spring-Mass System

EXAMPLE 5.8

Find the steady-state response of the system shown in Fig. 5.15 when the mass m_1 is excited by the force $F_1(t) = F_{10}\cos\omega t$. Also plot its frequency-response curve.

Solution: The equations of motion of the system can be expressed as

$$\begin{bmatrix} m & 0 \\ 0 & m \end{bmatrix}\begin{Bmatrix} \ddot{x}_1 \\ \ddot{x}_2 \end{Bmatrix} + \begin{bmatrix} 2k & -k \\ -k & 2k \end{bmatrix}\begin{Bmatrix} x_1 \\ x_2 \end{Bmatrix} = \begin{Bmatrix} F_{10}\cos\omega t \\ 0 \end{Bmatrix} \tag{E.1}$$

Comparison of Eq. (E.1) with Eq. (5.27) shows that

$$m_{11} = m_{22} = m, \qquad m_{12} = 0, \qquad c_{11} = c_{12} = c_{22} = 0,$$

$$k_{11} = k_{22} = 2k, \qquad k_{12} = -k, \qquad F_1 = F_{10}\cos\omega t, \qquad F_2 = 0$$

We assume the solution to be as follows:[2]

$$x_j(t) = X_j\cos\omega t, \qquad\qquad j = 1, 2 \tag{E.2}$$

Equation (5.31) gives

$$Z_{11}(\omega) = Z_{22}(\omega) = -m\omega^2 + 2k, \qquad Z_{12}(\omega) = -k \tag{E.3}$$

---

[2]Since $F_{10}\cos\omega t = \text{Re}(F_{10}e^{i\omega t})$, we shall assume the solution also to be $x_j = \text{Re}(X_j e^{i\omega t}) = X_j\cos\omega t$, $j = 1, 2$. It can be verified that X_j are real for an undamped system.

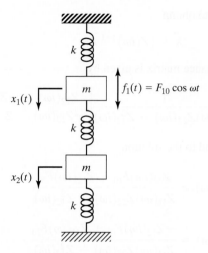

FIGURE 5.15   A two-mass system subjected to harmonic force.

Hence X_1 and X_2 are given by Eq. (5.35):

$$X_1(\omega) = \frac{(-\omega^2 m + 2k)\,F_{10}}{(-\omega^2 m + 2k)^2 - k^2} = \frac{(-\omega^2 m + 2k)\,F_{10}}{(-m\omega^2 + 3k)(-m\omega^2 + k)} \tag{E.4}$$

$$X_2(\omega) = \frac{kF_{10}}{(-m\omega^2 + 2k)^2 - k^2} = \frac{kF_{10}}{(-m\omega^2 + 3k)(-m\omega^2 + k)} \tag{E.5}$$

By defining $\omega_1^2 = k/m$ and $\omega_2^2 = 3k/m$, Eqs. (E.4) and (E.5) can be expressed as

$$X_1(\omega) = \frac{\left\{2 - \left(\dfrac{\omega}{\omega_1}\right)^2\right\}F_{10}}{k\left[\left(\dfrac{\omega_2}{\omega_1}\right)^2 - \left(\dfrac{\omega}{\omega_1}\right)^2\right]\left[1 - \left(\dfrac{\omega}{\omega_1}\right)^2\right]} \tag{E.6}$$

$$X_2(\omega) = \frac{F_{10}}{k\left[\left(\dfrac{\omega_2}{\omega_1}\right)^2 - \left(\dfrac{\omega}{\omega_1}\right)^2\right]\left[1 - \left(\dfrac{\omega}{\omega_1}\right)^2\right]} \tag{E.7}$$

The responses X_1 and X_2 are shown in Fig. 5.16 in terms of the dimensionless parameter ω/ω_1. In this parameter, ω_1 was selected arbitrarily; ω_2 could have been selected just as easily. It can be seen that the amplitudes X_1 and X_2 become infinite when $\omega^2 = \omega_1^2$ or $\omega^2 = \omega_2^2$. Thus there are two resonance conditions for the system: one at ω_1 and another at ω_2. At all other values of ω, the amplitudes of vibration are finite. It can be noted from Fig. 5.16 that there is a particular value of the frequency ω at

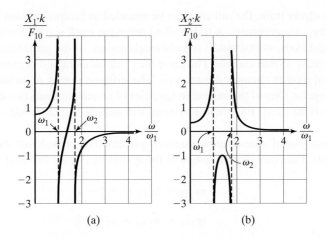

FIGURE 5.16   Frequency-response curves of Example 5.8.

which the vibration of the first mass m_1, to which the force $f_1(t)$ is applied, is reduced to zero. This characteristic forms the basis of the dynamic vibration absorber discussed in Chapter 9.

■

5.7   Semidefinite Systems

Semidefinite systems are also known as *unrestrained* or *degenerate systems*. Two examples of such systems are shown in Fig. 5.17. The arrangement in Fig. 5.17(a) may be considered to represent two railway cars of masses m_1 and m_2 with a coupling spring k. The arrangement in Fig. 5.17(c) may be considered to represent two rotors of mass moments of inertia J_1 and J_2 connected by a shaft of torsional stiffness k_t.

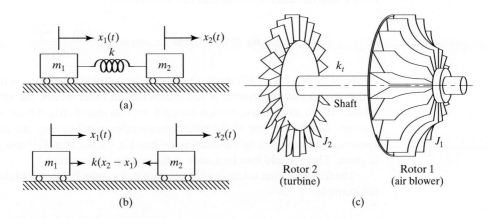

FIGURE 5.17   Semidefinite systems.

In a railway train, the rail cars can be modeled as lumped masses and the couplings between the cars as springs. A train rolling down the track can be considered as a system having rigid-body, unrestrained, translational motion. At the same time, the rail cars can vibrate relative to one another. The presence of an unrestrained degree of freedom in the equation of motion changes the analysis. The stiffness matrix of an unrestrained system will be singular. One of the natural frequencies of an unrestrained two-degree-of-freedom system will be zero. For such a system, the motion is composed of translation and vibration.

The analysis of unrestrained systems is presented by considering the system shown in Fig. 5.17(a). The equations of motion of the system can be written as (Fig. 5.17(b)):

$$m_1\ddot{x}_1 + k(x_1 - x_2) = 0$$

$$m_2\ddot{x}_2 + k(x_2 - x_1) = 0 \tag{5.36}$$

For free vibration, we assume the motion to be harmonic:

$$x_j(t) = X_j \cos(\omega t + \phi_j), \qquad j = 1, 2 \tag{5.37}$$

Substitution of Eq. (5.37) into Eq. (5.36) gives

$$(-m_1\omega^2 + k)X_1 - kX_2 = 0$$

$$-kX_1 + (-m_2\omega^2 + k)X_2 = 0 \tag{5.38}$$

By equating the determinant of the coefficients of X_1 and X_2 to zero, we obtain the frequency equation as

$$\omega^2[m_1m_2\omega^2 - k(m_1 + m_2)] = 0 \tag{5.39}$$

from which the natural frequencies can be obtained:

$$\omega_1 = 0 \qquad \text{and} \qquad \omega_2 = \sqrt{\frac{k(m_1 + m_2)}{m_1m_2}} \tag{5.40}$$

As stated earlier, Eq. (5.40) shows that one of the natural frequencies of the system is zero, which means that the system is not oscillating. In other words, the system moves as a whole without any relative motion between the two masses (rigid-body translation). Such systems, which have one of the natural frequencies equal to zero, are called *semidefinite systems*. We can verify, by substituting ω_2 into Eq. (5.38), that $X_1^{(2)}$ and $X_2^{(2)}$ are opposite in phase. There would thus be a node at the middle of the spring.

The free-vibration solution of an unrestrained system is illustrated through the following example.

EXAMPLE 5.9

Free Vibration of an Unrestrained System

Find the free-vibration solution of the unrestrained system shown in Fig. 5.17(a) for the following data: $m_1 = 1$ kg, $m_2 = 2$ kg, $k = 200$ N/m, $x_1(0) = 0.1$ m, and $x_2(0) = \dot{x}_1(0) = \dot{x}_2(0) = 0$.

Solution: The natural frequencies of the system can be computed, for the known data, from Eq. (5.40) as

$$\omega_1 = 0, \quad \omega_2 = \left\{\frac{200(1+2)}{1(2)}\right\}^{\frac{1}{2}} = 17.3205 \text{ rad/s} \tag{E.1}$$

To compute the mode shapes, Eq. (5.38) is written in matrix form as

$$\begin{bmatrix} (-m_1\omega^2 + k) & -k \\ -k & (-m_2\omega^2 + k) \end{bmatrix} \begin{Bmatrix} X_1 \\ X_2 \end{Bmatrix} = \begin{Bmatrix} 0 \\ 0 \end{Bmatrix} \tag{E.2}$$

For $\omega_1 = 0$, Eq. (E.2) becomes, for the known data,

$$\begin{bmatrix} 200 & -200 \\ -200 & 200 \end{bmatrix} \begin{Bmatrix} X_1 \\ X_2 \end{Bmatrix} = \begin{Bmatrix} 0 \\ 0 \end{Bmatrix} \tag{E.3}$$

The solution of Eq. (E.3) gives the first mode shape or modal vector as

$$\begin{Bmatrix} X_1 \\ X_2 \end{Bmatrix}^{(1)} = a_1 \begin{Bmatrix} 1 \\ 1 \end{Bmatrix} \tag{E.4}$$

where a_1 is a constant. For $\omega_2 = 17.3205$, Eq. (E.2) becomes, for the known data,

$$\begin{bmatrix} -100 & -200 \\ -200 & -400 \end{bmatrix} \begin{Bmatrix} X_1 \\ X_2 \end{Bmatrix} = \begin{Bmatrix} 0 \\ 0 \end{Bmatrix} \tag{E.5}$$

The solution of Eq. (E.5) gives the second mode shape or modal vector as

$$\begin{Bmatrix} X_1 \\ X_2 \end{Bmatrix}^{(2)} = a_2 \begin{Bmatrix} 1 \\ -0.5 \end{Bmatrix} \tag{E.6}$$

where a_2 is a constant. The free-vibration solution in each mode can be expressed as

$$\vec{x}^{(1)}(t) = \begin{Bmatrix} x_1^{(1)} \\ x_2^{(1')} \end{Bmatrix} = \begin{Bmatrix} X_1 \\ X_2 \end{Bmatrix}^{(1)} \cos(\omega_1 t + \phi_1) = a_1 \begin{Bmatrix} 1 \\ 1 \end{Bmatrix} \cos\phi_1 \tag{E.7}$$

$$\vec{x}^{(2)}(t) = \begin{Bmatrix} x_1^{(2)} \\ x_2^{(2')} \end{Bmatrix} = \begin{Bmatrix} X_1 \\ X_2 \end{Bmatrix}^{(2)} \cos(\omega_2 t + \phi_2) = a_2 \begin{Bmatrix} 1 \\ -0.5 \end{Bmatrix} \cos(17.3205t + \phi_2) \tag{E.8}$$

The free-vibration solution for any specified initial conditions can be expressed as a linear combination of the mode shapes as

$$\vec{x}(t) = \begin{Bmatrix} x_1(t) \\ x_2(t) \end{Bmatrix} = b_1 \vec{x}^{(1)}(t) + b_2 \vec{x}^{(2)}(t)$$

$$= c_1 \begin{Bmatrix} 1 \\ 1 \end{Bmatrix} \cos \phi_1 + c_2 \begin{Bmatrix} 1 \\ -0.5 \end{Bmatrix} \cos(17.3205t + \phi_2) \qquad (E.9)$$

where $b_1, b_2, c_1 = a_1 b_1,$ and $c_2 = a_2 b_2$ are (unknown) constants. The velocities of the masses can be determined by differentiating Eq. (E.9) as

$$\dot{\vec{x}}(t) = -c_2 \begin{Bmatrix} 1 \\ -0.5 \end{Bmatrix} (17.3205) \sin(17.3205t + \phi_2) \qquad (E.10)$$

Using the given initial conditions, Eqs. (E.9) and (E.10) give

$$x_1(0) = c_1 \cos \phi_1 + c_2 \cos \phi_2 = 0.1 \qquad (E.11)$$

$$x_2(0) = c_1 \cos \phi_1 + 0.5 c_2 \cos \phi_2 = 0 \qquad (E.12)$$

$$\dot{x}_1(0) = -17.3205 c_2 \sin \phi_2 = 0 \qquad (E.13)$$

$$\dot{x}_2(0) = -8.66025 c_2 \sin \phi_2 = 0 \qquad (E.14)$$

The solution of Eqs. (E.11)–(E.14) gives

$$c_2 = \pm 0.06666, \qquad \phi_2 = 0 \text{ or } \pi, \qquad c_1 \cos \phi_1 = 0.03333 \qquad (E.15)$$

Using Eq. (E.15), the free-vibration solution given by Eq. (E.9) can be expressed as

$$x_1(t) = 0.03333 \pm 0.06666 \cos(17.3205t + \phi_2) \qquad (E.16)$$

$$x_2(t) = 0.03333 \pm 0.03333 \cos(17.3205t + \phi_2) \qquad (E.17)$$

where plus (minus) is to be used when ϕ_2 is taken as 0 (π) in Eqs. (E.16) and (E.17).
Note: It can be seen from Eqs. (E.16) and (E.17) that the free-vibration response (or solution) is composed of a constant (translation) and a harmonic term (vibration).

■

5.8   Self-Excitation and Stability Analysis

In Section 3.11, the stability conditions of a single-degree-of-freedom system have been expressed in terms of the physical constants of the system. The procedure is extended to a two-degree-of-freedom system in this section. When the system is subjected to self-exciting forces, the force terms can be combined with the damping/stiffness terms, and the resulting equations of motion can be expressed in matrix notation as

$$\begin{bmatrix} m_{11} & m_{12} \\ m_{21} & m_{22} \end{bmatrix} \begin{Bmatrix} \ddot{x}_1 \\ \ddot{x}_2 \end{Bmatrix} + \begin{bmatrix} c_{11} & c_{12} \\ c_{21} & c_{22} \end{bmatrix} \begin{Bmatrix} \dot{x}_1 \\ \dot{x}_2 \end{Bmatrix}$$

$$+ \begin{bmatrix} k_{11} & k_{12} \\ k_{21} & k_{22} \end{bmatrix} \begin{Bmatrix} x_2 \\ x_2 \end{Bmatrix} = \begin{Bmatrix} 0 \\ 0 \end{Bmatrix} \tag{5.41}$$

By substituting the solution

$$x_j(t) = X_j e^{st}, \qquad j = 1, 2 \tag{5.42}$$

in Eq. (5.41) and setting the determinant of the coefficient matrix to zero, we obtain the characteristic equation of the form

$$a_0 s^4 + a_1 s^3 + a_2 s^2 + a_3 s + a_4 = 0 \tag{5.43}$$

The coefficients a_0, a_1, a_2, a_3, and a_4 are real numbers, since they are derived from the physical parameters of the system. If s_1, s_2, s_3, and s_4 denote the roots of Eq. (5.43), we have

$$(s - s_1)(s - s_2)(s - s_3)(s - s_4) = 0$$

or

$$s^4 - (s_1 + s_2 + s_3 + s_4)s^3$$
$$+ (s_1 s_2 + s_1 s_3 + s_1 s_4 + s_2 s_3 + s_2 s_4 + s_3 s_4)s^2$$
$$- (s_1 s_2 s_3 + s_1 s_2 s_4 + s_1 s_3 s_4 + s_2 s_3 s_4)s + (s_1 s_2 s_3 s_4) = 0 \tag{5.44}$$

A comparison of Eqs. (5.43) and (5.44) yields

$$a_0 = 1$$
$$a_1 = -(s_1 + s_2 + s_3 + s_4)$$
$$a_2 = s_1 s_2 + s_1 s_3 + s_1 s_4 + s_2 s_3 + s_2 s_4 + s_3 s_4$$
$$a_3 = -(s_1 s_2 s_3 + s_1 s_2 s_4 + s_1 s_3 s_4 + s_2 s_3 s_4)$$
$$a_4 = s_1 s_2 s_3 s_4 \tag{5.45}$$

The criterion for stability is that the real parts of s_i $(i = 1, 2, 3, 4)$ must be negative to avoid increasing exponentials in Eq. (5.42). Using the properties of a quartic equation, it can be derived that a necessary and sufficient condition for stability is that all the coefficients of the equation (a_0, a_1, a_2, a_3, and a_4) be positive and that the condition

$$a_1 a_2 a_3 > a_0 a_3^2 + a_4 a_1^2 \tag{5.46}$$

be fulfilled [5.8, 5.9]. A more general technique, which can be used to investigate the stability of an n-degree-of-freedom system, is known as the *Routh-Hurwitz criterion* [5.10]. For the system under consideration, Eq. (5.43), the Routh-Hurwitz criterion states that the system will be stable if all the coefficients $a_0, a_1, \ldots, a_4$ are positive and the determinants defined below are positive:

$$T_1 = |a_1| > 0 \tag{5.47}$$

$$T_2 = \begin{vmatrix} a_1 & a_3 \\ a_0 & a_2 \end{vmatrix} = a_1 a_2 - a_0 a_3 > 0 \tag{5.48}$$

$$T_3 = \begin{vmatrix} a_1 & a_3 & 0 \\ a_0 & a_2 & a_4 \\ 0 & a_1 & a_3 \end{vmatrix} = a_1 a_2 a_3 - a_1^2 a_4 - a_0 a_3^2 > 0 \tag{5.49}$$

Equation (5.47) simply states that the coefficient a_1 must be positive, while the satisfaction of Eq. (5.49), coupled with the satisfaction of the conditions $a_3 > 0$ and $a_4 > 0$, implies the satisfaction of Eq. (5.48). Thus the necessary and sufficient condition for the stability of the system is that all the coefficients a_0, a_1, a_2, a_3, and a_4 be positive and that the inequality stated in Eq. (5.46) be satisfied.

5.9   Transfer-Function Approach

As stated in Section 3.12, the transfer function of a differential equation denotes the ratio of the Laplace transform of the response (output) function to the Laplace transform of the forcing (input) function, assuming zero initial conditions. For the two-degree-of-freedom system shown in Fig. 5.5, the equations of motion are (Eqs. (5.1) and (5.2)):

$$m_1 \ddot{x}_1 + (c_1 + c_2)\dot{x}_1 - c_2 \dot{x}_2 + (k_1 + k_2)x_1 - k_2 x_2 = f_1 \tag{5.50}$$

$$m_2 \ddot{x}_2 + (c_2 + c_3)\dot{x}_2 - c_2 \dot{x}_1 + (k_2 + k_3)x_2 - k_2 x_1 = f_2 \tag{5.51}$$

By taking the Laplace transforms of Eqs. (5.50) and (5.51), assuming zero initial conditions, we obtain

$$m_1 s^2 X_1(s) + (c_1 + c_2)s X_1(s) - c_2 s X_2(s)$$
$$+ (k_1 + k_2)X_1(s) - k_2 X_2(s) = F_1(s) \tag{5.52}$$

$$m_2 s^2 X_2(s) + (c_2 + c_3)s X_2(s) - c_2 s X_1(s)$$
$$+ (k_2 + k_3)X_2(s) - k_2 X_1(s) = F_2(s) \tag{5.53}$$

Equations (5.52) and (5.53) can be rearranged to obtain

$$[m_1 s^2 + (c_1 + c_2)s + (k_1 + k_2)]X_1(s) - (c_2 s + k_2)X_2(s) = F_1(s) \quad (5.54)$$

$$[m_2 s^2 + (c_2 + c_3)s + (k_2 + k_3)]X_2(s) - (c_2 s + k_2)X_1(s) = F_2(s) \quad (5.55)$$

Equations (5.54) and (5.55) indicate two simultaneous linear algebraic equations in $X_1(s)$ and $X_2(s)$. These can be solved using Cramer's rule [5.11] as

$$X_1(s) = \frac{D_1(s)}{D(s)} \qquad (5.56)$$

$$X_2(s) = \frac{D_2(s)}{D(s)} \qquad (5.57)$$

where

$$D_1(s) = \begin{vmatrix} F_1(s) & -(c_2 s + k_2) \\ F_2(s) & m_2 s^2 + (c_2 + c_3)s + (k_2 + k_3) \end{vmatrix}$$

$$= [m_2 s^2 + (c_2 + c_3)s + (k_2 + k_3)]F_1(s) + (c_2 s + k_2)F_2(s) \quad (5.58)$$

$$D_2(s) = \begin{vmatrix} m_1 s^2 + (c_1 + c_2)s + (k_1 + k_2) & F_1(s) \\ -(c_2 s + k_2) & F_2(s) \end{vmatrix}$$

$$= [m_1 s^2 + (c_1 + c_2)s + (k_1 + k_2)]F_2(s) + (c_2 s + k_2)F_1(s) \quad (5.59)$$

$$D(s) = \begin{vmatrix} m_1 s^2 + (c_1 + c_2)s + (k_1 + k_2) & -(c_2 s + k_2) \\ -(c_2 s + k_2) & m_2 s^2 + (c_2 + c_3)s + (k_2 + k_3) \end{vmatrix}$$

$$\begin{aligned} = \; & m_1 m_2 s^4 + [m_2(c_1 + c_2) + m_1(c_2 + c_3)]s^3 \\ & + [m_2(k_1 + k_2) + m_1(k_2 + k_3) + c_1 c_2 + c_2 c_3 + c_3 c_1]s^2 \\ & + [(k_1 + k_2)(c_2 + c_3) + c_1 k_2 + c_1 k_3 - c_2 k_2 + c_2 k_3]s \\ & + (k_1 k_2 + k_2 k_3 + k_3 k_1) \end{aligned} \qquad (5.60)$$

Notes:

1. The denominator, $D(s)$, in the expressions of $X_1(s)$ and $X_2(s)$ given by Eq. (5.60), is a fourth-order polynomial in s and denotes the characteristic polynomial of the system. Because the characteristic polynomial is of order four, the model (or system) is said to be a fourth-order model (or system).

2. Equations (5.56) and (5.57) permit us to apply inverse Laplace transforms to obtain the fourth-order differential equations for $x_1(t)$ and $x_2(t)$, respectively (Problem 5.79).

3. Equations (5.56) and (5.57) can be used to derive the transfer functions of $x_1(t)$ and $x_2(t)$ corresponding to any specified forcing function.

5.10   Solutions Using Laplace Transform

The computation of responses of two-degree-of-freedom systems using the Laplace transform is illustrated through the following examples.

████████████████   **Free-Vibration Response of an Undamped System**

EXAMPLE 5.10

Find the free-vibration response of the system shown in Fig. 5.5(a) using the Laplace transform approach for the following data: $m_1 = 2, m_2 = 4, k_1 = 8, k_2 = 4, k_3 = 0, c_1 = 0, c_2 = 0, c_3 = 0$. Assume the initial conditions as $x_1(0) = 0, x_2(0) = 1$, and $\dot{x}_1(0) = \dot{x}_2(0) = 0$.

Solution: For the given data, for free vibration with $f_1(t) = f_2(t) = 0$, the equations of motion of the system, Eqs. (5.1) and (5.2), take the form

$$2\ddot{x}_1 + 12x_1 - 4x_2 = 0 \tag{E.1}$$

$$4\ddot{x}_2 - 4x_1 + 4x_2 = 0 \tag{E.2}$$

By taking the Laplace transform of Eqs. (E.1) and (E.2), we obtain

$$2[s^2 X_1(s) - s x_1(0) - \dot{x}_1(0)] + 12X_1(s) - 4X_2(s) = 0 \tag{E.3}$$

$$4[s^2 X_2(s) - s x_2(0) - \dot{x}_2(0)] - 4X_1(s) + 4X_2(s) = 0 \tag{E.4}$$

For the known initial conditions, $x_1(0) = 0, x_2(0) = 1$, and $\dot{x}_1(0) = \dot{x}_2(0) = 0$, Eqs. (E.3) and (E.4) become

$$(2s^2 + 12)X_1(s) - 4X_2(s) = 0 \tag{E.5}$$

$$(4s^2 + 4)X_2(s) - 4X_1(s) = 4s \tag{E.6}$$

By introducing

$$D_1(s) = \begin{vmatrix} 0 & -4 \\ 4s & 4s^2 + 4 \end{vmatrix} = 16s \tag{E.7}$$

$$D_2(s) = \begin{vmatrix} 2s^2 + 12 & 0 \\ -4 & 4s \end{vmatrix} = 8s^3 + 48s \tag{E.8}$$

$$D(s) = \begin{vmatrix} 2s^2 + 12 & -4 \\ -4 & 4s^2 + 4 \end{vmatrix} = 8s^4 + 56s^2 + 32 \tag{E.9}$$

the solution of Eqs. (E.5) and (E.6) for $X_1(s)$ and $X_2(s)$, based on Cramer's rule, can be expressed as

$$X_1(s) = \frac{D_1(s)}{D(s)} = \frac{16s}{8s^4 + 56s^2 + 32} = \frac{2s}{s^4 + 7s^2 + 4} \tag{E.10}$$

$$X_2(s) = \frac{D_2(s)}{D(s)} = \frac{8s^3 + 48s}{8s^4 + 56s^2 + 32} = \frac{s^3 + 6s}{s^4 + 7s^2 + 4} \tag{E.11}$$

As can be observed from Eqs. (E.10) and (E.11), the denominator is a quadratic in s^2 (true for all undamped two-degree-of-freedom systems). Since the roots of the denominator, $s^4 + 7s^2 + 4 = 0$, are given by

$$s^2 = -0.6277 \ (\text{or} -0.7923^2), \ -6.3723 \ (\text{or} -2.5243^2) \tag{E.12}$$

$X_1(s)$ and $X_2(s)$ can be expressed in factored form as

$$X_1(s) = \frac{2s}{(s^2 + 0.6277)(s^2 + 6.3723)} \tag{E.13}$$

$$X_2(s) = \frac{s^3 + 6s}{(s^2 + 0.6277)(s^2 + 6.3723)} \tag{E.14}$$

Using partial fractions, $X_1(s)$ and $X_2(s)$ can be written as

$$X_1(s) = \frac{0.7923C_1}{s^2 + 0.6277} + \frac{C_2 s}{s^2 + 0.6277} + \frac{2.5243C_3}{s^2 + 6.3723} + \frac{C_4 s}{s^2 + 6.3723} \tag{E.15}$$

$$X_2(s) = \frac{0.7923C_5}{s^2 + 0.6277} + \frac{C_6 s}{s^2 + 0.6277} + \frac{2.5243C_7}{s^2 + 6.3723} + \frac{C_8 s}{s^2 + 6.3723} \tag{E.16}$$

To determine $x_1(t)$, we equate Eqs. (E.13) and (E.15) to obtain (from the numerators)

$$0.7923C_1(s^2 + 6.3723) + C_2 s(s^2 + 6.3723) + 2.5243C_3(s^2 + 0.6277) + C_4 s(s^2 + 0.6277)$$
$$= 2s$$

or

$$s^3(C_2 + C_4) + s^2(0.7923C_1 + 2.5243C_3) + s(6.3723C_2 + 0.6277C_4)$$
$$+ (5.0488C_1 + 3.4676C_3) = 2s \tag{E.17}$$

By equating the corresponding terms on both sides of Eq. (E.17), we obtain

$$C_2 + C_4 = 0, \ 0.7923C_1 + 2.5243C_3 = 0, \ 6.3723C_2 + 0.6277C_4 = 2,$$
$$5.0488C_1 + 1.5845C_3 = 0 \tag{E.18}$$

The solution of Eqs. (E.18) yields $C_1 = 0, C_2 = 0.3481, C_3 = 0, C_4 = -0.3481$, and hence $X_1(s)$ of Eq. (E.15) becomes

$$X_1(s) = 0.3481 \frac{s}{s^2 + 0.6277} - 0.3481 \frac{s}{s^2 + 6.3723} \qquad \text{(E.19)}$$

The inverse Laplace transform of Eq. (E.19) gives

$$x_1(t) = 0.3481 \cos 0.7923t - 0.3481 \cos 2.5243t \qquad \text{(E.20)}$$

To determine $x_2(t)$, we equate Eqs. (E.14) and (E.16) to obtain (from the numerators)

$$0.7923 C_5 (s^2 + 6.3723) + C_6 s(s^2 + 6.3723)$$
$$+ 2.5243 C_7 (s^2 + 0.6277) + C_8 s(s^2 + 0.6277) = s^3 + 6s$$

or

$$s^3 (C_6 + C_8) + s^2 (0.7923 C_5 + 2.5243 C_7) + s(6.3723 C_6 + 0.6277 C_8)$$
$$+ (5.0488 C_5 + 1.5845 C_7) = s^3 + 6s \quad \text{(E.21)}$$

By equating the corresponding terms on both sides of Eq. (E.21), we obtain

$$C_6 + C_8 = 1, 0.7923 C_5 + 2.5243 C_7 = 0, 6.3723 C_6 + 0.6277 C_8 = 6,$$
$$5.0488 C_5 + 1.5845 C_7 = 0 \qquad \text{(E.22)}$$

The solution of Eqs. (E.22) yields $C_5 = 0, C_6 = 0.9352, C_7 = 0, C_8 = 0.0648$, and hence $X_2(s)$ of Eq. (E.16) becomes

$$X_2(s) = 0.9352 \frac{s}{s^2 + 0.6277} + 0.0648 \frac{s}{s^2 + 6.3723} \qquad \text{(E.23)}$$

The inverse Laplace transform of Eq. (E.23) gives

$$x_2(t) = 0.9352 \cos 0.7923t + 0.0648 \cos 2.5243t \qquad \text{(E.24)}$$

The free-vibration response of the system, $x_1(t)$ and $x_2(t)$, given by Eqs. (E.20) and (E.24), is shown graphically in Fig. 5.18.

(a)

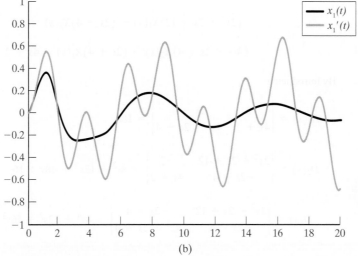

(b)

FIGURE 5.18

■

EXAMPLE 5.11

Free-Vibration Response of a Damped System

Find the free-vibration response of the system shown in Fig. 5.5(a) using the Laplace transform approach for the following data: $m_1 = 2, m_2 = 4, k_1 = 8, k_2 = 4, k_3 = 0, c_1 = 0, c_2 = 2, c_3 = 0$.

Assume the initial conditions as $x_1(0) = 0, x_2(0) = 1$, and $\dot{x}_1(0) = \dot{x}_2(0) = 0$.

Solution: For the given data, for free vibration with $f_1(t) = f_2(t) = 0$, the equations of motion of the system, Eqs. (5.1) and (5.2), take the form

$$2\ddot{x}_1 + 2\dot{x}_1 - 2\dot{x}_2 + 12x_1 - 4x_2 = 0 \qquad \text{(E.1)}$$

$$4\ddot{x}_1 - 2\dot{x}_1 + 2\dot{x}_2 - 4x_1 + 4x_2 = 0 \qquad \text{(E.2)}$$

By taking the Laplace transform of Eqs. (E.1) and (E.2), we obtain

$$2[s^2X_1(s) - sx_1(0) - \dot{x}_1(0)] + 2[sX_1(s) - x_1(0)] - 2[sX_2(s) - x_2(0)]$$
$$+ 12X_1(s) - 4X_2(s) = 0 \qquad \text{(E.3)}$$

$$4[s^2X_2(s) - sx_2(0) - \dot{x}_2(0)] - 2[sX_1(s) - x_1(0)]$$
$$+ 2[sX_2(s) - x_2(0)] - 4X_1(s) + 4X_2(s) = 0 \qquad \text{(E.4)}$$

For the known initial conditions, $x_1(0) = 0, x_2(0) = 1$, and $\dot{x}_1(0) = \dot{x}_2(0) = 0$, Eqs. (E.3) and (E.4) become

$$(2s^2 + 2s + 12)X_1(s) - (2s + 4)X_2(s) = -2 \qquad \text{(E.5)}$$

$$(4s^2 + 2s + 4)X_2(s) - (2s + 4)X_1(s) = 4s + 2 \qquad \text{(E.6)}$$

By introducing

$$D_1(s) = \begin{vmatrix} -2 & 2s - 4 \\ 4s + 2 & 4s^2 + 2s + 4 \end{vmatrix} = 16s \qquad \text{(E.7)}$$

$$D_2(s) = \begin{vmatrix} 2s^2 + 2s + 12 & -2 \\ -2s - 4 & 4s + 2 \end{vmatrix} = 8s^3 + 12s^2 + 48s + 16 \qquad \text{(E.8)}$$

$$D(s) = \begin{vmatrix} 2s^2 + 2s + 12 & -2s - 4 \\ -2s - 4 & 4s^2 + 2s + 4 \end{vmatrix} = 8s^4 + 12s^3 + 56s^2 + 16s + 32 \qquad \text{(E.9)}$$

the solution of Eqs. (E.5) and (E.6) for $X_1(s)$ and $X_2(s)$, based on Cramer's rule, can be expressed as

$$X_1(s) = \frac{D_1(s)}{D(s)} = \frac{16s}{8s^4 + 12s^3 + 56s^2 + 16s + 32} = \frac{2s}{s^4 + 1.5s^3 + 7s^2 + 2s + 4} \qquad \text{(E.10)}$$

$$X_2(s) = \frac{D_2(s)}{D(s)} = \frac{8s^3 + 12s^2 + 48s + 16}{8s^4 + 12s^3 + 56s^2 + 16s + 32} = \frac{s^3 + 1.5s^2 + 6s + 2}{s^4 + 1.5s^3 + 7s^2 + 2s + 4} \qquad \text{(E.11)}$$

As can be observed from Eqs. (E.10) and (E.11), the denominator is not a quadratic in s^2 (true for all damped two-degree-of-freedom systems). The roots of the denominator (characteristic roots of the system), $s^4 + 1.5s^3 + 7s^2 + 2s + 4 = 0$, can be found (for example, using MATLAB), as

$$s_{1,2} = -0.6567 \pm 2.3807i \equiv a \pm bi$$
$$s_{3,4} = -0.0933 \pm 0.8044i \equiv c \pm di \tag{E.12}$$

It can be seen that the roots are complex (true for all damped systems) instead of simply imaginary values (true for undamped systems). In view of the characteristic roots given in Eq. (E.12), $X_1(s)$ of Eq. (E.10) can be expressed as

$$X_1(s) = \frac{2s}{[(s+a)^2 + b^2][(s+c)^2 + d^2]} = \frac{C_1b + C_2(s+a)}{[(s+a)^2 + b^2]} + \frac{C_3d + C_4(s+c)}{[(s+c)^2 + d^2]} \tag{E.13}$$

where $a = 0.6567$, $b = 2.3807$, $c = 0.0933$, $d = 0.8044$, and C_i, $i = 1, 2, 3, 4$ are unknown constants. By writing the right-hand-side expression in Eq. (E.13) as

$$\frac{[C_1b + C_2(s+a)][(s+c)^2 + d^2]}{[(s+a)^2 + b^2]} + \frac{[C_3d + C_4(s+c)][(s+a)^2 + b^2]}{[(s+c)^2 + d^2]} \tag{E.14}$$

and equating the numerator of Eq. (E.14) to the numerator in the middle expression of Eq. (E.13), we obtain

$$(C_1b + C_2s + C_2a)(s^2 + 2sc + c^2 + d^2) + (C_3d + C_4s + C_4c)(s^2 + 2sa + a^2 + b^2) = 2s$$

or

$$s^3(C_2 + C_4) + s^2(2cC_2 + bC_1 + aC_2 + 2aC_4 + dC_3 + cC_4) + s[(c^2 + d^2)C_2$$
$$2c(bC_1 + aC_2) + (a^2 + b^2)C_4 + 2a(dC_3 + cC_4)] + [(bC_1 + aC_2)(c^2 + d^2)$$
$$+ (dC_3 + cC_4)(a^2 + b^2)] = 2s \tag{E.15}$$

By equating the coefficients of the corresponding terms on both sides of Eq. (E.15), we obtain

$$C_2 + C_4 = 0$$

$$bC_1 + (2c + a)C_2 + dC_3 + (2a + c)C_4 = 0$$

$$2cbC_1 + (2ac + c^2 + d^2)C_2 + 2adC_3 + (2ac + 2a^2 + b^2)C_4 = 0$$

$$(bc^2 + bd^2)C_1 + (ac^2 + ad^2)C_2 + (da^2 + db^2)C_3 + (ca^2 + cb^2)C_4 = 0 \tag{E.16}$$

where the values of a, b, c, and d are defined in Eq. (E.12). The solution of Eqs. (E.16), for example by MATLAB, gives $C_1 = -0.0945$, $C_2 = -0.3713$, $C_3 = 0.0196$, $C_4 = 0.3713$. Thus $X_1(s)$, in Eq. (E.13), becomes

$$X_1(s) = -0.0945\frac{b}{(s+a)^2 + b^2} - 0.3713\frac{s+a}{(s+a)^2 + b^2}$$

$$+ 0.0196\frac{d}{(s+c)^2 + d^2} + 0.3713\frac{s+c}{(s+c)^2 + d^2} \tag{E.17}$$

By taking the inverse Laplace transform of Eq. (E.17), we obtain

$$x_1(t) = e^{-0.6567t}(0.0945 \sin 2.3807t - 0.3713 \cos 2.3807t)$$

$$+ e^{-0.0933t}(0.0196 \sin 0.8044t + 0.3713 \cos 0.8044t) \qquad \text{(E.18)}$$

Similarly, based on the characteristic roots given in Eq. (E.12), $X_2(s)$ of Eq. (E.11) can be expressed as

$$X_2(s) = \frac{s^3 + 1.5s^2 + 6s + 2}{[(s + a)^2 + b^2][(s + c)^2 + d^2]} = \frac{C_5 b + C_6(s + a)}{[(s + a)^2 + b^2]} + \frac{C_7 d + C_8(s + c)}{[(s + c)^2 + d^2]} \qquad \text{(E.19)}$$

where $a = 0.6567$, $b = 2.3807$, $c = 0.0933$, $d = 0.8044$, and C_i, $i = 5, 6, 7, 8$, are unknown constants. By writing the right-hand-side expression in Eq. (E.19) as

$$\frac{[C_5 b + C_6(s + a)][(s + c)^2 + d^2]}{[(s + a)^2 + b^2]} + \frac{[C_7 d + C_8(s + c)][(s + a)^2 + b^2]}{[(s + c)^2 + d^2]} \qquad \text{(E.20)}$$

and equating the numerator of Eq. (E.20) to the numerator in the middle expression of Eq. (E.19), we obtain

$$(C_5 b + C_6 s + C_6 a)(s^2 + 2sc + c^2 + d^2) + (C_7 d + C_8 s + C_8 c)(s^2 + 2sa + a^2 + b^2)$$

$$= s^3 + 1.5s^2 + 6s + 2$$

or

$$s^3(C_6 + C_8) + s^2(2cC_6 + bC_5 + aC_6 + 2aC_8 + dC_7 + cC_8) + s[(c^2 + d^2)C_6$$
$$+ 2c(bC_5 + aC_6) + (a^2 + b^2)C_8 + 2a(dC_7 + cC_8)] + [(bC_5 + aC_6)(c^2 + d^2)$$
$$+ dC_7 + cC_8](a^2 + b^2) = s^3 + 1.5s^2 + 6s + 2 \qquad \text{(E.21)}$$

By equating the coefficients of the corresponding terms on both sides of Eq. (E.21), we obtain

$$C_6 + C_8 = 1$$

$$bC_5 + (2c + a)C_6 + dC_7 + (2a + c)C_8 = 1.5$$

$$2cbC_5 + (2ac + c^2 + d^2)C_6 + 2adC_7 + (2ac + a^2 + b^2)C_8 = 6$$

$$(bc^2 + bd^2)C_5 + (ac^2 + ad^2)C_6 + (da^2 + db^2)C_7 + (ca^2 + cb^2)C_8 = 2 \qquad \text{(E.22)}$$

where the values of a, b, c, and d are defined in Eq. (E.12). The solution of Eqs. (E.22), for example by MATLAB, gives $C_5 = -0.0418$, $C_6 = 0.0970$, $C_7 = 0.3077$, $C_8 = 0.9030$. Thus Eq. (E.19) can be written as

$$X_2(s) = -0.0418 \frac{b}{(s + a)^2 + b^2} + 0.0970 \frac{s + a}{(s + a)^2 + b^2}$$

$$+ 0.3077 \frac{d}{(s + c)^2 + d^2} + 0.9030 \frac{s + c}{(s + c)^2 + d^2} \tag{E.23}$$

By taking the inverse Laplace transform of Eq. (E.23), we obtain

$$x_2(t) = e^{-0.6567t}(-0.0418 \sin 2.3807t + 0.0970 \cos 2.3807t)$$
$$+ e^{-0.0933t}(0.3077 \sin 0.8044t + 0.9030 \cos 0.8044t) \tag{E.24}$$

The free-vibration response of the system, $x_1(t)$ and $x_2(t)$, given by Eqs. (E.18) and (E.24), is shown graphically in Fig. 5.18.

■

EXAMPLE 5.12

Response Under Impulse Using Laplace Transform Method

Two railway cars, of masses $m_1 = M$ and $m_2 = m$, are connected by a spring of stiffness k, as shown in Fig. 5.17(a). If the car of mass M is subjected to an impulse $F_0 \delta(t)$, determine the time responses of the cars using the Laplace transform method.

Solution: The responses of the cars can be determined using either of the following approaches:

a. Consider the system to be undergoing free vibration due to the initial velocity caused by the impulse applied to car M.

b. Consider the system to be undergoing forced vibration due to the force $f(t) = F_0 \delta(t)$ applied to car M (with the displacements and velocities of cars M and m considered to be zero initially).

Using the second approach, the equations of motion of the cars can be expressed as

$$M \ddot{x}_1 + k(x_1 - x_2) = F_0 \delta(t) \tag{E.1}$$

$$m \ddot{x}_2 + k(x_2 - x_1) = 0 \tag{E.2}$$

Using the Laplace transforms, Eqs. (E.1) and (E.2) can be written as

$$(Ms^2 + k)X_1(s) - kX_2(s) = F_0 \tag{E.3}$$

$$-kX_1(s) + (ms^2 + k)X_2(s) = 0 \tag{E.4}$$

Equations (E.3) and (E.4) can be solved for $X_1(s)$ and $X_2(s)$ as

$$X_1(s) = \frac{F_0(ms^2 + k)}{s^2\{Mms^2 + k(M + m)\}} \tag{E.5}$$

$$X_2(s) = \frac{F_0 k}{s^2\{Mms^2 + k(M + m)\}} \tag{E.6}$$

Using partial fractions, Eqs. (E.5) and (E.6) can be rewritten as

$$X_1(s) = \frac{F_0}{M + m}\left(\frac{1}{s^2} + \frac{m}{\omega M}\frac{\omega}{s^2 + \omega^2}\right) \tag{E.7}$$

$$X_2(s) = \frac{F_0}{M + m}\left(\frac{1}{s^2} - \frac{1}{\omega}\frac{\omega}{s^2 + \omega^2}\right) \tag{E.8}$$

where

$$\omega^2 = k\left(\frac{1}{M} + \frac{1}{m}\right) \tag{E.9}$$

The inverse transforms of Eqs. (E.7) and (E.8), using the results of Appendix D, yield the time responses of the cars as

$$x_1(t) = \frac{F_0}{M + m}\left(t + \frac{m}{\omega M}\sin \omega t\right) \tag{E.10}$$

$$x_2(t) = \frac{F_0}{M + m}\left(t - \frac{1}{\omega}\sin \omega t\right) \tag{E.11}$$

Note: Equations (E.10) and (E.11) are plotted in Example 5.18.

■

5.11   Solutions Using Frequency Transfer Functions

The frequency transfer function can be obtained by substituting $i\omega$ in place of s in the general transfer function of the system. The generation of the frequency transfer function and the method of finding the response of a system using this function is illustrated through the following examples.

Derivation of Frequency Transfer Functions

EXAMPLE 5.13

Derive the frequency transfer functions of $x_1(t)$ and $x_2(t)$ for the system shown in Fig. 5.19(a).

Solution: From the free-body diagrams of the masses m_1 and m_2 shown in Fig. 5.19(b), the equations of motion of the system can be obtained as

$$m_1\ddot{x}_1 + c_1\dot{x}_1 + k_1x_1 + c_2(\dot{x}_1 - \dot{x}_2) + k_2(x_1 - x_2) = p_1 = P_0 \sin \omega t \tag{E.1}$$

$$m_2\ddot{x}_2 + c_2(\dot{x}_2 - \dot{x}_1) + k_2(x_2 - x_1) = p_2 = 0 \tag{E.2}$$

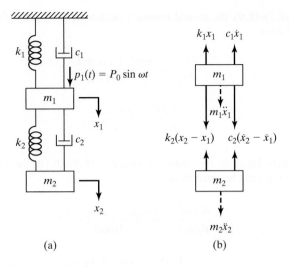

(a)   (b)

FIGURE 5.19

It can be observed that Eqs. (E.1) and (E.2) can be obtained from Eqs. (5.50) and (5.51) by setting $k_3 = c_3 = 0$, $f_1(t) = p_1(t)$, and $f_2(t) = 0$. By taking the Laplace transforms of Eqs. (E.1) and (E.2), assuming zero initial conditions, we obtain

$$m_1 s^2 X_1(s) + c_1 s X_1(s) + k_1 X_1(s) + c_2 s [X_1(s) - X_2(s)]$$
$$+ k_2 [X_1(s) - X_2(s)] = P_1(s) \tag{E.3}$$

$$m_2 s^2 X_2(s) + c_2 s [X_2(s) - X_1(s)] + k_2 [X_2(s) - X_1(x)] = 0 \tag{E.4}$$

The expressions for $X_1(s)$ and $X_2(s)$ can be found from the solution of Eqs. (E.3) and (E.4) by setting $k_3 = c_3 = 0$, $F_1(s) = P_1(s)$, and $F_2(s) = 0$ in Eqs. (5.56) – (5.60):

$$X_1(s) = \frac{D_1(s)}{D(s)} \tag{E.5}$$

$$X_2(s) = \frac{D_2(s)}{D(s)} \tag{E.6}$$

where

$$D_1(s) = (m_2 s^2 + c_2 s + k_2) P_1(s) \tag{E.7}$$

$$D_2(s) = (c_2 s + k_2) P_1(s) \tag{E.8}$$

$$D(s) = (m_1 m_2) s^4 + [m_1 c_2 + m_2 c_1 + m_2 c_2] s^3$$
$$+ [m_1 k_2 + m_2 k_1 + m_2 k_2 + c_1 c_2] s^2 + [c_1 k_2 + c_2 k_1] s + (k_1 k_2) \tag{E.9}$$

In view of Eqs. (E.7)–(E.9), the general transfer functions of $x_1(t)$ and $x_2(t)$ can be found from Eqs. (E.5) and (E.6) as

$$\frac{X_1(s)}{P_1(s)} = \frac{m_2 s^2 + c_2 s + k_2}{D(s)} \tag{E.10}$$

$$\frac{X_2(s)}{P_1(s)} = \frac{c_2 s + k_2}{D(s)} \tag{E.11}$$

where $D(s)$ is given by Eq. (E.9). By setting $s = i\omega$ in Eqs. (E.9)–(E.11), the frequency transfer functions of $x_1(t)$ and $x_2(t)$ can be obtained as

$$\frac{X_1(i\omega)}{P_1(i\omega)} = \frac{-m_2 \omega^2 + i\omega c_2 + k_2}{D(i\omega)} \tag{E.12}$$

$$\frac{X_2(i\omega)}{P_1(i\omega)} = \frac{i\omega c_2 + k_2}{D(i\omega)} \tag{E.13}$$

where

$$D(i\omega) = \omega^4 (m_1 m_2)\omega^4 - i\omega^3 [m_1 \, c_2 + m_2 \, c_1 + m_2 \, c_2]$$
$$- \omega^2 [m_1 k_2 + m_2 k_1 + m_2 k_2 + c_1 c_2] + i\omega [c_1 k_2 + c_2 k_1] + (k_1 k_2) \tag{E.14}$$

■

Steady-State Response of a System

EXAMPLE 5.14

Find the steady-state response of the system considered in Example 5.13 by neglecting damping.

Solution: By setting $c_1 = c_2 = 0$ in Eqs. (E.12) and (E.13) of Example 5.13, we obtain the frequency transfer functions

$$T_1(i\omega) = \frac{X_1(i\omega)}{P_1(i\omega)} = \frac{k_2 - m_2 \omega^2}{m_1 m_2 \omega^4 - (m_1 k_2 + m_2 k_1 + m_2 k_2)\,\omega^2 + k_1 k_2} \tag{E.1}$$

$$T_2(i\omega) = \frac{X_2(i\omega)}{P_1(i\omega)} = \frac{k_2}{m_1 m_2 \omega^4 - (m_1 k_2 + m_2 k_1 + m_2 k_2)\omega^2 + k_1 k_2} \tag{E.2}$$

and hence

$$\frac{X_2(i\omega)}{X_1(i\omega)} = \frac{k_2}{k_2 - m_2 \, \omega^2} \tag{E.3}$$

From Eq. (E.1), the steady-state solution $x_1(t)$ can be obtained, using $P_1(i\omega) = P_0 \sin \omega t$, as

$$x_1(t) = |X_1(i\omega)| \sin \omega t = \frac{(k_2 - m_2\omega^2)P_0}{[m_1 m_2 \omega^4 - (m_1 k_2 + m_2 k_1 + m_2 k_2)\omega^2 + k_1 k_2]} \sin(\omega t + \phi_1) \quad (E.4)$$

where

$$\phi_1 = \frac{X_1(i\omega)}{P_1(i\omega)} = 0 \text{ or } \pi \quad (E.5)$$

From Eqs. (E.3) and (E.4), the steady-state solution $x_2(t)$ can be determined as

$$x_2(t) = |X_2(i\omega)| \sin(\omega t + \phi_2) = \left| \frac{X_2(i\omega)}{X_1(i\omega)} \right| |X_1(i\omega)| \sin(\omega t + \phi_2)$$

$$= \frac{k_2}{(k_2 - m_2\omega^2)} \frac{(k_2 - m_2\omega^2)P_0}{[m_1 m_2 \omega^4 - (m_1 k_2 + m_2 k_1 + m_2 k_2)\omega^2 + k_1 k_2]} \sin(\omega t + \phi_2)$$

$$= \frac{k_2 P_0}{[m_1 m_2 \omega^4 - (m_1 k_2 + m_2 k_1 + m_2 k_2)\omega^2 + k_1 k_2]} \sin(\omega t + \phi_2) \quad (E.6)$$

where

$$\phi_2 = \frac{X_2(i\omega)}{P_1(i\omega)} = \frac{X_2(i\omega)}{X_1(i\omega)} \frac{X_1(i\omega)}{P_1(i\omega)} = 0 \text{ or } \pi \quad (E.7)$$

It can be seen that ϕ_1 and ϕ_2 are either 0 or π. Hence the masses m_1 and m_2 move either in phase $(\phi = 0)$ or out of phase $(\phi = \pi)$ with the applied force $P_1(i\omega)$. Thus the masses m_1 and m_2 will move in the same direction if $\omega < \sqrt{\dfrac{k_2}{m_2}}$ and in the opposite direction if $\omega > \sqrt{\dfrac{k_2}{m_2}}$. If $\omega = \sqrt{\dfrac{k_2}{m_2}}$, the mass m_1 will not move while the mass m_2 will have sinusoidal motion.

■

5.12   Examples Using MATLAB

Solution of the Eigenvalue Problem

EXAMPLE 5.15

Using MATLAB, determine the natural frequencies and mode shapes of the following problem:

$$\left[-\omega^2 m \begin{bmatrix} 1 & 0 \\ 0 & 1 \end{bmatrix} + k \begin{bmatrix} 2 & -1 \\ -1 & 2 \end{bmatrix} \right] \vec{X} = \vec{0} \quad (E.1)$$

Solution: The eigenvalue problem, Eq. (E.1), can be rewritten as

$$\begin{bmatrix} 2 & -1 \\ -1 & 2 \end{bmatrix} \vec{X} = \lambda \begin{bmatrix} 1 & 0 \\ 0 & 1 \end{bmatrix} \vec{X} \qquad\qquad \text{(E.2)}$$

where $\lambda = m\omega^2/k$ is the eigenvalue, ω is the natural frequency, and $\vec{X}$ is the eigenvector or mode shape. The solution of Eq. (E.2) can be found using MATLAB as follows:

```
>> A=[2 -1; -1 2]
A =

     2        -1
    -1         2

>> [V, D] = eig(A)

V =

   -0.7071      -0.7071
    0.7071      -0.7071

D =

    3.0000         0
         0    1.0000
```

Thus the eigenvalues are $\lambda_1 = 1.0$ and $\lambda_2 = 3.0$, and the corresponding eigenvectors are

$$\vec{X}_1 = \begin{Bmatrix} -0.7071 \\ -0.7071 \end{Bmatrix} \qquad \text{and} \qquad \vec{X}_2 = \begin{Bmatrix} -0.7071 \\ 0.7071 \end{Bmatrix}$$

$\blacksquare$

EXAMPLE 5.16

Roots of a Quartic Equation

Using MATLAB, find the roots of the quartic equation

$$f(x) = x^4 - 8x + 12 = 0$$

Solution: The MATLAB command **roots** is used to obtain the roots of the fourth-degree polynomial as

$$x_{1,2} = -1.37091 \pm 1.82709i$$

$$x_{3,4} = 1.37091 \pm 0.648457i$$

```
>> roots ([1 0 0 -8 12])
ans =
  -1.3709 + 1.8271i
  -1.3709 - 1.8271i
   1.3709 + 0.645-i
   1.3709 - 0.645-i
>>
```

$\blacksquare$

EXAMPLE 5.17

Plotting a Free-Vibration Response

Using MATLAB, plot the free-vibration response of the masses m_1 and m_2 of Example 5.3.

Solution: The time responses of the masses m_1 and m_2 are given by Eqs. (E.15) and (E.16) of Example 5.3. The MATLAB program to plot the responses is given below.

```
% E-_3.m
for i = 1: 501
    t(i) = 20 * (i-1)/500;
    x1(i) = (5/7) * cos(1.5811*t(i)) + (2/7) * cos(2.4495*t(i));
    x2(i) = (10/7) * cos(1.5811*t(i)) - (10/7) * cos(2.4495*t(i));
end
subplot(211);
plot(t, x1);
xlabel('t');
ylabel('x1(t)');
subplot(212);
plot(t, x2);
xlabel('t');
ylabel('x2(t)');
```

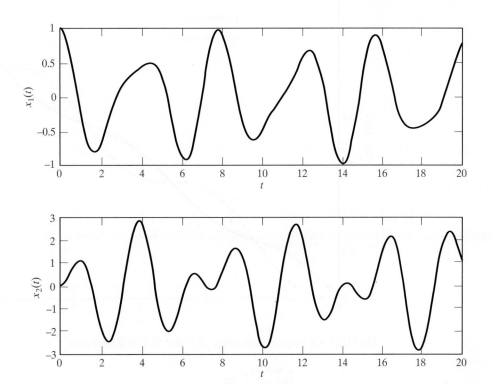

EXAMPLE 5.18

Time Response of Railway Cars

Using MATLAB, plot the time responses of the two railway cars considered in Example 5.12 for the following data: $F_0 = 1,500$ N, $M = 5,000$ kg, $m = 2,500$ kg, $k = 10^4$ N/m.

Solution: For the given data, the time responses of the railway cars can be expressed as (from Eqs. (E.10) and (E.11) of Example 5.12):

$$x_1(t) = 0.2(t + 0.204124 \sin 2.44949t) \tag{E.1}$$

$$x_2(t) = 0.2(t - 0.408248 \sin 2.44949t) \tag{E.2}$$

where

$$\omega^2 = 10^4 \left(\frac{1}{5,000} + \frac{1}{2,500} \right) \quad \text{or} \quad \omega = 2.44949 \text{ rad/s} \tag{E.3}$$

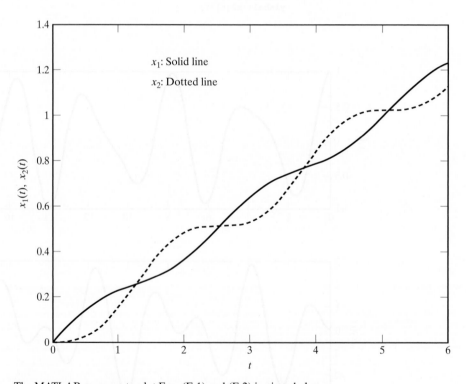

The MATLAB program to plot Eqs. (E.1) and (E.2) is given below.

```
% Ex5_18.m
for i=1 : 101
    t(i) = 6* (i - 1) / 100;
    x1(i) = 0.2* (t(i) + 0.204124*sin(2.44949*t(i)));
    x2(i) = 0.2* (t(i) - 0.408248*sin(2.44949*t(i)));
end
```

```
plot (t, x1);
xlabel ('t');
ylabel ('x1(t), x2(t)');
hold on;
plot (t, x2, '—');
gtext ('x1: Solid line');
gtext ('x2: Dotted line');
```

■

EXAMPLE 5.19

Plotting of Frequency Response of a Two-Degree-of-Freedom System

Using MATLAB, plot the frequency-response functions of the system considered in Example 5.8.

Solution: The frequency-response functions $X_1(\omega)$ and $X_2(\omega)$, given by Eqs. (E.6) and (E.7) of Example 5.8, are

$$\frac{X_1(\omega)k}{F_{10}} = \frac{(2 - \lambda^2)}{(\lambda_2^2 - \lambda^2)(1 - \lambda^2)} \tag{E.1}$$

$$\frac{X_2(\omega)k}{F_{10}} = \frac{1}{(\lambda_2^2 - \lambda^2)(1 - \lambda^2)} \tag{E.2}$$

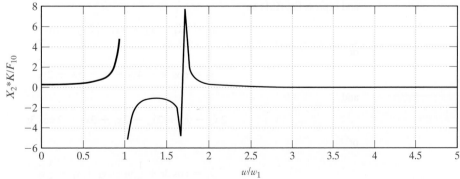

where $\lambda = \omega/\omega_1$ and $\lambda_2 = \omega_2/\omega_1$. From the results of Example 5.8, we find that $\lambda_2 = \omega_2/\omega_1 = (3k/m)/(k/m) = 3$. The MATLAB program to plot Eqs. (E.1) and (E.2) is given below.

```
% Ex-514.m
for i = 1: 101
    w_w1 (i) = 5 * (i - 1) / 100; % 0 to 5
    x1 (i) = (2-w_w1 (i) \^2) / ( (3-w_w1 (i) \^2) * (1-w_w1 (i) \^2) );
    x2 (i) = 1 / ( (3-w_w1 (i) \^2) * (1-w_w1 (i) \^2) );
end
subplot (211);
plot (w_w1, x1);
xlabel ('w/w_1');
ylabel ('X_1*K/F_1_0');
grid on;
subplot (212);
plot (w_w1, x2);
xlabel ('w/w_1');
ylabel ('X_2*K/F_1_0');
grid on
```

■

Forced Response of a Two-Degree-of-Freedom System

EXAMPLE 5.20

Determine and plot the time response of a two-degree-of-freedom system with equations of motion

$$\begin{bmatrix} 1 & 0 \\ 0 & 2 \end{bmatrix} \begin{Bmatrix} \ddot{x}_1 \\ \ddot{x}_2 \end{Bmatrix} + \begin{bmatrix} 4 & -1 \\ -1 & 2 \end{bmatrix} \begin{Bmatrix} \dot{x}_1 \\ \dot{x}_2 \end{Bmatrix} + \begin{bmatrix} 5 & -2 \\ -2 & 3 \end{bmatrix} \begin{Bmatrix} x_1 \\ x_2 \end{Bmatrix} = \begin{Bmatrix} 1 \\ 2 \end{Bmatrix} \cos 3t \qquad (E.1)$$

with the initial conditions

$$x_1(0) = 0.2, \qquad \dot{x}_1(0) = 1.0, \qquad x_2(0) = 0, \qquad \dot{x}_2(0) = 0 \qquad (E.2)$$

Solution: In order to use the MATLAB program **ode23**, the two coupled second-order differential equations, Eq. (E.1), are to be expressed as a system of coupled first-order differential equations. For this, we introduce new variables y_1, y_2, y_3, and y_4 as

$$y_1 = x_1, \qquad y_2 = \dot{x}_1, \qquad y_3 = x_2, \qquad y_4 = \dot{x}_2$$

and express Eq. (E.1) as

$$\ddot{x}_1 + 4\dot{x}_1 - \dot{x}_2 + 5x_1 - 2x_2 = \cos 3t \qquad (E.3)$$

or

$$\dot{y}_2 = \cos 3t - 4y_2 + y_4 - 5y_1 + 2y_3 \qquad (E.4)$$

and

$$2\ddot{x}_2 - \dot{x}_1 + 2\dot{x}_2 - 2x_1 + 3x_2 = 2 \cos 3t \qquad (E.5)$$

or

$$\dot{y}_4 = \cos 3t + \frac{1}{2} y_2 - y_4 + y_1 - \frac{3}{2} y_3 \qquad (E.6)$$

Thus Eq. (E.1) can be restated as

$$\begin{Bmatrix} \dot{y}_1 \\ \dot{y}_2 \\ \dot{y}_3 \\ \dot{y}_4 \end{Bmatrix} = \begin{Bmatrix} y_2 \\ \cos 3t - 4y_2 + y_4 - 5y_1 + 2y_3 \\ y_4 \\ \cos 3t + \dfrac{1}{2}y_2 - y_4 + y_1 - \dfrac{3}{2}y_3 \end{Bmatrix} \qquad (E.7)$$

with the initial conditions

$$\vec{y}(0) = \begin{Bmatrix} y_1(0) \\ y_2(0) \\ y_3(0) \\ y_4(0) \end{Bmatrix} = \begin{Bmatrix} 0.2 \\ 1.0 \\ 0.0 \\ 0.0 \end{Bmatrix} \qquad (E.8)$$

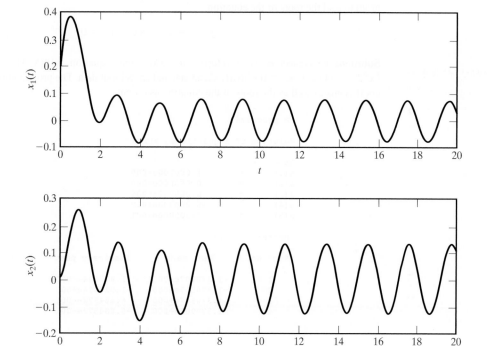

The MATLAB program to solve Eq. (E.7) with the initial conditions of Eq. (E.8) is given below.

```
% Ex5_20.m
tspan = [0: 0.01: 20];
y0 = [0.2; 1.0; 0.0; 0.0];
[t,y] = ode23('dfunc5_15', tspan, y0);
subplot (211)
plot (t,y (:, 1));
xlabel ('t');
ylabel ('x1 (t)');
subplot (212)
```

```
plot (t,y (:, 3));
xlabel ('t');
ylabel ('x2 (t)');

%dfunc5_15.m
function f = dfunc5_15(t,y)
f = zeros(4, 1);
f(1) = y(2);
f(2) = cos(3*t) - 4*y(2) + y(4) - 5*y(1) + 2*y(3);
f(3) = y(4);
f(4) = cos(3*t) + 0.5*y(2) - y(4) + y(1) - 1.5*y(3);
```

∎

Program to Find the Roots of a Quartic Equation

EXAMPLE 5.21

Develop a general program, called `Program6.m`, to find the roots of a quartic equation. Use the program to find the roots of the equation

$$f(x) = x^4 - 8x + 12 = 0$$

Solution: `Program6.m` is developed to solve the equation al*(x^4) + a2*(x^3) + a3* (x^2) + a4*x + a5 = 0 with al, a2, a3, a4, and a5 as input data. The program gives the polynomial coefficients as well as the roots of the equation as output.

```
>> program6

Solution of a quartic equation

Data:
a(1)      =        1.000000e+000
a(2)      =        0.000000e+000
a(3)      =        0.000000e+000
a(4)      =       -8.000000e+000
a(5)      =        1.200000e+001

    Roots:

Root No.       Real part        Imaginary part

    1        -1.370907e+000      1.827094e+000
    2        -1.370907e+000     -1.827094e+000
    3         1.370907e+000      6.484572e-001
    4         1.370907e+000     -6.484572e-001
```

∎

CHAPTER SUMMARY

We considered the determination of the coupled equations of motion of two-degree-of-freedom systems. We determined the eigenvalues or natural frequencies of vibration, modal vectors, and the free-vibration solutions. We presented the concepts of coordinate coupling, generalized coordinates, and principal coordinates. We studied the forced-vibration analysis of the system under a harmonic force.

We considered the transfer-function approach, Laplace transform method of solution, and the frequency transfer-function method. Finally, we presented the free- and forced-vibration solutions of two-degree-of-freedom systems using MATLAB.

Now that you have finished this chapter, you should be able to answer the review questions and solve the problems given below.

REFERENCES

5.1   H. Sato, Y. Kuroda, and M. Sagara, "Development of the finite element method for vibration analysis of machine tool structure and its application," *Proceedings of the Fourteenth International Machine Tool Design and Research Conference*, Macmillan, London, 1974, pp. 545–552.

5.2   F. Koenigsberger and J. Tlusty, *Machine Tool Structures*, Pergamon Press, Oxford, 1970.

5.3   C. P. Reddy and S. S. Rao, "Automated optimum design of machine tool structures for static rigidity, natural frequencies and regenerative chatter stability," *Journal of Engineering for Industry*, Vol. 100, 1978, pp. 137–146.

5.4   M. S. Hundal, "Effect of damping on impact response of a two degree of freedom system," *Journal of Sound and Vibration*, Vol. 68, 1980, pp. 407–412.

5.5   J. A. Linnett, "The effect of rotation on the steady-state response of a spring-mass system under harmonic excitation," *Journal of Sound and Vibration*, Vol. 35, 1974, pp. 1–11.

5.6   A. Hurwitz, "On the conditions under which an equation has only roots with negative real parts," in *Selected Papers on Mathematical Trends in Control Theory*, Dover Publications, New York, 1964, pp. 70–82.

5.7   R. C. Dorf, *Modern Control Systems* (6th ed.), Addison-Wesley, Reading, MA, 1992.

5.8   J. P. Den Hartog, *Mechanical Vibrations* (4th ed.), McGraw-Hill, New York, 1956.

5.9   R. H. Scanlan and R. Rosenbaum, *Introduction to the Study of Aircraft Vibration and Flutter*, Macmillan, New York, 1951.

5.10 L. A. Pipes and L. R. Harvill, *Applied Mathematics for Engineers and Physicists* (3rd ed.), McGraw-Hill, New York, 1970.

5.11 S. S. Rao, *Applied Numerical Methods for Engineers and Scientists*, Prentice Hall, Upper Saddle River, NJ, 2002.

REVIEW QUESTIONS

5.1   Give brief answers to the following:

1. How do you determine the number of degrees of freedom of a lumped-mass system?
2. Define these terms: *mass coupling*, *velocity coupling*, *elastic coupling*.
3. Is the nature of the coupling dependent on the coordinates used?
4. How many degrees of freedom does an airplane in flight have if it is treated as (a) a rigid body, and (b) an elastic body?
5. What are principal coordinates? What is their use?
6. Why are the mass, damping, and stiffness matrices symmetrical?
7. What is a node?
8. What is meant by static and dynamic coupling? How can you eliminate coupling of the equations of motion?

   9. Define the impedance matrix.

   10. How can we make a system vibrate in one of its natural modes?

   11. What is a degenerate system? Give two examples of physical systems that are degenerate.

   12. How many degenerate modes can a vibrating system have?

   13. What is the difference between a general transfer function and a frequency transfer function?

   14. How many natural frequencies can be zero for an unrestrained two-degree-of-freedom system?

5.2 Indicate whether each of the following statements is true or false:

   1. The normal modes can also be called *principal modes*.

   2. The generalized coordinates are linearly dependent.

   3. Principal coordinates can be considered as generalized coordinates.

   4. The vibration of a system depends on the coordinate system.

   5. The nature of coupling depends on the coordinate system.

   6. The principal coordinates avoid both static and dynamic coupling.

   7. The use of principal coordinates helps in finding the response of the system.

   8. The mass, stiffness, and damping matrices of a two-degree-of-freedom system are symmetric.

   9. The characteristics of a two-degree-of-freedom system are used in the design of a dynamic vibration absorber.

   10. Semidefinite systems are also known as *degenerate systems*.

   11. A semidefinite system cannot have nonzero natural frequencies.

   12. The generalized coordinates are always measured from the equilibrium position of the body.

   13. During free vibration, different degrees of freedom oscillate with different phase angles.

   14. During free vibration, different degrees of freedom oscillate at different frequencies.

   15. During free vibration, different degrees of freedom oscillate with different amplitudes.

   16. The relative amplitudes of different degrees of freedom in a two-degree-of-freedom system depend on the natural frequency.

   17. The modal vectors of a system denote the normal modes of vibration.

   18. The characteristic polynomial of a two-degree-of-freedom undamped system will be a quadratic in s^2.

   19. The characteristic polynomial of a two-degree-of-freedom damped system can be a quadratic in s^2.

   20. The equations of motion of a two-degree-of-freedom system can be expressed in terms of the displacement of either of the two masses.

5.3 Fill in each of the following blanks with the appropriate word:

   1. The free vibration of a two-degree-of-freedom system under arbitrary initial excitation can be found by superposing the two _____ modes of vibration.

   2. The motion of a two-degree-of-freedom system is described by two _____ coordinates.

   3. When the forcing frequency is equal to one of the natural frequencies of the system, a phenomenon known as _____ occurs.

   4. The amplitudes and phase angles are determined from the _____ conditions of the system.

   5. For a torsional system, _____ and _____ are analogous to the masses and linear springs, respectively, of a mass-spring system.

   6. The use of different generalized coordinates lead to different types of _____.

   7. A semidefinite system has at least one _____ body motion.

   8. The elastic coupling is also known as _____ *coupling*.

   9. The inertia coupling is also known as _____ *coupling*.

   10. The damping coupling is also known as _____ *coupling*.

11. The equations of motion of a system will be _____ when principal coordinates are used.
12. The Routh-Hurwitz criterion can be used to investigate the _____ of a system.
13. The equations of motion of a two-degree-of-freedom system are uncoupled only when the two masses are not _____ connected.
14. The vibration of a system under initial conditions only is called _____ *vibration.*
15. The vibration of a system under external forces is called _____ *vibration.*
16. The order of a system is the same as the order of the _____ polynomial of the system.
17. The response of an unrestrained system is composed of rigid-body motion and _____ motion.

5.4 Select the most appropriate answer out of the choices given:

1. When a two-degree-of-freedom system is subjected to a harmonic force, the system vibrates at the
  a. frequency of applied force
  b. smaller natural frequency
  c. larger natural frequency

2. The number of degrees of freedom of a vibrating system depends on
  a. number of masses
  b. number of masses and degrees of freedom of each mass
  c. number of coordinates used to describe the position of each mass

3. A two-degree-of-freedom system has
  a. one normal mode
  b. two normal modes
  c. many normal modes

4. The equations of motion of a two-degree-of-freedom system are in general
  a. coupled
  b. uncoupled
  c. linear

5. Mechanical impedance $Z_{rs}(i\omega)$ is
  a. $[m_{rs}]\ddot{\vec{x}} + [c_{rs}]\dot{\vec{x}} + [k_{rs}]\vec{x}$
  b. $\begin{Bmatrix} X_r(i\omega) \\ X_s(i\omega) \end{Bmatrix}$
  c. $-\omega^2 m_{rs} + i\omega c_{rs} + k_{rs}$

6. The impedance matrix, $[Z(i\omega)]$, can be used to find the solution as
  a. $\vec{X} = [Z(i\omega)]^{-1}\vec{F}_0$
  b. $\vec{X} = [Z(i\omega)]\vec{F}_0$
  c. $\vec{X} = [Z(i\omega)]\vec{X}_0$

7. The configuration of a system vibrating at one of its natural frequencies is called
  a. *natural mode*   **b.** *natural frequency*   **c.** *solution*
8. The equations of motion of a two-degree-of-freedom system are in general in the form of
  a. coupled algebraic equations
  b. coupled differential equations
  c. uncoupled equations

5.5   Match the items in the two columns below:

1. Static coupling	**a.** Only the mass matrix is nondiagonal
2. Inertia coupling	**b.** The mass and damping matrices are nondiagonal
3. Velocity coupling	**c.** Only the stiffness matrix is nondiagonal
4. Dynamic coupling	**d.** Only the damping matrix is nondiagonal

5.6   Match the data given in the left column with the frequency equations given in the right column for a two-degree-of-freedom system governed by the equations of motion:

$$J_0\ddot{\theta}_1 - 2k_t\theta_1 - k_t\theta_2 = 0$$

$$2J_0\ddot{\theta}_2 - k_t\theta_1 + k_t\theta_2 = 0$$

1. $J_0 = 1, k_t = 2$	**a.** $32\omega^4 - 20\omega^2 + 1 = 0$
2. $J_0 = 2, k_t = 1$	**b.** $\omega^4 - 5\omega^2 + 2 = 0$
3. $J_0 = 2, k_t = 2$	**c.** $\omega^4 - 10\omega^2 + 8 = 0$
4. $J_0 = 1, k_t = 4$	**d.** $8\omega^4 - 10\omega^2 + 1 = 0$
5. $J_0 = 4, k_t = 1$	**e.** $2\omega^4 - 5\omega^2 + 1 = 0$

PROBLEMS

Section 5.2 Equations of Motion for Forced Vibration

5.1   Derive the equations of motion of the system shown in Fig. 5.20.

5.2   Derive the equations of motion of the system shown in Fig. 5.21.

FIGURE 5.20

FIGURE 5.21

5.3 Two masses m_1 and m_2, each connected by two springs of stiffness k, are connected by a rigid massless horizontal rod of length l as shown in Fig. 5.22. (a) Derive the equations of motion of the system in terms of the vertical displacement of the C.G. of the system, $x(t)$, and the rotation about the C.G. of the system, $\theta(t)$. (b) Find the natural frequencies of vibration of the system for $m_1 = 50$ kg, $m_2 = 200$ kg, and $k = 1{,}000$ N/m. Assume $l = 1$ m.

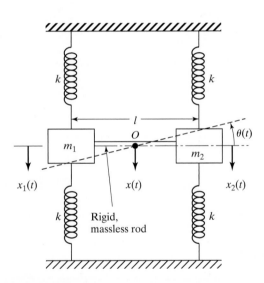

FIGURE 5.22

5.4 A two-mass system consists of a piston of mass m_1, connected by two elastic springs, that moves inside a tube as shown in Fig. 5.23. A pendulum of length l and end mass m_2 is connected to the piston as shown in Fig. 5.23. (a) Derive the equations of motion of the system in terms of $x_1(t)$ and $\theta(t)$. (b) Derive the equations of motion of the system in terms of $x_1(t)$ and $x_2(t)$. (c) Find the natural frequencies of vibration of the system.

Section 5.3 Free-Vibration Analysis of an Undamped System

5.5 Find the natural frequencies of the system shown in Fig. 5.24, with $m_1 = m$, $m_2 = 2m$, $k_1 = k$, and $k_2 = 2k$. Determine the response of the system when $k = 1{,}000$ N/m, $m = 20$ kg, and the initial values of the displacements of the masses m_1 and m_2 are 1 and -1, respectively.

FIGURE 5.23

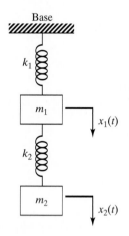

FIGURE 5.24

5.6   Set up the differential equations of motion for the double pendulum shown in Fig. 5.25, using the coordinates x_1 and x_2 and assuming small amplitudes. Find the natural frequencies, the ratios of amplitudes, and the locations of nodes for the two modes of vibration when $m_1 = m_2 = m$ and $l_1 = l_2 = l$.

5.7   Determine the natural modes of the system shown in Fig. 5.26 when $k_1 = k_2 = k_3 = k$.

5.8   A machine tool, having a mass of $m = 1,000$ kg and a mass moment of inertia of $J_0 = 300$ kg-m^2, is supported on elastic supports, as shown in Fig. 5.27. If the stiffnesses of the supports are given by $k_1 = 3,000$ N/mm and $k_2 = 2,000$ N/mm, and the supports are located at $l_1 = 0.5$ m and $l_2 = 0.8$ m, find the natural frequencies and mode shapes of the machine tool.

FIGURE 5.25

FIGURE 5.26

FIGURE 5.27

5.9   An electric overhead traveling crane, consisting of a girder, trolley, and wire rope, is shown in Fig. 5.28. The girder has a flexural rigidity (EI) of 17×10^9 N-m^2 and a span (L) of 9 m. The rope is made of steel and has a length (l) of 6 m. The weights of the trolley and the load lifted are 3,600 kg and 900 kg, respectively. Find the area of cross section of the rope such that the fundamental natural frequency is greater than 20 Hz.

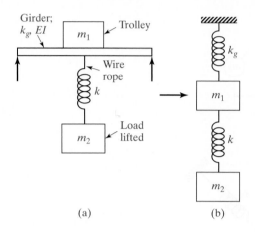

(a)   (b)

FIGURE 5.28

5.10   An overhead traveling crane can be modeled as indicated in Fig. 5.28. Assuming that the girder has a span of 40 m, an area moment of inertia (I) of 0.02 m^4, and a modulus of elasticity (E) of 2.06×10^{11} N/m^2, the trolley has a mass (m_1) of 1,000 kg, the load being lifted has a mass (m_2) of 5,000 kg, and the cable through which the mass (m_2) is lifted has a stiffness (k) of 3.0×10^5 N/m, determine the natural frequencies and mode shapes of the system.

5.11   The drilling machine shown in Fig. 5.29(a) can be modeled as a two-degree-of-freedom system as indicated in Fig. 5.29(b). Since a transverse force applied to mass m_1 or mass m_2 causes both the masses to deflect, the system exhibits elastic coupling. The bending stiffnesses of the column are given by (see Section 6.4 for the definition of stiffness influence coefficients)

$$k_{11} = \frac{768}{7}\frac{EI}{l^3}, \qquad k_{12} = k_{21} = -\frac{240}{7}\frac{EI}{l^3}, \qquad k_{22} = \frac{96}{7}\frac{EI}{l^3}$$

Determine the natural frequencies of the drilling machine.

5.12   One of the wheels and leaf springs of an automobile, traveling over a rough road, is shown in Fig. 5.30. For simplicity, all the wheels can be assumed to be identical and the system can be idealized as shown in Fig. 5.31. The automobile has a mass of $m_1 = 1,000$ kg and the leaf springs have a total stiffness of $k_1 = 400$ kN/m. The wheels and axles have a mass of $m_2 = 300$ kg and the tires have a stiffness of $k_2 = 500$ kN/m. If the road surface varies sinusoidally with an amplitude of $Y = 0.1$ m and a period of $l = 6$ m, find the critical velocities of the automobile.

(a)   (b)

FIGURE 5.29   (Photo courtesy of Atlas-Clausing division.)

FIGURE 5.30

5.13 Derive the equations of motion of the double pendulum shown in Fig. 5.25, using the coordinates θ_1 and θ_2. Also find the natural frequencies and mode shapes of the system for $m_1 = m_2 = m$ and $l_1 = l_2 = l$.

5.14 Find the natural frequencies and mode shapes of the system shown in Fig. 5.24 for $m_1 = m_2 = m$ and $k_1 = k_2 = k$.

5.15 The normal modes of a two-degree-of-freedom system are orthogonal if $X^{(1)^T}[m]\vec{X}^{(2)} = 0$. Prove that the mode shapes of the system shown in Fig. 5.5(a) are orthogonal.

5.16 Find the natural frequencies of the system shown in Fig. 5.6 for $k_1 = 300$ N/m, $k_2 = 500$ N/m, $k_3 = 200$ N/m, $m_1 = 2$ kg, and $m_2 = 1$ kg.

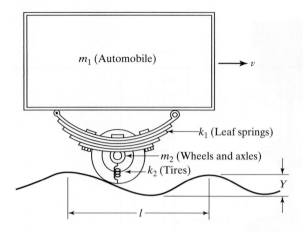

FIGURE 5.31

5.17 Find the natural frequencies and mode shapes of the system shown in Fig. 5.24 for $m_1 = m_2 = 1$ kg, $k_1 = 2,000$ N/m, and $k_2 = 6,000$ N/m.

5.18 Derive expressions for the displacements of the masses in Fig. 5.6 when $m_i = 4,400$ kg, $i = 1, 2$, and $k_i = 9 \times 10^6$ N/m, $i = 1, 2, 3$.

5.19 For the system shown in Fig. 5.6, $m_1 = 1$ kg, $m_2 = 2$ kg, $k_1 = 2,000$ N/m, $k_2 = 1,000$ N/m, $k_3 = 3,000$ N/m, and an initial velocity of 20 m/s is imparted to mass m_1. Find the resulting motion of the two masses.

5.20 For Problem 5.17, calculate $x_1(t)$ and $x_2(t)$ for the following initial conditions:

   a. $x_1(0) = 0.2, \dot{x}_1(0) = x_2(0) = \dot{x}_2(0) = 0.$
   b. $x_1(0) = 0.2, \dot{x}_1(0) = x_2(0) = 0, \dot{x}_2(0) = 5.0.$

5.21 A two-story building frame is modeled as shown in Fig. 5.32. The girders are assumed to be rigid, and the columns have flexural rigidities EI_1 and EI_2, with negligible masses. The stiffness of each column can be computed as

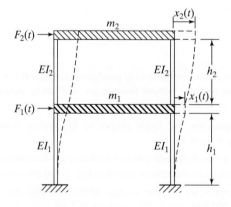

FIGURE 5.32

$$\frac{24EI_i}{h_i^3}, \qquad i = 1, 2$$

For $m_1 = 2m$, $m_2 = m$, $h_1 = h_2 = h$, and $EI_1 = EI_2 = EI$, determine the natural frequencies and mode shapes of the frame.

5.22 Figure 5.33 shows a system of two masses attached to a tightly stretched string, fixed at both ends. Determine the natural frequencies and mode shapes of the system for $m_1 = m_2 = m$ and $l_1 = l_2 = l_3 = l$.

FIGURE 5.33

5.23 Find the normal modes of the two-story building shown in Fig. 5.32 when $m_1 = 3m$, $m_2 = m$, $k_1 = 3k$, and $k_2 = k$, where k_1 and k_2 represent the total equivalent stiffnesses of the lower and upper columns, respectively.

5.24 A hoisting drum, having a weight W_1, is mounted at the end of a steel cantilever beam of thickness t, width a, and length b, as shown in Fig. 5.34. The wire rope is made of steel and has a diameter of d and a suspended length of l. If the load hanging at the end of the rope is W_2, derive expressions for the natural frequencies of the system.

(a)   (b)

FIGURE 5.34

5.25 Determine the initial conditions of the system shown in Fig. 5.24 for which the system vibrates only at its lowest natural frequency for the following data: $k_1 = k$, $k_2 = 2k$, $m_1 = m$, $m_2 = 2m$.

5.26 The system shown in Fig. 5.24 is initially disturbed by holding the mass m_1 stationary and giving the mass m_2 a downward displacement of 0.1 m. Discuss the nature of the resulting motion of the system.

5.27 Design the cantilever beam supporting the hoisting drum and the wire rope carrying the load in Problem 5.24 in order to have the natural frequencies of the system greater than 10 Hz when $m_1 = 450$ kg and $m_2 = 225$ kg, $b = 75$ cm, and $l = 150$ cm.

5.28 Find the free-vibration response of the two-degree-of-freedom system shown in Fig. 5.6 with $n = 1$, $k = 8$, and $m = 2$ for the initial conditions $x_1(0) = 1$, $x_2(0) = \dot{x}_1(0) = 0$, and $\dot{x}_2(0) = 1$.

5.29 Find the free-vibration response of the two-degree-of-freedom system shown in Fig. 5.6 with $n = 1$, $k = 8$, and $m = 2$ for the initial conditions $x_1(0) = 1$ and $x_2(0) = \dot{x}_1(0) = \dot{x}_2(0) = 0$.

5.30 Using the results of Example 5.1, verify that the mode shapes satisfy the following relations, known as *orthogonality relations*:

$$\vec{X}^{(1)^T} \vec{X}^{(2)} = 0, \qquad \vec{X}^{(1)^T}[m]\vec{X}^{(2)} = 0, \qquad \vec{X}^{(1)^T}[m]\vec{X}^{(1)} = c_1 = \text{constant}$$

$$\vec{X}^{(2)^T}[m]\vec{X}^{(2)} = c_2 = \text{constant}$$

$$\vec{X}^{(1)^T}[k]\vec{X}^{(1)} = c_1\omega_1^2, \qquad \vec{X}^{(2)^T}[k]\vec{X}^{(2)} = c_2\omega_2^2$$

5.31 Two identical pendulums, each with mass m and length l, are connected by a spring of stiffness k at a distance d from the fixed end, as shown in Fig. 5.35.
  a. Derive the equations of motion of the two masses.
  b. Find the natural frequencies and mode shapes of the system.
  c. Find the free-vibration response of the system for the initial conditions $\theta_1(0) = a$, $\theta_2(0) = 0$, $\dot{\theta}_1(0) = 0$, and $\dot{\theta}_2(0) = 0$.
  d. Determine the condition(s) under which the system exhibits a beating phenomenon.

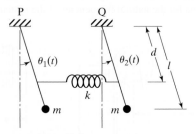

FIGURE 5.35

5.32 The motor-pump system shown in Fig. 5.36(a) is modeled as a rigid bar of mass $m = 50$ kg and mass moment of inertia $J_0 = 100$ kg-m^2. The foundation of the system can be replaced by two springs of stiffness $k_1 = 500$ N/m and $k_2 = 200$ N/m. Determine the natural frequencies of the system. Assume the distances between C.G. and spring, (k_1 and k_2) as $l_1 = 0.4$ m and $l_2 = 0.6$ m.

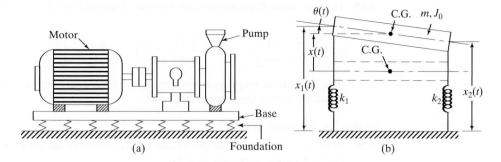

FIGURE 5.36   Motor-pump system on springs.

5.33   An airplane standing on a runway is shown in Fig. 5.37. The airplane has a mass $m = 20{,}000$ kg and a mass moment of inertia $J_0 = 50 \times 10^6$ kg-m^2. If the values of stiffness and damping constant are $k_1 = 10$ kN/m and $c_1 = 2$ kN-s/m for the main landing gear and $k_2 = 5$ kN/m and $c_2 = 5$ kN-s/m for the nose landing gear, (a) derive the equations of motion of the airplane, and (b) find the undamped natural frequencies of the system. Assume $l_1 = 20$ m and $l_2 = 30$ m.

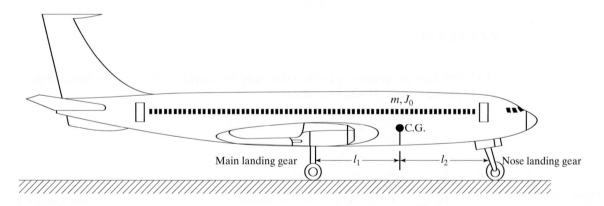

FIGURE 5.37

5.34   The mass and stiffness matrices and the mode shapes of a two-degree-of-freedom system are given by

$$[m] = \begin{bmatrix} 1 & 0 \\ 0 & 4 \end{bmatrix}, \qquad [k] = \begin{bmatrix} 12 & -k_{12} \\ -k_{12} & k_{22} \end{bmatrix}, \qquad \overline{X}^{(1)} = \begin{Bmatrix} 1 \\ 9.1109 \end{Bmatrix}, \qquad \overline{X}^{(2)} = \begin{Bmatrix} -9.1109 \\ 1 \end{Bmatrix}$$

If the first natural frequency is given by $\omega_1 = 1.7000$, determine the stiffness coefficients k_{12} and k_{22} and the second natural frequency of vibration, ω_2.

5.35   The mass and stiffness matrices and the mode shapes of a two-degree-of-freedom system are given by

$$[m] = \begin{bmatrix} m_1 & 0 \\ 0 & m_2 \end{bmatrix}, \qquad [k] = \begin{bmatrix} 27 & -3 \\ -3 & 3 \end{bmatrix}, \qquad \overline{X}^{(1)} = \begin{Bmatrix} 1 \\ 3 \end{Bmatrix}, \qquad \overline{X}^{(2)} = \begin{Bmatrix} -1 \\ 3 \end{Bmatrix}$$

If the first natural frequency is given by $\omega_1 = 1.4142$, determine the masses m_1 and m_2 and the second natural frequency of the system.

Section 5.4 Torsional System

5.36   Determine the natural frequencies and normal modes of the torsional system shown in Fig. 5.38 for $k_{t2} = 2k_{t1}$ and $J_2 = 2J_1$.

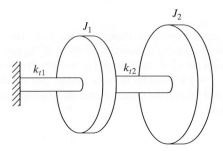

FIGURE 5.38

5.37   Determine the natural frequencies of the system shown in Fig. 5.39 by assuming that the rope passing over the cylinder does not slip.

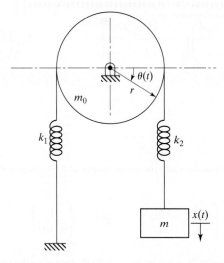

FIGURE 5.39

5.38 Find the natural frequencies and mode shapes of the system shown in Fig. 5.8(a) by assuming that $J_1 = J_0$, $J_2 = 2J_0$, and $k_{t1} = k_{t2} = k_{t3} = k_t$.

5.39 Determine the normal modes of the torsional system shown in Fig. 5.9 when $k_{t1} = k_t$, $k_{t2} = 5k_t$, $J_1 = J_0$, and $J_2 = 5J_0$.

Section 5.5 Coordinate Coupling and Principal Coordinates

5.40 A simplified ride model of the military vehicle in Fig. 5.40(a) is shown in Fig. 5.40(b). This model can be used to obtain information about the bounce and pitch modes of the vehicle. If the total mass of the vehicle is m and the mass moment of inertia about its C.G. is J_0, derive the equations of motion of the vehicle using two different sets of coordinates, as indicated in Section 5.5.

(a) Military vehicle   (b) Simplified ride model

FIGURE 5.40

5.41 Find the natural frequencies and the amplitude ratios of the system shown in Fig. 5.41.

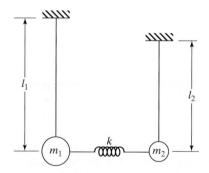

FIGURE 5.41

5.42 A rigid rod of negligible mass and length $2l$ is pivoted at the middle point and is constrained to move in the vertical plane by springs and masses, as shown in Fig. 5.42. Find the natural frequencies and mode shapes of the system.

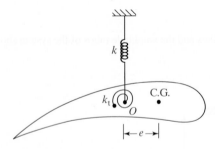

FIGURE 5.42

5.43   An airfoil of mass m is suspended by a linear spring of stiffness k and a torsional spring of stiffness k_t in a wind tunnel, as shown in Fig. 5.43. The C.G. is located at a distance of e from point O. The mass moment of inertia of the airfoil about an axis passing through point O is J_0. Find the natural frequencies of the airfoil.

FIGURE 5.43

5.44   The expansion joints of a concrete highway, which are located at 15-m intervals, cause a series of impulses to affect cars running at a constant speed. Determine the speeds at which bounce motion and pitch motion are most likely to arise for the automobile of Example 5.7.

5.45   Consider the overhead traveling crane described in Problem 5.9 (Fig. 5.28). If the rails on both sides of the girder have a sinusoidally varying surface in the z direction (perpendicular to the page), as shown in Fig. 5.44, set up the equations and the initial conditions for finding the vibration response of the load lifted (m) in the vertical direction. Assume that the velocity of the crane is 15 cm/s in the z direction.

FIGURE 5.44

5.46 An automobile is modeled with a capability of pitch and bounce motions, as shown in Fig. 5.45. It travels on a rough road whose surface varies sinusoidally with an amplitude of 0.05 m and a wavelength of 10 m. Derive the equations of motion of the automobile for the following data: mass = 1,000 kg, radius of gyration = 0.9 m, l_1 = 1.0 m, l_2 = 1.5 m, k_f = 18 kN/m, k_r = 22 kN/m, velocity = 50 km/h.

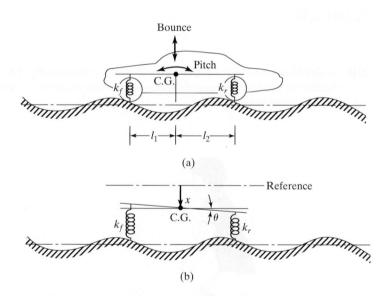

FIGURE 5.45

5.47 A steel shaft, of diameter 5 cm, is supported on two bearings and carries a pulley and a motor, as shown in Fig. 5.46. The masses of the pulley and the motor are 90 kg and 225 kg, respectively. A transverse load applied at any point along the length of the shaft results in the deflection of all points on the shaft, hence the system exhibits elastic coupling. The stiffness coefficients are given by (see Section 6.4 for the definition of stiffness influence coefficients)

$$k_{11} = \frac{1{,}296}{5}\frac{EI}{l^3}, \qquad k_{12} = k_{21} = \frac{324}{5}\frac{EI}{l^3}, \qquad k_{22} = \frac{216}{5}\frac{EI}{l^3}$$

Determine the natural frequencies of the system in bending vibration for l = 225 cm.

FIGURE 5.46

5.48 A simplified model of a mountain bike with a rider is shown in Fig. 5.47. Discuss methods of finding the vibratory response of the bicycle due to the unevenness of the terrain using a two-degree-of-freedom model.

FIGURE 5.47

5.49   A uniform rigid bar of length l and mass m is supported on two springs and is subjected to a force $F(t) = F_0 \sin \omega t$, as shown in Fig. 5.48. (a) Derive the equations of motion of the bar for small displacements. (b) Discuss the nature of coupling in the system.

FIGURE 5.48

5.50   A trailer of mass M, connected to a wall through a spring of stiffness k and a damper of damping coefficient c, slides on a frictionless surface, as shown in Fig. 5.49. A uniform rigid bar, pin-connected to the trailer, can oscillate about the hinge point, O. Derive the equations of motion of the system under the applied forces $F(t)$ and $M_t(t)$ indicated in Fig. 5.49.

FIGURE 5.49

5.51   A trailer of mass M is connected to a wall through a spring of stiffness k_1 and can move on a frictionless horizontal surface, as shown in Fig. 5.50. A uniform cylinder of mass m, connected to the wall of the trailer by a spring of stiffness k_2, can roll on the floor of the trailer without slipping. Derive the equations of motion of the system and discuss the nature of coupling present in the system.

Section 5.6 Forced-Vibration Analysis

5.52   The masses of the tup, frame, anvil (along with the workpiece), and foundation block in a forging hammer (Fig. 5.51) are 2,000 kg, 20,000 kg, 30,000 kg, and 65,000 kg, respectively.

FIGURE 5.50

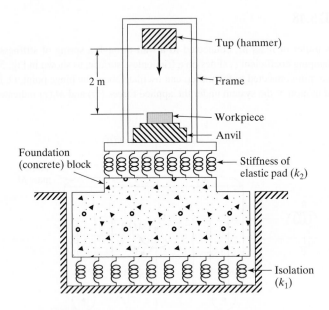

FIGURE 5.51

The stiffnesses of the elastic pad placed between the anvil and the foundation block and the isolation placed underneath the foundation (including the elasticity of the soil) are 1 GPa and 0.5 GPa, respectively. If the velocity of the tup before it strikes the anvil is 4.5 m/s, find (a) the natural frequencies of the system, and (b) the magnitudes of displacement of the anvil and the foundation block. Assume the coefficient of restitution as 0.5 and damping to be negligible in the system.

5.53   Find (a) the natural frequencies of the system, and (b) the responses of the anvil and the foundation block of the forging hammer shown in Fig. 5.51 when the time history of the force applied to the anvil is as shown in Fig. 5.52. Assume the following data:

Mass of anvil and frame = 200 Mg

Mass of foundation block = 250 Mg

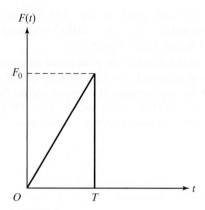

FIGURE 5.52

Stiffness of the elastic pad $= 150$ MN/m

Stiffness of the soil $= 75$ MN/m

$F_0 = 10^5$ N and $T = 0.5$ s

5.54 Derive the equations of motion for the free vibration of the system shown in Fig. 5.53. Assuming the solution as $x_i(t) = C_i e^{st}$, $i = 1, 2$, express the characteristic equation in the form

$$a_0 s^4 + a_1 s^3 + a_2 s^2 + a_3 s + a_4 = 0$$

Discuss the nature of possible solutions, $x_1(t)$ and $x_2(t)$.

FIGURE 5.53

5.55 Find the displacements $x_1(t)$ and $x_2(t)$ in Fig. 5.53 for $m_1 = 1$ kg, $m_2 = 2$ kg, $k_1 = k_2 = k_3 = 10{,}000$ N/m, and $c_1 = c_2 = c_3 = 2{,}000$ N-s/m using the initial conditions $x_1(0) = 0.2$ m, $x_2(0) = 0.1$ m, and $\dot{x}_1(0) = \dot{x}_2(0) = 0$.

5.56 A centrifugal pump, having an unbalance of me, is supported on a rigid foundation of mass m_2 through isolator springs of stiffness k_1, as shown in Fig. 5.54. If the soil stiffness and damping are k_2 and c_2, find the displacements of the pump and the foundation for the following data: $m = 0.2$ kg, $e = 0.15$ m, $m_1 = 400$ kg, $k_1 = 400$ kN/m, $m_2 = 900$ kg, $k_2 = 200$ kN/m, $c_2 = 40$ N-s/m, and speed of pump $= 1{,}200$ rpm.

FIGURE 5.54

5.57 A reciprocating engine of mass m_1 is mounted on a fixed-fixed beam of length l, width a, thickness t, and Young's modulus E, as shown in Fig. 5.55. A spring-mass system (k_2, m_2) is suspended from the beam as indicated in the figure. Find the relation between m_2 and k_2 that leads to no steady-state vibration of the beam when a harmonic force, $F_1(t) = F_0 \cos \omega t$, is developed in the engine during its operation.[3]

5.58 Find the steady-state response of the system shown in Fig. 5.24 by using the mechanical impedance method, when the mass m_1 is subjected to the force $F(t) = F_0 \sin \omega t$ in the direction of $x_1(t)$.

5.59 Find the steady-state response of the system shown in Fig. 5.24 when the base is subjected to a displacement $y(t) = Y_0 \cos \omega t$.

5.60 The mass m_1 of the two-degree-of-freedom system shown in Fig. 5.24 is subjected to a force $F_0 \cos \omega t$. Assuming that the surrounding air damping is equivalent to $c = 200$ N-s/m, find the steady-state response of the two masses. Assume $m_1 = m_2 = 1$ kg, $k_1 = k_2 = 500$ N/m, and $\omega = 1$ rad/s.

5.61 Determine the steady-state vibration of the system shown in Fig. 5.5(a), assuming that $c_1 = c_2 = c_3 = 0$, $F_1(t) = F_{10} \cos \omega t$, and $F_2(t) = F_{20} \cos \omega t$.

---

[3]The spring-mass system (k_2, m_2) added to make the amplitude of the first mass zero is known as a "vibration absorber." A detailed discussion of vibration absorbers is given in Section 9.11.

$F_1(t) = F_0 \cos \omega t$

m_1

k_2 $x_1(t)$

m_2

$x_2(t)$

t

$\dfrac{l}{2}$

$\dfrac{l}{2}$

FIGURE 5.55

5.62 In the system shown in Fig. 5.24, the mass m_1 is excited by a harmonic force having a maximum value of 50 N and a frequency of 2 Hz. Find the forced amplitude of each mass for $m_1 = 10$ kg, $m_2 = 5$ kg, $k_1 = 8{,}000$ N/m, and $k_2 = 2{,}000$ N/m.

5.63 Find the response of the two masses of the two-story building frame shown in Fig. 5.32 under the ground displacement $y(t) = 0.2 \sin \pi t$ m. Assume the equivalent stiffness of the lower and upper columns to be 800 N/m and 600 N/m, respectively, and $m_1 = m_2 = 50$ kg.

5.64 Find the forced-vibration response of the system shown in Fig. 5.15 when $F_1(t)$ is a step force of magnitude 5 N using the Laplace transform method. Assume $x_1(0) = \dot{x}_1(0) = x_2(0) = \dot{x}_2(0) = 0$, $m = 1$ kg, and $k = 100$ N/m.

Section 5.7 Semidefinite Systems

5.65 Determine the equations of motion and the natural frequencies of the system shown in Fig. 5.56.

5.66 Two identical circular cylinders, of radius r and mass m each, are connected by a spring, as shown in Fig. 5.57. Determine the natural frequencies of oscillation of the system.

5.67 The differential equations of motion for a two-degree-of-freedom system are given by

$$a_1 \ddot{x}_1 + b_1 x_1 + c_1 x_2 = 0$$

$$a_2 \ddot{x}_2 + b_2 x_1 + c_2 x_2 = 0$$

Derive the condition to be satisfied for the system to be degenerate.

FIGURE 5.56

FIGURE 5.57

5.68 Find the angular displacements $\theta_1(t)$ and $\theta_2(t)$ of the system shown in Fig. 5.58 for the initial conditions $\theta_1(t = 0) = \theta_1(0)$, $\theta_2(t = 0) = \theta_2(0)$, and $\dot{\theta}_1(t = 0) = \dot{\theta}_2(t = 0) = 0$.

FIGURE 5.58

5.69 Determine the normal modes of the system shown in Fig. 5.9 with $k_{t1} = 0$. Show that the system with $k_{t1} = 0$ can be treated as a single-degree-of-freedom system by using the coordinate $\alpha = \theta_1 - \theta_2$.

5.70 A turbine is connected to an electric generator through gears, as shown in Fig. 5.59. The mass moments of inertia of the turbine, generator, gear 1, and gear 2 are given, respectively, by 3,000, 2,000, 500, and 1,000 kg-m^2. Shafts 1 and 2 are made of steel and have diameters 30 cm and 10 cm and lengths 2 cm and 1.0 m, respectively. Find the natural frequencies of the system.

5.71 A hot-air balloon of mass m is used to lift a load, Mg, by means of 12 equally spaced elastic ropes, each of stiffness k (see Fig. 5.60). Find the natural frequencies of vibration of the balloon in the vertical direction. State the assumptions made in your solution and discuss their validity.

5.72 A turbine of mass moment of inertia of 0.5 kg-m^2 is connected to an electric generator of mass moment of inertia of 0.25 kg-m^2 by a hollow steel shaft of inner diameter 0.02 m, outer

FIGURE 5.59

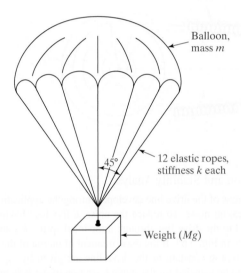

FIGURE 5.60

diameter 0.04 m, and length 0.4 m (similar to the system in Fig. 5.17(c)). If the turbine is suddenly stopped while delivering 70 kW of power at a speed of 6,000 rpm, the transmitted torque drops to zero. Find the resulting angular displacements of the turbine and the generator. Assume damping to be negligible in the system.

5.73 Figure 5.61 shows a 1,400-kg car connected to a 900-kg trailer by a flexible hitch having a stiffness of 1,750 N/cm. Assuming that both the car and the trailer can move freely on the roadway, determine the natural frequencies and mode shapes of vibration of the system.

5.74 Find the response of the car-trailer system described in Problem 5.73 if the values of initial displacement and velocity are 15 cm and 0 cm/s for the car and −7.5 cm and 0 cm/s for the trailer.

FIGURE 5.61

5.75 Two pulleys are driven by a belt as shown in Fig. 5.62. If the pulleys have radii r_1 and r_2 and mass moments of inertia J_1 and J_2, respectively, determine the natural frequencies of the pulley drive system. Assume the stiffness of the belt on each side as k as indicated in Fig. 5.62.

FIGURE 5.62

Section 5.8 Self-Excitation and Stability Analysis

5.76 The transient vibrations of the drive line developed during the application of a cone (friction) clutch lead to unpleasant noise. To reduce the noise, a flywheel having a mass moment of inertia J_2 is attached to the drive line through a torsional spring k_{t2} and a viscous torsional damper c_{t2}, as shown in Fig. 5.63. If the mass moment of inertia of the cone clutch is J_1 and the stiffness and damping constant of the drive line are given by k_{t1} and c_{t1}, respectively, derive the relations to be satisfied for the stable operation of the system.

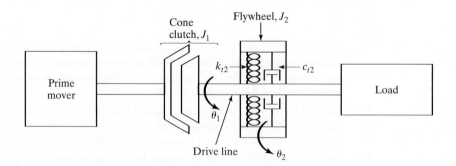

FIGURE 5.63

5.77 A uniform rigid bar of mass m is connected to the wall of a trailer by a spring of stiffness k (see Fig. 5.64). The trailer has a mass $5m$, is connected to a spring of stiffness $2k$, and moves on a frictionless surface. Derive the conditions necessary for the stability of the system.

FIGURE 5.64

5.78 A two-degree-of-freedom system consists of masses m_1 and m_2 connected to a damper and a spring as shown in Fig. 5.65. If the mass m_1 is subjected to a force proportional to its velocity, $f_1(t) = a\dot{x}_1(t)$, determine the conditions for the stability of the system.

FIGURE 5.65

Section 5.9 Transfer-Function Approach

5.79 Derive the fourth-order differential equations of motion of the two-degree-of-freedom system shown in Fig. 5.5(a) in terms of $x_1(t)$ and $x_2(t)$ separately.
Hint: Take the inverse Laplace transforms of Eqs. (5.56) and (5.57).

5.80 **a.** Suggest a method of solving the fourth-order differential equations derived in Problem 5.79 (in terms of $x_1(t)$ or $x_2(t)$).
b. How can we apply the known initial conditions $x_1(0)$, $x_2(0)$, $\dot{x}_1(0)$, and $\dot{x}_2(0)$ while solving the fourth-order differential equation in terms of $x_1(t)$?

5.81 Derive expressions for the Laplace transform of $x_1(t)$ and $x_2(t)$ for the system shown in Fig. 5.5(a) for the following data: $m_1 = 1$, $m_2 = 2$, $k_1 = 4$, $k_2 = 2$, $k_3 = 0$, $c_1 = 1$, $c_2 = 2$, $c_3 = 0$, $f_1(t) = F_0 u(t) =$ step function, and $f_2(t) = 0$. Assume the initial conditions of both $x_1(t)$ and $x_2(t)$ as zero.

5.82 Derive expressions for the Laplace transform of $x_1(t)$ and $x_2(t)$ for the system shown in Fig. 5.5(a) for the following data: $m_1 = 1$, $m_2 = 2$, $k_1 = 4$, $k_2 = 2$, $k_3 = 0$, $c_1 = 1$, $c_2 = 2$, $c_3 = 0$, $f_1(t) = 0$, $f_2(t) = F_0 u(t) =$ step function. Assume the initial conditions of both $x_1(t)$ and $x_2(t)$ as zero.

Section 5.10 Solutions Using Laplace Transform

5.83 Find the free-vibration response of the system shown in Fig. 5.5(a) using the Laplace transform approach for the following data: $m_1 = 2$, $m_2 = 4$, $k_1 = 8$, $k_2 = 4$, $k_3 = 0$, $c_1 = 0$, $c_2 = 0$, $c_3 = 0$. Assume the initial conditions as $x_1(0) = 1$, $x_2(0) = 0$, and $\dot{x}_1(0) = \dot{x}_2(0) = 0$. Plot the responses $x_1(t)$ and $x_2(t)$.

5.84 Find the free-vibration response of the system shown in Fig. 5.5(a) using the Laplace transform approach for the following data: $m_1 = 2$, $m_2 = 4$, $k_1 = 8$, $k_2 = 4$, $k_3 = 0$, $c_1 = 0$, $c_2 = 2$, $c_3 = 0$. Assume the initial conditions as $x_1(0) = 1$, $x_2(0) = 0$, and $\dot{x}_1(0) = \dot{x}_2(0) = 0$. Plot the responses $x_1(t)$ and $x_2(t)$.

5.85 Find the free-vibration response of the system shown in Fig. 5.5(a) using the Laplace transform approach for the following data: $m_1 = 2$, $m_2 = 8$, $k_1 = 8$, $k_2 = 4$, $k_3 = 0$, $c_1 = 0$, $c_2 = 0$, $c_3 = 0$. Assume the initial conditions as $x_1(0) = 1$, $x_2(0) = 0$, and $\dot{x}_1(0) = \dot{x}_2(0) = 0$. Plot the responses $x_1(t)$ and $x_2(t)$.

5.86 Find the free-vibration response of the system shown in Fig. 5.5(a) using the Laplace transform approach for the following data: $m_1 = 1$, $m_2 = 8$, $k_1 = 8$, $k_2 = 4$, $k_3 = 0$, $c_1 = 0$, $c_2 = 0$, $c_3 = 0$. Assume the initial conditions as $x_1(0) = 1$, $x_2(0) = 0$, and $\dot{x}_1(0) = \dot{x}_2(0) = 0$. Plot the responses $x_1(t)$ and $x_2(t)$.

5.87 Find the response of the system shown in Fig. 5.66 with $m_1 = 2$, $m_2 = 1$, $k_1 = 40$, and $k_2 = 20$ for the following initial conditions using the Laplace transform:

i. $x_1(0) = 0.05$, $x_2(0) = 0.10$, $\dot{x}_1(0) = 0$, $\dot{x}_2(0) = 0$

ii. $x_1(0) = 0.10$, $x_2(0) = -0.05$, $\dot{x}_1(0) = 0$, $\dot{x}_2(0) = 0$

FIGURE 5.66

Section 5.11 Solutions Using Frequency Transfer Functions

5.88 Find the steady-state response of the system considered in Example 5.13 by assuming $p_1(t) = 0$ and $p_2(t) = P_0 \sin \omega t$ and neglecting damping.

5.89 Find the steady-state response of the system considered in Example 5.13 by assuming $p_1(t) = P_{01} \sin \omega t$ and $p_2(t) = P_{02} \sin \omega t$ and neglecting damping.

Section 5.12 Examples Using MATLAB

5.90 Find the response of the system shown in Fig. 5.5(a) using a numerical procedure when $k_1 = k, k_2 = 2k, k_3 = k, m_1 = 2m, m_2 = m, F_2(t) = 0$, and $F_1(t)$ is a rectangular pulse of magnitude 500 N and duration 0.5 s. Assume $m = 10$ kg, $c_1 = c_2 = c_3 = 0$, $k = 2{,}000$ N/m, and zero initial conditions.

5.91 (a) Find the roots of the frequency equation of the system shown in Fig. 5.5 with the following data: $m_1 = m_2 = 40$ kg, $k_1 = k_2 = 3{,}000$ N/m, $k_3 = 0$, $c_1 = c_2 = c_3 = 0$. (b) If the initial conditions are $x_1(0) = x_2(0) = 0.05$ m, $\dot{x}_1(0) = \dot{x}_2(0) = 0$, determine the displacements $x_1(t)$ and $x_2(t)$ of the masses.

5.92 Write a computer program for finding the steady-state response of a two-degree-of-freedom system under the harmonic excitation $F_j(t) = F_{j0}e^{i\omega t}$ and $j = 1, 2$ using Eqs. (5.29) and (5.35). Use this program to find the response of a system with $m_{11} = m_{22} = 20$ kg, $m_{12} = 0$, $c_{11} = 200$ N-s/m, $c_{12} = c_{22} = 0$, $k_{11} = 7{,}000$ N/m, $k_{22} = 3{,}500$ N/m, $k_{12} = -3{,}500$ N/m, $F_{10} = 5$ N, $F_{20} = 10$ N, and $\omega = 5$ rad/s.

5.93 Find and plot the free-vibration response of the system shown in Fig. 5.24 for the following data: $k_1 = 1{,}000$ N/m, $k_2 = 500$ N/m, $m_1 = 2$ kg, $m_2 = 1$ kg, $x_1(0) = 1$, $x_2(0) = 0$, $\dot{x}_2(0) = 0$, $\dot{x}_1(0) = -1$.

5.94 Find and plot the free-vibration response of the system shown in Fig. 5.24 for the following data: $k_1 = 1{,}000$ N/m, $k_2 = 500$ N/m, $m_1 = 2$ kg, $m_2 = 1$ kg, $x_1(0) = 1$, $x_2(0) = 2$, $\dot{x}_1(0) = 1$, $\dot{x}_2(0) = -2$.

5.95 Solve the following eigenvalue problem using MATLAB:

$$\begin{bmatrix} 25 \times 10^6 & -5 \times 10^6 \\ -5 \times 10^6 & 5 \times 10^6 \end{bmatrix} \begin{Bmatrix} x_1 \\ x_2 \end{Bmatrix} = \omega^2 \begin{bmatrix} 10{,}000 & 0 \\ 0 & 5{,}000 \end{bmatrix} \begin{Bmatrix} x_1 \\ x_2 \end{Bmatrix}$$

5.96 Find and plot the response of the following two-degree-of-freedom system using MATLAB:

$$\begin{bmatrix} 2 & 0 \\ 0 & 10 \end{bmatrix} \begin{Bmatrix} \ddot{x}_1 \\ \ddot{x}_2 \end{Bmatrix} + \begin{bmatrix} 20 & -5 \\ -5 & 5 \end{bmatrix} \begin{Bmatrix} \dot{x}_1 \\ \dot{x}_2 \end{Bmatrix} + \begin{bmatrix} 50 & -10 \\ -10 & 10 \end{bmatrix} \begin{Bmatrix} x_1 \\ x_2 \end{Bmatrix} = \begin{Bmatrix} 2 \sin 3t \\ 5 \cos 5t \end{Bmatrix}$$

The initial conditions are $x_1(0) = 1$, $\dot{x}_1(0) = 0$, $x_2(0) = -1$, and $\dot{x}_2(0) = 0$.

5.97 Using MATLAB, solve Problem 5.90. Use the MATLAB function **stepfun** for the rectangular pulse.

5.98 Using MATLAB, solve Problem 5.91(a).

5.99 Using MATLAB, solve Problem 5.92. Plot the steady-state responses of masses m_{11} and m_{22}.

5.100 Using MATLAB, find the roots of the equation $x^4 - 32x^3 + 244x^2 - 20x - 1{,}200 = 0$.

DESIGN PROJECTS

5.101 A step-cone pulley with a belt drive (Fig. 5.67) is used to change the cutting speeds in a lathe. The speed of the driving shaft is 350 rpm and the speeds of the output shaft are 150, 250, 450, and 750 rpm. The diameters of the driving and the driven pulleys, corresponding to 150 rpm output speed, are 250 mm and 1,000 mm, respectively. The center distance between the shafts is 5 m. The mass moments of inertia of the driving and driven step cones are 0.1 and 0.2 kg-m^2, respectively. Find the cross-sectional area of the belt to avoid resonance with any of the input/output speeds of the system. Assume the Young's modulus of the belt material as 10^{10} N/m^2.

FIGURE 5.67

5.102 The masses of the tup, frame (along with the anvil and the workpiece), and concrete block in the forging hammer shown in Fig. 5.51 are 1,000 kg, 5,000 kg, and 25,000 kg, respectively. The tup drops onto the workpiece from a height of 2 m. Design suitable springs k_1 and k_2 for the following conditions: (a) The impact is inelastic—that is, the tup will not rebound after striking the workpiece. (b) The natural frequencies of vibration of the forging hammer should be greater than 5 Hz. (c) The stresses in the springs should be smaller than the yield stress of the material with a factor of safety of at least 1.5. Assume that the elasticity of the soil is negligible.

Joseph Louis Lagrange (1736–1813) was an Italian-born mathematician famous for his work on theoretical mechanics. He was made professor of mathematics in 1755 at the Artillery School in Turin. Lagrange's masterpiece, *Méchanique*, contains what are now known as "Lagrange's equations," which are very useful in the study of vibrations. His work on elasticity and strength of materials, where he considered the strength and deflection of struts, is less well known.
(Courtesy of Dirk J. Struik, *A Concise History of Mathematics* (2nd ed.), Dover Publications, New York, 1948.)

C H A P T E R 6

Multidegree-of-Freedom Systems

Chapter Outline

553

Multidegree-of-freedom systems are the topic of this chapter. The modeling of continuous systems as multidegree-of-freedom systems is presented. The equations of a general n-degree-of-freedom system are derived using Newton's second law of motion. Because the solution of the equations of motion in scalar form involve complicated algebraic manipulations, we use matrix representation for multidegree-of-freedom systems. By expressing the coupled set of n equations in matrix form, the mass, damping, and stiffness matrices are identified. The derivation of equations using influence coefficients is also presented. The stiffness, flexibility, and inertia influence coefficients are presented from first principles. The expressions for potential and kinetic energies and their use in deriving the equations of motion based on Lagrange's equations are presented. The concepts of generalized coordinates and generalized forces are presented. After expressing the free-vibration equations in matrix form, the eigenvalue problem is derived in matrix form. The solution of the eigenvalue problem using the solution of the characteristic (polynomial) equation is outlined to determine the natural frequencies and mode shapes (or normal modes) of the system. The concepts of orthogonality of normal modes, modal matrix, and orthonormalization of the mass and stiffness matrices are introduced. The expansion theorem and the unrestrained or semidefinite systems are also presented. The free vibration of undamped systems using modal vectors and the forced vibration of undamped systems using modal analysis are considered with illustrative examples. The equations of motion for the forced vibration of viscously damped systems are considered through the introduction of Rayleigh's dissipation function. The equations of motion are uncoupled for proportionally damped systems, and the solution of each of the uncoupled equations is outlined through the Duhamel integral. The self-excitation and stability analysis of multidegree-of-freedom systems is considered using Routh-Hurwitz stability criterion. Finally, MATLAB solutions are presented for the free and forced vibration of multidegree-of-freedom systems.

Learning Objectives

After you have finished studying this chapter, you should be able to do the following:

- Formulate the equations of motion of multidegree-of-freedom systems using Newton's second law, influence coefficients, or Lagrange's equations.
- Express the equation of motion in matrix form.
- Find the natural frequencies of vibration and the modal vectors by solving the eigenvalue problem.
- Determine the free- and forced-vibration response of undamped systems using modal analysis.
- Use proportional damping to find the response of damped systems.
- Analyze the stability characteristics of multidegree-of-freedom systems using the Routh-Hurwitz criterion.
- Solve free- and forced-vibration problems using MATLAB.

6.1 Introduction

As stated in Chapter 1, most engineering systems are continuous and have an infinite number of degrees of freedom. The vibration analysis of continuous systems requires the solution of partial differential equations, which is quite difficult. For many partial differential equations, in fact, analytical solutions do not exist. The analysis of a multidegree-of-freedom system, on the other hand, requires the solution of a set of ordinary differential equations, which is relatively simple. Hence, for simplicity of analysis, continuous systems are often approximated as multidegree-of-freedom systems.

All the concepts introduced in the preceding chapter can be directly extended to the case of multidegree-of-freedom systems. For example, there is one equation of motion for each degree of freedom; if generalized coordinates are used, there is one generalized coordinate for each degree of freedom. The equations of motion can be obtained from Newton's second law of motion or by using the influence coefficients defined in Section 6.4. However, it is often more convenient to derive the equations of motion of a multidegree-of-freedom system by using Lagrange's equations.

There are n natural frequencies, each associated with its own mode shape, for a system having n degrees of freedom. The method of determining the natural frequencies from the characteristic equation obtained by equating the determinant to zero also applies to these systems. However, as the number of degrees of freedom increases, the solution of the characteristic equation becomes more complex. The mode shapes exhibit a property known as *orthogonality*, which can be utilized for the solution of undamped forced-vibration problems using a procedure known as *modal analysis*. The solution of forced-vibration problems associated with viscously damped systems can also be found conveniently by using a concept called *proportional damping*.

6.2 Modeling of Continuous Systems as Multidegree-of-Freedom Systems

Different methods can be used to approximate a continuous system as a multidegree-of-freedom system. A simple method involves replacing the distributed mass or inertia of the system by a finite number of lumped masses or rigid bodies. The lumped masses are assumed to be connected by massless elastic and damping members. Linear (or angular) coordinates are used to describe the motion of the lumped masses (or rigid bodies). Such models are called *lumped-parameter* or *lumped-mass* or *discrete-mass* systems. The minimum number of coordinates necessary to describe the motion of the lumped masses and rigid bodies defines the number of degrees of freedom of the system. Naturally, the larger the number of lumped masses used in the model, the higher the accuracy of the resulting analysis.

Some problems automatically indicate the type of lumped-parameter model to be used. For example, the three-story building shown in Fig. 6.1(a) automatically suggests using a three-lumped-mass model, as indicated in Fig. 6.1(b). In this model, the inertia of the system is assumed to be concentrated as three point masses located at the floor levels, and the elasticities of the columns are replaced by the springs. Similarly, the radial drilling

(a)   (b)

FIGURE 6.1   Three-story building.

(a)   (b)

FIGURE 6.2   Radial drilling machine. (Photo courtesy of South Bend Lathe Corp.)

machine shown in Fig. 6.2(a) can be modeled using four lumped masses and four spring elements (elastic beams), as shown in Fig. 6.2(b).

Another popular method of approximating a continuous system as a multidegree-of-freedom system involves replacing the geometry of the system by a large number of small elements. By assuming a simple solution within each element, the principles of compatibility and equilibrium are used to find an approximate solution to the original system. This method, known as the *finite element method*, is considered in detail in Chapter 12.

6.3   Using Newton's Second Law to Derive Equations of Motion

The following procedure can be adopted to derive the equations of motion of a multidegree-of-freedom system using Newton's second law of motion:

1. Set up suitable coordinates to describe the positions of the various point masses and rigid bodies in the system. Assume suitable positive directions for the displacements, velocities, and accelerations of the masses and rigid bodies.

2. Determine the static equilibrium configuration of the system and measure the displacements of the masses and rigid bodies from their respective static equilibrium positions.

3. Draw the free-body diagram of each mass or rigid body in the system. Indicate the spring, damping, and external forces acting on each mass or rigid body when positive displacement and velocity are given to that mass or rigid body.

4. Apply Newton's second law of motion to each mass or rigid body shown by the free-body diagram as

$$m_i \ddot{x}_i = \sum_j F_{ij} \text{ (for mass } m_i)$$ (6.1)

or

$$J_i \ddot{\theta}_i = \sum_j M_{ij} \text{ (for rigid body of inertia } J_i)$$ (6.2)

where $\sum_j F_{ij}$ denotes the sum of all forces acting on mass m_i and $\sum_j M_{ij}$ indicates the sum of moments of all forces (about a suitable axis) acting on the rigid body of mass moment of inertia J_i.

The procedure is illustrated in the following examples.

EXAMPLE 6.1

Equations of Motion of a Spring-Mass-Damper System

Derive the equations of motion of the spring-mass-damper system shown in Fig. 6.3(a).

Solution:

Approach: Draw free-body diagrams of masses and apply Newton's second law of motion. The coordinates describing the positions of the masses, $x_i(t)$, are measured from their respective static equilibrium positions, as indicated in Fig. 6.3(a). The free-body diagram of a typical interior mass m_i is shown in Fig. 6.3(b) along with the assumed positive directions for its displacement, velocity, and acceleration. The application of Newton's second law of motion to mass m_i gives

$$m_i \ddot{x}_i = -k_i(x_i - x_{i-1}) + k_{i+1}(x_{i+1} - x_i) - c_i(\dot{x}_i - \dot{x}_{i-1})$$
$$+ c_{i+1}(\dot{x}_{i+1} - \dot{x}_i) + F_i; \qquad i = 2, 3, \ldots, n - 1$$

or

$$m_i \ddot{x}_i - c_i \dot{x}_{i-1} + (c_i + c_{i+1})\dot{x}_i - c_{i+1}\dot{x}_{i+1} - k_i x_{i-1}$$
$$+ (k_i + k_{i+1})x_i - k_{i+1}x_{i+1} = F_i; \qquad i = 2, 3, \ldots, n - 1 \qquad \text{(E.1)}$$

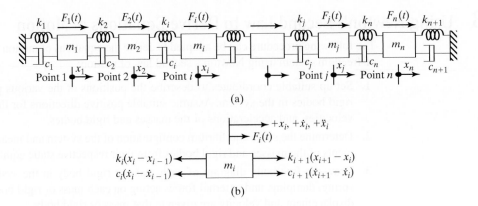

(a)

(b)

FIGURE 6.3   Spring-mass-damper system.

The equations of motion of the masses m_1 and m_n can be derived from Eq. (E.1) by setting $i = 1$ along with $x_0 = 0$ and $i = n$ along with $x_{n+1} = 0$, respectively:

$$m_1\ddot{x}_1 + (c_1 + c_2)\dot{x}_1 - c_2\dot{x}_2 + (k_1 + k_2)x_1 - k_2 x_2 = F_1 \tag{E.2}$$

$$m_n\ddot{x}_n - c_n\dot{x}_{n-1} + (c_n + c_{n+1})\dot{x}_n - k_n x_{n-1} + (k_n + k_{n+1})x_n = F_n \tag{E.3}$$

Notes:

1. The equations of motion, Eqs. (E.1) to (E.3), of Example 6.1 can be expressed in matrix form as

$$[m]\ddot{\vec{x}} + [c]\dot{\vec{x}} + [k]\vec{x} = \vec{F} \tag{6.3}$$

where $[m]$, $[c]$, and $[k]$ are called the *mass*, *damping*, and *stiffness matrices*, respectively, and are given by

$$[m] = \begin{bmatrix} m_1 & 0 & 0 & \cdots & 0 & 0 \\ 0 & m_2 & 0 & \cdots & 0 & 0 \\ 0 & 0 & m_3 & \cdots & 0 & 0 \\ \vdots & & & & & \\ 0 & 0 & 0 & \cdots & 0 & m_n \end{bmatrix} \tag{6.4}$$

$$[c] = \begin{bmatrix} (c_1 + c_2) & -c_2 & 0 & \cdots & 0 & 0 \\ -c_2 & (c_2 + c_3) & -c_3 & \cdots & 0 & 0 \\ 0 & -c_3 & (c_3 + c_4) & \cdots & 0 & 0 \\ \cdot & \cdot & \cdot & \cdots & \cdot & \cdot \\ \cdot & \cdot & \cdot & \cdots & \cdot & \cdot \\ \cdot & \cdot & \cdot & \cdots & \cdot & \cdot \\ 0 & 0 & 0 & \cdots & -c_n & (c_n + c_{n+1}) \end{bmatrix} \tag{6.5}$$

$$[k] = \begin{bmatrix} (k_1 + k_2) & -k_2 & 0 & \cdots & 0 & 0 \\ -k_2 & (k_2 + k_3) & -k_3 & \cdots & 0 & 0 \\ 0 & -k_3 & (k_3 + k_4) & \cdots & 0 & 0 \\ \cdot & & & & & \\ \cdot & & & & & \\ \cdot & & & & & \\ 0 & 0 & 0 & \cdots & -k_n & (k_n + k_{n+1}) \end{bmatrix} \tag{6.6}$$

and $\vec{x}, \dot{\vec{x}}, \ddot{\vec{x}}$, and $\vec{F}$ are the displacement, velocity, acceleration, and force vectors, given by

$$\vec{x} = \begin{Bmatrix} x_1(t) \\ x_2(t) \\ \cdot \\ \cdot \\ \cdot \\ x_n(t) \end{Bmatrix}, \qquad \dot{\vec{x}} = \begin{Bmatrix} \dot{x}_1(t) \\ \dot{x}_2(t) \\ \cdot \\ \cdot \\ \cdot \\ \dot{x}_n(t) \end{Bmatrix},$$

$$\ddot{\vec{x}} = \begin{Bmatrix} \ddot{x}_1(t) \\ \ddot{x}_2(t) \\ \cdot \\ \cdot \\ \cdot \\ \ddot{x}_n(t) \end{Bmatrix}, \qquad \vec{F} = \begin{Bmatrix} F_1(t) \\ F_2(t) \\ \cdot \\ \cdot \\ \cdot \\ F_n(t) \end{Bmatrix} \tag{6.7}$$

2. For an undamped system (with all $c_i = 0, i = 1, 2, \ldots, n + 1$), the equations of motion reduce to

$$[m]\ddot{\vec{x}} + [k]\vec{x} = \vec{F} \tag{6.8}$$

3. The spring-mass-damper system considered above is a particular case of a general n-degree-of-freedom spring-mass-damper system. In their most general form, the mass, damping, and stiffness matrices are given by

$$[m] = \begin{bmatrix} m_{11} & m_{12} & m_{13} & \cdots & m_{1n} \\ m_{12} & m_{22} & m_{23} & \cdots & m_{2n} \\ \cdot & & & & \\ \cdot & & & & \\ \cdot & & & & \\ m_{1n} & m_{2n} & m_{3n} & \cdots & m_{nn} \end{bmatrix} \tag{6.9}$$

$$[c] = \begin{bmatrix} c_{11} & c_{12} & c_{13} & \cdots & c_{1n} \\ c_{12} & c_{22} & c_{23} & \cdots & c_{2n} \\ \cdot & & & & \\ \cdot & & & & \\ \cdot & & & & \\ c_{1n} & c_{2n} & c_{3n} & \cdots & c_{nn} \end{bmatrix} \tag{6.10}$$

and

$$[k] = \begin{bmatrix} k_{11} & k_{12} & k_{13} & \cdots & k_{1n} \\ k_{12} & k_{22} & k_{23} & \cdots & k_{2n} \\ \cdot & & & & \\ \cdot & & & & \\ \cdot & & & & \\ k_{1n} & k_{2n} & k_{3n} & \cdots & k_{nn} \end{bmatrix} \tag{6.11}$$

As stated in Section 5.5, if the mass matrix is not diagonal, the system is said to have mass or inertia coupling. If the damping matrix is not diagonal, the system is said to have damping or velocity coupling. Finally, if the stiffness matrix is not diagonal, the system is said to have elastic or static coupling. Both mass and damping coupling are also known as *dynamic coupling*.

4. The differential equations of the spring-mass system considered in Example 6.1 (Fig. 6.3(a)) can be seen to be coupled; each equation involves more than one coordinate. This means that the equations cannot be solved one at a time; they can only be solved simultaneously. In addition, the system can be seen to be statically coupled, since stiffnesses are coupled—that is, the stiffness matrix has at least one nonzero off-diagonal term. On the other hand, if the mass matrix has at least one nonzero off-diagonal term, the system is said to be *dynamically coupled*. Further, if both the stiffness and mass matrices have nonzero off-diagonal terms, the system is said to be *coupled* both *statically* and *dynamically*.

■

Equations of Motion of a Trailer–Compound Pendulum System

EXAMPLE 6.2

Derive the equations of motion of the trailer-compound pendulum system shown in Fig. 6.4(a).

Solution:

Approach: Draw the free-body diagrams and apply Newton's second law of motion.

The coordinates $x(t)$ and $\theta(t)$ are used to describe, respectively, the linear displacement of the trailer and the angular displacement of the compound pendulum from their respective static equilibrium positions. When positive values are assumed for the displacements $x(t)$ and $\theta(t)$, velocities $\dot{x}(t)$ and $\dot{\theta}(t)$, and accelerations $\ddot{x}(t)$ and $\ddot{\theta}(t)$, the external forces on the trailer will be the applied force $F(t)$, the spring forces $k_1 x$ and $k_2 x$, and the damping forces $c_1 \dot{x}$ and $c_2 \dot{x}$, as shown in Fig. 6.4(b). The external forces on the compound pendulum will be the applied torque $M_t(t)$ and the gravitational force mg, as shown in Fig. 6.4(b). The inertia forces that act on the trailer and the compound pendulum are indicated by the dashed lines in Fig. 6.4(b). Note that the rotational motion of the compound pendulum about the hinge O induces a radially inward force (toward O) $m\frac{l}{2}\dot{\theta}^2$ and a normal force (perpendicular to OC) $m\frac{l}{2}\ddot{\theta}$ as shown in Fig. 6.4(b). The application of Newton's second law for translatory motion in the horizontal direction gives

$$M\ddot{x} + m\ddot{x} + m\frac{l}{2}\ddot{\theta}\cos\theta - m\frac{l}{2}\dot{\theta}^2\sin\theta = -k_1 x - k_2 x - c_1\dot{x} - c_2\dot{x} + F(t) \tag{E.1}$$

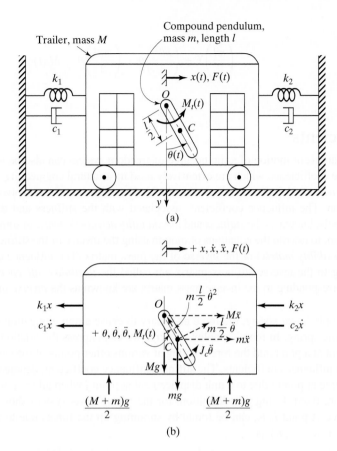

FIGURE 6.4   Compound pendulum and trailer system.

Similarly the application of Newton's second law for rotational motion about hinge O yields

$$\left(m\frac{l}{2}\ddot{\theta}\right)\frac{l}{2} + \left(m\frac{l^2}{12}\right)\ddot{\theta} + (m\ddot{x})\frac{l}{2}\cos\theta = -(mg)\frac{l}{2}\sin\theta + M_t(t) \qquad \text{(E.2)}$$

Notes:

1. The equations of motion, Eqs. (E.1) and (E.2), can be seen to be nonlinear due to the presence of the terms involving $\sin\theta$, $\cos\theta$, and $(\dot{\theta})^2\sin\theta$.

2. Equations (E.1) and (E.2) can be linearized if the term involving $(\dot{\theta})^2\sin\theta$ is assumed negligibly small and the displacements are assumed small so that $\cos\theta \approx 1$ and $\sin\theta \approx \theta$. The linearized equations can be derived as

$$(M + m)\ddot{x} + \left(m\frac{l}{2}\right)\ddot{\theta} + (k_1 + k_2)x$$

$$+ (c_1 + c_2)\dot{x} = F(t) \qquad \text{(E.3)}$$

and

$$\left(\frac{ml}{2}\right)\ddot{x} + \left(\frac{ml^2}{3}\right)\ddot{\theta} + \left(\frac{mgl}{2}\right)\theta = M_t(t) \qquad \text{(E.4)}$$

∎

6.4   Influence Coefficients

The equations of motion of a multidegree-of-freedom system can also be written in terms of influence coefficients, which are extensively used in structural engineering. Basically, one set of influence coefficients can be associated with each of the matrices involved in the equations of motion. The influence coefficients associated with the stiffness and mass matrices are, respectively, known as the *stiffness* and *inertia influence coefficients*. In some cases, it is more convenient to rewrite the equations of motion using the inverse of the stiffness matrix (known as the *flexibility matrix*) or the inverse of the mass matrix. The influence coefficients corresponding to the inverse stiffness matrix are called the *flexibility influence coefficients*, and those corresponding to the inverse mass matrix are known as the *inverse inertia coefficients*.

6.4.1
Stiffness
Influence
Coefficients

For a simple linear spring, the force necessary to cause a unit elongation is called the *stiffness* of the spring. In more complex systems, we can express the relation between the displacement at a point and the forces acting at various other points of the system by means of stiffness influence coefficients. The stiffness influence coefficient, denoted as k_{ij}, is defined as the force at point i due to a unit displacement at point j when all the points other than the point j are fixed. Using this definition, for the spring-mass system shown in Fig. 6.5, the total force at point i, F_i, can be found by summing up the forces due to all displacements $x_j (j = 1, 2, \ldots, n)$ as

$$F_i = \sum_{j=1}^{n} k_{ij} x_j, \qquad i = 1, 2, \ldots, n \qquad \text{(6.12)}$$

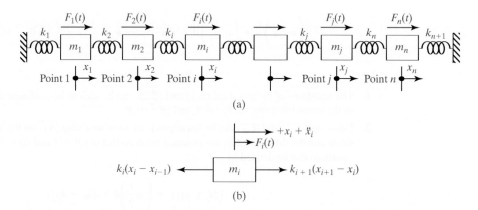

FIGURE 6.5   Multidegree-of-freedom spring-mass system.

Equation (6.12) can be stated in matrix form as

$$\vec{F} = [k]\vec{x} \tag{6.13}$$

where $\vec{x}$ and $\vec{F}$ are the displacement and force vectors defined in Eq. (6.7) and $[k]$ is the stiffness matrix given by

$$[k] = \begin{bmatrix} k_{11} & k_{12} & \cdots & k_{1n} \\ k_{21} & k_{22} & \cdots & k_{2n} \\ \vdots & & & \\ k_{n1} & k_{n2} & \cdots & k_{nn} \end{bmatrix} \tag{6.14}$$

The following aspects of stiffness influence coefficients are to be noted:

1. Since the force required at point i to cause a unit deflection at point j and zero deflection at all other points is the same as the force required at point j to cause a unit deflection at point i and zero deflection at all other points (Maxwell's reciprocity theorem [6.1]), we have $k_{ij} = k_{ji}$.

2. The stiffness influence coefficients can be calculated by applying the principles of statics and solid mechanics.

3. The stiffness influence coefficients for torsional systems can be defined in terms of unit angular displacement and the torque that causes the angular displacement. For example, in a multirotor torsional system, k_{ij} can be defined as the torque at point i (rotor i) due to a unit angular displacement at point j and zero angular displacement at all other points.

The stiffness influence coefficients of a multidegree-of-freedom system can be determined as follows:

1. Assume a value of one for the displacement x_j ($j = 1$ to start with) and a value of zero for all other displacements $x_1, x_2, \ldots, x_{j-1}, x_{j+1}, \ldots, x_n$. By definition, the set of forces k_{ij} ($i = 1, 2, \ldots, n$) will maintain the system in the assumed configuration ($x_j = 1, x_1 = x_2 = \cdots = x_{j-1} = x_{j+1} = \cdots = x_n = 0$). Then the static equilibrium equations are written for each mass and the resulting set of n equations solved to find the n influence coefficients k_{ij} ($i = 1, 2, \ldots, n$).

2. After completing step 1 for $j = 1$, the procedure is repeated for $j = 2, 3, \ldots, n$.

The following examples illustrate the procedure.

EXAMPLE 6.3

Stiffness Influence Coefficients

Find the stiffness influence coefficients of the system shown in Fig. 6.6(a).

Solution:

Approach: Use the definition of k_{ij} and static equilibrium equations.

Let $x_1, x_2,$ and x_3 denote the displacements of the masses $m_1, m_2,$ and m_3, respectively. The stiffness influence coefficients k_{ij} of the system can be determined in terms of the spring stiffnesses

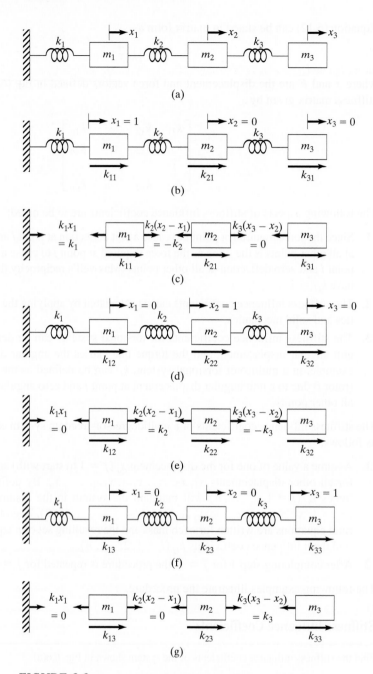

FIGURE 6.6 Determination of stiffness influence coefficients.

k_1, k_2, and k_3 as follows. First, we set the displacement of m_1 equal to one ($x_1 = 1$) and the displacements of m_2 and m_3 equal to zero ($x_2 = x_3 = 0$), as shown in Fig. 6.6(b). The set of forces k_{i1} ($i = 1, 2, 3$) is assumed to maintain the system in this configuration. The free-body diagrams of the masses corresponding to the configuration of Fig. 6.6(b) are indicated in Fig. 6.6(c). The equilibrium of forces for the masses m_1, m_2, and m_3 in the horizontal direction yields

$$\text{Mass } m_1: k_1 = -k_2 + k_{11} \tag{E.1}$$

$$\text{Mass } m_2: k_{21} = -k_2 \tag{E.2}$$

$$\text{Mass } m_3: k_{31} = 0 \tag{E.3}$$

The solution of Eqs. (E.1) to (E.3) gives

$$k_{11} = k_1 + k_2, \qquad k_{21} = -k_2, \qquad k_{31} = 0 \tag{E.4}$$

Next, the displacements of the masses are assumed as $x_1 = 0$, $x_2 = 1$, and $x_3 = 0$, as shown in Fig. 6.6(d). Since the forces k_{i2} ($i = 1, 2, 3$) are assumed to maintain the system in this configuration, the free-body diagrams of the masses can be developed as indicated in Fig. 6.6(e). The force equilibrium equations of the masses are:

$$\text{Mass } m_1: k_{12} + k_2 = 0 \tag{E.5}$$

$$\text{Mass } m_2: k_{22} - k_3 = k_2 \tag{E.6}$$

$$\text{Mass } m_3: k_{32} = -k_3 \tag{E.7}$$

The solution of Eqs. (E.5) to (E.7) yields

$$k_{12} = -k_2, \qquad k_{22} = k_2 + k_3, \qquad k_{32} = -k_3 \tag{E.8}$$

Finally, the set of forces k_{i3} ($i = 1, 2, 3$) is assumed to maintain the system with $x_1 = 0$, $x_2 = 0$, and $x_3 = 1$ (Fig. 6.6(f)). The free-body diagrams of the various masses in this configuration are shown in Fig. 6.6(g), and the force equilibrium equations lead to

$$\text{Mass } m_1: k_{13} = 0 \tag{E.9}$$

$$\text{Mass } m_2: k_{23} + k_3 = 0 \tag{E.10}$$

$$\text{Mass } m_3: k_{33} = k_3 \tag{E.11}$$

The solution of Eqs. (E.9) to (E.11) yields

$$k_{13} = 0, \qquad k_{23} = -k_3, \qquad k_{33} = k_3 \tag{E.12}$$

Thus the stiffness matrix of the system is given by

$$[k] = \begin{bmatrix} (k_1 + k_2) & -k_2 & 0 \\ -k_2 & (k_2 + k_3) & -k_3 \\ 0 & -k_3 & k_3 \end{bmatrix} \tag{E.13}$$

∎

Stiffness Matrix of a Frame

EXAMPLE 6.4

Determine the stiffness matrix of the frame shown in Fig. 6.7(a). Neglect the effect of axial stiffness of the members AB and BC.

Solution: Since the segments AB and BC of the frame can be considered as beams, the beam force-deflection formulas can be used to generate the stiffness matrix of the frame. The forces necessary to cause a displacement along one coordinate while maintaining zero displacements along other coordinates of a beam are indicated in Fig. 6.7(b) [6.1, 6.8]. In Fig. 6.7(a), the ends A and C are fixed and hence the joint B will have three possible displacements—x, y, and θ, as indicated. The forces necessary to maintain a unit displacement along the x direction and zero displacement along y and θ directions at the joint B are given by (from Fig. 6.7(b))

$$F_x = \left(\frac{12EI}{l^3}\right)_{BC} = \frac{3EI}{2\,l^3}, \qquad F_y = 0, \qquad M_\theta = \left(\frac{6EI}{l^2}\right)_{BC} = \frac{3EI}{2\,l^2}$$

Similarly, when a unit displacement is given along the y direction at joint B with zero displacements along x and θ directions, the forces required to maintain the configuration can be found from Fig. 6.7(b) as

$$F_x = 0, \qquad F_y = \left(\frac{12EI}{l^3}\right)_{BA} = \frac{24EI}{l^3}, \qquad M_\theta = -\left(\frac{6EI}{l^2}\right)_{BA} = -\frac{12EI}{l^2}$$

Finally, the forces necessary to maintain a unit displacement along the θ direction and zero displacements along x and y directions at joint B can be seen, from Fig. 6.7(b), as

$$F_x = \left(\frac{6EI}{l^2}\right)_{BC} = \frac{3EI}{2\,l^2}, \qquad F_y = -\left(\frac{6EI}{l^2}\right)_{BA} = -\frac{12EI}{l^3}$$

$$M_\theta = \left(\frac{4EI}{l}\right)_{BC} + \left(\frac{4EI}{l}\right)_{BA} = \frac{2EI}{l} + \frac{8EI}{l} = \frac{10EI}{l}$$

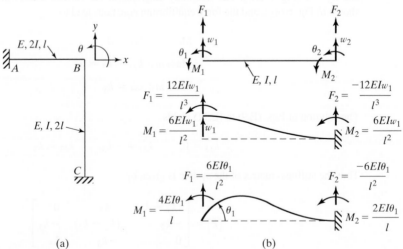

(a)   (b)

FIGURE 6.7   Stiffness matrix of a frame.

Thus the stiffness matrix, $[k]$, is given by

$$\vec{F} = [k]\vec{x}$$

where

$$\vec{F} = \begin{Bmatrix} F_x \\ F_y \\ M_\theta \end{Bmatrix}, \qquad \vec{x} = \begin{Bmatrix} x \\ y \\ \theta \end{Bmatrix}, \qquad [k] = \frac{EI}{l^3} \begin{bmatrix} \dfrac{3}{2} & 0 & \dfrac{3l}{2} \\ 0 & 24 & -12l \\ \dfrac{3l}{2} & -12l & 10l^2 \end{bmatrix}$$

$\blacksquare$

6.4.2 Flexibility Influence Coefficients

As seen in Examples 6.3 and 6.4, the computation of stiffness influence coefficients requires the application of the principles of statics and some algebraic manipulation. In fact, the generation of n stiffness influence coefficients $k_{1j}, k_{2j}, \ldots, k_{nj}$ for any specific j requires the solution of n simultaneous linear equations. Thus n sets of linear equations (n equations in each set) are to be solved to generate all the stiffness influence coefficients of an n-degree-of-freedom system. This implies a significant computational effort for large values of n. The generation of the flexibility influence coefficients, on the other hand, proves to be simpler and more convenient. To illustrate the concept, consider again the spring-mass system shown in Fig. 6.5.

Let the system be acted on by just one force F_j, and let the displacement at point i (i.e., mass m_i) due to F_j be x_{ij}. The flexibility influence coefficient, denoted by a_{ij}, is defined as the deflection at point i due to a unit load at point j. Since the deflection increases proportionately with the load for a linear system, we have

$$x_{ij} = a_{ij}F_j \tag{6.15}$$

If several forces F_j ($j = 1, 2, \ldots, n$) act at different points of the system, the total deflection at any point i can be found by summing up the contributions of all forces F_j:

$$x_i = \sum_{j=1}^{n} x_{ij} = \sum_{j=1}^{n} a_{ij}F_j, \quad i = 1, 2, \ldots, n \tag{6.16}$$

Equation (6.16) can be expressed in matrix form as

$$\vec{x} = [a]\vec{F} \tag{6.17}$$

where $\vec{x}$ and $\vec{F}$ are the displacement and force vectors defined in Eq. (6.7) and $[a]$ is the flexibility matrix given by

$$[a] = \begin{bmatrix} a_{11} & a_{12} & \cdots & a_{1n} \\ a_{21} & a_{22} & \cdots & a_{2n} \\ \vdots & & & \\ a_{n1} & a_{n2} & \cdots & a_{nn} \end{bmatrix} \tag{6.18}$$

The following characteristics of flexibility influence coefficients can be noted:

1. An examination of Eqs. (6.13) and (6.17) indicates that the flexibility and stiffness matrices are related. If we substitute Eq. (6.13) into Eq. (6.17), we obtain

$$\vec{x} = [a]\vec{F} = [a][k]\vec{x} \tag{6.19}$$

from which we can obtain the relation

$$[a][k] = [I] \tag{6.20}$$

where $[I]$ denotes the unit matrix. Equation (6.20) is equivalent to

$$[k] = [a]^{-1}, \qquad [a] = [k]^{-1} \tag{6.21}$$

That is, the stiffness and flexibility matrices are the inverse of one another. The use of dynamic stiffness influence coefficients in the vibration of nonuniform beams is discussed in [6.10].

2. Since the deflection at point i due to a unit load at point j is the same as the deflection at point j due to a unit load at point i for a linear system (Maxwell's reciprocity theorem [6.1]), we have $a_{ij} = a_{ji}$.

3. The flexibility influence coefficients of a torsional system can be defined in terms of unit torque and the angular deflection it causes. For example, in a multirotor torsional system, a_{ij} can be defined as the angular deflection of point i (rotor i) due to a unit torque at point j (rotor j).

The flexibility influence coefficients of a multidegree-of-freedom system can be determined as follows:

1. Assume a unit load at point j ($j = 1$ to start with). By definition, the displacements of the various points i ($i = 1, 2, \ldots, n$) resulting from this load give the flexibility influence coefficients, $a_{ij}, i = 1, 2, \ldots, n$. Thus a_{ij} can be found by applying the simple principles of statics and solid mechanics.

2. After completing Step 1 for $j = 1$, the procedure is repeated for $j = 2, 3, \ldots, n$.

3. Instead of applying Steps 1 and 2, the flexibility matrix, $[a]$, can be determined by finding the inverse of the stiffness matrix, $[k]$, if the stiffness matrix is available.

The following examples illustrate the procedure.

EXAMPLE 6.5

Flexibility Influence Coefficients

Find the flexibility influence coefficients of the system shown in Fig. 6.8(a).

Solution: Let x_1, x_2, and x_3 denote the displacements of the masses m_1, m_2, and m_3, respectively. The flexibility influence coefficients a_{ij} of the system can be determined in terms of the spring stiffnesses k_1, k_2, and k_3 as follows. Apply a unit force at mass m_1 and no force at other masses ($F_1 = 1, F_2 = F_3 = 0$), as shown in Fig. 6.8(b). The resulting deflections of the masses m_1, m_2, and

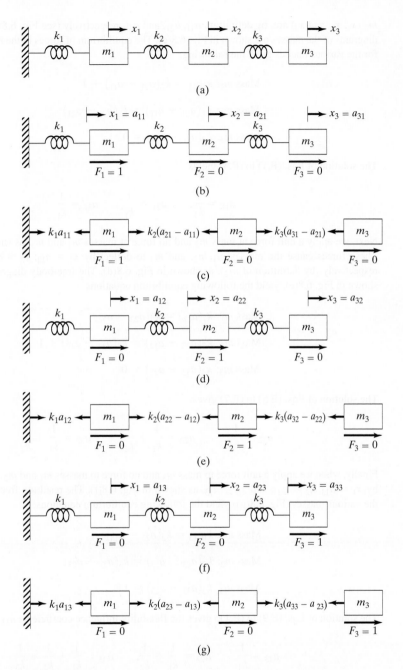

FIGURE 6.8   Determination of flexibility influence coefficients.

m_3 (x_1, x_2, and x_3) are, by definition, a_{11}, a_{21}, and a_{31}, respectively (see Fig. 6.8(b)). The free-body diagrams of the masses are shown in Fig. 6.8(c). The equilibrium of forces in the horizontal direction for the various masses gives the following:

$$\text{Mass } m_1: k_1 a_{11} = k_2(a_{21} - a_{11}) + 1 \tag{E.1}$$

$$\text{Mass } m_2: k_2(a_{21} - a_{11}) = k_3(a_{31} - a_{21}) \tag{E.2}$$

$$\text{Mass } m_3: k_3(a_{31} - a_{21}) = 0 \tag{E.3}$$

The solution of Eqs. (E.1) to (E.3) yields

$$a_{11} = \frac{1}{k_1}, \qquad a_{21} = \frac{1}{k_1}, \qquad a_{31} = \frac{1}{k_1} \tag{E.4}$$

Next, we apply a unit force at mass m_2 and no force at masses m_1 and m_3, as shown in Fig. 6.8(d). These forces cause the masses m_1, m_2, and m_3 to deflect by $x_1 = a_{12}$, $x_2 = a_{22}$, and $x_3 = a_{32}$, respectively (by definition of a_{i2}), as shown in Fig. 6.8(d). The free-body diagrams of the masses, shown in Fig. 6.8(e), yield the following equilibrium equations:

$$\text{Mass } m_1: k_1(a_{12}) = k_2(a_{22} - a_{12}) \tag{E.5}$$

$$\text{Mass } m_2: k_2(a_{22} - a_{12}) = k_3(a_{32} - a_{22}) + 1 \tag{E.6}$$

$$\text{Mass } m_3: k_3(a_{32} - a_{22}) = 0 \tag{E.7}$$

The solution of Eqs. (E.5) to (E.7) gives

$$a_{12} = \frac{1}{k_1}, \qquad a_{22} = \frac{1}{k_1} + \frac{1}{k_2}, \qquad a_{32} = \frac{1}{k_1} + \frac{1}{k_2} \tag{E.8}$$

Finally, when we apply a unit force to mass m_3 and no force to masses m_1 and m_2, the masses deflect by $x_1 = a_{13}$, $x_2 = a_{23}$, and $x_3 = a_{33}$, as shown in Fig. 6.8(f). The resulting free-body diagrams of the various masses (Fig. 6.8(g)) yield the following equilibrium equations:

$$\text{Mass } m_1: k_1 a_{13} = k_2(a_{23} - a_{13}) \tag{E.9}$$

$$\text{Mass } m_2: k_2(a_{23} - a_{13}) = k_3(a_{33} - a_{23}) \tag{E.10}$$

$$\text{Mass } m_3: k_3(a_{33} - a_{23}) = 1 \tag{E.11}$$

The solution of Eqs. (E.9) to (E.11) gives the flexibility influence coefficients a_{i3} as

$$a_{13} = \frac{1}{k_1}, \qquad a_{23} = \frac{1}{k_1} + \frac{1}{k_2}, \qquad a_{33} = \frac{1}{k_1} + \frac{1}{k_2} + \frac{1}{k_3} \tag{E.12}$$

It can be verified that the stiffness matrix of the system, given by Eq. (E.13) of Example 6.3, can also be found from the relation $[k] = [a]^{-1}$.

∎

EXAMPLE 6.6

Flexibility Matrix of a Beam

Derive the flexibility matrix of the weightless beam shown in Fig. 6.9(a). The beam is simply supported at both ends, and the three masses are placed at equal intervals. Assume the beam to be uniform with stiffness EI.

Solution: Let x_1, x_2, and x_3 denote the total transverse deflection of the masses m_1, m_2, and m_3, respectively. From the known formula for the deflection of a pinned-pinned beam [6.2], the influence coefficients $a_{1j}(j = 1, 2, 3)$ can be found by applying a unit load at the location of m_1 and zero load at the locations of m_2 and m_3 (see Fig. 6.9(b)):

$$a_{11} = \frac{9}{768}\frac{l^3}{EI}, \qquad a_{12} = \frac{11}{768}\frac{l^3}{EI}, \qquad a_{13} = \frac{7}{768}\frac{l^3}{EI} \tag{E.1}$$

Similarly, by applying a unit load at the locations of m_2 and m_3 separately (with zero load at other locations), we obtain

$$a_{21} = a_{12} = \frac{11}{768}\frac{l^3}{EI}, \qquad a_{22} = \frac{1}{48}\frac{l^3}{EI}, \qquad a_{23} = \frac{11}{768}\frac{l^3}{EI} \tag{E.2}$$

and

$$a_{31} = a_{13} = \frac{7}{768}\frac{l^3}{EI}, \qquad a_{32} = a_{23} = \frac{11}{768}\frac{l^3}{EI}, \qquad a_{33} = \frac{9}{768}\frac{l^3}{EI} \tag{E.3}$$

Thus the flexibility matrix of the system is given by

$$[a] = \frac{l^3}{768EI}\begin{bmatrix} 9 & 11 & 7 \\ 11 & 16 & 11 \\ 7 & 11 & 9 \end{bmatrix} \tag{E.4}$$

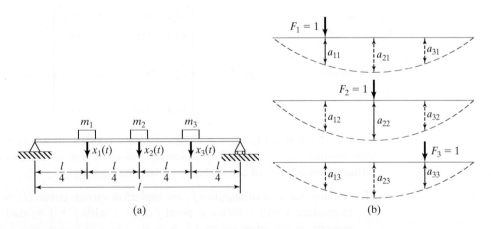

FIGURE 6.9   Beam deflections.

**6.4.3
Inertia Influence
Coefficients**

The elements of the mass matrix, m_{ij}, are known as the *inertia influence coefficients*. Although it is more convenient to derive the inertia influence coefficients from the expression for kinetic energy of the system (see Section 6.5), the coefficients m_{ij} can be computed using the impulse-momentum relations. The inertia influence coefficients $m_{1j}, m_{2j}, \ldots, m_{nj}$ are defined as the set of impulses applied at points $1, 2, \ldots, n$, respectively, to produce a unit velocity at point j and zero velocity at every other point (that is, $\dot{x}_j = 1, \dot{x}_1 = \dot{x}_2 = \cdots = \dot{x}_{j-1} = \dot{x}_{j+1} = \cdots = \dot{x}_n = 0$). Thus, for a multidegree-of-freedom system, the total impulse at point i, F_i, can be found by summing up the impulses causing the velocities $\dot{x}_j (j = 1, 2, \ldots, n)$ as

$$F_i = \sum_{j=1}^{n} m_{ij}\,\dot{x}_j \tag{6.22}$$

Equation (6.22) can be stated in matrix form as

$$\vec{F} = [m]\vec{x} \tag{6.23}$$

where $\vec{x}$ and $\vec{F}$ are the velocity and impulse vectors given by

$$\vec{x} = \begin{Bmatrix} \dot{x}_1 \\ \dot{x}_2 \\ \cdot \\ \cdot \\ \cdot \\ \dot{x}_n \end{Bmatrix}, \qquad \vec{F} = \begin{Bmatrix} F_1 \\ F_2 \\ \cdot \\ \cdot \\ \cdot \\ F_n \end{Bmatrix} \tag{6.24}$$

and $[m]$ is the mass matrix given by

$$[m] = \begin{bmatrix} m_{11} & m_{12} & \cdots & m_{1n} \\ m_{21} & m_{22} & \cdots & m_{2n} \\ \cdot \\ \cdot \\ \cdot \\ m_{n1} & m_{n2} & \cdots & m_{nn} \end{bmatrix} \tag{6.25}$$

It can be verified easily that the inertia influence coefficients are symmetric for a linear system—that is, $m_{ij} = m_{ji}$. The following procedure can be used to derive the inertia influence coefficients of a multidegree-of-freedom system.

1. Assume that a set of impulses f_{ij} are applied at various points i $(i = 1, 2, \ldots, n)$ so as to produce a unit velocity at point j ($\dot{x}_j = 1$ with $j = 1$ to start with) and a zero velocity at all other points ($\dot{x}_1 = \dot{x}_2 = \cdots \dot{x}_{j-1} = \dot{x}_{j+1} = \cdots = \dot{x}_n = 0$). By definition, the set of impulses f_{ij} ($i = 1, 2, \ldots, n$) denote the inertia influence coefficients m_{ij} ($i = 1, 2, \ldots, n$).

2. After completing step 1 for $j = 1$, the procedure is repeated for $j = 2, 3, \ldots, n$.

Note that if x_j denotes an angular coordinate, then $\dot{x}_j$ represents an angular velocity and F_j indicates an angular impulse. The following example illustrates the procedure of generating m_{ij}.

EXAMPLE 6.7

Inertia Influence Coefficients

Find the inertia influence coefficients of the system shown in Fig. 6.4(a).

Solution:

Approach: Use the definition of m_{ij} along with impulse-momentum relations.

Let $x(t)$ and $\theta(t)$ denote the coordinates to define the linear and angular positions of the trailer (M) and the compound pendulum (m). To derive the inertia influence coefficients, impulses of magnitudes m_{11} and m_{21} are applied along the directions $x(t)$ and $\theta(t)$ to result in the velocities $\dot{x} = 1$ and $\dot{\theta} = 0$. Then the linear impulse-linear momentum equation gives

$$m_{11} = (M + m)(1) \tag{E.1}$$

and the angular impulse-angular momentum equation (about O) yields

$$m_{21} = m(1)\frac{l}{2} \tag{E.2}$$

Next, impulses of magnitudes m_{12} and m_{22} are applied along the directions $x(t)$ and $\theta(t)$ to obtain the velocities $\dot{x} = 0$ and $\dot{\theta} = 1$. Then the linear impulse-linear momentum relation provides

$$m_{12} = m(1)\left(\frac{l}{2}\right) \tag{E.3}$$

and the angular impulse-angular momentum equation (about O) gives

$$m_{22} = \left(\frac{ml^2}{3}\right)(1) \tag{E.4}$$

Thus the mass or inertia matrix of the system is given by

$$[m] = \begin{bmatrix} M + m & \dfrac{ml}{2} \\[2mm] \dfrac{ml}{2} & \dfrac{ml^2}{3} \end{bmatrix} \tag{E.5}$$

■

6.5   Potential and Kinetic Energy Expressions in Matrix Form

Let x_i denote the displacement of mass m_i and F_i the force applied in the direction of x_i at mass m_i in an n-degree-of-freedom system similar to the one shown in Fig. 6.5.

The elastic potential energy (also known as *strain energy* or *energy of deformation*) of the ith spring is given by

$$V_i = \frac{1}{2} F_i x_i \tag{6.26}$$

The total potential energy can be expressed as

$$V = \sum_{i=1}^{n} V_i = \frac{1}{2} \sum_{i=1}^{n} F_i x_i \tag{6.27}$$

Since

$$F_i = \sum_{j=1}^{n} k_{ij} x_j \tag{6.28}$$

Eq. (6.27) becomes

$$V = \frac{1}{2} \sum_{i=1}^{n} \left(\sum_{j=1}^{n} k_{ij} x_j \right) x_i = \frac{1}{2} \sum_{i=1}^{n} \sum_{j=1}^{n} k_{ij} x_i x_j \tag{6.29}$$

Equation (6.29) can also be written in matrix form as[1]

$$V = \frac{1}{2} \vec{x}^T [k] \vec{x} \tag{6.30}$$

where the displacement vector is given by Eq. (6.7) and the stiffness matrix is given by

$$[k] = \begin{bmatrix} k_{11} & k_{12} & \cdots & k_{1n} \\ k_{21} & k_{22} & \cdots & k_{2n} \\ \vdots & & & \\ k_{n1} & k_{n2} & \cdots & k_{nn} \end{bmatrix} \tag{6.31}$$

The kinetic energy associated with mass m_i is, by definition, equal to

$$T_i = \frac{1}{2} m_i \dot{x}_i^2 \tag{6.32}$$

---

[1]Since the indices i and j can be interchanged in Eq. (6.29), we have the relation $k_{ij} = k_{ji}$.

The total kinetic energy of the system can be expressed as

$$T = \sum_{i=1}^{n} T_i = \frac{1}{2} \sum_{i=1}^{n} m_i \dot{x}_i^2 \tag{6.33}$$

which can be written in matrix form as

$$T = \frac{1}{2} \vec{x}^T [m] \vec{x} \tag{6.34}$$

where the velocity vector $\vec{x}$ is given by

$$\vec{x} = \begin{Bmatrix} \dot{x}_1 \\ \dot{x}_2 \\ \vdots \\ \dot{x}_n \end{Bmatrix}$$

and the mass matrix $[m]$ is a diagonal matrix given by

$$[m] = \begin{bmatrix} m_1 & & & 0 \\ & m_2 & & \\ & & \ddots & \\ 0 & & & m_n \end{bmatrix} \tag{6.35}$$

If generalized coordinates (q_i), discussed in Section 6.6, are used instead of the physical displacements (x_i), the kinetic energy can be expressed as

$$T = \frac{1}{2} \vec{q}^T [m] \vec{q} \tag{6.36}$$

where $\vec{q}$ is the vector of generalized velocities, given by

$$\vec{q} = \begin{Bmatrix} \dot{q}_1 \\ \dot{q}_2 \\ \vdots \\ \dot{q}_n \end{Bmatrix} \tag{6.37}$$

and $[m]$ is called the *generalized mass matrix*, given by

$$[m] = \begin{bmatrix} m_{11} & m_{12} & \cdots & m_{1n} \\ m_{21} & m_{22} & \cdots & m_{2n} \\ \vdots & & & \\ m_{n1} & m_{n2} & \cdots & m_{nn} \end{bmatrix} \tag{6.38}$$

with $m_{ij} = m_{ji}$. The generalized mass matrix given by Eq. (6.38) is full, as opposed to the diagonal mass matrix of Eq. (6.35).

It can be seen that the potential energy is a quadratic function of the displacements, and the kinetic energy is a quadratic function of the velocities. Hence they are said to be in quadratic form. Since kinetic energy, by definition, cannot be negative and vanishes only when all the velocities vanish, Eqs. (6.34) and (6.36) are called *positive definite quadratic forms* and the mass matrix [m] is called a *positive definite matrix*. On the other hand, the potential energy expression, Eq. (6.30), is a positive definite quadratic form, but the matrix [k] is positive definite only if the system is a stable one. There are systems for which the potential energy is zero without the displacements or coordinates $x_1, x_2, \ldots, x_n$ being zero. In these cases, the potential energy will be a positive quadratic function rather than positive definite; correspondingly, the matrix [k] is said to be positive. A system for which [k] is positive and [m] is positive definite is called a *semidefinite system* (see Section 6.12).

6.6   Generalized Coordinates and Generalized Forces

The equations of motion of a vibrating system can be formulated in a number of different coordinate systems. As stated earlier, *n* independent coordinates are necessary to describe the motion of a system having *n* degrees of freedom. Any set of *n* independent coordinates is called *generalized coordinates*, usually designated by $q_1, q_2, \ldots, q_n$. The generalized coordinates may be lengths, angles, or any other set of numbers that define the configuration of the system at any time uniquely. They are also independent of the conditions of constraint.

To illustrate the concept of generalized coordinates, consider the triple pendulum shown in Fig. 6.10. The configuration of the system can be specified by the six coordinates

FIGURE 6.10   Triple pendulum.

(x_j, y_j), $j = 1, 2, 3$. However, these coordinates are not independent but are constrained by the relations

$$x_1^2 + y_1^2 = l_1^2$$
$$(x_2 - x_1)^2 + (y_2 - y_1)^2 = l_2^2$$
$$(x_3 - x_2)^2 + (y_3 - y_2)^2 = l_3^2 \tag{6.39}$$

Since the coordinates (x_j, y_j), $j = 1, 2, 3$ are not independent, they cannot be called generalized coordinates. Without the constraints of Eq. (6.39), each of the masses m_1, m_2, and m_3 will be free to occupy any position in the x-, y- plane. The constraints eliminate three degrees of freedom from the six coordinates (two for each mass), and the system thus has only three degrees of freedom. If the angular displacements $\theta_j (j = 1, 2, 3)$ are used to specify the locations of the masses $m_j (j = 1, 2, 3)$ at any time, there will be no constraints on θ_j. Thus they form a set of generalized coordinates and are denoted as $q_j = \theta_j$, $j = 1, 2, 3$.

When external forces act on the system, its configuration changes. The new configuration of the system can be obtained by changing the generalized coordinates q_j by δq_j, $j = 1, 2, \ldots, n$, where n denotes the number of generalized coordinates (or degrees of freedom) of the system. If U_j denotes the work done in changing the generalized coordinate q_j by the amount δq_j, the corresponding generalized force Q_j can be defined as

$$Q_j = \frac{U_j}{\delta q_j}, \qquad j = 1, 2, \ldots, n \tag{6.40}$$

where Q_j will be a force (moment) when q_j is a linear (angular) displacement.

6.7   Using Lagrange's Equations to Derive Equations of Motion

The equations of motion of a vibrating system can often be derived in a simple manner in terms of generalized coordinates by the use of Lagrange's equations [6.3]. Lagrange's equations can be stated, for an n-degree-of-freedom system, as

$$\frac{d}{dt}\left(\frac{\partial T}{\partial \dot{q}_j}\right) - \frac{\partial T}{\partial q_j} + \frac{\partial V}{\partial q_j} = Q_j^{(n)}, \qquad j = 1, 2, \ldots, n \tag{6.41}$$

where $\dot{q}_j = \partial q_j/\partial t$ is the generalized velocity and $Q_j^{(n)}$ is the nonconservative generalized force corresponding to the generalized coordinate q_j. The forces represented by $Q_j^{(n)}$ may be dissipative (damping) forces or other external forces that are not derivable from a potential function. For example, if F_{xk}, F_{yk}, and F_{zk} represent the external forces acting on the kth mass of the system in the x, y, and z directions, respectively, then the generalized force $Q_j^{(n)}$ can be computed as follows:

$$Q_j^{(n)} = \sum_k \left(F_{xk}\frac{\partial x_k}{\partial q_j} + F_{yk}\frac{\partial y_k}{\partial q_j} + F_{zk}\frac{\partial z_k}{\partial q_j} \right) \tag{6.42}$$

where x_k, y_k, and z_k are the displacements of the kth mass in the x, y, and z directions, respectively. Note that for a torsional system, the force F_{xk}, for example, is to be replaced by the moment acting about the x-axis (M_{xk}), and the displacement x_k by the angular displacement about the x-axis (θ_{xk}) in Eq. (6.42). For a conservative system, $Q_j^{(n)} = 0$, so Eq. (6.41) takes the form

$$\frac{d}{dt}\left(\frac{\partial T}{\partial \dot{q}_j}\right) - \frac{\partial T}{\partial q_j} + \frac{\partial V}{\partial q_j} = 0, \qquad j = 1, 2, \ldots, n \tag{6.43}$$

Equation (6.41) or (6.43) represents a system of n differential equations, one corresponding to each of the n generalized coordinates. Thus the equations of motion of the vibrating system can be derived, provided the energy expressions are available.

EXAMPLE 6.8

Equations of Motion of a Torsional System

The arrangement of the compressor, turbine, and generator in a thermal power plant is shown in Fig. 6.11. This arrangement can be considered as a torsional system where J_i denotes the mass moments of inertia of the three components (compressor, turbine, and generator), M_{ti} indicates the external moments acting on the components, and k_{ti} represents the torsional spring constants of the shaft between the components, as indicated in Fig. 6.11. Derive the equations of motion of the system using Lagrange's equations by treating the angular displacements of the components θ_i as generalized coordinates.

Solution: Here $q_1 = \theta_1$, $q_2 = \theta_2$, and $q_3 = \theta_3$, and the kinetic energy of the system is given by

$$T = \frac{1}{2}J_1\dot{\theta}_1^2 + \frac{1}{2}J_2\dot{\theta}_2^2 + \frac{1}{2}J_3\dot{\theta}_3^2 \tag{E.1}$$

For the shaft, the potential energy is equal to the work done by the shaft as it returns from the dynamic configuration to the reference equilibrium position. Thus if θ denotes the angular displacement, for a shaft having a torsional spring constant k_t, the potential energy is equal to the work done in causing an angular displacement θ of the shaft:

$$V = \int_0^\theta (k_t\theta)\, d\theta = \frac{1}{2}k_t\theta^2 \tag{E.2}$$

FIGURE 6.11 Torsional system.

Thus the total potential energy of the system can be expressed as

$$V = \frac{1}{2}k_{t1}\theta_1^2 + \frac{1}{2}k_{t2}(\theta_2 - \theta_1)^2 + \frac{1}{2}k_{t3}(\theta_3 - \theta_2)^2 \qquad \text{(E.3)}$$

There are external moments applied to the components, so Eq. (6.42) gives

$$Q_j^{(n)} = \sum_{k=1}^{3} M_{tk}\frac{\partial\theta_k}{\partial q_j} = \sum_{k=1}^{3} M_{tk}\frac{\partial\theta_k}{\partial\theta_j} \qquad \text{(E.4)}$$

from which we can obtain

$$Q_1^{(n)} = M_{t1}\frac{\partial\theta_1}{\partial\theta_1} + M_{t2}\frac{\partial\theta_2}{\partial\theta_1} + M_{t3}\frac{\partial\theta_3}{\partial\theta_1} = M_{t1}$$

$$Q_2^{(n)} = M_{t1}\frac{\partial\theta_1}{\partial\theta_2} + M_{t2}\frac{\partial\theta_2}{\partial\theta_2} + M_{t3}\frac{\partial\theta_3}{\partial\theta_2} = M_{t2}$$

$$Q_3^{(n)} = M_{t1}\frac{\partial\theta_1}{\partial\theta_3} + M_{t2}\frac{\partial\theta_2}{\partial\theta_3} + M_{t3}\frac{\partial\theta_3}{\partial\theta_3} = M_{t3} \qquad \text{(E.5)}$$

Substituting Eqs. (E.1), (E.3), and (E.5) in Lagrange's equations, Eq. (6.41), we obtain for $j = 1, 2, 3$ the equations of motion

$$J_1\ddot{\theta}_1 + (k_{t1} + k_{t2})\theta_1 - k_{t2}\theta_2 = M_{t1}$$

$$J_2\ddot{\theta}_2 + (k_{t2} + k_{t3})\theta_2 - k_{t2}\theta_1 - k_{t3}\theta_3 = M_{t2}$$

$$J_3\ddot{\theta}_3 + k_{t3}\theta_3 - k_{t3}\theta_2 = M_{t3} \qquad \text{(E.6)}$$

which can be expressed in matrix form as

$$\begin{bmatrix} J_1 & 0 & 0 \\ 0 & J_2 & 0 \\ 0 & 0 & J_3 \end{bmatrix}\begin{Bmatrix} \ddot{\theta}_1 \\ \ddot{\theta}_2 \\ \ddot{\theta}_3 \end{Bmatrix} + \begin{bmatrix} (k_{t1} + k_{t2}) & -k_{t2} & 0 \\ -k_{t2} & (k_{t2} + k_{t3}) & -k_{t3} \\ 0 & -k_{t3} & k_{t3} \end{bmatrix}\begin{Bmatrix} \theta_1 \\ \theta_2 \\ \theta_3 \end{Bmatrix} = \begin{Bmatrix} M_{t1} \\ M_{t2} \\ M_{t3} \end{Bmatrix} \qquad \text{(E.7)}$$

■

EXAMPLE 6.9

Lagrange's Equations

Derive the equations of motion of the trailer-compound pendulum system shown in Fig. 6.4(a).

Solution: The coordinates $x(t)$ and $\theta(t)$ can be used as generalized coordinates to describe, respectively, the linear displacement of the trailer and the angular displacement of the compound pendulum. If a y-coordinate is introduced, for convenience, as shown in Fig. 6.4(a), the displacement components of point C can be expressed as

$$x_C = x + \frac{l}{2}\sin\theta \qquad \text{(E.1)}$$

$$y_C = \frac{l}{2}\cos\theta \qquad \text{(E.2)}$$

Differentiation of Eqs. (E.1) and (E.2) with respect to time gives the velocities of point C as

$$\dot{x}_C = \dot{x} + \frac{l}{2}\dot{\theta}\cos\theta \tag{E.3}$$

$$\dot{y}_C = -\frac{l}{2}\dot{\theta}\sin\theta \tag{E.4}$$

The kinetic energy of the system, T, can be expressed as

$$T = \frac{1}{2}M\dot{x}^2 + \frac{1}{2}m(\dot{x}_C^2 + \dot{y}_C^2) + \frac{1}{2}J_C\dot{\theta}^2 \tag{E.5}$$

where $J_C = \frac{1}{12}ml^2$. Using Eqs. (E.3) and (E.4), Eq. (E.5) can be rewritten as

$$T = \frac{1}{2}M\dot{x}^2 + \frac{1}{2}m\left(\dot{x}^2 + \frac{l^2\dot{\theta}^2}{4} + \dot{x}\dot{\theta}l\cos\theta\right) + \frac{1}{2}\left(\frac{ml^2}{12}\right)\dot{\theta}^2$$

$$= \frac{1}{2}(M + m)\dot{x}^2 + \frac{1}{2}\left(\frac{ml^2}{3}\right)\dot{\theta}^2 + \frac{1}{2}(ml\cos\theta)\dot{x}\dot{\theta} \tag{E.6}$$

The potential energy of the system, V, due to the strain energy of the springs and the gravitational potential, can be expressed as

$$V = \frac{1}{2}k_1x^2 + \frac{1}{2}k_2x^2 + mg\frac{l}{2}(1 - \cos\theta) \tag{E.7}$$

where the lowest position of point C is taken as the datum. Since there are nonconservative forces acting on the system, the generalized forces corresponding to $x(t)$ and $\theta(t)$ are to be computed. The force, $X(t)$, acting in the direction of $x(t)$ can be found from Eq. (6.42) as

$$X(t) = Q_1^{(n)} = F(t) - c_1\dot{x}(t) - c_2\dot{x}(t) \tag{E.8}$$

where the negative sign for the terms $c_1\dot{x}$ and $c_2\dot{x}$ indicates that the damping forces oppose the motion. Similarly, the force $\Theta(t)$ acting in the direction of $\theta(t)$ can be determined as

$$\Theta(t) = Q_2^{(n)} = M_t(t) \tag{E.9}$$

where $q_1 = x$ and $q_2 = \theta$. By differentiating the expressions of T and V as required by Eq. (6.41) and substituting the resulting expressions, along with Eqs. (E.8) and (E.9), we obtain the equations of motion of the system as

$$(M + m)\ddot{x} + \frac{1}{2}(ml\cos\theta)\ddot{\theta} - \frac{1}{2}ml\sin\theta\dot{\theta}^2 + k_1x + k_2x$$

$$= F(t) - c_1\dot{x} - c_2\dot{x} \tag{E.10}$$

$$\left(\frac{1}{3}ml^2\right)\ddot{\theta} + \frac{1}{2}(ml\cos\theta)\ddot{x} - \frac{1}{2}ml\sin\theta\dot{\theta}\dot{x} + \frac{1}{2}ml\sin\theta\dot{\theta}\dot{x}$$

$$+ \frac{1}{2}mgl\sin\theta = M_t(t) \tag{E.11}$$

Equations (E.10) and (E.11) can be seen to be identical to those obtained using Newton's second law of motion (Eqs. (E.1) and (E.2) in Example 6.2).

∎

6.8 Equations of Motion of Undamped Systems in Matrix Form

We can derive the equations of motion of a multidegree-of-freedom system in matrix form from Lagrange's equations[2]

$$\frac{d}{dt}\left(\frac{\partial T}{\partial \dot{x}_i}\right) - \frac{\partial T}{\partial x_i} + \frac{\partial V}{\partial x_i} = F_i, \quad i = 1, 2, \ldots n \tag{6.44}$$

where F_i is the nonconservative generalized force corresponding to the ith generalized coordinate x_i and $\dot{x}_i$ is the time derivative of x_i (generalized velocity). The kinetic and potential energies of a multidegree-of-freedom system can be expressed in matrix form as indicated in Section 6.5:

$$T = \frac{1}{2}\vec{x}^T[m]\vec{x} \tag{6.45}$$

$$V = \frac{1}{2}\vec{x}^T[k]\vec{x} \tag{6.46}$$

where $\vec{x}$ is the column vector of the generalized coordinates

$$\vec{x} = \begin{Bmatrix} x_1 \\ x_2 \\ \cdot \\ \cdot \\ \cdot \\ x_n \end{Bmatrix} \tag{6.47}$$

From the theory of matrices, we obtain, by taking note of the symmetry of $[m]$,

$$\frac{\partial T}{\partial \dot{x}_i} = \frac{1}{2}\vec{\delta}^T[m]\vec{x} + \frac{1}{2}\vec{x}^T[m]\vec{\delta} = \vec{\delta}^T[m]\vec{x}$$

$$= \vec{m}_i^T\vec{x}, \quad i = 1, 2, \ldots, n \tag{6.48}$$

where δ_{ji} is the Kronecker delta ($\delta_{ji} = 1$ if $j = i$ and $\delta_{ji} = 0$ if $j \neq i$), $\vec{\delta}$ is the column vector of Kronecker deltas whose elements in the rows for which $j \neq i$ are equal to zero and

---

[2]The generalized coordinates are denoted as x_i instead of q_i and the generalized forces as F_i instead of $Q_i^{(n)}$ in Eq. (6.44).

whose element in the row $i = j$ is equal to 1, and $\vec{m}_i^T$ is a row vector which is identical to the ith row of the matrix $[m]$. All the relations represented by Eq. (6.48) can be expressed as

$$\frac{\partial T}{\partial \dot{x}_i} = \vec{m}_i^T \, \vec{\dot{x}} \tag{6.49}$$

Differentiation of Eq. (6.49) with respect to time gives

$$\frac{d}{dt}\left(\frac{\partial T}{\partial \dot{x}_i}\right) = \vec{m}_i^T \, \vec{\ddot{x}}, \qquad i = 1, 2, \ldots, n \tag{6.50}$$

since the mass matrix is not a function of time. Further, the kinetic energy is a function of only the velocities $\dot{x}_i$, and so

$$\frac{\partial T}{\partial x_i} = 0, \qquad i = 1, 2, \ldots, n \tag{6.51}$$

Similarly, we can differentiate Eq. (6.46), taking note of the symmetry of $[k]$,

$$\frac{\partial V}{\partial x_i} = \frac{1}{2}\vec{\delta}^T[k]\vec{x} + \frac{1}{2}\vec{x}^T[k]\vec{\delta} = \vec{\delta}^T[k]\vec{x}$$

$$= \vec{k}_i^T \vec{x}, \qquad i = 1, 2, \ldots, n \tag{6.52}$$

where $\vec{k}_i^T$ is a row vector identical to the ith row of the matrix $[k]$. By substituting Eqs. (6.50) to (6.52) into Eq. (6.44), we obtain the desired equations of motion in matrix form

$$[m]\vec{\ddot{x}} + [k]\vec{x} = \vec{F} \tag{6.53}$$

where

$$\vec{F} = \begin{Bmatrix} F_1 \\ F_2 \\ \cdot \\ \cdot \\ \cdot \\ F_n \end{Bmatrix} \tag{6.54}$$

Note that if the system is conservative, there are no nonconservative forces F_i, so the equations of motion become

$$[m]\vec{\ddot{x}} + [k]\vec{x} = \vec{0} \tag{6.55}$$

Note also that if the generalized coordinates x_i are the same as the actual (physical) displacements, the mass matrix $[m]$ is a diagonal matrix.

6.9   Eigenvalue Problem

The solution of Eq. (6.55) corresponds to the undamped free vibration of the system. In this case, if the system is given some energy in the form of initial displacements or initial velocities or both, it vibrates indefinitely, because there is no dissipation of energy. We can find the solution of Eq. (6.55) by assuming a solution of the form

$$x_i(t) = X_i T(t), \qquad i = 1, 2, \ldots, n \tag{6.56}$$

where X_i is a constant and T is a function of time t. Equation (6.56) shows that the amplitude ratio of two coordinates

$$\left\{ \frac{x_i(t)}{x_j(t)} \right\}$$

is independent of time. Physically, this means that all coordinates have synchronous motions. The configuration of the system does not change its shape during motion, but its amplitude does. The configuration of the system, given by the vector

$$\vec{X} = \begin{Bmatrix} X_1 \\ X_2 \\ \cdot \\ \cdot \\ \cdot \\ X_n \end{Bmatrix}$$

is known as the *mode shape* of the system. Substituting Eq. (6.56) into Eq. (6.55), we obtain

$$[m]\vec{X}\ddot{T}(t) + [k]\vec{X}T(t) = \vec{0} \tag{6.57}$$

Equation (6.57) can be written in scalar form as n separate equations

$$\left(\sum_{j=1}^{n} m_{ij}X_j \right) \ddot{T}(t) + \left(\sum_{j=1}^{n} k_{ij}X_j \right) T(t) = 0, \qquad i = 1, 2, \ldots, n \tag{6.58}$$

from which we can obtain the relations

$$-\frac{\ddot{T}(t)}{T(t)} = \frac{\left(\displaystyle\sum_{j=1}^{n} k_{ij}X_j \right)}{\left(\displaystyle\sum_{j=1}^{n} m_{ij}X_j \right)}, \qquad i = 1, 2, \ldots, n \tag{6.59}$$

Since the left side of Eq. (6.59) is independent of the index i, and the right side is independent of t, both sides must be equal to a constant. By assuming this constant[3] as ω^2, we can write Eq. (6.59) as

$$\ddot{T}(t) + \omega^2 T(t) = 0 \tag{6.60}$$

$$\sum_{j=1}^{n} \left(k_{ij} - \omega^2 m_{ij} \right) X_j = 0, \qquad i = 1, 2, \ldots, n$$

or

$$\left[[k] - \omega^2 [m] \right] \vec{X} = \vec{0} \tag{6.61}$$

The solution of Eq. (6.60) can be expressed as

$$T(t) = C_1 \cos \left(\omega t + \phi \right) \tag{6.62}$$

where C_1 and ϕ are constants, known as the *amplitude* and the *phase angle*, respectively. Equation (6.62) shows that all the coordinates can perform a harmonic motion with the same frequency ω and the same phase angle ϕ. However, the frequency ω cannot take any arbitrary value; it has to satisfy Eq. (6.61). Since Eq. (6.61) represents a set of n linear homogeneous equations in the unknowns $X_i (i = 1, 2, \ldots, n)$, the trivial solution is $X_1 = X_2 = \cdots = X_n = 0$. For a nontrivial solution of Eq. (6.61), the determinant Δ of the coefficient matrix must be zero. That is,

$$\Delta = |k_{ij} - \omega^2 m_{ij}| = |[k] - \omega^2 [m]| = 0 \tag{6.63}$$

Equation (6.61) represents what is known as the *eigenvalue* or *characteristic value problem*, Eq. (6.63) is called the *characteristic equation*, ω^2 is known as the *eigenvalue* or the *characteristic value*, and ω is called the *natural frequency* of the system.

The expansion of Eq. (6.63) leads to an nth-order polynomial equation in ω^2. The solution (roots) of this polynomial or characteristic equation gives n values of ω^2. It can be shown that all the n roots are real and positive when the matrices $[k]$ and $[m]$ are symmetric and positive definite [6.4], as in the present case. If $\omega_1^2, \omega_2^2, \ldots, \omega_n^2$ denote the n roots in ascending order of magnitude, their positive square roots give the n natural frequencies of the system $\omega_1 \leq \omega_2 \leq \cdots \leq \omega_n$. The lowest value (ω_1) is called the *fundamental* or *first natural frequency*. In general, all the natural frequencies ω_i are distinct, although in some cases, two natural frequencies might possess the same value.

---

[3]The constant is assumed to be a positive number, ω^2, so as to obtain a harmonic solution to the resulting Eq. (6.60). Otherwise, the solution of $T(t)$ and hence that of $x(t)$ becomes exponential, which violates the physical limitations of finite total energy.

6.10   Solution of the Eigenvalue Problem

Several methods are available to solve an eigenvalue problem. We shall consider an elementary method in this section.

6.10.1 Solution of the Characteristic (Polynomial) Equation

Equation (6.61) can also be expressed as

$$\left[\lambda[k] - [m]\right]\vec{X} = \vec{0} \qquad (6.64)$$

where

$$\lambda = \frac{1}{\omega^2} \qquad (6.65)$$

By premultiplying Eq. (6.64) by $[k]^{-1}$, we obtain

$$\left[\lambda[I] - [D]\right]\vec{X} = \vec{0}$$

or

$$\lambda[I]\vec{X} = [D]\vec{X} \qquad (6.66)$$

where $[I]$ is the identity matrix and

$$[D] = [k]^{-1}[m] \qquad (6.67)$$

is called the *dynamical matrix*. The eigenvalue problem of Eq. (6.66) is known as the *standard eigenvalue problem*. For a nontrivial solution of $\vec{X}$, the characteristic determinant must be zero—that is,[4]

$$\Delta = \left|\lambda[I] - [D]\right| = 0 \qquad (6.68)$$

On expansion, Eq. (6.68) gives an nth-degree polynomial in λ, known as the *characteristic* or *frequency equation*. If the degree of freedom of the system (n) is large, the solution of this polynomial equation becomes quite tedious. We must use some numerical method, several of which are available to find the roots of a polynomial equation [6.5].

---

[4]By premultiplying Eq. (6.61) by $[m]^{-1}$, the characteristic determinantal equation can be expressed as $\Delta = \left|\omega^2[I] - [m]^{-1}[k]\right| = 0$.

EXAMPLE 6.10

Equations of Motion of a Three-Degree-of-Freedom System

State the free-vibration equations of motion of the system shown in Fig. 6.12.

Solution: By setting $n = 3$ along with $k_{n+1} = 0$ and $c_i = 0$ for $i = 1, 2, \ldots, n, n + 1$ in Fig. 6.3(a), the equations of motion of the system shown in Fig. 6.12, for forced vibration, can be obtained from Eq. (6.3) as

$$[m]\ddot{\overline{x}} + [k]\overline{x} = \overline{F} \tag{E.1}$$

where

$$[m] = \begin{bmatrix} m_1 & 0 & 0 \\ 0 & m_2 & 0 \\ 0 & 0 & m_3 \end{bmatrix}, \qquad [k] = \begin{bmatrix} k_1 + k_2 & -k_2 & 0 \\ -k_2 & k_2 + k_3 & -k_3 \\ 0 & -k_3 & k_3 \end{bmatrix}, \qquad \overline{F} = \begin{Bmatrix} F_1(t) \\ F_2(t) \\ F_3(t) \end{Bmatrix} \tag{E.2}$$

By setting $\vec{F} = \vec{0}$, the free-vibration equations can be obtained as

$$[m]\ddot{\overline{x}} + [k]\overline{x} = \overline{0} \tag{E.3}$$

FIGURE 6.12   A three-degree-of-freedom spring-mass system.

■

EXAMPLE 6.11

Natural Frequencies of a Three-Degree-of-Freedom System

Find the natural frequencies and mode shapes of the system shown in Fig. 6.12 for $k_1 = k_2 = k_3 = k$ and $m_1 = m_2 = m_3 = m$.

Solution: The dynamical matrix is given by

$$[D] = [k]^{-1}[m] \equiv [a][m] \tag{E.1}$$

where the elements of the flexibility matrix can be obtained by setting $k_i = k, i = 1, 2, 3$ in Eqs. (E.8) and (E.12) of Example 6.5 and the mass matrix can be obtained by setting $m_i = m, i = 1, 2, 3$ in the $[m]$ shown in Eq. (E.2) of Example 6.10 so that:

$$[a] = [k]^{-1} = \frac{1}{k}\begin{bmatrix} 1 & 1 & 1 \\ 1 & 2 & 2 \\ 1 & 2 & 3 \end{bmatrix} \tag{E.2}$$

and

$$[m] = m\begin{bmatrix} 1 & 0 & 0 \\ 0 & 1 & 0 \\ 0 & 0 & 1 \end{bmatrix} \tag{E.3}$$

Thus

$$[D] = [k]^{-1}[m] = \frac{m}{k}\begin{bmatrix} 1 & 1 & 1 \\ 1 & 2 & 2 \\ 1 & 2 & 3 \end{bmatrix} \tag{E.4}$$

By setting the characteristic determinant equal to zero, we obtain the frequency equation

$$\Delta = |\lambda[I] - [D]| = \begin{vmatrix} \begin{bmatrix} \lambda & 0 & 0 \\ 0 & \lambda & 0 \\ 0 & 0 & \lambda \end{bmatrix} - \frac{m}{k}\begin{bmatrix} 1 & 1 & 1 \\ 1 & 2 & 2 \\ 1 & 2 & 3 \end{bmatrix} \end{vmatrix} = 0 \tag{E.5}$$

where

$$\lambda = \frac{1}{\omega^2} \tag{E.6}$$

By dividing throughout by λ, Eq. (E.5) gives

$$\begin{vmatrix} 1 - \alpha & -\alpha & -\alpha \\ -\alpha & 1 - 2\alpha & -2\alpha \\ -\alpha & -2\alpha & 1 - 3\alpha \end{vmatrix} = \alpha^3 - 5\alpha^2 + 6\alpha - 1 = 0 \tag{E.7}$$

where

$$\alpha = \frac{m}{k\lambda} = \frac{m\omega^2}{k} \tag{E.8}$$

The roots of the cubic equation (E.7) are given by

$$\alpha_1 = \frac{m\omega_1^2}{k} = 0.19806, \qquad \omega_1 = 0.44504\sqrt{\frac{k}{m}} \qquad \text{(E.9)}$$

$$\alpha_2 = \frac{m\omega_2^2}{k} = 1.5553, \qquad \omega_2 = 1.2471\sqrt{\frac{k}{m}} \qquad \text{(E.10)}$$

$$\alpha_3 = \frac{m\omega_3^2}{k} = 3.2490, \qquad \omega_3 = 1.8025\sqrt{\frac{k}{m}} \qquad \text{(E.11)}$$

Once the natural frequencies are known, the mode shapes or eigenvectors can be calculated using Eq. (6.66):

$$[\lambda_i[I] - [D]]\,\vec{X}^{(i)} = \vec{0}, \qquad i = 1, 2, 3 \qquad \text{(E.12)}$$

where

$$\vec{X}^{(i)} = \begin{Bmatrix} X_1^{(i)} \\ X_2^{(i)} \\ X_3^{(i)} \end{Bmatrix}$$

denotes the ith mode shape. The procedure is outlined below.

First Mode: By substituting the value of ω_1 $\left(\text{i.e., } \lambda_1 = 5.0489\frac{m}{k}\right)$ in Eq. (E.12), we obtain

$$\left[5.0489\frac{m}{k}\begin{bmatrix} 1 & 0 & 0 \\ 0 & 1 & 0 \\ 0 & 0 & 1 \end{bmatrix} - \frac{m}{k}\begin{bmatrix} 1 & 1 & 1 \\ 1 & 2 & 2 \\ 1 & 2 & 3 \end{bmatrix} \right]\begin{Bmatrix} X_1^{(1)} \\ X_2^{(1)} \\ X_3^{(1)} \end{Bmatrix} = \begin{Bmatrix} 0 \\ 0 \\ 0 \end{Bmatrix}$$

—that is,

$$\begin{bmatrix} 4.0489 & -1.0 & -1.0 \\ -1.0 & 3.0489 & -2.0 \\ -1.0 & -2.0 & 2.0489 \end{bmatrix}\begin{Bmatrix} X_1^{(1)} \\ X_2^{(1)} \\ X_3^{(1)} \end{Bmatrix} = \begin{Bmatrix} 0 \\ 0 \\ 0 \end{Bmatrix} \qquad \text{(E.13)}$$

Equation (E.13) denotes a system of three homogeneous linear equations in the three unknowns $X_1^{(1)}$, $X_2^{(1)}$, and $X_3^{(1)}$. Any two of these unknowns can be expressed in terms of the remaining one. If we choose, arbitrarily, to express $X_2^{(1)}$ and $X_3^{(1)}$ in terms of $X_1^{(1)}$, we obtain from the first two rows of Eq. (E.13)

$$X_2^{(1)} + X_3^{(1)} = 4.0489X_1^{(1)}$$

$$3.0489X_2^{(1)} - 2.0\,X_3^{(1)} = X_1^{(1)} \qquad \text{(E.14)}$$

Once Eqs. (E.14) are satisfied, the third row of Eq. (E.13) is satisfied automatically. The solution of Eqs. (E.14) can be obtained:

$$X_2^{(1)} = 1.8019X_1^{(1)} \qquad \text{and} \qquad X_3^{(1)} = 2.2470X_1^{(1)} \qquad \text{(E.15)}$$

Thus the first mode shape is given by

$$\vec{X}^{(1)} = X_1^{(1)} \begin{Bmatrix} 1.0 \\ 1.8019 \\ 2.2470 \end{Bmatrix} \tag{E.16}$$

where the value of $X_1^{(1)}$ can be chosen arbitrarily.

Second Mode: The substitution of the value of $\omega_2 \left(\text{i.e., } \lambda_2 = 0.6430 \frac{m}{k} \right)$ in Eq. (E.12) leads to

$$\left[0.6430 \frac{m}{k} \begin{bmatrix} 1 & 0 & 0 \\ 0 & 1 & 0 \\ 0 & 0 & 1 \end{bmatrix} - \frac{m}{k} \begin{bmatrix} 1 & 1 & 1 \\ 1 & 2 & 2 \\ 1 & 2 & 3 \end{bmatrix} \right] \begin{Bmatrix} X_1^{(2)} \\ X_2^{(2)} \\ X_3^{(2)} \end{Bmatrix} = \begin{Bmatrix} 0 \\ 0 \\ 0 \end{Bmatrix}$$

—that is,

$$\begin{bmatrix} -0.3570 & -1.0 & -1.0 \\ -1.0 & -1.3570 & -2.0 \\ -1.0 & -2.0 & -2.3570 \end{bmatrix} \begin{Bmatrix} X_1^{(2)} \\ X_2^{(2)} \\ X_3^{(2)} \end{Bmatrix} = \begin{Bmatrix} 0 \\ 0 \\ 0 \end{Bmatrix} \tag{E.17}$$

As before, the first two rows of Eq. (E.17) can be used to obtain

$$-X_2^{(2)} - X_3^{(2)} = 0.3570 X_1^{(2)}$$
$$-1.3570 X_2^{(2)} - 2.0 X_3^{(2)} = X_1^{(2)} \tag{E.18}$$

The solution of Eqs. (E.18) leads to

$$X_2^{(2)} = 0.4450 X_1^{(2)} \quad \text{and} \quad X_3^{(2)} = -0.8020 X_1^{(2)} \tag{E.19}$$

Thus the second mode shape can be expressed as

$$\vec{X}^{(2)} = X_1^{(2)} \begin{Bmatrix} 1.0 \\ 0.4450 \\ -0.8020 \end{Bmatrix} \tag{E.20}$$

where the value of $X_1^{(2)}$ can be chosen arbitrarily.

Third Mode: To find the third mode, we substitute the value of $\omega_3 \left(\text{i.e., } \lambda_3 = 0.3078 \frac{m}{k} \right)$ in Eq. (E.12) and obtain

$$\left[0.3078 \frac{m}{k} \begin{bmatrix} 1 & 0 & 0 \\ 0 & 1 & 0 \\ 0 & 0 & 1 \end{bmatrix} - \frac{m}{k} \begin{bmatrix} 1 & 1 & 1 \\ 1 & 2 & 2 \\ 1 & 2 & 3 \end{bmatrix} \right] \begin{Bmatrix} X_1^{(3)} \\ X_2^{(3)} \\ X_3^{(3)} \end{Bmatrix} = \begin{Bmatrix} 0 \\ 0 \\ 0 \end{Bmatrix}$$

—that is,

$$\begin{bmatrix} -0.6922 & -1.0 & -1.0 \\ -1.0 & -1.6922 & -2.0 \\ -1.0 & -2.0 & -2.6922 \end{bmatrix} \begin{Bmatrix} X_1^{(3)} \\ X_2^{(3)} \\ X_3^{(3)} \end{Bmatrix} = \begin{Bmatrix} 0 \\ 0 \\ 0 \end{Bmatrix} \tag{E.21}$$

The first two rows of Eq. (E.21) can be written as

$$-X_2^{(3)} - X_3^{(3)} = 0.6922X_1^{(3)}$$

$$-1.6922X_2^{(3)} - 2.0X_3^{(3)} = X_1^{(3)} \tag{E.22}$$

Equations (E.22) give

$$X_2^{(3)} = -1.2468X_1^{(3)} \qquad \text{and} \qquad X_3^{(3)} = 0.5544X_1^{(3)} \tag{E.23}$$

Hence the third mode shape can be written as

$$\vec{X}^{(3)} = X_1^{(3)} \begin{Bmatrix} 1.0 \\ -1.2468 \\ 0.5544 \end{Bmatrix} \tag{E.24}$$

where the value of $X_1^{(3)}$ is arbitrary. The values of $X_1^{(1)}$, $X_1^{(2)}$, and $X_1^{(3)}$ are usually taken as 1, and the mode shapes are shown in Fig. 6.13.

First mode

Second mode

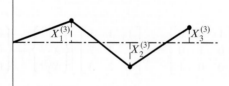

Third mode

FIGURE 6.13   Mode shapes of a three-degree-of-freedom system.

∎

6.10.2 Orthogonality of Normal Modes

In the previous section, we considered a method of finding the n natural frequencies ω_i and the corresponding normal modes or modal vectors $\vec{X}^{(i)}$. We shall now see an important property of the normal modes—*orthogonality*.[5] The natural frequency ω_i and the corresponding modal vector $\vec{X}^{(i)}$ satisfy Eq. (6.61) so that

$$\omega_i^2[m]\vec{X}^{(i)} = [k]\vec{X}^{(i)} \tag{6.69}$$

If we consider another natural frequency ω_j and the corresponding modal vector $\vec{X}^{(j)}$, they also satisfy Eq. (6.61) so that

$$\omega_j^2[m]\vec{X}^{(j)} = [k]\vec{X}^{(j)} \tag{6.70}$$

By premultiplying Eqs. (6.69) and (6.70) by $\vec{X}^{(j)^T}$ and $\vec{X}^{(i)^T}$, respectively, we obtain, by considering the symmetry of the matrices $[k]$ and $[m]$,

$$\omega_i^2 \, \vec{X}^{(j)T}[m]\vec{X}^{(i)} = \vec{X}^{(j)T}[k]\vec{X}^{(i)} \equiv \vec{X}^{(i)T}[k]\vec{X}^{(j)} \tag{6.71}$$

$$\omega_j^2 \, \vec{X}^{(i)T}[m]\vec{X}^{(j)} \equiv \omega_j^2\vec{X}^{(j)T}[m]\vec{X}^{(i)} = \vec{X}^{(i)T}[k]\vec{X}^{(j)} \tag{6.72}$$

By subtracting Eq. (6.72) from Eq. (6.71), we obtain

$$(\omega_i^2 - \omega_j^2)\vec{X}^{(j)T}[m]\vec{X}^{(i)} = 0 \tag{6.73}$$

In general, $\omega_i^2 \neq \omega_j^2$, so Eq. (6.73) leads to[6]

$$\vec{X}^{(j)T}[m]\vec{X}^{(i)} = 0, \qquad i \neq j \tag{6.74}$$

From Eqs. (6.71) and (6.72), we obtain, in view of Eq. (6.74),

$$\vec{X}^{(j)T}[k]\vec{X}^{(i)} = 0, \qquad i \neq j \tag{6.75}$$

Equations (6.74) and (6.75) indicate that the modal vectors $\vec{X}^{(i)}$ and $\vec{X}^{(j)}$ are orthogonal with respect to both mass and stiffness matrices.

---

[5]Two vectors $\vec{X}^{(i)}$ and $\vec{X}^{(j)}$ are said to be orthogonal (perpendicular to one another if the vectors are defined in two- or three-dimensional space) if the following relation is satisfied:

$$\vec{X}^{(i)^T} \vec{X}^{(j)} = 0$$

A vector $\vec{X}^{(i)}$ is said to be normal if its magnitude is unity—that is,

$$|\vec{X}^{(i)}|^2 = \vec{X}^{(i)^T} \vec{X}^{(i)} = 1$$

Thus the vectors $\vec{X}^{(i)}$ and $\vec{X}^{(j)}$ are said to be orthonormal if they satisfy the orthogonality and normality relations:

$$\vec{X}^{(i)^T} \vec{X}^{(j)} = 0, \quad |\vec{X}^{(i)}|^2 = \vec{X}^{(i)^T} \vec{X}^{(i)} = 1, \quad |\vec{X}^{(j)}|^2 = \vec{X}^{(j)^T} \vec{X}^{(j)} = 1$$

[6]In the case of repeated eigenvalues, $\omega_i = \omega_j$, the associated modal vectors are orthogonal to all the remaining modal vectors but are not usually orthogonal to each other.

When $i = j$, the left sides of Eqs. (6.74) and (6.75) are not equal to zero, but they do yield the generalized mass and stiffness coefficients of the ith mode:

$$M_{ii} = \vec{X}^{(i)T}[m]\vec{X}^{(i)}, \qquad i = 1, 2, \ldots, n \tag{6.76}$$

$$K_{ii} = \vec{X}^{(i)T}[k]\vec{X}^{(i)}, \qquad i = 1, 2, \ldots, n \tag{6.77}$$

Equations (6.76) and (6.77) can be written in matrix form as

$$\begin{bmatrix} \diagdown M \diagdown \end{bmatrix} = \begin{bmatrix} M_{11} & & & 0 \\ & M_{22} & & \\ & & \ddots & \\ 0 & & & M_{nn} \end{bmatrix} = [X]^T[m][X] \tag{6.78}$$

$$\begin{bmatrix} \diagdown K \diagdown \end{bmatrix} = \begin{bmatrix} K_{11} & & & 0 \\ & K_{22} & & \\ & & \ddots & \\ 0 & & & K_{nn} \end{bmatrix} = [X]^T[k][X] \tag{6.79}$$

where $[X]$ is called the *modal matrix*, in which the ith column corresponds to the ith modal vector:

$$[X] = \begin{bmatrix} \vec{X}^{(1)}\vec{X}^{(2)} \cdots \vec{X}^{(n)} \end{bmatrix} \tag{6.80}$$

In many cases, we normalize the modal vectors $\vec{X}^{(i)}$ such that $\begin{bmatrix} \diagdown M \diagdown \end{bmatrix} = [I]$—that is,

$$\vec{X}^{(i)T}[m]\vec{X}^{(i)} = 1, \qquad i = 1, 2, \ldots, n \tag{6.81}$$

In this case, the matrix $\begin{bmatrix} \diagdown K \diagdown \end{bmatrix}$ reduces to

$$\begin{bmatrix} \diagdown K \diagdown \end{bmatrix} = \begin{bmatrix} \diagdown \omega_i^2 \diagdown \end{bmatrix} = \begin{bmatrix} \omega_1^2 & & & 0 \\ & \omega_2^2 & & \\ & & \ddots & \\ 0 & & & \omega_n^2 \end{bmatrix} \tag{6.82}$$

Note: If an eigenvector $\vec{X}^{(i)}$ satisfies Eq. (6.81), it is said to be orthonormal with respect to the mass matrix $[m]$.

EXAMPLE 6.12

Orthonormalization of Eigenvectors

Orthonormalize the eigenvectors of Example 6.11 with respect to the mass matrix.

Solution:

Approach: Multiply each eigenvector by a constant and find its value from the relation $\vec{X}^{(i)T}[m]\vec{X}^{(i)} = 1, i = 1, 2, 3$.

The eigenvectors of Example 6.11 are given by

$$\vec{X}^{(1)} = X_1^{(1)} \begin{Bmatrix} 1.0 \\ 1.8019 \\ 2.2470 \end{Bmatrix}$$

$$\vec{X}^{(2)} = X_1^{(2)} \begin{Bmatrix} 1.0 \\ 0.4450 \\ -0.8020 \end{Bmatrix}$$

$$\vec{X}^{(3)} = X_1^{(3)} \begin{Bmatrix} 1.0 \\ -1.2468 \\ 0.5544 \end{Bmatrix}$$

The mass matrix is given by

$$[m] = m \begin{bmatrix} 1 & 0 & 0 \\ 0 & 1 & 0 \\ 0 & 0 & 1 \end{bmatrix}$$

The eigenvector $\vec{X}^{(i)}$ is said to be $[m]$-orthonormal if the following condition is satisfied:

$$\vec{X}^{(i)T}[m]\vec{X}^{(i)} = 1 \qquad\qquad\qquad\text{(E.1)}$$

Thus for $i = 1$, Eq. (E.1) leads to

$$m(X_1^{(1)})^2(1.0^2 + 1.8019^2 + 2.2470^2) = 1$$

or

$$X_1^{(1)} = \frac{1}{\sqrt{m(9.2959)}} = \frac{0.3280}{\sqrt{m}}$$

Similarly, for $i = 2$ and $i = 3$, Eq. (E.1) gives

$$m(X_1^{(2)})^2(1.0^2 + 0.4450^2 + \{-0.8020\}^2) = 1 \qquad \text{or} \qquad X_1^{(2)} = \frac{0.7370}{\sqrt{m}}$$

and

$$m(X_1^{(3)})^2(1.0^2 + \{-1.2468\}^2 + 0.5544^2) = 1 \qquad \text{or} \qquad X_1^{(3)} = \frac{0.5911}{\sqrt{m}}$$

∎

6.10.3
Repeated
Eigenvalues

When the characteristic equation possesses repeated roots, the corresponding mode shapes are not unique. To see this, let $\vec{X}^{(1)}$ and $\vec{X}^{(2)}$ be the mode shapes corresponding to the repeated eigenvalue $\lambda_1 = \lambda_2 = \lambda$ and let $\vec{X}^{(3)}$ be the mode shape corresponding to a different eigenvalue λ_3. Equation (6.66) can be written as

$$[D]\vec{X}^{(1)} = \lambda\vec{X}^{(1)} \tag{6.83}$$

$$[D]\vec{X}^{(2)} = \lambda\vec{X}^{(2)} \tag{6.84}$$

$$[D]\vec{X}^{(3)} = \lambda_3\vec{X}^{(3)} \tag{6.85}$$

By multiplying Eq. (6.83) by a constant p and adding to Eq. (6.84), we obtain

$$[D](p\vec{X}^{(1)} + \vec{X}^{(2)}) = \lambda(p\vec{X}^{(1)} + \vec{X}^{(2)}) \tag{6.86}$$

This shows that the new mode shape, $(p\vec{X}^{(1)} + \vec{X}^{(2)})$, which is a linear combination of the first two, also satisfies Eq. (6.66), so the mode shape corresponding to λ is not unique. Any $\vec{X}$ corresponding to λ must be orthogonal to $\vec{X}^{(3)}$ if it is to be a normal mode. If all three modes are orthogonal, they will be linearly independent and can be used to describe the free vibration resulting from any initial conditions.

The response of a multidegree-of-freedom system with repeated natural frequencies to force and displacement excitation was presented by Mahalingam and Bishop [6.16].

EXAMPLE 6.13

Repeated Eigenvalues

Determine the eigenvalues and eigenvectors of a vibrating system for which

$$[m] = \begin{bmatrix} 1 & 0 & 0 \\ 0 & 2 & 0 \\ 0 & 0 & 1 \end{bmatrix} \quad \text{and} \quad [k] = \begin{bmatrix} 1 & -2 & 1 \\ -2 & 4 & -2 \\ 1 & -2 & 1 \end{bmatrix}$$

Solution: The eigenvalue equation $[[k] - \lambda[m]]\vec{X} = \vec{0}$ can be written in the form

$$\begin{bmatrix} (1 - \lambda) & -2 & 1 \\ -2 & 2(2 - \lambda) & -2 \\ 1 & -2 & (1 - \lambda) \end{bmatrix} \begin{Bmatrix} X_1 \\ X_2 \\ X_3 \end{Bmatrix} = \begin{Bmatrix} 0 \\ 0 \\ 0 \end{Bmatrix} \tag{E.1}$$

where $\lambda = \omega^2$. The characteristic equation gives

$$|[k] - \lambda[m]| = \lambda^2(\lambda - 4) = 0$$

so

$$\lambda_1 = 0, \quad \lambda_2 = 0, \quad \lambda_3 = 4 \tag{E.2}$$

Eigenvector for $\lambda_3 = 4$: Using $\lambda_3 = 4$, Eq. (E.1) gives

$$-3X_1^{(3)} - 2X_2^{(3)} + X_3^{(3)} = 0$$

$$-2X_1^{(3)} - 4X_2^{(3)} - 2X_3^{(3)} = 0$$

$$X_1^{(3)} - 2X_2^{(3)} - 3X_3^{(3)} = 0 \qquad \text{(E.3)}$$

If $X_1^{(3)}$ is set equal to 1, Eqs. (E.3) give the eigenvector $\vec{X}^{(3)}$:

$$\vec{X}^{(3)} = \begin{Bmatrix} 1 \\ -1 \\ 1 \end{Bmatrix} \qquad \text{(E.4)}$$

Eigenvector for $\lambda_1 = \lambda_2 = 0$: The value $\lambda_1 = 0$ or $\lambda_2 = 0$ indicates that the system is degenerate (see Section 6.12). Using $\lambda_1 = 0$ in Eq. (E.1), we obtain

$$X_1^{(1)} - 2X_2^{(1)} + X_3^{(1)} = 0$$

$$-2X_1^{(1)} + 4X_2^{(1)} - 2X_3^{(1)} = 0$$

$$X_1^{(1)} - 2X_2^{(1)} + X_3^{(1)} = 0 \qquad \text{(E.5)}$$

All these equations are of the form

$$X_1^{(1)} = 2X_2^{(1)} - X_3^{(1)}$$

Thus the eigenvector corresponding to $\lambda_1 = \lambda_2 = 0$ can be written as

$$\vec{X}^{(1)} = \begin{Bmatrix} 2X_2^{(1)} - X_3^{(1)} \\ X_2^{(1)} \\ X_3^{(1)} \end{Bmatrix} \qquad \text{(E.6)}$$

If we choose $X_2^{(1)} = 1$ and $X_3^{(1)} = 1$, we obtain

$$\vec{X}^{(1)} = \begin{Bmatrix} 1 \\ 1 \\ 1 \end{Bmatrix} \qquad \text{(E.7)}$$

If we select $X_2^{(1)} = 1$ and $X_3^{(1)} = -1$, Eq. (E.6) gives

$$\vec{X}^{(1)} = \begin{Bmatrix} 3 \\ 1 \\ -1 \end{Bmatrix} \qquad \text{(E.8)}$$

As shown earlier in Eq. (6.86), $\vec{X}^{(1)}$ and $\vec{X}^{(2)}$ are not unique: Any linear combination of $\vec{X}^{(1)}$ and $\vec{X}^{(2)}$ will also satisfy the original Eq. (E.1). Note that $\vec{X}^{(1)}$ given by Eq. (E.6) is orthogonal to $\vec{X}^{(3)}$ of Eq. (E.4) for all values of $X_2^{(1)}$ and $X_3^{(1)}$, since

$$\vec{X}^{(3)T}[m]\vec{X}^{(1)} = \begin{pmatrix} 1 & -1 & 1 \end{pmatrix} \begin{bmatrix} 1 & 0 & 0 \\ 0 & 2 & 0 \\ 0 & 0 & 1 \end{bmatrix} \begin{Bmatrix} 2X_2^{(1)} - X_3^{(1)} \\ X_2^{(1)} \\ X_3^{(1)} \end{Bmatrix} = 0$$

∎

6.11 Expansion Theorem

The eigenvectors, due to their property of orthogonality, are linearly independent.[7] Hence they form a basis in the n-dimensional space.[8] This means that any vector in the n-dimensional space can be expressed by a linear combination of the n linearly independent vectors. If $\vec{x}$ is an arbitrary vector in n-dimensional space, it can be expressed as

$$\vec{x} = \sum_{i=1}^{n} c_i \vec{X}^{(i)} \tag{6.87}$$

where c_i are constants. By premultiplying Eq. (6.87) throughout by $\vec{X}^{(i)T}[m]$, the value of the constant c_i can be determined as

$$c_i = \frac{\vec{X}^{(i)T}[m]\,\vec{x}}{\vec{X}^{(i)T}[m]\,\vec{X}^{(i)}} = \frac{\vec{X}^{(i)T}[m]\,\vec{x}}{M_{ii}}, \qquad i = 1, 2, \ldots, n \tag{6.88}$$

where M_{ii} is the generalized mass in the ith normal mode. If the modal vectors $\vec{X}^{(i)}$ are normalized according to Eq. (6.81), c_i is given by

$$c_i = \vec{X}^{(i)T}[m]\vec{x}, \qquad i = 1, 2, \ldots, n \tag{6.89}$$

Equation (6.89) represents what is known as the *expansion theorem* [6.6]. It is very useful in finding the response of multidegree-of-freedom systems subjected to arbitrary forcing conditions according to a procedure called *modal analysis*.

6.12 Unrestrained Systems

As stated in Section 5.7, an unrestrained system is one that has no restraints or supports and that can move as a rigid body. It is not uncommon to see, in practice, systems that are not attached to any stationary frame. A common example is the motion of two railway cars

---

[7]A set of vectors is called *linearly independent* if no vector in the set can be obtained by a linear combination of the remaining ones.
[8]Any set of n linearly independent vectors in an n-dimensional space is called a *basis* in that space.

with masses m_1 and m_2 and a coupling spring k. Such systems are capable of moving as rigid bodies, which can be considered as modes of oscillation with zero frequency. For a conservative system, the kinetic and potential energies are given by Eqs. (6.30) and (6.34), respectively. By definition, the kinetic energy is always positive, so the mass matrix $[m]$ is a positive definite matrix. However, the stiffness matrix $[k]$ is a semidefinite matrix: V is zero without the displacement vector $\vec{x}$ being zero for unrestrained systems. To see this, consider the equation of motion for free vibration in normal coordinates:

$$\ddot{q}(t) + \omega^2 q(t) = 0 \tag{6.90}$$

For $\omega = 0$, the solution of Eq. (6.90) can be expressed as

$$q(t) = \alpha + \beta t \tag{6.91}$$

where α and β are constants. Equation (6.91) represents a rigid-body translation. Let the modal vector of a multidegree-of-freedom system corresponding to the rigid-body mode be denoted by $\vec{X}^{(0)}$. The eigenvalue problem, Eq. (6.64) can be expressed as

$$\omega^2 [m] \vec{X}^{(0)} = [k] \vec{X}^{(0)} \tag{6.92}$$

With $\omega = 0$, Eq. (6.92) gives

$$[k] \vec{X}^{(0)} = \vec{0}$$

—that is,

$$k_{11} X_1^{(0)} + k_{12} X_2^{(0)} + \cdots + k_{1n} X_n^{(0)} = 0$$
$$k_{21} X_1^{(0)} + k_{22} X_2^{(0)} + \cdots + k_{2n} X_n^{(0)} = 0$$
$$\vdots$$
$$k_{n1} X_1^{(0)} + k_{n2} X_2^{(0)} + \cdots + k_{nn} X_n^{(0)} = 0 \tag{6.93}$$

If the system undergoes rigid-body translation, not all the components $X_i^{(0)}, i = 1, 2, \ldots, n$, are zero—that is, the vector $\vec{X}^{(0)}$ is not equal to $\vec{0}$. Hence, in order to satisfy Eq. (6.93), the determinant of $[k]$ must be zero. Thus the stiffness matrix of an unrestrained system (having zero natural frequency) is singular. If $[k]$ is singular, the potential energy is given by

$$V = \frac{1}{2} \vec{X}^{(0)^T} [k] \vec{X}^{(0)} \tag{6.94}$$

by virtue of Eq. (6.93). The mode $\vec{X}^{(0)}$ is called a *zero mode* or *rigid-body mode*. If we substitute any vector $\vec{X}$ other than $\vec{X}^{(0)}$ and $\vec{0}$ for $\vec{x}$ in Eq. (6.30), the potential energy V

becomes a positive quantity. The matrix $[k]$ is then a positive semidefinite matrix. This is why an unrestrained system is also called a *semidefinite system*.

Note that a multidegree-of-freedom system can have at most six rigid-body modes with the corresponding frequencies equal to zero. There can be three modes for rigid-body translation, one for translation along each of the three Cartesian coordinates, and three modes for rigid-body rotation, one for rotation about each of the three Cartesian coordinates. We can determine the mode shapes and natural frequencies of a semidefinite system by the procedures outlined in Section 6.10.

Natural Frequencies of a Free System

EXAMPLE 6.14

Three freight cars are coupled by two springs, as shown in Fig. 6.14. Find the natural frequencies and mode shapes of the system for $m_1 = m_2 = m_3 = m$ and $k_1 = k_2 = k$.

Solution: The kinetic energy of the system can be written as

$$T = \frac{1}{2}(m_1 \dot{x}_1^2 + m_2 \dot{x}_2^2 + m_3 \dot{x}_3^2) = \frac{1}{2}\vec{x}^T[m]\vec{x} \tag{E.1}$$

where

$$\vec{x} = \begin{Bmatrix} x_1 \\ x_2 \\ x_3 \end{Bmatrix}, \qquad \dot{\vec{x}} = \begin{Bmatrix} \dot{x}_1 \\ \dot{x}_2 \\ \dot{x}_3 \end{Bmatrix}$$

and

$$[m] = \begin{bmatrix} m_1 & 0 & 0 \\ 0 & m_2 & 0 \\ 0 & 0 & m_3 \end{bmatrix} \tag{E.2}$$

The elongations of the springs k_1 and k_2 are $(x_2 - x_1)$ and $(x_3 - x_2)$, respectively, so the potential energy of the system is given by

$$V = \frac{1}{2}\{k_1(x_2 - x_1)^2 + k_2(x_3 - x_2)^2\} = \frac{1}{2}\vec{x}^T[k]\vec{x} \tag{E.3}$$

where

$$[k] = \begin{bmatrix} k_1 & -k_1 & 0 \\ -k_1 & k_1 + k_2 & -k_2 \\ 0 & -k_2 & k_2 \end{bmatrix} \tag{E.4}$$

FIGURE 6.14   Semidefinite system.

It can be verified that the stiffness matrix $[k]$ is singular. Furthermore, if we take all the displacement components to be the same as $x_1 = x_2 = x_3 = c$ (rigid-body motion), the potential energy V can be seen to be equal to zero.

To find the natural frequencies and the mode shapes of the system, we express the eigenvalue problem as

$$[[k] - \omega^2[m]]\vec{X} = \vec{0} \tag{E.5}$$

Since $[k]$ is singular, we cannot find its inverse $[k]^{-1}$ and the dynamical matrix $[D] = [k]^{-1}[m]$. Hence we set the determinant of the coefficient matrix of $\vec{X}$ in Eq. (E.5) equal to zero. For $k_1 = k_2 = k$ and $m_1 = m_2 = m_3 = m$, this yields

$$\begin{vmatrix} (k - \omega^2 m) & -k & 0 \\ -k & (2k - \omega^2 m) & -k \\ 0 & -k & (k - \omega^2 m) \end{vmatrix} = 0 \tag{E.6}$$

The expansion of the determinant in Eq. (E.6) leads to

$$m^3\omega^6 - 4m^2k\omega^4 + 3mk^2\omega^2 = 0 \tag{E.7}$$

By setting

$$\lambda = \omega^2 \tag{E.8}$$

Eq. (E.7) can be rewritten as

$$m\lambda\left(\lambda - \frac{k}{m}\right)\left(\lambda - \frac{3k}{m}\right) = 0 \tag{E.9}$$

As $m \neq 0$, the roots of Eq. (E.9) are

$$\lambda_1 = \omega_1^2 = 0$$

$$\lambda_2 = \omega_2^2 = \frac{k}{m}$$

$$\lambda_3 = \omega_3^2 = \frac{3k}{m} \tag{E.10}$$

The first natural frequency ω_1 can be observed to be zero in Eq. (E.10). To find the mode shapes, we substitute the values of ω_1, ω_2, and ω_3 into Eq. (E.5) and solve for $\vec{X}^{(1)}$, $\vec{X}^{(2)}$, and $\vec{X}^{(3)}$, respectively. For $\omega_1 = 0$, Eq. (E.5) gives

$$kX_1^{(1)} - kX_2^{(1)} = 0$$

$$-kX_1^{(1)} + 2kX_2^{(1)} - kX_3^{(1)} = 0$$

$$-kX_2^{(1)} + kX_3^{(1)} = 0 \tag{E.11}$$

By fixing the value of one component of $\vec{X}^{(1)}$—say, $X_1^{(1)}$ as 1—Eq. (E.11) can be solved to obtain

$$X_2^{(1)} = X_1^{(1)} = 1 \quad \text{and} \quad X_3^{(1)} = X_2^{(1)} = 1$$

Thus the first (rigid-body) mode $\vec{X}^{(1)}$ corresponding to $\omega_1 = 0$ is given by

$$\vec{X}^{(1)} = \begin{Bmatrix} 1 \\ 1 \\ 1 \end{Bmatrix} \tag{E.12}$$

Note that the mode shape $\vec{X}^{(1)}$ represents the rigid-body translation of the system (all masses undergo the same displacement). Although the natural frequency ω_1 (or the eigenvalue ω_1^2) is zero, the corresponding mode shape (or eigenvector) $\vec{X}^{(1)}$ is not zero.

For $\omega_2 = (k/m)^{1/2}$, Eq. (E.5) yields

$$-kX_2^{(2)} = 0$$

$$-kX_1^{(2)} + kX_2^{(2)} - kX_3^{(2)} = 0$$

$$-kX_2^{(2)} = 0 \tag{E.13}$$

By fixing the value of one component of $\vec{X}^{(2)}$—say, $X_1^{(2)}$ as 1—Eq. (E.13) can be solved to obtain

$$X_2^{(2)} = 0 \qquad \text{and} \qquad X_3^{(2)} = -X_1^{(2)} = -1$$

Thus the second mode $\vec{X}^{(2)}$ corresponding to $\omega_2 = (k/m)^{1/2}$ is given by

$$\vec{X}^{(2)} = \begin{Bmatrix} 1 \\ 0 \\ -1 \end{Bmatrix} \tag{E.14}$$

For $\omega_3 = (3k/m)^{1/2}$, Eq. (E.5) gives

$$-2kX_1^{(3)} - kX_2^{(3)} = 0$$

$$-kX_1^{(3)} - kX_2^{(3)} - kX_3^{(3)} = 0$$

$$-kX_2^{(3)} - 2kX_3^{(3)} = 0 \tag{E.15}$$

By fixing the value of one component of $\vec{X}^{(3)}$—say, $X_1^{(3)}$ as 1—Eq. (E.15) can be solved to obtain

$$X_2^{(3)} = -2X_1^{(3)} = -2 \qquad \text{and} \qquad X_3^{(3)} = -\frac{1}{2}X_2^{(3)} = 1$$

Thus the third mode $\vec{X}^{(3)}$ corresponding to $\omega_3 = (3k/m)^{1/2}$ is given by

$$\vec{X}^{(3)} = \begin{Bmatrix} 1 \\ -2 \\ 1 \end{Bmatrix} \tag{E.16}$$

∎

6.13   Free Vibration of Undamped Systems

The equation of motion for the free vibration of an undamped system can be expressed in matrix form as

$$[m]\overset{\cdot\cdot}{\vec{x}} + [k]\vec{x} = \vec{0} \tag{6.95}$$

The most general solution of Eq. (6.95) can be expressed as a linear combination of all possible solutions given by Eqs. (6.56) and (6.62) as

$$\vec{x}(t) = \sum_{i=1}^{n} \vec{X}^{(i)} A_i \cos(\omega_i t + \phi_i) \tag{6.96}$$

where $\vec{X}^{(i)}$ is the ith modal vector and ω_i is the corresponding natural frequency, and A_i and ϕ_i are constants. The constants A_i and $\phi_i (i = 1, 2, \ldots, n)$ can be evaluated from the specified initial conditions of the system. If

$$\vec{x}(0) = \begin{Bmatrix} x_1(0) \\ x_2(0) \\ \cdot \\ \cdot \\ \cdot \\ x_n(0) \end{Bmatrix} \quad \text{and} \quad \overset{\cdot}{\vec{x}}(0) = \begin{Bmatrix} \dot{x}_1(0) \\ \dot{x}_2(0) \\ \cdot \\ \cdot \\ \cdot \\ \dot{x}_n(0) \end{Bmatrix} \tag{6.97}$$

denote the initial displacements and velocities given to the system, Eq. (6.96) give

$$\vec{x}(0) = \sum_{i=1}^{n} \vec{X}^{(i)} A_i \cos \phi_i \tag{6.98}$$

$$\overset{\cdot}{\vec{x}}(0) = -\sum_{i=1}^{n} \vec{X}^{(i)} A_i \omega_i \sin \phi_i \tag{6.99}$$

Equations (6.98) and (6.99) represent, in scalar form, $2n$ simultaneous equations which can be solved to find the n values of $A_i (i = 1, 2, \ldots, n)$ and n values of $\phi_i (i = 1, 2, \ldots, n)$.

Free-Vibration Analysis of a Spring-Mass System

EXAMPLE 6.15

Find the free-vibration response of the spring-mass system shown in Fig. 6.12 corresponding to the initial conditions $\dot{x}_i(0) = 0 \ (i = 1, 2, 3)$, $x_1(0) = x_{10}, x_2(0) = x_3(0) = 0$. Assume that $k_i = k$ and $m_i = m$ for $i = 1, 2, 3$.

Solution:

Approach: Assume free-vibration response as a sum of natural modes.

The natural frequencies and mode shapes of the system are given by (see Example 6.11):

$$\omega_1 = 0.44504 \sqrt{\frac{k}{m}}, \qquad \omega_2 = 1.2471 \sqrt{\frac{k}{m}}, \qquad \omega_3 = 1.8025 \sqrt{\frac{k}{m}}$$

$$\vec{X}^{(1)} = \begin{Bmatrix} 1.0 \\ 1.8019 \\ 2.2470 \end{Bmatrix}, \qquad \vec{X}^{(2)} = \begin{Bmatrix} 1.0 \\ 0.4450 \\ -0.8020 \end{Bmatrix}, \qquad \vec{X}^{(3)} = \begin{Bmatrix} 1.0 \\ -1.2468 \\ 0.5544 \end{Bmatrix}$$

where the first component of each mode shape is assumed as unity for simplicity. The application of the initial conditions, Eqs. (6.98) and (6.99), leads to

$$A_1 \cos \phi_1 + A_2 \cos \phi_2 + A_3 \cos \phi_3 = x_{10} \tag{E.1}$$

$$1.8019 A_1 \cos \phi_1 + 0.4450 A_2 \cos \phi_2 - 1.2468 A_3 \cos \phi_3 = 0 \tag{E.2}$$

$$2.2470 A_1 \cos \phi_1 - 0.8020 A_2 \cos \phi_2 + 0.5544 A_3 \cos \phi_3 = 0 \tag{E.3}$$

$$-0.44504 \sqrt{\frac{k}{m}} A_1 \sin \phi_1 - 1.2471 \sqrt{\frac{k}{m}} A_2 \sin \phi_2 - 1.8025 \sqrt{\frac{k}{m}} A_3 \sin \phi_3 = 0 \tag{E.4}$$

$$-0.80192 \sqrt{\frac{k}{m}} A_1 \sin \phi_1 - 0.55496 \sqrt{\frac{k}{m}} A_2 \sin \phi_2 + 2.2474 \sqrt{\frac{k}{m}} A_3 \sin \phi_3 = 0 \tag{E.5}$$

$$-1.0 \sqrt{\frac{k}{m}} A_1 \sin \phi_1 + 1.0 \sqrt{\frac{k}{m}} A_2 \sin \phi_2 - 1.0 \sqrt{\frac{k}{m}} A_3 \sin \phi_3 = 0 \tag{E.6}$$

The solution of Eqs. (E.1) to (E.6) is given by[9] $A_1 = 0.1076 x_{10}$, $A_2 = 0.5431 x_{10}$, $A_3 = 0.3493 x_{10}$, $\phi_1 = 0$, $\phi_2 = 0$, and $\phi_3 = 0$. Thus the free-vibration solution of the system can be expressed as

$$x_1(t) = x_{10} \left[0.1076 \cos \left(0.44504 \sqrt{\frac{k}{m}} \, t \right) \right.$$

$$+ \, 0.5431 \cos \left(1.2471 \sqrt{\frac{k}{m}} \, t \right)$$

$$\left. + \, 0.3493 \cos \left(1.8025 \sqrt{\frac{k}{m}} \, t \right) \right] \tag{E.7}$$

-----

[9]Note that Eqs. (E.1) to (E.3) can be considered as a system of linear equations in the unknowns $A_1 \cos \phi_1$, $A_2 \cos \phi_2$, and $A_3 \cos \phi_3$, while Eqs. (E.4) to (E.6) can be considered as a set of linear equations in the unknowns

$$\sqrt{\frac{k}{m}} A_1 \sin \phi_1, \quad \sqrt{\frac{k}{m}} A_2 \sin \phi_2, \quad \text{and} \quad \sqrt{\frac{k}{m}} A_3 \sin \phi_3.$$

$$x_2(t) = x_{10} \left[0.1939 \cos\left(0.44504 \sqrt{\frac{k}{m}}\, t \right) \right.$$

$$+ 0.2417 \cos\left(1.2471 \sqrt{\frac{k}{m}}\, t \right)$$

$$\left. - 0.4355 \cos\left(1.8025 \sqrt{\frac{k}{m}}\, t \right) \right] \tag{E.8}$$

$$x_3(t) = x_{10} \left[0.2418 \cos\left(0.44504 \sqrt{\frac{k}{m}}\, t \right) \right.$$

$$- 0.4356 \cos\left(1.2471 \sqrt{\frac{k}{m}}\, t \right)$$

$$\left. + 0.1937 \cos\left(1.8025 \sqrt{\frac{k}{m}}\, t \right) \right] \tag{E.9}$$

■

6.14   Forced Vibration of Undamped Systems Using Modal Analysis

When external forces act on a multidegree-of-freedom system, the system undergoes forced vibration. For a system with n coordinates or degrees of freedom, the governing equations of motion are a set of n coupled ordinary differential equations of second order. The solution of these equations becomes more complex when the degree of freedom of the system (n) is large and/or when the forcing functions are nonperiodic.[10] In such cases, a more convenient method known as *modal analysis* can be used to solve the problem. In this method, the expansion theorem is used, and the displacements of the masses are expressed as a linear combination of the normal modes of the system. This linear transformation uncouples the equations of motion so that we obtain a set of n uncoupled differential equations of second order. The solution of these equations, which is equivalent to the solution of the equations of n single-degree-of-freedom systems, can be readily obtained. We shall now consider the procedure of modal analysis.

Modal Analysis.   The equations of motion of a multidegree-of-freedom system under external forces are given by

$$[m]\ddot{\vec{x}} + [k]\vec{x} = \vec{F} \tag{6.100}$$

---

[10]The dynamic response of multidegree-of-freedom systems with statistical properties is considered in [6.15].

where $\vec{F}$ is the vector of arbitrary external forces. To solve Eq. (6.100) by modal analysis, it is necessary first to solve the eigenvalue problem.

$$\omega^2[m]\vec{X} = [k]\vec{X} \tag{6.101}$$

and find the natural frequencies $\omega_1, \omega_2, \ldots, \omega_n$ and the corresponding normal modes $\vec{X}^{(1)}, \vec{X}^{(2)}, \ldots, \vec{X}^{(n)}$. According to the expansion theorem, the solution vector of Eq. (6.100) can be expressed by a linear combination of the normal modes

$$\vec{x}(t) = q_1(t)\vec{X}^{(1)} + q_2(t)\vec{X}^{(2)} + \cdots + q_n(t)\vec{X}^{(n)} \tag{6.102}$$

where $q_1(t), q_2(t), \ldots, q_n(t)$ are time-dependent generalized coordinates, also known as the *principal coordinates* or *modal participation coefficients*. By defining a modal matrix $[X]$ in which the jth column is the vector $\vec{X}^{(j)}$—that is,

$$[X] = [\vec{X}^{(1)}\vec{X}^{(2)} \cdots \vec{X}^{(n)}] \tag{6.103}$$

Eq. (6.102) can be rewritten as

$$\vec{x}(t) = [X]\vec{q}(t) \tag{6.104}$$

where

$$\vec{q}(t) = \begin{Bmatrix} q_1(t) \\ q_2(t) \\ \vdots \\ q_n(t) \end{Bmatrix} \tag{6.105}$$

Since $[X]$ is not a function of time, we obtain from Eq. (6.104)

$$\ddot{\vec{x}}(t) = [X]\ddot{\vec{q}}(t) \tag{6.106}$$

Using Eqs. (6.104) and (6.106), we can write Eq. (6.100) as

$$[m][X]\ddot{\vec{q}} + [k][X]\vec{q} = \vec{F} \tag{6.107}$$

Premultiplying Eq. (6.107) throughout by $[X]^T$, we obtain

$$[X]^T[m][X]\ddot{\vec{q}} + [X]^T[k][X]\vec{q} = [X]^T\vec{F} \tag{6.108}$$

If the normal modes are normalized according to Eqs. (6.74) and (6.75), we have

$$[X]^T[m][X] = [I] \tag{6.109}$$

$$[X]^T[k][X] = [\diagdown\omega^2\diagdown] \tag{6.110}$$

By defining the vector of generalized forces $\vec{Q}(t)$ associated with the generalized coordinates $\vec{q}(t)$ as

$$\vec{Q}(t) = [X]^T \vec{F}(t) \qquad (6.111)$$

Eq. (6.108) can be expressed, using Eqs. (6.109) and (6.110), as

$$\ddot{\vec{q}}(t) + [\diagdown\omega^2\diagdown]\vec{q}(t) = \vec{Q}(t) \qquad (6.112)$$

Equation (6.112) denotes a set of n uncoupled differential equations of second order[11]

$$\ddot{q}_i(t) + \omega_i^2 q_i(t) = Q_i(t), \qquad i = 1, 2, \ldots, n \qquad (6.113)$$

It can be seen that Eqs. (6.113) have precisely the form of the differential equation describing the motion of an undamped single-degree-of-freedom system. The solution of Eqs. (6.113) can be expressed (see Eq. (4.31)) as

$$q_i(t) = q_i(0) \cos \omega_i t + \left(\frac{\dot{q}(0)}{\omega_i}\right) \sin \omega_i t$$
$$+ \frac{1}{\omega_i} \int_0^t Q_i(\tau) \sin \omega_i (t - \tau) d\tau,$$
$$i = 1, 2, \ldots, n \qquad (6.114)$$

The initial generalized displacements $q_i(0)$ and the initial generalized velocities $\dot{q}_i(0)$ can be obtained from the initial values of the physical displacements $x_i(0)$ and physical velocities $\dot{x}_i(0)$ as (see Problem 6.94):

$$\vec{q}(0) = [X]^T[m]\vec{x}(0) \qquad (6.115)$$

$$\dot{\vec{q}}(0) = [X]^T[m]\dot{\vec{x}}(0) \qquad (6.116)$$

---

[11]It is possible to approximate the solution vector $\vec{x}(t)$ by only the first r $(r < n)$ modal vectors (instead of n vectors as in Eq. (6.102)):

$$\underset{n\times 1}{\vec{x}(t)} = \underset{n\times r}{[X]}\underset{r\times 1}{\vec{q}(t)}$$

where

$$[X] = [\vec{X}^{(1)} \vec{X}^{(2)} \cdots \vec{X}^{(r)}] \qquad \text{and} \qquad \vec{q}(t) = \begin{Bmatrix} q_1(t) \\ q_2(t) \\ \vdots \\ q_r(t) \end{Bmatrix}$$

This leads to only r uncoupled differential equations

$$\ddot{q}_i(t) + \omega_i^2 q_i(t) = Q_i(t), \qquad i = 1, 2, \ldots, r$$

instead of n equations. The resulting solution $\vec{x}(t)$ will be an approximate solution. This procedure is called the *mode displacement method*. An alternate procedure, *mode acceleration method*, for finding an approximate solution is indicated in Problem 6.92.

where

$$\vec{q}(0) = \begin{Bmatrix} q_1(0) \\ q_2(0) \\ \vdots \\ q_n(0) \end{Bmatrix},$$

$$\dot{\vec{q}}(0) = \begin{Bmatrix} \dot{q}_1(0) \\ \dot{q}_2(0) \\ \vdots \\ \dot{q}_n(0) \end{Bmatrix},$$

$$\vec{x}(0) = \begin{Bmatrix} x_1(0) \\ x_2(0) \\ \vdots \\ x_n(0) \end{Bmatrix},$$

$$\dot{\vec{x}}(0) = \begin{Bmatrix} \dot{x}_1(0) \\ \dot{x}_2(0) \\ \vdots \\ \dot{x}_n(0) \end{Bmatrix}$$

Once the generalized displacements $q_i(t)$ are found, using Eqs. (6.114) to (6.116), the physical displacements $x_i(t)$ can be found with the help of Eq. (6.104).

Free-Vibration Response Using Modal Analysis

EXAMPLE 6.16

Using modal analysis, find the free-vibration response of a two-degree-of-freedom system with equations of motion

$$\begin{bmatrix} m_1 & 0 \\ 0 & m_2 \end{bmatrix} \begin{Bmatrix} \ddot{x}_1 \\ \ddot{x}_2 \end{Bmatrix} + \begin{bmatrix} k_1 + k_2 & -k_2 \\ -k_2 & k_2 + k_3 \end{bmatrix} \begin{Bmatrix} x_1 \\ x_2 \end{Bmatrix} = \vec{F} = \begin{Bmatrix} 0 \\ 0 \end{Bmatrix} \tag{E.1}$$

Assume the following data: $m_1 = 10$, $m_2 = 1$, $k_1 = 30$, $k_2 = 5$, $k_3 = 0$, and

$$\vec{x}(0) = \begin{Bmatrix} x_1(0) \\ x_2(0) \end{Bmatrix} = \begin{Bmatrix} 1 \\ 0 \end{Bmatrix}, \qquad \dot{\vec{x}}(0) = \begin{Bmatrix} \dot{x}_1(0) \\ \dot{x}_2(0) \end{Bmatrix} = \begin{Bmatrix} 0 \\ 0 \end{Bmatrix} \tag{E.2}$$

Solution: The natural frequencies and normal modes of the system are given by (see Example 5.3)

$$\omega_1 = 1.5811, \qquad \vec{X}^{(1)} = \begin{Bmatrix} 1 \\ 2 \end{Bmatrix} X_1^{(1)}$$

$$\omega_2 = 2.4495, \qquad \vec{X}^{(2)} = \begin{Bmatrix} 1 \\ -5 \end{Bmatrix} X_1^{(2)}$$

where $X_1^{(1)}$ and $X_1^{(2)}$ are arbitrary constants. By orthogonalizing the normal modes with respect to the mass matrix, we can find the values of $X_1^{(1)}$ and $X_1^{(2)}$ as

$$\vec{X}^{(1)^T}[m]\vec{X}^{(1)} = 1 \Rightarrow (X_1^{(1)})^2 \{1 \quad 2\} \begin{bmatrix} 10 & 0 \\ 0 & 1 \end{bmatrix} \begin{Bmatrix} 1 \\ 2 \end{Bmatrix} = 1$$

or $X_1^{(1)} = 0.2673$

$$\vec{X}^{(2)^T}[m]\vec{X}^{(2)} = 1 \Rightarrow (X_1^{(2)})^2 \{1 \quad -5\} \begin{bmatrix} 10 & 0 \\ 0 & 1 \end{bmatrix} \begin{Bmatrix} 1 \\ -5 \end{Bmatrix} = 1$$

or $X_1^{(2)} = 0.1690$

Thus the modal matrix becomes

$$[X] = \begin{bmatrix} \vec{X}^{(1)} & \vec{X}^{(2)} \end{bmatrix} = \begin{bmatrix} 0.2673 & 0.1690 \\ 0.5346 & -0.8450 \end{bmatrix} \tag{E.3}$$

Using

$$\vec{x}(t) = [X]\vec{q}(t) \tag{E.4}$$

Eq. (E.1) can be expressed as (see Eq. (6.112)):

$$\ddot{\vec{q}}(t) + [\diagdown\omega^2\diagdown]\vec{q}(t) = \vec{Q}(t) = \vec{0} \tag{E.5}$$

where $\vec{Q}(t) = [X]^T\vec{F} = \vec{0}$. Equation (E.5) can be written in scalar form as

$$\ddot{q}_i(t) + \omega_i^2 q_i(t) = 0, \qquad i = 1, 2 \tag{E.6}$$

The solution of Eq. (E.6) is given by (see Eq. (2.18)):

$$q_i(t) = q_{i0} \cos \omega_i t + \frac{\dot{q}_{i0}}{\omega_i} \sin \omega_i t \tag{E.7}$$

where q_{i0} and $\dot{q}_{i0}$ denote the initial values of $q_i(t)$ and $\dot{q}_i(t)$, respectively. Using the initial conditions of Eq. (E.2), we can find (see Eqs. (6.115) and (6.116)):

$$\vec{q}(0) = \begin{Bmatrix} q_{10}(0) \\ q_{20}(0) \end{Bmatrix} = [X]^T[m]\,\vec{x}(0)$$

$$= \begin{bmatrix} 0.2673 & 0.5346 \\ 0.1690 & -0.8450 \end{bmatrix} \begin{bmatrix} 10 & 0 \\ 0 & 1 \end{bmatrix} \begin{Bmatrix} 1 \\ 0 \end{Bmatrix} = \begin{Bmatrix} 2.673 \\ 1.690 \end{Bmatrix} \tag{E.8}$$

$$\dot{\vec{q}}(0) = \begin{Bmatrix} \dot{q}_{10}(0) \\ \dot{q}_{20}(0) \end{Bmatrix} = [X]^T[m]\dot{\vec{x}}(0) = \begin{Bmatrix} 0 \\ 0 \end{Bmatrix} \tag{E.9}$$

Equations (E.7) to (E.9) lead to

$$q_1(t) = 2.673 \cos 1.5811t \qquad \text{(E.10)}$$

$$q_2(t) = 1.690 \cos 2.4495t \qquad \text{(E.11)}$$

Using Eq. (E.4), we obtain the displacements of the masses m_1 and m_2 as

$$\vec{x}(t) = \begin{bmatrix} 0.2673 & 0.1690 \\ 0.5346 & -0.8450 \end{bmatrix} \begin{Bmatrix} 2.673 \cos 1.5811t \\ 1.690 \cos 2.4495t \end{Bmatrix}$$

or

$$\begin{Bmatrix} x_1(t) \\ x_2(t) \end{Bmatrix} = \begin{Bmatrix} 0.7145 \cos 1.5811t + 0.2856 \cos 2.4495t \\ 1.4280 \cos 1.5811t - 1.4280 \cos 2.4495t \end{Bmatrix} \qquad \text{(E.12)}$$

It can be seen that this solution is identical to the one obtained in Example 5.3 and plotted in Example 5.17.

∎

Forced-Vibration Response of a Forging Hammer

EXAMPLE 6.17

The force acting on the workpiece of the forging hammer shown in Fig. 5.51 due to impact by the hammer can be approximated as a rectangular pulse, as shown in Fig. 6.15(a). Find the resulting vibration of the system for the following data: mass of the workpiece, anvil, and frame (m_1) = 200 Mg, mass of the foundation block (m_2) = 250 Mg, stiffness of the elastic pad (k_1) = 150 MN/m, and stiffness of the soil (k_2) = 75 MN/m. Assume the initial displacements and initial velocities of the masses as zero.

Solution: The forging hammer can be modeled as a two-degree-of-freedom system as indicated in Fig. 6.15(b). The equations of motion of the system can be expressed as

$$[m]\ddot{\vec{x}} + [k]\vec{x} = \vec{F}(t) \qquad \text{(E.1)}$$

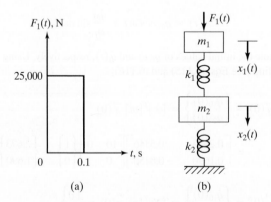

(a)   (b)

FIGURE 6.15   Impact caused by forging hammer.

where

$$[m] = \begin{bmatrix} m_1 & 0 \\ 0 & m_2 \end{bmatrix} = \begin{bmatrix} 200 & 0 \\ 0 & 250 \end{bmatrix} \text{Mg}$$

$$[k] = \begin{bmatrix} k_1 & -k_1 \\ -k_1 & k_1 + k_2 \end{bmatrix} = \begin{bmatrix} 150 & -150 \\ -150 & 225 \end{bmatrix} \text{MN/m}$$

$$\vec{F}(t) = \begin{Bmatrix} F_1(t) \\ 0 \end{Bmatrix}$$

Natural Frequencies and Mode Shapes: The natural frequencies of the system can be found by solving the frequency equation

$$\left| -\omega^2 [m] + [k] \right| = \left| -\omega^2 \begin{bmatrix} 2 & 0 \\ 0 & 2.5 \end{bmatrix} 10^5 + \begin{bmatrix} 150 & -150 \\ -150 & 225 \end{bmatrix} 10^6 \right| = 0 \qquad \text{(E.2)}$$

as

$$\omega_1 = 12.2474 \text{ rad/s} \qquad \text{and} \qquad \omega_2 = 38.7298 \text{ rad/s}$$

The mode shapes can be found as

$$\vec{X}^{(1)} = \begin{Bmatrix} 1 \\ 0.8 \end{Bmatrix} \qquad \text{and} \qquad \vec{X}^{(2)} = \begin{Bmatrix} 1 \\ -1 \end{Bmatrix}$$

Orthonormalization of Mode Shapes: The mode shapes are assumed as

$$\vec{X}^{(1)} = a \begin{Bmatrix} 1 \\ 0.8 \end{Bmatrix} \qquad \text{and} \qquad \vec{X}^{(2)} = b \begin{Bmatrix} 1 \\ -1 \end{Bmatrix}$$

where a and b are constants. The constants a and b can be determined by normalizing the vectors $\vec{X}^{(1)}$ and $\vec{X}^{(2)}$ as

$$[X]^T [m] [X] = [I] \qquad \text{(E.3)}$$

where $[X] = [\vec{X}^{(1)} \vec{X}^{(2)}]$ denotes the modal matrix. Equation (E.3) gives $a = 1.6667 \times 10^{-3}$ and $b = 1.4907 \times 10^{-3}$, which means that the new modal matrix (with normalized mode shapes) becomes

$$[X] = [\vec{X}^{(1)} \vec{X}^{(2)}] = \begin{bmatrix} 1.6667 & 1.4907 \\ 1.3334 & -1.4907 \end{bmatrix} \times 10^{-3}$$

Response in Terms of Generalized Coordinates: Since the two masses m_1 and m_2 are at rest at $t = 0$, the initial conditions are $x_1(0) = x_2(0) = \dot{x}_1(0) = \dot{x}_2(0) = 0$, hence Eqs. (6.115) and (6.116) give $q_1(0) = q_2(0) = \dot{q}_1(0) = \dot{q}_2(0) = 0$. Thus the generalized coordinates are given by the solution of the equations

$$q_i(t) = \frac{1}{\omega_i} \int_0^t Q_i(\tau) \sin \omega_i(t - \tau) \, d\tau, \, i = 1, 2 \qquad \text{(E.4)}$$

where

$$\vec{Q}(t) = [X]^T \vec{F}(t) \tag{E.5}$$

or

$$\begin{Bmatrix} Q_1(t) \\ Q_2(t) \end{Bmatrix} = \begin{bmatrix} 1.6667 & 1.3334 \\ 1.4907 & -1.4907 \end{bmatrix} 10^{-3} \begin{Bmatrix} F_1(t) \\ 0 \end{Bmatrix}$$

$$= \begin{Bmatrix} 1.6667 \times 10^{-3} F_1(t) \\ 1.4907 \times 10^{-3} F_1(t) \end{Bmatrix} \tag{E.6}$$

with $F_1(t) = 25{,}000$ N for $0 \le t \le 0.1$ s and 0 for $t > 0.1$ s. Using Eq. (6.104), the displacements of the masses can be found as

$$\begin{Bmatrix} x_1(t) \\ x_2(t) \end{Bmatrix} = [X] \vec{q}(t) = \begin{Bmatrix} 1.6667 q_1(t) + 1.4907 q_2(t) \\ 1.3334 q_1(t) - 1.4907 q_2(t) \end{Bmatrix} 10^{-3} \text{ m} \tag{E.7}$$

where

$$q_1(t) = 3.4021 \int_0^t \sin 12.2474 \, (t - \tau) \, d\tau = 0.2778 \, (1 - \cos 12.2474 \, t)$$

$$q_2(t) = 0.9622 \int_0^t \sin 38.7298 \, (t - \tau) \, d\tau = 0.02484 \, (1 - \cos 38.7298 \, t) \tag{E.8}$$

Note that the solution given by Eqs. (E.8) is valid for $t \le 0.1$ s. For $t > 0.1$ s, there is no applied force, hence the response is given by the free-vibration solution of an undamped single-degree-of-freedom system (Eq. (2.18)) for $q_1(t)$ and $q_2(t)$ with $q_1(0.1)$ and $\dot{q}_1(0.1)$, and $q_2(0.1)$ and $\dot{q}_2(0.1)$ as initial conditions for $q_1(t)$ and $q_2(t)$, respectively.

■

6.15 Forced Vibration of Viscously Damped Systems

Modal analysis, as presented in Section 6.14, applies only to undamped systems. In many cases, the influence of damping upon the response of a vibratory system is minor and can be disregarded. However, it must be considered if the response of the system is required for a relatively long period of time compared to the natural periods of the system. Further, if the frequency of excitation (in the case of a periodic force) is at or near one of the natural frequencies of the system, damping is of primary importance and must be taken into account. In general, since the effects are not known in advance, damping must be considered in the vibration analysis of any system. In this section, we shall consider the equations of motion of a damped multidegree-of-freedom system and their solution using Lagrange's equations. If the system has viscous damping, its motion will be resisted by a force whose magnitude is proportional to that of the velocity but in the opposite direction. It is convenient to introduce a function R, known as *Rayleigh's dissipation function*, in deriving the equations of motion by means of Lagrange's equations [6.7]. This function is defined as

$$R = \frac{1}{2}\dot{\vec{x}}^T[c]\dot{\vec{x}} \qquad (6.117)$$

where the matrix $[c]$ is called the *damping matrix* and is positive definite, like the mass and stiffness matrices. Lagrange's equations, in this case [6.8], can be written as

$$\frac{d}{dt}\left(\frac{\partial T}{\partial \dot{x}_i}\right) - \frac{\partial T}{\partial x_i} + \frac{\partial R}{\partial \dot{x}_i} + \frac{\partial V}{\partial x_i} = F_i, \qquad i = 1, 2, \ldots, n \qquad (6.118)$$

where F_i is the force applied to mass m_i. By substituting Eqs. (6.30), (6.34), and (6.117) into Eq. (6.118), we obtain the equations of motion of a damped multidegree-of-freedom system in matrix form:

$$[m]\ddot{\vec{x}} + [c]\dot{\vec{x}} + [k]\vec{x} = \vec{F} \qquad (6.119)$$

For simplicity, we shall consider a special system for which the damping matrix can be expressed as a linear combination of the mass and stiffness matrices:

$$[c] = \alpha[m] + \beta[k] \qquad (6.120)$$

where α and β are constants. This is known as *proportional damping* because $[c]$ is proportional to a linear combination of $[m]$ and $[k]$. By substituting Eq. (6.120) into Eq. (6.119), we obtain

$$[m]\ddot{\vec{x}} + [\alpha[m] + \beta[k]]\dot{\vec{x}} + [k]\vec{x} = \vec{F} \qquad (6.121)$$

By expressing the solution vector $\vec{x}$ as a linear combination of the natural modes of the undamped system, as in the case of Eq. (6.104),

$$\vec{x}(t) = [X]\vec{q}(t) \qquad (6.122)$$

Eq. (6.121) can be rewritten as

$$[m][X]\ddot{\vec{q}}(t) + [\alpha[m] + \beta[k]][X]\dot{\vec{q}}(t)$$
$$+ [k][X]\vec{q}(t) = \vec{F}(t) \qquad (6.123)$$

Premultiplication of Eq. (6.123) by $[X]^T$ leads to

$$[X]^T[m][X]\ddot{\vec{q}} + [\alpha[X]^T[m][X] + \beta[X]^T[k][X]]\dot{\vec{q}}$$
$$+ [X]^T[k][X]\vec{q} = [X]^T\vec{F} \qquad (6.124)$$

If the eigenvectors $\vec{X}^{(j)}$ are normalized according to Eqs. (6.74) and (6.75), Eq. (6.124) reduces to

$$[I]\ddot{\vec{q}}(t) + \left[\alpha[I] + \beta\left[\diagdown\omega^2\diagdown\right]\right]\dot{\vec{q}}(t) + \left[\diagdown\omega^2\diagdown\right]\vec{q}(t) = \vec{Q}(t)$$

—that is,

$$\ddot{q}_i(t) + (\alpha + \omega_i^2\beta)\dot{q}_i(t) + \omega_i^2 q_i(t) = Q_i(t),$$

$$i = 1, 2, \ldots, n \tag{6.125}$$

where ω_i is the ith natural frequency of the undamped system and

$$\vec{Q}(t) = [X]^T \vec{F}(t) \tag{6.126}$$

By writing

$$\alpha + \omega_i^2\beta = 2\zeta_i\omega_i \tag{6.127}$$

where ζ_i is called the *modal damping ratio* for the ith normal mode, Eqs. (6.125) can be rewritten as

$$\ddot{q}_i(t) + 2\zeta_i\omega_i\dot{q}_i(t) + \omega_i^2 q_i(t) = Q_i(t), \qquad i = 1, 2, \ldots, n \tag{6.128}$$

It can be seen that each of the n equations represented by this expression is uncoupled from all of the others. Hence we can find the response of the ith mode in the same manner as that of a viscously damped single-degree-of-freedom system. The solution of Eq. (6.128), when $\zeta_i < 1$, can be expressed as

$$q_i(t) = e^{-\zeta_i\omega_i t}\left\{\cos\omega_{di}t + \frac{\zeta_i}{\sqrt{1 - \zeta_i^2}}\sin\omega_{di}t\right\}q_i(0)$$

$$+ \left\{\frac{1}{\omega_{di}}e^{-\zeta_i\omega_i t}\sin\omega_{di}t\right\}\dot{q}_0(0)$$

$$+ \frac{1}{\omega_{di}}\int_0^t Q_i(\tau)e^{-\zeta_i\omega_i(t-\tau)}\sin\omega_{di}(t - \tau)\,d\tau,$$

$$i = 1, 2, \ldots, n \tag{6.129}$$

where

$$\omega_{di} = \omega_i\sqrt{1 - \zeta_i^2} \tag{6.130}$$

Note the following aspects of these systems:

1. The identification of the sources and magnitude of damping is difficult in most practical problems. More than one type of damping—Coulomb, viscous, and hysteretic—may be present in the system. In addition, the exact nature of damping, such as linear, quadratic, cubic, or other type of variation, is not known. Even when the source and nature of damping are known, obtaining the precise magnitude is very difficult. For some practical systems, experimentally determined damping values may be available for use in vibration analysis. Some damping, in the form of structural damping, is present in automobile, aerospace, and machine structures. Damping is introduced deliberately in certain practical applications such as vehicle suspension systems, aircraft landing gear, and machine isolation systems. Because the analysis of damped systems involves lengthy mathematical manipulations, in many vibration studies, damping is either neglected or assumed to be proportional.

2. It has been shown by Caughey [6.9] that the condition given by Eq. (6.120) is sufficient but not necessary for the existence of normal modes in damped systems. The necessary condition is that the transformation that diagonalizes the damping matrix also uncouples the coupled equations of motion. This condition is less restrictive than Eq. (6.120) and covers more possibilities.

3. In the general case of damping, the damping matrix cannot be diagonalized simultaneously with the mass and stiffness matrices. In this case, the eigenvalues of the system are either real and negative or complex with negative real parts. The complex eigenvalues exist as conjugate pairs: the associated eigenvectors also consist of complex conjugate pairs. A common procedure for finding the solution of the eigenvalue problem of a damped system involves the transformation of the n coupled second-order equations of motion into $2n$ uncoupled first-order equations [6.6].

4. The error bounds and numerical methods in the modal analysis of dynamic systems are discussed in [6.11, 6.12].

Equations of Motion of a Dynamic System

EXAMPLE 6.18

Derive the equations of motion of the system shown in Fig. 6.16.

Solution:

Approach: Use Lagrange's equations in conjunction with Rayleigh's dissipation function.
The kinetic energy of the system is

$$T = \frac{1}{2}(m_1\dot{x}_1^2 + m_2\dot{x}_2^2 + m_3\dot{x}_3^2) \tag{E.1}$$

The potential energy has the form

$$V = \frac{1}{2}[k_1x_1^2 + k_2(x_2 - x_1)^2 + k_3(x_3 - x_2)^2] \tag{E.2}$$

FIGURE 6.16 Three-degree-of-freedom dynamic system.

and Rayleigh's dissipation function is

$$R = \frac{1}{2}[c_1\dot{x}_1^2 + c_2(\dot{x}_2 - \dot{x}_1)^2 + c_3(\dot{x}_3 - \dot{x}_2)^2 + c_4\dot{x}_2^2 + c_5(\dot{x}_3 - \dot{x}_1)^2] \tag{E.3}$$

Lagrange's equations can be written as

$$\frac{d}{dt}\left(\frac{\partial T}{\partial \dot{x}_i}\right) - \frac{\partial T}{\partial x_i} + \frac{\partial R}{\partial \dot{x}_i} + \frac{\partial V}{\partial x_i} = F_i, \qquad i = 1, 2, 3 \tag{E.4}$$

By substituting Eqs. (E.1) to (E.3) into Eq. (E.4), we obtain the differential equations of motion

$$[m]\ddot{\vec{x}} + [c]\dot{\vec{x}} + [k]\vec{x} = \vec{F} \tag{E.5}$$

where

$$[m] = \begin{bmatrix} m_1 & 0 & 0 \\ 0 & m_2 & 0 \\ 0 & 0 & m_3 \end{bmatrix} \tag{E.6}$$

$$[c] = \begin{bmatrix} c_1 + c_2 + c_5 & -c_2 & -c_5 \\ -c_2 & c_2 + c_3 + c_4 & -c_3 \\ -c_5 & -c_3 & c_3 + c_5 \end{bmatrix} \tag{E.7}$$

$$[k] = \begin{bmatrix} k_1 + k_2 & -k_2 & 0 \\ -k_2 & k_2 + k_3 & -k_3 \\ 0 & -k_3 & k_3 \end{bmatrix} \tag{E.8}$$

$$\vec{x} = \begin{Bmatrix} x_1(t) \\ x_2(t) \\ x_3(t) \end{Bmatrix} \quad \text{and} \quad \vec{F} = \begin{Bmatrix} F_1(t) \\ F_2(t) \\ F_3(t) \end{Bmatrix} \tag{E.9}$$

■

EXAMPLE 6.19

Steady-State Response of a Forced System

Find the steady-state response of the system shown in Fig. 6.16 when the masses are subjected to the simple harmonic forces $F_1 = F_2 = F_3 = F_0 \cos \omega t$, where $\omega = 1.75\sqrt{k/m}$. Assume that $m_1 = m_2 = m_3 = m$, $k_1 = k_2 = k_3 = k$, $c_4 = c_5 = 0$, and the damping ratio in each normal mode is given by $\zeta_i = 0.01$, $i = 1, 2, 3$.

Solution: The (undamped) natural frequencies of the system (see Example 6.11) are given by

$$\omega_1 = 0.44504\sqrt{\frac{k}{m}}$$

$$\omega_2 = 1.2471\sqrt{\frac{k}{m}}$$

$$\omega_3 = 1.8025\sqrt{\frac{k}{m}} \tag{E.1}$$

and the corresponding $[m]$-orthonormal mode shapes (see Example 6.12) are given by

$$\vec{X}^{(1)} = \frac{0.3280}{\sqrt{m}} \begin{Bmatrix} 1.0 \\ 1.8019 \\ 2.2470 \end{Bmatrix}, \quad \vec{X}^{(2)} = \frac{0.7370}{\sqrt{m}} \begin{Bmatrix} 1.0 \\ 1.4450 \\ -0.8020 \end{Bmatrix},$$

$$\vec{X}^{(3)} = \frac{0.5911}{\sqrt{m}} \begin{Bmatrix} 1.0 \\ -1.2468 \\ 0.5544 \end{Bmatrix} \tag{E.2}$$

Thus the modal vector can be expressed as

$$[X] = [\vec{X}^{(1)}\vec{X}^{(2)}\vec{X}^{(3)}] = \frac{1}{\sqrt{m}} \begin{bmatrix} 0.3280 & 0.7370 & 0.5911 \\ 0.5911 & 0.3280 & -0.7370 \\ 0.7370 & -0.5911 & 0.3280 \end{bmatrix} \tag{E.3}$$

The generalized force vector

$$\vec{Q}(t) = [X]^T \vec{F}(t) = \frac{1}{\sqrt{m}} \begin{bmatrix} 0.3280 & 0.5911 & 0.7370 \\ 0.7370 & 0.3280 & -0.5911 \\ 0.5911 & -0.7370 & 0.3280 \end{bmatrix} \begin{Bmatrix} F_0 \cos \omega t \\ F_0 \cos \omega t \\ F_0 \cos \omega t \end{Bmatrix}$$

$$= \begin{Bmatrix} Q_{10} \\ Q_{20} \\ Q_{30} \end{Bmatrix} \cos \omega t \tag{E.4}$$

can be obtained where

$$Q_{10} = 1.6561 \frac{F_0}{\sqrt{m}}, \qquad Q_{20} = 0.4739 \frac{F_0}{\sqrt{m}}, \qquad Q_{30} = 0.1821 \frac{F_0}{\sqrt{m}} \tag{E.5}$$

If the generalized coordinates or the modal participation factors for the three principal modes are denoted as $q_1(t)$, $q_2(t)$, and $q_3(t)$, the equations of motion can be expressed as

$$\ddot{q}_i(t) + 2\zeta_i \omega_i \dot{q}_i(t) + \omega_i^2 q_i(t) = Q_i(t), \qquad i = 1, 2, 3 \tag{E.6}$$

The steady-state solution of Eqs. (E.6) can be written as

$$q_i(t) = q_{i0} \cos(\omega t - \phi), \qquad i = 1, 2, 3 \tag{E.7}$$

where

$$q_{i0} = \frac{Q_{i0}}{\omega_i^2} \frac{1}{\left[\left\{ 1 - \left(\frac{\omega}{\omega_i} \right)^2 \right\}^2 + \left(2\zeta_i \frac{\omega}{\omega_i} \right)^2 \right]^{1/2}} \tag{E.8}$$

and

$$\phi_i = \tan^{-1} \left\{ \frac{2\zeta_i \dfrac{\omega}{\omega_i}}{1 - \left(\dfrac{\omega}{\omega_i} \right)^2} \right\} \tag{E.9}$$

By substituting the values given in Eqs. (E.1) and (E.5) into Eqs. (E.8) and (E.9), we obtain

$$q_{10} = 0.57815 \frac{F_0 \sqrt{m}}{k}, \qquad \phi_1 = \tan^{-1}(-0.00544)$$

$$q_{20} = 0.31429 \frac{F_0 \sqrt{m}}{k}, \qquad \phi_2 = \tan^{-1}(-0.02988)$$

$$q_{30} = 0.92493 \frac{F_0 \sqrt{m}}{k}, \qquad \phi_3 = \tan^{-1}(0.33827) \tag{E.10}$$

Finally, the steady-state response can be found using Eq. (6.122).

∎

6.16   Self-Excitation and Stability Analysis

In a number of damped vibratory systems, friction leads to negative damping instead of positive damping. This leads to the instability (or self-excited vibration) of the system. In general, for an n-degree-of-freedom system shown in Fig. 6.17, the equations of motion will be a set of second-order linear differential equations (as given by Eqs. (6.119) or (6.128)):

$$[m]\ddot{\vec{x}} + [c]\dot{\vec{x}} + [k]\vec{x} = \vec{F} \tag{6.131}$$

The method presented in Section 5.8 can be extended to study the stability of the system governed by Eq. (6.131). Accordingly, we assume a solution of the form

$$x_j(t) = C_j e^{st}, \qquad j = 1, 2, \ldots, n$$

or

$$\vec{x}(t) = \vec{C} e^{st} \tag{6.132}$$

where s is a complex number to be determined, C_j is the amplitude of x_j, and

$$\vec{C} = \begin{Bmatrix} C_1 \\ C_2 \\ \vdots \\ C_n \end{Bmatrix}$$

The real part of s determines the damping, and its imaginary part gives the natural frequency of the system. The substitution of Eq. (6.132) into the free-vibration equations (obtained by setting $\vec{F} = \vec{0}$ in Eq. (6.131)) leads to

$$([m]s^2 + [c]s + [k])\vec{C} e^{st} = \vec{0} \tag{6.133}$$

For a nontrivial solution of C_j, the determinant of the coefficients of C_j is set equal to zero, which leads to the "characteristic equation," similar to Eq. (6.63):

$$D(s) = |[m]s^2 + [c]s + [k]| = 0 \tag{6.134}$$

FIGURE 6.17   Multidegree-of-freedom system.

The expansion of Eq. (6.134) leads to a polynomial in s of order $m = 2n$, which can be expressed in the form

$$D(s) = a_0 s^m + a_1 s^{m-1} + a_2 s^{m-2} + \cdots + a_{m-1} s + a_m = 0 \qquad (6.135)$$

The stability or instability of the system depends on the roots of the polynomial equation, $D(s) = 0$. Let the roots of Eq. (6.135) be denoted as

$$s_j = b_j + i\omega_j, \qquad j = 1, 2, \ldots, m \qquad (6.136)$$

If the real parts of all the roots b_j are negative numbers, there will be decaying time functions $e^{b_j t}$, in Eq. (6.132), hence the solution (system) will be stable. On the other hand, if one or more roots s_j have a positive real part, then the solution of Eq. (6.132) will contain one or more exponentially increasing time functions $e^{b_j t}$, hence the solution (system) will be unstable. If there is a purely imaginary root of the form $s_j = i\omega_j$, it will lead to an oscillatory solution $e^{i\omega_j t}$, which represents a borderline case between stability and instability. If s_j is a multiple root, the above conclusion still holds unless it is a pure imaginary number, such as $s_j = i\omega_j$. In this case, the solution contains functions of the type $e^{i\omega_j t}, te^{i\omega_j t}, t^2 e^{i\omega_j t}, \ldots$, which increase with time. Thus the multiple roots with purely imaginary values indicate the instability of the system. As such, in order for a linear system governed by Eq. (6.131) to be stable, it is necessary and sufficient that the roots of Eq. (6.135) should have nonpositive real parts, and that, if any purely imaginary root exists, it should not appear as a multiple root.

Since finding the roots of the polynomial equation (6.135) is a lengthy procedure, a simplified procedure, known as the *Routh-Hurwitz stability criterion* [6.13, 6.14], can be used to investigate the stability of the system. In order to apply this procedure, the following mth-order determinant T_m is defined in terms of the coefficients of the polynomial equation (6.135) as

$$T_m = \begin{vmatrix} a_1 & a_3 & a_5 & a_7 & \cdots & a_{2m-1} \\ a_0 & a_2 & a_4 & a_6 & \cdots & a_{2m-2} \\ 0 & a_1 & a_3 & a_5 & \cdots & a_{2m-3} \\ 0 & a_0 & a_2 & a_4 & \cdots & a_{2m-4} \\ 0 & 0 & a_1 & a_3 & \cdots & a_{2m-5} \\ \vdots \\ \cdot & \cdot & \cdot & \cdot & \cdots & a_m \end{vmatrix} \qquad (6.137)$$

Then the following subdeterminants, indicated by the dashed lines in Eq. (6.137), are defined:

$$T_1 = a_1 \qquad (6.138)$$

$$T_2 = \begin{vmatrix} a_1 & a_3 \\ a_0 & a_2 \end{vmatrix} \qquad (6.139)$$

$$T_3 = \begin{vmatrix} a_1 & a_3 & a_5 \\ a_0 & a_2 & a_4 \\ 0 & a_1 & a_3 \end{vmatrix}$$

$$\vdots \qquad (6.140)$$

In constructing these subdeterminants, all the coefficients a_i with $i > m$ or $i < 0$ are to be replaced by zeros. According to the Routh-Hurwitz criterion, a necessary and sufficient condition for the stability of the system is that all the coefficients $a_0, a_1, \ldots, a_m$ must be positive and all the determinants $T_1, T_2, \ldots, T_m$ must be positive.

6.17   Examples Using MATLAB

Solution of Eigenvalue Problem

EXAMPLE 6.20

Find the eigenvalues and eigenvectors of the matrix (see Example 6.11):

$$[A] = \begin{bmatrix} 1 & 1 & 1 \\ 1 & 2 & 2 \\ 1 & 2 & 3 \end{bmatrix}$$

Solution:

```
% Ex 6.20
>> A = [1 1 1; 1 2 2; 1 2 3]

A =

        1        1        1
        1        2        2
        1        2        3
>> [V, D] = eig(A)

V =

      0.5910      0.7370      0.3280
     -0.7370      0.3280      0.5910
      0.3280     -0.5910      0.7370

D =

      0.3080           0           0
           0      0.6431           0
           0           0      5.0489
```

Free-Vibration Response of a Multidegree-of-Freedom System

EXAMPLE 6.21

Plot the free-vibration response, $x_1(t)$, $x_2(t)$, and $x_3(t)$, of the system considered in Example 6.15 for the following data: $x_{10} = 1.0$, $k = 4,000$, and $m = 10$.

Solution: The free-vibration response of the masses, $x_1(t)$, $x_2(t)$, and $x_3(t)$, is given by Eqs. (E.7) to (E.9) of Example 6.15.

```
% Ex6_21.m
x10 = 1.0;
k = 4000;
m = 10;
for i = 1: 1001
    t(i) = 5* (i-1) / 1000;
    x1 (i) = x10 * ( 0.1076 * cos (0.44504 * sqrt (k/m) * t(i)) +
        0.5431 * cos (1.2471*sqrt(k/m) *t(i)) + 0.3493 *
        cos (1.8025*sqrt (k/m) *t(i)) );
```

```
    x2 (i) = x10 * ( 0.1939 * cos (0.44504 * sqrt (k/m) * t(i)) +
           0.2417 * cos (1.2471*sqrt (k/m) *t(i)) - 0.4355 *
           cos (1.8025*sqrt (k/m) *t(i)) );
    x3 (i) = x10 * ( 0.2418 * cos (0.44504 * sqrt(k/m) * t(i)) -
           0.4356 * cos (1.2471*sqrt (k/m) *t(i)) + 0.1937 *
           cos (1.8025*sqrt (k/m) *t(i)) );
end
subplot (311);
plot (t, x1);
ylabel ('x1 (t) ');
subplot (312);
plot (t, x2);
ylabel ('x2 (t) ');
subplot (313);
plot (t, x3);
ylabel ('x3 (t) ');
xlabel ('t');
```

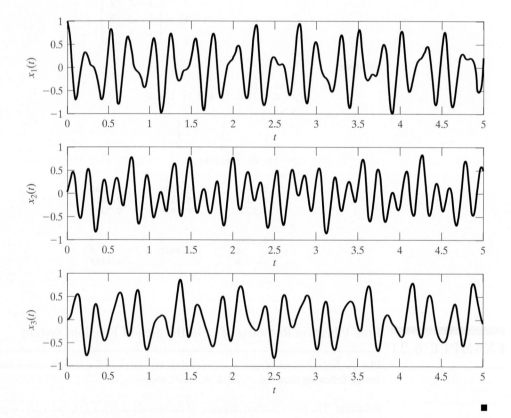

EXAMPLE 6.22

Forced-Vibration Response of a Multidegree-of-Freedom System

Find and plot the forced-vibration response of the forging hammer considered in Example 6.17 by solving the governing differential equations. Assume the initial conditions to be zero.

Solution: The governing equations are given by

$$[m]\ddot{\vec{x}}(t) + [k]\vec{x}(t) = \vec{F}(t) \tag{E.1}$$

with

$$[m] = 10^5 \begin{bmatrix} 2 & 0 \\ 0 & 2.5 \end{bmatrix}, \qquad [k] = 10^6 \begin{bmatrix} 150 & -150 \\ -150 & 225 \end{bmatrix}, \qquad \vec{F}(t) = \begin{Bmatrix} F_1(t) \\ 0 \end{Bmatrix}$$

where $F_1(t)$ is a step function of magnitude 25,000 N and duration $0 \le t \le 0.1$ s.

Equations (E.1) can be expressed as a set of four coupled first-order differential equations as

$$\dot{y}_1 = y_2$$

$$\dot{y}_2 = \frac{F_1}{m_1} - \frac{k_1}{m_1} y_1 + \frac{k_1}{m_1} y_3$$

$$\dot{y}_3 = y_4$$

$$\dot{y}_4 = \frac{k_1}{m_2} y_1 - \frac{k_2}{m_2} y_3$$

where $y_1 = x_1, y_2 = \dot{x}_1, y_3 = x_2, y_4 = \dot{x}_2, m_1 = 2 \times 10^5, m_2 = 2.5 \times 10^5, k_1 = 150 \times 10^6,$ and $k_2 = 225 \times 10^6$.

Using the initial values of all $y_i = 0$, the following results can be obtained.

```
% Ex6_22.m
% This program will use the function dfunc6_21.m, they should
% be in the same folder
tspan = [0: 0.001: 10];
y0 = [0; 0; 0; 0];
[t, y] = ode23 ('dfunc6_21', tspan, y0);
subplot (211);
plot (t, y(:, 1));
xlabel ('t');
ylabel ('x1 (t) ');
subplot (212);
plot (t, y(:, 3));
xlabel ('t');
ylabel ('x2 (t) ');

% dfunc6_21.m
function f = dfunc6_21 (t, y)
f = zeros (4, 1);
m1 = 2*1e5;
m2 = 2.5*1e5;
k1 = 150 * 1e6;
k2 = 225 * 1e6;
F1 = 25000 * (stepfun (t, 0) - stepfun (t, 0.1));
f(1) = y(2);
f(2) = F1/m1 + k1 * y(3) /m1 - k1 * y(1) /m1;
f(3) = y(4);
f(4) = -k2 * y(3) /m2 + k1 * y(1) /m2;
```

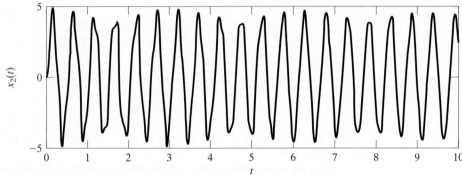

■

EXAMPLE 6.23

Roots of a Polynomial Equation

Using MATLAB, find the roots of the polynomial

$$f(x) = x^3 - 6x^2 + 11x - 6 = 0$$

Solution:

```
>> roots ([1 -6 11 -6])

ans =

    3.0000
    2.0000
    1.0000

>>
```

■

Forced-Vibration Response of a Damped System

EXAMPLE 6.24

Find the forced-vibration response of a damped multidegree-of-freedom system with equations of motion

$$[m]\ddot{\vec{x}} + [c]\dot{\vec{x}} + [k]\vec{x} = \vec{f} \qquad (E.1)$$

with

$$[m] = \begin{bmatrix} 100 & 0 & 0 \\ 0 & 10 & 0 \\ 0 & 0 & 10 \end{bmatrix}, \quad [c] = 100\begin{bmatrix} 4 & -2 & 0 \\ -2 & 4 & -2 \\ 0 & -2 & 2 \end{bmatrix}, \quad [k] = 1,000\begin{bmatrix} 8 & -4 & 0 \\ -4 & 8 & -4 \\ 0 & -4 & 4 \end{bmatrix}$$

$$\vec{f} = \begin{Bmatrix} 1 \\ 1 \\ 1 \end{Bmatrix} F_0 \cos \omega t$$

with $F_0 = 50$ and $\omega = 50$. Assume zero initial conditions.

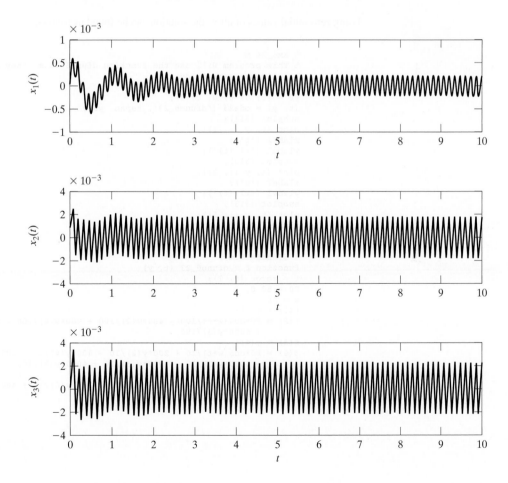

Solution: Equations (E.1) can be rewritten as a set of six first-order differential equations

$$\dot{y}_1 = y_2$$

$$\dot{y}_2 = \frac{F_0}{10}\cos\omega t - \frac{400}{10}y_2 + \frac{200}{10}y_4 - \frac{8{,}000}{10}y_1 + \frac{4{,}000}{10}y_3$$

$$\dot{y}_3 = y_4$$

$$\dot{y}_4 = \frac{F_0}{10}\cos\omega t + \frac{200}{10}y_2 - \frac{400}{10}y_4 + \frac{200}{10}y_6 + \frac{4{,}000}{10}y_1 - \frac{8{,}000}{10}y_3 + \frac{4{,}000}{10}y_5$$

$$\dot{y}_5 = y_6$$

$$\dot{y}_6 = \frac{F_0}{10}\cos\omega t + \frac{200}{10}y_4 - \frac{200}{10}y_6 + \frac{4{,}000}{10}y_3 - \frac{4{,}000}{10}y_5$$

where $y_1 = x_1, y_2 = \dot{x}_1, y_3 = x_2, y_4 = \dot{x}_2, y_5 = x_3$, and $y_6 = \dot{x}_3$.

Using zero initial values of all y_i, the solution can be found as follows.

```
% Ex6_24.m
% This program will use the function dfunc6_23.m, they should
% be in the same folder
tspan = [0: 0.01: 10];
y0 = [0; 0; 0; 0; 0; 0];
[t, y] = ode23 ('dfunc6_23', tspan, y0);
subplot (311);
plot (t, y (:, 1));
xlabel ('t');
ylabel ('x1 (t)');
subplot (312);
plot (t, y (:, 3));
xlabel ('t');
ylabel ('x2 (t)');
subplot (313);
plot (t, y (:, 5));
xlabel ('t');
ylabel ('x3 (t)');

% dfunc6_23.m
function f = dfunc6_23 (t, y)
f = zeros (6, 1);
F0 = 50.0;
w = 50.0;
f(1) = y(2);
f(2) = F0*cos(w*t)/100 - 400*y(2)/100 + 200*y(4)/100 - 8000*y(1)/100
       + 4000*y(3)/100;
f(3) = y(4);
f(4) = F0*cos(w*t)/10 + 200*y(2)/10 - 400*y(4)/10 + 200*y(6)/10
       + 4000*y(1)/10 - 8000*y(3)/10 + 4000*y(5)/10;
f(5) = y(6);
f(6) = F0*cos(w*t)/10 + 200*y(4)/10 - 200*y(6)/10 + 4000*y(3)/10
       - 4000*y(5)/10;
```

EXAMPLE 6.25

Program to Generate Characteristic Polynomial

Develop a general computer program, called `Program7.m`, to generate the characteristic polynomial corresponding to a given square matrix. Use the program to generate the characteristic polynomial corresponding to the matrix

$$[A] = \begin{bmatrix} 2 & -1 & 0 \\ -1 & 2 & -1 \\ 0 & -1 & 2 \end{bmatrix}$$

Solution: `Program7.m` is developed to accept the following input data:

n = order of the matrix [A]

$[a]$ = given matrix [A]

The following output is generated by the program:

pcf = vector of polynomial coefficients starting from the constant term.

```
>> program7

polynomial expansion of a determinantal equation

data: determinant A:

    2.000000e+000   -1.000000e+000    0.000000e+000
   -1.000000e+000    2.000000e+000   -1.000000e+000
    0.000000e+000   -1.000000e+000    2.000000e+000

result: polynomial coefficients in
pcf(np)*(x^n)+pcf(n)*(x^(n-1))+...+pcf(2)+pcf(1)=0

   -4.000000e+000    1.000000e+001   -6.000000e+000    1.000000e+000
```

■

EXAMPLE 6.26

Program for Modal Analysis of Multidegree-of-Freedom Systems

Develop a MATLAB program, called `Program8.m`, to find the response of a multidegree-of-freedom system using modal analysis. Use the program to find the solution of a system with the following data:

Mass matrix:

$$[m] = \begin{bmatrix} 1 & 0 & 0 \\ 0 & 1 & 0 \\ 0 & 0 & 1 \end{bmatrix}$$

Modal matrix (with modes as columns; modes are not made m-orthogonal):

$$[ev] = \begin{bmatrix} 1.0000 & 1.0000 & 1.0000 \\ 1.8019 & 0.4450 & -1.2468 \\ 2.2470 & -0.8020 & 0.5544 \end{bmatrix}$$

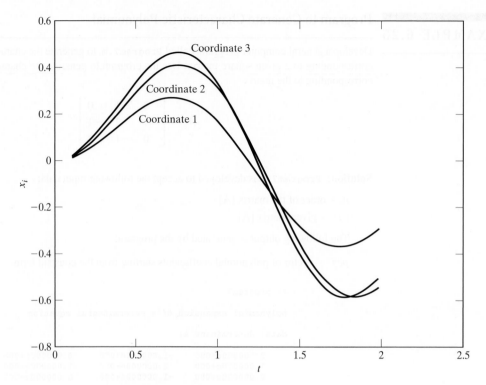

Natural frequencies: $\omega_1 = 0.89008$, $\omega_2 = 1.4942$, $\omega_3 = 3.6050$
Modal damping ratios: $\zeta_i = 0.01$, $i = 1, 2, 3$
Vector of forces applied to different masses:

$$\vec{F}(t) = \begin{Bmatrix} F_0 \\ F_0 \\ F_0 \end{Bmatrix} \cos \omega t; \qquad F_0 = 2.0, \qquad \omega = 3.5$$

Initial conditions: $\vec{x}(0) = \vec{0}$, $\dot{\vec{x}}(0) = \vec{0}$

Solution: `Program8.m` is developed to accept the following input data:

n = degree of freedom of the system

nvec = number of modes to be used in the modal analysis

xm = mass matrix of size $n \times n$

ev = modal matrix of size $n \times nvec$

z = vector of size $nvec$ = modal damping ratios vector

om = vector of size $nvec$ = natural frequencies vector

f = vector of forces applied to masses, of size n

x_0 = initial displacements of masses, vector of size n

xd_0 = initial velocities of masses, vector of size n

nstep = number of time stations or integration points $t_1, t_2, \ldots, t_{nstep}$

delt = interval between consecutive time stations

t = array of size nstep containing times $t_1, t_2, \ldots, t_{nstep}$

The program gives the following output:

x = matrix of size $n \times$ nstep = displacements of masses $m_1, m_2, \ldots, m_n$ at various time stations $t_1, t_2, \ldots, t_{nstep}$

```
>> program8

Response of system using modal analysis

Coordinate 1
 1.21920e-002  4.62431e-002  9.57629e-002  1.52151e-001  2.05732e-001
 2.47032e-001  2.68028e-001  2.63214e-001  2.30339e-001  1.70727e-001
 8.91432e-002 -6.79439e-003 -1.07562e-001 -2.02928e-001 -2.83237e-001
-3.40630e-001 -3.70023e-001 -3.69745e-001 -3.41725e-001 -2.91231e-001

Coordinate 2
 1.67985e-002  6.40135e-002  1.33611e-001  2.14742e-001  2.94996e-001
 3.61844e-001  4.04095e-001  4.13212e-001  3.84326e-001  3.16843e-001
 2.14565e-001  8.53051e-002 -5.99475e-002 -2.08242e-001 -3.46109e-001
-4.61071e-001 -5.43061e-001 -5.85566e-001 -5.86381e-001 -5.47871e-001

Coordinate 3
 1.99158e-002  7.57273e-002  1.57485e-001  2.51794e-001  3.43491e-001
 4.17552e-001  4.60976e-001  4.64416e-001  4.23358e-001  3.38709e-001
 2.16699e-001  6.81361e-002 -9.29091e-002 -2.50823e-001 -3.90355e-001
-4.98474e-001 -5.65957e-001 -5.88490e-001 -5.67173e-001 -5.08346e-001
```

∎

CHAPTER SUMMARY

The analysis of multidegree-of-freedom systems requires tedious algebraic manipulations. Matrix representation can be used to simplify the manipulations. We derived the equations of motion using three different approaches—Newton's second law of motion, influence coefficients, and Lagrange's equations. We presented the computation of the natural frequencies by solving the eigenvalue problem. We used modal analysis procedure for the free and forced vibration of undamped and proportionately damped systems. Finally, we presented the free- and forced-vibration solution of multidegree-of-freedom problems using MATLAB.

Now that you have finished this chapter, you should be able to answer the review questions and solve the problems given below.

REFERENCES

6.1 F. W. Beaufait, *Basic Concepts of Structural Analysis*, Prentice Hall, Englewood Cliffs, NJ, 1977.

6.2 R. J. Roark and W. C. Young, *Formulas for Stress and Strain* (5th ed.), McGraw-Hill, New York, 1975.

6.3 D. A. Wells, *Theory and Problems of Lagrangian Dynamics*, Schaum's Outline Series, McGraw-Hill, New York, 1967.

6.4 J. H. Wilkinson, *The Algebraic Eigenvalue Problem*, Clarendon Press, Oxford, 1965.

6.5 A. Ralston, *A First Course in Numerical Analysis*, McGraw-Hill, New York, 1965.

6.6 L. Meirovitch, *Analytical Methods in Vibrations*, Macmillan, New York, 1967.

6.7 J. W. Strutt, Lord Rayleigh, *The Theory of Sound*, Macmillan, London, 1877 (reprinted by Dover Publications, New York, in 1945).

6.8 W. C. Hurty and M. F. Rubinstein, *Dynamics of Structures*, Prentice Hall, Englewood Cliffs, NJ, 1964.

6.9 T. K. Caughey, "Classical normal modes in damped linear dynamic systems," *Journal of Applied Mechanics*, Vol. 27, 1960, pp. 269–271.

6.10 A. Avakian and D. E. Beskos, "Use of dynamic stiffness influence coefficients in vibrations of non-uniform beams," letter to the editor, *Journal of Sound and Vibration*, Vol. 47, 1976, pp. 292–295.

6.11 I. Gladwell and P. M. Hanson, "Some error bounds and numerical experiments in modal methods for dynamics of systems," *Earthquake Engineering and Structural Dynamics*, Vol. 12, 1984, pp. 9–36.

6.12 R. Bajan, A. R. Kukreti, and C. C. Feng, "Method for improving incomplete modal coupling," *Journal of Engineering Mechanics*, Vol. 109, 1983, pp. 937–949.

6.13 E. J. Routh, *Advanced Rigid Dynamics*, Macmillan, New York, 1905.

6.14 D. W. Nicholson and D. J. Inman, "Stable response of damped linear systems," *Shock and Vibration Digest*, Vol. 15, November 1983, pp. 19–25.

6.15 P. C. Chen and W. W. Soroka, "Multidegree dynamic response of a system with statistical properties," *Journal of Sound and Vibration*, Vol. 37, 1974, pp. 547–556.

6.16 S. Mahalingam and R. E. D. Bishop, "The response of a system with repeated natural frequencies to force and displacement excitation," *Journal of Sound and Vibration*, Vol. 36, 1974, pp. 285–295.

6.17 S. G. Kelly, *Fundamentals of Mechanical Vibrations*, McGraw-Hill, New York, 1993.

6.18 S. S. Rao, *Applied Numerical Methods for Engineers and Scientists*, Prentice Hall, Upper Saddle River, NJ, 2002.

REVIEW QUESTIONS

6.1 Give brief answers to the following:

1. Define the flexibility and stiffness influence coefficients. What is the relation between them?
2. Write the equations of motion of a multidegree-of-freedom system in matrix form using
   a. the flexibility matrix and
   b. the stiffness matrix.
3. Express the potential and kinetic energies of an *n*-degree-of-freedom system using matrix notation.
4. What is a generalized mass matrix?
5. Why is the mass matrix $[m]$ always positive definite?
6. Is the stiffness matrix $[k]$ always positive definite? Why?

7. What is the difference between generalized coordinates and Cartesian coordinates?
8. State Lagrange's equations.
9. What is an eigenvalue problem?
10. What is a mode shape? How is it computed?
11. How many distinct natural frequencies can exist for an n-degree-of-freedom system?
12. What is a dynamical matrix? What is its use?
13. How is the frequency equation derived for a multidegree-of-freedom system?
14. What is meant by the orthogonality of normal modes? What are orthonormal modal vectors?
15. What is a basis in n-dimensional space?
16. What is the expansion theorem? What is its importance?
17. Explain the modal analysis procedure.
18. What is a rigid-body mode? How is it determined?
19. What is a degenerate system?
20. How can we find the response of a multidegree-of-freedom system using the first few modes only?
21. Define Rayleigh's dissipation function.
22. Define these terms: *proportional damping, modal damping ratio, modal participation factor.*
23. When do we get complex eigenvalues?
24. What is the use of the Routh-Hurwitz criterion?

6.2 Indicate whether each of the following statements is true or false:

1. For a multidegree-of-freedom system, one equation of motion can be written for each degree of freedom.
2. Lagrange's equation cannot be used to derive the equations of motion of a multidegree-of-freedom system.
3. The mass, stiffness, and damping matrices of a multidegree-of-freedom system are always symmetric.
4. The product of stiffness and flexibility matrices of a system is always an identity matrix.
5. The modal analysis of a n-degree-of-freedom system can be conducted using r modes with $r < n$.
6. For a damped multidegree-of-freedom system, all the eigenvalues can be complex.
7. The modal damping ratio denotes damping in a particular normal mode.
8. A multidegree-of-freedom system can have six of the natural frequencies equal to zero.
9. The generalized coordinates will always have the unit of length.
10. The generalized coordinates are independent of the conditions of constraint of the system.
11. The generalized mass matrix of a multidegree-of-freedom system is always diagonal.
12. The potential and kinetic energies of a multidegree-of-freedom system are always quadratic functions.
13. The mass matrix of a system is always symmetric and positive definite.
14. The stiffness matrix of a system is always symmetric and positive definite.
15. The rigid-body mode is also called the *zero mode.*
16. An unrestrained system is also known as a *semidefinite system.*
17. Newton's second law of motion can always be used to derive the equations of motion of a vibrating system.

6.3 Fill in each of the following blanks with the appropriate word:

1. The spring constant denotes the _____ necessary to cause a unit elongation.
2. The flexibility influence coefficient a_{ij} denotes the deflection at point _____ due to a unit load at point _____ .

3. The force at point i due to a unit displacement at point j, when all the points other than the point j are fixed, is known as _____ *influence coefficient.*

4. The mode shapes of a multidegree-of-freedom system are _____ .

5. The equations of motion of a multidegree-of-freedom system can be expressed in terms of _____ coefficients.

6. Lagrange's equations are expressed in terms of _____ coordinates.

7. The value of the Kronecker delta (δ_{ij}) is 1 for $i = j$ and _____ for $i \neq j$.

8. The stiffness matrix of a semidefinite system is _____ .

9. A multidegree-of-freedom system can have at most _____ rigid-body modes.

10. When the solution vector is denoted as a linear combination of the normal modes as $\vec{x}(t) = \sum_{i=1}^{n} q_i(t)\vec{X}^{(i)}$, the generalized coordinates $q_i(t)$ are also known as the _____ *participation coefficients.*

11. Any set of n linearly independent vectors in an n-dimensional space is called a _____ .

12. The representation of an arbitrary n-dimensional vector as a linear combination of n linearly independent vectors is known as _____ *theorem.*

13. The _____ analysis is based on the expansion theorem.

14. The modal analysis basically _____ the equations of motion.

15. The eigenvalues of an n-degree-of-freedom system form a _____ in the n-dimensional space.

16. The application of Lagrange's equations requires the availability of _____ expressions.

17. The determinantal equation, $|[k] - \omega^2[m]| = 0$, is known as the _____ *equation.*

18. The symmetry of stiffness and flexibility matrices is due to the _____ reciprocity theorem.

19. Maxwell's reciprocity theorem states that the influence coefficients are _____ .

20. The stiffness matrix is positive definite only if the system is _____ .

21. During free vibration of an undamped system, all coordinates will have _____ motion.

22. In proportional damping, the damping matrix is assumed to be a linear combination of the _____ and _____ matrices.

6.4 Select the most appropriate answer out of the choices given:

1. The number of distinct natural frequencies for an n-degree-of-freedom system can be

   a. 1   **b.** ∞ **c.** n

2. The dynamical matrix, $[D]$, is given by

   a. $[k]^{-1}[m]$ **b.** $[m]^{-1}[k]$ **c.** $[k][m]$

3. The orthognolity of modes implies

   a. $\vec{X}^{(i)^T}[m]\vec{X}^{(j)} = 0$ only

   b. $\vec{X}^{(i)^T}[k]\vec{X}^{(j)} = 0$ only

   c. $\vec{X}^{(i)^T}[m]\vec{X}^{(j)} = 0$ and $\vec{X}^{(i)^T}[k]\vec{X}^{(j)} = 0$

4. The modal matrix, $[X]$, is given by

   a. $[X] = \begin{bmatrix} \vec{X}^{(1)} & \vec{X}^{(2)} & \cdots & \vec{X}^{(n)} \end{bmatrix}$

   b. $[X] = \begin{bmatrix} \vec{X}^{(1)^T} \\ \vec{X}^{(2)^T} \\ \vdots \\ \vec{X}^{(n)^T} \end{bmatrix}$

   c. $[X] = [k]^{-1}[m]$

5. Rayleigh's dissipation function is used to generate a
   a. stiffness matrix   **b.** damping matrix   **c.** mass matrix

6. The characteristic equation of an *n*-degree-of-freedom system is a
   a. transcendental equation
   b. polynomial of degree *n*
   c. differential equation of order *n*

7. The fundamental natural frequency of a system is
   a. the largest value   **b.** the smallest value   **c.** any value

8. Negative damping leads to
   a. instability   **b.** fast convergence   **c.** oscillations

9. The Routh-Hurwitz criterion can be used to investigate the
   a. convergence of a system
   b. oscillations of a system
   c. stability of a system

10. The stiffness and flexibility matrices are related as
    a. $[k] = [a]$ **b.** $[k] = [a]^{-1}$ **c.** $[k] = [a]^T$

11. A system for which $[k]$ is positive and $[m]$ is positive definite is called a
    a. *semidefinite system*
    b. *positive-definite system*
    c. *indefinite system*

12. $[m]$-orthogonality of modal vectors implies
    a. $\vec{X}^{(i)^T}[m]\vec{X}^{(i)} = 0$
    b. $\vec{X}^{(i)^T}[m]\vec{X}^{(j)} = 0$
    c. $[X]^T[m][X] = [\omega_i^2]$

13. Modal analysis can be used conveniently to find the response of a multidegree-of-freedom system
    a. under arbitrary forcing conditions
    b. under free-vibration conditions
    c. involving several modes

6.5 Match the items in the two columns below:

1. $\frac{1}{2}\dot{\vec{X}}^T[m]\dot{\vec{X}}$ **a.** equal to zero yields the characteristic values

2. $\frac{1}{2}\vec{X}^T[m]\vec{X}$ **b.** equal to $[\omega_i^2]$ when modes are normalized

3. $\vec{X}^{(i)^T}[m]\vec{X}^{(j)}$ **c.** kinetic energy of the system

4. $\vec{X}^{(i)^T}[m]\vec{X}^{(i)}$ **d.** equal to zero when modes are orthogonal

5. $[X]^T[k][X]$ **e.** equal to the dynamical matrix $[D]$

6. $[m]\ddot{\vec{x}} + [k]\vec{x}$ **f.** strain energy of the system

7. $|[k] - \omega^2[m]|$ **g.** equal to the applied force vector $\vec{F}$

8. $[k]^{-1}[m]$ **h.** equal to one when modes are orthonormal

PROBLEMS

Section 6.3 Using Newton's Second Law to Derive Equations of Motions

6.1–6.5   Derive the equations of motion, using Newton's second law of motion, for each of the systems shown in Figs. 6.18 to 6.22.

FIGURE 6.18

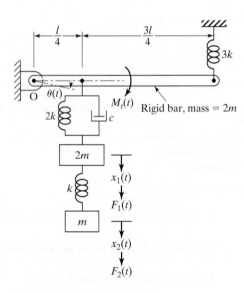

FIGURE 6.19

6.6   An automobile is modeled as shown in Fig. 6.23. Derive the equations of motion using Newton's second law of motion.

FIGURE 6.20

FIGURE 6.21

Number of teeth on gear $G_i = n_i$ ($i = 1$ to 6)
Mass moment of inertia of gear $G_i = I_i$ ($i = 1$ to 6)

FIGURE 6.22

6.7   The equations of motion derived using the displacements of the masses, x_1, x_2, and x_3 as degrees of freedom in Fig. 6.12 (Example 6.10) lead to symmetric mass and stiffness matrices in Eq. (E.3) of Example 6.10. Express the equations of motion (E.3) of Example 6.10, using $x_1, x_2 - x_1$, and $x_3 - x_2$ as degrees of freedom in the form:

$$[\overline{m}]\ddot{\vec{y}} + [\overline{k}]\vec{y} = \vec{0}$$

where

$$\vec{y} = \begin{Bmatrix} y_1 \\ y_2 \\ y_3 \end{Bmatrix}$$

Show that the resulting mass and stiffness matrices $[\overline{m}]$ and $[\overline{k}]$ are nonsymmetric.

FIGURE 6.23

6.8 A simplified vibration analysis of an airplane considers bounce and pitch motions (Fig. 6.24(a)). For this, a model consisting of a rigid bar (corresponding to the body of the airplane) supported on two springs (corresponding to the stiffnesses of the main and nose landing gears) as shown in Fig. 6.24(b) is used. The analysis can be conducted using three different coordinate systems as shown in Figs. 6.24(c)–(e). Derive the equations of motion in the three coordinate systems and identify the type of coupling associated with each coordinate system.

(a)

(b)

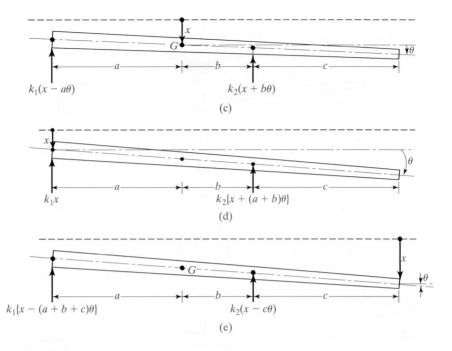

$k_1(x - a\theta)$ $k_2(x + b\theta)$

(c)

$k_1 x$ $k_2\{x + (a + b)\theta\}$

(d)

$k_1\{x - (a + b + c)\theta\}$ $k_2(x - c\theta)$

(e)

FIGURE 6.24

6.9   Consider the two-degree-of-freedom system shown in Fig. 6.25 with $m_1 = m_2 = 1$ and $k_1 = k_2 = 4$. The masses m_1 and m_2 move on a rough surface for which the equivalent viscous damping constants can be assumed as $c_1 = c_2 = 2$.

a. Derive the equations of motion of the system.

b. Find the natural frequencies and mode shapes of the undamped system.

6.10 For a simplified analysis of the vibration of an airplane in the vertical direction, a three-degree-of-freedom model, as shown in Fig. 6.26, can be used. The three masses indicate the

FIGURE 6.25

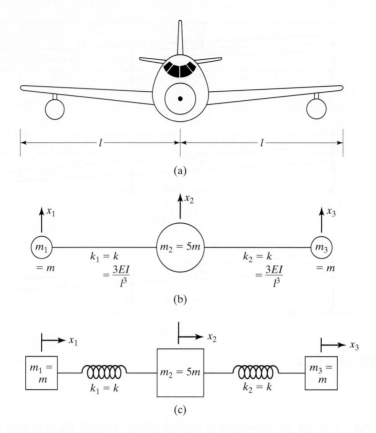

FIGURE 6.26

masses of the two wings ($m_1 = m_3 = m$) and the fuselage ($m_2 = 5m$). The stiffnesses $k_1 = k_2 = k$ correspond to the bending stiffnesses of the two wings, which can be modeled as cantilever beams so that $k_1 = k_2 = k = \dfrac{3EI}{l^3}$.

a. Derive the equations of motion of the airplane using the three-degree-of-freedom model.
b. Using the equations of motion derived in part (a), find the natural frequencies and mode shapes of the airplane. Give an interpretation of the results.

6.11 A simplified model of the main landing gear system of a small airplane is shown in Fig. 6.27 with $m_1 = 100$ kg, $m_2 = 5,000$ kg, $k_1 = 10^4$ N/m, and $k_2 = 10^6$ N/m.

a. Find the equations of motion of the system.
b. Find the natural frequencies and the mode shapes of the system.

Section 6.4 Influence Coefficients

6.12–6.17 Derive the stiffness matrix of each of the systems shown in Figs. 6.18 to 6.23 using the indicated coordinates.

6.18 Derive the flexibility matrix of the system shown in Fig. 5.39.

6.19 Derive the stiffness matrix of the system shown in Fig. 5.39.

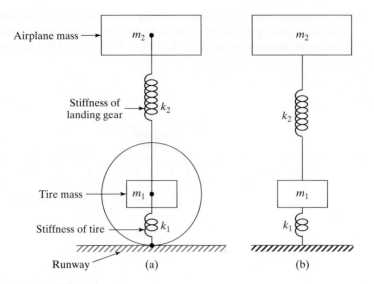

Airplane mass → m_2

Stiffness of landing gear → k_2

Tire mass → m_1

Stiffness of tire → k_1

Runway   (a)   (b)

FIGURE 6.27

6.20 Derive the flexibility matrix of the system shown in Fig. 5.42.

6.21 Derive the stiffness matrix of the system shown in Fig. 5.42.

6.22 Derive the mass matrix of the system shown in Fig. 5.42.

6.23 Find the flexibility and stiffness influence coefficients of the torsional system shown in Fig. 6.28. Also write the equations of motion of the system.

θ_1 θ_2 θ_3

$(GJ)_1$ $(GJ)_2$ $(GJ)_3$ $(GJ)_4$

l_1 l_2 l_3 l_4

Compressor   Turbine   Generator
(J_{d1}) (J_{d2}) (J_{d3})

FIGURE 6.28

6.24 Find the flexibility and stiffness influence coefficients of the system shown in Fig. 6.29. Also derive the equations of motion of the system.

k_1 k_2 k_3 k_4

m_1 m_2 m_3

FIGURE 6.29

6.25 An airplane wing, Fig. 6.30(a), is modeled as a three-degree-of-freedom lumped-mass system, as shown in Fig. 6.30(b). Derive the flexibility matrix and the equations of motion of the wing by assuming that all $A_i = A$, $(EI)_i = EI$, $l_i = l$ and that the root is fixed.

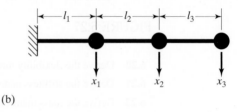

(a)

(b)

FIGURE 6.30

6.26 Determine the flexibility matrix of the uniform beam shown in Fig. 6.31. Disregard the mass of the beam compared to the concentrated masses placed on the beam and assume all $l_i = l$.

6.27 Derive the flexibility and stiffness matrices of the spring-mass system shown in Fig. 6.32 assuming that all the contacting surfaces are frictionless.

FIGURE 6.31   **FIGURE 6.32**

6.28 Derive the equations of motion for the tightly stretched string carrying three masses, as shown in Fig. 6.33. Assume the ends of the string to be fixed.

6.29 Derive the equations of motion of the system shown in Fig. 6.34.

6.30 Four identical springs, each having a stiffness k, are arranged symmetrically at 90° from each other, as shown in Fig. 2.65. Find the influence coefficient of the junction point in an arbitrary direction.

FIGURE 6.33   **FIGURE 6.34**

6.31 Show that the stiffness matrix of the spring-mass system shown in Fig. 6.3(a) is a band matrix along the diagonal.

6.32–6.36 Derive the mass matrix of each of the systems shown in Figs. 6.18 to 6.22 using the indicated coordinates.

6.37 The inverse mass influence coefficient b_{ij} is defined as the velocity induced at point i due to a unit impulse applied at point j. Using this definition, derive the inverse mass matrix of the system shown in Fig. 6.4(a).

Section 6.6 Generalized Coordinates and Generalized Forces

6.38 For the four-story shear building shown in Fig. 6.35, there is no rotation of the horizontal section at the levels of floors. Assuming that the floors are rigid and the total mass is concentrated at the levels of the floors, derive the equations of motion of the building using (a) Newton's second law of motion and (b) Lagrange's equations.

Section 6.7 Using Lagrange's Equations to Derive Equations of Motion

6.39 Derive the equations of motion of the system shown in Fig. 6.36, using Lagrange's equations with x and θ as generalized coordinates.

6.40 Derive the equations of motion of the system shown in Fig. 5.12(a), using Lagrange's equations with (a) x_1 and x_2 as generalized coordinates and (b) x and θ as generalized coordinates.

6.41 Derive the equations of motion of the system shown in Fig. 6.29, using Lagrange's equations.

6.42 Derive the equations of motion of the triple pendulum shown in Fig. 6.10, using Lagrange's equations.

FIGURE 6.35   **FIGURE 6.36**

6.43 When an airplane (see Fig. 6.37(a)) undergoes symmetric vibrations, the fuselage can be idealized as a concentrated central mass M_0 and the wings can be modeled as rigid bars carrying end masses M, as shown in Fig. 6.37(b). The flexibility between the wings and the fuselage can be represented by two torsional springs of stiffness k_t each. (a) Derive the equations of motion of the airplane, using Lagrange's equations with x and θ as generalized coordinates.

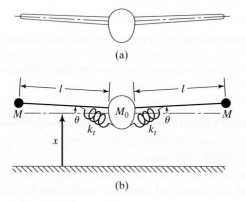

FIGURE 6.37

(b) Find the natural frequencies and mode shapes of the airplane. (c) Find the torsional spring constant in order to have the natural frequency of vibration, in torsional mode, greater than 2 Hz when $M_0 = 1,000$ kg, $M = 500$ kg, and $l = 6$ m.

6.44–6.48 Use Lagrange's equations to derive the equations of motion of each of the systems shown in Figs. 6.18 to 6.22.

Section 6.9 Eigenvalue Problem

6.49 Set up the eigenvalue problem of Example 6.11 in terms of the coordinates $q_1 = x_1, q_2 = x_2 - x_1$, and $q_3 = x_3 - x_2$, and solve the resulting problem. Compare the results obtained with those of Example 6.11 and draw conclusions.

6.50 Derive the frequency equation of the system shown in Fig. 6.29.

Section 6.10 Solution of Eigenvalue Problem

6.51 Find the natural frequencies and mode shapes of the system shown in Fig. 6.6(a) when $k_1 = k, k_2 = 2k, k_3 = 3k, m_1 = m, m_2 = 2m$, and $m_3 = 3m$. Plot the mode shapes.

6.52 Set up the matrix equation of motion and determine the three principal modes of vibration for the system shown in Fig. 6.6(a) with $k_1 = 3k, k_2 = k_3 = k, m_1 = 3m$, and $m_2 = m_3 = m$. Check the orthogonality of the modes found.

6.53 Find the natural frequencies of the system shown in Fig. 6.10 with $l_1 = 20$ cm, $l_2 = 30$ cm, $l_3 = 40$ cm, $m_1 = 1$ kg, $m_2 = 2$ kg, and $m_3 = 3$ kg.

6.54* (a) Find the natural frequencies of the system shown in Fig. 6.31 with $m_1 = m_2 = m_3 = m$ and $l_1 = l_2 = l_3 = l_4 = l/4$. (b) Find the natural frequencies of the beam when $m = 10$ kg, $l = 0.5$ m, the cross section is a solid circular section with diameter 2.5 cm, and the material is steel. (c) Consider using a hollow circular, solid rectangular, or hollow rectangular cross section for the beam to achieve the same natural frequencies as in (b). Identify the cross section corresponding to the least weight of the beam.

6.55 The frequency equation of a three-degree-of-freedom system is given by

$$\begin{vmatrix} \lambda - 5 & -3 & -2 \\ -3 & \lambda - 6 & -4 \\ -1 & -2 & \lambda - 6 \end{vmatrix} = 0$$

Find the roots of this equation.

6.56 Determine the eigenvalues and eigenvectors of the system shown in Fig. 6.29, taking $k_1 = k_2 = k_3 = k_4 = k$ and $m_1 = m_2 = m_3 = m$.

6.57 Find the natural frequencies and mode shapes of the system shown in Fig. 6.29 for $k_1 = k_2 = k_3 = k_4 = k, m_1 = 2m, m_2 = 3m$, and $m_3 = 2m$.

6.58 Find the natural frequencies and principal modes of the triple pendulum shown in Fig. 6.10, assuming that $l_1 = l_2 = l_3 = l$ and $m_1 = m_2 = m_3 = m$.

---

*The asterisk denotes a problem with no unique answer.

6.59   Find the natural frequencies and mode shapes of the system considered in Problem 6.27 with $m_1 = m$, $m_2 = 2m$, $m_3 = m$, $k_1 = k_2 = k$, and $k_3 = 2k$.

6.60   Show that the natural frequencies of the system shown in Fig. 6.6(a), with $k_1 = 3k$, $k_2 = k_3 = k$, $m_1 = 4m$, $m_2 = 2m$, and $m_3 = m$, are given by $\omega_1 = 0.46\sqrt{k/m}$, $\omega_2 = \sqrt{k/m}$, and $\omega_3 = 1.34\sqrt{k/m}$. Find the eigenvectors of the system.

6.61   Find the natural frequencies of the system considered in Problem 6.28 with $m_1 = 2m$, $m_2 = m$, $m_3 = 3m$, and $l_1 = l_2 = l_3 = l_4 = l$.

6.62   Find the natural frequencies and principal modes of the torsional system shown in Fig. 6.28 for $(GJ)_i = GJ$, $i = 1, 2, 3, 4$, $J_{d1} = J_{d2} = J_{d3} = J_0$, and $l_1 = l_2 = l_3 = l_4 = l$.

6.63   The mass matrix $[m]$ and the stiffness matrix $[k]$ of a uniform bar are

$$[m] = \frac{\rho Al}{4}\begin{bmatrix} 1 & 0 & 0 \\ 0 & 2 & 0 \\ 0 & 0 & 1 \end{bmatrix} \quad \text{and} \quad [k] = \frac{2AE}{l}\begin{bmatrix} 1 & -1 & 0 \\ -1 & 2 & -1 \\ 0 & -1 & 1 \end{bmatrix}$$

where ρ is the density, A is the cross-sectional area, E is Young's modulus, and l is the length of the bar. Find the natural frequencies of the system by finding the roots of the characteristic equation. Also find the principal modes.

6.64   The mass matrix of a vibrating system is given by

$$[m] = \begin{bmatrix} 1 & 0 & 0 \\ 0 & 2 & 0 \\ 0 & 0 & 1 \end{bmatrix}$$

and the eigenvectors by

$$\begin{Bmatrix} 1 \\ -1 \\ 1 \end{Bmatrix}, \quad \begin{Bmatrix} 1 \\ 1 \\ 1 \end{Bmatrix}, \quad \text{and} \quad \begin{Bmatrix} 0 \\ 1 \\ 2 \end{Bmatrix}$$

Find the $[m]$-orthonormal modal matrix of the system.

6.65   For the system shown in Fig. 6.38, (a) determine the characteristic polynomial $\Delta(\omega^2) = \det|[k] - \omega^2[m]|$, (b) plot $\Delta(\omega^2)$ from $\omega^2 = 0$ to $\omega^2 = 4.0$ (using increments $\Delta\omega^2 = 0.2$), and (c) find ω_1^2, ω_2^2, and ω_3^2.

6.66   (a) Two of the eigenvectors of a vibrating system are known to be

$$\begin{Bmatrix} 0.2754946 \\ 0.3994672 \\ 0.4490562 \end{Bmatrix} \quad \text{and} \quad \begin{Bmatrix} 0.6916979 \\ 0.2974301 \\ -0.3389320 \end{Bmatrix}$$

Prove that these are orthogonal with respect to the mass matrix

$$[m] = \begin{bmatrix} 1 & 0 & 0 \\ 0 & 2 & 0 \\ 0 & 0 & 3 \end{bmatrix}$$

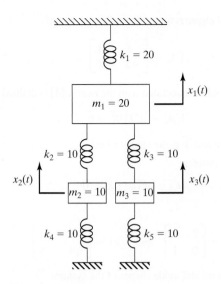

FIGURE 6.38

Find the remaining $[m]$-orthogonal eigenvector. (b) If the stiffness matrix of the system is given by

$$\begin{bmatrix} 6 & -4 & 0 \\ -4 & 10 & 0 \\ 0 & 0 & 6 \end{bmatrix}$$

determine all the natural frequencies of the system, using the eigenvectors of part (a).

6.67 Find the natural frequencies of the system shown in Fig. 6.18 for $m_i = m, i = 1, 2, 3$.

6.68 Find the natural frequencies of the system shown in Fig. 6.19 with $m = 1$ kg, $l = 1$ m, $k = 1,000$ N/m, and $c = 100$ N-s/m.

6.69 Consider the eigenvalue problem

$$[[k] - \omega^2[m]]\vec{X} = \vec{0}$$

where

$$[k] = k\begin{bmatrix} 2 & -1 & 0 \\ -1 & 2 & -1 \\ 0 & -1 & 1 \end{bmatrix} \quad \text{and} \quad [m] = m\begin{bmatrix} 1 & 0 & 0 \\ 0 & 1 & 0 \\ 0 & 0 & 1 \end{bmatrix}$$

Find the natural frequencies by finding the roots of the characteristic equation

$$|[m]^{-1}[k] - \omega^2[I]| = 0$$

Compare your results with the ones obtained in Example 6.11.

6.70 Find the eigenvalues and eigenvectors of the following matrix:

$$[A] = \begin{bmatrix} 8 & -1 \\ -4 & 4 \end{bmatrix}$$

Hint: The eigenvalue problem associated with the matrix [A] is defined as

$$[[A] - \lambda[I]]\vec{X} = \vec{0}$$

where λ is the eigenvalue and $\vec{X}$ is the eigenvector.

6.71 Consider the eigenvalue problem

$$[[k] - \omega^2[m]]\vec{X} = \vec{0}$$

where

$$[m] = \begin{bmatrix} 2 & 0 \\ 0 & 1 \end{bmatrix} \quad \text{and} \quad [k] = \begin{bmatrix} 8 & -4 \\ -4 & 4 \end{bmatrix}$$

Find the natural frequencies and mode shapes of the system:

a. by solving the equation

$$[[m]^{-1}[k] - \omega^2[I]]\vec{X} = \vec{0}$$

b. by solving the equation

$$[-\omega^2[k]^{-1}[m] + [I]]\vec{X} = \vec{0}$$

Compare the two sets of results and give your observations.

6.72 Consider the eigenvalue problem:

$$\omega^2 \begin{bmatrix} 1 & 0 \\ 0 & 2 \end{bmatrix} \begin{Bmatrix} X_1 \\ X_2 \end{Bmatrix} = \begin{bmatrix} 6 & -2 \\ -2 & 2 \end{bmatrix} \begin{Bmatrix} X_1 \\ X_2 \end{Bmatrix} \tag{E.1}$$

a. Find the natural frequencies and mode shapes of the system.

b. Change the coordinates in Eq. (E.1) as $X_1 = Y_1$ and $X_2 = 3Y_2$ and express the eigenvalue problem in terms of the eigenvector $\vec{Y} = \begin{Bmatrix} Y_1 \\ Y_2 \end{Bmatrix}$, solve it, and find the natural frequencies and mode shapes.

c. Compare the results found in parts (a) and (b) and give your observations.

6.73 Consider the eigenvalue problem:

$$\lambda[m]\vec{X} = [k]\vec{X} \tag{E.1}$$

where

$$[m] = \begin{bmatrix} 1 & 0 \\ 0 & 4 \end{bmatrix}, \quad [k] = \begin{bmatrix} 8 & -2 \\ -2 & 2 \end{bmatrix}, \quad \text{and} \quad \lambda = \omega^2$$

Equation (E.1) can be expressed as

$$[D]\vec{X} = \lambda\vec{X}$$

where

$$[D] = \left([m]^{\frac{1}{2}}\right)^{-1}[k]\left([m]^{\frac{1}{2}}\right)^{-1}$$

is called the *mass normalized stiffness matrix*. Determine the mass normalized stiffness matrix and use it to find the eigenvalues and orthonormal eigenvectors of the problem stated in Eq. (E.1).

Hint: The square root of a diagonal matrix $[m]$, of order n, is given by

$$[m]^{\frac{1}{2}} = \begin{bmatrix} \sqrt{m_{11}} & \cdots & 0 \\ & \cdot & \\ & \cdot & \\ & \cdot & \\ 0 & \cdots & \sqrt{m_{nn}} \end{bmatrix}$$

6.74 A symmetric positive definite matrix, such as the mass matrix of a multidegree-of-freedom system, $[m]$, can be expressed as the product of a lower triangular matrix, $[L]$, and an upper triangular matrix, $[L]^T$, as

$$[m] = [L][L]^T \tag{E.1}$$

using a procedure known as the *Choleski method* [6.18]. For a mass matrix of order 3×3, Eq. (E.1) becomes

$$\begin{bmatrix} m_{11} & m_{12} & m_{13} \\ m_{12} & m_{22} & m_{23} \\ m_{13} & m_{23} & m_{33} \end{bmatrix} = \begin{bmatrix} L_{11} & 0 & 0 \\ L_{21} & L_{22} & 0 \\ L_{31} & L_{32} & L_{33} \end{bmatrix}\begin{bmatrix} L_{11} & L_{21} & L_{31} \\ 0 & L_{22} & L_{32} \\ 0 & 0 & L_{33} \end{bmatrix} \tag{E.2}$$

By carrying out the multiplication of the matrices on the right-hand side of Eq. (E.2) and equating each of the elements of the resulting 3×3 matrix to the corresponding element of the matrix on the left-hand side of Eq. (E.2), the matrix [L] can be identified. Using this procedure, decompose the matrix

$$[m] = \begin{bmatrix} 4 & 2 & 1 \\ 2 & 6 & 2 \\ 1 & 2 & 8 \end{bmatrix}$$

in the form $[L][L]^T$.

Section 6.12 Unrestrained Systems

6.75 Find the natural frequencies and mode shapes of the system shown in Fig. 6.14 with $m_1 = m, m_2 = 2m, m_3 = 3m$, and $k_1 = k_2 = k$.

6.76   Find the modal matrix for the semidefinite system shown in Fig. 6.39 for $J_1 = J_2 = J_3 = J_0$, $k_{t1} = k_t$, and $k_{t2} = 2k_t$.

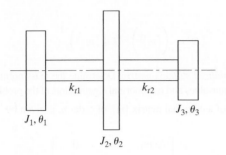

FIGURE 6.39

Section 6.13 Free Vibration of Undamped Systems

6.77   Find the free-vibration response of the spring-mass system shown in Fig. 6.29 for $k_i = k \, (i = 1, 2, 3, 4)$, $m_1 = 2m$, $m_2 = 3m$, $m_3 = 2m$ for the initial conditions $x_1(0) = x_{10}$ and $x_2(0) = x_3(0) = \dot{x}_1(0) = \dot{x}_2(0) = \dot{x}_3(0) = 0$.

6.78   Find the free-vibration response of the triple pendulum shown in Fig. 6.10 for $l_i = l \, (i = 1, 2, 3)$ and $m_i = m \, (i = 1, 2, 3)$ for the initial conditions $\theta_1(0) = \theta_2(0) = 0$, $\theta_3(0) = \theta_{30}$, $\dot{\theta}_i(0) = 0 \, (i = 1, 2, 3)$.

6.79   Find the free-vibration response of the tightly stretched string shown in Fig. 6.33 for $m_1 = 2m$, $m_2 = m$, $m_3 = 3m$, and $l_i = l \, (i = 1, 2, 3, 4)$. Assume the initial conditions as $x_1(0) = x_3(0) = 0$, $x_2(0) = x_{20}$, $\dot{x}_i(0) = 0 \, (i = 1, 2, 3)$.

6.80   Find the free-vibration response of the spring-mass system shown in Fig. 6.6(a) for $k_1 = k$, $k_2 = 2k$, $k_3 = 3k$, $m_1 = m$, $m_2 = 2m$, and $m_3 = 3m$ corresponding to the initial conditions $\dot{x}_1(0) = \dot{x}_{10}$, $x_i(0) = 0 \, (i = 1, 2, 3)$, and $\dot{x}_2(0) = \dot{x}_3(0) = 0$.

6.81   Find the free-vibration response of the spring-mass system shown in Fig. 6.32 for $m_1 = m$, $m_2 = 2m$, $m_3 = m$, $k_1 = k_2 = k$, and $k_3 = 2k$ corresponding to the initial conditions $\dot{x}_3(0) = \dot{x}_{30}$ and $x_1(0) = x_2(0) = x_3(0) = \dot{x}_1(0) = \dot{x}_2(0) = 0$.

6.82   In the freight car system shown in Fig. 6.14, the first car acquires a velocity of $\dot{x}_0$ due to an impact. Find the resulting free vibration of the system. Assume $m_i = m \, (i = 1, 2, 3)$ and $k_1 = k_2 = k$.

6.83   Find the free-vibration response of a three-degree-of-freedom system governed by the equation

$$10 \begin{bmatrix} 1 & 0 & 0 \\ 0 & 1 & 0 \\ 0 & 0 & 1 \end{bmatrix} \ddot{\vec{x}}(t) + 100 \begin{bmatrix} 2 & -1 & 0 \\ -1 & 2 & -1 \\ 0 & -1 & 1 \end{bmatrix} \vec{x}(t) = \vec{0}$$

Assume the initial conditions as $x_i(0) = 0.1$ and $\dot{x}_i(0) = 0$; $i = 1, 2, 3$.

Note: The natural frequencies and mode shapes of the system are given in Examples 6.11 and 6.12.

6.84 Using modal analysis, determine the free-vibration response of a two-degree-of-freedom system with equations of motion

$$2\begin{bmatrix} 1 & 0 \\ 0 & 1 \end{bmatrix}\ddot{\vec{x}}(t) + 8\begin{bmatrix} 2 & -1 \\ -1 & 2 \end{bmatrix}\vec{x}(t) = \vec{0}$$

with initial conditions

$$\vec{x}(0) = \begin{Bmatrix} 1 \\ 0 \end{Bmatrix}, \quad \dot{\vec{x}}(0) = \begin{Bmatrix} 0 \\ 1 \end{Bmatrix}$$

6.85 Consider the free-vibration equations of an undamped two-degree-of-freedom system:

$$[m]\ddot{\vec{x}} + [k]\vec{x} = \vec{0}$$

with

$$[m] = \begin{bmatrix} 1 & 0 \\ 0 & 4 \end{bmatrix} \quad \text{and} \quad [k] = \begin{bmatrix} 8 & -2 \\ -2 & 2 \end{bmatrix}$$

a. Find the orthonormal eigenvectors using the mass normalized stiffness matrix.
b. Determine the principal coordinates of the system and obtain the modal equations.

6.86 For the two-degree-of-freedom system considered in Problem 6.85, find the free-vibration response, $x_1(t)$ and $x_2(t)$, using the modal equations derived in Problem 6.85 for the following initial conditions: $x_1(0) = 2$, $x_2(0) = 3$, $\dot{x}_1(0) = \dot{x}_2(0) = 0$.

6.87 Find the free-vibration response of the three-degree-of-freedom airplane model considered in Problem 6.10 for the following data: $m = 5{,}000$ kg, $l = 5$ m, $E = 7$ GPa, $I = 8 \times 10^{-6}$ m^4. Assume that the initial conditions correspond to that of a gust which results in $x_1(0) = 0$, $x_2(0) = 0.1$ m, $x_3(0) = 0$, $\dot{x}_1(0) = \dot{x}_2(0) = \dot{x}_3(0) = 0$.

6.88 The free-vibration solution of a two-degree-of-freedom system can be determined by solving the equations

$$[m]\ddot{\vec{x}} + [k]\vec{x} = \vec{0} \tag{E.1}$$

with $\vec{x} = \begin{Bmatrix} x_1(t) \\ x_2(t) \end{Bmatrix}$, using the initial conditions

$$\vec{x}(t = 0) = \vec{x}_0 = \begin{Bmatrix} x_{10} \\ x_{20} \end{Bmatrix} \quad \text{and} \quad \dot{\vec{x}}(t = 0) = \dot{\vec{x}}_0 = \begin{Bmatrix} \dot{x}_{10} \\ \dot{x}_{20} \end{Bmatrix}$$

If ω_1 and ω_2 are the natural frequencies and $\vec{u}_1$ and $\vec{u}_2$ are the mode shapes of the system obtained from the solution of the characteristic equation

$$[[m]s^2 + [k]]\vec{u} = \vec{0} \tag{E.2}$$

with $s = \pm\omega_1, \pm\omega_2$ (characteristic roots), the solution of Eq. (E.1), $\vec{x}(t)$, can be found as a linear combination of different solutions as:

$$\vec{x}(t): \quad C_1\vec{u}_1 e^{-i\omega_1 t} + C_2\vec{u}_1 e^{+i\omega_1 t} + C_3\vec{u}_2 e^{-i\omega_2 t} + C_4\vec{u}_2 e^{+i\omega_2 t} \tag{E.3}$$

where C_i, $i = 1, 2, 3, 4$, are constants. Show that the solution in Eq. (E.3) can be expressed, in equivalent form, as

$$\vec{x}(t) = A_1 \sin(\omega_1 t + \phi_1)\vec{u}_1 + A_2 \sin(\omega_2 t + \phi_2)\vec{u}_2 \tag{E.4}$$

where A_1, A_2, ϕ_1, and ϕ_2 are constants.

Section 6.14 Forced Vibration of Undamped Systems Using Modal Analysis

6.89   Determine the amplitudes of motion of the three masses in Fig. 6.40 when a harmonic force $F(t) = F_0 \sin \omega t$ is applied to the lower left mass with $m = 1$ kg, $k = 1,000$ N/m, $F_0 = 5$ N, and $\omega = 10$ rad/s using the mode superposition method.

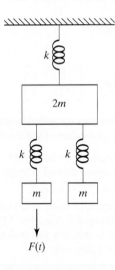

FIGURE 6.40

6.90   (a) Determine the natural frequencies and mode shapes of the torsional system shown in Fig. 6.11 for $k_{t1} = k_{t2} = k_{t3} = k_t$ and $J_1 = J_2 = J_3 = J_0$. (b) If a torque $M_{t3}(t) = M_{t0} \cos \omega t$, with $M_{t0} = 500$ N-m and $\omega = 100$ rad/s, acts on the generator (J_3), find the amplitude of each component. Assume $M_{t1} = M_{t2} = 0$, $k_t = 100$ N-m/rad, and $J_0 = 1$ kg-m^2.

6.91   Using the results of Problems 6.24 and 6.56, determine the modal matrix $[X]$ of the system shown in Fig. 6.29 and derive the uncoupled equations of motion.

6.92   An approximate solution of a multidegree-of-freedom system can be obtained using the mode acceleration method. According to this method, the equations of motion of an undamped system, for example, are expressed as

$$\vec{x} = [k]^{-1}(\vec{F} - [m]\ddot{\vec{x}}) \tag{E.1}$$

and $\ddot{\vec{x}}$ is approximated using the first r modes $(r < n)$ as

$$\underset{n \times 1}{\ddot{\vec{x}}} = \underset{n \times r}{[X]} \underset{r \times 1}{\ddot{\vec{q}}} \tag{E.2}$$

Since $([k] - \omega_i^2[m])\vec{X}^{(i)} = \vec{0}$, Eq. (E.1) can be written as

$$\vec{x}(t) = [k]^{-1}\vec{F}(t) - \sum_{i=1}^{r}\frac{1}{\omega_i^2}\vec{X}^{(i)}\ddot{q}_i(t) \tag{E.3}$$

Find the approximate response of the system described in Example 6.19 (without damping), using the mode acceleration method with $r = 1$.

6.93 Determine the response of the system in Problem 6.51 to the initial conditions $x_1(0) = 1$, $\dot{x}_1(0) = 0$, $x_2(0) = 2$, $\dot{x}_2(0) = 1$, $x_3(0) = 1$, and $\dot{x}_3(0) = -1$. Assume $k/m = 1$.

6.94 Show that the initial conditions of the generalized coordinates $q_i(t)$ can be expressed in terms of those of the physical coordinates $x_i(t)$ in modal analysis as

$$\vec{q}(0) = [X]^T[m]\vec{x}(0), \qquad \dot{\vec{q}}(0) = [X]^T[m]\dot{\vec{x}}(0)$$

6.95 A simplified model of a bicycle with its rider is shown in Fig. 6.41. Find the vertical motion of the rider when the bicycle hits an elevated road, as shown in the figure.

6.96 Find the response of the triple pendulum shown in Fig. 6.10 for $l_i = 0.5$ m $(i = 1, 2, 3)$ and $m_i = 1$ kg $(i = 1, 2, 3)$ when a moment, in the form of a rectangular pulse of magnitude 0.1 N-m and duration 0.1 s, is applied along the direction of θ_3. Assume that the pendulum is at rest at $t = 0$.

6.97 Find the response of the spring-mass system shown in Fig. 6.6(a) for $k_1 = k$, $k_2 = 2k$, $k_3 = 3k$, $m_1 = m$, $m_2 = 2m$, and $m_3 = 3m$ with $k = 10^4$ N/m and $m = 2$ kg when a force, in the form of a rectangular pulse of magnitude 1,000 N and duration 0.25 s, is applied to mass m_1 in the direction of x_1.

6.98 Consider a two-degree-of-freedom system with the equations of motion $[m]\ddot{\vec{x}} + [k]\vec{x} = \vec{f}(t)$ with

$$[m] = \begin{bmatrix} 1 & 0 \\ 0 & 4 \end{bmatrix}, \quad [k] = \begin{bmatrix} 8 & -2 \\ -2 & 2 \end{bmatrix}, \quad \text{and} \quad \vec{f}(t) = \begin{Bmatrix} f_1(t) \\ f_2(t) \end{Bmatrix}$$

FIGURE 6.41

a. Derive the modal equations for the forced-vibration response of the system.

b. Determine the conditions to be satisfied by $f_1(t)$ and $f_2(t)$ in order to influence both the modes.

Section 6.15 Forced Vibration of Viscously Damped Systems

6.99 Find the steady-state response of the system shown in Fig. 6.17 with $k_1 = k_2 = k_3 = k_4 = 100$ N/m, $c_1 = c_2 = c_3 = c_4 = 1$ N-s/m, $m_1 = m_2 = m_3 = 1$ kg, $F_1(t) = F_0 \cos \omega t$, $F_0 = 10$ N, and $\omega = 1$ rad/s. Assume that the spring k_4 and the damper c_4 are connected to a rigid wall at the right end. Use the mechanical impedance method described in Section 5.6 for solution.

6.100 An airplane wing, Fig. 6.42(a), is modeled as a 12-degree-of-freedom lumped-mass system as shown in Fig. 6.42(b). The first three natural mode shapes, obtained experimentally, are given below.

	Degrees of Freedom												
Mode Shape	0	1	2	3	4	5	6	7	8	9	10	11	12
$\vec{X}^{(1)}$	0.0	0.126	0.249	0.369	0.483	0.589	0.686	0.772	0.846	0.907	0.953	0.984	1.000
$\vec{X}^{(2)}$	0.0	-0.375	-0.697	-0.922	-1.017	-0.969	-0.785	-0.491	-0.127	0.254	0.599	0.860	1.000
$\vec{X}^{(3)}$	0.0	0.618	1.000	1.000	0.618	0.000	-0.618	-1.000	-1.000	-0.618	0.000	0.618	1.000

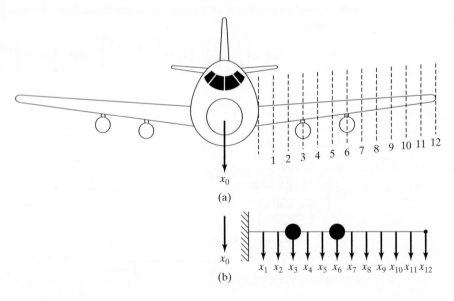

FIGURE 6.42

The natural frequencies corresponding to the three mode shapes are given by $\omega_1 = 225$ rad/s, $\omega_2 = 660$ rad/s, and $\omega_3 = 1,100$ rad/s. If the fuselage of the airplane is subjected to a known vertical motion $x_0(t)$, derive the uncoupled equations for determining the dynamic response of the wing by approximating it as a linear combination of the first three normal modes. *Hint:* The equation of motion of the airplane wing can be written, similar to Eq. (3.64), as

$$[m]\ddot{\vec{x}} + [c](\dot{\vec{x}} - \dot{x}_0\vec{u}_1) + [k](\vec{x} - x_0\vec{u}_1) = \vec{0}$$

or

$$[m]\ddot{\vec{x}} + [c]\dot{\vec{x}} + [k]\vec{x} = -x_0[m]\vec{u}_1$$

where $\vec{u}_1 = \{1, 0, 0, \ldots, 0\}^T$ is a unit vector.

Section 6.17 MATLAB Problems

6.101 Using MATLAB, find the eigenvalues and eigenvectors of a system with mass and stiffness matrices given in Example 6.13.

6.102 Using MATLAB, find and plot the free-vibration response of the system described in Problem 6.79 for the following data: $x_{20} = 0.5$, $P = 100$, $l = 5$, $m = 2$.

6.103 Using the MATLAB function `ode23`, find and plot the forced-vibration response of the system described in Problem 6.89.

6.104 Using the MATLAB function `roots`, find the roots of the following equation:

$$f(x) = x^{12} - 2 = 0$$

6.105 Find the forced-vibration response of a viscously damped three-degree-of-freedom system with equations of motion:

$$10\begin{bmatrix} 1 & 0 & 0 \\ 0 & 2 & 0 \\ 0 & 0 & 3 \end{bmatrix}\ddot{\vec{x}}(t) + 5\begin{bmatrix} 3 & -1 & 0 \\ -1 & 4 & -3 \\ 0 & -3 & 3 \end{bmatrix}\dot{\vec{x}}(t) + 20\begin{bmatrix} 7 & -3 & 0 \\ -3 & 5 & -2 \\ 0 & -2 & 2 \end{bmatrix}\vec{x}(t) = \begin{Bmatrix} 5\cos 2t \\ 0 \\ 0 \end{Bmatrix}$$

Assume zero initial conditions.

6.106 Using the MATLAB function `ode23`, solve Problem 6.99 and plot $x_1(t)$, $x_2(t)$, and $x_3(t)$.

6.107 Using `Program7.m`, generate the characteristic polynomial corresponding to the matrix

$$[A] = \begin{bmatrix} 5 & 3 & 2 \\ 3 & 6 & 4 \\ 1 & 2 & 6 \end{bmatrix}$$

6.108 Using `Program8.m`, find the steady-state response of a three-degree-of-freedom system with the following data:

$$\omega_1 = 25.076 \text{ rad/s}, \qquad \omega_2 = 53.578 \text{ rad/s}, \qquad \omega_3 = 110.907 \text{ rad/s}$$

$$\zeta_i = 0.001, \qquad i = 1, 2, 3$$

$$[m] = \begin{bmatrix} 41.4 & 0 & 0 \\ 0 & 38.8 & 0 \\ 0 & 0 & 25.88 \end{bmatrix}, \qquad [ev] = \begin{bmatrix} 1 & 1.0 & 1.0 \\ 1.303 & 0.860 & -1.000 \\ 1.947 & -1.685 & 0.183 \end{bmatrix}$$

$$\vec{F}(t) = \begin{Bmatrix} F_1(t) \\ F_2(t) \\ F_3(t) \end{Bmatrix} = \begin{Bmatrix} 5{,}000 \cos 5\,t \\ 10{,}000 \cos 10\,t \\ 20{,}000 \cos 20\,t \end{Bmatrix}$$

6.109 Find and plot the response, $x_1(t)$ and $x_2(t)$, of a system with the following equations of motion:

$$\begin{bmatrix} 5 & 0 \\ 0 & 2 \end{bmatrix} \begin{Bmatrix} \ddot{x}_1 \\ \ddot{x}_2 \end{Bmatrix} + \begin{bmatrix} 0.5 & -0.6 \\ -0.6 & 0.8 \end{bmatrix} \begin{Bmatrix} \dot{x}_1 \\ \dot{x}_2 \end{Bmatrix} + \begin{bmatrix} 20 & -2 \\ -2 & 2 \end{bmatrix} \begin{Bmatrix} x_1 \\ x_2 \end{Bmatrix} = \begin{Bmatrix} 1 \\ 0 \end{Bmatrix} \sin 2t \quad (E.1)$$

using the initial conditions:

$$\vec{x}(t = 0) = \begin{Bmatrix} 0.1 \\ 0 \end{Bmatrix} \text{ m} \quad \text{and} \quad \dot{\vec{x}}(t = 0) = \begin{Bmatrix} 0 \\ 1 \end{Bmatrix} \text{m/s}$$

Solve the differential equations, (E.1), numerically using a suitable MATLAB function.

6.110 Write a computer program for finding the eigenvectors using the known eigenvalues in Eq. (6.61). Find the mode shapes of Problem 6.57 using this program.

6.111 Write a computer program for generating the $[m]$-orthonormal modal matrix $[X]$. The program should accept the number of degrees of freedom, the normal modes, and the mass matrix as input. Solve Problem 6.64 using this program.

6.112 The equations of motion of an undamped system in SI units are given by

$$\begin{bmatrix} 2 & 0 & 0 \\ 0 & 2 & 0 \\ 0 & 0 & 2 \end{bmatrix} \ddot{\vec{x}} + \begin{bmatrix} 16 & -8 & 0 \\ -8 & 16 & -8 \\ 0 & -8 & 16 \end{bmatrix} \vec{x} = \begin{Bmatrix} 10 \sin \omega t \\ 0 \\ 0 \end{Bmatrix}$$

Using subroutine **MODAL**, find the steady-state response of the system when $\omega = 5$ rad/s.

6.113 Find the response of the system in Problem 6.112 by varying ω between 1 and 10 rad/s in increments of 1 rad/s. Plot the graphs showing the variations of magnitudes of the first peaks of $x_i(t), i = 1, 2, 3$, with respect to ω.

6.114 Find the natural frequencies of vibration and the corresponding mode shapes of the beam shown in Fig. 6.9 using the mass matrix

$$[m] = \begin{bmatrix} m_1 & 0 & 0 \\ 0 & m_2 & 0 \\ 0 & 0 & m_3 \end{bmatrix} \equiv m \begin{bmatrix} 1 & 0 & 0 \\ 0 & 1 & 0 \\ 0 & 0 & 1 \end{bmatrix}$$

and the flexibility matrix given by Eq. (E.4) of Example 6.6.

DESIGN PROJECT

6.115 A heavy machine tool mounted on the first floor of a building, Fig. 6.43(a), has been modeled as a three-degree-of-freedom system as indicated in Fig. 6.43(b). (a) For $k_1 = 900$ kN/m, $k_2 = 900$ kN/m, $k_3 = 3.5$ MN/m, $c_1 = c_2 = c_3 = 2,000$ N-s/m, $m_f = 9,000$ kg, $m_b = 2,000$ kg, $m_h = 400$ kg, and $F(t) = 200 \cos 60t$ N, find the steady-state vibration of the system using the mechanical impedance method described in Section 5.6. (b) If the maximum response of the machine tool head (x_3) has to be reduced by 25 percent, how should the stiffness of the mounting (k_2) be changed? (c) Is there any better way of achieving the goal stated in (b)? Provide details.

FIGURE 6.43

C H A P T E R   7

Determination of Natural Frequencies and Mode Shapes

John William Strutt, Lord Rayleigh (1842–1919), was an English physicist who held the positions of professor of experimental physics at Cambridge University, professor of natural philosophy at the Royal Institution in London, president of the Royal Society, and chancellor of Cambridge University. His works in optics and acoustics are well known, with *Theory of Sound* (1877) considered as a standard reference even today. The method of computing approximate natural frequencies of vibrating bodies using an energy approach has become known as "Rayleigh's method." (Courtesy of *Applied Mechanics Reviews*.)

Chapter Outline

Several methods of determining the natural frequencies and mode shapes of multidegree-of-freedom systems are outlined in this chapter. Specifically, Dunkerley's formula, Rayleigh's method, Holzer's method, the matrix iteration method, and Jacobi's method are presented. Derivation of Dunkerley's formula is based on the fact that higher natural frequencies of most systems are large compared to their fundamental frequencies. It gives an approximate value, always smaller than the exact value, of the fundamental natural frequency. Rayleigh's method, which is based on Rayleigh's principle, also gives an approximate value of the fundamental natural frequency, which is always larger than the exact value. Proof is given of Rayleigh's quotient and its stationariness in the neighborhood of an eigenvalue. It is also shown that Rayleigh's quotient is never lower than the first eigenvalue and never higher than the highest eigenvalue. Use of the static deflection curve in estimating the fundamental natural frequencies of beams and shafts using Rayleigh's method is presented. Holzer's method, based on a trial-and-error scheme, is presented to find the natural frequencies of undamped, damped, semidefinite, or branched translational and torsional systems. The matrix iteration method and its extensions for finding the smallest, highest, and intermediate natural frequencies are presented. A proof for the convergence of the method to the smallest frequency is given. Jacobi's method, which finds all the eigenvalues and eigenvectors of real symmetric matrices, is outlined. The standard eigenvalue problem is defined and the method of deriving it from the general eigenvalue problem, based on the Choleski decomposition method, is presented. Finally, the use of MATLAB in finding the eigenvalues and eigenvectors of multidegree-of-freedom systems is illustrated with several numerical examples.

Learning Objectives

After you have finished studying this chapter, you should be able to do the following:

- Find the approximate fundamental frequency of a composite system in terms of the natural frequencies of component parts using Dunkerley's formula.
- Understand Rayleigh's principle, and the properties of Rayleigh's quotient, and compute the fundamental natural frequency of a system using Rayleigh's method.
- Find the approximate natural frequencies of vibration and the modal vectors by using Holzer's method.
- Determine the smallest, intermediate, and highest natural frequencies of a system by using the matrix iteration method and its extensions (using the matrix deflation procedure).
- Find all the eigenvalues and eigenvectors of a multidegree-of-freedom system using Jacobi's method.
- Convert a general eigenvalue problem into a standard eigenvalue problem based on the Choleski decomposition method.
- Solve eigenvalue problems using MATLAB.

7.1 Introduction

In the preceding chapter, the natural frequencies (eigenvalues) and the natural modes (eigenvectors) of a multidegree-of-freedom system were found by setting the characteristic determinant equal to zero. Although this is an exact method, the expansion of the characteristic determinant and the solution of the resulting nth-degree polynomial equation to obtain the natural frequencies can become quite tedious for large values of n. Several analytical and numerical methods have been developed to compute the natural frequencies and mode shapes of multidegree-of-freedom systems. In this chapter, we shall consider Dunkerley's formula, Rayleigh's method, Holzer's method, the matrix iteration method, and Jacobi's method. Dunkerley's formula and Rayleigh's method are useful only for estimating the fundamental natural frequency. Holzer's method is essentially a tabular method that can be used to find partial or full solutions to eigenvalue problems. The matrix iteration method finds one natural frequency at a time, usually starting from the lowest value. The method can thus be terminated after finding the required number of natural frequencies and mode shapes. When all the natural frequencies and mode shapes are required, Jacobi's method can be used; it finds all the eigenvalues and eigenvectors simultaneously.

7.2   Dunkerley's Formula

Dunkerley's formula gives the approximate value of the fundamental frequency of a composite system in terms of the natural frequencies of its component parts. It is derived by making use of the fact that the higher natural frequencies of most vibratory systems are large compared to their fundamental frequencies [7.1–7.3]. To derive Dunkerley's formula, consider a general n-degree-of-freedom system whose eigenvalues can be determined by solving the frequency equation, Eq. (6.63):

$$| - [k] + \omega^2[m]| = 0$$

or

$$\left| -\frac{1}{\omega^2}[I] + [a][m] \right| = 0 \tag{7.1}$$

For a lumped-mass system with a diagonal mass matrix, Eq. (7.1) becomes

$$\left| -\frac{1}{\omega^2} \begin{bmatrix} 1 & 0 & \cdots & 0 \\ 0 & 1 & \cdots & 0 \\ \vdots & & & \\ 0 & 0 & \cdots & 1 \end{bmatrix} + \begin{bmatrix} a_{11} & a_{12} & \cdots & a_{1n} \\ a_{21} & a_{22} & \cdots & a_{2n} \\ \vdots & & & \\ a_{n1} & a_{n2} & \cdots & a_{nn} \end{bmatrix} \begin{bmatrix} m_1 & 0 & \cdots & 0 \\ 0 & m_2 & \cdots & 0 \\ \vdots & & & \\ 0 & 0 & \cdots & m_n \end{bmatrix} \right| = 0$$

—that is,

$$\left| \begin{matrix} \left(-\frac{1}{\omega^2} + a_{11}m_1 \right) & a_{12}m_2 & \cdots & a_{1n}m_n \\ a_{21}m_1 & \left(-\frac{1}{\omega^2} + a_{22}m_2 \right) & \cdots & a_{2n}m_n \\ \vdots & \vdots & & \vdots \\ a_{n1}m_1 & a_{n2}m_2 & \cdots & \left(-\frac{1}{\omega^2} + a_{nn}m_n \right) \end{matrix} \right| = 0 \tag{7.2}$$

The expansion of Eq. (7.2) leads to

$$\left(\frac{1}{\omega^2}\right)^n - (a_{11}m_1 + a_{22}m_2 + \cdots + a_{nn}m_n)\left(\frac{1}{\omega^2}\right)^{n-1}$$

$$+ (a_{11}a_{22}m_1m_2 + a_{11}a_{33}m_1m_3 + \cdots + a_{n-1,n-1}a_{nn}m_{n-1}m_n$$

$$- a_{12}a_{21}m_1m_2 - \cdots - a_{n-1,n}a_{n,n-1}m_{n-1}m_n)\left(\frac{1}{\omega^2}\right)^{n-2}$$

$$- \cdots = 0 \tag{7.3}$$

This is a polynomial equation of nth degree in $(1/\omega^2)$. Let the roots of Eq. (7.3) be denoted as $1/\omega_1^2, 1/\omega_2^2, \ldots, 1/\omega_n^2$. Thus

$$\left(\frac{1}{\omega^2} - \frac{1}{\omega_1^2}\right)\left(\frac{1}{\omega^2} - \frac{1}{\omega_2^2}\right) \cdots \left(\frac{1}{\omega^2} - \frac{1}{\omega_n^2}\right)$$

$$= \left(\frac{1}{\omega^2}\right)^n - \left(\frac{1}{\omega_1^2} + \frac{1}{\omega_2^2} + \cdots + \frac{1}{\omega_n^2}\right)\left(\frac{1}{\omega^2}\right)^{n-1} - \cdots = 0 \tag{7.4}$$

Equating the coefficient of $(1/\omega^2)^{n-1}$ in Eqs. (7.3) and (7.4) gives

$$\frac{1}{\omega_1^2} + \frac{1}{\omega_2^2} + \cdots + \frac{1}{\omega_n^2} = a_{11}m_1 + a_{22}m_2 + \cdots + a_{nn}m_n \tag{7.5}$$

In most cases, the higher frequencies $\omega_2, \omega_3, \ldots, \omega_n$ are considerably larger than the fundamental frequency ω_1, and so

$$\frac{1}{\omega_i^2} \ll \frac{1}{\omega_1^2}, \qquad i = 2, 3, \ldots, n$$

Thus, Eq. (7.5) can be approximately written as

$$\frac{1}{\omega_1^2} \simeq a_{11}m_1 + a_{22}m_2 + \cdots + a_{nn}m_n \tag{7.6}$$

This equation is known as *Dunkerley's formula*. The fundamental frequency given by Eq. (7.6) will always be smaller than the exact value. In some cases, it will be more convenient to rewrite Eq. (7.6) as

$$\frac{1}{\omega_1^2} \simeq \frac{1}{\omega_{1n}^2} + \frac{1}{\omega_{2n}^2} + \cdots + \frac{1}{\omega_{nn}^2} \tag{7.7}$$

where $\omega_{in} = (1/a_{ii}m_i)^{1/2} = (k_{ii}/m_i)^{1/2}$ denotes the natural frequency of a single-degree-of-freedom system consisting of mass m_i and spring of stiffness k_{ii}, $i = 1, 2, \ldots, n$. The use of Dunkerley's formula for finding the lowest frequency of elastic systems is presented in [7.4, 7.5].

EXAMPLE 7.1

Fundamental Frequency of a Beam

Estimate the fundamental natural frequency of a simply supported beam carrying three identical equally spaced masses, as shown in Fig. 7.1.

Solution: The flexibility influence coefficients (see Example 6.6) required for the application of Dunkerley's formula are given by

$$a_{11} = a_{33} = \frac{3}{256}\frac{l^3}{EI}, \qquad a_{22} = \frac{1}{48}\frac{l^3}{EI} \tag{E.1}$$

Using $m_1 = m_2 = m_3 = m$, Eq. (7.6) thus gives

$$\frac{1}{\omega_1^2} \simeq \left(\frac{3}{256} + \frac{1}{48} + \frac{3}{256}\right)\frac{ml^3}{EI} = 0.04427\frac{ml^3}{EI}$$

$$\omega_1 \simeq 4.75375\sqrt{\frac{EI}{ml^3}}$$

This value can be compared with the exact value of the fundamental frequency, $4.9326\sqrt{\dfrac{EI}{ml^3}}$ (see Problem 6.54).

FIGURE 7.1   Beam carrying masses.

7.3 Rayleigh's Method

Rayleigh's method was presented in Section 2.5 to find the natural frequencies of single-degree-of-freedom systems. The method can be extended to find the approximate value of the fundamental natural frequency of a discrete system.[1] The method is based on *Rayleigh's principle*, which can be stated as follows [7.6]:

> *The frequency of vibration of a conservative system vibrating about an equilibrium position has a stationary value in the neighborhood of a natural mode. This stationary value, in fact, is a minimum value in the neighborhood of the fundamental natural mode.*

We shall now derive an expression for the approximate value of the first natural frequency of a multidegree-of-freedom system according to Rayleigh's method.

---

[1]Rayleigh's method for continuous systems is presented in Section 8.7

The kinetic and potential energies of an *n*-degree-of-freedom discrete system can be expressed as

$$T = \frac{1}{2}\dot{\vec{x}}^T[m]\dot{\vec{x}} \tag{7.8}$$

$$V = \frac{1}{2}\vec{x}^T[k]\vec{x} \tag{7.9}$$

To find the natural frequencies, we assume harmonic motion to be

$$\vec{x} = \vec{X}\cos\omega t \tag{7.10}$$

where $\vec{X}$ denotes the vector of amplitudes (mode shape) and ω represents the natural frequency of vibration. If the system is conservative, the maximum kinetic energy is equal to the maximum potential energy:

$$T_{max} = V_{max} \tag{7.11}$$

By substituting Eq. (7.10) into Eqs. (7.8) and (7.9), we find

$$T_{max} = \frac{1}{2}\vec{X}^T[m]\vec{X}\omega^2 \tag{7.12}$$

$$V_{max} = \frac{1}{2}\vec{X}^T[k]\vec{X} \tag{7.13}$$

By equating T_{max} and V_{max}, we obtain[2]

$$\omega^2 = \frac{\vec{X}^T[k]\vec{X}}{\vec{X}^T[m]\vec{X}} \tag{7.14}$$

The right-hand side of Eq. (7.14) is known as *Rayleigh's quotient* and is denoted as $R(\vec{X})$.

7.3.1
Properties
of Rayleigh's
Quotient

As stated earlier, $R(\vec{X})$ has a stationary value when the arbitrary vector $\vec{X}$ is in the neighborhood of any eigenvector $\vec{X}^{(r)}$. To prove this, we express the arbitrary vector $\vec{X}$ in terms of the normal modes of the system, $\vec{X}^{(i)}$, as

$$\vec{X} = c_1\vec{X}^{(1)} + c_2\vec{X}^{(2)} + c_3\vec{X}^{(3)} + \cdots \tag{7.15}$$

Then

$$\vec{X}^T[k]\vec{X} = c_1^2\,\vec{X}^{(1)^T}[k]\vec{X}^{(1)} + c_2^2\vec{X}^{(2)^T}[k]\vec{X}^{(2)}$$
$$+ c_3^2\vec{X}^{(3)^T}[k]\vec{X}^{(3)} + \cdots \tag{7.16}$$

---

[2]Equation (7.14) can also be obtained from the relation $[k]\vec{X} = \omega^2[m]\vec{X}$. Premultiplying this equation by $\vec{X}^T$ and solving the resulting equation gives Eq. (7.14).

and

$$\vec{X}^T[m]\vec{X} = c_1^2\vec{X}^{(1)^T}[m]\vec{X}^{(1)} + c_2^2\vec{X}^{(2)^T}[m]\vec{X}^{(2)}$$
$$+ c_3^2\vec{X}^{(3)^T}[m]\vec{X}^{(3)} + \cdots \tag{7.17}$$

as the cross terms of the form $c_ic_j\vec{X}^{(i)^T}[k]\vec{X}^{(j)}$ and $c_ic_j\vec{X}^{(i)^T}[m]\vec{X}^{(j)}, i \neq j$, are zero by the orthogonality property. Using Eqs. (7.16) and (7.17) and the relation

$$\vec{X}^{(i)^T}[k]\vec{X}^{(i)} = \omega_i^2\vec{X}^{(i)^T}[m]\vec{X}^{(i)} \tag{7.18}$$

the Rayleigh's quotient of Eq. (7.14) can be expressed as

$$\omega^2 = R(\vec{X}) = \frac{c_1^2\,\omega_1^2\vec{X}^{(1)^T}[m]\vec{X}^{(1)} + c_2^2\omega_2^2\vec{X}^{(2)^T}[m]\vec{X}^{(2)} + \cdots}{c_1^2\vec{X}^{(1)^T}[m]\vec{X}^{(1)} + c_2^2\,\vec{X}^{(2)^T}[m]\vec{X}^{(2)} + \cdots} \tag{7.19}$$

If the normal modes are normalized, this equation becomes

$$\omega^2 = R(\vec{X}) = \frac{c_1^2\omega_1^2 + c_2^2\omega_2^2 + \cdots}{c_1^2 + c_2^2 + \cdots} \tag{7.20}$$

If $\vec{X}$ differs little from the eigenvector $\vec{X}^{(r)}$, the coefficient c_r will be much larger than the remaining coefficients $c_i\,(i \neq r)$, and Eq. (7.20) can be written as

$$R(\vec{X}) = \frac{c_r^2\omega_r^2 + c_r^2 \displaystyle\sum_{\substack{i=1,2,\ldots \\ i\neq r}} \left(\frac{c_i}{c_r}\right)^2 \omega_i^2}{c_r^2 + c_r^2 \displaystyle\sum_{\substack{i=1,2,\ldots \\ i\neq r}} \left(\frac{c_i}{c_r}\right)^2} \tag{7.21}$$

Since $|c_i/c_r| = \varepsilon_i \ll 1$, where ε_i is a small number for all $i \neq r$, Eq. (7.21) gives

$$R(\vec{X}) = \omega_r^2\{1 + 0(\varepsilon^2)\} \tag{7.22}$$

where $0(\varepsilon^2)$ represents an expression in ε of the second order or higher. Equation (7.22) indicates that if the arbitrary vector $\vec{X}$ differs from the eigenvector $\vec{X}^{(r)}$ by a small quantity of the first order, $R(\vec{X})$ differs from the eigenvalue ω_r^2 by a small quantity of the second order. This means that Rayleigh's quotient has a stationary value in the neighborhood of an eigenvector.

The stationary value is actually a minimum value in the neighborhood of the fundamental mode, $\vec{X}^{(1)}$. To see this, let $r = 1$ in Eq. (7.21) and write

$$R(\vec{X}) = \frac{\omega_1^2 + \displaystyle\sum_{i=2,3,\ldots} \left(\frac{c_i}{c_1}\right)^2 \omega_i^2}{\left\{ 1 + \displaystyle\sum_{i=2,3,\ldots} \left(\frac{c_i}{c_1}\right)^2 \right\}}$$

$$\simeq \omega_1^2 + \sum_{i=2,3,\ldots} \varepsilon_i^2 \omega_i^2 - \omega_1^2 \sum_{i=2,3,\ldots} \varepsilon_i^2$$

$$\simeq \omega_1^2 + \sum_{i=2,3,\ldots} (\omega_i^2 - \omega_1^2)\varepsilon_i^2 \tag{7.23}$$

Since, in general, $\omega_i^2 > \omega_1^2$ for $i = 2, 3, \ldots$, Eq. (7.23) leads to

$$R(\vec{X}) \geq \omega_1^2 \tag{7.24}$$

which shows that Rayleigh's quotient is never lower than the first eigenvalue. By proceeding in a similar manner, we can show that

$$R(\vec{X}) \leq \omega_n^2 \tag{7.25}$$

which means that Rayleigh's quotient is never higher than the highest eigenvalue. Thus Rayleigh's quotient provides an upper bound for ω_1^2 and a lower bound for ω_n^2.

7.3.2 Computation of the Fundamental Natural Frequency

Equation (7.14) can be used to find an approximate value of the first natural frequency (ω_1) of the system. For this, we select a trial vector $\vec{X}$ to represent the first natural mode $\vec{X}^{(1)}$ and substitute it on the right-hand side of Eq. (7.14). This yields the approximate value of ω_1^2. Because Rayleigh's quotient is stationary, remarkably good estimates of ω_1^2 can be obtained even if the trial vector $\vec{X}$ deviates greatly from the true natural mode $\vec{X}^{(1)}$. Obviously, the estimated value of the fundamental frequency ω_1 is more accurate if the trial vector $(\vec{X})$ chosen resembles the true natural mode $\vec{X}^{(1)}$ closely. Rayleigh's method is compared with Dunkerley's and other methods in [7.7–7.9].

EXAMPLE 7.2

Fundamental Frequency of a Three-Degree-of-Freedom System

Estimate the fundamental frequency of vibration of the system shown in Fig. 7.2. Assume that $m_1 = m_2 = m_3 = m$, $k_1 = k_2 = k_3 = k$, and the mode shape is

$$\vec{X} = \begin{Bmatrix} 1 \\ 2 \\ 3 \end{Bmatrix}$$

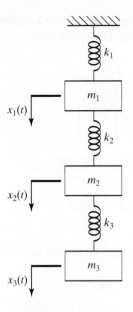

FIGURE 7.2   Three-degree-of-freedom spring-mass system.

Solution: The stiffness and mass matrices of the system are

$$[k] = k \begin{bmatrix} 2 & -1 & 0 \\ -1 & 2 & -1 \\ 0 & -1 & 1 \end{bmatrix} \tag{E.1}$$

$$[m] = m \begin{bmatrix} 1 & 0 & 0 \\ 0 & 1 & 0 \\ 0 & 0 & 1 \end{bmatrix} \tag{E.2}$$

By substituting the assumed mode shape in the expression for Rayleigh's quotient, we obtain

$$R(\vec{X}) = \omega^2 = \frac{\vec{X}^T[k]\vec{X}}{\vec{X}^T[m]\vec{X}} = \frac{(1\ 2\ 3)k \begin{bmatrix} 2 & -1 & 0 \\ -1 & 2 & -1 \\ 0 & -1 & 1 \end{bmatrix} \begin{Bmatrix} 1 \\ 2 \\ 3 \end{Bmatrix}}{(1\ 2\ 3)m \begin{bmatrix} 1 & 0 & 0 \\ 0 & 1 & 0 \\ 0 & 0 & 1 \end{bmatrix} \begin{Bmatrix} 1 \\ 2 \\ 3 \end{Bmatrix}} = 0.2143 \frac{k}{m} \tag{E.3}$$

$$\omega_1 = 0.4629 \sqrt{\frac{k}{m}} \tag{E.4}$$

This value is 4.0225 percent larger than the exact value of $0.4450\sqrt{k/m}$. The exact fundamental mode shape (see Example 6.10) in this case is

$$\vec{X}^{(1)} = \begin{Bmatrix} 1.0000 \\ 1.8019 \\ 2.2470 \end{Bmatrix} \tag{E.5}$$

∎

7.3.3 Fundamental Frequency of Beams and Shafts

Although the procedure outlined above is applicable to all discrete systems, a simpler equation can be derived for the fundamental frequency of the lateral vibration of a beam or a shaft carrying several masses such as pulleys, gears, or flywheels. In these cases, the static deflection curve is used as an approximation of the dynamic deflection curve.

Consider a shaft carrying several masses, as shown in Fig. 7.3. The shaft is assumed to have negligible mass. The potential energy of the system is the strain energy of the deflected shaft, which is equal to the work done by the static loads. Thus

$$V_{\max} = \frac{1}{2}\left(m_1 g w_1 + m_2 g w_2 + \cdots\right) \tag{7.26}$$

where $m_i g$ is the static load due to the mass m_i, and w_i is the total static deflection of mass m_i due to all the masses. For harmonic oscillation (free vibration), the maximum kinetic energy due to the masses is

$$T_{\max} = \frac{\omega^2}{2}\left(m_1 w_1^2 + m_2 w_2^2 + \cdots\right) \tag{7.27}$$

where ω is the frequency of oscillation. Equating $V_{\max}$ and $T_{\max}$, we obtain

$$\omega = \left\{\frac{g(m_1 w_1 + m_2 w_2 + \cdots)}{(m_1 w_1^2 + m_2 w_2^2 + \cdots)}\right\}^{1/2} \tag{7.28}$$

FIGURE 7.3   Shaft carrying masses.

███████████████
EXAMPLE 7.3
─────────────────

Fundamental Frequency of a Shaft with Rotors

Estimate the fundamental frequency of the lateral vibration of a shaft carrying three rotors (masses), as shown in Fig. 7.3, with $m_1 = 20$ kg, $m_2 = 50$ kg, $m_3 = 40$ kg, $l_1 = 1$ m, $l_2 = 3$ m, $l_3 = 4$ m, and $l_4 = 2$ m. The shaft is made of steel with a solid circular cross section of diameter 10 cm.

Solution: From strength of materials, the deflection of the beam shown in Fig. 7.4 due to a static load P [7.10] is given by

$$w(x) = \begin{cases} \dfrac{Pbx}{6EIl}(l^2 - b^2 - x^2); & 0 \leq x \leq a \quad \text{(E.1)} \\[2mm] -\dfrac{Pa(l-x)}{6EIl}[a^2 + x^2 - 2lx]; & a \leq x \leq l \quad \text{(E.2)} \end{cases}$$

Deflection Due to the Weight of m_1: At the location of mass m_1 (with $x = 1$ m, $b = 9$ m, and $l = 10$ m in Eq. (E.1)):

$$w_1' = \frac{(20 \times 9.81)(9)(1)}{6EI(10)}(100 - 81 - 1) = \frac{529.74}{EI} \tag{E.3}$$

At the location of m_2 (with $a = 1$ m, $x = 4$ m, and $l = 10$ m in Eq. (E.2)):

$$w_2' = -\frac{(20 \times 9.81)(1)(6)}{6EI(10)}[1 + 16 - 2(10)(4)] = \frac{1,236.06}{EI} \tag{E.4}$$

At the location of m_3 (with $a = 1$ m, $x = 8$ m, and $l = 10$ m in Eq. (E.2)):

$$w_3' = -\frac{(20 \times 9.81)(1)(2)}{6EI(10)}[1 + 64 - 2(10)(8)] = \frac{621.3}{EI} \tag{E.5}$$

Deflection Due to the Weight of m_2: At the location of m_1 (with $x = 1$ m, $b = 6$ m, and $l = 10$ m in Eq. (E.1)):

$$w_1'' = \frac{(50 \times 9.81)(6)(1)}{6EI(10)}(100 - 36 - 1) = \frac{3,090.15}{EI} \tag{E.6}$$

FIGURE 7.4   Beam under static load.

At the location of m_2 (with $x = 4$ m, $b = 6$ m, and $l = 10$ m in Eq. (E.1)):

$$w_2'' = \frac{(50 \times 9.81)(6)(4)}{6EI(10)}(100 - 36 - 16) = \frac{9{,}417.6}{EI} \tag{E.7}$$

At the location of m_3 (with $a = 4$ m, $x = 8$ m, and $l = 10$ m in Eq. (E.2)):

$$w_3'' = -\frac{(50 \times 9.81)(4)(2)}{6EI(10)}[16 + 64 - 2(10)(8)] = \frac{5{,}232.0}{EI} \tag{E.8}$$

Deflection Due to the Weight of m_3: At the location of m_1 (with $x = 1$ m, $b = 2$ m, and $l = 10$ m in Eq. (E.1)):

$$w_1''' = \frac{(40 \times 9.81)(2)(1)}{6EI(10)}(100 - 4 - 1) = \frac{1{,}242.6}{EI} \tag{E.9}$$

At the location of m_2 (with $x = 4$ m, $b = 2$ m, and $l = 10$ m in Eq. (E.1)):

$$w_2''' = \frac{(40 \times 9.81)(2)(4)}{6EI(10)}(100 - 4 - 16) = \frac{4{,}185.6}{EI} \tag{E.10}$$

At the location of m_3 (with $x = 8$ m, $b = 2$ m, and $l = 10$ m in Eq. (E.1)):

$$w_3''' = \frac{(40 \times 9.81)(2)(8)}{6EI(10)}(100 - 4 - 64) = \frac{3{,}348.48}{EI} \tag{E.11}$$

The total deflections of the masses m_1, m_2, and m_3 are

$$w_1 = w_1' + w_1'' + w_1''' = \frac{4{,}862.49}{EI}$$

$$w_2 = w_2' + w_2'' + w_2''' = \frac{14{,}839.26}{EI}$$

$$w_3 = w_3' + w_3'' + w_3''' = \frac{9{,}201.78}{EI}$$

Substituting into Eq. (7.28), we find the fundamental natural frequency:

$$\omega = \left\{ \frac{9.81(20 \times 4{,}862.49 + 50 \times 14{,}839.26 + 40 \times 9{,}201.78)EI}{20 \times (4{,}862.49)^2 + 50 \times (14{,}839.26)^2 + 40 \times (9{,}201.78)^2} \right\}^{1/2}$$

$$= 0.028222\sqrt{EI} \tag{E.12}$$

For the shaft, $E = 2.07 \times 10^{11}$ N/m^2 and $I = \pi(0.1)^4/64 = 4.90875 \times 10^{-6}$ m^4 and hence Eq. (E.12) gives

$$\omega = 28.4482 \text{ rad/s}$$

■

7.4   Holzer's Method

Holzer's method is essentially a trial-and-error scheme to find the natural frequencies of undamped, damped, semidefinite, fixed, or branched vibrating systems involving linear and angular displacements [7.11, 7.12]. The method can also be programmed for computer applications. A trial frequency of the system is first assumed, and a solution is found when the assumed frequency satisfies the constraints of the system. This generally requires several trials. Depending on the trial frequency used, the fundamental as well as the higher frequencies of the system can be determined. The method also gives the mode shapes.

7.4.1
Torsional
Systems

Consider the undamped torsional semidefinite system shown in Fig. 7.5. The equations of motion of the discs can be derived as follows:

$$J_1\ddot{\theta}_1 + k_{t1}(\theta_1 - \theta_2) = 0 \tag{7.29}$$

$$J_2\ddot{\theta}_2 + k_{t1}(\theta_2 - \theta_1) + k_{t2}(\theta_2 - \theta_3) = 0 \tag{7.30}$$

$$J_3\ddot{\theta}_3 + k_{t2}(\theta_3 - \theta_2) = 0 \tag{7.31}$$

Since the motion is harmonic in a natural mode of vibration, we assume that $\theta_i = \Theta_i \cos(\omega t + \phi)$ in Eqs. (7.29) to (7.31) and obtain

$$\omega^2 J_1\Theta_1 = k_{t1}(\Theta_1 - \Theta_2) \tag{7.32}$$

$$\omega^2 J_2\Theta_2 = k_{t1}(\Theta_2 - \Theta_1) + k_{t2}(\Theta_2 - \Theta_3) \tag{7.33}$$

$$\omega^2 J_3\Theta_3 = k_{t2}(\Theta_3 - \Theta_2) \tag{7.34}$$

Summing these equations gives

$$\sum_{i=1}^{3} \omega^2 J_i\Theta_i = 0 \tag{7.35}$$

Equation (7.35) states that the sum of the inertia torques of the semidefinite system must be zero. This equation can be treated as another form of the frequency equation, and the trial frequency must satisfy this requirement.

FIGURE 7.5   Torsional semidefinite system.

In Holzer's method, a trial frequency ω is assumed, and Θ_1 is arbitrarily chosen as unity. Next, Θ_2 is computed from Eq. (7.32), and then Θ_3 is found from Eq. (7.33). Thus we obtain

$$\Theta_1 = 1 \tag{7.36}$$

$$\Theta_2 = \Theta_1 - \frac{\omega^2 J_1 \Theta_1}{k_{t1}} \tag{7.37}$$

$$\Theta_3 = \Theta_2 - \frac{\omega^2}{k_{t2}}(J_1 \Theta_1 + J_2 \Theta_2) \tag{7.38}$$

These values are substituted in Eq. (7.35) to verify whether the constraint is satisfied. If Eq. (7.35) is not satisfied, a new trial value of ω is assumed and the process repeated. Equations (7.35), (7.37), and (7.38) can be generalized for an n-disc system as follows:

$$\sum_{i=1}^{n} \omega^2 J_i \Theta_i = 0 \tag{7.39}$$

$$\Theta_i = \Theta_{i-1} - \frac{\omega^2}{k_{ti-1}}\left(\sum_{k=1}^{i-1} J_k \Theta_k\right), \qquad i = 2, 3, \cdots, n \tag{7.40}$$

Thus the method uses Eqs. (7.39) and (7.40) repeatedly for different trial frequencies. If the assumed trial frequency is not a natural frequency of the system, Eq. (7.39) is not satisfied. The resultant torque in Eq. (7.39) represents a torque applied at the last disc. This torque M_t is then plotted for the chosen ω. When the calculation is repeated with other values of ω, the resulting graph appears as shown in Fig. 7.6. From this graph, the natural frequencies of the system can be identified as the values of ω at which $M_t = 0$. The amplitudes Θ_i ($i = 1, 2, \cdots, n$) corresponding to the natural frequencies are the mode shapes of the system.

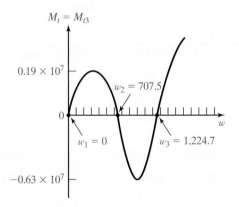

FIGURE 7.6   Resultant torque versus frequency.

Holzer's method can also be applied to systems with fixed ends. At a fixed end, the amplitude of vibration must be zero. In this case, the natural frequencies can be found by plotting the resulting amplitude (instead of the resultant torque) against the assumed frequencies. For a system with one end free and the other end fixed, Eq. (7.40) can be used for checking the amplitude at the fixed end. An improvement of Holzer's method is presented in [7.13, 7.14].

Natural Frequencies of a Torsional System

EXAMPLE 7.4

The arrangement of the compressor, turbine, and generator in a thermal power plant is shown in Fig. 7.7. Find the natural frequencies and mode shapes of the system.

Solution: This system represents an unrestrained or free-free torsional system. Table 7.1 shows its parameters and the sequence of computations. The calculations for the trial frequencies $\omega = 0, 10, 20, 700,$ and 710 are shown in this table. The quantity M_{t3} denotes the torque to the right of Station 3

FIGURE 7.7   Free-free torsional system.

TABLE 7.1

Parameters of the System	Quantity	Trial					
		1	2	3	. . .	71	72
		0	10	20		700	710
	ω^2	0	100	400		490,000	504,100
Station 1:							
$J_1 = 8$	Θ_1	1.0	1.0	1.0		1.0	1.0
$k_{t1} = 4 \times 10^6$	$M_{t1} = \omega^2 J_1 \Theta_1$	0	800	3,200		0.392E7	0.403E7
Station 2:							
$J_2 = 6$	$\Theta_2 = 1 - \dfrac{M_{t1}}{k_{t1}}$	1.0	0.9998	0.9992		0.0200	−0.0082
$k_{t2} = 2 \times 10^6$	$M_{t2} = M_{t1} + \omega^2 J_2 \Theta_2$	0	1,400	5,598		0.398E7	0.401E7
Station 3:							
$J_3 = 4$	$\Theta_3 = \Theta_2 - \dfrac{M_{t2}}{k_{t2}}$	1.0	0.9991	0.9964		−1.9690	−2.0120
$K_{t3} = 0$	$M_{t3} = M_{t2} + \omega^2 J_3 \Theta_3$	0	1,800	7,192		0.119E6	−0.494E5

FIGURE 7.8 First two flexible modes.

(generator), which must be zero at the natural frequencies. Figure 7.6 shows the graph of M_{t3} versus ω. Closely spaced trial values of ω are used in the vicinity of $M_{t3} = 0$ to obtain accurate values of the first two flexible mode shapes, shown in Fig. 7.8. Note that the value $\omega = 0$ corresponds to the rigid-body rotation.

∎

7.4.2 Spring-Mass Systems

Although Holzer's method has been extensively applied to torsional systems, the procedure is equally applicable to the vibration analysis of spring-mass systems. The equations of motion of a spring-mass system (see Fig. 7.9) can be expressed as

$$m_1\ddot{x}_1 + k_1(x_1 - x_2) = 0 \tag{7.41}$$

$$m_2\ddot{x}_2 + k_1(x_2 - x_1) + k_2(x_2 - x_3) = 0$$
$$\cdots \tag{7.42}$$

For harmonic motion, $x_i(t) = X_i \cos \omega t$, where X_i is the amplitude of mass m_i, and Eqs. (7.41) and (7.42) can be written as

$$\omega^2 m_1 X_1 = k_1(X_1 - X_2) \tag{7.43}$$

$$\omega^2 m_2 X_2 = k_1(X_2 - X_1) + k_2(X_2 - X_3)$$
$$= -\omega^2 m_1 X_1 + k_2(X_2 - X_3)$$
$$\cdots \tag{7.44}$$

FIGURE 7.9 Free-free spring-mass system.

The procedure for Holzer's method starts with a trial frequency ω and the amplitude of mass m_1 as $X_1 = 1$. Equations (7.43) and (7.44) can then be used to obtain the amplitudes of the masses $m_2, m_3, \cdots, m_i$:

$$X_2 = X_1 - \frac{\omega^2 m_1 X_1}{k_1} \tag{7.45}$$

$$X_3 = X_2 - \frac{\omega^2}{k_2}(m_1 X_1 + m_2 X_2) \tag{7.46}$$

$$X_i = X_{i-1} - \frac{\omega^2}{k_{i-1}}\left(\sum_{k=1}^{i-1} m_k X_k\right), \qquad i = 2, 3, \ldots, n \tag{7.47}$$

As in the case of torsional systems, the resultant force applied to the last (nth) mass can be computed as follows:

$$F = \sum_{i=1}^{n} \omega^2 m_i X_i \tag{7.48}$$

The calculations are repeated with several other trial frequencies ω. The natural frequencies are identified as those values of ω that give $F = 0$ for a free-free system. For this, it is convenient to plot a graph between F and ω, using the same procedure for spring-mass systems as for torsional systems.

7.5   Matrix Iteration Method

The matrix iteration method assumes that the natural frequencies are distinct and well separated such that $\omega_1 < \omega_2 < \cdots < \omega_n$. The iteration is started by selecting a trial vector $\vec{X}_1$, which is then premultiplied by the dynamical matrix $[D]$. The resulting column vector is then normalized, usually by making one of its components equal to unity. The normalized column vector is premultiplied by $[D]$ to obtain a third column vector, which is normalized in the same way as before and becomes still another trial column vector. The process is repeated until the successive normalized column vectors converge to a common vector: the fundamental eigenvector. The normalizing factor gives the largest value of $\lambda = 1/\omega^2$— that is, the smallest or the fundamental natural frequency [7.15]. The convergence of the process can be explained as follows.

According to the expansion theorem, any arbitrary n-dimensional vector $\vec{X}_1$ can be expressed as a linear combination of the n orthogonal eigenvectors of the system $\vec{X}^{(i)}, i = 1, 2, \ldots, n$:

$$\vec{X}_1 = c_1\vec{X}^{(1)} + c_2\vec{X}^{(2)} + \cdots + c_n\vec{X}^{(n)} \tag{7.49}$$

where $c_1, c_2, \ldots, c_n$ are constants. In the iteration method, the trial vector $\vec{X}_1$ is selected arbitrarily and is therefore a known vector. The modal vectors $\vec{X}^{(i)}$, although unknown, are constant vectors because they depend upon the properties of the system. The constants c_i

are unknown numbers to be determined. According to the iteration method, we premultiply $\vec{X}_1$ by the matrix $[D]$. In view of Eq. (7.49), this gives

$$[D]\vec{X}_1 = c_1[D]\vec{X}^{(1)} + c_2[D]\vec{X}^{(2)} + \cdots + c_n[D]\vec{X}^{(n)} \tag{7.50}$$

Now, according to Eq. (6.66), we have

$$[D]\vec{X}^{(i)} = \lambda_i[I]\vec{X}^{(i)} = \frac{1}{\omega_i^2}\vec{X}^{(i)}; \qquad i = 1, 2, \ldots, n \tag{7.51}$$

Substitution of Eq. (7.51) into Eq. (7.50) yields

$$\begin{aligned} [D]\vec{X}_1 &= \vec{X}_2 \\ &= \frac{c_1}{\omega_1^2}\vec{X}^{(1)} + \frac{c_2}{\omega_2^2}\vec{X}^{(2)} + \cdots + \frac{c_n}{\omega_n^2}\vec{X}^{(n)} \end{aligned} \tag{7.52}$$

where $\vec{X}_2$ is the second trial vector. We now repeat the process and premultiply $\vec{X}_2$ by $[D]$ to obtain, by Eqs. (6.66) and (7.49),

$$\begin{aligned} [D]\vec{X}_2 &= \vec{X}_3 \\ &= \frac{c_1}{\omega_1^4}\vec{X}^{(1)} + \frac{c_2}{\omega_2^4}\vec{X}^{(2)} + \cdots + \frac{c_n}{\omega_n^4}\vec{X}^{(n)} \end{aligned} \tag{7.53}$$

By repeating the process we obtain, after the rth iteration,

$$\begin{aligned} [D]\vec{X}_r &= \vec{X}_{r+1} \\ &= \frac{c_1}{\omega_1^{2r}}\vec{X}^{(1)} + \frac{c_2}{\omega_2^{2r}}\vec{X}^{(2)} + \cdots + \frac{c_n}{\omega_n^{2r}}\vec{X}^{(n)} \end{aligned} \tag{7.54}$$

Since the natural frequencies are assumed to be $\omega_1 < \omega_2 < \cdots < \omega_n$, a sufficiently large value of r yields

$$\frac{1}{\omega_1^{2r}} \gg \frac{1}{\omega_2^{2r}} \gg \cdots \gg \frac{1}{\omega_n^{2r}} \tag{7.55}$$

Thus the first term on the right-hand side of Eq. (7.54) becomes the only significant one. Hence we have

$$\vec{X}_{r+1} = \frac{c_1}{\omega_1^{2r}}\vec{X}^{(1)} \tag{7.56}$$

which means that the $(r + 1)$th trial vector becomes identical to the fundamental modal vector to within a multiplicative constant. Since

$$\vec{X}_r = \frac{c_1}{\omega_1^{2(r-1)}}\vec{X}^{(1)} \tag{7.57}$$

the fundamental natural frequency ω_1 can be found by taking the ratio of any two corresponding components in the vectors $\vec{X}_r$ and $\vec{X}_{r+1}$:

$$\omega_1^2 \simeq \frac{X_{i,r}}{X_{i,r+1}}, \qquad \text{for any } i = 1, 2, \ldots, n \tag{7.58}$$

where $X_{i,r}$ and $X_{i,r+1}$ are the ith elements of the vectors $\vec{X}_r$ and $\vec{X}_{r+1}$, respectively.

Discussion

1. In the above proof, nothing has been said about the normalization of the successive trial vectors $\vec{X}_i$. Actually, it is not necessary to establish the proof of convergence of the method. The normalization amounts to a readjustment of the constants $c_1, c_2, \ldots, c_n$ in each iteration.

2. Although it is theoretically necessary to have $r \rightarrow \infty$ for the convergence of the method, in practice only a finite number of iterations suffices to obtain a reasonably good estimate of ω_1.

3. The actual number of iterations necessary to find the value of ω_1 to within a desired degree of accuracy depends on how closely the arbitrary trial vector $\vec{X}_1$ resembles the fundamental mode $\vec{X}^{(1)}$ and on how well ω_1 and ω_2 are separated. The required number of iterations is less if ω_2 is very large compared to ω_1.

4. The method has a distinct advantage in that any computational errors made do not yield incorrect results. Any error made in premultiplying $\vec{X}_i$ by $[D]$ results in a vector other than the desired one, $\vec{X}_{i+1}$. But this wrong vector can be considered as a new trial vector. This may delay the convergence but does not produce wrong results.

5. One can take any set of n numbers for the first trial vector $\vec{X}_1$ and still achieve convergence to the fundamental modal vector. Only in the unusual case in which the trial vector $\vec{X}_1$ is exactly proportional to one of the modes $\vec{X}^{(i)}$ ($i \neq 1$) does the method fail to converge to the first mode. In such a case, the premultiplication of $\vec{X}^{(i)}$ by $[D]$ results in a vector proportional to $\vec{X}^{(i)}$ itself.

7.5.1 Convergence to the Highest Natural Frequency

To obtain the highest natural frequency ω_n and the corresponding mode shape or eigenvector $\vec{X}^{(n)}$ by the matrix iteration method, we first rewrite Eq. (6.66) as

$$[D]^{-1} \vec{X} = \omega^2 [I] \vec{X} = \omega^2 \vec{X} \tag{7.59}$$

where $[D]^{-1}$ is the inverse of the dynamical matrix $[D]$ given by

$$[D]^{-1} = [m]^{-1}[k] \tag{7.60}$$

Now we select any arbitrary trial vector $\vec{X}_1$ and premultiply it by $[D]^{-1}$ to obtain an improved trial vector $\vec{X}_2$. The sequence of trial vectors $\vec{X}_{i+1}$ ($i = 1, 2, \ldots$) obtained by premultiplying by $[D]^{-1}$ converges to the highest normal mode $\vec{X}^{(n)}$. It can be seen that the procedure is similar to the one already described. The constant of proportionality in this case is ω^2 instead of $1/\omega^2$.

7.5.2
Computation
of Intermediate
Natural
Frequencies

Once the first natural frequency ω_1 (or the largest eigenvalue $\lambda_1 = 1/\omega_1^2$) and the corresponding eigenvector $\vec{X}^{(1)}$ are determined, we can proceed to find the higher natural frequencies and the corresponding mode shapes by the matrix iteration method. Before we proceed, it should be remembered that any arbitrary trial vector premultiplied by $[D]$ would lead again to the largest eigenvalue. It is thus necessary to remove the largest eigenvalue from the matrix $[D]$. The succeeding eigenvalues and eigenvectors can be obtained by eliminating the root λ_1 from the characteristic or frequency equation

$$\left| [D] - \lambda[I] \right| = 0 \tag{7.61}$$

A procedure known as *matrix deflation* can be used for this purpose [7.16]. To find the eigenvector $\vec{X}^{(i)}$ by this procedure, the previous eigenvector $\vec{X}^{(i-1)}$ is normalized with respect to the mass matrix such that

$$\vec{X}^{(i-1)T}[m]\vec{X}^{(i-1)} = 1 \tag{7.62}$$

The deflated matrix $[D_i]$ is then constructed as

$$[D_i] = [D_{i-1}] - \lambda_{i-1}\vec{X}^{(i-1)}\vec{X}^{(i-1)T}[m], \qquad i = 2, 3, \ldots, n \tag{7.63}$$

where $[D_1] = [D]$. Once $[D_i]$ is constructed, the iterative scheme

$$\vec{X}_{r+1} = [D_i]\vec{X}_r \tag{7.64}$$

is used, where $\vec{X}_1$ is an arbitrary trial eigenvector.

EXAMPLE 7.5

Natural Frequencies of a Three-Degree-of-Freedom System

Find the natural frequencies and mode shapes of the system shown in Fig. 7.2 for $k_1 = k_2 = k_3 = k$ and $m_1 = m_2 = m_3 = m$ by the matrix iteration method.

Solution: The mass and stiffness matrices of the system are given in Example 7.2. The flexibility matrix is

$$[a] = [k]^{-1} = \frac{1}{k}\begin{bmatrix} 1 & 1 & 1 \\ 1 & 2 & 2 \\ 1 & 2 & 3 \end{bmatrix} \tag{E.1}$$

and so the dynamical matrix is

$$[k]^{-1}[m] = \frac{m}{k}\begin{bmatrix} 1 & 1 & 1 \\ 1 & 2 & 2 \\ 1 & 2 & 3 \end{bmatrix} \tag{E.2}$$

The eigenvalue problem can be stated as

$$[D]\vec{X} = \lambda\vec{X} \tag{E.3}$$

where

$$[D] = \begin{bmatrix} 1 & 1 & 1 \\ 1 & 2 & 2 \\ 1 & 2 & 3 \end{bmatrix} \tag{E.4}$$

and

$$\lambda = \frac{k}{m} \cdot \frac{1}{\omega^2} \tag{E.5}$$

First Natural Frequency: By assuming the first trial eigenvector or mode shape to be

$$\vec{X}_1 = \begin{Bmatrix} 1 \\ 1 \\ 1 \end{Bmatrix} \tag{E.6}$$

the second trial eigenvector can be obtained:

$$\vec{X}_2 = [D]\vec{X}_1 = \begin{Bmatrix} 3 \\ 5 \\ 6 \end{Bmatrix} \tag{E.7}$$

By making the first element equal to unity, we obtain

$$\vec{X}_2 = 3.0 \begin{Bmatrix} 1.0000 \\ 1.6667 \\ 2.0000 \end{Bmatrix} \tag{E.8}$$

and the corresponding eigenvalue is given by

$$\lambda_1 \simeq 3.0 \quad \text{or} \quad \omega_1 \simeq 0.5773\sqrt{\frac{k}{m}} \tag{E.9}$$

The subsequent trial eigenvector can be obtained from the relation

$$\vec{X}_{i+1} = [D]\vec{X}_i \tag{E.10}$$

and the corresponding eigenvalues are given by

$$\lambda_1 \simeq X_{1,\,i+1} \tag{E.11}$$

where $X_{1,i+1}$ is the first component of the vector $\vec{X}_{i+1}$ before normalization. The various trial eigenvectors and eigenvalues obtained by using Eqs. (E.10) and (E.11) are shown in the table below.

i	$\vec{X}_i$ with $X_{1,i} = 1$	$\vec{X}_{i+1} = [D]\vec{X}_i$	$\lambda_I \simeq X_{1,i+1}$	ω_1
1	$\begin{Bmatrix} 1 \\ 1 \\ 1 \end{Bmatrix}$	$\begin{Bmatrix} 3 \\ 5 \\ 6 \end{Bmatrix}$	3.0	$0.5773\sqrt{\dfrac{k}{m}}$
2	$\begin{Bmatrix} 1.00000 \\ 1.66667 \\ 2.00000 \end{Bmatrix}$	$\begin{Bmatrix} 4.66667 \\ 8.33333 \\ 10.33333 \end{Bmatrix}$	4.66667	$0.4629\sqrt{\dfrac{k}{m}}$
3	$\begin{Bmatrix} 1.0000 \\ 1.7857 \\ 2.2143 \end{Bmatrix}$	$\begin{Bmatrix} 5.00000 \\ 9.00000 \\ 11.2143 \end{Bmatrix}$	5.00000	$0.4472\sqrt{\dfrac{k}{m}}$
.				
.				
.				
7	$\begin{Bmatrix} 1.00000 \\ 1.80193 \\ 2.24697 \end{Bmatrix}$	$\begin{Bmatrix} 5.04891 \\ 9.09781 \\ 11.34478 \end{Bmatrix}$	5.04891	$0.44504\sqrt{\dfrac{k}{m}}$
8	$\begin{Bmatrix} 1.00000 \\ 1.80194 \\ 2.24698 \end{Bmatrix}$	$\begin{Bmatrix} 5.04892 \\ 9.09783 \\ 11.34481 \end{Bmatrix}$	5.04892	$0.44504\sqrt{\dfrac{k}{m}}$

It can be seen that the mode shape and the natural frequency converged (to the fourth decimal place) in eight iterations. Thus the first eigenvalue and the corresponding natural frequency and mode shape are given by

$$\lambda_1 = 5.04892, \qquad \omega_1 = 0.44504\sqrt{\frac{k}{m}}$$

$$\vec{X}^{(1)} = \begin{Bmatrix} 1.00000 \\ 1.80194 \\ 2.24698 \end{Bmatrix} \tag{E.12}$$

Second Natural Frequency: To compute the second eigenvalue and the eigenvector, we must first produce a deflated matrix:

$$[D_2] = [D_1] - \lambda_1 \vec{X}^{(1)}\vec{X}^{(1)T}[m] \tag{E.13}$$

This equation, however, calls for a normalized vector $\vec{X}^{(1)}$ satisfying $\vec{X}^{(1)T}[m]\vec{X}^{(1)} = 1$. Let the normalized vector be denoted as

$$\vec{X}^{(1)} = \alpha \begin{Bmatrix} 1.00000 \\ 1.80194 \\ 2.24698 \end{Bmatrix}$$

where α is a constant whose value must be such that

$$\vec{X}^{(1)T}[m]\vec{X}^{(1)} = \alpha^2 m \begin{Bmatrix} 1.00000 \\ 1.80194 \\ 2.24698 \end{Bmatrix}^T \begin{bmatrix} 1 & 0 & 0 \\ 0 & 1 & 0 \\ 0 & 0 & 1 \end{bmatrix} \begin{Bmatrix} 1.00000 \\ 1.80194 \\ 2.24698 \end{Bmatrix}$$

$$= \alpha^2 m(9.29591) = 1 \tag{E.14}$$

from which we obtain $\alpha = 0.32799 m^{-1/2}$. Hence the first normalized eigenvector is

$$\vec{X}^{(1)} = m^{-1/2} \begin{Bmatrix} 0.32799 \\ 0.59102 \\ 0.73699 \end{Bmatrix} \tag{E.15}$$

Next, we use Eq. (E.13) and form the first deflated matrix:

$$[D_2] = \begin{bmatrix} 1 & 1 & 1 \\ 1 & 2 & 2 \\ 1 & 2 & 3 \end{bmatrix} - 5.04892 \begin{Bmatrix} 0.32799 \\ 0.59102 \\ 0.73699 \end{Bmatrix} \begin{Bmatrix} 0.32799 \\ 0.59102 \\ 0.73699 \end{Bmatrix}^T \begin{bmatrix} 1 & 0 & 0 \\ 0 & 1 & 0 \\ 0 & 0 & 1 \end{bmatrix}$$

$$= \begin{bmatrix} 0.45684 & 0.02127 & -0.22048 \\ 0.02127 & 0.23641 & -0.19921 \\ -0.22048 & -0.19921 & 0.25768 \end{bmatrix} \tag{E.16}$$

Since the trial vector can be chosen arbitrarily, we again take

$$\vec{X}_1 = \begin{Bmatrix} 1 \\ 1 \\ 1 \end{Bmatrix} \tag{E.17}$$

By using the iterative scheme

$$\vec{X}_{i+1} = [D_2]\vec{X}_i \tag{E.18}$$

we obtain $\vec{X}_2$

$$\vec{X}_2 = \begin{Bmatrix} 0.25763 \\ 0.05847 \\ -0.16201 \end{Bmatrix} = 0.25763 \begin{Bmatrix} 1.00000 \\ 0.22695 \\ -0.62885 \end{Bmatrix} \tag{E.19}$$

Hence λ_2 can be found from the general relation

$$\lambda_2 \simeq X_{1,i+1} \tag{E.20}$$

as 0.25763. Continuation of this procedure gives the results shown in the table below.

i	$\vec{X}_i$ with $X_{1,i} = 1$	$\vec{X}_{i+1} = [D_2]\vec{X}_i$	$\lambda_2 \simeq X_{1,i+1}$	ω_2
1	$\begin{Bmatrix} 1 \\ 1 \\ 1 \end{Bmatrix}$	$\begin{Bmatrix} 0.25763 \\ 0.05847 \\ -0.16201 \end{Bmatrix}$	0.25763	$1.97016\sqrt{\dfrac{k}{m}}$
2	$\begin{Bmatrix} 1.00000 \\ 0.22695 \\ -0.62885 \end{Bmatrix}$	$\begin{Bmatrix} 0.60032 \\ 0.20020 \\ -0.42773 \end{Bmatrix}$	0.60032	$1.29065\sqrt{\dfrac{k}{m}}$
$\vdots$				
10	$\begin{Bmatrix} 1.00000 \\ 0.44443 \\ -0.80149 \end{Bmatrix}$	$\begin{Bmatrix} 0.64300 \\ 0.28600 \\ -0.51554 \end{Bmatrix}$	0.64300	$1.24708\sqrt{\dfrac{k}{m}}$
11	$\begin{Bmatrix} 1.00000 \\ 0.44479 \\ -0.80177 \end{Bmatrix}$	$\begin{Bmatrix} 0.64307 \\ 0.28614 \\ -0.51569 \end{Bmatrix}$	0.64307	$1.24701\sqrt{\dfrac{k}{m}}$

Thus the converged second eigenvalue and the eigenvector are

$$\lambda_2 = 0.64307, \qquad \omega_2 = 1.24701\sqrt{\frac{k}{m}}$$

$$\vec{X}^{(2)} = \begin{Bmatrix} 1.00000 \\ 0.44496 \\ -0.80192 \end{Bmatrix} \tag{E.21}$$

Third Natural Frequency: For the third eigenvalue and the eigenvector, we use a similar procedure. The detailed calculations are left as an exercise to the reader. Note that before computing the deflated matrix $[D_3]$, we need to normalize $\vec{X}^{(2)}$ by using Eq. (7.62), which gives

$$\vec{X}^{(2)} = m^{-1/2}\begin{Bmatrix} 0.73700 \\ 0.32794 \\ -0.59102 \end{Bmatrix} \tag{E.22}$$

∎

7.6   Jacobi's Method

The matrix iteration method described in the preceding section produces the eigenvalues and eigenvectors of matrix $[D]$ one at a time. Jacobi's method is also an iterative method but produces all the eigenvalues and eigenvectors of $[D]$ simultaneously, where $[D] = [d_{ij}]$ is a real symmetric matrix of order $n \times n$. The method is based on a theorem in linear algebra stating that a real symmetric matrix $[D]$ has only real eigenvalues and that there exists a real orthogonal matrix $[R]$ such that $[R]^T[D][R]$ is diagonal [7.17]. The diagonal elements are the eigenvalues, and the columns of the matrix $[R]$ are the eigenvectors. According to Jacobi's method, the matrix $[R]$ is generated as a product of several rotation matrices [7.18] of the form

$$
[R_1] \underset{n \times n}{=}
\begin{bmatrix}
1 & 0 & & & & & \\
0 & 1 & & & & & \\
& & \ddots & & & & \\
& & & \cos\theta & & -\sin\theta & \\
& & & & \ddots & & \\
& & & \sin\theta & & \cos\theta & \\
& & & & & & \ddots \\
& & & & & & & 1
\end{bmatrix}
\begin{matrix}
\\ \\ \\ i\text{th row} \\ \\ j\text{th row} \\ \\ \\
\end{matrix}
\tag{7.65}
$$

$$
\overset{i\text{th column} \qquad j\text{th column}}{}
$$

where all elements other than those appearing in columns and rows i and j are identical with those of the identity matrix $[I]$. If the sine and cosine entries appear in positions (i, i), (i, j), (j, i), and (j, j), then the corresponding elements of $[R_1]^T[D][R_1]$ can be computed as follows:

$$
\underline{d}_{ii} = d_{ii}\cos^2\theta + 2d_{ij}\sin\theta\cos\theta + d_{jj}\sin^2\theta \tag{7.66}
$$

$$
\underline{d}_{ij} = \underline{d}_{ji} = (d_{jj} - d_{ii})\sin\theta\cos\theta + d_{ij}(\cos^2\theta - \sin^2\theta) \tag{7.67}
$$

$$
\underline{d}_{jj} = d_{ii}\sin^2\theta - 2d_{ij}\sin\theta\cos\theta + d_{jj}\cos^2\theta \tag{7.68}
$$

If θ is chosen to be

$$
\tan 2\theta = \left(\frac{2d_{ij}}{d_{ii} - d_{jj}}\right) \tag{7.69}
$$

then it makes $\underline{d}_{ij} = \underline{d}_{ji} = 0$. Thus each step of Jacobi's method reduces a pair of off-diagonal elements to zero. Unfortunately, in the next step, while the method reduces a new pair of zeros, it introduces nonzero contributions to formerly zero positions. However, successive matrices of the form

$$
[R_2]^T[R_1]^T[D][R_1][R_2], \qquad [R_3]^T[R_2]^T[R_1]^T[D][R_1][R_2][R_3], \ \cdots
$$

converge to the required diagonal form; the final matrix $[R]$, whose columns give the eigenvectors, then becomes

$$[R] = [R_1][R_2][R_3]\cdots \tag{7.70}$$

EXAMPLE 7.6

Eigenvalue Solution Using Jacobi Method

Find the eigenvalues and eigenvectors of the matrix

$$[D] = \begin{bmatrix} 1 & 1 & 1 \\ 1 & 2 & 2 \\ 1 & 2 & 3 \end{bmatrix}$$

using Jacobi's method.

Solution: We start with the largest off-diagonal term, $d_{23} = 2$, in the matrix $[D]$ and try to reduce it to zero. From Eq. (7.69),

$$\theta_1 = \frac{1}{2}\tan^{-1}\left(\frac{2d_{23}}{d_{22}-d_{33}}\right) = \frac{1}{2}\tan^{-1}\left(\frac{4}{2-3}\right) = -37.981878°$$

$$[R_1] = \begin{bmatrix} 1.0 & 0.0 & 0.0 \\ 0.0 & 0.7882054 & 0.6154122 \\ 0.0 & -0.6154122 & 0.7882054 \end{bmatrix}$$

$$[D'] = [R_1]^T[D][R_1] = \begin{bmatrix} 1.0 & 0.1727932 & 1.4036176 \\ 0.1727932 & 0.4384472 & 0.0 \\ 1.4036176 & 0.0 & 4.5615525 \end{bmatrix}$$

Next, we try to reduce the largest off-diagonal term of $[D']$—namely, $d'_{13} = 1.4036176$—to zero. Equation (7.69) gives

$$\theta_2 = \frac{1}{2}\tan^{-1}\left(\frac{2d'_{13}}{d'_{11}-d'_{33}}\right) = \frac{1}{2}\tan^{-1}\left(\frac{2.8072352}{1.0-4.5615525}\right) = -19.122686°$$

$$[R_2] = \begin{bmatrix} 0.9448193 & 0.0 & 0.3275920 \\ 0.0 & 1.0 & 0.0 \\ -0.3275920 & 0.0 & 0.9448193 \end{bmatrix}$$

$$[D''] = [R_2]^T[D'][R_2] = \begin{bmatrix} 0.5133313 & 0.1632584 & 0.0 \\ 0.1632584 & 0.4384472 & 0.0566057 \\ 0.0 & 0.0566057 & 5.0482211 \end{bmatrix}$$

The largest off-diagonal element in $[D'']$ is $d''_{12} = 0.1632584$. θ_3 can be obtained from Eq. (7.69) as

$$\theta_3 = \frac{1}{2} \tan^{-1} \left(\frac{2d''_{12}}{d''_{11} - d''_{22}} \right) = \frac{1}{2} \tan^{-1} \left(\frac{0.3265167}{0.5133313 - 0.4384472} \right) = 38.541515°$$

$$[R_3] = \begin{bmatrix} 0.7821569 & -0.6230815 & 0.0 \\ 0.6230815 & 0.7821569 & 0.0 \\ 0.0 & 0.0 & 1.0 \end{bmatrix}$$

$$[D'''] = [R_3]^T [D''][R_3] = \begin{bmatrix} 0.6433861 & 0.0 & 0.0352699 \\ 0.0 & 0.3083924 & 0.0442745 \\ 0.0352699 & 0.0442745 & 5.0482211 \end{bmatrix}$$

Assuming that all the off-diagonal terms in $[D''']$ are close to zero, we can stop the process here. The diagonal elements of $[D''']$ give the eigenvalues (values of $1/\omega^2$) as 0.6433861, 0.3083924, and 5.0482211. The corresponding eigenvectors are given by the columns of the matrix $[R]$, where

$$[R] = [R_1][R_2][R_3] = \begin{bmatrix} 0.7389969 & -0.5886994 & 0.3275920 \\ 0.3334301 & 0.7421160 & 0.5814533 \\ -0.5854125 & -0.3204631 & 0.7447116 \end{bmatrix}$$

The iterative process can be continued for obtaining a more accurate solution. The present eigenvalues can be compared with the exact values: 0.6431041, 0.3079786, and 5.0489173.

∎

7.7 Standard Eigenvalue Problem

In the preceding chapter, the eigenvalue problem was stated as

$$[k]\vec{X} = \omega^2 [m]\vec{X} \tag{7.71}$$

which can be rewritten in the form of a standard eigenvalue problem [7.19] as

$$[D]\vec{X} = \lambda \vec{X} \tag{7.72}$$

where

$$[D] = [k]^{-1}[m] \tag{7.73}$$

and

$$\lambda = \frac{1}{\omega^2} \tag{7.74}$$

In general, the matrix $[D]$ is nonsymmetric, although the matrices $[k]$ and $[m]$ are both symmetric. Since Jacobi's method (described in Section 7.6) is applicable only to symmetric matrices $[D]$, we can adopt the following procedure [7.18] to derive a standard eigenvalue problem with a symmetric matrix $[D]$.

Assuming that the matrix $[k]$ is symmetric and positive definite, we can use Choleski decomposition (see Section 7.7.1) and express $[k]$ as

$$[k] = [U]^T[U] \tag{7.75}$$

where $[U]$ is an upper triangular matrix. Using this relation, the eigenvalue problem of Eq. (7.71) can be stated as

$$\lambda[U]^T[U]\vec{X} = [m]\vec{X} \tag{7.76}$$

Premultiplying this equation by $([U]^T)^{-1}$, we obtain

$$\lambda[U]\vec{X} = ([U]^T)^{-1}[m]\vec{X} = ([U]^T)^{-1}[m][U]^{-1}[U]\vec{X} \tag{7.77}$$

By defining a new vector $\vec{Y}$ as

$$\vec{Y} = [U]\vec{X} \tag{7.78}$$

Eq. (7.77) can be written as a standard eigenvalue problem

$$[D]\vec{Y} = \lambda\vec{Y} \tag{7.79}$$

where

$$[D] = ([U]^T)^{-1}[m][U]^{-1} \tag{7.80}$$

Thus, to formulate $[D]$ according to Eq. (7.80), we first decompose the symmetric matrix $[k]$ as shown in Eq. (7.75), find $[U]^{-1}$ and $([U]^T)^{-1} = ([U]^{-1})^T$ as outlined in the next section, and then carry out the matrix multiplication as stated in Eq. (7.80). The solution of the eigenvalue problem stated in Eq. (7.79) yields λ_i and $\vec{Y}^{(i)}$. We then apply inverse transformation and find the desired eigenvectors:

$$\vec{X}^{(i)} = [U]^{-1}\vec{Y}^{(i)} \tag{7.81}$$

7.7.1
Choleski
Decomposition

Any symmetric and positive definite matrix $[A]$ of order $n \times n$ can be decomposed uniquely [7.20]:

$$[A] = [U]^T[U] \tag{7.82}$$

where $[U]$ is an upper triangular matrix given by

$$[U] = \begin{bmatrix} u_{11} & u_{12} & u_{13} & \cdots & u_{1n} \\ 0 & u_{22} & u_{23} & \cdots & u_{2n} \\ 0 & 0 & u_{33} & \cdots & u_{3n} \\ \vdots & & & & \\ 0 & 0 & 0 & \cdots & u_{nn} \end{bmatrix} \qquad (7.83)$$

with

$$u_{11} = (a_{11})^{1/2}$$

$$u_{1j} = \frac{a_{1j}}{u_{11}}, \qquad j = 2, 3, \ldots, n$$

$$u_{ij} = \frac{1}{u_{ii}}\left(a_{ij} - \sum_{k=1}^{i-1} u_{ki}u_{kj} \right), \qquad i = 2, 3, \ldots, n \quad \text{and} \quad j = i+1, i+2, \ldots, n$$

$$u_{ii} = \left(a_{ii} - \sum_{k=1}^{i-1} u_{ki}^2 \right)^{1/2}, \qquad i = 2, 3, \ldots, n$$

$$u_{ij} = 0, \qquad i > j \qquad\qquad (7.84)$$

Inverse of the Matrix [U].   If the inverse of the upper triangular matrix $[U]$ is denoted as $[\alpha_{ij}]$, the elements α_{ij} can be determined from the relation

$$[U][U]^{-1} = [I] \qquad\qquad (7.85)$$

which gives

$$\alpha_{ii} = \frac{1}{u_{ii}}$$

$$\alpha_{ij} = \frac{-1}{u_{ii}}\left(\sum_{k=i+1}^{j} u_{ik}\alpha_{kj} \right), \qquad i < j$$

$$\alpha_{ij} = 0, \qquad i > j \qquad\qquad (7.86)$$

Thus the inverse of $[U]$ is also an upper triangular matrix.

Decomposition of a Symmetric Matrix

EXAMPLE 7.7

Decompose the matrix

$$[A] = \begin{bmatrix} 5 & 1 & 0 \\ 1 & 3 & 2 \\ 0 & 2 & 8 \end{bmatrix}$$

into the form of Eq. (7.82).

Solution: Equation (7.84) gives

$$u_{11} = \sqrt{a_{11}} = \sqrt{5} = 2.2360680$$

$$u_{12} = a_{12}/u_{11} = 1/2.236068 = 0.4472136$$

$$u_{13} = a_{13}/u_{11} = 0$$

$$u_{22} = [a_{22} - u_{12}^2]^{1/2} = (3 - 0.4472136^2)^{1/2} = 1.6733201$$

$$u_{33} = [a_{33} - u_{13}^2 - u_{23}^2]^{1/2}$$

where

$$u_{23} = (a_{23} - u_{12}u_{13})/u_{22} = (2 - 0.4472136 \times 0)/1.6733201 = 1.1952286$$

$$u_{33} = (8 - 0^2 - 1.1952286^2)^{1/2} = 2.5634799$$

Since $u_{ij} = 0$ for $i > j$, we have

$$[U] = \begin{bmatrix} 2.2360680 & 0.4472136 & 0.0 \\ 0.0 & 1.6733201 & 1.1952286 \\ 0.0 & 0.0 & 2.5634799 \end{bmatrix}$$

■

7.7.2 Other Solution Methods

Several other methods have been developed for finding the numerical solution of an eigenvalue problem [7.18, 7.21]. Bathe and Wilson [7.22] have done a comparative study of some of these methods. Recent emphasis has been on the economical solution of large eigenproblems [7.23, 7.24]. The estimation of natural frequencies by the use of Sturm sequences is presented in [7.25] and [7.26]. An alternative way to solve a class of lumped mechanical vibration problems using topological methods is presented in [7.27].

7.8   Examples Using MATLAB

Solution of an Eigenvalue Problem

EXAMPLE 7.8

Using MATLAB, find the eigenvalues and eigenvectors of the matrix

$$[A] = \begin{bmatrix} 3 & -1 & 0 \\ -2 & 4 & -3 \\ 0 & -1 & 1 \end{bmatrix}$$

Solution:

```
>> A=[3 -1 0; -2 4 -3; 0 -1 1]

A =

     3       -1        0
    -2        4       -3
     0       -1        1
>> [V, D] = eig (A)
```

```
V =

        -0.3665        -0.8305         0.2262
         0.9080        -0.4584         0.6616
        -0.2028         0.3165         0.7149

D =

         5.4774              0              0
              0         2.4481              0
              0              0         0.0746

>>
```

■

EXAMPLE 7.9

Using a Program for Jacobi's Method to Solve an Eigenvalue Problem

Develop a general program, called **Program9.m**, to implement Jacobi's method to find the eigenvalues and eigenvectors of a symmetric matrix. Use the program to find the eigenvalues and eigenvectors of the matrix

$$[A] = \begin{bmatrix} 1 & 1 & 1 \\ 1 & 2 & 2 \\ 1 & 2 & 3 \end{bmatrix}$$

Solution: **Program9.m** is developed to accept the following data:

n = order of the matrix

d = given matrix of order $n \times n$

eps = convergence specification, a small quantity on the order of 10^{-5}

itmax = maximum number of iterations permitted

The program gives the eigenvalues and eigenvectors of the matrix d.

```
>> program9
Eigenvalue solution by Jacobi Method

Given matrix
1.00000000e+000        1.00000000e+000        1.00000000e+000
1.00000000e+000        2.00000000e+000        2.00000000e+000
1.00000000e+000        2.00000000e+000        3.00000000e+000

Eigen values are
5.04891734e+000        6.43104132e-001        3.07978528e-001

Eigen vectors are
    First                  Second                 Third
3.27984948e-001        -7.36976229e-001        5.91009231e-001
5.91009458e-001        -3.27985278e-001        -7.36975900e-001
7.36976047e-001         5.91009048e-001        3.27985688e-001
```

■

EXAMPLE 7.10

Program for an Eigenvalue Solution Using the Matrix Iteration Method

Develop a general computer program, called **Program10.m**, to implement the matrix iteration method. Use the program to find the eigenvalues and eigenvectors of the matrix $[A]$ given in Example 7.9.

Solution: Program10.m is developed to accept the following input data:

n = order of the matrix d

d = given matrix of order $n \times n$

xs = initial guess vector of order n

nvec = number of eigenvalues and eigenvectors to be determined

xm = mass matrix of order $n \times n$

eps = convergence requirement, a small quantity on the order of 10^{-5}

The program gives the following output:

freq = array of size nvec, containing the computed natural frequencies

eig = array of size $n \times$ nvec, containing the computed eigenvectors (columns)

```
>> program10
Solution of eigenvalue problem by
matrix iteration method

Natural frequencies:

      4.450424e-001        1.246983e+000        1.801938e+000

Mode shapes (Columnwise):

      1.000000e+000        1.000000e+000        1.000000e+000
      1.801937e+000        4.450328e-001       -1.247007e+000
      2.246979e+000       -8.019327e-001        5.549798e-001
```

∎

EXAMPLE 7.11

Program for Solving a General Eigenvalue Problem

Develop a general program, called **Program11.m**, to solve a general eigenvalue problem. Use the program to find the solution of the general eigenvalue problem

$$[k]\vec{X} = \omega^2[m]\vec{X}$$

where

$$[k] = \begin{bmatrix} 2 & -1 & 0 \\ -1 & 2 & -1 \\ 0 & -1 & 1 \end{bmatrix}, \quad [m] = \begin{bmatrix} 1 & 0 & 0 \\ 0 & 1 & 0 \\ 0 & 0 & 1 \end{bmatrix}$$

Solution: Program11.m is developed to solve the problem $[k]\vec{X} = \omega^2[m]\vec{X}$ by first converting it to the form of a special eigenvalue problem $[D]\vec{Y} = \frac{1}{\omega^2}[I]\vec{Y}$, where $[D]$ is equal to $([U]^T)^{-1}[m][U]^{-1}$ and $[k] = [U]^T[U]$. The program is developed to accept the following input data:

nd = size of the problem (size of mass and stiffness matrices)

bk = stiffness matrix of size $nd \times nd$

bm = mass matrix of size $nd \times nd$

The program gives the upper triangular matrix of $[bk]$, the inverse of the upper triangular matrix $[ui]$, the matrix $[uti]$ $[bm]$ $[ui]$ where $[uti]$ is the transpose of $[ui]$, and the eigenvalues and eigenvectors of the problem.

```
>> program11
Upper triangular matrix [U]:

   1.414214e+000       -7.071068e-001        0.000000e+000
   0.000000e+000        1.224745e+000       -8.164966e-001
   0.000000e+000        0.000000e+000        5.773503e-001

Inverse of the upper triangular matrix:

   7.071068e-001        4.082483e-001        5.773503e-001
   0.000000e+000        8.164966e-001        1.154701e+000
   0.000000e+000        0.000000e+000        1.732051e+000

Matrix [UMU] = [UTI] [M] [UI]:

   5.000000e-001        2.886751e-001        4.082483e-001
   2.886751e-001        8.333333e-001        1.178511e+000
   4.082483e-001        1.178511e+000        4.666667e+000

Eigenvectors:

   5.048917e+000        6.431041e-001        3.079785e-001

Eigenvectors (Columnwise):

   7.369762e-001       -5.910090e-001        3.279853e-001
   1.327985e+000       -2.630237e-001       -4.089910e-001
   1.655971e+000        4.739525e-001        1.820181e-001
```

■

CHAPTER SUMMARY

The determination of natural frequencies (eigenvalues) and mode shapes (eigenvectors) of multidegree-of-freedom systems is a tedious procedure. Because the fundamental (lowest) natural frequency and the corresponding mode shape are most important in many applications, we presented several methods of finding the approximate value of the fundamental frequency and the mode shape. We discussed Dunkerley's formula, Rayleigh's method, Holzer's method, and the matrix iteration method. We also outlined the extension on the matrix iteration method to find the intermediate as well as the highest natural frequency and the mode shapes. For finding all the eigenvalues and eigenvectors simultaneously, we discussed Jacobi's method. Since most mathematical methods require the eigenvalue problem in standard form, we outlined a method of converting a general eigenvalue problem into the standard form. Finally, we presented the solution of an eigenvalue problem of multidegree-of- freedom systems using MATLAB.

Now that you have finished this chapter, you should be able to answer the review questions and solve the problems given below.

REFERENCES

7.1 S. Dunkerley, "On the whirling and vibration of shafts," *Philosophical Transactions of the Royal Society of London*, 1894, Series A, Vol. 185, Part I, pp. 279–360.

7.2 B. Atzori, "Dunkerley's formula for finding the lowest frequency of vibration of elastic systems," letter to the editor, *Journal of Sound and Vibration*, 1974, Vol. 36, pp. 563–564.

7.3 H. H. Jeffcott, "The periods of lateral vibration of loaded shafts—The rational derivation of Dunkerley's empirical rule for determining whirling speeds," *Proceedings of the Royal Society of London*, 1919, Series A, Vol. 95, No. A666, pp. 106–115.

7.4 M. Endo and O. Taniguchi, "An extension of the Southwell-Dunkerley methods for synthesizing frequencies," *Journal of Sound and Vibration*, 1976, "Part I: Principles," Vol. 49, pp. 501–516; "Part II: Applications," Vol. 49, pp. 517–533.

7.5 A. Rutenberg, "A lower bound for Dunkerley's formula in continuous elastic systems," *Journal of Sound and Vibration*, 1976, Vol. 45, pp. 249–252.

7.6 G. Temple and W. G. Bickley, *Rayleigh's Principle and Its Applications to Engineering*, Dover, New York, 1956.

7.7 N. G. Stephen, "Rayleigh's, Dunkerley's, and Southwell's methods," *International Journal of Mechanical Engineering Education*, January 1983, Vol. 11, pp. 45–51.

7.8 A. Rutenberg, "Dunkerley's formula and alternative approximations," letter to the editor, *Journal of Sound and Vibration*, 1975, Vol. 39, pp. 530–531.

7.9 R. Jones, "Approximate expressions for the fundamental frequency of vibration of several dynamic systems," *Journal of Sound and Vibration*, 1976, Vol. 44, pp. 475–478.

7.10 R. W. Fitzgerald, *Mechanics of Materials* (2nd ed.), Addison-Wesley, Reading, Mass., 1982.

7.11 H. Holzer, *Die Berechnung der Drehschwingungen*, Julius Springer, Berlin, 1921.

7.12 H. E. Fettis, "A modification of the Holzer method for computing uncoupled torsion and bending modes," *Journal of the Aeronautical Sciences*, October 1949, pp. 625–634; May 1954, pp. 359–360.

7.13 S. H. Crandall and W. G. Strang, "An improvement of the Holzer table based on a suggestion of Rayleigh's," *Journal of Applied Mechanics*, 1957, Vol. 24, p. 228.

7.14 S. Mahalingam, "An improvement of the Holzer method," *Journal of Applied Mechanics*, 1958, Vol. 25, p. 618.

7.15 S. Mahalingam, "Iterative procedures for torsional vibration analysis and their relationships," *Journal of Sound and Vibration*, 1980, Vol. 68, pp. 465–467.

7.16 L. Meirovitch, *Computational Methods in Structural Dynamics*, Sijthoff and Noordhoff, The Netherlands, 1980.

7.17 J. H. Wilkinson and G. Reinsch, *Linear Algebra*, Springer Verlag, New York, 1971.

7.18 J. W. Wilkinson, *The Algebraic Eigenvalue Problem*, Oxford University Press, London, 1965.

7.19 R. S. Martin and J. H. Wilkinson, "Reduction of a symmetric eigenproblem $Ax = \lambda Bx$ and related problems to standard form," *Numerical Mathematics*, 1968, Vol. 11, pp. 99–110.

7.20 G. B. Haggerty, *Elementary Numerical Analysis with Programming*, Allyn and Bacon, Boston, 1972.

7.21 A. Jennings, "Eigenvalue methods for vibration analysis," *Shock and Vibration Digest*, Part I, February 1980, Vol. 12, pp. 3–16; Part II, January 1984, Vol. 16, pp. 25–33.

7.22 K. Bathe and E. L. Wilson, "Solution methods for eigenvalue problems in structural mechanics," *International Journal for Numerical Methods in Engineering*, 1973, Vol. 6, pp. 213–226.

7.23 E. Cohen and H. McCallion, "Economical methods for finding eigenvalues and eigenvectors," *Journal of Sound and Vibration*, 1967, Vol. 5, pp. 397–406.

7.24 A. J. Fricker, "A method for solving high-order real symmetric eigenvalue problems," *International Journal for Numerical Methods in Engineering*, 1983, Vol. 19, pp. 1131–1138.

7.25 G. Longbottom and K. F. Gill, "The estimation of natural frequencies by use of Sturm sequences," *International Journal of Mechanical Engineering Education*, 1976, Vol. 4, pp. 319–329.

7.26 K. K. Gupta, "Solution of eigenvalue problems by Sturm sequence method," *International Journal for Numerical Methods in Engineering*, 1972, Vol. 4, pp. 379–404.

7.27 W. K. Chen and F. Y. Chen, "Topological analysis of a class of lumped vibrational systems," *Journal of Sound and Vibration*, 1969, Vol. 10, pp. 198–207.

REVIEW QUESTIONS

7.1 Give brief answers to the following:

1. Name a few methods for finding the fundamental natural frequency of a multidegree-of-freedom system.
2. What is the basic assumption made in deriving Dunkerley's formula?
3. What is Rayleigh's principle?
4. State whether we get a lower bound or an upper bound to the fundamental natural frequency if we use (a) Dunkerley's formula and (b) Rayleigh's method.
5. What is Rayleigh's quotient?
6. What is the basic principle used in Holzer's method?
7. What is the matrix iteration method?
8. Can we use any trial vector $\vec{X}_1$ in the matrix iteration method to find the largest natural frequency?
9. Using the matrix iteration method, how do you find the intermediate natural frequencies?
10. What is the difference between the matrix iteration method and Jacobi's method?
11. What is a rotation matrix? What is its purpose in Jacobi's method?
12. What is a standard eigenvalue problem?
13. What is the role of Choleski decomposition in deriving a standard eigenvalue problem?
14. How do you find the inverse of an upper triangular matrix?

7.2 Indicate whether each of the following statements is true or false:

1. The fundamental frequency given by Durkerley's formula will always be larger than the exact value.
2. The fundamental frequency given by Rayleigh's method will always be larger than the exact value.
3. $[A]\vec{X} = \lambda[B]\vec{X}$ is a standard eigenvalue problem.
4. $[A]\vec{X} = \lambda[I][B]\vec{X}$ is a standard eigenvalue problem.
5. Jacobi's method can find the eigenvalues of only symmetric matrices.
6. Jacobi's method uses rotation matrices.
7. The matrix iteration method requires the natural frequencies to be distinct and well separated.
8. In the matrix iteration method, any computational error will not yield incorrect results.
9. The matrix iteration method will never fail to converge to higher frequencies.
10. When Rayleigh's method is used for a shaft carrying several rotors, the static deflection curve can be used as the appropriate mode shape.
11. Rayleigh's method can be considered to be the same as the conservation of energy for a vibrating system.

7.3 Fill in each of the following blanks with the appropriate word:

1. Any symmetric positive definite matrix $[A]$ can be decomposed as $[A] = [U]^T[U]$, where $[U]$ is _____ triangular matrix.
2. The method of decomposing a symmetric positive definite matrix $[A]$ as $[A] = [U]^T[U]$ is known as _____ *method*.
3. Each step of Jacobi's method reduces a pair of off-diagonal elements to _____.
4. The _____ theorem permits the representation of any vector as a linear combination of the eigenvectors of the system.
5. If the matrix iteration method converges to the smallest eigenvalue with $[D]\vec{X} = \lambda\vec{X}$, the method converges to the _____ eigenvalue with $[D]^{-1}\vec{X} = \mu\vec{X}$.
6. Rayleigh's quotient provides _____ bound to ω_1^2 and _____ bound to ω_n^2.
7. Rayleigh's quotient has a stationary value in the neighborhood of an _____.
8. For a shaft carrying masses $m_1, m_2, \ldots$, Rayleigh's method gives the natural frequency as

$$\omega = \left\{ \frac{g(m_1 w_1 + m_2 w_2 + \cdots)}{m_1 w_1^2 + m_2 w_2^2 + \cdots} \right\}^{1/2}$$

   where $w_1, w_2, \ldots$ denote the _____ deflections of $m_1, m_2, \ldots$, respectively.

9. Holzer's method is basically a _____ method.
10. _____ method is more extensively applied to torsional systems, although the method is equally applicable to linear systems.
11. The computation of higher natural frequencies, based on the matrix iteration method, involves a process known as *matrix* _____.

7.4 Select the most appropriate answer out of the choices given:

1. When the trial vector

$$\vec{X}^{(1)} = \begin{Bmatrix} 1 \\ 1 \\ 1 \end{Bmatrix}$$

   is used for the solution of the eigenvalue problem,

$$\begin{bmatrix} 1 & 1 & 2 \\ 1 & 2 & 2 \\ 1 & 2 & 3 \end{bmatrix} \vec{X} = \lambda\vec{X}$$

   the next trial vector, $\vec{X}^{(2)}$, given by the matrix iteration method is

   a. $\begin{Bmatrix} 3 \\ 5 \\ 6 \end{Bmatrix}$ **b.** $\begin{Bmatrix} 1 \\ 1 \\ 1 \end{Bmatrix}$ **c.** $\begin{Bmatrix} 3 \\ 3 \\ 3 \end{Bmatrix}$

2. For a semidefinite system, the final equation in Holzer's method denotes the
   a. amplitude at the end as zero
   b. sum of inertia forces as zero
   c. equation of motion

3. Dunkerley's formula is given by

   a. $\omega_1^2 \approx a_{11}m_1 + a_{22}m_2 + \cdots + a_{nn}m_n$

   b. $\dfrac{1}{\omega_1^2} \approx a_{11}m_1 + a_{22}m_2 + \cdots + a_{nn}m_n$

   c. $\dfrac{1}{\omega_1^2} \approx k_{11}m_1 + k_{22}m_2 + \cdots + k_{nn}m_n$

4. Rayleigh's quotient is given by

   a. $\dfrac{\vec{X}^T[k]\vec{X}}{\vec{X}^T[m]\vec{X}}$ **b.** $\dfrac{\vec{X}^T[m]\vec{X}}{\vec{X}^T[k]\vec{X}}$ **c.** $\dfrac{\vec{X}^T[k]\vec{X}}{\dot{\vec{X}}^T[m]\dot{\vec{X}}}$

5. Rayleigh's quotient satisfies the following relation:

   a. $R(\vec{X}) \leq \omega_1^2$ **b.** $R(\vec{X}) \geq \omega_n^2$ **c.** $R(\vec{X}) \geq \omega_1^2$

6. For a vibrating system with $[k] = \begin{bmatrix} 2 & -1 \\ -1 & 2 \end{bmatrix}$ and $[m] = \begin{bmatrix} 1 & 0 \\ 0 & 1 \end{bmatrix}$, the mode shape closest to the fundamental mode, according to Rayleigh's quotient, $R(\vec{X}) = \dfrac{\vec{X}^T[k]\vec{X}}{\vec{X}^T[m]\vec{X}}$, is given by $\vec{X} =$

   a. $\begin{Bmatrix} 1 \\ 1 \end{Bmatrix}$ **b.** $\begin{Bmatrix} 1 \\ -1 \end{Bmatrix}$ **c.** $\begin{Bmatrix} -1 \\ 1 \end{Bmatrix}$

7.5   Match the items in the two columns below:

1. Dunkerley's formula	**a.** Finds the natural frequencies and mode shapes of the system, one at a time, using several trial values for each frequency.
2. Rayleigh's method	**b.** Finds all the natural frequencies using trial vectors and the matrix deflation procedure.
3. Holzer's method	**c.** Finds all the eigenvalues and eigenvectors simultaneously without using trial vectors.
4. Matrix iteration method	**d.** Finds the approximate value of the fundamental frequency of a composite system.
5. Jacobi's method	**e.** Finds the approximate value of the fundamental frequency of a system that is always larger than the true value.

PROBLEMS

Section 7.2 Dunkerley's Formula

7.1 Estimate the fundamental frequency of the beam shown in Fig. 6.9, using Dunkerley's formula for the following data: (a) $m_1 = m_3 = 5m$, $m_2 = m$ and (b) $m_1 = m_3 = m$, $m_2 = 5m$.

7.2 Find the fundamental frequency of the torsional system shown in Fig. 6.11, using Dunkerley's formula for the following data: (a) $J_1 = J_2 = J_3 = J_0$; $k_{t1} = k_{t2} = k_{t3} = k_t$; and (b) $J_1 = J_0$, $J_2 = 2J_0$, $J_3 = 3J_0$; $k_{t1} = k_t$, $k_{t2} = 2k_t$, $k_{t3} = 3k_t$.

7.3 Estimate the fundamental frequency of the shaft shown in Fig. 7.3, using Dunkerley's formula for the following data: $m_1 = m$, $m_2 = 2m$, $m_3 = 3m$, $l_1 = l_2 = l_3 = l_4 = l/4$.

7.4 The natural frequency of vibration, in bending, of the wing of a military aircraft is found to be 20 Hz. Find the new frequency of bending vibration of the wing when a weapon, weighing 900 kg, is attached at the tip of the wing, as shown in Fig. 7.10. The stiffness of the wing tip, in bending, is known to be 700 kN/m.

FIGURE 7.10

7.5 In an overhead crane (see Fig. 7.11), the trolley weighs 10 times the weight of the girder. Using Dunkerley's formula, estimate the fundamental frequency of the system.

FIGURE 7.11

7.6 Using Dunkerley's formula, determine the fundamental natural frequency of the stretched string system shown in Fig. 5.33 with $m_1 = m_2 = m$ and $l_1 = l_2 = l_3 = l$.

7.7* Design a minimum-weight tubular section for the shaft shown in Fig. 7.3 to achieve a fundamental frequency of vibration of 0.5 Hz. Assume $m_1 = 20$ kg, $m_2 = 50$ kg, $m_3 = 40$ kg, $l_1 = 1$ m, $l_2 = 3$ m, $l_3 = 4$ m, $l_4 = 2$ m, and $E = 2.07 \times 10^{11}$ N/m^2.

7.8 A uniform simply supported beam carries two masses m_1 and m_2 with $m_2 = 3m_1$ as shown in Fig. 7.12. Find the fundamental natural frequency of the beam using Dunkerley's method.

---

*The asterisk denotes a problem with no unique answer.

FIGURE 7.12

7.9 A uniform fixed-fixed beam carries two masses m_1 and m_2 with $m_2 = m_1$ as shown in Fig. 7.13. Find the fundamental natural frequency of the beam using Dunkerley's method.

FIGURE 7.13

Section 7.3 Rayleigh's Method

7.10 Using Rayleigh's method, determine the first natural frequency of vibration of the system shown in Fig. 7.2. Assume $k_1 = k$, $k_2 = 2k$, $k_3 = 3k$, and $m_1 = m$, $m_2 = 2m$, $m_3 = 3m$.

7.11 Using Rayleigh's method, find the fundamental natural frequency of the torsional system shown in Fig. 6.11. Assume that $J_1 = J_0$, $J_2 = 2J_0$, $J_3 = 3J_0$, and $k_{t1} = k_{t2} = k_{t3} = k_t$.

7.12 Using Rayleigh's method, solve Problem 7.6.

7.13 Using Rayleigh's method, determine the fundamental natural frequency of the system shown in Fig. 5.33 when $m_1 = m$, $m_2 = 5m$, $l_1 = l_2 = l_3 = l$.

7.14 A two-story shear building is shown in Fig. 7.14 in which the floors are assumed to be rigid. Using Rayleigh's method, compute the first natural frequency of the building for $m_1 = 2m$,

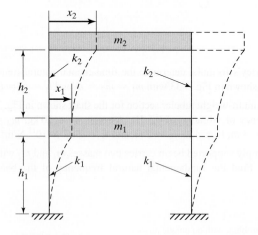

FIGURE 7.14

$m_2 = m$, $h_1 = h_2 = h$, and $k_1 = k_2 = 3EI/h^3$. Assume the first mode configuration to be the same as the static equilibrium shape due to loads proportional to the floor weights.

7.15 Prove that Rayleigh's quotient is never higher than the highest eigenvalue.

7.16 Figure 7.15 shows a steel stepped cantilever beam. The steps have square cross sections of size 10 cm × 10 cm and 5 cm × 5 cm each with a length of 130 cm. Assuming the Young's modulus as E = 205 GPa and the unit mass as $\gamma = 7.83$ g/cm^3 for the material of the beam, determine the fundamental natural frequency of bending vibration of the beam using Rayleigh's method. Assume the deflection of the beam as

$$y(x) = C\left(1 - \cos\frac{\pi x}{2l}\right)$$

where C is a constant.

FIGURE 7.15

7.17 A uniform simply supported beam of length 250 cm with a hollow rectangular section is shown in Fig. 7.16. Assuming a deflection shape of

$$y(x) = C \sin\frac{\pi x}{l}$$

find the natural frequency of transverse vibration of the beam. The material of the beam has a Young's modulus of $E = 205$ GPa and a unit mass of $\gamma = 7.83$ g/cm^3.

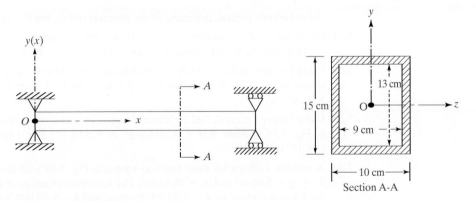

FIGURE 7.16

7.18 A uniform fixed-fixed beam of length l with a rectangular cross section $w \times h$ is shown in Fig. 7.17. Assuming the Young's modulus as E and unit mass as γ for the material of the beam and the deflection shape as

$$y(x) = C \left(1 - \cos \frac{2\pi x}{l} \right)$$

Determine the natural frequency of the beam.

FIGURE 7.17

Section 7.4 Holzer's Method

7.19 Using Holzer's method, find the natural frequencies and mode shapes of the system shown in Fig. 6.14, with $m_1 = 100$ kg, $m_2 = 20$ kg, $m_3 = 200$ kg, $k_1 = 8{,}000$ N/m, and $k_2 = 4{,}000$ N/m.

7.20 The stiffness and mass matrices of a vibrating system are given by

$$[k] = k \begin{bmatrix} 2 & -1 & 0 \\ -1 & 2 & -1 \\ 0 & -1 & 3 \end{bmatrix}, \quad [m] = m \begin{bmatrix} 1 & 0 & 0 \\ 0 & 1 & 0 \\ 0 & 0 & 2 \end{bmatrix}$$

Using Holzer's method, determine all the principal modes and the natural frequencies.

7.21 For the torsional system shown in Fig. 6.11, determine a principal mode and the corresponding frequency by Holzer's method. Assume that $k_{t1} = k_{t2} = k_{t3} = k_t$ and $J_1 = J_2 = J_3 = J_0$.

7.22 Using Holzer's method, find the natural frequencies and mode shapes of the shear building shown in Fig. 7.14. Assume that $m_1 = 2m$, $m_2 = m$, $h_1 = h_2 = h$, $k_1 = 2k$, $k_2 = k$, and $k = 3EI/h^3$.

7.23 Using Holzer's method, find the natural frequencies and mode shapes of the system shown in Fig. 6.39. Assume that $J_1 = 10$ kg-m^2, $J_2 = 5$ kg-m^2, $J_3 = 1$ kg-m^2, and $k_{t1} = k_{t2} = 1 \times 10^6$ N-m/rad.

7.24 A uniform shaft carries three rotors as shown in Fig. 7.18 with mass moments of inertia $J_1 = J_2 = 5$ kg-m^2 and $J_3 = 10$ kg-m^2. The torsional stiffnesses of the segments between the rotors are given by $k_{t1} = 20{,}000$ N-m/rad and $k_{t2} = 10{,}000$ N-m/rad. Determine the natural frequencies and mode shapes of the system using Holzer's method.

FIGURE 7.18

7.25 A uniform shaft carries three rotors as shown in Fig. 7.18 with mass moments of inertia $J_1 = 5$ kg-m^2, $J_2 = 15$ kg-m^2, and $J_3 = 25$ kg-m^2. The torsional stiffnesses of the segments between the rotors are given by $k_{t1} = 20{,}000$ N-m/rad and $k_{t2} = 60{,}000$ N-m/rad. Determine the natural frequencies and mode shapes of the system using Holzer's method.

7.26 The mass and stiffness matrices of a three-degree-of-freedom spring-mass system are given by

$$[m] = \begin{bmatrix} 3 & 0 & 0 \\ 0 & 2 & 0 \\ 0 & 0 & 1 \end{bmatrix} \quad \text{and} \quad [k] = \begin{bmatrix} 2 & -1 & 0 \\ -1 & 2 & -1 \\ 0 & -1 & 1 \end{bmatrix}$$

Determine the natural frequencies and mode shapes of the system using Holzer's method.

Section 7.5 Matrix Iteration Method

7.27 The largest eigenvalue of the matrix

$$[D] = \begin{bmatrix} 2.5 & -1 & 0 \\ -1 & 5 & -\sqrt{2} \\ 0 & -\sqrt{2} & 10 \end{bmatrix}$$

is given by $\lambda_1 = 10.38068$. Using the matrix iteration method, find the other eigenvalues and all the eigenvectors of the matrix. Assume $[m] = [I]$.

7.28 The mass and stiffness matrices of a spring-mass system are known to be

$$[m] = m \begin{bmatrix} 1 & 0 & 0 \\ 0 & 1 & 0 \\ 0 & 0 & 2 \end{bmatrix} \quad \text{and} \quad [k] = k \begin{bmatrix} 2 & -1 & 0 \\ -1 & 3 & -2 \\ 0 & -2 & 2 \end{bmatrix}$$

Using the matrix iteration method, find the natural frequencies and mode shapes of the system.

7.29 Using the matrix iteration method, find the natural frequencies and mode shapes of the system shown in Fig. 6.6 with $k_1 = k$, $k_2 = 2k$, $k_3 = 3k$, and $m_1 = m_2 = m_3 = m$.

7.30 Using the matrix iteration method, find the natural frequencies of the system shown in Fig. 6.28. Assume that $J_{d1} = J_{d2} = J_{d3} = J_0$, $l_i = l$, and $(GJ)_i = GJ$ for $i = 1$ to 4.

7.31 Using the matrix iteration method, solve Problem 7.6.

7.32 The stiffness and mass matrices of a vibrating system are given by

$$[k] = k \begin{bmatrix} 4 & -2 & 0 & 0 \\ -2 & 3 & -1 & 0 \\ 0 & -1 & 2 & -1 \\ 0 & 0 & -1 & 1 \end{bmatrix} \quad \text{and} \quad [m] = m \begin{bmatrix} 3 & 0 & 0 & 0 \\ 0 & 2 & 0 & 0 \\ 0 & 0 & 1 & 0 \\ 0 & 0 & 0 & 1 \end{bmatrix}$$

Using the matrix iteration method, find the fundamental frequency and the mode shape of the system.

7.33 The mass and stiffness matrices of an airplane in flight, with a three-degree-of-freedom model for vertical motion (similar to Fig. 6.26) are given by

$$[m] = \begin{bmatrix} 1 & 0 & 0 \\ 0 & 4 & 0 \\ 0 & 0 & 1 \end{bmatrix} \quad \text{and} \quad [k] = \begin{bmatrix} 3 & -3 & 0 \\ -3 & 6 & -3 \\ 0 & -3 & 3 \end{bmatrix}$$

Determine the highest natural frequency of vibration of the airplane using the matrix iteration method.

7.34 The mass and flexibility matrices of a three-degree-of-freedom system are given by

$$[m] = \begin{bmatrix} 1 & 0 & 0 \\ 0 & 2 & 0 \\ 0 & 0 & 1 \end{bmatrix} \quad \text{and} \quad [a] = [k]^{-1} = \begin{bmatrix} 1 & 1 & 1 \\ 1 & 2 & 2 \\ 1 & 2 & 3 \end{bmatrix}$$

Find the lowest natural frequency of vibration of the system using the matrix iteration method.

7.35 For the system considered in Problem 7.34, determine the highest natural frequency of vibration of the system using the matrix iteration method.

7.36 Find the middle natural frequency of vibration of the system considered in Problems 7.34 and 7.35 using the matrix iteration method.

Section 7.6 Jacobi's Method

7.37 Using Jacobi's method, find the eigenvalues and eigenvectors of the matrix

$$[D] = \begin{bmatrix} 3 & -2 & 0 \\ -2 & 5 & -3 \\ 0 & -3 & 3 \end{bmatrix}$$

7.38 Using Jacobi's method, find the eigenvalues and eigenvectors of the matrix

$$[D] = \begin{bmatrix} 3 & 2 & 1 \\ 2 & 2 & 1 \\ 1 & 1 & 1 \end{bmatrix}$$

7.39 Using Jacobi's method, find the eigenvalues of the matrix

$$[A] = \begin{bmatrix} 4 & -2 & 6 & 4 \\ -2 & 2 & -1 & 3 \\ 6 & -1 & 22 & 13 \\ 4 & 3 & 13 & 46 \end{bmatrix}$$

Section 7.7 Standard Eigenvalue Problem

7.40 Using the Choleski decomposition technique, decompose the matrix given in Problem 7.39.

7.41 Using the decomposition $[A] = [U]^T[U]$, find the inverse of the following matrix:

$$[A] = \begin{bmatrix} 5 & -1 & 1 \\ -1 & 6 & -4 \\ 1 & -4 & 3 \end{bmatrix}$$

7.42 Using the Choleski decomposition technique, find the inverse of the following matrix:

$$[A] = \begin{bmatrix} 2 & 5 & 8 \\ 5 & 16 & 28 \\ 8 & 28 & 54 \end{bmatrix}$$

7.43 Convert Problem 7.32 to a standard eigenvalue problem with a symmetric matrix.

7.44 Using the Choleski decomposition technique, express the following matrix as the product of two triangular matrices:

$$[A] = \begin{bmatrix} 16 & -20 & -24 \\ -20 & 89 & -50 \\ -24 & -50 & 280 \end{bmatrix}$$

Section 7.8 MATLAB Problems

7.45 Using MATLAB, find the eigenvalues and eigenvectors of the following matrix:

$$[A] = \begin{bmatrix} 3 & -2 & 0 \\ -2 & 5 & -3 \\ 0 & -1 & 1 \end{bmatrix}$$

7.46 Using MATLAB, find the eigenvalues and eigenvectors of the following matrix:

$$[A] = \begin{bmatrix} -5 & 2 & 1 \\ 1 & -9 & -1 \\ 2 & -1 & 7 \end{bmatrix}$$

7.47 Using **Program9.m**, find the eigenvalues and eigenvectors of the matrix $[D]$ given in Problem 7.27.

7.48 Using **Program10.m**, determine the eigenvalues and eigenvectors of the matrix $[D]$ given in Problem 7.38.

7.49 Using **Program11.m**, find the solution of the general eigenvalue problem given in Problem 7.32 with $k = m = 1$.

7.50 Find the eigenvalues and eigenvectors of the following matrix using MATLAB:

$$[A] = \begin{bmatrix} 2 & 2 & 2 \\ 2 & 5 & 5 \\ 2 & 5 & 12 \end{bmatrix}$$

7.51   Solve the following eigenvalue problem using MATLAB:

$$\omega^2 \begin{bmatrix} 3 & 0 & 0 \\ 0 & 2 & 0 \\ 0 & 0 & 1 \end{bmatrix} \vec{X} = \begin{bmatrix} 10 & -4 & 0 \\ -4 & 6 & -2 \\ 0 & -2 & 2 \end{bmatrix} \vec{X}$$

DESIGN PROJECTS

7.52   A flywheel of mass $m_1 = 100$ kg and a pulley of mass $m_2 = 50$ kg are to be mounted on a shaft of length $l = 2$ m, as shown in Fig. 7.19. Determine their locations l_1 and l_2 to maximize the fundamental frequency of vibration of the system.

FIGURE 7.19

7.53   A simplified diagram of an overhead traveling crane is shown in Fig. 7.20. The girder, with square cross section, and the wire rope, with circular cross section, are made up of steel. Design the girders and the wire rope such that the natural frequencies of the system are greater than the operating speed, 1,500 rpm, of an electric motor located in the trolley.

FIGURE 7.20

Stephen Prokf'yevich Timoshenko (1878–1972), a Russian-born engineer who emigrated to the United States, was one of the most widely known authors of books in the field of elasticity, strength of materials, and vibrations. He held the chair of mechanics at the University of Michigan and later at Stanford University, and he is regarded as the father of engineering mechanics in the United States. The improved theory he presented in 1921 for the vibration of beams has become known as the Timoshenko beam theory. (Courtesy of *Applied Mechanics Reviews*.)

C H A P T E R 8

Continuous Systems

Chapter Outline

The vibration analysis of continuous systems, which are also known as *distributed systems*, is considered in this chapter. The equations of motion of continuous systems will be partial differential equations. The equations of motion of several continuous systems, including the transverse vibration of a tightly stretched string or cable, longitudinal vibration of a bar, torsional vibration of a shaft or rod, the lateral vibration of beams, and transverse vibration of a membrane are derived by considering the free-body diagram of an infinitesimally small element of the particular system and applying Newton's second law of motion. The free-vibration solution of the system is found by assuming harmonic motion and applying the relevant boundary conditions. The solution gives an infinite number of natural frequencies and the corresponding mode shapes. The free-vibration displacement of the system is found as a linear superposition of the mode shapes, the constants involved being determined from the known initial conditions of the system. In the case of transverse vibration of a string of infinite length, the traveling-wave solution is presented. In the case of the longitudinal vibration of a bar, the vibration response under an initial force is also found. In the case of the transverse vibration of beams, all the common boundary conditions are summarized and the orthogonality of normal modes is proved. The forced vibration of beams is presented using the mode superposition method. The effect of axial force on the natural frequencies and mode shapes of beams is considered. The *thick beam theory*, also called the *Timoshenko beam theory*, is presented by considering the effects of rotary inertia and shear deformation. The free vibration of rectangular membranes is presented. *Rayleigh's method*, based on Rayleigh's quotient, for finding the approximate fundamental frequencies of continuous systems is outlined. The extension of the method, known as the *Rayleigh-Ritz* method, is outlined for determining approximate values of several frequencies. Finally, MATLAB solutions are presented for the free and forced vibration of typical continuous systems.

Learning Objectives

After you have finished studying this chapter, you should be able to do the following:

- Derive the equation of motion of a continuous system from the free-body diagram of an infinitesimally small element of the system and Newton's second law.
- Find the natural frequencies and mode shapes of the system using harmonic solution.
- Determine the free-vibration solution using a linear superposition of the mode shapes and the initial conditions.
- Find the free-vibration solutions of string, bar, shaft, beam, and membrane problems.
- Express the vibration of an infinite string in the form of traveling waves.
- Determine the forced-vibration solution of continuous systems using mode superposition method.
- Find the effects of axial force, rotary inertia, and shear deformation on the vibration of beams.
- Apply the Rayleigh and Rayleigh-Ritz methods to find the approximate natural frequencies of continuous systems.
- Use MATLAB to find the natural frequencies, mode shapes, and forced response of continuous systems.

8.1   Introduction

We have so far dealt with discrete systems where mass, damping, and elasticity were assumed to be present only at certain discrete points in the system. In many cases, known as *distributed* or *continuous systems*, it is not possible to identify discrete masses, dampers, or springs. We must then consider the continuous distribution of the mass, damping, and elasticity and assume that each of the infinite number of points of the system can vibrate. This is why a continuous system is also called a *system of infinite degrees of freedom*.

If a system is modeled as a discrete one, the governing equations are ordinary differential equations, which are relatively easy to solve. On the other hand, if the system is modeled as a continuous one, the governing equations are partial differential equations, which are more difficult. However, the information obtained from a discrete model of a system may not be as accurate as that obtained from a continuous model. The choice between the two models must be made carefully, with due consideration of factors such as the purpose of the analysis, the influence of the analysis on design, and the computational time available.

In this chapter, we shall consider the vibration of simple continuous systems—strings, bars, shafts, beams, and membranes. A more specialized treatment of the vibration of continuous structural elements is given in [8.1–8.3]. In general, the frequency equation of a continuous system is a transcendental equation that yields an infinite number of natural frequencies and normal modes. This is in contrast to the behavior of discrete systems, which yield a finite number of such frequencies and modes. We need to apply boundary conditions to find the natural frequencies of a continuous system. The question of boundary conditions does not arise in the case of discrete systems except in an indirect way, because the influence coefficients depend on the manner in which the system is supported.

8.2   Transverse Vibration of a String or Cable

8.2.1 Equation of Motion

Consider a tightly stretched elastic string or cable of length l subjected to a transverse force $f(x, t)$ per unit length, as shown in Fig. 8.1(a). The transverse displacement of the string, $w(x, t)$, is assumed to be small. Equilibrium of the forces in the z direction gives (see Fig. 8.1(b)):

The net force acting on an element is equal to the inertia force acting on the element, or

$$(P + dP) \sin(\theta + d\theta) + f\,dx - P \sin \theta = \rho\,dx \frac{\partial^2 w}{\partial t^2} \tag{8.1}$$

where P is the tension, ρ is the mass per unit length, and θ is the angle the deflected string makes with the x-axis. For an elemental length dx,

$$dP = \frac{\partial P}{\partial x}\,dx \tag{8.2}$$

$$\sin \theta \simeq \tan \theta = \frac{\partial w}{\partial x} \tag{8.3}$$

and

$$\sin(\theta + d\theta) \simeq \tan(\theta + d\theta) = \frac{\partial w}{\partial x} + \frac{\partial^2 w}{\partial x^2}\,dx \tag{8.4}$$

(a)

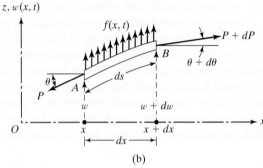

(b)

FIGURE 8.1   A vibrating string.

Hence the forced-vibration equation of the nonuniform string, Eq. (8.1), can be simplified to

$$\frac{\partial}{\partial x}\left[P\frac{\partial w(x,t)}{\partial x}\right] + f(x,t) = \rho(x)\frac{\partial^2 w(x,t)}{\partial t^2} \qquad (8.5)$$

If the string is uniform and the tension is constant, Eq. (8.5) reduces to

$$P\frac{\partial^2 w(x,t)}{\partial x^2} + f(x,t) = \rho\frac{\partial^2 w(x,t)}{\partial t^2} \qquad (8.6)$$

If $f(x,t) = 0$, we obtain the free-vibration equation

$$P\frac{\partial^2 w(x,t)}{\partial x^2} = \rho\frac{\partial^2 w(x,t)}{\partial t^2} \qquad (8.7)$$

or

$$c^2\frac{\partial^2 w}{\partial x^2} = \frac{\partial^2 w}{\partial t^2} \qquad (8.8)$$

where

$$c = \left(\frac{P}{\rho}\right)^{1/2} \tag{8.9}$$

Equation (8.8) is also known as the *wave equation*.

8.2.2
Initial and
Boundary
Conditions

The equation of motion, Eq. (8.5) or its special forms Eqs. (8.6) and (8.7), is a partial differential equation of the second order. Since the order of the highest derivative of w with respect to x and t in this equation is two, we need to specify two boundary and two initial conditions in finding the solution $w(x, t)$. If the string has a known deflection $w_0(x)$ and velocity $\dot{w}_0(x)$ at time $t = 0$, the initial conditions are specified as

$$w(x, t = 0) = w_0(x)$$

$$\frac{\partial w}{\partial t}(x, t = 0) = \dot{w}_0(x) \tag{8.10}$$

If the string is fixed at an end, say $x = 0$, the displacement w must always be zero, and so the boundary condition is

$$w(x = 0, t) = 0, \qquad t \ge 0 \tag{8.11}$$

If the string or cable is connected to a pin that can move in a perpendicular direction as shown in Fig. 8.2, the end cannot support a transverse force. Hence the boundary condition becomes

$$P(x)\frac{\partial w(x, t)}{\partial x} = 0 \tag{8.12}$$

FIGURE 8.2   String connected to pins at the ends.

FIGURE 8.3   String with elastic constraint.

If the end $x = 0$ is free and P is a constant, then Eq. (8.12) becomes

$$\frac{\partial w(0, t)}{\partial x} = 0, \qquad t \geq 0 \qquad (8.13)$$

If the end $x = l$ is constrained elastically as shown in Fig. 8.3, the boundary condition becomes

$$P(x)\frac{\partial w(x, t)}{\partial x}\bigg|_{x=l} = -\underset{\sim}{k}w(x, t)|_{x=l}, \qquad t \geq 0 \qquad (8.14)$$

where $\underset{\sim}{k}$ is the spring constant.

8.2.3
Free Vibration
of a Uniform
String

The free-vibration equation, Eq. (8.8), can be solved by the method of separation of variables. In this method, the solution is written as the product of a function $W(x)$ (which depends only on x) and a function $T(t)$ (which depends only on t) [8.4]:

$$w(x, t) = W(x)T(t) \qquad (8.15)$$

Substitution of Eq. (8.15) into Eq. (8.8) leads to

$$\frac{c^2}{W}\frac{d^2W}{dx^2} = \frac{1}{T}\frac{d^2T}{dt^2} \qquad (8.16)$$

Since the left-hand side of this equation depends only on x and the right-hand side depends only on t, their common value must be a constant—say, a—so that

$$\frac{c^2}{W}\frac{d^2W}{dx^2} = \frac{1}{T}\frac{d^2T}{dt^2} = a \qquad (8.17)$$

The equations implied in Eq. (8.17) can be written as

$$\frac{d^2W}{dx^2} - \frac{a}{c^2}W = 0 \qquad (8.18)$$

$$\frac{d^2T}{dt^2} - aT = 0 \tag{8.19}$$

Since the constant a is generally negative (see Problem 8.9), we can set $a = -\omega^2$ and write Eqs. (8.18) and (8.19) as

$$\frac{d^2W}{dx^2} + \frac{\omega^2}{c^2}W = 0 \tag{8.20}$$

$$\frac{d^2T}{dt^2} + \omega^2 T = 0 \tag{8.21}$$

The solutions of these equations are given by

$$W(x) = A \cos\frac{\omega x}{c} + B \sin\frac{\omega x}{c} \tag{8.22}$$

$$T(t) = C \cos\omega t + D \sin\omega t \tag{8.23}$$

where ω is the frequency of vibration and the constants A, B, C, and D can be evaluated from the boundary and initial conditions.

8.2.4
Free Vibration
of a String with
Both Ends Fixed

If the string is fixed at both ends, the boundary conditions are $w(0, t) = w(l, t) = 0$ for all time $t \geq 0$. Hence, from Eq. (8.15), we obtain

$$W(0) = 0 \tag{8.24}$$

$$W(l) = 0 \tag{8.25}$$

In order to satisfy Eq. (8.24), A must be zero in Eq. (8.22). Equation (8.25) requires that

$$B \sin\frac{\omega l}{c} = 0 \tag{8.26}$$

Since B cannot be zero for a nontrivial solution, we have

$$\sin\frac{\omega l}{c} = 0 \tag{8.27}$$

Equation (8.27) is called the *frequency* or *characteristic equation* and is satisfied by several values of ω. The values of ω are called the *eigenvalues* (or *natural frequencies* or *characteristic values*) of the problem. The nth natural frequency is given by

$$\frac{\omega_n l}{c} = n\pi, \qquad n = 1, 2, \ldots$$

or

$$\omega_n = \frac{nc\pi}{l}, \qquad n = 1, 2, \ldots \tag{8.28}$$

The solution $w_n(x, t)$ corresponding to ω_n can be expressed as

$$w_n(x, t) = W_n(x)T_n(t) = \sin\frac{n\pi x}{l}\left[C_n \cos\frac{nc\pi t}{l} + D_n \sin\frac{nc\pi t}{l}\right] \tag{8.29}$$

where C_n and D_n are arbitrary constants. The solution $w_n(x, t)$ is called the *nth mode of vibration* or *nth harmonic* or *nth normal mode* of the string. In this mode, each point of the string vibrates with an amplitude proportional to the value of W_n at that point, with the circular frequency $\omega_n = (nc\pi)/l$. The function $W_n(x)$ is called the *nth normal mode*, or *characteristic function*. The first three modes of vibration are shown in Fig. 8.4. The mode corresponding to $n = 1$ is called the *fundamental mode*, and ω_1 is called the *fundamental frequency*. The fundamental period is

$$\tau_1 = \frac{2\pi}{\omega_1} = \frac{2l}{c}$$

The points at which $w_n = 0$ for all times are called *nodes*. Thus the fundamental mode has two nodes, at $x = 0$ and $x = l$; the second mode has three nodes, at $x = 0$, $x = l/2$, and $x = l$; etc.

The general solution of Eq. (8.8), which satisfies the boundary conditions of Eqs. (8.24) and (8.25), is given by the superposition of all $w_n(x, t)$:

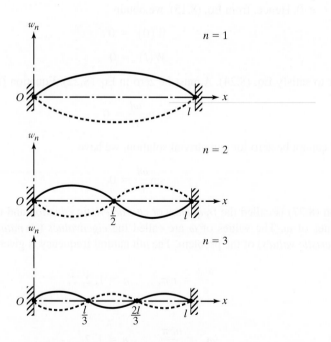

FIGURE 8.4   Mode shapes of a string.

$$w(x, t) = \sum_{n=1}^{\infty} w_n(x, t)$$

$$= \sum_{n=1}^{\infty} \sin\frac{n\pi x}{l}\left[C_n \cos\frac{nc\pi t}{l} + D_n \sin\frac{nc\pi t}{l} \right] \tag{8.30}$$

This equation gives all possible vibrations of the string; the particular vibration that occurs is uniquely determined by the specified initial conditions. The initial conditions give unique values of the constants C_n and D_n. If the initial conditions are specified as in Eq. (8.10), we obtain

$$\sum_{n=1}^{\infty} C_n \sin\frac{n\pi x}{l} = w_0(x) \tag{8.31}$$

$$\sum_{n=1}^{\infty} \frac{nc\pi}{l} D_n \sin\frac{n\pi x}{l} = \dot{w}_0(x) \tag{8.32}$$

which can be seen to be Fourier sine series expansions of $w_0(x)$ and $\dot{w}_0(x)$ in the interval $0 \le x \le l$. The values of C_n and D_n can be determined by multiplying Eqs. (8.31) and (8.32) by $\sin(n\pi x / l)$ and integrating with respect to x from 0 to l:

$$C_n = \frac{2}{l}\int_0^l w_0(x) \sin\frac{n\pi x}{l} \, dx \tag{8.33}$$

$$D_n = \frac{2}{nc\pi}\int_0^l \dot{w}_0(x) \sin\frac{n\pi x}{l} \, dx \tag{8.34}$$

Note: The solution given by Eq. (8.30) can be identified as the *mode superposition method* since the response is expressed as a superposition of the normal modes. The procedure is applicable in finding not only the free-vibration solution but also the forced-vibration solution of continuous systems.

EXAMPLE   8.1

Dynamic Response of a Plucked String

If a string of length l, fixed at both ends, is plucked at its midpoint as shown in Fig. 8.5 and then released, determine its subsequent motion.

Solution: The solution is given by Eq. (8.30) with C_n and D_n given by Eqs. (8.33) and (8.34), respectively. Since there is no initial velocity, $\dot{w}_0(x) = 0$, and so $D_n = 0$. Thus the solution of Eq. (8.30) reduces to

$$w(x, t) = \sum_{n=1}^{\infty} C_n \sin\frac{n\pi x}{l} \cos\frac{nc\pi t}{l} \tag{E.1}$$

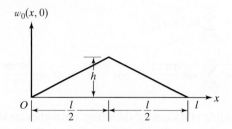

FIGURE 8.5   Initial deflection of the string.

where

$$C_n = \frac{2}{l} \int_0^l w_0(x) \sin \frac{n\pi x}{l} \, dx \tag{E.2}$$

The initial deflection $w_0(x)$ is given by

$$w_0(x) = \begin{cases} \dfrac{2hx}{l} & \text{for } 0 \le x \le \dfrac{l}{2} \\ \dfrac{2h(l-x)}{l} & \text{for } \dfrac{l}{2} \le x \le l \end{cases} \tag{E.3}$$

By substituting Eq. (E.3) into Eq. (E.2), C_n can be evaluated:

$$C_n = \frac{2}{l} \left\{ \int_0^{l/2} \frac{2hx}{l} \sin \frac{n\pi x}{l} \, dx + \int_{l/2}^l \frac{2h}{l} (l-x) \sin \frac{n\pi x}{l} \, dx \right\}$$

$$= \begin{cases} \dfrac{8h}{\pi^2 n^2} \sin \dfrac{n\pi}{2} & \text{for } n = 1, 3, 5, \ldots \\ 0 & \text{for } n = 2, 4, 6, \ldots \end{cases} \tag{E.4}$$

By using the relation

$$\sin \frac{n\pi}{2} = (-1)^{(n-1)/2}, \qquad n = 1, 3, 5, \ldots \tag{E.5}$$

the desired solution can be expressed as

$$w(x, t) = \frac{8h}{\pi^2} \left\{ \sin \frac{\pi x}{l} \cos \frac{\pi ct}{l} - \frac{1}{9} \sin \frac{3\pi x}{l} \cos \frac{3\pi ct}{l} + \cdots \right\} \tag{E.6}$$

In this case, no even harmonics are excited.

■

**8.2.5
Traveling-Wave
Solution**

The solution of the wave equation, Eq. (8.8), for a string of infinite length can be expressed as [8.5]

$$w(x, t) = w_1(x - ct) + w_2(x + ct) \tag{8.35}$$

where w_1 and w_2 are arbitrary functions of $(x - ct)$ and $(x + ct)$, respectively. To show that Eq. (8.35) is the correct solution of Eq. (8.8), we first differentiate Eq. (8.35):

$$\frac{\partial^2 w(x, t)}{\partial x^2} = w_1''(x - ct) + w_2''(x + ct) \tag{8.36}$$

$$\frac{\partial^2 w(x, t)}{\partial t^2} = c^2 w_1''(x - ct) + c^2 w_2''(x + ct) \tag{8.37}$$

Substitution of these equations into Eq. (8.8) reveals that the wave equation is satisfied. In Eq. (8.35), $w_1(x - ct)$ and $w_2(x + ct)$ represent waves that propagate in the positive and negative directions of the x-axis, respectively, with a velocity c.

For a given problem, the arbitrary functions w_1 and w_2 are determined from the initial conditions, Eq. (8.10). Substitution of Eq. (8.35) into Eq. (8.10) gives, at $t = 0$,

$$w_1(x) + w_2(x) = w_0(x) \tag{8.38}$$

$$-cw_1'(x) + cw_2'(x) = \dot{w}_0(x) \tag{8.39}$$

where the prime indicates differentiation with respect to the respective argument at $t = 0$ (that is, with respect to x). Integration of Eq. (8.39) yields

$$-w_1(x) + w_2(x) = \frac{1}{c} \int_{x_0}^{x} \dot{w}_0(x') \, dx' \tag{8.40}$$

where x_0 is a constant. Solution of Eqs. (8.38) and (8.40) gives w_1 and w_2:

$$w_1(x) = \frac{1}{2} \left[w_0(x) - \frac{1}{c} \int_{x_0}^{x} \dot{w}_0(x') \, dx' \right] \tag{8.41}$$

$$w_2(x) = \frac{1}{2} \left[w_0(x) + \frac{1}{c} \int_{x_0}^{x} \dot{w}_0(x') \, dx' \right] \tag{8.42}$$

By replacing x with $(x - ct)$ and $(x + ct)$, respectively, in Eqs. (8.41) and (8.42), we obtain the total solution:

$$w(x, t) = w_1(x - ct) + w_2(x + ct)$$

$$= \frac{1}{2} [w_0(x - ct) + w_0(x + ct)] + \frac{1}{2c} \int_{x-ct}^{x+ct} \dot{w}_0(x') \, dx' \tag{8.43}$$

The following points should be noted:

1. As can be seen from Eq. (8.43), there is no need to apply boundary conditions to the problem.

2. The solution given by Eq. (8.43) can be expressed as

$$w(x, t) = w_D(x, t) + w_V(x, t) \tag{8.44}$$

where $w_D(x, t)$ denotes the waves propagating due to the known initial displacement $w_0(x)$ with zero initial velocity, and $w_V(x, t)$ represents waves traveling due only to the known initial velocity $\dot{w}_0(x)$ with zero initial displacement.

The transverse vibration of a string fixed at both ends excited by the transverse impact of an elastic load at an intermediate point was considered in [8.6]. A review of the literature on the dynamics of cables and chains was given by Triantafyllou [8.7].

8.3   Longitudinal Vibration of a Bar or Rod

**8.3.1
Equation of
Motion and
Solution**

Consider an elastic bar of length l with varying cross-sectional area $A(x)$, as shown in Fig. 8.6. The forces acting on the cross sections of a small element of the bar are given by P and $P + dP$ with

$$P = \sigma A = EA \frac{\partial u}{\partial x} \tag{8.45}$$

where σ is the axial stress, E is Young's modulus, u is the axial displacement, and $\partial u / \partial x$ is the axial strain. If $f(x, t)$ denotes the external force per unit length, the summation of the forces in the x direction gives the equation of motion

$$(P + dP) + f \, dx - P = \rho A \, dx \frac{\partial^2 u}{\partial t^2} \tag{8.46}$$

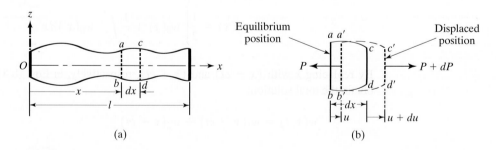

(a)   (b)

FIGURE 8.6   Longitudinal vibration of a bar.

where ρ is the mass density of the bar. By using the relation $dP = (\partial P/\partial x)\, dx$ and Eq. (8.45), the equation of motion for the forced longitudinal vibration of a nonuniform bar, Eq. (8.46), can be expressed as

$$\frac{\partial}{\partial x}\left[EA(x)\frac{\partial u(x,t)}{\partial x}\right] + f(x,t) = \rho(x)A(x)\frac{\partial^2 u}{\partial t^2}(x,t) \tag{8.47}$$

For a uniform bar, Eq. (8.47) reduces to

$$EA\frac{\partial^2 u}{\partial x^2}(x,t) + f(x,t) = \rho A\frac{\partial^2 u}{\partial t^2}(x,t) \tag{8.48}$$

The free-vibration equation can be obtained from Eq. (8.48), by setting $f = 0$, as

$$c^2\frac{\partial^2 u}{\partial x^2}(x,t) = \frac{\partial^2 u}{\partial t^2}(x,t) \tag{8.49}$$

where

$$c = \sqrt{\frac{E}{\rho}} \tag{8.50}$$

Note that Eqs. (8.47) to (8.50) can be seen to be similar to Eqs. (8.5), (8.6), (8.8), and (8.9), respectively. The solution of Eq. (8.49), which can be obtained as in the case of Eq. (8.8), can thus be written as

$$u(x,t) = U(x)T(t) \equiv \left(\underset{\sim}{A}\cos\frac{\omega x}{c} + \underset{\sim}{B}\sin\frac{\omega x}{c}\right)(C\cos\omega t + D\sin\omega t)^{1} \tag{8.51}$$

End Conditions of Bar	Boundary Conditions	Frequency Equation	Mode Shape (Normal Function)	Natural Frequencies
Fixed-free	$u(0,t) = 0$ $\dfrac{\partial u}{\partial x}(l,t) = 0$	$\cos\dfrac{\omega l}{c} = 0$	$U_n(x) = C_n \sin\dfrac{(2n+1)\pi x}{2l}$	$\omega_n = \dfrac{(2n+1)\pi c}{2l}$; $n = 0,1,2,\ldots$
Free-free	$\dfrac{\partial u}{\partial x}(0,t) = 0$ $\dfrac{\partial u}{\partial x}(l,t) = 0$	$\sin\dfrac{\omega l}{c} = 0$	$U_n(x) = C_n \cos\dfrac{n\pi x}{l}$	$\omega_n = \dfrac{n\pi c}{l}$; $n = 0,1,2,\ldots$
Fixed-fixed	$u(0,t) = 0$ $u(l,t) = 0$	$\sin\dfrac{\omega l}{c} = 0$	$U_n(x) = C_n \cos\dfrac{n\pi x}{l}$	$\omega_n = \dfrac{n\pi c}{l}$; $n = 1,2,3,\ldots$

FIGURE 8.7   Common boundary conditions for a bar in longitudinal vibration.

---

[1]We use $\underset{\sim}{A}$ and $\underset{\sim}{B}$ in this section; A is used to denote the cross-sectional area of the bar.

where the function $U(x)$ represents the normal mode and depends only on x and the function $T(t)$ depends only on t. If the bar has known initial axial displacement $u_0(x)$ and initial velocity $\dot{u}_0(x)$, the initial conditions can be stated as

$$u(x, t = 0) = u_0(x)$$

$$\frac{\partial u}{\partial t}(x, t = 0) = \dot{u}_0(x) \tag{8.52}$$

The common boundary conditions and the corresponding frequency equations for the longitudinal vibration of uniform bars are shown in Fig. 8.7.

Boundary Conditions for a Bar

EXAMPLE 8.2

A uniform bar of cross-sectional area A, length l, and Young's modulus E is connected at both ends by springs, dampers, and masses, as shown in Fig. 8.8(a). State the boundary conditions.

Solution: The free-body diagrams of the masses m_1 and m_2 are shown in Fig. 8.8(b). From this, we find that at the left end $(x = 0)$, the force developed in the bar due to positive u and $\partial u/\partial x$ must be equal to the sum of spring, damper, and inertia forces:

$$AE\frac{\partial u}{\partial x}(0, t) = k_1 u(0, t) + c_1\frac{\partial u}{\partial t}(0, t) + m_1\frac{\partial^2 u}{\partial t^2}(0, t) \tag{E.1}$$

Similarly at the right end $(x = l)$, the force developed in the bar due to positive u and $\partial u/\partial x$ must be equal to the negative sum of spring, damper, and inertia forces:

$$AE\frac{\partial u}{\partial x}(l, t) = -k_2 u(l, t) - c_2\frac{\partial u}{\partial t}(l, t) - m_2\frac{\partial^2 u}{\partial t^2}(l, t) \tag{E.2}$$

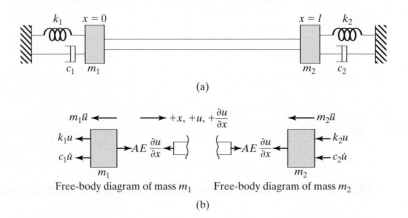

(a)

Free-body diagram of mass m_1 Free-body diagram of mass m_2

(b)

FIGURE 8.8   Bar connected to springs-masses-dampers at ends.

8.3.2
Orthogonality
of Normal
Functions

The normal functions for the longitudinal vibration of bars satisfy the orthogonality relation

$$\int_0^l U_i(x)U_j(x)\,dx = 0 \qquad (8.53)$$

where $U_i(x)$ and $U_j(x)$ denote the normal functions corresponding to the ith and jth natural frequencies ω_i and ω_j, respectively. When $u(x,t) = U_i(x)T(t)$ and $u(x,t) = U_j(x)T(t)$ are assumed as solutions, Eq. (8.49) gives

$$c^2 \frac{d^2 U_i(x)}{dx^2} + \omega_i^2 U_i(x) = 0 \qquad \text{or} \qquad c^2 U_i''(x) + \omega_i^2 U_i(x) = 0 \qquad (8.54)$$

and

$$c^2 \frac{d^2 U_j(x)}{dx^2} + \omega_j^2 U_j(x) = 0 \qquad \text{or} \qquad c^2 U_j''(x) + \omega_j^2 U_j(x) = 0 \qquad (8.55)$$

where $U_i'' = \dfrac{d^2 U_i}{dx^2}$ and $U_j'' = \dfrac{d^2 U_j}{dx^2}$. Multiplication of Eq. (8.54) by U_j and Eq. (8.55) by U_i gives

$$c^2 U_i'' U_j + \omega_i^2 U_i U_j = 0 \qquad (8.56)$$

$$c^2 U_j'' U_i + \omega_j^2 U_j U_i = 0 \qquad (8.57)$$

Subtraction of Eq. (8.57) from Eq. (8.56) and integration from 0 to l results in

$$\int_0^l U_i U_j\,dx = -\frac{c^2}{\omega_i^2 - \omega_j^2} \int_0^l (U_i'' U_j - U_j'' U_i)\,dx$$

$$= -\frac{c^2}{\omega_i^2 - \omega_j^2} \left[U_i' U_j - U_j' U_i \right]\Big|_0^l \qquad (8.58)$$

The right-hand side of Eq. (8.58) can be proved to be zero for any combination of boundary conditions. For example, if the bar is fixed at $x = 0$ and free at $x = l$,

$$u(0,t) = 0, \qquad t \geq 0 \qquad \text{or} \qquad U(0) = 0 \qquad (8.59)$$

$$\frac{\partial u}{\partial x}(l,t) = 0, \qquad t \geq 0 \qquad \text{or} \qquad U'(l) = 0 \qquad (8.60)$$

Thus $(U_i' U_j - U_j' U_i)\big|_{x=l} = 0$ due to U' being zero (Eq. (8.60)) and $(U_i' U_j - U_j' U_i)\big|_{x=0} = 0$ due to U being zero (Eq. (8.59)). Equation (8.58) thus reduces to Eq. (8.53), which is also known as the *orthogonality principle for the normal functions*.

EXAMPLE 8.3

Free Vibrations of a Fixed-Free Bar

Find the natural frequencies and the free-vibration solution of a bar fixed at one end and free at the other.

Solution: Let the bar be fixed at $x = 0$ and free at $x = l$, so that the boundary conditions can be expressed as

$$u(0, t) = 0, \qquad t \geq 0 \tag{E.1}$$

$$\frac{\partial u}{\partial x}(l, t) = 0, \qquad t \geq 0 \tag{E.2}$$

The use of Eq. (E.1) in Eq. (8.51) gives $\underset{\sim}{A} = 0$, while the use of Eq. (E.2) gives the frequency equation

$$B \frac{\omega}{\sim c} \cos \frac{\omega l}{c} = 0 \qquad \text{or} \qquad \cos \frac{\omega l}{c} = 0 \tag{E.3}$$

The eigenvalues or natural frequencies are given by

$$\frac{\omega_n l}{c} = (2n + 1)\frac{\pi}{2}, \qquad n = 0, 1, 2, \ldots$$

or

$$\omega_n = \frac{(2n + 1)\pi c}{2l}, \qquad n = 0, 1, 2, \ldots \tag{E.4}$$

Thus the total (free-vibration) solution of Eq. (8.49) can be written, using the mode superposition method, as

$$u(x, t) = \sum_{n=0}^{\infty} u_n(x, t)$$

$$= \sum_{n=0}^{\infty} \sin\frac{(2n + 1)\pi x}{2l} \left[C_n \cos\frac{(2n + 1)\pi c t}{2l} + D_n \sin\frac{(2n + 1)\pi c t}{2l} \right] \tag{E.5}$$

where the values of the constants C_n and D_n can be determined from the initial conditions, as in Eqs. (8.33) and (8.34):

$$C_n = \frac{2}{l} \int_0^l u_0(x) \sin\frac{(2n + 1)\pi x}{2l} dx \tag{E.6}$$

$$D_n = \frac{4}{(2n + 1)\pi c} \int_0^l \dot{u}_0(x) \sin\frac{(2n + 1)\pi x}{2l} dx \tag{E.7}$$

∎

Natural Frequencies of a Bar Carrying a Mass

EXAMPLE 8.4

Find the natural frequencies of a bar with one end fixed and a mass attached at the other end, as in Fig. 8.9.

Solution: The equation governing the axial vibration of the bar is given by Eq. (8.49) and the solution by Eq. (8.51). The boundary condition at the fixed end $(x = 0)$

$$u(0, t) = 0 \tag{E.1}$$

leads to $\underset{\sim}{A} = 0$ in Eq. (8.51). At the end $x = l$, the tensile force in the bar must be equal to the inertia force of the vibrating mass M, and so

$$AE\frac{\partial u}{\partial x}(l, t) = -M\frac{\partial^2 u}{\partial t^2}(l, t) \tag{E.2}$$

With the help of Eq. (8.51), this equation can be expressed as

$$AE\frac{\omega}{c}\cos\frac{\omega l}{c}(C\cos\omega t + D\sin\omega t) = M\omega^2 \sin\frac{\omega l}{c}(C\cos\omega t + D\sin\omega t)$$

—that is,

$$\frac{AE\omega}{c}\cos\frac{\omega l}{c} = M\omega^2 \sin\frac{\omega l}{c}$$

or

$$\alpha \tan \alpha = \beta \tag{E.3}$$

where

$$\alpha = \frac{\omega l}{c} \tag{E.4}$$

and

$$\beta = \frac{AEl}{c^2 M} = \frac{A\rho l}{M} = \frac{m}{M} \tag{E.5}$$

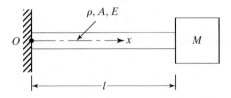

FIGURE 8.9 Bar carrying an end mass.

TABLE 8.1

| | Values of the Mass Ratio β | | | | |
	0.01	0.1	1.0	10.0	100.0
Value of α_1 $\left(\omega_1 = \dfrac{\alpha_1 c}{l}\right)$	0.1000	0.3113	0.8602	1.4291	1.5549
Value of α_2 $\left(\omega_2 = \dfrac{\alpha_2 c}{l}\right)$	3.1448	3.1736	3.4267	4.3063	4.6658

where m is the mass of the bar. Equation (E.3) is the frequency equation (in the form of a transcendental equation) whose solution gives the natural frequencies of the system. The first two natural frequencies are given in Table 8.1 for different values of the parameter β.

Note: If the mass of the bar is negligible compared to the mass attached, $m \simeq 0$,

$$c = \left(\frac{E}{\rho}\right)^{1/2} = \left(\frac{EAl}{m}\right)^{1/2} \to \infty \quad \text{and} \quad \alpha = \frac{\omega l}{c} \to 0$$

In this case

$$\tan\frac{\omega l}{c} \simeq \frac{\omega l}{c}$$

and the frequency equation (E.3) can be taken as

$$\left(\frac{\omega l}{c}\right)^2 = \beta$$

This gives the approximate value of the fundamental frequency

$$\omega_1 = \frac{c}{l}\beta^{1/2} = \frac{c}{l}\left(\frac{\rho Al}{M}\right)^{1/2} = \left(\frac{EA}{lM}\right)^{1/2} = \left(\frac{g}{\delta_s}\right)^{1/2}$$

where

$$\delta_s = \frac{Mgl}{EA}$$

represents the static elongation of the bar under the action of the load Mg.

∎

EXAMPLE 8.5

Vibrations of a Bar Subjected to Initial Force

A bar of uniform cross-sectional area A, density ρ, modulus of elasticity E, and length l is fixed at one end and free at the other end. It is subjected to an axial force F_0 at its free end, as shown in Fig. 8.10(a). Study the resulting vibrations if the force F_0 is suddenly removed.

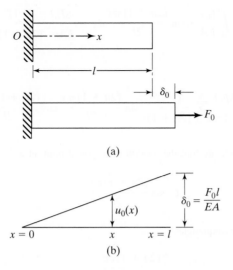

(a)

(b)

FIGURE 8.10   Bar subjected to an axial force at end.

Solution: The tensile strain induced in the bar due to F_0 is

$$\varepsilon = \frac{F_0}{EA}$$

Thus the displacement of the bar just before the force F_0 is removed (initial displacement) is given by (see Fig. (8.10)(b))

$$u_0 = u(x, 0) = \varepsilon x = \frac{F_0 x}{EA}, \qquad 0 \le x \le l \tag{E.1}$$

Since the initial velocity is zero, we have

$$\dot{u}_0 = \frac{\partial u}{\partial t}(x, 0) = 0, \qquad 0 \le x \le l \tag{E.2}$$

The general solution of a bar fixed at one end and free at the other end is given by Eq. (E.5) of Example 8.3:

$$u(x, t) = \sum_{n=0}^{\infty} u_n(x, t)$$

$$= \sum_{n=0}^{\infty} \sin \frac{(2n + 1)\pi x}{2l} \left[C_n \cos \frac{(2n + 1)\pi ct}{2l} + D_n \sin \frac{(2n + 1)\pi ct}{2l} \right] \tag{E.3}$$

where C_n and D_n are given by Eqs. (E.6) and (E.7) of Example 8.3. Since $\dot{u}_0 = 0$, we obtain $D_n = 0$. By using the initial displacement of Eq. (E.1) in Eq. (E.6) of Example 8.3, we obtain

$$C_n = \frac{2}{l} \int_0^l \frac{F_0 x}{EA} \cdot \sin \frac{(2n+1)\pi x}{2l} dx = \frac{8F_0 l}{EA\pi^2} \frac{(-1)^n}{(2n+1)^2} \tag{E.4}$$

Thus the solution becomes

$$u(x, t) = \frac{8F_0 l}{EA\pi^2} \sum_{n=0}^{\infty} \frac{(-1)^n}{(2n+1)^2} \sin \frac{(2n+1)\pi x}{2l} \cos \frac{(2n+1)\pi ct}{2l} \tag{E.5}$$

Equations (E.3) and (E.5) indicate that the motion of a typical point at $x = x_0$ on the bar is composed of the amplitudes

$$C_n \sin \frac{(2n+1)\pi x_0}{2l}$$

corresponding to the circular frequencies

$$\frac{(2n+1)\pi c}{2l}$$

∎

8.4 Torsional Vibration of a Shaft or Rod

Figure 8.11 represents a nonuniform shaft subjected to an external torque $f(x, t)$ per unit length. If $\theta(x, t)$ denotes the angle of twist of the cross section, the relation between the torsional deflection and the twisting moment $M_t(x, t)$ is given by [8.8]

$$M_t(x, t) = GJ(x) \frac{\partial \theta}{\partial x}(x, t) \tag{8.61}$$

where G is the shear modulus and $GJ(x)$ is the torsional stiffness, with $J(x)$ denoting the polar moment of inertia of the cross section in the case of a circular section. If the mass polar moment of inertia of the shaft per unit length is I_0, the inertia torque acting on an element of length dx becomes

$$I_0 dx \frac{\partial^2 \theta}{\partial t^2}$$

If an external torque $f(x, t)$ acts on the shaft per unit length, the application of Newton's second law yields the equation of motion:

$$(M_t + dM_t) + f \, dx - M_t = I_0 \, dx \frac{\partial^2 \theta}{\partial t^2} \tag{8.62}$$

By expressing dM_t as

$$\frac{\partial M_t}{\partial x} dx$$

(a)

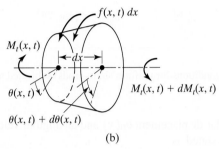

(b)

FIGURE 8.11   Torsional vibration of a shaft.

and using Eq. (8.61), the forced torsional vibration equation for a nonuniform shaft can be obtained:

$$\frac{\partial}{\partial x}\left[GJ(x)\frac{\partial \theta}{\partial x}(x, t)\right] + f(x, t) = I_0(x)\frac{\partial^2 \theta}{\partial t^2}(x, t) \qquad (8.63)$$

For a uniform shaft, Eq. (8.63) takes the form

$$GJ\frac{\partial^2 \theta}{\partial x^2}(x, t) + f(x, t) = I_0\frac{\partial^2 \theta}{\partial t^2}(x, t) \qquad (8.64)$$

which, in the case of free vibration, reduces to

$$c^2\frac{\partial^2 \theta}{\partial x^2}(x, t) = \frac{\partial^2 \theta}{\partial t^2}(x, t) \qquad (8.65)$$

where

$$c = \sqrt{\frac{GJ}{I_0}} \qquad (8.66)$$

Notice that Eqs. (8.63) to (8.66) are similar to the equations derived in the cases of transverse vibration of a string and longitudinal vibration of a bar. If the shaft has a uniform cross section, $I_0 = \rho J$, Eq. (8.66) becomes

$$c = \sqrt{\frac{G}{\rho}} \qquad (8.67)$$

End Conditions of Shaft	Boundary Conditions	Frequency Equation	Mode Shape (Normal Function)	Natural Frequencies
Fixed-free	$\theta(0, t) = 0$ $\dfrac{\partial \theta}{\partial x}(l, t) = 0$	$\cos \dfrac{\omega l}{c} = 0$	$\theta(x) = C_n \sin \dfrac{(2n + 1)\,\pi x}{2l}$	$\omega_n = \dfrac{(2n + 1)\,\pi c}{2l}$; $n = 0, 1, 2, \ldots$
Free-free	$\dfrac{\partial \theta}{\partial x}(0, t) = 0$ $\dfrac{\partial \theta}{\partial x}(l, t) = 0$	$\sin \dfrac{\omega l}{c} = 0$?	$\theta(x) = C_n \cos \dfrac{n\pi x}{l}$	$\omega_n = \dfrac{n\pi c}{l}$; $n = 0, 1, 2, \ldots$
Fixed-fixed	$\theta(0, t) = 0$ $\theta(l, t) = 0$	$\sin \dfrac{\omega l}{c} = 0$	$\theta(x) = C_n \cos \dfrac{n\pi x}{l}$	$\omega_n = \dfrac{n\pi c}{l}$; $n = 1, 2, 3, \ldots$

FIGURE 8.12   Boundary conditions for uniform shafts (rods) subjected to torsional vibration.

If the shaft is given an angular displacement $\theta_0(x)$ and an angular velocity $\dot{\theta}_0(x)$ at $t = 0$, the initial conditions can be stated as

$$\theta(x, t = 0) = \theta_0(x)$$

$$\frac{\partial \theta}{\partial t}(x, t = 0) = \dot{\theta}_0(x) \tag{8.68}$$

The general solution of Eq. (8.65) can be expressed as

$$\theta(x, t) = \left(A \cos \frac{\omega x}{c} + B \sin \frac{\omega x}{c} \right) (C \cos \omega t + D \sin \omega t) \tag{8.69}$$

The common boundary conditions for the torsional vibration of uniform shafts are indicated in Fig. 8.12 along with the corresponding frequency equations and the normal functions.

EXAMPLE 8.6

Natural Frequencies of a Milling Cutter

Find the natural frequencies of the plane milling cutter shown in Fig. 8.13 when the free end of the shank is fixed. Assume the torsional rigidity of the shank as GJ and the mass moment of inertia of the cutter as I_0.

Solution: The general solution is given by Eq. (8.69). From this equation, by using the fixed boundary condition $\theta(0, t) = 0$, we obtain $A = 0$. The boundary condition at $x = l$ can be stated as

$$GJ \frac{\partial \theta}{\partial x}(l, t) = -I_0 \frac{\partial^2 \theta}{\partial t^2}(l, t) \tag{E.1}$$

—that is,

$$BGJ \frac{\omega}{c} \cos \frac{\omega l}{c} = BI_0 \omega^2 \sin \frac{\omega l}{c}$$

FIGURE 8.13   Plane milling cutter.

or

$$\frac{\omega l}{c} \tan \frac{\omega l}{c} = \frac{J\rho l}{I_0} = \frac{J_{\text{rod}}}{I_0} \tag{E.2}$$

where $J_{\text{rod}} = J\rho l$. Equation (E.2) can be expressed as

$$\alpha \tan \alpha = \beta \qquad \text{where } \alpha = \frac{\omega l}{c} \qquad \text{and} \qquad \beta = \frac{J_{\text{rod}}}{I_0} \tag{E.3}$$

The solution of Eq. (E.3), and thus the natural frequencies of the system, can be obtained as in the case of Example 8.4.

∎

8.5   Lateral Vibration of Beams

8.5.1
Equation of Motion

Consider the free-body diagram of an element of a beam shown in Fig. 8.14, where $M(x, t)$ is the bending moment, $V(x, t)$ is the shear force, and $f(x, t)$ is the external force per unit length of the beam. Since the inertia force acting on the element of the beam is

$$\rho A(x)\, dx\, \frac{\partial^2 w}{\partial t^2}(x, t)$$

the force equation of motion in the z direction gives

$$-(V + dV) + f(x, t)\, dx + V = \rho A(x)\, dx \frac{\partial^2 w}{\partial t^2}(x, t) \tag{8.70}$$

where ρ is the mass density and $A(x)$ is the cross-sectional area of the beam. The moment equation of motion about the y-axis passing through point O in Fig. 8.14 leads to

$$(M + dM) - (V + dV)\, dx + f(x, t)\, dx \frac{dx}{2} - M = 0 \tag{8.71}$$

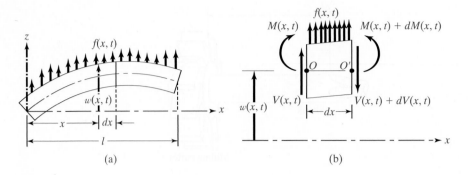

FIGURE 8.14   A beam in bending.

By writing

$$dV = \frac{\partial V}{\partial x} dx \qquad \text{and} \qquad dM = \frac{\partial M}{\partial x} dx$$

and disregarding terms involving second powers in dx, Eqs. (8.70) and (8.71) can be written as

$$-\frac{\partial V}{\partial x}(x, t) + f(x, t) = \rho A(x) \frac{\partial^2 w}{\partial t^2}(x, t) \tag{8.72}$$

$$\frac{\partial M}{\partial x}(x, t) - V(x, t) = 0 \tag{8.73}$$

By using the relation $V = \partial M/\partial x$ from Eq. (8.73), Eq. (8.72) becomes

$$-\frac{\partial^2 M}{\partial x^2}(x, t) + f(x, t) = \rho A(x) \frac{\partial^2 w}{\partial t^2}(x, t) \tag{8.74}$$

From the elementary theory of bending of beams (also known as the *Euler-Bernoulli* or *thin beam theory*), the relationship between bending moment and deflection can be expressed as [8.8]

$$M(x, t) = EI(x) \frac{\partial^2 w}{\partial x^2}(x, t) \tag{8.75}$$

where E is Young's modulus and $I(x)$ is the moment of inertia of the beam cross section about the y-axis. Inserting Eq. (8.75) into Eq. (8.74), we obtain the equation of motion for the forced lateral vibration of a nonuniform beam:

$$\frac{\partial^2}{\partial x^2}\left[EI(x) \frac{\partial^2 w}{\partial x^2}(x, t)\right] + \rho A(x) \frac{\partial^2 w}{\partial t^2}(x, t) = f(x, t) \tag{8.76}$$

For a uniform beam, Eq. (8.76) reduces to

$$EI\frac{\partial^4 w}{\partial x^4}(x, t) + \rho A \frac{\partial^2 w}{\partial t^2}(x, t) = f(x, t) \tag{8.77}$$

For free vibration, $f(x, t) = 0$, and so the equation of motion becomes

$$c^2 \frac{\partial^4 w}{\partial x^4}(x, t) + \frac{\partial^2 w}{\partial t^2}(x, t) = 0 \tag{8.78}$$

where

$$c = \sqrt{\frac{EI}{\rho A}} \tag{8.79}$$

8.5.2
Initial
Conditions

Since the equation of motion involves a second-order derivative with respect to time and a fourth-order derivative with respect to x, two initial conditions and four boundary conditions are needed for finding a unique solution for $w(x, t)$. Usually, the values of lateral displacement and velocity are specified as $w_0(x)$ and $\dot{w}_0(x)$ at $t = 0$, so that the initial conditions become

$$w(x, t = 0) = w_0(x)$$

$$\frac{\partial w}{\partial t}(x, t = 0) = \dot{w}_0(x) \tag{8.80}$$

8.5.3
Free Vibration

The free-vibration solution can be found using the method of separation of variables as

$$w(x, t) = W(x)T(t) \tag{8.81}$$

Substituting Eq. (8.81) into Eq. (8.78) and rearranging leads to

$$\frac{c^2}{W(x)}\frac{d^4 W(x)}{dx^4} = -\frac{1}{T(t)}\frac{d^2 T(t)}{dt^2} = a = \omega^2 \tag{8.82}$$

where $a = \omega^2$ is a positive constant (see Problem 8.45). Equation (8.82) can be written as two equations:

$$\frac{d^4 W(x)}{dx^4} - \beta^4 W(x) = 0 \tag{8.83}$$

$$\frac{d^2 T(t)}{dt^2} + \omega^2 T(t) = 0 \tag{8.84}$$

where

$$\beta^4 = \frac{\omega^2}{c^2} = \frac{\rho A \omega^2}{EI} \tag{8.85}$$

The solution of Eq. (8.84) can be expressed as

$$T(t) = A \cos \omega t + B \sin \omega t \tag{8.86}$$

where A and B are constants that can be found from the initial conditions. For the solution of Eq. (8.83), we assume

$$W(x) = Ce^{sx} \tag{8.87}$$

where C and s are constants, and derive the auxiliary equation as

$$s^4 - \beta^4 = 0 \tag{8.88}$$

The roots of this equation are

$$s_{1,2} = \pm\beta, \qquad s_{3,4} = \pm i\beta \tag{8.89}$$

Hence the solution of Eq. (8.83) becomes

$$W(x) = C_1 e^{\beta x} + C_2 e^{-\beta x} + C_3 e^{i\beta x} + C_4 e^{-i\beta x} \tag{8.90}$$

where $C_1, C_2, C_3,$ and C_4 are constants. Equation (8.90) can also be expressed as

$$W(x) = C_1 \cos \beta x + C_2 \sin \beta x + C_3 \cosh \beta x + C_4 \sinh \beta x \tag{8.91}$$

or

$$W(x) = C_1(\cos \beta x + \cosh \beta x) + C_2(\cos \beta x - \cosh \beta x)$$
$$+ C_3(\sin \beta x + \sinh \beta x) + C_4(\sin \beta x - \sinh \beta x) \tag{8.92}$$

where $C_1, C_2, C_3,$ and C_4, in each case, are different constants. The constants $C_1, C_2, C_3,$ and C_4 can be found from the boundary conditions. The natural frequencies of the beam are computed from Eq. (8.85) as

$$\omega = \beta^2 \sqrt{\frac{EI}{\rho A}} = (\beta l)^2 \sqrt{\frac{EI}{\rho A l^4}} \tag{8.93}$$

The function $W(x)$ is known as the *normal mode* or *characteristic function* of the beam and ω is called the *natural frequency of vibration*. For any beam, there will be an infinite number of normal modes with one natural frequency associated with each normal mode. The unknown constants C_1 to C_4 in Eq. (8.91) or (8.92) and the value of β in Eq. (8.93) can be determined from the boundary conditions of the beam as indicated below.

8.5.4 Boundary Conditions

The common boundary conditions are as follows:

1. *Free end:*

$$\text{Bending moment} = EI\frac{\partial^2 w}{\partial x^2} = 0$$

$$\text{Shear force} = \frac{\partial}{\partial x}\left(EI\frac{\partial^2 w}{\partial x^2}\right) = 0 \tag{8.94}$$

2. *Simply supported (pinned) end:*

$$\text{Deflection} = w = 0, \qquad \text{Bending moment} = EI\frac{\partial^2 w}{\partial x^2} = 0 \tag{8.95}$$

3. *Fixed (clamped) end:*

$$\text{Deflection} = 0, \qquad \text{Slope} = \frac{\partial w}{\partial x} = 0 \tag{8.96}$$

The frequency equations, the mode shapes (normal functions), and the natural frequencies for beams with common boundary conditions are given in Fig. 8.15 [8.13, 8.17]. We shall now consider some other possible boundary conditions for a beam.

4. *End connected to a linear spring, damper, and mass* (Fig. 8.16(a)): When the end of a beam undergoes a transverse displacement w and slope $\partial w/\partial x$ with velocity $\partial w/\partial t$ and acceleration $\partial^2 w/\partial t^2$, the resisting forces due to the spring, damper, and mass are proportional to w, $\partial w/\partial t$, and $\partial^2 w/\partial t^2$, respectively. This resisting force is balanced by the shear force at the end. Thus

$$\frac{\partial}{\partial x}\left(EI\frac{\partial^2 w}{\partial x^2}\right) = a\left[kw + c\frac{\partial w}{\partial t} + m\frac{\partial^2 w}{\partial t^2}\right] \tag{8.97}$$

where $a = -1$ for the left end and $+1$ for the right end of the beam. In addition, the bending moment must be zero; hence

$$EI\frac{\partial^2 w}{\partial x^2} = 0 \tag{8.98}$$

5. *End connected to a torsional spring, torsional damper, and rotational inertia* (Fig. 8.16(b)): In this case, the boundary conditions are

$$EI\frac{\partial^2 w}{\partial x^2} = a\left[k_t\frac{\partial w}{\partial x} + c_t\frac{\partial^2 w}{\partial x\partial t} + I_0\frac{\partial^3 w}{\partial x\partial t^2}\right] \tag{8.99}$$

where $a = +1$ for the left end and -1 for the right end of the beam, and

$$\frac{\partial}{\partial x}\left[EI\frac{\partial^2 w}{\partial x^2}\right] = 0 \tag{8.100}$$

End Conditions of Beam	Frequency Equation	Mode Shape (Normal Function)	Value of $\beta_n l$
Pinned-pinned	$\sin \beta_n l = 0$	$W_n(x) = C_n[\sin \beta_n x]$	$\beta_1 l = \pi$ $\beta_2 l = 2\pi$ $\beta_3 l = 3\pi$ $\beta_4 l = 4\pi$
Free-free	$\cos \beta_n l \cdot \cosh \beta_n l = 1$	$W_n(x) = C_n[\sin \beta_n x + \sinh \beta_n x$ $+ \alpha_n (\cos \beta_n x + \cosh \beta_n x)]$ where $\alpha_n = \left(\dfrac{\sin \beta_n l - \sinh \beta_n l}{\cosh \beta_n l - \cos \beta_n l} \right)$	$\beta_1 l = 4.730041$ $\beta_2 l = 7.853205$ $\beta_3 l = 10.995608$ $\beta_4 l = 14.137165$ ($\beta l = 0$ for rigid-body mode)
Fixed-fixed	$\cos \beta_n l \cdot \cosh \beta_n l = 1$	$W_n(x) = C_n[\sinh \beta_n x - \sin \beta_n x$ $+ \alpha_n (\cosh \beta_n x - \cos \beta_n x)]$ where $\alpha_n = \left(\dfrac{\sinh \beta_n l - \sin \beta_n l}{\cos \beta_n l - \cosh \beta_n l} \right)$	$\beta_1 l = 4.730041$ $\beta_2 l = 7.853205$ $\beta_3 l = 10.995608$ $\beta_4 l = 14.137165$
Fixed-free	$\cos \beta_n l \cdot \cosh \beta_n l = -1$	$W_n(x) = C_n[\sin \beta_n x - \sinh \beta_n x$ $- \alpha_n (\cos \beta_n x - \cosh \beta_n x)]$ where $\alpha_n = \left(\dfrac{\sin \beta_n l + \sinh \beta_n l}{\cos \beta_n l + \cosh \beta_n l} \right)$	$\beta_1 l = 1.875104$ $\beta_2 l = 4.694091$ $\beta_3 l = 7.854757$ $\beta_4 l = 10.995541$
Fixed-pinned	$\tan \beta_n l - \tanh \beta_n l = 0$	$W_n(x) = C_n[\sin \beta_n x - \sinh \beta_n x$ $+ \alpha_n (\cosh \beta_n x - \cos \beta_n x)]$ where $\alpha_n = \left(\dfrac{\sin \beta_n l - \sinh \beta_n l}{\cos \beta_n l - \cosh \beta_n l} \right)$	$\beta_1 l = 3.926602$ $\beta_2 l = 7.068583$ $\beta_3 l = 10.210176$ $\beta_4 l = 13.351768$
Pinned-free	$\tan \beta_n l - \tanh \beta_n l = 0$	$W_n(x) = C_n[\sin \beta_n x + \alpha_n \sinh \beta_n x]$ where $\alpha_n = \left(\dfrac{\sin \beta_n l}{\sinh \beta_n l} \right)$	$\beta_1 l = 3.926602$ $\beta_2 l = 7.068583$ $\beta_3 l = 10.210176$ $\beta_4 l = 13.351768$ ($\beta l = 0$ for rigid-body mode)

FIGURE 8.15   Common boundary conditions for the transverse vibration of a beam.

8.5.5 Orthogonality of Normal Functions

The normal functions $W(x)$ satisfy Eq. (8.83):

$$c^2 \frac{d^4 W}{dx^4}(x) - \omega^2 W(x) = 0 \tag{8.101}$$

Let $W_i(x)$ and $W_j(x)$ be the normal functions corresponding to the natural frequencies ω_i and $\omega_j (i \neq j)$, so that

$$c^2 \frac{d^4 W_i}{dx^4} - \omega_i^2 W_i = 0 \tag{8.102}$$

FIGURE 8.16   Beams connected with springs-dampers-masses at ends.

and

$$c^2 \frac{d^4 W_j}{dx^4} - \omega_j^2 W_j = 0 \tag{8.103}$$

Multiplying Eq. (8.102) by W_j and Eq. (8.103) by W_i, subtracting the resulting equations one from the other, and integrating from 0 to l gives

$$\int_0^l \left[c^2 \frac{d^4 W_i}{dx^4} W_j - \omega_i^2 W_i W_j \right] dx - \int_0^l \left[c^2 \frac{d^4 W_j}{dx^4} W_i - \omega_j^2 W_j W_i \right] dx = 0$$

or

$$\int_0^l W_i W_j \, dx = -\frac{c^2}{\omega_i^2 - \omega_j^2} \int_0^l (W_i'''' W_j - W_i W_j'''') \, dx \tag{8.104}$$

where a prime indicates differentiation with respect to x. The right-hand side of Eq. (8.104) can be evaluated using integration by parts to obtain

$$\int_0^l W_i W_j \, dx = -\frac{c^2}{\omega_i^2 - \omega_j^2} \left[W_i W_j''' - W_j W_i''' + W_j' W_i'' - W_i' W_j'' \right] \Big|_0^l \tag{8.105}$$

The right-hand side of Eq. (8.105) can be shown to be zero for any combination of free, fixed, or simply supported end conditions. At a free end, the bending moment and shear force are equal to zero so that

$$W'' = 0, \qquad W''' = 0 \tag{8.106}$$

For a fixed end, the deflection and slope are zero:

$$W = 0, \qquad W' = 0 \tag{8.107}$$

At a simply supported end, the bending moment and deflection are zero:

$$W'' = 0, \qquad W = 0 \tag{8.108}$$

Since each term on the right-hand side of Eq. (8.105) is zero at $x = 0$ or $x = l$ for any combination of the boundary conditions in Eqs. (8.106) to (8.108), Eq. (8.105) reduces to

$$\int_0^l W_i W_j \, dx = 0 \tag{8.109}$$

which proves the orthogonality of normal functions for the transverse vibration of beams.

Natural Frequencies of a Fixed-Pinned Beam

EXAMPLE 8.7

Determine the natural frequencies of vibration of a uniform beam fixed at $x = 0$ and simply supported at $x = l$.

Solution: The boundary conditions can be stated as

$$W(0) = 0 \tag{E.1}$$

$$\frac{dW}{dx}(0) = 0 \tag{E.2}$$

$$W(l) = 0 \tag{E.3}$$

$$EI\frac{d^2W}{dx^2}(l) = 0 \qquad \text{or} \qquad \frac{d^2W}{dx^2}(l) = 0 \tag{E.4}$$

Condition (E.1) leads to

$$C_1 + C_3 = 0 \tag{E.5}$$

in Eq. (8.91), while Eqs. (E.2) and (8.91) give

$$\frac{dW}{dx}\bigg|_{x=0} = \beta[-C_1 \sin \beta x + C_2 \cos \beta x + C_3 \sinh \beta x + C_4 \cosh \beta x]_{x=0} = 0$$

or

$$\beta[C_2 + C_4] = 0 \tag{E.6}$$

Thus the solution, Eq. (8.91), becomes

$$W(x) = C_1(\cos \beta x - \cosh \beta x) + C_2(\sin \beta x - \sinh \beta x) \qquad \text{(E.7)}$$

Applying conditions (E.3) and (E.4) to Eq. (E.7) yields

$$C_1(\cos \beta l - \cosh \beta l) + C_2(\sin \beta l - \sinh \beta l) = 0 \qquad \text{(E.8)}$$

$$-C_1(\cos \beta l + \cosh \beta l) - C_2(\sin \beta l + \sinh \beta l) = 0 \qquad \text{(E.9)}$$

For a nontrivial solution of C_1 and C_2, the determinant of their coefficients must be zero—that is,

$$\begin{vmatrix} (\cos \beta l - \cosh \beta l) & (\sin \beta l - \sinh \beta l) \\ -(\cos \beta l + \cosh \beta l) & -(\sin \beta l + \sinh \beta l) \end{vmatrix} = 0 \qquad \text{(E.10)}$$

Expanding the determinant gives the frequency equation

$$\cos \beta l \sinh \beta l - \sin \beta l \cosh \beta l = 0$$

or

$$\tan \beta l = \tanh \beta l \qquad \text{(E.11)}$$

The roots of this equation, $\beta_n l$, give the natural frequencies of vibration

$$\omega_n = (\beta_n l)^2 \left(\frac{EI}{\rho A l^4} \right)^{1/2}, \qquad n = 1, 2, \ldots \qquad \text{(E.12)}$$

where the values of $\beta_n l$, $n = 1, 2, \ldots$ satisfying Eq. (E.11) are given in Fig. 8.15. If the value of C_2 corresponding to β_n is denoted as C_{2n}, it can be expressed in terms of C_{1n} from Eq. (E.8) as

$$C_{2n} = -C_{1n} \left(\frac{\cos \beta_n l - \cosh \beta_n l}{\sin \beta_n l - \sinh \beta_n l} \right) \qquad \text{(E.13)}$$

Hence Eq. (E.7) can be written as

$$W_n(x) = C_{1n} \left[(\cos \beta_n x - \cosh \beta_n x) - \left(\frac{\cos \beta_n l - \cosh \beta_n l}{\sin \beta_n l - \sinh \beta_n l} \right)(\sin \beta_n x - \sinh \beta_n x) \right] \qquad \text{(E.14)}$$

The normal modes of vibration can be obtained by the use of Eq. (8.81)

$$w_n(x, t) = W_n(x)(A_n \cos \omega_n t + B_n \sin \omega_n t) \qquad \text{(E.15)}$$

with $W_n(x)$ given by Eq. (E.14). The general or total solution of the fixed-simply supported beam can be expressed by the sum of the normal modes:

$$w(x, t) = \sum_{n=1}^{\infty} w_n(x, t) \qquad \text{(E.16)}$$

■

8.5.6
Forced Vibration

The forced-vibration solution of a beam can be determined using the mode superposition principle. For this, the deflection of the beam is assumed as

$$w(x, t) = \sum_{n=1}^{\infty} W_n(x) q_n(t) \tag{8.110}$$

where $W_n(x)$ is the nth normal mode or characteristic function satisfying the differential equation (Eq. (8.101))

$$EI \frac{d^4 W_n(x)}{dx^4} - \omega_n^2 \rho A W_n(x) = 0; \qquad n = 1, 2, \ldots \tag{8.111}$$

and $q_n(t)$ is the generalized coordinate in the nth mode. By substituting Eq. (8.110) into the forced-vibration equation, Eq. (8.77), we obtain

$$EI \sum_{n=1}^{\infty} \frac{d^4 W_n(x)}{dx^4} q_n(t) + \rho A \sum_{n=1}^{\infty} W_n(x) \frac{d^2 q_n(t)}{dt^2} = f(x, t) \tag{8.112}$$

In view of Eq. (8.111), Eq. (8.112) can be written as

$$\sum_{n=1}^{\infty} \omega_n^2 W_n(x) q_n(t) + \sum_{n=1}^{\infty} W_n(x) \frac{d^2 q_n(t)}{dt^2} = \frac{1}{\rho A} f(x, t) \tag{8.113}$$

By multiplying Eq. (8.113) throughout by $W_m(x)$, integrating from 0 to l, and using the orthogonality condition, Eq. (8.109), we obtain

$$\frac{d^2 q_n(t)}{dt^2} + \omega_n^2 q_n(t) = \frac{1}{\rho A b} Q_n(t) \tag{8.114}$$

where $Q_n(t)$ is called the *generalized force corresponding to* $q_n(t)$

$$Q_n(t) = \int_0^l f(x, t) W_n(x) \, dx \tag{8.115}$$

and the constant b is given by

$$b = \int_0^l W_n^2(x) \, dx \tag{8.116}$$

Equation (8.114) can be identified to be, essentially, the same as the equation of motion of an undamped single-degree-of-freedom system. Using the Duhamel integral, the solution of Eq. (8.114) can be expressed as

$$q_n(t) = A_n \cos \omega_n t + B_n \sin \omega_n t$$

$$+ \frac{1}{\rho A b \omega_n} \int_0^t Q_n(\tau) \sin \omega_n(t - \tau) \, d\tau \tag{8.117}$$

where the first two terms on the right-hand side of Eq. (8.117) represent the transient or free vibration (resulting from the initial conditions) and the third term denotes the steady-state vibration (resulting from the forcing function). Once Eq. (8.117) is solved for $n = 1, 2, \ldots$, the total solution can be determined from Eq. (8.110).

Forced Vibration of a Simply Supported Beam

EXAMPLE 8.8

Find the steady-state response of a pinned-pinned beam subject to a harmonic force $f(x, t) = f_0 \sin \omega t$ applied at $x = a$, as shown in Fig. 8.17.

Solution: *Approach:* Mode superposition method.
The normal mode functions of a pinned-pinned beam are given by (see Fig. 8.15; also Problem 8.33)

$$W_n(x) = \sin \beta_n x = \sin \frac{n\pi x}{l} \tag{E.1}$$

where

$$\beta_n l = n\pi \tag{E.2}$$

The generalized force $Q_n(t)$, given by Eq. (8.115), becomes

$$Q_n(t) = \int_0^l f(x, t) \sin \beta_n x \, dx = f_0 \sin \frac{n\pi a}{l} \sin \omega t \tag{E.3}$$

The steady-state response of the beam is given by Eq. (8.117)

$$q_n(t) = \frac{1}{\rho Ab\omega_n} \int_0^l Q_n(\tau) \sin \omega_n (t - \tau) \, d\tau \tag{E.4}$$

where

$$b = \int_0^l W_n^2(x) \, dx = \int_0^l \sin^2 \beta_n x \, dx = \frac{l}{2} \tag{E.5}$$

The solution of Eq. (E.4) can be expressed as

$$q_n(t) = \frac{2f_0}{\rho Al} \frac{\sin \frac{n\pi a}{l}}{\omega_n^2 - \omega^2} \sin \omega t \tag{E.6}$$

Thus the response of the beam is given by Eq. (8.110):

$$w(x, t) = \frac{2f_0}{\rho Al} \sum_{n=1}^{\infty} \frac{1}{\omega_n^2 - \omega^2} \sin \frac{n\pi a}{l} \sin \frac{n\pi x}{l} \sin \omega t \tag{E.7}$$

FIGURE 8.17 Pinned-pinned beam under harmonic force.

8.5.7
Effect of Axial
Force

The problem of vibrations of a beam under the action of axial force finds application in the study of vibrations of cables and guy wires. For example, although the vibrations of a cable can be found by treating it as an equivalent string, many cables have failed due to fatigue caused by alternating flexure. The alternating flexure is produced by the regular shedding of vortices from the cable in a light wind. We must therefore consider the effects of axial force and bending stiffness on lateral vibrations in the study of fatigue failure of cables.

To find the effect of an axial force $P(x, t)$ on the bending vibrations of a beam, consider the equation of motion of an element of the beam, as shown in Fig. 8.18. For the vertical motion, we have

$$-(V + dV) + f\,dx + V + (P + dP)\sin(\theta + d\theta) - P\sin\theta = \rho A\,dx\,\frac{\partial^2 w}{\partial t^2} \quad (8.118)$$

and for the rotational motion about 0,

$$(M + dM) - (V + dV)\,dx + f\,dx\,\frac{dx}{2} - M = 0 \quad (8.119)$$

For small deflections,

$$\sin(\theta + d\theta) \simeq \theta + d\theta = \theta + \frac{\partial\theta}{\partial x}\,dx = \frac{\partial w}{\partial x} + \frac{\partial^2 w}{\partial x^2}\,dx$$

FIGURE 8.18   An element of a beam under axial load.

With this, Eqs. (8.75), (8.118), and (8.119) can be combined to obtain a single differential equation of motion:

$$\frac{\partial^2}{\partial x^2}\left[EI\frac{\partial^2 w}{\partial x^2} \right] + \rho A\frac{\partial^2 w}{\partial t^2} - P\frac{\partial^2 w}{\partial x^2} = f \tag{8.120}$$

For the free vibration of a uniform beam, Eq. (8.120) reduces to

$$EI\frac{\partial^4 w}{\partial x^4} + \rho A\frac{\partial^2 w}{\partial t^2} - P\frac{\partial^2 w}{\partial x^2} = 0 \tag{8.121}$$

The solution of Eq. (8.121) can be obtained using the method of separation of variables as

$$w(x, t) = W(x)(A \cos \omega t + B \sin \omega t) \tag{8.122}$$

Substitution of Eq. (8.122) into Eq. (8.121) gives

$$EI\frac{d^4 W}{dx^4} - P\frac{d^2 W}{dx^2} - \rho A\omega^2 W = 0 \tag{8.123}$$

By assuming the solution $W(x)$ to be

$$W(x) = Ce^{sx} \tag{8.124}$$

in Eq. (8.123), the auxiliary equation can be obtained:

$$s^4 - \frac{P}{EI}s^2 - \frac{\rho A\omega^2}{EI} = 0 \tag{8.125}$$

The roots of Eq. (8.125) are

$$s_1^2, s_2^2 = \frac{P}{2EI} \pm \left(\frac{P^2}{4E^2 I^2} + \frac{\rho A\omega^2}{EI} \right)^{1/2} \tag{8.126}$$

and so the solution can be expressed as (with absolute value of s_2)

$$W(x) = C_1 \cosh s_1 x + C_2 \sinh s_1 x + C_3 \cos s_2 x + C_4 \sin s_2 x \tag{8.127}$$

where the constants C_1 to C_4 are to be determined from the boundary conditions.

Beam Subjected to an Axial Compressive Force

EXAMPLE 8.9

Find the natural frequencies of a simply supported beam subjected to an axial compressive force.

Solution: The boundary conditions are

$$W(0) = 0 \tag{E.1}$$

$$\frac{d^2 W}{dx^2}(0) = 0 \tag{E.2}$$

$$W(l) = 0 \tag{E.3}$$

$$\frac{d^2W}{dx^2}(l) = 0 \tag{E.4}$$

Equations (E.1) and (E.2) require that $C_1 = C_3 = 0$ in Eq. (8.127), and so

$$W(x) = C_2 \sinh s_1 x + C_4 \sin s_2 x \tag{E.5}$$

The application of Eqs. (E.3) and (E.4) to Eq. (E.5) leads to

$$\sinh s_1 l \cdot \sin s_2 l = 0 \tag{E.6}$$

Since $\sinh s_1 l > 0$ for all values of $s_1 l \neq 0$, the only roots to this equation are

$$s_2 l = n\pi, \qquad n = 0, 1, 2, \ldots \tag{E.7}$$

Thus Eqs. (E.7) and (8.126) give the natural frequencies of vibration:

$$\omega_n = \frac{\pi^2}{l^2} \sqrt{\frac{EI}{\rho A}} \left(n^4 + \frac{n^2 P l^2}{\pi^2 EI} \right)^{1/2} \tag{E.8}$$

Since the axial force P is compressive, P is negative. Further, from the strength of materials, the smallest Euler buckling load for a simply supported beam is given by [8.9]

$$P_{\text{cri}} = \frac{\pi^2 EI}{l^2} \tag{E.9}$$

Thus Eq. (E.8) can be written as

$$\omega_n = \frac{\pi^2}{l^2} \left(\frac{EI}{\rho A} \right)^{1/2} \left(n^4 - n^2 \frac{P}{P_{\text{cri}}} \right)^{1/2} \tag{E.10}$$

The following observations can be made from the present example:

1. If $P = 0$, the natural frequency will be the same as that of a simply supported beam given in Fig. 8.15.
2. If $EI = 0$, the natural frequency (see Eq. (E.8)) reduces to that of a taut string.
3. If $P > 0$, the natural frequency increases as the tensile force stiffens the beam.
4. As $P \rightarrow P_{\text{cri}}$, the natural frequency approaches zero for $n = 1$.

∎

8.5.8
Effects of Rotary Inertia and Shear Deformation

If the cross-sectional dimensions are not small compared to the length of the beam, we need to consider the effects of rotary inertia and shear deformation. The procedure, presented by Timoshenko [8.10], is known as the *thick beam theory* or *Timoshenko beam theory*. Consider the element of the beam shown in Fig. 8.19. If the effect of shear deformation is disregarded, the tangent to the deflected center line $O'T$ coincides with the normal to the face $Q'R'$ (since cross sections normal to the center line remain normal even

FIGURE 8.19   An element of Timoshenko beam.

after deformation). Due to shear deformation, the tangent to the deformed center line $O'T$ will not be perpendicular to the face $Q'R'$. The angle γ between the tangent to the deformed center line $(O'T)$ and the normal to the face $(O'N)$ denotes the shear deformation of the element. Since positive shear on the right face $Q'R'$ acts downward, we have, from Fig. 8.19,

$$\gamma = \phi - \frac{\partial w}{\partial x} \qquad (8.128)$$

where ϕ denotes the slope of the deflection curve due to bending deformation alone. Note that because of shear alone, the element undergoes distortion but no rotation.

The bending moment M and the shear force V are related to ϕ and w by the formulas[2]

$$M = EI \frac{\partial \phi}{\partial x} \qquad (8.129)$$

and

$$V = kAG\gamma = kAG\left(\phi - \frac{\partial w}{\partial x}\right) \qquad (8.130)$$

where G denotes the modulus of rigidity of the material of the beam and k is a constant, also known as *Timoshenko's shear coefficient*, which depends on the shape of

---

[2]Equation (8.129) is similar to Eq. (8.75). Equation (8.130) can be obtained as follows:

Shear force = Shear stress × Area = Shear strain × Shear modulus × Area

or

$$V = \gamma G A$$

This equation is modified as $V = kAG\gamma$ by introducing a factor k on the right-hand side to take care of the shape of the cross section.

the cross section. For a rectangular section, the value of k is 5/6; for a circular section, it is 9/10 [8.11].

The equations of motion for the element shown in Fig. 8.19 can be derived as follows:

1. For translation in the z direction:

$$- [V(x, t) + dV(x, t)] + f(x, t) \, dx + V(x, t)$$

$$= \rho A(x) \, dx \, \frac{\partial^2 w}{\partial t^2}(x, t)$$

$$\equiv \text{Translational inertia of the element} \qquad (8.131)$$

2. For rotation about a line passing through point D and parallel to the y-axis:

$$[M(x, t) + dM(x, t)] + [V(x, t) + dV(x, t)] \, dx$$

$$+ f(x, t) \, dx \, \frac{dx}{2} - M(x, t)$$

$$= \rho I(x) \, dx \, \frac{\partial^2 \phi}{\partial t^2} \equiv \text{Rotary inertia of the element} \qquad (8.132)$$

Using the relations

$$dV = \frac{\partial V}{\partial x} \, dx \qquad \text{and} \qquad dM = \frac{\partial M}{\partial x} \, dx$$

along with Eqs. (8.129) and (8.130) and disregarding terms involving second powers in dx, Eqs. (8.131) and (8.132) can be expressed as

$$-kAG \left(\frac{\partial \phi}{\partial x} - \frac{\partial^2 w}{\partial x^2} \right) + f(x, t) = \rho A \frac{\partial^2 w}{\partial t^2} \qquad (8.133)$$

$$EI \frac{\partial^2 \phi}{\partial x^2} - kAG \left(\phi - \frac{\partial w}{\partial x} \right) = \rho I \frac{\partial^2 \phi}{\partial t^2} \qquad (8.134)$$

By solving Eq. (8.133) for $\partial \phi / \partial x$ and substituting the result in Eq. (8.134), we obtain the desired equation of motion for the forced vibration of a uniform beam:

$$EI \frac{\partial^4 w}{\partial x^4} + \rho A \frac{\partial^2 w}{\partial t^2} - \rho I \left(1 + \frac{E}{kG} \right) \frac{\partial^4 w}{\partial x^2 \partial t^2} + \frac{\rho^2 I}{kG} \frac{\partial^4 w}{\partial t^4}$$

$$+ \frac{EI}{kAG} \frac{\partial^2 f}{\partial x^2} - \frac{\rho I}{kAG} \frac{\partial^2 f}{\partial t^2} - f = 0 \tag{8.135}$$

For free vibration, $f = 0$, and Eq. (8.135) reduces to

$$EI \frac{\partial^4 w}{\partial x^4} + \rho A \frac{\partial^2 w}{\partial t^2} - \rho I \left(1 + \frac{E}{kG} \right) \frac{\partial^4 w}{\partial x^2 \partial t^2} + \frac{\rho^2 I}{kG} \frac{\partial^4 w}{\partial t^4} = 0 \tag{8.136}$$

The following boundary conditions are to be applied in the solution of Eq. (8.135) or (8.136):

1. **Fixed end:**

$$\phi = w = 0$$

2. **Simply supported end:**

$$EI \frac{\partial \phi}{\partial x} = w = 0$$

3. **Free end:**

$$kAG \left(\frac{\partial w}{\partial x} - \phi \right) = EI \frac{\partial \phi}{\partial x} = 0$$

▬▬▬▬▬▬▬▬▬▬

EXAMPLE 8.10

Natural Frequencies of a Simply Supported Beam

Determine the effects of rotary inertia and shear deformation on the natural frequencies of a simply supported uniform beam.

Solution: By defining

$$\alpha^2 = \frac{EI}{\rho A} \quad \text{and} \quad r^2 = \frac{I}{A} \tag{E.1}$$

Eq. (8.136) can be written as

$$\alpha^2 \frac{\partial^4 w}{\partial x^4} + \frac{\partial^2 w}{\partial t^2} - r^2 \left(1 + \frac{E}{kG} \right) \frac{\partial^4 w}{\partial x^2 \partial t^2} + \frac{\rho r^2}{kG} \frac{\partial^4 w}{\partial t^4} = 0 \tag{E.2}$$

We can express the solution of Eq. (E.2) as

$$w(x, t) = C \sin \frac{n \pi x}{l} \cos \omega_n t \tag{E.3}$$

which satisfies the necessary boundary conditions at $x = 0$ and $x = l$. Here, C is a constant and ω_n is the nth natural frequency. By substituting Eq. (E.3) into Eq. (E.2), we obtain the frequency equation:

$$\omega_n^4\left(\frac{\rho r^2}{kG}\right) - \omega_n^2\left(1 + \frac{n^2\pi^2 r^2}{l^2} + \frac{n^2\pi^2 r^2}{l^2}\frac{E}{kG}\right) + \left(\frac{\alpha^2 n^4 \pi^4}{l^4}\right) = 0 \qquad \text{(E.4)}$$

It can be seen that Eq. (E.4) is a quadratic equation in ω_n^2, and for any given n there are two values of ω_n that satisfy Eq. (E.4). The smaller value corresponds to the bending deformation mode, while the larger one corresponds to the shear deformation mode.

The values of the ratio of ω_n given by Eq. (E.4) to the natural frequency given by the classical theory (in Fig. 8.15) are plotted for three values of E/kG in Fig. 8.20 [8.22].[3]

Note the following aspects of rotary inertia and shear deformation:

1. If the effect of rotary inertia alone is considered, the resulting equation of motion does not contain any term involving the shear coefficient k. Hence we obtain (from Eq. (8.136)):

$$EI\frac{\partial^4 w}{\partial x^4} + \rho A\frac{\partial^2 w}{\partial t^2} - \rho I\frac{\partial^4 w}{\partial x^2 \partial t^2} = 0 \qquad \text{(E.5)}$$

In this case, the frequency equation (E.4) reduces to

$$\omega_n^2 = \frac{\alpha^2 n^4 \pi^4}{l^4\left(1 + \frac{n^2\pi^2 r^2}{l^2}\right)} \qquad \text{(E.6)}$$

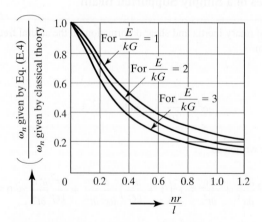

FIGURE 8.20   Variation of frequency.

---

[3]The theory used for the derivation of the equation of motion (8.76), which disregards the effects of rotary inertia and shear deformation, is called the *classical* or *Euler-Bernoulli* or *thin beam theory*.

2. If the effect of shear deformation alone is considered, the resulting equation of motion does not contain the terms originating from $\rho I (\partial^2 \phi / \partial t^2)$ in Eq. (8.134). Thus we obtain the equation of motion

$$EI \frac{\partial^4 w}{\partial x^4} + \rho A \frac{\partial^2 w}{\partial t^2} - \frac{EI\rho}{kG} \frac{\partial^4 w}{\partial x^2 \partial t^2} = 0 \qquad \text{(E.7)}$$

and the corresponding frequency equation

$$\omega_n^2 = \frac{\alpha^2 n^4 \pi^4}{l^4 \left(1 + \dfrac{n^2 \pi^2 r^2}{l^2} \dfrac{E}{kG} \right)} \qquad \text{(E.8)}$$

3. If both the effects of rotary inertia and shear deformation are disregarded, Eq. (8.136) reduces to the classical equation of motion, Eq. (8.78),

$$EI \frac{\partial^4 w}{\partial x^4} + \rho A \frac{\partial^2 w}{\partial t^2} = 0 \qquad \text{(E.9)}$$

and Eq. (E.4) to

$$\omega_n^2 = \frac{\alpha^2 n^4 \pi^4}{l^4} \qquad \text{(E.10)}$$

∎

8.5.9 Other Effects

The transverse vibration of tapered beams is presented in [8.12, 8.14]. The natural frequencies of continuous beams are discussed by Wang [8.15]. The dynamic response of beams resting on elastic foundation is considered in [8.16]. The effect of support flexibility on the natural frequencies of beams is presented in [8.18, 8.19]. A treatment of the problem of natural vibrations of a system of elastically connected Timoshenko beams is given in [8.20]. A comparison of the exact and approximate solutions of vibrating beams is made by Hutchinson [8.30]. The steady-state vibration of damped beams is considered in [8.21].

8.6 Vibration of Membranes

A membrane is a plate that is subjected to tension and has negligible bending resistance. Thus a membrane bears the same relationship to a plate as a string bears to a beam. A drumhead is an example of a membrane.

8.6.1 Equation of Motion

To derive the equation of motion of a membrane, consider the membrane to be bounded by a plane curve S in the xy-plane, as shown in Fig. 8.21. Let $f(x, y, t)$ denote the pressure loading acting in the z direction and P the intensity of tension at a point that is equal to the product of the tensile stress and the thickness of the membrane. The magnitude of P is usually

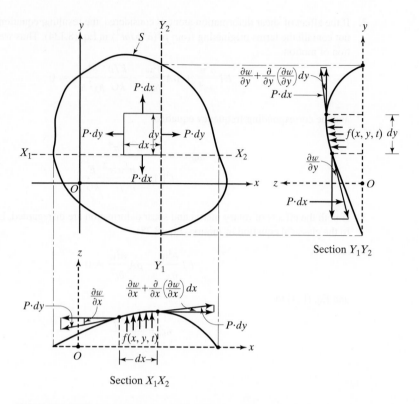

FIGURE 8.21 A membrane under uniform tension.

constant throughout the membrane, as in a drumhead. If we consider an elemental area $dx\,dy$, forces of magnitude $P\,dx$ and $P\,dy$ act on the sides parallel to the y- and x-axes, respectively, as shown in Fig. 8.21. The net forces acting along the z direction due to these forces are

$$\left(P\frac{\partial^2 w}{\partial y^2}\,dx\,dy\right) \qquad \text{and} \qquad \left(P\frac{\partial^2 w}{\partial x^2}\,dx\,dy\right)$$

The pressure force along the z direction is $f(x, y, t)\,dx\,dy$, and the inertia force is

$$\rho(x, y)\frac{\partial^2 w}{\partial t^2}\,dx\,dy$$

where $\rho(x, y)$ is the mass per unit area. The equation of motion for the forced transverse vibration of the membrane can be obtained as

$$P\left(\frac{\partial^2 w}{\partial x^2} + \frac{\partial^2 w}{\partial y^2}\right) + f = \rho\frac{\partial^2 w}{\partial t^2} \qquad (8.137)$$

If the external force $f(x, y, t) = 0$, Eq. (8.137) gives the free-vibration equation

$$c^2 \left(\frac{\partial^2 w}{\partial x^2} + \frac{\partial^2 w}{\partial y^2} \right) = \frac{\partial^2 w}{\partial t^2} \tag{8.138}$$

where

$$c = \left(\frac{P}{\rho} \right)^{1/2} \tag{8.139}$$

Equations (8.137) and (8.138) can be expressed as

$$P \, \nabla^2 w + f = \rho \frac{\partial^2 w}{\partial t^2} \tag{8.140}$$

and

$$c^2 \, \nabla^2 w = \frac{\partial^2 w}{\partial t^2} \tag{8.141}$$

where

$$\nabla^2 = \frac{\partial^2}{\partial x^2} + \frac{\partial^2}{\partial y^2} \tag{8.142}$$

is the Laplacian operator.

8.6.2 Initial and Boundary Conditions

Since the equation of motion, Eq. (8.137) or (8.138), involves second-order partial derivatives with respect to each of t, x, and y, we need to specify two initial conditions and four boundary conditions to find a unique solution of the problem. Usually, the displacement and velocity of the membrane at $t = 0$ are specified as $w_0(x, y)$ and $\dot{w}_0(x, y)$. Hence the initial conditions are given by

$$w(x, y, 0) = w_0(x, y)$$

$$\frac{\partial w}{\partial t}(x, y, 0) = \dot{w}_0(x, y) \tag{8.143}$$

The boundary conditions are of the following types:

1. If the membrane is fixed at any point (x_1, y_1) on a segment of the boundary, we have

$$w(x_1, y_1, t) = 0, \qquad t \geq 0 \tag{8.144}$$

2. If the membrane is free to deflect transversely (in the z direction) at a different point (x_2, y_2) of the boundary, then the force component in the z direction must be zero. Thus

$$P \frac{\partial w}{\partial n}(x_2, y_2, t) = 0, \qquad t \geq 0 \tag{8.145}$$

where $\partial w / \partial n$ represents the derivative of w with respect to a direction n normal to the boundary at the point (x_2, y_2).

The solution of the equation of motion of the vibrating membrane was presented in [8.23–8.25].

EXAMPLE 8.11

Free Vibrations of a Rectangular Membrane

Find the free-vibration solution of a rectangular membrane of sides a and b along the x- and y-axes, respectively.

Solution: By using the method of separation of variables, $w(x, y, t)$ can be assumed to be

$$w(x, y, t) = W(x, y) \, T(t) = X(x) \, Y(y) \, T(t) \tag{E.1}$$

By using Eqs. (E.1) and (8.138), we obtain

$$\frac{d^2 X(x)}{dx^2} + \alpha^2 X(x) = 0 \tag{E.2}$$

$$\frac{d^2 Y(y)}{dy^2} + \beta^2 Y(y) = 0 \tag{E.3}$$

$$\frac{d^2 T(t)}{dt^2} + \omega^2 T(t) = 0 \tag{E.4}$$

where α^2 and β^2 are constants related to ω^2 as follows:

$$\beta^2 = \frac{\omega^2}{c^2} - \alpha^2 \tag{E.5}$$

The solutions of Eqs. (E.2) to (E.4) are given by

$$X(x) = C_1 \cos \alpha x + C_2 \sin \alpha x \tag{E.6}$$

$$Y(y) = C_3 \cos \beta y + C_4 \sin \beta y \tag{E.7}$$

$$T(t) = A \cos \omega t + B \sin \omega t \tag{E.8}$$

where the constants C_1 to C_4, A, and B can be determined from the boundary and initial conditions.

∎

8.7   Rayleigh's Method

Rayleigh's method can be applied to find the fundamental natural frequency of continuous systems. This method is much simpler than exact analysis for systems with varying distributions of mass and stiffness. Although the method is applicable to all continuous systems, we shall apply it only to beams in this section.[4] Consider the beam shown in Fig. 8.14. In order to apply Rayleigh's method, we need to derive expressions for the maximum kinetic and potential energies and Rayleigh's quotient. The kinetic energy of the beam can be expressed as

$$T = \frac{1}{2}\int_0^l \dot{w}^2 \, dm = \frac{1}{2}\int_0^l \dot{w}^2 \rho A(x) \, dx \qquad (8.146)$$

The maximum kinetic energy can be found by assuming a harmonic variation $w(x,t) = W(x)\cos\omega t$:

$$T_{max} = \frac{\omega^2}{2}\int_0^l \rho A(x) W^2(x) \, dx \qquad (8.147)$$

The potential energy of the beam V is the same as the work done in deforming the beam. By disregarding the work done by the shear forces, we have

$$V = \frac{1}{2}\int_0^l M \, d\theta \qquad (8.148)$$

where M is the bending moment given by Eq. (8.75) and θ is the slope of the deformed beam given by $\theta = \dfrac{\partial w}{\partial x}$. Thus Eq. (8.148) can be rewritten as

$$V = \frac{1}{2}\int_0^l \left(EI\frac{\partial^2 w}{\partial x^2}\right)\frac{\partial^2 w}{\partial x^2}\, dx = \frac{1}{2}\int_0^l EI\left(\frac{\partial^2 w}{\partial x^2}\right)^2 dx \qquad (8.149)$$

Since the maximum value of $w(x,t)$ is $W(x)$, the maximum value of V is given by

$$V_{max} = \frac{1}{2}\int_0^l EI(x)\left(\frac{d^2 W(x)}{dx^2}\right)^2 dx \qquad (8.150)$$

By equating T_{max} to V_{max}, we obtain Rayleigh's quotient:

$$R(\omega) = \omega^2 = \frac{\int_0^l EI\left(\dfrac{d^2 W(x)}{dx^2}\right)^2 dx}{\int_0^l \rho A(W(x))^2 \, dx} \qquad (8.151)$$

[4]An integral equation approach for the determination of the fundamental frequency of vibrating beams is presented by Penny and Reed [8.26].

Thus the natural frequency of the beam can be found once the deflection $W(x)$ is known. In general, $W(x)$ is not known and must therefore be assumed. Generally, the static equilibrium shape is assumed for $W(x)$ to obtain the fundamental frequency. It is to be noted that the assumed shape $W(x)$ unintentionally introduces a constraint on the system (which amounts to adding additional stiffness to the system), and so the frequency given by Eq. (8.151) is higher than the exact value [8.27].

For a stepped beam, Eq. (8.151) can be more conveniently written as

$$R(\omega) = \omega^2$$

$$= \frac{E_1 I_1 \displaystyle\int_0^{l_1} \left(\frac{d^2W}{dx^2}\right)^2 dx + E_2 I_2 \displaystyle\int_{l_1}^{l_2} \left(\frac{d^2W}{dx^2}\right)^2 dx + \cdots}{\rho A_1 \displaystyle\int_0^{l_1} W^2 \, dx + \rho A_2 \displaystyle\int_{l_1}^{l_2} W^2 \, dx + \cdots} \tag{8.152}$$

where E_i, I_i, A_i, and l_i correspond to the ith step ($i = 1, 2, \ldots$).

EXAMPLE 8.12
Fundamental Frequency of a Tapered Beam

Find the fundamental frequency of transverse vibration of the nonuniform cantilever beam shown in Fig. 8.22, using the deflection shape $W(x) = (1 - x/l)^2$.

Solution: The given deflection shape can be verified to satisfy the boundary conditions of the beam. The cross-sectional area A and the moment of inertia I of the beam can be expressed as

$$A(x) = \frac{hx}{l}, \qquad I(x) = \frac{1}{12}\left(\frac{hx}{l}\right)^3 \tag{E.1}$$

FIGURE 8.22   Tapered cantilever beam.

Thus Rayleigh's quotient gives

$$\omega^2 = \frac{\int_0^l E\left(\frac{h^3 x^3}{12l^3}\right)\left(\frac{2}{l^2}\right)^2 dx}{\int_0^l \rho\left(\frac{hx}{l}\right)\left(1 - \frac{x}{l}\right)^4 dx} = 2.5 \frac{Eh^2}{\rho l^4}$$

or

$$\omega = 1.5811\left(\frac{Eh^2}{\rho l^4}\right)^{1/2} \tag{E.2}$$

The exact value of the frequency for this case [8.2] is known to be

$$\omega_1 = 1.5343\left(\frac{Eh^2}{\rho l^4}\right)^{1/2} \tag{E.3}$$

Thus the value of ω_1 given by Rayleigh's method can be seen to be 3.0503 percent higher than the exact value.

■

8.8   The Rayleigh-Ritz Method

The Rayleigh-Ritz method can be considered an extension of Rayleigh's method. It is based on the premise that a closer approximation to the exact natural mode can be obtained by superposing a number of assumed functions than by using a single assumed function, as in Rayleigh's method. If the assumed functions are suitably chosen, this method provides not only the approximate value of the fundamental frequency but also the approximate values of the higher natural frequencies and the mode shapes. An arbitrary number of functions can be used, and the number of frequencies that can be obtained is equal to the number of functions used. A large number of functions, although it involves more computational work, leads to more accurate results.

In the case of transverse vibration of beams, if n functions are chosen for approximating the deflection $W(x)$, we can write

$$W(x) = c_1 w_1(x) + c_2 w_2(x) + \cdots + c_n w_n(x) \tag{8.153}$$

where $w_1(x), w_2(x), \ldots, w_n(x)$ are known linearly independent functions of the spatial coordinate x, which satisfy all the boundary conditions of the problem, and $c_1, c_2, \ldots, c_n$ are coefficients to be found. The coefficients c_i are to be determined so that the assumed functions

$w_i(x)$ provide the best possible approximation to the natural modes. To obtain such approximations, the coefficients c_i are adjusted and the natural frequency is made stationary at the natural modes. For this, we substitute Eq. (8.153) in Rayleigh's quotient, Eq. (8.151), and the resulting expression is partially differentiated with respect to each of the coefficients c_i. To make the natural frequency stationary, we set each of the partial derivatives equal to zero and obtain

$$\frac{\partial(\omega^2)}{\partial c_i} = 0, \qquad i = 1, 2, \ldots, n \tag{8.154}$$

Equation (8.154) denotes a set of n linear algebraic equations in the coefficients $c_1, c_2, \ldots, c_n$ and also contains the undetermined quantity ω^2. This defines an algebraic eigenvalue problem similar to the ones that arise in multidegree-of-freedom systems. The solution of this eigenvalue problem generally gives n natural frequencies $\omega_i^2, i = 1, 2, \ldots, n$, and n eigenvectors, each containing a set of numbers for $c_1, c_2, \ldots, c_n$. For example, the ith eigenvector corresponding to ω_i may be expressed as

$$\vec{C}^{(i)} = \begin{Bmatrix} c_1^{(i)} \\ c_2^{(i)} \\ \vdots \\ c_n^{(i)} \end{Bmatrix} \tag{8.155}$$

When this eigenvector—the values of $c_1^{(i)}, c_2^{(i)}, \ldots, c_n^{(i)}$—is substituted into Eq. (8.153), we obtain the best possible approximation to the ith mode of the beam. A method of reducing the size of the eigenproblem in the Rayleigh-Ritz method is presented in [8.28]. A new approach, which combines the advantages of the Rayleigh-Ritz analysis and the finite element method is given in [8.29]. The basic Rayleigh-Ritz procedure is illustrated with the help of the following example.

First Two Frequencies of a Tapered Beam

EXAMPLE 8.13

Find the natural frequencies of the tapered cantilever beam of Example 8.12 by using the Rayleigh-Ritz method.

Solution: We assume the deflection functions $w_i(x)$ to be

$$w_1(x) = \left(1 - \frac{x}{l}\right)^2 \tag{E.1}$$

$$w_2(x) = \frac{x}{l}\left(1 - \frac{x}{l}\right)^2 \tag{E.2}$$

$$w_3(x) = \frac{x^2}{l^2}\left(1 - \frac{x}{l}\right)^2 \tag{E.3}$$

$$\vdots$$

If we use the one-term approximation

$$W(x) = c_1\left(1 - \frac{x}{l}\right)^2 \tag{E.4}$$

the fundamental frequency will be the same as the one found in Example 8.12. Now we use the two-term approximation

$$W(x) = c_1\left(1 - \frac{x}{l}\right)^2 + c_2\frac{x}{l}\left(1 - \frac{x}{l}\right)^2 \tag{E.5}$$

Rayleigh's quotient is given by

$$R[W(x)] = \omega^2 = \frac{X}{Y} \tag{E.6}$$

where

$$X = \int_0^l EI(x)\left(\frac{d^2W(x)}{dx^2}\right)^2 dx \tag{E.7}$$

and

$$Y = \int_0^l \rho A(x)[W(x)]^2\, dx \tag{E.8}$$

If Eq. (E.5) is substituted, Eq. (E.6) becomes a function of c_1 and c_2. The conditions that make ω^2 or $R[W(x)]$ stationary are

$$\frac{\partial(\omega^2)}{\partial c_1} = \frac{Y\frac{\partial X}{\partial c_1} - X\frac{\partial Y}{\partial c_1}}{Y^2} = 0 \tag{E.9}$$

$$\frac{\partial(\omega^2)}{\partial c_2} = \frac{Y\frac{\partial X}{\partial c_2} - X\frac{\partial Y}{\partial c_2}}{Y^2} = 0 \tag{E.10}$$

These equations can be rewritten as

$$\frac{\partial X}{\partial c_1} - \frac{X}{Y}\frac{\partial Y}{\partial c_1} = \frac{\partial X}{\partial c_1} - \omega^2\frac{\partial Y}{\partial c_1} = 0 \tag{E.11}$$

$$\frac{\partial X}{\partial c_2} - \frac{X}{Y}\frac{\partial Y}{\partial c_2} = \frac{\partial X}{\partial c_2} - \omega^2\frac{\partial Y}{\partial c_2} = 0 \tag{E.12}$$

By substituting Eq. (E.5) into Eqs. (E.7) and (E.8), we obtain

$$X = \frac{Eh^3}{3l^3}\left(\frac{c_1^2}{4} + \frac{c_2^2}{10} + \frac{c_1 c_2}{5}\right) \tag{E.13}$$

$$Y = \rho h l\left(\frac{c_1^2}{30} + \frac{c_2^2}{280} + \frac{2c_1 c_2}{105}\right) \tag{E.14}$$

With the help of Eqs. (E.13) and (E.14), Eqs. (E.11) and (E.12) can be expressed as

$$\begin{bmatrix} \left(\dfrac{1}{2} - \underset{\sim}{\omega}^2 \cdot \dfrac{1}{15}\right) & \left(\dfrac{1}{5} - \underset{\sim}{\omega}^2 \cdot \dfrac{2}{105}\right) \\ \left(\dfrac{1}{5} - \underset{\sim}{\omega}^2 \cdot \dfrac{2}{105}\right) & \left(\dfrac{1}{5} - \underset{\sim}{\omega}^2 \cdot \dfrac{1}{140}\right) \end{bmatrix} \begin{Bmatrix} c_1 \\ c_2 \end{Bmatrix} = \begin{Bmatrix} 0 \\ 0 \end{Bmatrix} \tag{E.15}$$

where

$$\underset{\sim}{\omega}^2 = \frac{3\omega^2 \rho l^4}{Eh^2} \tag{E.16}$$

By setting the determinant of the matrix in Eq. (E.15) equal to zero, we obtain the frequency equation

$$\frac{1}{8,820}\underset{\sim}{\omega}^4 - \frac{13}{1,400}\underset{\sim}{\omega}^2 + \frac{3}{50} = 0 \tag{E.17}$$

The roots of Eq. (E.17) are given by $\underset{\sim}{\omega}_1 = 2.6599$ and $\underset{\sim}{\omega}_2 = 8.6492$. Thus the natural frequencies of the tapered beam are

$$\omega_1 \simeq 1.5367 \left(\frac{Eh^2}{\rho l^4}\right)^{1/2} \tag{E.18}$$

and

$$\omega_2 \simeq 4.9936 \left(\frac{Eh^2}{\rho l^4}\right)^{1/2} \tag{E.19}$$

∎

8.9 Examples Using MATLAB

Plotting the Forced-Vibration Response of a Simply Supported Beam

EXAMPLE 8.14

Using MATLAB, plot the steady-state response of the pinned-pinned beam considered in Example 8.8, Eq. (E.7), for $n = 1, 2,$ and 5.

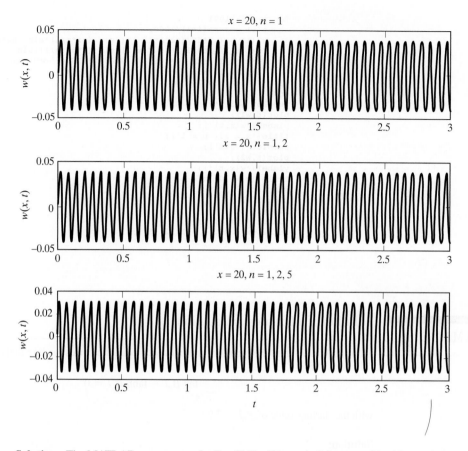

Solution: The MATLAB program to plot Eq. (E.7) of Example 8.8 at $x = 20$ with $n = 1, 2,$ and 5 is given below.

```
%Ex8_14.m
x = 20;
f0 = 100;
a = 10;
A = 1;
l = 40;
ro = 0.283/386.4;
w = 100;
n = 1;
wn = (n^2) * 360.393674;
for i = 1: 1001
    t(i) = 3 * (i-1)/1000;
    w1(i) = ( 2*f0/(ro*A*1) )*sin (n*pi*a/1)*sin (n*pi*x/1)*sin
             (w*t(i))/(wn^2-w^2);
end
n = 2;
for i = 1: 1001
    t(i) = 3 * (i-1)/1000;
    w2(i) = ( 2*f0/(ro*A*1) )*( sin (pi*a/1)*sin (pi*x/1)*sin
    (w*t(i))/(360.393674^2-w^2)+sin (2*pi*a/1)*sin
    (2*pi*x/1)*sin (w*t(i))/((2*360.393674)^2-w^2) );
end
```

```
for i = 1: 1001
    t(i) = 3 * (i-1)/1000;
    w3(i) = ( 2*f0/(ro*A*1) )*( sin (pi*a/1)*sin (pi*x/1)*sin
    (w*t(i))/(360.393674^2-w^2)+sin (2*pi*a/1)+sin
    (2*pi*x/1)*sin (w*t(i))/((2*360.393674)^2-w^2)+sin
    (5*pi*a/1)*sin (5*pi*x/1)*sin (w*t(i))/
    ((5*360.393674)^2-w^2) );
end
subplot ('311');
plot(t,w1);
ylabel('w(x,t)');
title('x = 20, n = 1');
subplot ('312');
plot(t,w2);
ylabel('w(x,t)');
title('x = 20, n = 1, 2');
subplot('313');
plot(t,w3);
xlabel('t');
ylabel('w(x,t)');
title('x = 20, n = 1, 2, 5');
```

∎

EXAMPLE 8.15

Solution of a Frequency Equation

Using MATLAB, find the root(s) of the frequency equation corresponding to the fixed-pinned beam

$$\tan \beta_n l - \tanh \beta_n l = 0$$

with the starting value of $\beta_n l = 3.0$.

Solution:

```
>> x = fzero(inline('tan(y)-tanh(y)'), 3.0)

x =

    3.92660231204792

>> tan (x) - tanh (x)

ans =

    -4.440892098500626e-016
```

∎

EXAMPLE 8.16

Program to Find the Roots of Transcendental and Nonlinear Equations

Develop a general MATLAB program, called `Program12.m`, for finding the roots of nonlinear and transcendental equations. Use the program to find the root of the equation

$$\tan \beta l - \tanh \beta l = 0 \qquad \text{(E.1)}$$

Solution: `Program12.m` is developed to accept the following input data:

n = number of roots to be determined

xs = initial guess for the first root

xinc = initial increment to be used in searching for the root

nint = maximum number of subintervals to be used (usual value: 50)

iter = maximum number of iterations permitted in finding a root (usual value: 100)

eps = convergence requirement (usual value: 10^{-6})

The given nonlinear equation is to be defined in a subprogram called `function.m`. The program gives the computed roots as output.

```
>> programs12
Roots of nonlinear equation
Data:
n    = 5
xs   = 2.000000e+000
xinc = 1.000000e-001
nint = 50
iter = 100
eps  = 1.000000e-006
Roots
  3.926602e+000
  7.068583e+000
  1.021018e+001
  1.335177e+001
  1.649336e+001
```

■

CHAPTER SUMMARY

We have studied the method of deriving the equation of motion of continuous systems such as a string, bar, shaft, beam, and membrane. We presented methods of finding the natural frequencies, mode shapes, and free-vibration solutions using the relevant boundary and initial conditions. We outlined a method of analyzing the forced vibration of beams using the mode superposition method. In addition, we presented the effects of axial force, rotary inertia, and shear deformation on the vibration of beams. We studied the Rayleigh and Rayleigh-Ritz methods to find the approximate natural frequencies of continuous systems. Finally, we presented the use of MATLAB in finding the free- and forced-vibration solutions of continuous systems.

Now that you have finished this chapter, you should be able to answer the review questions and solve the problems given below.

REFERENCES

8.1 S. K. Clark, *Dynamics of Continuous Elements*, Prentice Hall, Englewood Cliffs, N.J, 1972.

8.2 S. Timoshenko, D. H. Young, and W. Weaver, Jr., *Vibration Problems in Engineering* (4th ed.), Wiley, New York, 1974.

8.3   A. Leissa, *Vibration of Plates*, NASA SP-160, Washington, DC, 1969.

8.4   I. S. Habib, *Engineering Analysis Methods*, Lexington Books, Lexington, MA., 1975.

8.5   J. D. Achenbach, *Wave Propagation in Elastic Solids*, North-Holland Publishing, Amsterdam, 1973.

8.6   K. K. Deb, "Dynamics of a string and an elastic hammer," *Journal of Sound and Vibration*, Vol. 40, 1975, pp. 243–248.

8.7   M. S. Triantafyllou, "Linear dynamics of cables and chains," *Shock and Vibration Digest*, Vol. 16, March 1984, pp. 9–17.

8.8   R. W. Fitzgerald, *Mechanics of Materials* (2nd ed.), Addison-Wesley, Reading, MA, 1982.

8.9   S. P. Timoshenko and J. Gere, *Theory of Elastic Stability* (2nd ed.), McGraw-Hill, New York, 1961.

8.10   S. P. Timoshenko, "On the correction for shear of the differential equation for transverse vibration of prismatic bars," *Philosophical Magazine*, Series 6, Vol. 41, 1921, pp. 744–746.

8.11   G. R. Cowper, "The shear coefficient in Timoshenko's beam theory," *Journal of Applied Mechanics*, Vol. 33, 1966, pp. 335–340.

8.12   G. W. Housner and W. O. Keightley, "Vibrations of linearly tapered beams," Part I, *Transactions of ASCE*, Vol. 128, 1963, pp. 1020–1048.

8.13   C. M. Harris (ed.), *Shock and Vibration Handbook* (3rd ed.), McGraw-Hill, New York, 1988.

8.14   J. H. Gaines and E. Volterra, "Transverse vibrations of cantilever bars of variable cross section," *Journal of the Acoustical Society of America*, Vol. 39, 1966, pp. 674–679.

8.15   T. M. Wang, "Natural frequencies of continuous Timoshenko beams," *Journal of Sound and Vibration*, Vol. 13, 1970, pp. 409–414.

8.16   S. L. Grassie, R. W. Gregory, D. Harrison, and K. L. Johnson, "The dynamic response of railway track to high frequency vertical excitation," *Journal of Mechanical Engineering Science*, Vol. 24, June 1982, pp. 77–90.

8.17   A. Dimarogonas, *Vibration Engineering*, West Publishing, Saint Paul, MN, 1976.

8.18   T. Justine and A. Krishnan, "Effect of support flexibility on fundamental frequency of beams," *Journal of Sound and Vibration*, Vol. 68, 1980, pp. 310–312.

8.19   K. A. R. Perkins, "The effect of support flexibility on the natural frequencies of a uniform cantilever," *Journal of Sound and Vibration*, Vol. 4, 1966, pp. 1–8.

8.20   S. S. Rao, "Natural frequencies of systems of elastically connected Timoshenko beams," *Journal of the Acoustical Society of America*, Vol. 55, 1974, pp. 1232–1237.

8.21   A. M. Ebner and D. P. Billington, "Steady-state vibration of damped Timoshenko beams," *Journal of Structural Division* (ASCE), Vol. 3, 1968, p. 737.

8.22   M. Levinson and D. W. Cooke, "On the frequency spectra of Timoshenko beams," *Journal of Sound and Vibration*, Vol. 84, 1982, pp. 319–326.

8.23   N. Y. Olcer, "General solution to the equation of the vibrating membrane," *Journal of Sound and Vibration*, Vol. 6, 1967, pp. 365–374.

8.24   G. R. Sharp, "Finite transform solution of the vibrating annular membrane," *Journal of Sound and Vibration*, Vol. 6, 1967, pp. 117–128.

8.25   J. Mazumdar, "A review of approximate methods for determining the vibrational modes of membranes," *Shock and Vibration Digest*, Vol. 14, February 1982, pp. 11–17.

8.26   J. E. Penny and J. R. Reed, "An integral equation approach to the fundamental frequency of vibrating beams," *Journal of Sound and Vibration*, Vol. 19, 1971, pp. 393–400.

8.27 G. Temple and W. G. Bickley, *Rayleigh's Principle and Its Application to Engineering*, Dover, New York, 1956.

8.28 W. L. Craver, Jr. and D. M. Egle, "A method for selection of significant terms in the assumed solution in a Rayleigh-Ritz analysis," *Journal of Sound and Vibration*, Vol. 22, 1972, pp. 133–142.

8.29 L. Klein, "Transverse vibrations of non-uniform beams," *Journal of Sound and Vibration*, Vol. 37, 1974, pp. 491–505.

8.30 J. R. Hutchinson, "Transverse vibrations of beams: Exact versus approximate solutions," *Journal of Applied Mechanics*, Vol. 48, 1981, pp. 923–928.

REVIEW QUESTIONS

8.1 Give brief answers to the following:

1. How does a continuous system differ from a discrete system in the nature of its equation of motion?
2. How many natural frequencies does a continuous system have?
3. Are the boundary conditions important in a discrete system? Why?
4. What is a wave equation? What is a traveling-wave solution?
5. What is the significance of wave velocity?
6. State the boundary conditions to be specified at the simply supported end of a beam if (a) thin-beam theory is used and (b) Timoshenko beam theory is used.
7. State the possible boundary conditions at the ends of a string.
8. What is the main difference in the nature of the frequency equations of a discrete system and a continuous system?
9. What is the effect of a tensile force on the natural frequencies of a beam?
10. Under what circumstances does the frequency of vibration of a beam subjected to an axial load become zero?
11. Why does the natural frequency of a beam become lower if the effects of shear deformation and rotary inertia are considered?
12. Give two practical examples of the vibration of membranes.
13. What is the basic principle used in Rayleigh's method?
14. Why is the natural frequency given by Rayleigh's method always larger than the true value of ω_1?
15. What is the difference between Rayleigh's method and the Rayleigh-Ritz method?
16. What is Rayleigh's quotient?

8.2 Indicate whether each of the following statements is true or false:

1. Continuous systems are the same as distributed systems.
2. Continuous systems can be considered to have an infinite number of degrees of freedom.
3. The governing equation of a continuous system is an ordinary differential equation.
4. The free-vibration equations corresponding to the transverse motion of a string, the longitudinal motion of a bar, and the torsional motion of a shaft have the same form.
5. The normal modes of a continuous system are orthogonal.
6. A membrane has zero bending resistance.
7. Rayleigh's method can be considered as a method of conservation of energy.

8. The Rayleigh-Ritz method assumes that the solution is a series of functions that satisfy the boundary conditions of the problem.

9. For a discrete system, the boundary conditions are to be applied explicitly.

10. The Euler-Bernoulli beam theory is more accurate than the Timoshenko theory.

8.3 Fill in each of the following blanks with appropriate words:

1. The free-vibration equation of a string is also called a _____ equation.

2. The frequency equation is also known as the _____ equation.

3. The method of separation of variables is used to express the free-vibration solution of a string as a _____ of function of x and function of t.

4. Both boundary and _____ conditions are to be specified to find the solution of a vibrating continuous system.

5. In the wave-solution $w(x, t) = w_1(x - ct) + w_2(x + ct)$, the first term represents the wave that propagates in the _____ directions of x.

6. The quantities EI and GJ are called the _____ and _____ stiffnesses, respectively.

7. The thin beam theory is also known as the _____ theory.

8. The lateral vibration of a thin beam is governed by a _____ order partial differential equation in spatial variable.

9. When a beam is subjected to an axial force (tension), it _____ the natural frequency.

10. The Timoshenko beam theory can be considered as _____ beam theory.

11. A drumhead can be considered as a _____.

12. A string has the same relationship to a beam as a membrane bears to a _____.

13. Rayleigh's method can be used to estimate the _____ natural frequency of a continuous system.

14. $EI \dfrac{\partial^2 w}{\partial x^2}$ denotes the _____ in a beam.

15. For a discrete system, the governing equations are _____ differential equations.

16. An axial tensile load increases the bending _____ of a beam.

17. The _____ energy of a beam is denoted by $\dfrac{1}{2} \displaystyle\int_0^l \rho A \left(\dfrac{\partial w}{\partial t} \right)^2 dx$.

18. The _____ energy of a beam is denoted by $\dfrac{1}{2} \displaystyle\int_0^l EI \left(\dfrac{\partial^2 w}{\partial x^2} \right)^2 dx$.

8.4 Select the most appropriate answer out of the choices given:

1. The frequency equation of a continuous system is a
   a. polynomial equation
   b. transcendental equation
   c. differential equation

2. The number of natural frequencies of a continuous system is
   a. infinite   b. one   c. finite

3. When the axial force approaches the Euler buckling load, the fundamental frequency of the beam reaches
   a. infinity   b. the frequency of a taut string   c. zero

4. The value of the Timoshenko shear coefficient depends on the following:
   a. shape of the cross section
   b. size of the cross section
   c. length of the beam

5. A Laplacian operator is given by

   a. $\dfrac{\partial^2}{\partial x\,\partial y}$

   b. $\dfrac{\partial^2}{\partial x^2} + \dfrac{\partial^2}{\partial y^2} + 2\,\dfrac{\partial^2}{\partial x\,\partial y}$

   c. $\dfrac{\partial^2}{\partial x^2} + \dfrac{\partial^2}{\partial y^2}$

6. The boundary condition corresponding to the free end of a bar in longitudinal vibration is given by
   a. $u(0, t) = 0$

   b. $\dfrac{\partial u}{\partial x}(0, t) = 0$

   c. $AE\,\dfrac{\partial u}{\partial x}(0, t) - u(0, t) = 0$

7. The orthogonality of normal functions of the longitudinal vibration of a bar is given by

   a. $\displaystyle\int_0^l U_i(x)U_j(x)\,dx = 0$

   b. $\displaystyle\int_0^l \left(U_i'U_j - U_j'U_i\right)\,dx = 0$

   c. $\displaystyle\int_0^l \left(U_i(x) + U_j(x)\right)\,dx = 0$

8.5 Match the items in the two columns below regarding boundary conditions for a thin beam:

   1. Free end
   2. Pinned end
   3. Fixed end
   4. Elastically restrained end

   a. Bending moment = 0; shear force equals the spring force
   b. Deflection = 0; slope = 0
   c. Deflection = 0; bending moment = 0
   d. Bending moment = 0; shear force = 0

8.6 Match the items in the two columns below regarding a uniform beam:

   1. $W = 0$
   2. $W' = 0$
   3. $W'' = 0$
   4. $W''' = 0$

   a. Zero bending moment
   b. Zero transverse displacement
   c. Zero shear force
   d. Zero slope

8.7   Match the items in the two columns below regarding the wave equation $c^2 \dfrac{\partial^2 w}{\partial x^2} = \dfrac{\partial^2 w}{\partial t^2}$:

1. $c = \left(\dfrac{P}{\rho}\right)^{1/2}$ **a.** Longitudinal vibration of a bar

2. $c = \left(\dfrac{E}{\rho}\right)^{1/2}$ **b.** Torsional vibration of a shaft

3. $c = \left(\dfrac{G}{\rho}\right)^{1/2}$ **c.** Transverse vibration of a string

PROBLEMS

Section 8.2 Transverse Vibration of a String or Cable

8.1   Determine the velocity of wave propagation in a cable of mass $\rho = 5$ kg/m when stretched by a tension $P = 4{,}000$ N.

8.2   A steel wire of 2 mm diameter is fixed between two points located 2 m apart. The tensile force in the wire is 250 N. Determine (a) the fundamental frequency of vibration and (b) the velocity of wave propagation in the wire.

8.3   A stretched cable of length 2 m has a fundamental frequency of 3,000 Hz. Find the frequency of the third mode. How are the fundamental and third mode frequencies changed if the tension is increased by 20 percent?

8.4   Find the time it takes for a transverse wave to travel along a transmission line from one tower to another one 300 m away. Assume the horizontal component of the cable tension as 30,000 N and the mass of the cable as 2 kg/m of length.

8.5   A cable of length l and mass ρ per unit length is stretched under a tension P. One end of the cable is connected to a mass m, which can move in a frictionless slot, and the other end is fastened to a spring of stiffness k, as shown in Fig. 8.23. Derive the frequency equation for the transverse vibration of the cable.

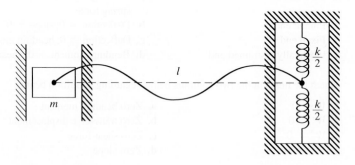

FIGURE 8.23

8.6 The cord of a musical instrument is fixed at both ends and has a length 2 m, diameter 0.5 mm, and density 7,800 kg/m³. Find the tension required in order to have a fundamental frequency of (a) 1 Hz and (b) 5 Hz.

8.7 A cable of length l and mass ρ per unit length is stretched under a tension P. One end of the cable is fixed and the other end is connected to a pin, which can move in a frictionless slot. Find the natural frequencies of vibration of the cable.

8.8 Find the free-vibration solution of a cord fixed at both ends when its initial conditions are given by

$$w(x, 0) = 0, \qquad \frac{\partial w}{\partial t}(x, 0) = \frac{2ax}{l} \qquad \text{for} \qquad 0 \leq x \leq \frac{l}{2}$$

and

$$\frac{\partial w}{\partial t}(x, 0) = 2a\left(1 - \frac{x}{l}\right) \qquad \text{for} \qquad \frac{l}{2} \leq x \leq l$$

8.9 Prove that the constant a in Eqs. (8.18) and (8.19) is negative for common boundary conditions. *Hint:* Multiply Eq. (8.18) by $W(x)$ and integrate with respect to x from 0 to l.

8.10* The cable between two electric transmission towers has a length of 2,000 m. It is clamped at its ends under a tension P (Fig. 8.24). The density of the cable material is 8,890 kg/m³. If the first four natural frequencies are required to lie between 0 and 20 Hz, determine the necessary cross-sectional area of the cable and the initial tension.

FIGURE 8.24

8.11 If a string of length l, fixed at both ends, is given an initial transverse displacement of h at $x = l/3$ and then released, determine its subsequent motion. Compare the deflection shapes of the string at times $t = 0, l/(4c), l/(3c), l/(2c)$, and l/c by considering the first four terms of the series solution.

8.12 A cord of length l is made to vibrate in a viscous medium. Derive the equation of motion considering the viscous damping force.

8.13 Determine the free-vibration solution of a string fixed at both ends under the initial conditions $w(x, 0) = w_0 \sin(\pi x/l)$ and $(\partial w/\partial t)(x, 0) = 0$.

---

*The asterisk denotes a problem with no unique answer.

8.14 The strings of a guitar (Fig. 8.25) are made of music wire with diameter 0.05 mm, weight density 76.5 kN/m^3, and Young's modulus 207 GPa. If the lengths of two of the strings are given by 0.60 m and 0.65 m, determine the fundamental natural frequencies of the strings if the tension in each string is 5×10^4 N.

FIGURE 8.25

8.15 The vertical and horizontal forces (reactions) at joints A and B of a typical cable of the suspension bridge shown in Fig. 8.26 are given by $F_x = 2.8 \times 10^6$ N and $F_y = 1.1 \times 10^6$ N. The cables are made of steel with a weight density of 76.5 kN/m^3. If the effective diameter of the cable is 25 cm, find the first two natural frequencies of vibration of the cable in the vertical direction. State the assumptions involved in the solution.

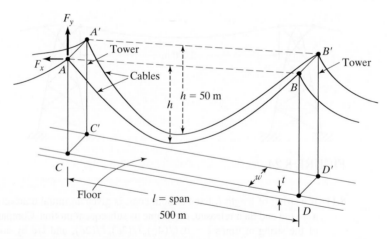

FIGURE 8.26   A suspension bridge.

Section 8.3 Longitudinal Vibration of a Bar or Rod

8.16 Derive an equation for the principal modes of longitudinal vibration of a uniform bar having both ends free.

8.17 Derive the frequency equation for the longitudinal vibration of the systems shown in Fig. 8.27.

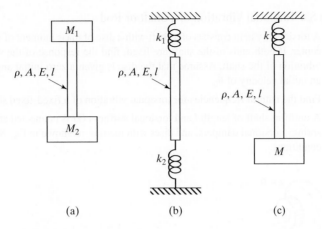

(a)   (b)   (c)

FIGURE 8.27

8.18* A thin bar of length l and mass m is clamped at one end and free at the other. What mass M must be attached to the free end in order to decrease the fundamental frequency of longitudinal vibration by 50 percent from its fixed-free value?

8.19 Show that the normal functions corresponding to the longitudinal vibration of the bar shown in Fig. 8.28 are orthogonal.

$x = 0$ $x = l$

FIGURE 8.28

8.20 Derive the frequency equation for the longitudinal vibration of a stepped bar having two different cross-sectional areas A_1 and A_2 over lengths l_1 and l_2, respectively. Assume fixed-free end conditions.

8.21 A steel shaft of diameter d and length l is fixed at one end and carries a propeller of mass m and mass moment of inertia J_0 at the other end (Fig. 8.29). Determine the fundamental natural frequency of vibration of the shaft in (a) axial vibration and (b) torsional vibration. Data: $d = 5$ cm, $l = 1$ m, $m = 100$ kg, $J_0 = 10$ kg-m^2.

Propeller; mass m,
mass moment of inertia J_0

d

x

l

FIGURE 8.29

Section 8.4 Torsional Vibration of a Shaft or Rod

8.22 A torsional system consists of a shaft with a disc of mass moment of inertia I_0 mounted at its center. If both ends of the shaft are fixed, find the response of the system in free torsional vibration of the shaft. Assume that the disc is given a zero initial angular displacement and an initial velocity of $\dot{\theta}_0$.

8.23 Find the natural frequencies for torsional vibration of a fixed-fixed shaft.

8.24 A uniform shaft of length l and torsional stiffness GJ is connected at both ends by torsional springs, torsional dampers, and discs with inertias, as shown in Fig. 8.30. State the boundary conditions.

FIGURE 8.30

8.25 Solve Problem 8.23 if one end of the shaft is fixed and the other free.

8.26 Derive the frequency equation for the torsional vibration of a uniform shaft carrying rotors of mass moment of inertia I_{01} and I_{02} one at each end.

8.27 An external torque $M_t(t) = M_{t0} \cos \omega t$ is applied at the free end of a fixed-free uniform shaft. Find the steady-state vibration of the shaft.

8.28 Find the fundamental frequency for torsional vibration of a shaft of length 2 m and diameter 50 mm when both the ends are fixed. The density of the material is 7,800 kg/m³ and the modulus of rigidity is 0.8×10^{11} N/m².

8.29 A uniform shaft, supported at $x = 0$ and rotating at an angular velocity ω, is suddenly stopped at the end $x = 0$. If the end $x = l$ is free, determine the subsequent angular displacement response of the shaft.

Section 8.5 Lateral Vibration of Beams

8.30 Compute the first three natural frequencies and the corresponding mode shapes of the transverse vibrations of a uniform beam of rectangular cross section (100 mm × 300 mm) with $l = 2$ m, $E = 20.5 \times 10^{10}$ N/m², and $\rho = 7.83 \times 10^3$ kg/m³ for the following cases: (a) when both ends are simply supported; (b) when both ends are built-in (clamped); (c) when one end is fixed and the other end is free; and (d) when both ends are free. Plot the mode shapes.

8.31 Derive an expression for the natural frequencies for the lateral vibration of a uniform fixed-free beam.

8.32 Prove that the normal functions of a uniform beam, whose ends are connected by springs as shown in Fig. 8.31, are orthogonal.

FIGURE 8.31

8.33 Derive an expression for the natural frequencies for the transverse vibration of a uniform beam with both ends simply supported.

8.34 Derive the expression for the natural frequencies for the lateral vibration of a uniform beam suspended as a pendulum, neglecting the effect of dead weight.

8.35 Find the cross-sectional area (A) and the area moment of inertia (I) of a simply supported steel beam of length 1 m for which the first three natural frequencies lie in the range 1,500–5,000 Hz.

8.36 A uniform beam, simply supported at both ends, is found to vibrate in its first mode with an amplitude of 10 mm at its center. If $A = 120$ mm^2, $I = 1,000$ mm^4, $E = 20.5 \times 10^{10}$ N/m^2, $\rho = 7.83 \times 10^3$ kg/m^3, and $l = 1$ m, determine the maximum bending moment in the beam.

8.37 Derive the frequency equation for the transverse vibration of a uniform beam resting on springs at both ends, as shown in Fig. 8.32. The springs can deflect vertically only, and the beam is horizontal in the equilibrium position.

FIGURE 8.32

8.38 A simply supported uniform beam of length l carries a mass M at the center of the beam. Assuming M to be a point mass, obtain the frequency equation of the system.

8.39 A uniform fixed-fixed beam of length $2l$ is simply supported at the middle point. Derive the frequency equation for the transverse vibration of the beam.

8.40 A simply supported beam carries initially a uniformly distributed load of intensity f_0. Find the vibration response of the beam if the load is suddenly removed.

8.41 Estimate the fundamental frequency of a cantilever beam whose cross-sectional area and moment of inertia vary as

$$A(x) = A_0 \frac{x}{l} \qquad \text{and} \qquad I(x) = \overline{I} \frac{x}{l}$$

where x is measured from the free end.

8.42 (a) Derive a general expression for the response of a uniform beam subjected to an arbitrary force. (b) Use the result of part (a) to find the response of a uniform simply supported beam under the harmonic force $F_0 \sin \omega t$ applied at $x = a$. Assume the initial conditions as $w(x, 0) = (\partial w / \partial t)(x, 0) = 0$.

8.43 Derive Eqs. (E.5) and (E.6) of Example 8.10.

8.44 Derive Eqs. (E.7) and (E.8) of Example 8.10.

8.45 Prove that the constant a in Eq. (8.82) is positive for common boundary conditions. *Hint:* Multiply Eq. (8.83) by $W(x)$ and integrate with respect to x from 0 to l.

8.46 Find the response of a simply supported beam subject to a uniformly distributed harmonically varying load.

8.47 A fixed-fixed beam carries an electric motor of mass 100 kg and operational speed 3,000 rpm at its midspan, as shown in Fig. 8.33. If the motor has a rotational unbalance of 0.5 kg-m, determine the steady-state response of the beam. Assume the length of the beam as $l = 2$ m, cross section as 10 cm $\times$ 10 cm, and the material as steel.

FIGURE 8.33

8.48 A steel cantilever beam of diameter 2 cm and length 1 m is subjected to an exponentially decaying force $100e^{-0.1t}$ N at the free end, as shown in Fig. 8.34. Determine the steady-state response of the beam. Assume the density and Young's modulus of steel as 7,500 kg/m^3 and 210×10^9 N/m^2, respectively.

FIGURE 8.34

8.49 Find the steady-state response of a cantilever beam that is subjected to a suddenly applied step bending moment of magnitude M_0 at its free end.

8.50 A cantilever beam of length l, density ρ, Young's modulus E, area of cross section A, and area moment of inertia I carries a concentrated mass M at its free end. Derive the frequency equation for the transverse vibration of the beam.

8.51 Consider a railway car moving on a railroad track as shown in Fig. 8.35(a). The track can be modeled as an infinite beam resting on an elastic foundation and the car can be idealized as a

(a)

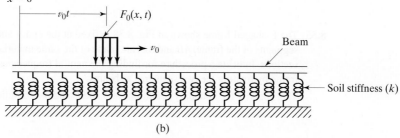

(b)

FIGURE 8.35

moving load $F_0(x, t)$ (see Fig. 8.35(b)). If the soil stiffness per unit length is k, and the constant velocity of the car is v_0, show that the equation of motion of the beam can be expressed as

$$EI \frac{\partial^4 w(x, t)}{\partial x^4} + \rho A \frac{\partial^2 w(x, t)}{\partial t^2} + k w(x, t) = F_0(x - v_0 t)$$

Indicate a method of solving the equation of motion if the moving load is assumed to be constant in magnitude.

8.52 Find the first two natural frequencies of vibration in the vertical direction of the floor of the suspension bridge shown in Fig. 8.26 under the following assumptions:

1. The floor can be considered as a uniform beam with simple supports at both ends C and D.
2. The floor has a width (w) 12 m, thickness (t) 0.75 m, and weight, including the supporting girders, 3,000 N/m.
3. The Young's modulus of the floor is 175 GPa.

8.53 A uniform beam of length $2l$ is fixed at the left end, supported on a simple support at the middle, and free at the right end as shown in Fig. 8.36. Derive the frequency equation for determining the natural frequencies of vibration of the continuous beam.

FIGURE 8.36

8.54 A uniform fixed-fixed beam of length $2l$ is supported on a pin joint at the midpoint as shown in Fig. 8.37. Derive the frequency equation for determining the natural frequencies of vibration of the continuous beam.

FIGURE 8.37

8.55 The L-shaped frame shown in Fig. 8.38 is fixed at the end A and free at the end C. The two segments of the frame, AB and BC, are made of the same material with identical square cross sections. Indicate a procedure for finding the natural frequencies of in-plane vibration of the frame by treating the two segments as beams.

Hint: Identify the boundary conditions at A and C and the conditions to be satisfied at point B.

FIGURE 8.38

Section 8.6 Vibration of Membranes

8.56 Starting from fundamentals, show that the equation for the lateral vibration of a circular membrane is given by

$$\frac{\partial^2 w}{\partial r^2} + \frac{1}{r}\frac{\partial w}{\partial r} + \frac{1}{r^2}\frac{\partial^2 w}{\partial \theta^2} = \frac{\rho}{P}\frac{\partial^2 w}{\partial t^2}$$

8.57 Consider a rectangular membrane of sides a and b supported along all the edges. (a) Derive an expression for the deflection $w(x, y, t)$ under an arbitrary pressure $f(x, y, t)$. (b) Find the response when a uniformly distributed pressure f_0 is applied to a membrane that is initially at rest.

8.58 Find the free-vibration solution and the natural frequencies of a rectangular membrane that is clamped along all the sides. The membrane has dimensions a and b along the x and y directions, respectively.

8.59 Find the free-vibration response of a rectangular membrane of sides a and b subject to the following initial conditions:

$$w(x, y, 0) = w_0 \sin\frac{\pi x}{a} \sin\frac{\pi y}{b}, \qquad 0 \le x \le a, \qquad 0 \le y \le b$$

$$\frac{\partial w}{\partial t}(x, y, 0) = 0, \qquad 0 \le x \le a, \qquad 0 \le y \le b$$

8.60 Find the free-vibration response of a rectangular membrane of sides a and b subjected to the following initial conditions:

$$\left.\begin{array}{l} w(x, y, 0) = 0 \\[2mm] \dfrac{\partial w}{\partial t}(x, y, 0) = \dot{w}_0 \sin\dfrac{\pi x}{a} \sin\dfrac{2\pi y}{b} \end{array}\right\}, \qquad \begin{array}{l} 0 \le x \le a \\[2mm] 0 \le y \le b \end{array}$$

Assume that the edges of the membrane are fixed.

8.61 Compare the fundamental natural frequencies of transverse vibration of membranes of the following shapes: (a) square; (b) circular; and (c) rectangular with sides in the ratio of 2:1. Assume that all the membranes are clamped around their edges and have the same area, material, and tension.

8.62 Using the equation of motion given in Problem 8.56, find the natural frequencies of a circular membrane of radius R clamped around the boundary at $r = R$.

Section 8.7 Rayleigh's Method

8.63 Find the fundamental natural frequency of a fixed-fixed beam using the static deflection curve

$$W(x) = \frac{c_0 x^2}{24EI}(l - x)^2$$

where c_0 is a constant.

8.64 Solve Problem 8.63 using the deflection shape $W(x) = c_0\left(1 - \cos\dfrac{2\pi x}{l}\right)$, where c_0 is a constant.

8.65 Find the fundamental natural frequency of vibration of a uniform beam of length l that is fixed at one end and simply supported at the other end. Assume the deflection shape of the beam to be the same as the static deflection curve under its self weight. *Hint:* The static deflection of a uniform beam under self weight is governed by

$$EI\frac{d^4 W(x)}{dx^4} = \rho g A$$

where ρ is the density, g is the acceleration due to gravity, and A is the area of cross section of the beam. This equation can be integrated for any known boundary conditions of the beam.

8.66 Determine the fundamental frequency of a uniform fixed-fixed beam carrying a mass M at the middle by applying Rayleigh's method. Use the static deflection curve for $W(x)$.

8.67 Applying Rayleigh's method, determine the fundamental frequency of a cantilever beam (fixed at $x = l$) whose cross-sectional area $A(x)$ and moment of inertia $I(x)$ vary as $A(x) = A_0 x / l$ and $I(x) = I_0 x / l$.

8.68 Using Rayleigh's method, find the fundamental frequency for the lateral vibration of the beam shown in Fig. 8.39. The restoring force in the spring k is proportional to the deflection, and the restoring moment in the spring k_t is proportional to the angular deflection.

FIGURE 8.39

8.69 Using Rayleigh's method, estimate the fundamental frequency for the lateral vibration of a uniform beam fixed at both ends. Assume the deflection curve to be

$$W(x) = c_1 \left(1 - \cos \frac{2\pi x}{l} \right)$$

8.70 Find the fundamental frequency of longitudinal vibration of the tapered bar shown in Fig. 8.40, using Rayleigh's method with the mode shape

$$U(x) = c_1 \sin \frac{\pi x}{2l}$$

The mass per unit length is given by

$$m(x) = 2m_0 \left(1 - \frac{x}{l} \right)$$

and the stiffness by

$$EA(x) = 2EA_0 \left(1 - \frac{x}{l} \right)$$

FIGURE 8.40

8.71 Approximate the fundamental frequency of a rectangular membrane supported along all the edges by using Rayleigh's method with

$$W(x, y) = c_1 xy(x - a)(y - b).$$

Hint:

$$V = \frac{P}{2} \iint \left[\left(\frac{\partial w}{\partial x} \right)^2 + \left(\frac{\partial w}{\partial y} \right)^2 \right] dx\, dy \quad \text{and} \quad T = \frac{\rho}{2} \iint \left(\frac{\partial w}{\partial t} \right)^2 dx\, dy$$

8.72 Using Rayleigh's method, determine the fundamental natural frequency of the system shown in Fig. 8.41.

Disc, mass moment of inertia 5 kg-m^2

Steel shaft, dia. 5 cm

0.8 m   0.2 m

FIGURE 8.41

Section 8.8 The Rayleigh-Ritz Method

8.73 Estimate the fundamental frequency of a fixed-fixed string, assuming the mode shape (a) $W(x) = c_1 x(l - x)$ and (b) $W(x) = c_1 x(l - x) + c_2 x^2(l - x)^2$.

8.74 Estimate the fundamental frequency for the longitudinal vibration of a uniform bar fixed at $x = 0$ and free at $x = l$ by assuming the mode shapes as (a) $U(x) = c_1(x/l)$ and (b) $U(x) = c_1(x/l) + c_2(x/l)^2$.

8.75 A stepped bar, fixed at $x = 0$ and free at $x = l$, has a cross-sectional area of $2A$ for $0 \le x < l/3$ and A for $l/3 \le x \le l$. Assuming the mode shape

$$U(x) = c_1 \sin\frac{\pi x}{2l} + c_2 \sin\frac{3\pi x}{2l}$$

estimate the first two natural frequencies of longitudinal vibration.

8.76 Solve Problem 8.70 using the Rayleigh-Ritz method with the mode shape

$$U(x) = c_1 \sin\frac{\pi x}{2l} + c_2 \sin\frac{3\pi x}{2l}$$

8.77 Find the first two natural frequencies of a fixed-fixed uniform string of mass density ρ per unit length stretched between $x = 0$ and $x = l$ with an initial tension P. Assume the deflection functions

$$w_1(x) = x(l - x)$$
$$w_2(x) = x^2(l - x)^2$$

Section 8.9 MATLAB Problems

8.78 Using `Program12.m`, solve Example 8.4.

8.79 Using `Program12.m`, find the first five natural frequencies of a thin fixed-fixed beam.

8.80 Using MATLAB, plot the dynamic response of the plucked string, Eq. (E.6) of Example 8.1, at $x = l/2$. Data: $h = 0.1$ m, $l = 1.0$ m, $c = 100$ m/s.

8.81 Write a computer program for finding numerically the mode shapes of thin fixed-simply supported beams by using the known values of the natural frequencies.

DESIGN PROJECT

8.82 A vehicle, of weight F_0, moving at a constant speed on a bridge (Fig. 8.42(a)) can be modeled as a concentrated load traveling on a simply supported beam as shown in Fig. 8.42(b). The concentrated load F_0 can be considered as a uniformly distributed load over an infinitesimal length 2Δ and can be expressed as a sum of sine terms using Fourier sine series expansion (of the distributed load). Find the transverse displacement of the bridge as a sum of the responses due to each of the moving harmonic load components. Assume the initial conditions of the bridge as $w(x, 0) = \partial w/\partial t(x, 0) = 0$.

FIGURE 8.42

Leonhard Euler (1707–1783) was a Swiss mathematician who became a court mathematician and later a professor of mathematics in Saint Petersburg, Russia. He produced many works in algebra and geometry and was interested in the geometrical form of deflection curves in strength of materials. Euler's column buckling load is quite familiar to mechanical and civil engineers, and Euler's constant and Euler's coordinate system are well known to mathematicians. He derived the equation of motion for the bending vibrations of a rod (Euler-Bernoulli theory) and presented a series form of solution, as well as studied the dynamics of a vibrating ring. (Courtesy of Dirk J. Struik, *A Concise History of Mathematics* (2nd ed.), Dover Publications, New York, 1948.)

CHAPTER 9

Vibration Control

Chapter Outline

We studied all the aspects of modeling and analysis of vibrating systems in the previous chapters. We will now consider methods of eliminating or reducing unwanted vibration. The acceptable levels of vibration must be known before we can quantify the levels to be eliminated or reduced. The vibration nomograph and vibration criteria which indicate acceptable levels of vibration are outlined at the beginning. The vibration to be eliminated or reduced can be in one or more forms of disturbance—displacement, velocity, acceleration,

769

and transmitted force. The following methods are discussed to eliminate or reduce vibration at the source:

- Balancing of rotating machines—single- and two-plane balancing.
- Controlling the response and stability of rotating shafts.
- Balancing of reciprocating engines.
- Reducing vibration caused by impacts due to clearances in the joints of machines and mechanisms.

The following methods are discussed to reduce transmission of vibration from the source:

- Changing the natural frequency of the system when the forcing frequency cannot be altered.
- Introducing a power-dissipation mechanism by adding dashpots or viscoelastic materials.
- Designing an isolator which changes the stiffness or damping of the system.
- Using an active control technique.
- Designing a vibration absorber by adding an auxiliary mass to absorb the vibration energy of the original mass.

Finally, the solution of various vibration-control problems using MATLAB is presented with numerical examples.

Learning Objectives

After you have finished studying this chapter, you should be able to do the following:

- Use vibration nomographs and vibration criteria to determine the levels of vibration to be controlled or reduced.
- Apply single- and two-plane balancing techniques for eliminating vibration (unbalance).
- Control the vibration caused by the unbalance in rotating shafts.
- Reduce the unbalance in reciprocating engines.
- Design vibration and shock isolations for systems with a fixed base as well as a vibrating base.
- Design active vibration-control systems.
- Design undamped and damped vibration absorbers.
- Use MATLAB for solving vibration-control problems.

9.1   Introduction

There are numerous sources of vibration in an industrial environment: impact processes such as pile driving and blasting; rotating or reciprocating machinery such as engines, compressors, and motors; transportation vehicles such as trucks, trains, and aircraft; the flow of fluids; and many others. The presence of vibration often leads to excessive wear of bearings, formation of cracks, loosening of fasteners, structural and mechanical failures, frequent and costly maintenance of machines, electronic malfunctions through fracture of solder joints, and abrasion of insulation around electric conductors causing shorts. The occupational exposure of humans to vibration leads to pain, discomfort, and reduced efficiency. Vibration can sometimes be eliminated on the basis of theoretical analysis. However, the manufacturing costs involved in eliminating the vibration may be too high; a designer must compromise between an acceptable amount of vibration and a reasonable manufacturing cost. In some cases, the excitation or shaking force is inherent in the machine. As seen earlier, even a relatively small excitation force can cause an undesirably large response near resonance, especially in lightly damped systems. In these cases, the magnitude of the response can be significantly reduced by the use of isolators and auxiliary mass absorbers [9.1]. In this chapter, we shall consider various techniques of vibration control—that is, methods involving the elimination or reduction of vibration.

9.2   Vibration Nomograph and Vibration Criteria

The acceptable levels of vibration are often specified in terms of the response of an undamped single-degree-of-freedom system undergoing harmonic vibration. The bounds are shown in a graph, called the *vibration nomograph*, which displays the variations of displacement, velocity, and acceleration amplitudes with respect to the frequency of vibration. For the harmonic motion

$$x(t) = X \sin \omega t \tag{9.1}$$

the velocity and acceleration are given by

$$v(t) = \dot{x}(t) = \omega X \cos \omega t = 2\pi f X \cos \omega t \tag{9.2}$$

$$a(t) = \ddot{x}(t) = -\omega^2 X \sin \omega t = -4\pi^2 f^2 X \sin \omega t \tag{9.3}$$

where ω is the circular frequency (rad/s), f is the linear frequency (Hz), and X is the amplitude of displacement. The amplitudes of displacement (X), velocity (v_{max}), and acceleration (a_{max}) are related as

$$v_{max} = 2\pi f X \tag{9.4}$$

$$a_{max} = -4\pi^2 f^2 X = -2\pi f v_{max} \tag{9.5}$$

By taking logarithms of Eqs. (9.4) and (9.5), we obtain the following linear relations:

$$\ln v_{max} = \ln (2\pi f) + \ln X \tag{9.6}$$

$$\ln v_{max} = -\ln a_{max} - \ln (2\pi f) \tag{9.7}$$

It can be seen that for a constant value of the displacement amplitude (*X*), Eq. (9.6) shows that $\ln v_{max}$ varies with $\ln(2\pi f)$ as a straight line with slope +1. Similarly, for a constant value of the acceleration amplitude (a_{max}), Eq. (9.7) indicates that $\ln v_{max}$ varies with $\ln(2\pi f)$ as a straight line with slope −1. These variations are shown as a nomograph in Fig. 9.1. Thus every point on the nomograph denotes a specific sinusoidal (harmonic) vibration.

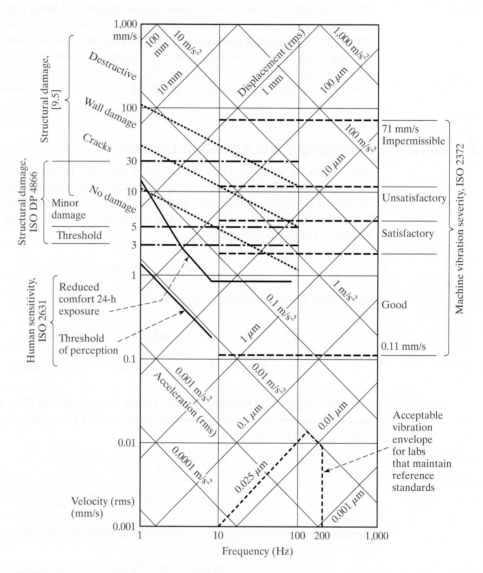

FIGURE 9.1 Vibration nomograph and vibration criteria [9.2].

Since the vibration imparted to a human or machine is composed of many frequencies—rarely of just one frequency—the root mean square values of $x(t)$, $v(t)$, and $a(t)$ are used in the specification of vibration levels.

The usual ranges of vibration encountered in different scientific and engineering applications are given below [9.2]:

1. Atomic vibrations: Frequency = 10^{12} Hz, displacement amplitude = 10^{-8} to 10^{-6} mm.

2. Microseisms or minor tremors of earth's crust: Frequency = 0.1 to 1 Hz, displacement amplitude = 10^{-5} to 10^{-3} mm. This vibration also denotes the threshold of disturbance of optical, electronic, and computer equipment.

3. Machinery and building vibration: Frequency = 10 to 100 Hz, displacement amplitude = 0.01 to 1 mm. The threshold of human perception falls in the frequency range 1 to 8 Hz.

4. Swaying of tall buildings: Frequency range = 0.1 to 5 Hz, displacement amplitude = 10 to 1,000 mm.

Vibration severity of machinery is defined in terms of the rms value of the vibration velocity in ISO 2372 [9.3]. The ISO definition identifies 15 vibration severity ranges in the velocity range 0.11–71 mm/s for four classes of machines: (1) small, (2) medium, (3) large, and (4) turbomachine. The vibration severity of class 3 machines, including large prime movers, is shown in Fig. 9.1. In order to apply these criteria, the vibration is to be measured on machine surfaces such as bearing caps in the frequency range 10–1,000 Hz.

ISO DP 4866 [9.4] gives the vibration severity for whole-building vibration under blasting and steady-state vibration in the frequency range 1–100 Hz. For the vibration from blasting, the velocity is to be measured at the building foundation nearest the blast, and for the steady-state vibration, the peak velocity is to be measured on the top floor. The limits given are 3–5 mm/s for threshold of damage and 5–30 mm/s for minor damage. The vibration results reported by Steffens [9.5] on structural damage are also shown in Fig. 9.1.

The vibration limits recommended in ISO 2631 [9.6] on human sensitivity to vibration are also shown in Fig. 9.1. In the United States, an estimated 8 million workers are exposed to either whole-body vibration or segmented vibration to specific body parts. The whole-body vibration may be due to transmission through a supporting structure such as the seat of a helicopter, and the vibration to specific body parts may be due to work processes such as compacting, drilling, and chain-saw operations. Human tolerance of whole-body vibration is found to be lowest in the 4–8 Hz frequency range. The segmental vibration is found to cause localized stress injuries to different body parts at different frequencies, as indicated in Fig. 9.2. In addition, the following effects have been observed at different frequencies [9.7]: motion sickness (0.1–1 Hz), blurring vision (2–20 Hz), speech disturbance (1–20 Hz), interference with tasks (0.5–20 Hz), and after-fatigue (0.2–15 Hz).

The acceptable vibration levels for laboratories that maintain reference standards are also shown in Fig. 9.1.

Head (25 Hz)
Eyeballs (30–60 Hz)

Chest wall (60 Hz)

Spine
(10–12 Hz)

Arm (16–30 Hz)

Pelvic mass
Buttocks
(4–8 Hz)

Hand
(50–150 Hz)

Legs (2–20 Hz)

FIGURE 9.2   Vibration frequency sensitivity of different parts of
the human body.

Helicopter Seat Vibration Reduction

EXAMPLE   9.1

The seat of a helicopter, with the pilot, weighs 1,000 N and is found to have a static deflection of 10 mm under self weight. The vibration of the rotor is transmitted to the base of the seat as harmonic motion with frequency 4 Hz and amplitude 0.2 mm.

a.   What is the level of vibration felt by the pilot?

b.   How can the seat be redesigned to reduce the effect of vibration?

Solution:

a.   By modeling the seat as an undamped single-degree-of-freedom system, we can compute the following:

$$\text{Mass} = m = 1,000/9.81 = 101.9368 \text{ kg}$$

$$\text{Stiffness} = k = \frac{W}{\delta_{\text{st}}} = \frac{1,000}{0.01} = 10^5 \text{ N/m}$$

$$\text{Natural frequency} = \omega_n = \sqrt{\frac{k}{m}} = \sqrt{\frac{10^5}{101.9368}} = 31.3209 \text{ rad/s} = 4.9849 \text{ Hz}$$

$$\text{Frequency ratio} = r = \frac{\omega}{\omega_n} = \frac{4.9849}{4.0} = 1.2462$$

Since the seat is subject to harmonic base excitation, the amplitude of vibration felt by the pilot (mass of the seat) is given by Eq. (3.68) with $\zeta = 0$:

$$X = \pm \frac{Y}{1 - r^2} \tag{E.1}$$

where Y is the amplitude of base displacement. Equation (E.1) yields

$$X = \frac{0.2}{1 - 1.2462^2} = 0.3616 \text{ mm}$$

The amplitudes of velocity and acceleration felt by the pilot are given by $\omega X = 2\pi f X = 2(\pi)(5)(0.3616) = 9.0887$ mm/s, and $\omega^2 X = (2\pi f)^2 X = 228.4074$ mm/s^2 = 0.2284 m/s^2. Corresponding to the frequency 4 Hz, Fig. 9.1 shows that the amplitude of motion of 0.3616 mm may not cause much discomfort. However, the velocity and acceleration levels at the same frequency (4 Hz) are not acceptable for a comfortable ride.

b. To bring the vibration level to an acceptable level, let us try to bring the acceleration felt by the pilot from the level 0.2284 m/s^2 to 0.01 m/s^2. Using $a_{max} = 10$ mm/s^2 = $-(2\pi f)^2 X = -(8\pi)^2 X$, we obtain $X = 0.01583$ mm. This leads to

$$\frac{X}{Y} = \frac{0.01583}{0.2} = \pm \frac{1}{1 - r^2} \qquad \text{or} \qquad r = 3.6923$$

This gives the new natural frequency of the seat as

$$\omega_n = \frac{\omega}{3.6923} = \frac{8\pi}{3.6923} = 6.8068 \text{ rad/s}$$

Using the relation $\omega_n = \sqrt{k/m}$ with $m = 101.9368$ kg, the new stiffness is given by $k = 4{,}722.9837$ N/m. This implies that the stiffness of the seat is to be reduced from 10^5 N/m to 4,722.9837 N/m. This can be accomplished by using a softer material for the seat or by using a different spring design. Alternatively, the desired acceleration level can be achieved by increasing the mass of the seat. However, this solution is not usually acceptable, as it increases the weight of the helicopter.

■

9.3   Reduction of Vibration at the Source

The first thing to be explored to control vibrations is to try to alter the source so that it produces less vibration. This method may not always be feasible. Some examples of the sources of vibration that cannot be altered are earthquake excitation, atmospheric turbulence, road roughness, and engine combustion instability. On the other hand, certain sources such as unbalance in rotating or reciprocating machines can be altered to reduce the vibrations. This can be achieved, usually, by using either internal balancing or an increase in the precision of machine elements. The use of close tolerances and better surface finish for machine parts (which have relative motion with respect to one another) makes the machine less susceptible to vibration. Of course, there may be economic and manufacturing constraints on the degree of balancing that can be achieved or the precision with which the machine parts can be made. We shall consider the analysis of rotating and reciprocating machines in the presence of unbalance as well as the means of controlling the vibrations that result from unbalanced forces.

9.4   Balancing of Rotating Machines

The presence of an eccentric or unbalanced mass in a rotating disc causes vibration, which may be acceptable up to a certain level. If the vibration caused by an unbalanced mass is not acceptable, it can be eliminated either by removing the eccentric mass or by adding an equal mass in such a position that it cancels the effect of the unbalance. In order to use this procedure, we need to determine the amount and location of the eccentric mass experimentally. The unbalance in practical machines can be attributed to such irregularities as machining errors and variations in sizes of bolts, nuts, rivets, and welds. In this section, we shall consider two types of balancing: *single-plane* or *static balancing* and *two-plane* or *dynamic balancing* [9.9, 9.10].

9.4.1
Single-Plane
Balancing

Consider a machine element in the form of a thin circular disc, such as a fan, flywheel, gear, and a grinding wheel mounted on a shaft. When the center of mass is displaced from the axis of rotation due to manufacturing errors, the machine element is said to be statically unbalanced. To determine whether a disc is balanced or not, mount the shaft on two low-friction bearings, as shown in Fig. 9.3(a). Rotate the disc and permit it to come to rest. Mark the lowest point on the circumference of the disc with chalk. Repeat the process several times, each time marking the lowest point on the disc with chalk. If the disc is balanced, the chalk marks will be scattered randomly all over the circumference. On the other hand, if the disc is unbalanced, all the chalk marks will coincide.

The unbalance detected by this procedure is known as *static unbalance*. The static unbalance can be corrected by removing (drilling) metal at the chalk mark or by adding a weight at $180°$ from the chalk mark. Since the magnitude of unbalance is not known, the amount of material to be removed or added must be determined by trial and error. This procedure is called "single-plane balancing," since all the mass lies practically in a single plane. The amount of unbalance can be found by rotating the disc at a known speed ω and measuring the reactions at the two bearings (see Fig. 9.3(b)). If an unbalanced mass m is located at a radius r of the disc, the centrifugal force will be $mr\omega^2$. Thus the measured bearing reactions F_1 and F_2 give m and r:

$$F_1 = \frac{a_2}{l} mr\omega^2, \qquad F_2 = \frac{a_1}{l} mr\omega^2 \qquad (9.8)$$

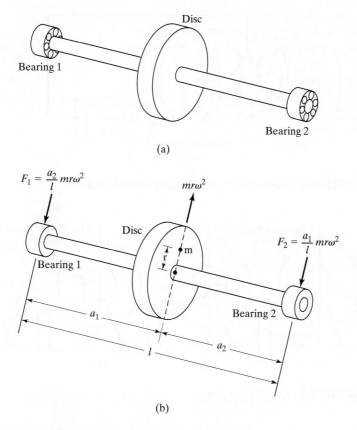

FIGURE 9.3   Single-plane balancing of a disc.

Another procedure for single-plane balancing, using a vibration analyzer, is illustrated in Fig. 9.4. Here, a grinding wheel (disc) is attached to a rotating shaft that has bearing at A and is driven by an electric motor rotating at an angular velocity ω.

Before starting the procedure, *reference marks*, also known as *phase marks*, are placed both on the rotor (wheel) and the stator, as shown in Fig. 9.5(a). A vibration pickup is placed in contact with the bearing, as shown in Fig. 9.4, and the vibration analyzer is set to a frequency corresponding to the angular velocity of the grinding wheel. The vibration signal (the displacement amplitude) produced by the unbalance can be read from the indicating meter of the vibration analyzer. A stroboscopic light is fired by the vibration analyzer at the frequency of the rotating wheel. When the rotor rotates at speed ω, the phase mark on the rotor appears stationary under the stroboscopic light but is positioned at an angle θ from the mark on the stator, as shown in Fig. 9.5(b), due to phase lag in the response. Both the angle θ and the amplitude A_u (read from the vibration analyzer) caused by the original unbalance are noted. The rotor is then stopped, and a known trial weight W is attached to the rotor, as shown in Fig. 9.5(b). When the rotor runs at speed ω, the new angular position

FIGURE 9.4   Single-plane balancing using vibration analyzer.

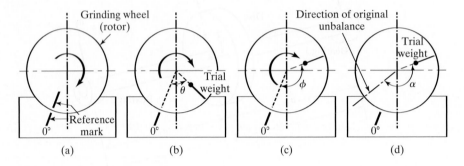

FIGURE 9.5   Use of phase marks.

of the rotor phase mark ϕ and the vibration amplitude A_{u+w}, caused by the combined unbalance of rotor and trial weight, are noted (see Fig. 9.5(c)).[1]

Now we construct a vector diagram to find the magnitude and location of the correction mass for balancing the wheel. The original unbalance vector $\vec{A}_u$ is drawn in an arbitrary direction, with its length equal to A_u, as shown in Fig. 9.6. Then the combined unbalance vector is drawn as $\vec{A}_{u+w}$ at an angle $\phi - \theta$ from the direction of $\vec{A}_u$ with a length of A_{u+w}. The difference vector $\vec{A}_w = \vec{A}_{u+w} - \vec{A}_u$ in Fig. 9.6 then represents the unbalance vector due to the trial weight W. The magnitude of $\vec{A}_w$ can be computed using the law of cosines:

$$A_w = [A_u^2 + A_{u+w}^2 - 2A_u A_{u+w} \cos(\phi - \theta)]^{1/2} \qquad (9.9)$$

Since the magnitude of the trial weight W and its direction relative to the original unbalance (α in Fig. 9.6) are known, the original unbalance itself must be at an angle α away

---

[1]Note that if the trial weight is placed in a position that shifts the net unbalance in a clockwise direction, the stationary position of the phase mark will be shifted by exactly the same amount in the counterclockwise direction, and vice versa.

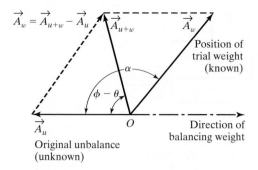

FIGURE 9.6   Unbalance due to trial weight W.

from the position of the trial weight, as shown in Fig. 9.5(d). The angle α can be obtained from the law of cosines:

$$\alpha = \cos^{-1}\left[\frac{A_u^2 + A_w^2 - A_{u+w}^2}{2A_uA_w}\right] \tag{9.10}$$

The magnitude of the original unbalance is $W_0 = (A_u/A_w) \cdot W$, located at the same radial distance from the rotation axis of the rotor as the weight W. Once the location and magnitude of the original unbalance are known, correction weight can be added to balance the wheel properly.

9.4.2
Two-Plane
Balancing

The single-plane balancing procedure can be used for balancing in one plane—that is, for rotors of the rigid disc type. If the rotor is an elongated rigid body, as shown in Fig. 9.7, the unbalance can be anywhere along the length of the rotor. In this case, the rotor can be balanced by adding balancing weights in any two planes [9.10, 9.11]. For convenience, the two planes are usually chosen as the end planes of the rotor (shown by dashed lines in Fig. 9.7).

To see that any unbalanced mass in the rotor can be replaced by two equivalent unbalanced masses (in any two planes), consider a rotor with an unbalanced mass m at a distance

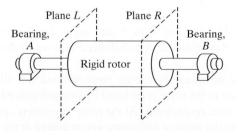

FIGURE 9.7   Two-plane balancing of a rotor.

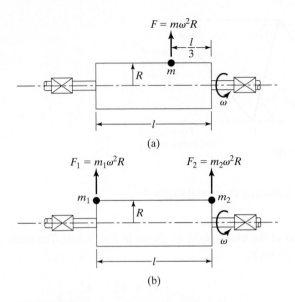

(a)

(b)

FIGURE 9.8   Representation of an unbalanced mass
as two equivalent unbalanced masses.

$l/3$ from the right end, as shown in Fig. 9.8(a). When the rotor rotates at a speed of ω, the
force due to the unbalance will be $F = m\omega^2 R$, where R is the radius of the rotor. The unbalanced mass m can be replaced by two masses m_1 and m_2, located at the ends of the rotor, as
shown in Fig. 9.8(b). The forces exerted on the rotor by these masses are $F_1 = m_1\omega^2 R$ and
$F_2 = m_2\omega^2 R$. For the equivalence of force in Figs. 9.8(a) and (b), we have

$$m\omega^2 R = m_1\omega^2 R + m_2\omega^2 R \qquad \text{or} \qquad m = m_1 + m_2 \qquad (9.11)$$

For the equivalence of moments in the two cases, we consider moments about the right end
so that

$$m\omega^2 R\frac{l}{3} = m_1\omega^2 Rl \qquad \text{or} \qquad m = 3m_1 \qquad (9.12)$$

Equations (9.11) and (9.12) give $m_1 = m/3$ and $m_2 = 2m/3$. Thus any unbalanced
mass can be replaced by two equivalent unbalanced masses in the end planes of the
rotor.

We now consider the two-plane balancing procedure using a vibration analyzer. In
Fig. 9.9, the total unbalance in the rotor is replaced by two unbalanced weights U_L and U_R
in the left and the right planes, respectively. At the rotor's operating speed ω, the vibration
amplitude and phase due to the original unbalance are measured at the two bearings A and
B, and the results are recorded as vectors $\vec{V}_A$ and $\vec{V}_B$. The magnitude of the vibration vector
is taken as the vibration amplitude, while the direction of the vector is taken as the negative

FIGURE 9.9   Two-plane balancing.

of the phase angle observed under stroboscopic light with reference to the stator reference line. The measured vectors $\vec{V}_A$ and $\vec{V}_B$ can be expressed as

$$\vec{V}_A = \vec{A}_{AL}\vec{U}_L + \vec{A}_{AR}\vec{U}_R \tag{9.13}$$

$$\vec{V}_B = \vec{A}_{BL}\vec{U}_L + \vec{A}_{BR}\vec{U}_R \tag{9.14}$$

where $\vec{A}_{ij}$ can be considered as a vector, reflecting the effect of the unbalance in plane j $(j = L, R)$ on the vibration at bearing i $(i = A, B)$. Note that $\vec{U}_L$, $\vec{U}_R$, and all the vectors $\vec{A}_{ij}$ are unknown in Eqs. (9.13) and (9.14).

As in the case of single-plane balancing, we add known trial weights and take measurements to obtain information about the unbalanced masses. We first add a known weight $\vec{W}_L$ in the left plane at a known angular position and measure the displacement and phase of vibration at the two bearings while the rotor is rotating at speed ω. We denote these measured vibrations as vectors as

$$\vec{V}_A' = \vec{A}_{AL}(\vec{U}_L + \vec{W}_L) + \vec{A}_{AR}\vec{U}_R \tag{9.15}$$

$$\vec{V}_B' = \vec{A}_{BL}(\vec{U}_L + \vec{W}_L) + \vec{A}_{BR}\vec{U}_R \tag{9.16}$$

By subtracting Eqs. (9.13) and (9.14) from Eqs. (9.15) and (9.16), respectively, and solving, we obtain[2]

$$\vec{A}_{AL} = \frac{\vec{V}_A' - \vec{V}_A}{\vec{W}_L} \tag{9.17}$$

---

[2]It can be seen that complex subtraction, division, and multiplication are often used in the computation of the balancing weights. If

$$\vec{A} = a\underline{/\theta_A} \qquad \text{and} \qquad \vec{B} = b\underline{/\theta_B}$$

we can rewrite $\vec{A}$ and $\vec{B}$ as $\vec{A} = a_1 + ia_2$ and $\vec{B} = b_1 + ib_2$, where $a_1 = a\cos\theta_A$, $a_2 = a\sin\theta_A$, $b_1 = b\cos\theta_B$, and $b_2 = b\sin\theta_B$. Then the formulas for complex subtraction, division, and multiplication are [9.12]:

$$\vec{A} - \vec{B} = (a_1 - b_1) + i(a_2 - b_2)$$

$$\frac{\vec{A}}{\vec{B}} = \frac{(a_1 b_1 + a_2 b_2) + i(a_2 b_1 - a_1 b_2)}{(b_1^2 + b_2^2)}$$

$$\vec{A} \cdot \vec{B} = (a_1 b_1 - a_2 b_2) + i(a_2 b_1 + a_1 b_2)$$

$$\vec{A}_{BL} = \frac{\vec{V}_B' - \vec{V}_B}{\vec{W}_L} \tag{9.18}$$

We then remove $\vec{W}_L$ and add a known weight $\vec{W}_R$ in the right plane at a known angular position and measure the resulting vibrations while the rotor is running at speed ω. The measured vibrations can be denoted as vectors:

$$\vec{V}_A'' = \vec{A}_{AR}(\vec{U}_R + \vec{W}_R) + \vec{A}_{AL}\vec{U}_L \tag{9.19}$$

$$\vec{V}_B'' = \vec{A}_{BR}(\vec{U}_R + \vec{W}_R) + \vec{A}_{BL}\vec{U}_L \tag{9.20}$$

As before, we subtract Eqs. (9.13) and (9.14) from Eqs. (9.19) and (9.20), respectively, to find

$$\vec{A}_{AR} = \frac{\vec{V}_A'' - \vec{V}_A}{\vec{W}_R} \tag{9.21}$$

$$\vec{A}_{BR} = \frac{\vec{V}_B'' - \vec{V}_B}{\vec{W}_R} \tag{9.22}$$

Once the vector operators $\vec{A}_{ij}$ are known, Eqs. (9.13) and (9.14) can be solved to find the unbalance vectors $\vec{U}_L$ and $\vec{U}_R$:

$$\vec{U}_L = \frac{\vec{A}_{BR}\vec{V}_A - \vec{A}_{AR}\vec{V}_B}{\vec{A}_{BR}\vec{A}_{AL} - \vec{A}_{AR}\vec{A}_{BL}} \tag{9.23}$$

$$\vec{U}_R = \frac{\vec{A}_{BL}\vec{V}_A - \vec{A}_{AL}\vec{V}_B}{\vec{A}_{BL}\vec{A}_{AR} - \vec{A}_{AL}\vec{A}_{BR}} \tag{9.24}$$

The rotor can now be balanced by adding equal and opposite balancing weights in each plane. The balancing weights in the left and right planes can be denoted vectorially as $\vec{B}_L = -\vec{U}_L$ and $\vec{B}_R = -\vec{U}_R$. It can be seen that the two-plane balancing procedure is a straightforward extension of the single-plane balancing procedure. Although high-speed rotors are balanced during manufacture, usually it becomes necessary to rebalance them in the field due to slight unbalances introduced due to creep, high-temperature operation, and the like. Figure 9.10 shows a practical example of two-plane balancing.

FIGURE 9.10 Two-plane balancing. (Courtesy of Bruel and Kjaer Instruments, Inc., Marlborough, MA.)

EXAMPLE 9.2

Two-Plane Balancing of Turbine Rotor

In the two-plane balancing of a turbine rotor, the data obtained from measurement of the original unbalance, the right-plane trial mass, and the left-plane trial mass are shown below. The displacement amplitudes are in mm. Determine the size and location of the balance masses required.

	Vibration (Displacement) Amplitude		Phase Angle	
Condition	At Bearing A	At Bearing B	At Bearing A	At Bearing B
Original unbalance	0.2	0.16	60°	205°
$W_L = 300$ g added at 270° from reference mark	0.15	0.1	125°	230°
$W_R = 340$ g added at 180° from reference mark	0.15	0.25	35°	160°

784   CHAPTER 9   VIBRATION CONTROL

Solution: The given data can be expressed in vector notation as

$$\vec{V_A} = 0.2\underline{/60°} = 0.1000 + i0.1732$$

$$\vec{V_B} = 0.16\underline{/205°} = -0.1450 - i0.0676$$

$$\vec{V_A'} = 0.15\underline{/125°} = -0.0860 + i0.1229$$

$$\vec{V_B'} = 0.1\underline{/230°} = -0.0643 - i0.0766$$

$$\vec{V_A''} = 0.15\underline{/35°} = 0.1229 + i0.0860$$

$$\vec{V_B''} = 0.25\underline{/160°} = -0.2349 + i0.0855$$

$$\vec{W_L} = 0.3000\underline{/270°} = 0.0000 - i0.3000$$

$$\vec{W_R} = 0.3400\underline{/180°} = -0.3400 + i0.0000$$

Equations (9.17) and (9.18) give

$$\vec{A_{AL}} = \frac{\vec{V_A'} - \vec{V_A}}{\vec{W_L}} = \frac{-0.1860 - i0.0503}{0.0000 - i0.3000} = 0.1677 - i0.6200$$

$$\vec{A_{BL}} = \frac{\vec{V_B'} - \vec{V_B}}{\vec{W_L}} = \frac{0.0807 - i0.009}{0.0000 - i0.3000} = 0.0300 + i0.2690$$

The use of Eqs. (9.21) and (9.22) leads to

$$\vec{A_{AR}} = \frac{\vec{V_A''} - \vec{V_A}}{\vec{W_R}} = \frac{0.0229 - i0.0872}{-0.3400 + i0.0000} = -0.0674 + i0.2565$$

$$\vec{A_{BR}} = \frac{\vec{V_B''} - \vec{V_B}}{\vec{W_R}} = \frac{-0.0899 + i0.1531}{-0.3400 + i0.0000} = 0.2644 - i0.4503$$

The unbalance masses can be determined from Eqs. (9.23) and (9.24):

$$\vec{U_L} = \frac{(0.1045 + i0.0008) - (0.0271 - i0.0327)}{(-0.2347 - i0.2397) - (-0.712 - i0.0104)}$$

$$= -0.2565 + i0.1546$$

$$\vec{U_R} = \frac{(-0.0436 + i0.0321) - (0.0663 + i0.0785)}{(-0.0712 + i0.0104) - (-0.2347 - i0.2397)}$$

$$= -0.0875 - i0.1612$$

Thus the required balance masses are given by

$$\vec{B_L} = -\vec{U_L} = (0.2565 - i0.1546) = 0.2994\underline{/-31.0772°}$$

$$\vec{B_R} = -\vec{U_R} = (0.0875 + i0.1612) = 0.1834\underline{/61.5242°}$$

This shows that the addition of a 0.2994-kg mass in the left plane at $-31.0772°$ and a 0.1834-kg mass in the right plane at $61.5242°$ from the reference position will balance the turbine rotor. It is implied that the balance masses are added at the same radial distance as the trial masses. If a balance mass is to be located at a different radial position, the required balance mass is to be modified in inverse proportion to the radial distance from the axis of rotation.

■

9.5 Whirling of Rotating Shafts

In the previous section, the rotor system—the shaft as well as the rotating body—was assumed to be rigid. However, in many practical applications, such as turbines, compressors, electric motors, and pumps, a heavy rotor is mounted on a lightweight, flexible shaft that is supported by bearings. There will be unbalance in all rotors due to manufacturing errors. These unbalances as well as other effects, such as the stiffness and damping of the shaft, gyroscopic effects, and fluid friction in bearings, will cause a shaft to bend in a complicated manner at certain rotational speeds, known as the *whirling*, *whipping*, or *critical speeds*. Whirling is defined as the rotation of the plane made by the line of centers of the bearings and the bent shaft. We consider the aspects of modeling the rotor system, critical speeds, response of the system, and stability in this section [9.13–9.14].

9.5.1 Equations of Motion

Consider a shaft supported by two bearings and carrying a rotor or disc of mass m at the middle, as shown in Fig. 9.11. We shall assume that the rotor is subjected to a steady-state excitation due to mass unbalance. The forces acting on the rotor are the inertia force due to the acceleration of the mass center, the spring force due to the elasticity of the shaft, and the external and internal damping forces.[3]

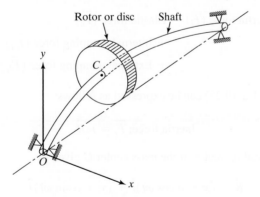

FIGURE 9.11   Shaft carrying a rotor.

---

[3]Any rotating system responds in two different ways to damping or friction forces, depending upon whether the forces rotate with the shaft or not. When the positions at which the forces act remain fixed in space, as in the case of damping forces (which cause energy losses) in the bearing support structure, the damping is called *stationary* or *external damping*. On the other hand, if the positions at which they act rotate with the shaft in space, as in the case of internal friction of the shaft material, the damping is called *rotary* or *internal damping*.

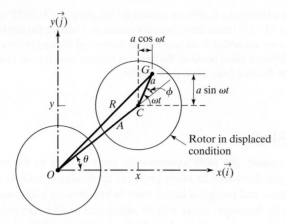

FIGURE 9.12   Rotor with eccentricity.

Let O denote the equilibrium position of the shaft when balanced perfectly, as shown in Fig. 9.12. The shaft (line CG) is assumed to rotate with a constant angular velocity ω. During rotation, the rotor deflects radially by a distance $A = OC$ (in steady state). The rotor (disc) is assumed to have an eccentricity a so that its mass center (center of gravity) G is at a distance a from the geometric center C. We use a fixed coordinate system (x and y fixed to the earth) with O as the origin for describing the motion of the system. The angular velocity of the line OC, $\dot{\theta} = d\theta/dt$, is known as the *whirling speed* and, in general, is not equal to ω. The equations of motion of the rotor (mass m) can be written as

$$\text{Inertia force } (\vec{F_i}) = \text{Elastic force } (\vec{F_e})$$
$$+ \text{ Internal damping force } (\vec{F_{di}})$$
$$+ \text{ External damping force } (\vec{F_{de}}) \qquad (9.25)$$

The various forces in Eq. (9.25) can be expressed as follows:

$$\text{Inertia force: } \vec{F_i} = m\ddot{\vec{R}} \qquad (9.26)$$

where $\vec{R}$ denotes the radius vector of the mass center G given by

$$\vec{R} = (x + a\cos\omega t)\vec{i} + (y + a\sin\omega t)\vec{j} \qquad (9.27)$$

with x and y representing the coordinates of the geometric center C and with $\vec{i}$ and $\vec{j}$ denoting the unit vectors along the x- and y- coordinates, respectively. Equations (9.26) and (9.27) lead to

$$\vec{F_i} = m[(\ddot{x} - a\omega^2\cos\omega t)\vec{i} + (\ddot{y} - a\omega^2\sin\omega t)\vec{j}] \qquad (9.28)$$

$$\text{Elastic force: } \vec{F_e} = -k(x\vec{i} + y\vec{j}) \qquad (9.29)$$

where k is the stiffness of the shaft.

$$\text{Internal damping force: } \vec{F}_{di} = -c_i\left[(\dot{x} + \omega y)\vec{i} + (\dot{y} + \omega x)\vec{j}\right] \quad (9.30)$$

where c_i is the internal or rotary damping coefficient:

$$\text{External damping force: } \vec{F}_{de} = -c(\dot{x}\vec{i} + \dot{y}\vec{j}) \quad (9.31)$$

where c is the external damping coefficient. By substituting Eqs. (9.28) to (9.31) into Eq. (9.25), we obtain the equations of motion in scalar form:

$$m\ddot{x} + (c_i + c)\dot{x} + kx - c_i\omega y = m\omega^2 a \cos \omega t \quad (9.32)$$

$$m\ddot{y} + (c_i + c)\dot{y} + ky - c_i\omega x = m\omega^2 a \sin \omega t \quad (9.33)$$

These equations of motion, which describe the lateral vibration of the rotor, are coupled and are dependent on the speed of the steady-state rotation of the shaft, ω. By defining a complex quantity w as

$$w = x + iy \quad (9.34)$$

where $i = (-1)^{1/2}$, and by adding Eq. (9.32) to Eq. (9.33) multiplied by i, we obtain a single equation of motion:

$$m\ddot{w} + (c_i + c)\dot{w} + kw - i\omega c_i w = m\omega^2 a\, e^{i\omega t} \quad (9.35)$$

9.5.2
Critical Speeds

A critical speed is said to exist when the frequency of the rotation of a shaft equals one of the natural frequencies of the shaft. The undamped natural frequency of the rotor system can be obtained by solving Eq. (9.32), (9.33), or (9.35), retaining only the homogeneous part with $c_i = c = 0$. This gives the natural frequency of the system (or critical speed of the undamped system):

$$\omega_n = \left(\frac{k}{m}\right)^{1/2} \quad (9.36)$$

When the rotational speed is equal to this critical speed, the rotor undergoes large deflections, and the force transmitted to the bearings can cause bearing failures. A rapid transition of the rotating shaft through a critical speed is expected to limit the whirl amplitudes, while a slow transition through the critical speed aids the development of large amplitudes. The behavior of the rotor during acceleration and deceleration through critical speeds is investigated in [9.15]. A FORTRAN computer program for calculating the critical speeds of rotating shafts is given in [9.16].

9.5.3
Response of the System

To determine the response of the rotor, we assume the excitation to be a harmonic force due to the unbalance of the rotor. In addition, we assume the internal damping to be negligible ($c_i = 0$). Then, we can solve Eqs. (9.32) and (9.33) (or equivalently, Eq. (9.35)) and find the rotor's dynamic whirl amplitudes resulting from the mass unbalance. With $c_i = 0$, Eq. (9.35) reduces to

$$m\ddot{w} + c\dot{w} + kw = m\omega^2 a e^{i\omega t} \tag{9.37}$$

The solution of Eq. (9.37) can be expressed as

$$w(t) = Ce^{-(\alpha t + \beta)} + Ae^{i(\omega t - \phi)} \tag{9.38}$$

where C, β, A, and ϕ are constants. Note that the first term on the right-hand side of Eq. (9.38) contains a decaying exponential term representing a transient solution and the second term denotes a steady-state circular motion (whirl). By substituting the steady-state part of Eq. (9.38) into Eq. (9.37), we can find the amplitude of the circular motion (whirl) as

$$A = \frac{m\omega^2 a}{[(k - m\omega^2)^2 + \omega^2 c^2]^{1/2}} = \frac{ar^2}{[(1 - r^2)^2 + (2\zeta r)^2]^{1/2}} \tag{9.39}$$

and the phase angle as

$$\phi = \tan^{-1}\left(\frac{c\omega}{k - m\omega^2}\right) = \tan^{-1}\left(\frac{2\zeta r}{1 - r^2}\right) \tag{9.40}$$

where

$$r = \frac{\omega}{\omega_n}, \qquad \omega_n = \sqrt{\frac{k}{m}}, \qquad \text{and} \qquad \zeta = \frac{c}{2\sqrt{km}}$$

By differentiating Eq. (9.39) with respect to ω and setting the result equal to zero, we can find the rotational speed ω at which the whirl amplitude becomes a maximum:

$$\omega \approx \frac{\omega_n}{\sqrt{1 - 2\zeta^2}} \tag{9.41}$$

where ω_n is given by Eq. (9.36). It can be seen that the critical speed corresponds exactly to the natural frequency ω_n only when the damping (c) is zero. Furthermore, Eq. (9.41) shows that the presence of damping, in general, increases the value of the critical speed compared to ω_n. A plot of Eqs. (9.39) and (9.40) is shown in Fig. 9.13 [9.14]. Since the forcing function is proportional to ω^2, we normally expect the vibration amplitude to increase with the speed ω. However, the actual amplitude appears as shown in Fig. 9.13. From Eq. (9.39), we note that the amplitude of circular whirl A at low speeds is determined

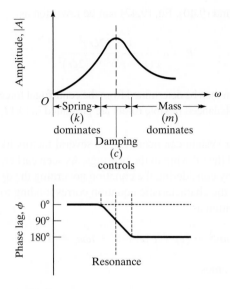

FIGURE 9.13 Plots of Eqs. (9.39) and (9.40).

by the spring constant k, since the other two terms, $m\omega^2$ and $c^2\omega^2$, are small. Also, the value of the phase angle ϕ can be seen to be 0° from Eq. (9.40) for small values of ω. As ω increases, the amplitude of the response reaches a peak, since resonance occurs at $k - m\omega^2 = 0$. Around resonance, the response is essentially limited by the damping term. The phase lag is 90° at resonance. As the speed ω increases beyond ω_n, the response is dominated by the mass term $m^2\omega^4$ in Eq. (9.39). Since this term is 180° out of phase with the unbalanced force, the shaft rotates in a direction opposite to that of the unbalanced force, hence the response of the shaft will be limited.

Notes

1. Equation (9.38) implicitly assumes a condition of forward synchronous whirl under steady state (that is, $\dot{\theta} = \omega$). As a general case, if the steady-state solution of Eq. (9.37) is assumed as $w(t) = Ae^{i(\gamma t - \phi)}$, the solution can be obtained as $\gamma = \pm\omega$, with $\gamma = +\omega$ representing the forward synchronous whirl and $\gamma = -\omega$ denoting a backward synchronous whirl. For simple rotors, such as the one shown in Fig. 9.11, only a forward synchronous whirl occurs in practice.

2. To determine the bearing reactions, we first find the deflection of the mass center of the disc from the bearing axis, R in Fig. 9.12, as

$$R^2 = A^2 + a^2 + 2Aa \cos \phi \qquad (9.42)$$

In view of Eqs. (9.39) and (9.40), Eq. (9.42) can be rewritten as

$$R = a \left[\frac{1 + (2\zeta r)^2}{(1 - r^2)^2 + (2\zeta r)^2} \right]^{1/2} \tag{9.43}$$

The bearing reactions can then be determined from the centrifugal force, $m\omega^2 R$. The vibration and balancing of unbalanced flexible rotors are presented in [9.17, 9.18].

9.5.4
Stability
Analysis

Instability in a flexible rotor system can occur due to several factors like internal friction, eccentricity of the rotor, and the oil whip in the bearings. As seen earlier, the stability of the system can be investigated by considering the equation governing the dynamics of the system. Assuming $w(t) = e^{st}$, the characteristic equation corresponding to the homogeneous part of Eq. (9.35) can be written as

$$ms^2 + (c_i + c)s + k - i\omega c_i = 0 \tag{9.44}$$

With $s = i\lambda$, Eq. (9.44) becomes

$$-m\lambda^2 + (c_i + c)i\lambda + k - i\omega c_i = 0 \tag{9.45}$$

This equation is a particular case of the more general equation

$$(p_2 + iq_2)\lambda^2 + (p_1 + iq_1)\lambda + (p_0 + iq_0) = 0 \tag{9.46}$$

A necessary and sufficient condition for the system governed by Eq. (9.46) to be stable, according to the Routh-Hurwitz criterion, is that the following inequalities are satisfied:

$$- \begin{vmatrix} p_2 & p_1 \\ q_2 & q_1 \end{vmatrix} > 0 \tag{9.47}$$

and

$$\begin{vmatrix} p_2 & p_1 & p_0 & 0 \\ q_2 & q_1 & q_0 & 0 \\ 0 & p_2 & p_1 & p_0 \\ 0 & q_2 & q_1 & q_0 \end{vmatrix} > 0 \tag{9.48}$$

Noting that $p_2 = -m$, $q_2 = 0$, $p_1 = 0$, $q_1 = c_i + c$, $p_0 = k$, and $q_0 = -\omega c_i$, from Eq. (9.45), the application of Eqs. (9.47) and (9.48) leads to

$$m(c_i + c) > 0 \tag{9.49}$$

and

$$km(c_i + c)^2 - m^2(\omega^2 c_i^2) > 0 \tag{9.50}$$

Equation (9.49) is automatically satisfied, while Eq. (9.50) yields the condition

$$\sqrt{\frac{k}{m}}\left(1 + \frac{c}{c_i}\right) - \omega > 0 \tag{9.51}$$

This equation also shows that internal and external friction can cause instability at rotating speeds above the first critical speed of $\omega = \sqrt{\frac{k}{m}}$.

EXAMPLE   9.3

Whirl Amplitude of a Shaft Carrying an Unbalanced Rotor

A shaft, carrying a rotor of mass 50 kg and eccentricity 2 mm, rotates at 12,000 rpm. Determine (a) the steady-state whirl amplitude and (b) the maximum whirl amplitude during start-up conditions of the system. Assume the stiffness of the shaft as 40 MN/m and the external damping ratio as 0.1.

Solution: The forcing frequency of the rotor (rotational speed of the shaft) is given by

$$\omega = \frac{12,000 \times 2\pi}{60} = 400\pi = 1,256.6371 \text{ rad/s}$$

The natural frequency of the system can be determined as

$$\omega_n = \sqrt{\frac{k}{m}} = \sqrt{\frac{40 \times 10^6}{50}} = 894.4272 \text{ rad/s}$$

and the frequency ratio as

$$r = \frac{\omega}{\omega_n} = \frac{1,256.6371}{894.4272} = 1.4049$$

(a)   The steady-state amplitude is given by Eq. (9.39):

$$A = \frac{ar^2}{\sqrt{(1 - r^2)^2 + (2\zeta r)^2}} \tag{E.1}$$

$$= \frac{(0.002)(1.4049)^2}{\sqrt{(1 - 1.4049^2)^2 + (2 \times 0.1 \times 1.4049)^2}} = 3.8950 \text{ m} \times 10^{-3} \text{ m} = 3.8950 \text{ mm} \tag{E.2}$$

(b)   During start-up conditions, the frequency (speed) of the rotor, ω, passes through the natural frequency of the system. Thus, using $r = 1$ in Eq. (E.1), we obtain the whirl amplitude as

$$A|_{r=1} = \frac{a}{2\zeta} = \frac{0.002}{2(0.1)} = 0.01 \text{ m} = 10 \text{ mm}$$

■

9.6   Balancing of Reciprocating Engines

The essential moving elements of a reciprocating engine are the piston, the crank, and the connecting rod. Vibrations in reciprocating engines arise due to (1) periodic variations of the gas pressure in the cylinder and (2) inertia forces associated with the moving parts [9.19]. We shall now analyze a reciprocating engine and find the unbalanced forces caused by these factors.

**9.6.1
Unbalanced
Forces Due to
Fluctuations in
Gas Pressure**

Figure 9.14(a) is a schematic diagram of a cylinder of a reciprocating engine. The engine is driven by the expanding gas in the cylinder. The expanding gas exerts on the piston a pressure force F, which is transmitted to the crankshaft through the connecting rod. The reaction to the force F can be resolved into two components: one of magnitude $F/\cos\phi$, acting along the connecting rod, and the other of magnitude $F\tan\phi$, acting in a horizontal direction. The force $F/\cos\phi$ induces a torque M_t, which tends to rotate the crankshaft. (In Fig. 9.14(b), M_t acts about an axis perpendicular to the plane of the paper and passes through point Q.)

$$M_t = \left(\frac{F}{\cos\phi}\right) r \cos\theta \qquad (9.52)$$

(a)   (b)   (c)

FIGURE 9.14   Forces in a reciprocating engine.

For force equilibrium of the overall system, the forces at the bearings of the crankshaft will be F in the vertical direction and $F \tan \phi$ in the horizontal direction.

Thus the forces transmitted to the stationary parts of the engine are as follows:

1. Force F acting upward at the cylinder head
2. Force $F \tan \phi$ acting toward the right at the cylinder head
3. Force F acting downward at the crankshaft bearing Q
4. Force $F \tan \phi$ acting toward the left at the crankshaft bearing

These forces are shown in Fig. 9.14(c). Although the total resultant force is zero, there is a resultant torque $M_Q = Fh \tan \phi$ on the body of the engine, where h can be found from the geometry of the system:

$$h = \frac{r \cos \theta}{\sin \phi} \tag{9.53}$$

Thus the resultant torque is given by

$$M_Q = \frac{Fr \cos \theta}{\cos \phi} \tag{9.54}$$

As expected, M_t and M_Q given by Eqs. (9.52) and (9.54) can be seen to be identical, which indicates that the torque induced on the crankshaft due to the gas pressure on the piston is felt at the support of the engine. Since the magnitude of the gas force F varies with time, the torque M_Q also varies with time. The magnitude of force F changes from a maximum to a minimum at a frequency governed by the number of cylinders in the engine, the type of the operating cycle, and the rotating speed of the engine.

9.6.2 Unbalanced Forces Due to Inertia of the Moving Parts

Acceleration of the Piston.   Figure 9.15 shows the crank (of length r), the connecting rod (of length l), and the piston of a reciprocating engine. The crank is assumed to rotate in an anticlockwise direction at a constant angular speed of ω, as shown in Fig. 9.15. If we consider the origin of the x-axis (O) as the uppermost position of the piston, the displacement of the piston P corresponding to an angular displacement of the crank of $\theta = \omega t$ can be expressed as in Fig. 9.15. The displacement of the piston P corresponding to an angular displacement of the crank $\theta = \omega t$ from its topmost position (origin O) can be expressed as

$$x_p = r + l - r \cos \theta - l \cos \phi$$

$$= r + l - r \cos \omega t - l \sqrt{1 - \sin^2 \phi} \tag{9.55}$$

But

$$l \sin \phi = r \sin \theta = r \sin \omega t \tag{9.56}$$

FIGURE 9.15 Motions of crank, connecting rod, and piston.

and hence

$$\cos \phi = \left(1 - \frac{r^2}{l^2} \sin^2 \omega t \right)^{1/2} \tag{9.57}$$

By substituting Eq. (9.57) into Eq. (9.55), we obtain

$$x_p = r + l - r \cos \omega t - l\sqrt{1 - \frac{r^2}{l^2} \sin^2 \omega t} \tag{9.58}$$

Due to the presence of the term involving the square root, Eq. (9.58) is not very convenient in further calculation. Equation (9.58) can be simplified by noting that, in general, $r/l < \frac{1}{4}$ and by using the expansion relation

$$\sqrt{1 - \varepsilon} \simeq 1 - \frac{\varepsilon}{2} \tag{9.59}$$

Hence Eq. (9.58) can be approximated as

$$x_p \simeq r(1 - \cos \omega t) + \frac{r^2}{2l} \sin^2 \omega t \qquad (9.60)$$

or, equivalently,

$$x_p = r\left(1 + \frac{r}{2l}\right) - r\left(\cos \omega t + \frac{r}{4l} \cos 2\omega t\right) \qquad (9.61)$$

Equation (9.61) can be differentiated with respect to time to obtain expressions for the velocity and the acceleration of the piston:

$$\dot{x}_p = r\omega\left(\sin \omega t + \frac{r}{2l} \sin 2\omega t\right) \qquad (9.62)$$

$$\ddot{x}_p = r\omega^2\left(\cos \omega t + \frac{r}{l} \cos 2\omega t\right) \qquad (9.63)$$

Acceleration of the Crankpin.   With respect to the xy-coordinate axes shown in Fig. 9.15, the vertical and horizontal displacements of the crankpin C are given by

$$x_c = OA + AB = l + r(1 - \cos \omega t) \qquad (9.64)$$

$$y_c = CB = r \sin \omega t \qquad (9.65)$$

Differentiation of Eqs. (9.64) and (9.65) with respect to time gives the velocity and acceleration components of the crankpin as

$$\dot{x}_c = r\omega \sin \omega t \qquad (9.66)$$

$$\dot{y}_c = r\omega \cos \omega t \qquad (9.67)$$

$$\ddot{x}_c = r\omega^2 \cos \omega t \qquad (9.68)$$

$$\ddot{y}_c = -r\omega^2 \sin \omega t \qquad (9.69)$$

Inertia Forces.   Although the mass of the connecting rod is distributed throughout its length, it is generally idealized as a massless link with two masses concentrated at its ends—the piston end and the crankpin end. If m_p and m_c denote the total mass of the piston and of the crankpin (including the concentrated mass of the connecting rod), respectively, the vertical component of the inertia force (F_x) for one cylinder is given by

$$F_x = m_p \ddot{x}_p + m_c \ddot{x}_c \qquad (9.70)$$

By substituting Eqs. (9.63) and (9.68) for the accelerations of P and C, Eq. (9.70) becomes

$$F_x = (m_p + m_c)r\omega^2 \cos \omega t + m_p \frac{r^2\omega^2}{l} \cos 2\omega t \qquad (9.71)$$

It can be observed that the vertical component of the inertia force consists of two parts. One part, known as the *primary part*, has a frequency equal to the rotational frequency of the crank ω. The other part, known as the *secondary part*, has a frequency equal to twice the rotational frequency of the crank.

Similarly, the horizontal component of inertia force for a cylinder can be obtained

$$F_y = m_p \ddot{y}_p + m_c \ddot{y}_c \qquad (9.72)$$

where $\ddot{y}_p = 0$ and $\ddot{y}_c$ is given by Eq. (9.69). Thus

$$F_y = -m_c r\omega^2 \sin \omega t \qquad (9.73)$$

The horizontal component of the inertia force can be observed to have only a primary part.

9.6.3 Balancing of Reciprocating Engines

The unbalanced or inertia forces on a single cylinder are given by Eqs. (9.71) and (9.73). In these equations, m_p and m_c represent the equivalent reciprocating and rotating masses, respectively. The mass m_p is always positive, but m_c can be made zero by counterbalancing the crank. It is therefore possible to reduce the horizontal inertia force F_y to zero, but the vertical unbalanced force always exists. Thus a single-cylinder engine is inherently unbalanced.

In a multicylinder engine, it is possible to balance some or all of the inertia forces and torques by proper arrangement of the cranks. Figure 9.16(a) shows the general arrangement of an N-cylinder engine (only six cylinders, $N = 6$, are shown in the figure). The lengths of all the cranks and connecting rods are assumed to be r and l, respectively, and the angular velocity of all the cranks is taken to be a constant, ω. The axial displacement and angular orientation of the ith cylinder from those of the first cylinder are assumed to be α_i and l_i, respectively; $i = 2, 3, \ldots, N$. For force balance, the total inertia force in the x and y directions must be zero. Thus

$$(F_x)_{\text{total}} = \sum_{i=1}^{N}(F_x)_i = 0 \qquad (9.74)$$

$$(F_y)_{\text{total}} = \sum_{i=1}^{N}(F_y)_i = 0 \qquad (9.75)$$

where $(F_x)_i$ and $(F_y)_i$ are the vertical and horizontal components of inertia force of cylinder i given by (see Eqs. (9.71) and (9.73)):

$$(F_x)_i = (m_p + m_c)_i r\omega^2 \cos(\omega t + \alpha_i) + (m_p)_i \frac{r^2\omega^2}{l} \cos(2\omega t + 2\alpha_i) \qquad (9.76)$$

$$(F_y)_i = -(m_c)_i r\omega^2 \sin(\omega t + \alpha_i) \qquad (9.77)$$

FIGURE 9.16   Arrangement of an N-cylinder engine.

For simplicity, we assume the reciprocating and rotating masses for each cylinder to be the same—that is, $(m_p)_i = m_p$ and $(m_c)_i = m_c$ for $i = 1, 2, \ldots, N$. Without loss of generality, Eqs. (9.74) and (9.75) can be applied at time $t = 0$. Thus the conditions necessary for the total force balance are given by

$$\sum_{i=1}^{N} \cos \alpha_i = 0 \quad \text{and} \quad \sum_{i=1}^{N} \cos 2\alpha_i = 0 \tag{9.78}$$

$$\sum_{i=1}^{N} \sin \alpha_i = 0 \tag{9.79}$$

The inertia forces $(F_x)_i$ and $(F_y)_i$ of the ith cylinder induce moments about the y- and x-axes, respectively, as shown in Fig. 9.16(b). The moments about the z- and x-axes are given by

$$M_z = \sum_{i=2}^{N} (F_x)_i l_i = 0 \tag{9.80}$$

$$M_x = \sum_{i=2}^{N} (F_y)_i l_i = 0 \tag{9.81}$$

By substituting Eqs. (9.76) and (9.77) into Eqs. (9.80) and (9.81) and assuming $t = 0$, we obtain the necessary conditions to be satisfied for the balancing of moments about the z- and x-axes as

$$\sum_{i=2}^{N} l_i \cos \alpha_i = 0 \qquad \text{and} \qquad \sum_{i=2}^{N} l_i \cos 2\alpha_i = 0 \qquad (9.82)$$

$$\sum_{i=2}^{N} l_i \sin \alpha_i = 0 \qquad (9.83)$$

Thus we can arrange the cylinders of a multicylinder reciprocating engine so as to satisfy Eqs. (9.78), (9.79), (9.82), and (9.83); it will be completely balanced against the inertia forces and moments.

9.7   Control of Vibration

In many practical situations, it is possible to reduce but not eliminate the dynamic forces that cause vibrations. Several methods can be used to control vibrations. Among them, the following are important:

1. Controlling the natural frequencies of the system and avoiding resonance under external excitations.
2. Preventing excessive response of the system, even at resonance, by introducing a damping or energy-dissipating mechanism.
3. Reducing the transmission of the excitation forces from one part of the machine to another by the use of vibration isolators.
4. Reducing the response of the system by the addition of an auxiliary mass neutralizer or vibration absorber.

We shall now consider the details of these methods.

9.8   Control of Natural Frequencies

It is well known that whenever the frequency of excitation coincides with one of the natural frequencies of the system, resonance occurs. The most prominent feature of resonance is a large displacement. In most mechanical and structural systems, large displacements indicate undesirably large strains and stresses, which can lead to the failure of the system. Hence in any system, resonance conditions must be avoided. In most cases, the excitation frequency cannot be controlled, because it is imposed by the functional requirements of the system or machine. We must concentrate on controlling the natural frequencies of the system to avoid resonance.

As indicated by Eq. (2.14), the natural frequency of a system can be changed by varying either the mass *m* or the stiffness *k*.[4] In many practical cases, however, the mass cannot be changed easily, since its value is determined by the functional requirements of the system. For example, the mass of a flywheel on a shaft is determined by the amount of energy it must store in one cycle. Therefore, the stiffness of the system is the factor that is most often changed to alter its natural frequencies. For example, the stiffness of a rotating shaft can be altered by varying one or more of its parameters, such as the materials or the number and location of support points (bearings).

9.9   Introduction of Damping

Although damping is disregarded so as to simplify the analysis, especially in finding the natural frequencies, most systems possess damping to some extent. The presence of damping is helpful in many cases. In systems such as automobile shock absorbers and many vibration-measuring instruments, damping must be introduced to fulfill the functional requirements [9.20–9.21].

If the system undergoes forced vibration, its response or amplitude of vibration tends to become large near resonance if there is no damping. The presence of damping always limits the amplitude of vibration. If the forcing frequency is known, it may be possible to avoid resonance by changing the natural frequency of the system. However, the system or the machine may be required to operate over a range of speeds, as in the case of a variable-speed electric motor or an internal combustion engine. It may not be possible to avoid resonance under all operating conditions. In such cases, we can introduce damping into the system to control its response, by the use of structural materials having high internal damping, such as cast iron or laminated or sandwich materials.

In some structural applications, damping is introduced through joints. For example, bolted and riveted joints, which permit slip between surfaces, dissipate more energy compared to welded joints, which do not permit slip. Hence a bolted or riveted joint is desirable to increase the damping of the structure. However, bolted and riveted joints reduce the stiffness of the structure, produce debris due to joint slip, and cause fretting corrosion. In spite of this, if a highly damped structure is desired, bolted or riveted joints should not be ignored.

Use of Viscoelastic Materials.   The equation of motion of a single-degree-of-freedom system with internal damping, under harmonic excitation $F(t) = F_0 e^{i\omega t}$, can be expressed as

$$m\ddot{x} + k(1 + i\eta)x = F_0 e^{i\omega t} \tag{9.84}$$

where η is called the *loss factor* (or *loss coefficient*), which is defined as (see Section 2.6.4)

$$\eta = \frac{(\Delta W/2\pi)}{W} = \left(\frac{\text{Energy dissipated during 1 cycle of harmonic displacement/radian}}{\text{Maximum strain energy in cycle}} \right) \tag{9.85}$$

---

[4]Although this statement is made with reference to a single-degree-of-freedom system, it is generally true even for multidegree-of-freedom and continuous systems.

The amplitude of the response of the system at resonance $(\omega = \omega_n)$ is given by

$$\frac{F_0}{k\eta} = \frac{F_0}{aE\eta} \tag{9.86}$$

since the stiffness is proportional to the Young's modulus $(k = aE; a = \text{constant})$.

The viscoelastic materials have larger values of the loss factor and hence are used to provide internal damping. When viscoelastic materials are used for vibration control, they are subjected to shear or direct strains. In the simplest arrangement, a layer of viscoelastic material is attached to an elastic one. In another arrangement, a viscoelastic layer is sandwiched between the elastic layers. This arrangement is known as *constrained layer damping*.[5] Damping tapes, consisting of thin metal foil covered with a viscoelastic adhesive, are used on existing vibrating structures. A disadvantage with the use of viscoelastic materials is that their properties change with temperature, frequency, and strain. Equation (9.86) shows that a material with the highest value of $(E\eta)$ gives the smallest resonance amplitude. Since the strain is proportional to the displacement x and the stress is proportional to Ex, the material with the largest value of the loss factor will be subjected to the smallest stresses. The values of loss coefficient for some materials are given below:

Material	Loss Factor (η)
Polystyrene	2.0
Hard rubber	1.0
Fiber mats with matrix	0.1
Cork	0.13–0.17
Aluminum	1×10^{-4}
Iron and steel	$2\text{–}6 \times 10^{-4}$

The damping ratios obtainable with different types of construction or arrangement are indicated below:

Type of Construction or Arrangement	Equivalent Viscous Damping Ratio (%)
Welded construction	1–4
Bolted construction	3–10
Steel frame	5–6
Unconstrained viscoelastic layer on steel-concrete girder	4–5
Constrained viscoelastic layer on steel-concrete girder	5–8

---

[5]It appears that constrained layer damping was used, possibly unknowingly, as far back as the seventeenth century, in the manufacture of violins [9.22]. Antonio Stradivari (1644–1737), the renowned Italian violin manufacturer, bought the wood necessary for the manufacture of violins from Venice. The varnish used for sealing the wood was made from a mixture of resin and ground gem stones. This varnish—stone particles in resin matrix—acted as a form of constrained layer (friction mechanism) that created enough damping to explain why many of his violins had a rich, full tone.

9.10   Vibration Isolation

Vibration isolation is a procedure by which the undesirable effects of vibration are reduced [9.22–9.24]. Basically, it involves the insertion of a resilient member (or isolator) between the vibrating mass (or equipment or payload) and the source of vibration so that a reduction in the dynamic response of the system is achieved under specified conditions of vibration excitation. An isolation system is said to be active or passive depending on whether or not external power is required for the isolator to perform its function. A passive isolator consists of a resilient member (stiffness) and an energy dissipator (damping). Examples of passive isolators include metal springs, cork, felt, pneumatic springs, and elastomer (rubber) springs. Figure 9.17 shows typical spring and pneumatic mounts that can be used as passive isolators, and Fig. 9.18 illustrates the use of passive isolators in the mounting of a high-speed punch press [9.25]. The optimal synthesis of vibration isolators is presented in [9.26–9.30]. An active isolator is comprised of a servomechanism with a sensor, signal processor, and actuator.

(a)

(b)

(c)

FIGURE 9.17   (a) Undamped spring mount; (b) damped spring mount; (c) pneumatic rubber mount. (Courtesy of *Sound and Vibration*.)

FIGURE 9.18   High-speed punch press mounted on pneumatic rubber mounts.   (Courtesy of *Sound and Vibration*.)

Vibration isolation can be used in two types of situations. In the first type, the foundation or base of a vibrating machine is protected against large unbalanced forces. In the second type, the system is protected against the motion of its foundation or base.

The first type of isolation is used when a mass (or a machine) is subjected to a force or excitation. For example, in forging and stamping presses, large impulsive forces act on the object to be formed or stamped. These impacts are transmitted to the base or foundation of the forging or stamping machine, which can damage not only the base or foundation but also the surrounding or nearby structures and machines. They can also cause discomfort to operators of these machines. Similarly, in the case of reciprocating and rotating machines, the inherent unbalanced forces are transmitted to the base or foundation of the machine. In such cases, the force transmitted to the base, $F_t(t)$, varies harmonically, and the resulting stresses in the foundation bolts also vary harmonically, which might lead to fatigue

FIGURE 9.19   Vibration isolation.

failure. Even if the force transmitted is not harmonic, its magnitude is to be limited to safe permissible values. In these applications, we can insert an isolator (in the form of stiffness and/or damping) between the mass being subjected to force or excitation and the base or foundation to reduce the force transmitted to the base or foundation. This is called *force isolation.* In many applications, the isolator is also intended to reduce the vibratory motion of the mass under the applied force (as in the case of forging or stamping machines). Thus both force and displacement transmissibilities become important for this type of isolators.

The second type of isolation is used when a mass is to be protected against the motion or excitation of its base or foundation. When the base is subjected to vibration, the mass m will experience not only a displacement $x(t)$ but also a force $F_t(t)$. The displacement of the mass $x(t)$ is expected to be smaller than the displacement of the base $y(t)$. For example, a delicate instrument or equipment is to be protected from the motion of its container or package (as when the vehicle carrying the package experiences vibration while moving on a rough road). The force transmitted to the mass also needs to be reduced. For example, the package or container is to be designed properly to avoid transmission of large forces to the delicate instrument inside to avoid damage. The force experienced by the instrument or mass m (same as the force transmitted to mass m) is given by

$$F_t(t) = m\ddot{x}(t) = k\{x(t) - y(t)\} + c\{\dot{x}(t) - \dot{y}(t)\} \qquad (9.87)$$

where $y(t)$ is the displacement of the base, $x(t) - y(t)$ is the relative displacement of the spring, and $\dot{x}(t) - \dot{y}(t)$ is the relative velocity of the damper. In such cases, we can insert an isolator (which provides stiffness and/or damping) between the base being subjected to force or excitation and the mass to reduce the motion and/or force transmitted to the mass. Thus both displacement isolation and force isolation become important in this case also.

It is to be noted that the effectiveness of an isolator depends on the nature of the force or excitation. For example, an isolator designed to reduce the force transmitted to the base or foundation due to impact forces of forging or stamping may not be effective if the disturbance is a harmonic unbalanced force. Similarly, an isolator designed to handle harmonic excitation at a particular frequency may not be effective for other frequencies or other types of excitation such as step-type excitation.

**9.10.1
Vibration
Isolation System
with Rigid
Foundation**

Reduction of the Force Transmitted to Foundation.   When a machine is bolted directly to a rigid foundation or floor, the foundation will be subjected to a harmonic load due to the unbalance in the machine in addition to the static load due to the weight of the machine. Hence an elastic or resilient member is placed between the machine and the rigid foundation to reduce the force transmitted to the foundation. The system can then be idealized as a single-degree-of-freedom system, as shown in Fig. 9.20(a). The resilient member is assumed to have both elasticity and damping and is modeled as a spring k and a dashpot c, as shown in Fig. 9.20(b). It is assumed that the operation of the machine gives rise to a harmonically varying force $F(t) = F_0 \cos \omega t$. The equation of motion of the machine (of mass m) is given by

$$m\ddot{x} + c\dot{x} + kx = F_0 \cos \omega t \tag{9.88}$$

Since the transient solution dies out after some time, only the steady-state solution will be left. The steady-state solution of Eq. (9.88) is given by (see Eq. (3.25))

$$x(t) = X \cos(\omega t - \phi) \tag{9.89}$$

where

$$X = \frac{F_0}{[(k - m\omega^2)^2 + \omega^2 c^2]^{1/2}} \tag{9.90}$$

and

$$\phi = \tan^{-1}\left(\frac{\omega c}{k - m\omega^2}\right) \tag{9.91}$$

The force transmitted to the foundation through the spring and the dashpot, $F_t(t)$, is given by

$$F_t(t) = kx(t) + c\dot{x}(t) = kX \cos(\omega t - \phi) - c\omega X \sin(\omega t - \phi) \tag{9.92}$$

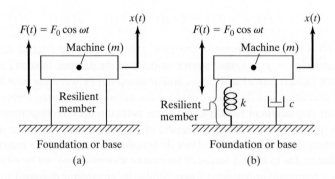

FIGURE 9.20   Machine and resilient member on rigid foundation.

The magnitude of the total transmitted force (F_T) is given by

$$F_T = [(kx)^2 + (c\dot{x})^2]^{1/2} = X\sqrt{k^2 + \omega^2 c^2}$$

$$= \frac{F_0(k^2 + \omega^2 c^2)^{1/2}}{[(k - m\omega^2)^2 + \omega^2 c^2]^{1/2}} \tag{9.93}$$

The transmissibility or transmission ratio of the isolator (T_f) is defined as the ratio of the magnitude of the force transmitted to that of the exciting force:

$$T_f = \frac{F_T}{F_0} = \left\{\frac{k^2 + \omega^2 c^2}{(k - m\omega^2)^2 + \omega^2 c^2}\right\}^{1/2}$$

$$= \left\{\frac{1 + (2\zeta r)^2}{[1 - r^2]^2 + (2\zeta r)^2}\right\}^{1/2} \tag{9.94}$$

where $r = \frac{\omega}{\omega_n}$ is the frequency ratio. The variation of T_f with the frequency ratio $r = \frac{\omega}{\omega_n}$ is shown in Fig. 9.21. In order to achieve isolation, the force transmitted to the foundation needs to be less than the excitation force. It can be seen, from Fig. 9.21, that the forcing frequency has to be greater than $\sqrt{2}$ times the natural frequency of the system in order to achieve isolation of vibration.

FIGURE 9.21   Variation of transmission ratio (T_f) with r.

For small values of damping ratio ζ and for frequency ratio $r > 1$, the force transmissibility, given by Eq. (9.94), can be approximated as

$$T_f = \frac{F_t}{F} \approx \frac{1}{r^2 - 1} \quad \text{or} \quad r^2 \approx \frac{1 + T_f}{T_f} \tag{9.95}$$

Notes

1. The magnitude of the force transmitted to the foundation can be reduced by decreasing the natural frequency of the system (ω_n).

2. The force transmitted to the foundation can also be reduced by decreasing the damping ratio. However, since vibration isolation requires $r > \sqrt{2}$, the machine should pass through resonance during start-up and stopping. Hence, some damping is essential to avoid infinitely large amplitudes at resonance.

3. Although damping reduces the amplitude of the mass (X) for all frequencies, it reduces the maximum force transmitted to the foundation (F_t) only if $r < \sqrt{2}$. Above that value, the addition of damping increases the force transmitted.

4. If the speed of the machine (forcing frequency) varies, we must compromise in choosing the amount of damping to minimize the force transmitted. The amount of damping should be sufficient to limit the amplitude X and the force transmitted F_t while passing through the resonance, but not so much to increase unnecessarily the force transmitted at the operating speed.

Reduction of the Vibratory Motion of the Mass.   In many applications, the isolation is required to reduce the motion of the mass (machine) under the applied force. The displacement amplitude of the mass m due to the force $F(t)$, given by Eq. (9.90), can be expressed as:

$$T_d = \frac{X}{\delta_{st}} = \frac{kX}{F_0} = \frac{1}{\sqrt{(1 - r^2)^2 + (2\zeta r)^2}} \tag{9.96}$$

where $\dfrac{X}{\delta_{st}}$ is called, in the present context, the *displacement transmissibility* or *amplitude ratio* and indicates the ratio of the amplitude of the mass, X, to the static deflection under the constant force F_0, $\delta_{st} = \dfrac{F_0}{k}$. The variation of the displacement transmissibility with the frequency ratio r for several values of the damping ratio ζ is shown in Fig. 9.22. The following observations can be made from Fig. 9.22:

1. The displacement transmissibility increases to a maximum value at (Eq. (3.33)):

$$r = \sqrt{1 - 2\zeta^2} \tag{9.97}$$

Equation (9.97) shows that, for small values of damping ratio ζ, the displacement transmissibility (or the amplitude of the mass) will be maximum at $r \approx 1$ or $\omega \approx \omega_n$.

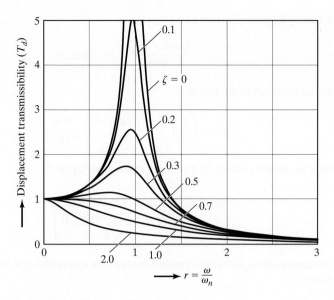

FIGURE 9.22   Variation of displacement transmissibility (T_d) with r.

Thus the value of $r \approx 1$ is to be avoided in practice. In most cases, the excitation frequency ω is fixed and hence we can avoid $r \approx 1$ by altering the value of the natural frequency $\omega_n = \sqrt{\dfrac{k}{m}}$ which can be accomplished by changing the value of either or both of m and k.

2. The amplitude of the mass, X, approaches zero as r increases to a large value. The reason is that at large values of r, the applied force $F(t)$ varies very rapidly and the inertia of the mass prevents it from following the fluctuating force.

EXAMPLE   9.4

Spring Support for Exhaust Fan

An exhaust fan, rotating at 1,000 rpm, is to be supported by four springs, each having a stiffness of K. If only 10 percent of the unbalanced force of the fan is to be transmitted to the base, what should be the value of K? Assume the mass of the exhaust fan to be 40 kg.

Solution: Since the transmissibility has to be 0.1, we have, from Eq. (9.94),

$$0.1 = \left[\frac{1 + \left(2\zeta \dfrac{\omega}{\omega_n} \right)^2}{\left\{ 1 - \left(\dfrac{\omega}{\omega_n} \right)^2 \right\}^2 + \left(2\zeta \dfrac{\omega}{\omega_n} \right)^2} \right]^{1/2} \tag{E.1}$$

where the forcing frequency is given by

$$\omega = \frac{1{,}000 \times 2\pi}{60} = 104.72 \text{ rad/s} \qquad (E.2)$$

and the natural frequency of the system by

$$\omega_n = \left(\frac{k}{m}\right)^{1/2} = \left(\frac{4K}{40}\right)^{1/2} = \frac{\sqrt{K}}{3.1623} \qquad (E.3)$$

By assuming the damping ratio to be $\zeta = 0$, we obtain from Eq. (E.1),

$$0.1 = \frac{\pm 1}{\left\{1 - \left(\dfrac{104.72 \times 3.1623}{\sqrt{K}}\right)^2\right\}} \qquad (E.4)$$

To avoid imaginary values, we need to consider the negative sign on the right-hand side of Eq. (E.4). This leads to

$$\frac{331.1561}{\sqrt{K}} = 3.3166$$

or

$$K = 9{,}969.6365 \text{ N/m}$$

∎

EXAMPLE 9.5

Design of an Undamped Isolator

A 50-kg mass is subjected to the harmonic force $F(t) = 1{,}000 \cos 120t$ N. Design an undamped isolator so that the force transmitted to the base does not exceed 5% of the applied force. Also find the displacement amplitude of the mass of the system with isolation.

Solution: By setting the value of force transmissibility as 0.05 and using $\zeta = 0$, Eq. (9.95) gives

$$r^2 \approx \frac{1 + T_f}{T_f} = \frac{1 + 0.05}{0.05} = 21 \qquad (E.1)$$

Using the definition of r, along with the values of $m = 50$ kg and $\omega = 120$ rad/s, Eq. (E.1) yields

$$r^2 = \frac{\omega^2}{\omega_n^2} = \frac{\omega^2 m}{k}$$

or

$$k = \frac{\omega^2 m}{r^2} = \frac{(120)^2(50)}{21} = 34.2857 \times 10^3 \text{ N/m} \qquad (E.2)$$

The displacement amplitude of the mass of the system with isolation can be found from Eq. (9.96), with $\zeta = 0$:

$$X = \frac{F_0}{k}\frac{1}{(r^2-1)} = \frac{1{,}000}{34.2857 \times 10^3}\frac{1}{(21-1)} = 1.4583 \times 10^{-3}\,\text{m} \qquad (\text{E.3})$$

■

Design Chart for Isolation:

The force transmitted to the base or ground by a source of vibration (vibrating mass) is given by Eq. (9.94) and is shown in Fig. 9.21 as a graph between $T_f = F_T/F_0$ and $r = \omega/\omega_n$. As noted earlier, vibration isolation—reduction of the force transmitted to the ground—can be achieved for $r > \sqrt{2}$. In the region $r > \sqrt{2}$, low values of damping are desired for more effective isolation. For large values of r and low values of ζ, the term $(2\zeta r)^2$ becomes very small and can be neglected in Eq. (9.94) for simplicity. Thus Eq. (9.94) can be approximated as shown in Eq. (9.95) for $r > \sqrt{2}$ and small ζ.

By defining the natural frequency of vibration of the undamped system as

$$\omega_n = \sqrt{\frac{k}{m}} = \sqrt{\frac{g}{\delta_{st}}} \qquad (9.98)$$

and the exciting frequency ω as

$$\omega = \frac{2\pi N}{60} \qquad (9.99)$$

where δ_{st} is the static deflection of the spring and N is the frequency in cycles per minute (cpm) or revolutions per minute (rpm) of rotating machines such as electric motors and turbines, Eqs. (9.95) to (9.99) can be combined to obtain

$$r = \frac{\omega}{\omega_n} = \frac{2\pi N}{60}\sqrt{\frac{\delta_{st}}{g}} = \sqrt{\frac{2-R}{1-R}} \qquad (9.100)$$

where $R = 1 - T_f$ is used to indicate the quality of the isolator and denotes the percent reduction achieved in the transmitted force. Equation (9.100) can be rewritten as

$$N = \frac{30}{\pi}\sqrt{\frac{g}{\delta_{st}}\left(\frac{2-R}{1-R}\right)} = 29.9092\sqrt{\frac{2-R}{\delta_{st}(1-R)}} \qquad (9.101)$$

Equation (9.101) can be used to generate a graph between $\log N$ and $\log \delta_{st}$ as a series of straight lines for different values of R, as shown in Fig. 9.23. This graph serves as a design chart for selecting a suitable spring for the isolation.

FIGURE 9.23   Isolation efficiency.

EXAMPLE 9.6

Isolator for Stereo Turntable

A stereo turntable, of mass 1 kg, generates an excitation force at a frequency of 3 Hz. If it is supported on a base through a rubber mount, determine the stiffness of the rubber mount to reduce the vibration transmitted to the base by 80 percent.

Solution: Using $N = 3 \times 60 = 180$ cpm and $R = 0.80$, Eq. (9.105) gives

$$180 = 29.9092 \sqrt{\frac{2 - 0.80}{\delta_{st}(1 - 0.80)}}$$

or

$$\delta_{st} = 0.1657 \text{ m}$$

The static deflection of the rubber mount can be expressed, in terms of its stiffness (k), as

$$\delta_{st} = \frac{mg}{k}$$

which gives the stiffness of the rubber mount as

$$0.1657 = \frac{1(9.81)}{k} \qquad \text{or} \qquad k = 59.2179 \text{ N/m}$$

∎

Isolation of systems with rotating unbalance:

A common source of forced harmonic force is imbalance in rotating machines such as turbines, centrifugal pumps, and turbogenerators. Imbalance in a rotating machine implies that the axis of rotation does not coincide with the center of mass of the whole system. Even a very small eccentricity can cause a large unbalanced force in high-speed machines such as turbines. A typical rotating system with an unbalance is shown in Fig. 9.24. Here the total mass of the system is assumed to be M and the unbalanced mass is considered as a point mass m located at the center of mass of the system (which has an eccentricity of e from the center of rotation) as shown in Fig. 9.24. If the unbalanced mass rotates at an angular velocity ω and the system is constrained to move in the vertical direction, the equation of motion of the system is given by

$$M\ddot{x} + c\dot{x} + kx = F_0 \sin \omega t \equiv me\omega^2 \sin \omega t \qquad (9.102)$$

FIGURE 9.24   A system with rotating unbalance.

Using $F_0 = me\omega^2$, the force transmissibility of the system can be found from Eq. (9.88). However, the presence of ω^2 in F_0 results in the following equation for the force transmissibility (T_f) due to rotating unbalance:

$$T_f = \frac{F_t}{F_0} = \frac{F_t}{me\omega^2} = \frac{F_t}{mer^2\omega_n^2}$$

or

$$\frac{F_t}{me\omega_n^2} = r^2 \left\{ \frac{1 + (2\zeta r)^2}{(1 - r^2)^2 + (2\zeta r)^2} \right\}^{\frac{1}{2}} \tag{9.103}$$

EXAMPLE 9.7

Centrifugal Pump with Rotating Unbalance—Rattle Space

A centrifugal pump, with a mass of 50 kg and rotational speed of 3,000 rpm, is mounted in the middle of a simply supported beam of length 100 cm, width 20 cm, and thickness 0.5 cm. The damping ratio of the system (beam) can be assumed as $\zeta = 0.05$. The impeller (rotating part) of the pump has a mass of 5 kg with an eccentricity of 1 mm. If the maximum deflection of the beam is constrained to be less than the available rattle space[6] of 3 mm, determine whether the support system of the pump is adequate.

Solution: The bending stiffness or spring constant of the simply supported beam is given by

$$k = \frac{48EI}{l^3}$$

where the moment of inertia of the beam cross section can be computed as

$$I = \frac{1}{12} wt^3 = \frac{(20)(0.05^3)}{12} = 0.208333 \text{ cm}^4 = 20.8333 \times 10^{-10} \text{ m}^4$$

Using $E = 207 \times 10^9$ Pa, the spring constant of the beam can be found as

$$k = \frac{48(207 \times 10^9)(20.8333 \times 10^{-10})}{(1.0^3)} = 206,999.6688 \text{ N/m}$$

Using the density of steel as 7.85 gram/cm^3, the mass of the beam (m_b) can be determined as

$$m_b = 7.85(100)(20)(0.5) = 7,850 \text{ gram} = 7.85 \text{ kg}$$

---

[6]The available clearance space that permits the system to undergo the induced deflection freely during vibration is called the *rattle space* or *clearance*. If the rattle space is too small to accommodate the deflection of the system, the system will undergo impacts (as it hits the surrounding or nearby surface or object) in each cycle of vibration.

The total mass of the system (M) is equal to the mass of the pump plus the effective mass of the beam at its center (equal to $\frac{17}{35}m_b$; see Problem 2.86):

$$M = m_{\text{pump}} + \frac{17}{35}m_b = 50 + \frac{17}{35}(7.85) = 53.8128 \text{ kg}$$

The natural frequency of the system is given by

$$\omega_n = \sqrt{\frac{k}{M}} = \sqrt{\frac{206,999.6688}{53.8128}} = 62.0215 \text{ rad/s}$$

The impeller (rotor) speed of 3,000 rpm gives $\omega = 2\pi(3,000)/60 = 314.16$ rad/s. Thus the frequency ratio (r) becomes

$$r = \frac{\omega}{\omega_n} = \frac{314.16}{62.0215} = 5.0653; \qquad r^2 = 25.6577$$

The amplitude of the forcing function is

$$me\omega^2 = 5(10^{-3})(314.16^2) = 493.4825 \text{ N}$$

Using $\zeta = 0.05$, the steady-state amplitude of the pump can be found from Eq. (9.96), using $me\omega^2$ for F_0, as

$$X = \frac{me\omega^2}{k} \frac{1}{\sqrt{(1-r^2)^2 + (2\zeta r)^2}} = \frac{493.4825}{206,999.6688} \frac{1}{\sqrt{(1-25.6577)^2 + \{2(0.05)(5.0653)\}^2}}$$

$$= \frac{493.4825}{206,999.6688} \frac{1}{24.6629} = 9.6662 \times 10^{-5} \text{ m}$$

The static deflection of the beam under the weight of the pump can be determined as

$$\delta_{\text{pump}} = \frac{W_{\text{pump}}}{k} = \frac{(50)(9.81)}{206,999.6688} = 236.9569 \times 10^{-5} \text{ m}$$

Thus the total deflection of the system is

$$\delta_{\text{total}} = X + \delta_{\text{pump}} = 9.662 \times 10^{-5} + 236.9569 \times 10^{-5} = 246.6231 \times 10^{-5} \text{ m} = 2.4662 \text{ mm}$$

This deflection is less than the rattle space of 3 mm. As such, the support system of the pump is adequate. In case the value of δ_{total} exceeds the rattle space, we need to redesign (modify) the support system. This can be achieved by changing the spring constant (dimensions) of the beam and/or by introducing a damper.

∎

9.10.2
Vibration
Isolation System
with Base
Motion

In some applications, the base of the system is subjected to a vibratory motion. For example, the base or foundation of a machine such as a turbine in a power plant may be subjected to ground motion during an earthquake. In the absence of a suitably designed isolation system, the motion of the base transmitted to the mass (turbine) might cause damage and power failure. Similarly, a delicate instrument (mass) may have to be protected from a force or shock when the package containing the instrument is dropped from a height accidentally. Also, if the instrument is to be transported, the vehicle carrying it may experience vibration as it travels on a rough road with potholes. In this case also, proper isolation is to be used to protect the instrument against excessive displacement or force transmitted from the base motion.

For a single-degree-of-freedom system with base excitation, shown in Fig. 9.19(b), the analysis was presented in Section 3.6. When the base of the system is subjected to a harmonic motion, $y(t) = Y \sin \omega t$, the equation of motion is given by Eq. (3.75):

$$m\ddot{z} + c\dot{z} + kz = -m\ddot{y} \tag{9.104}$$

where $z = x - y$ denotes the displacement of the mass relative to the base. If the base motion is harmonic, then the motion of the mass will also be harmonic. Hence the displacement transmissibility, $T_d = \dfrac{X}{Y}$, is given by Eq. (3.68):

$$T_d = \frac{X}{Y} = \left\{ \frac{1 + (2\zeta r)^2}{(1 - r^2)^2 + (2\zeta r)^2} \right\}^{1/2} \tag{9.105}$$

where X and Y denote the displacement amplitudes of the mass and the base, respectively, and the right-hand-side expression can be identified to be the same as that in Eq. (9.94). Note that Eq. (9.105) is also equal to the ratio of the maximum steady-state accelerations of the mass and the base. The variation of the displacement transmissibility with the frequency ratio (r) for different values of the damping ratio (ζ) is shown in Fig. 9.25. The following observations can be made from Fig. 9.25:

1. For an undamped system, the displacement transmissibility approaches infinity at resonance ($r = 1$). Thus the undamped isolator (stiffness) is to be designed to ensure that the natural frequency of the system (ω_n) is away from the excitation frequency (ω).

2. For a damped system, the displacement transmissibility (and hence the displacement amplitude) attains a maximum for frequency ratios close to 1. The maximum displacement amplitude of the mass can be larger than the amplitude of base motion—that is, the base motion can get amplified by a large factor.

3. The displacement transmissibility is close to 1 for small values of the frequency ratio (r) and is exactly equal to 1 at $r = \sqrt{2}$.

4. The displacement amplitude is larger than 1 for $r < \sqrt{2}$ and smaller than 1 for $r > \sqrt{2}$. Note that a smaller damping ratio corresponds to a larger T_d for $r < \sqrt{2}$ and a smaller T_d for $r > \sqrt{2}$. Thus, if the damping of the system cannot be altered, the natural frequency of the system (stiffness) can be changed to achieve a value of $r > \sqrt{2}$.

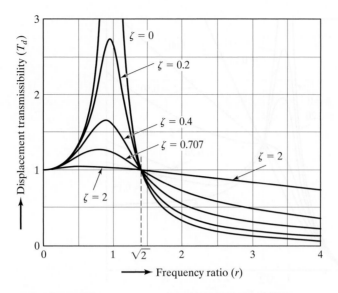

FIGURE 9.25   Variation of T_d with r (for base motion).

If F_t denotes the magnitude of the force transmitted to the mass by the spring and the damper, the force transmissibility (T_f) of the system is given by Eq. (3.74):

$$T_f = \frac{F_t}{kY} = r^2 \left\{ \frac{1 + (2\zeta r)^2}{(1 - r^2)^2 + (2\zeta r)^2} \right\}^{\frac{1}{2}}$$

(9.106)

where kY is used to make the force transmissibility dimensionless. Note that once the displacement transmissibility, T_d, or the displacement amplitude of the mass (X) is computed using Eq. (9.105), the force transmitted to the mass, F_t, can be determined using the relation

$$\frac{F_t}{kY} = r^2 \frac{X}{Y} \qquad \text{or} \qquad F_t = kr^2 X$$

(9.107)

The variation of the force transmissibility with the frequency ratio (r) for different values of the damping ratio (ζ) is shown in Fig. 9.26. The following observations can be made from Fig. 9.26:

1. The force transmissibility (T_f) will be 2 at the frequency ratio $r = \sqrt{2}$ for all values of the damping ratio (ζ).
2. For $r > \sqrt{2}$, a lower damping ratio corresponds to a lower value of force transmissibility.
3. For $r > \sqrt{2}$, for any specific value of the damping ratio, the force transmissibility increases with r. This behavior is opposite to that of the displacement transmissibility.
4. The force transmissibility is close to zero at small values of the frequency ratio r and attains a maximum at values of r close to 1.

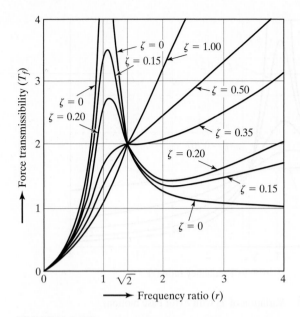

FIGURE 9.26   Variation of T_f with r (for base motion).

EXAMPLE 9.8

Isolation from Vibrating Base

A vibrating system is to be isolated from its vibrating base. Find the required damping ratio that must be achieved by the isolator to limit the displacement transmissibility to $T_d = 4$. Assume the system to have a single degree of freedom.

Solution: By setting $\omega = \omega_n$, Eq. (9.105) gives

$$T_d = \frac{\sqrt{1 + (2\zeta)^2}}{2\zeta}$$

or

$$\zeta = \frac{1}{2\sqrt{T_d^2 - 1}} = \frac{1}{2\sqrt{15}} = 0.1291$$

■

EXAMPLE 9.9

Design of Isolation for a Precision Machine with Base Motion

A precision machine used for the manufacture of integrated circuits, having a mass of 50 kg, is placed on a work bench (as a base). The ground vibration transmitted by a nearby internal combustion engine causes the base (all four corners of the bench) to vibrate at a frequency of 1,800 rpm.

Helical springs, with a damping ratio of $\zeta = 0.01$ and a relationship of bilinear load (P) to deflection (x) given by

$$P = \begin{cases} 50{,}000x; & 0 \le x \le 8 \times 10^{-3} \\ 10^5 x - 4 \times 10^5; & 8 \times 10^{-3} \le x \le 13 \times 10^{-3} \end{cases} \tag{E.1}$$

(P in Newtons and x in meters) are available for use as isolators at the four corners of the base. If no more than 10% of the vibration of the base is to be transmitted to the precision machine, determine a method of achieving the isolation.

Solution: Since the displacement transmissibility is required to be 0.1, Eq. (9.105), for $\zeta = 0.01$, gives

$$T_d = \frac{X}{Y} = 0.1 = \sqrt{\frac{1 + \{2(0.01)r\}^2}{(1 - r^2)^2 + \{2(0.01)r\}^2}} \tag{E.2}$$

The simplification of Eq. (E.2) yields a quadratic equation in r^2 as

$$r^4 - 2.0396r^2 - 99 = 0 \tag{E.3}$$

The solution of Eq. (E.3) gives

$$r^2 = 11.0218, \, -8.9822$$

which gives the positive value of r as 3.3199. Using the excitation frequency of

$$\omega = \frac{2\pi(1{,}800)}{60} = 188.496 \text{ rad/s}$$

and the frequency ratio of $r = 3.3199$, the required natural frequency of the system can be determined as

$$r = 3.3199 = \frac{\omega}{\omega_n} = \frac{188.496}{\omega_n} \tag{E.4}$$

Equation (E.4) gives $\omega_n = 56.7776$ rad/s.

We assume that one helical spring is installed at each corner of the base (under the four corners of the work bench). Because the expected deflection of the springs is unknown, the correct stiffness of the springs (out of the two possible values) is unknown. Hence we use the relation (see Eq. (2.28)):

$$\omega_n = \sqrt{\frac{g}{\delta_{st}}} \quad \text{or} \quad 56.7776 = \sqrt{\frac{9.81}{\delta_{st}}} \tag{E.5}$$

to find the static deflection of the system (δ_{st}) as

$$\delta_{st} = \frac{9.81}{56.7776^2} = 3.0431 \times 10^{-3} \text{ m}$$

Since all the four springs experience δ_{st}, the static load acting on each spring can be found from Eq. (E.1) as

$$P = 50,000(3.0431 \times 10^{-3}) = 152.155 \text{ N}$$

The total load on the four springs is $4 \times 152.155 = 608.62$ N. Because the weight of the machine is $50 g = 50(9.81) = 490.5$ N, in order to achieve the total load of 608.62 N, we need to add a weight of $608.62 - 490.50 = 118.12$ N to the system. This weight, in the form of a rectangular steel plate, is to be attached at the bottom of the machine, so that the total vibrating mass becomes 62.0408 kg (with a weight of 608.62 N).

∎

EXAMPLE 9.10

Isolation System for a System with Base Motion

A printed circuit board (PCB) made of fiber-reinforced plastic composite material is used for the computer control of an automobile engine. It is attached to the chassis of the computer, which is fixed to the frame of the automobile as shown in Fig. 9.27(a). The frame of the automobile and the chassis of the computer are subject to vibration at the engine speed of 3,000 rpm. If it is required to achieve a displacement transmissibility of no more than 10% at the PCB, design a suitable isolation system between the chassis of the computer and the frame of the automobile. Assume that the chassis of the computer is rigid with a mass of 0.25 kg.

Data of PCB: Length (l): 25 cm, width (w): 20 cm, thickness (t): 0.3 cm, mass per unit surface area: 0.005 kg/cm^2, Young's modulus (E): 15×10^9 N/m^2, damping ratio: 0.01.

Solution: The PCB is assumed to be fixed to the chassis of the computer as a cantilever beam. Its mass (m_{PCB}) is given by $25 \times 20 \times 0.005 = 2.5$ kg. The equivalent mass at the free end of the cantilever (m_b) is (see Example 2.9):

$$m_b = \frac{33}{140} m_{PCB} = \frac{33}{140}(2.5) = 0.5893 \text{ kg}$$

Using the moment of inertia of the cross section of the PCB

$$I = \frac{1}{12} wt^3 = \frac{1}{12}(0.20)(0.003)^3 = 45 \times 10^{-8} \text{ m}^4$$

the stiffness of the PCB as a cantilever beam can be computed as

$$k_b = \frac{3EI}{l^3} = \frac{3(15 \times 10^9)(45 \times 10^{-8})}{(0.25)^3} = 1.296 \times 10^6 \text{ N/m}$$

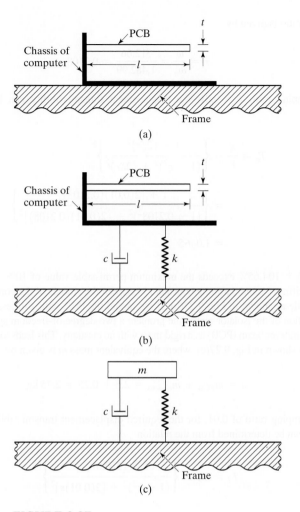

FIGURE 9.27

The natural frequency of the PCB is given by

$$\omega_n = \sqrt{\frac{k_b}{m_b}} = \sqrt{\frac{1.296 \times 10^6}{0.5893}} = 1,482.99 \text{ rad/s}$$

The frequency of vibration of the base (chassis of the computer) is

$$\omega = \frac{2\pi(3,000)}{60} = 312.66 \text{ rad/s}$$

The frequency ratio is given by

$$r = \frac{\omega}{\omega_n} = \frac{312.66}{1,482.99} = 0.2108$$

Using the damping ratio $\zeta = 0.01$, the displacement transmissibility can be determined from Eq. (9.105):

$$T_d = \frac{X}{Y} = \left\{ \frac{1 + (2\zeta r)^2}{(1 - r^2)^2 + (2\zeta r)^2} \right\}^{\frac{1}{2}}$$

$$= \left\{ \frac{1 + [2(0.01)(0.2108)]^2}{(1 - 0.2108^2)^2 + [2(0.01)(0.2108)]^2} \right\}$$

$$= 1.0465 \qquad\qquad\qquad (E.1)$$

This value of $T_d = 104.65\%$ exceeds the maximum permissible value of 10%. Hence we design an isolator (with stiffness k and damping constant c) between the chassis of the computer and the frame of the automobile as shown in Fig. 9.27(b). If we model the PCB with stiffness k_b and mass m_b as before, the addition of the isolator makes the problem a two-degree-of-freedom system. For simplicity, we model the cantilever beam (PCB) as a rigid mass with no elasticity. This leads to the single-degree-of-freedom system shown in Fig. 9.27(c), where the equivalent mass m is given by

$$m = m_{\text{PCB}} + m_{\text{chassis}} = 2.5 + 0.25 = 2.75 \text{ kg}$$

Assuming a damping ratio of 0.01, for the required displacement transmissibility of 10%, the frequency ratio r can be determined from the relation

$$T_d = 0.1 = \left\{ \frac{1 + [2(0.01)r]^2}{(1 - r^2)^2 + [2(0.01)r]^2} \right\}^{\frac{1}{2}} \qquad\qquad (E.2)$$

By squaring both sides of Eq. (E.2) and rearranging the terms, we obtain

$$r^4 - 2.0396r^2 - 99 = 0 \qquad\qquad\qquad (E.3)$$

The positive root of Eq. (E.3) is $r^2 = 11.0218$ or $r = 3.3199$. The stiffness of the isolator is given by

$$k = \frac{m\omega^2}{r^2} = \frac{(2.75)(312.66^2)}{11.0218} = 24,390.7309 \text{ N/m}$$

The damping constant of the isolator can be computed as

$$c = 2\zeta\sqrt{mk} = 2(0.01)\sqrt{(2.75)(24,390.7309)} = 5.1797 \text{ N-s/m}$$

∎

9.10.3
Vibration
Isolation System
with Flexible
Foundation

In many practical situations, the structure or foundation to which the isolator is connected moves when the machine mounted on the isolator operates. For example, in the case of a turbine supported on the hull of a ship or an aircraft engine mounted on the wing of an airplane, the area surrounding the point of support also moves with the isolator. In such cases, the system can be represented as having two degrees of freedom. In Fig. 9.28, m_1 and m_2 denote the masses of the machine and the supporting structure that moves with the isolator, respectively. The isolator is represented by a spring k, and the damping is disregarded for the sake of simplicity. The equations of motion of the masses m_1 and m_2 are

$$m_1\ddot{x}_1 + k(x_1 - x_2) = F_0 \cos \omega t$$
$$m_2\ddot{x}_2 + k(x_2 - x_1) = 0 \tag{9.108}$$

By assuming a harmonic solution of the form

$$x_j = X_j \cos \omega t, \qquad j = 1, 2$$

Eqs. (9.108) give

$$\left. \begin{array}{l} X_1(k - m_1\omega^2) - X_2 k = F_0 \\ -X_1 k + X_2(k - m_2\omega^2) = 0 \end{array} \right\} \tag{9.109}$$

The natural frequencies of the system are given by the roots of the equation

$$\begin{vmatrix} (k - m_1\omega^2) & -k \\ -k & (k - m_2\omega^2) \end{vmatrix} = 0 \tag{9.110}$$

The roots of Eq. (9.110) are given by

$$\omega_1^2 = 0, \qquad \omega_2^2 = \frac{(m_1 + m_2)k}{m_1 m_2} \tag{9.111}$$

FIGURE 9.28   Machine with isolator on a flexible foundation.

The value $\omega_1 = 0$ corresponds to rigid-body motion, since the system is unconstrained. In the steady state, the amplitudes of m_1 and m_2 are governed by Eq. (9.109), whose solution yields

$$X_1 = \frac{(k - m_2\omega^2)F_0}{[(k - m_1\omega^2)(k - m_2\omega^2) - k^2]}$$

$$X_2 = \frac{kF_0}{[(k - m_1\omega^2)(k - m_2\omega^2) - k^2]} \qquad (9.112)$$

The force transmitted to the supporting structure (F_t) is given by the amplitude of $m_2\ddot{x}_2$:

$$F_t = -m_2\omega^2 X_2 = \frac{-m_2k\omega^2 F_0}{[(k - m_1\omega^2)(k - m_2\omega^2) - k^2]} \qquad (9.113)$$

The transmissibility of the isolator (T_f) is given by

$$T_f = \frac{F_t}{F_0}$$

$$= \frac{-m_2k\omega^2}{[(k - m_1\omega^2)(k - m_2\omega^2) - k^2]}$$

$$= \frac{1}{\left(\dfrac{m_1 + m_2}{m_2} - \dfrac{m_1\omega^2}{k}\right)}$$

$$= \frac{m_2}{(m_1 + m_2)}\left(\dfrac{1}{1 - \dfrac{\omega^2}{\omega_2^2}}\right) \qquad (9.114)$$

where ω_2 is the natural frequency of the system given by Eq. (9.111). Equation (9.114) shows, as in the case of an isolator on a rigid base, that the force transmitted to the foundation becomes less as the natural frequency of the system ω_2 is reduced.

**9.10.4
Vibration
Isolation System
with Partially
Flexible
Foundation**

Figure 9.29 shows a more realistic situation in which the base of the isolator, instead of being completely rigid or completely flexible, is partially flexible [9.34]. We can define the mechanical impedance of the base structure, $Z(\omega)$, as the force at frequency ω required to produce a unit displacement of the base (as in Section 3.5):

$$Z(\omega) = \frac{\text{Applied force of frequency } \omega}{\text{Displacement}}$$

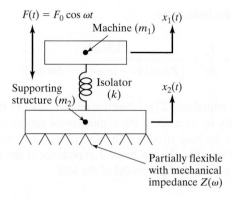

FIGURE 9.29 Machine with isolator on a partially flexible foundation.

The equations of motion are given by[7]

$$m_1\ddot{x}_1 + k(x_1 - x_2) = F_0 \cos \omega t \tag{9.115}$$

$$k(x_2 - x_1) = -x_2 Z(\omega) \tag{9.116}$$

By substituting the harmonic solution

$$x_j(t) = X_j \cos \omega t, \qquad j = 1, 2 \tag{9.117}$$

into Eqs. (9.115) and (9.116), X_1 and X_2 can be obtained as in the previous case:

$$X_1 = \frac{[k + Z(\omega)]X_2}{k} = \frac{[k + Z(\omega)]F_0}{[Z(\omega)(k - m_1\omega^2) - km_1\omega^2]}$$

$$X_2 = \frac{kF_0}{[Z(\omega)(k - m_1\omega^2) - km_1\omega^2]} \tag{9.118}$$

The amplitude of the force transmitted is given by

$$F_t = X_2 Z(\omega) = \frac{kZ(\omega)F_0}{[Z(\omega)(k - m_1\omega^2) - km_1\omega^2]} \tag{9.119}$$

---

[7]If the base is completely flexible with an unconstrained mass of m_2, $Z(\omega) = -\omega^2 m_2$, and Eqs. (9.115) to (9.117) lead to Eq. (9.109).

and the transmissibility of the isolator by

$$T_f = \frac{F_t}{F_0} = \frac{kZ(\omega)}{[Z(\omega)(k - m_1\omega^2) - km_1\omega^2]} \tag{9.120}$$

In practice, the mechanical impedance $Z(\omega)$ depends on the nature of the base structure. It can be found experimentally by measuring the displacement produced by a vibrator that applies a harmonic force on the base structure. In some cases—for example, if an isolator is resting on a concrete raft on soil—the mechanical impedance at any frequency ω can be found in terms of the spring-mass-dashpot model of the soil.

9.10.5
Shock Isolation

As stated earlier, a shock load involves the application of a force for a short duration, usually for less than one natural time period of the system. The forces involved in forge hammers, punch presses, blasts, and explosions are examples of shock loads. Shock isolation can be defined as a procedure by which the undesirable effects of shock are reduced. We noted that vibration isolation under a harmonic disturbance (input) occurs for the frequency ratio $r > \sqrt{2}$, with a smaller value of the damping ratio (ζ) leading to better isolation. On the other hand, shock isolation must occur over a wide range of frequencies, usually with large values of ζ. Thus a good vibration isolation design proves to be a poor shock isolation design and vice versa. In spite of the differences, the basic principles involved in shock isolation are similar to those of vibration isolation; however, the equations are different due to the transient nature of the shock.

A short-duration shock load $F(t)$, applied over a time period T, can be treated as an impulse F:

$$F = \int_0^T F(t)\, dt \tag{9.121}$$

Since this impulse acts on the mass m, the principle of impulse-momentum can be applied to find the velocity imparted to the mass (v) as

$$v = \frac{F}{m} \tag{9.122}$$

This indicates that the application of a short-duration shock load can be considered as equivalent to giving an initial velocity to the system. Thus the response of the system under the shock load can be determined as the free-vibration solution with a specified initial velocity. By assuming the initial conditions as $x(0) = x_0 = 0$ and $\dot{x}(0) = \dot{x}_0 = v$, the free-vibration solution of a viscously damped single-degree-of-freedom system (displacement of the mass m) can be found from Eq. (2.72) as

$$x(t) = \frac{ve^{-\zeta\omega_n t}}{\omega_d} \sin \omega_d t \tag{9.123}$$

where $\omega_d = \sqrt{1 - \zeta^2}\,\omega_n$ is the frequency of damped vibrations. The force transmitted to the foundation, $F_t(t)$, due to the spring and the damper is given by

$$F_t(t) = kx(t) + c\dot{x}(t) \tag{9.124}$$

Using Eq. (9.123), $F_t(t)$ can be expressed as

$$F_t(t) = \frac{v}{\omega_d}\sqrt{(k - c\zeta\omega_n)^2 + (c\omega_d)^2}\; e^{-\zeta\omega_n t}\sin(\omega_d t + \phi) \tag{9.125}$$

where

$$\phi = \tan^{-1}\left(\frac{c\omega_d}{k - c\zeta\omega_n}\right) \tag{9.126}$$

Equations (9.125) and (9.126) can be used to find the maximum value of the force transmitted to the foundation.

For longer-duration shock loads, the maximum transmitted force can occur while the shock is being applied. In such cases, the shock spectrum, discussed in Section 4.6, can be used to find the maximum force transmitted to the foundation.

The following examples illustrate different approaches that can be used for the design of shock isolators.

EXAMPLE 9.11

Isolation Under Shock

An electronic instrument of mass 20 kg is subject to a shock in the form of a step velocity of 2 m/s. If the maximum allowable values of deflection (due to clearance limit) and acceleration are specified as 20 mm and 25g, respectively, determine the spring constant of an undamped shock isolator.

Solution: The electronic instrument supported on the spring can be considered as an undamped system subject to base motion (in the form of step velocity). The mass vibrates at the natural frequency of the system with the magnitudes of velocity and acceleration given by

$$\dot{x}_{\text{max}} = X\omega_n \tag{E.1}$$

$$\ddot{x}_{\text{max}} = -X\omega_n^2 \tag{E.2}$$

where X is the amplitude of displacement of the mass. Since the maximum value of (step) velocity is specified as 2 m/s and the maximum allowable value of X is given to be 0.02 m, Eq. (E.1) yields

$$X = \frac{\dot{x}_{\text{max}}}{\omega_n} < 0.02 \quad\text{or}\quad \omega_n > \frac{\dot{x}_{\text{max}}}{X} = \frac{2}{0.02} = 100 \text{ rad/s} \tag{E.3}$$

Similarly, using the maximum specified value of $\ddot{x}_{max}$ as 25g, Eq. (E.2) gives

$$X\omega_n^2 \leq 25\,(9.81) = 245.25 \text{ m/s}^2 \quad \text{or} \quad \omega_n \leq \sqrt{\frac{\ddot{x}_{max}}{X}} = \sqrt{\frac{245.25}{0.02}} = 110.7362 \text{ rad/s} \quad \text{(E.4)}$$

Equations (E.3) and (E.4) give 100 rad/s $\leq \omega_n \leq$ 110.7362 rad/s. By selecting the value of ω_n in the middle of the permissible range as 105.3681 rad/s, the stiffness of the spring (isolator) can be found as

$$k = m\omega_n^2 = 20\,(105.3681)^2 = 2.2205 \times 10^5 \text{ N/m} \quad \text{(E.5)}$$

∎

EXAMPLE 9.12

Isolation Under Step Load

A sensitive electronic instrument of mass 100 kg is supported on springs and packaged for shipment. During shipping, the package is dropped from a height that effectively applied a shock load of intensity F_0 to the instrument, as shown in Fig. 9.30(a). Determine the stiffness of the springs used in the package if the maximum deflection of the instrument is required to be less than 2 mm. The response spectrum of the shock load is shown in Fig. 9.30(b) with $F_0 = 1,000$ N and $t_0 = 0.1$ s.

Solution: The response spectrum, indicating the maximum response of an undamped single-degree-of-freedom system subject to the given shock, is given by

$$\frac{x_{max}k}{F_0} = 1 + \frac{1}{\omega_n t_0}\sqrt{2\,(1 - \cos 2\omega_n t_0)} \quad \text{(E.1)}$$

where ω_n is the natural frequency of the system:

$$\omega_n = \sqrt{\frac{k}{m}} = \sqrt{\frac{k}{100}} = 0.1\sqrt{k} \quad \text{(E.2)}$$

$F_0 = 1,000$ N, $t_0 = 0.1$ s, and k is the stiffness of the springs used in the package. Using the known data, Eq. (E.1) can be expressed as

$$\frac{x_{max}k}{1,000} = 1 + \frac{1}{0.1\sqrt{k}\,(0.1)}\sqrt{2\,(1 - \cos 2(0.1\sqrt{k})\,(0.1))} \leq \frac{2}{1,000}\left(\frac{k}{1,000}\right) \quad \text{(E.3)}$$

By using the equality sign, Eq. (E.3) can be rearranged as

$$\frac{100}{\sqrt{k}}\sqrt{2(1 - \cos 0.02\sqrt{k})} - 2 \times 10^{-6}k + 1 = 0 \quad \text{(E.4)}$$

FIGURE 9.30   Shock load on electronic instrument.

The root of Eq. (E.4) gives the desired stiffness value as $k = 6.2615 \times 10^5$ N/m. The following MATLAB program can be used to find the root of Eq. (E.4):

```
>> x=1000:1:10000000;
>> f='(100/sqrt(x))*sqrt(2*(1-cos(0.02*sqrt(x))))-0.000002*x+1';
>> root=fzero(f,100000)

root =

 6.2615e+005

>>
```

■

9.10.6
Active Vibration
Control

A vibration isolation system is called *active* if it uses external power to perform its function. It consists of a servomechanism with a sensor, signal processor, and an actuator, as shown schematically in Fig. 9.31 [9.31–9.33]. This system maintains a constant distance (l) between the vibrating mass and the reference plane. As the force $F(t)$ applied to the system (mass) varies, the distance l tends to vary. This change in l is sensed by the sensor and a signal, proportional to the magnitude of the excitation (or response) of the vibrating

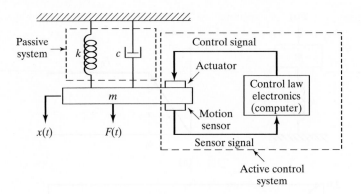

FIGURE 9.31   Active vibration isolation system.

body, is produced. The signal processor produces a command signal to the actuator based on the sensor signal it receives. The actuator develops a motion or force proportional to the command signal. The actuator motion or force will control the base displacement such that the distance l is maintained at the desired constant value.

Different types of sensors are available to create feedback signals based on the displacement, velocity, acceleration, jerk, or force. The signal processor may consist of a passive mechanism, such as a mechanical linkage, or an active electronic or fluidic network that can perform functions such as addition, integration, differentiation, attenuation, or amplification. The actuator may be a mechanical system such as a rack-and-pinion or ball screw mechanism, a fluidic system, or piezoelectric and electromagnetic force generating system. Depending on the types of sensor, signal processor, and actuator used, an active vibration control system can be called *electromechanical, electrofluidic, electromagnetic, piezoelectric,* or *fluidic*.

Analysis: Consider a single-degree-of-freedom system in which the mass m is subjected to an applied force $f(t)$ as shown in Fig. 9.31. If we use an active control system to control the vibration of the mass m, the actuator will be designed to exert a control force $f_c(t)$ so that the equation of motion of the system becomes

$$m\ddot{x} + c\dot{x} + kx = F(t) = f(t) + f_c(t) \tag{9.127}$$

Most commonly, the sensor (computer) measures the displacement x and the velocity $\dot{x}$ of the mass in real time (continuously). The computer computes the control force $f_c(t)$ necessary to control the motion and commands the actuator to exert the force $f_c(t)$ on the mass m.

Usually the computer is programmed to generate the control force proportional to the displacement $x(t)$ and the displacement derivative or velocity $\dot{x}(t)$ of the mass so that

$$f_c(t) = -g_p x - g_d \dot{x} \tag{9.128}$$

where g_p and g_d are constants whose values are to be determined and programmed into the computer by the designer. The constants g_p and g_d are known as *control gains*, with g_p denoting the proportional gain and g_d indicating the derivative or rate gain. The control algorithm in this case is known as the *proportional and derivative (PD) control*. By substituting Eq. (9.128) into Eq. (9.127), we obtain

$$m\ddot{x} + (c + g_d)\dot{x} + (k + g_p)x = f(t) \tag{9.129}$$

which shows that g_d acts like additional (or artificial) damping and g_p like additional (or artificial) stiffness. Equation (9.129), known as the *closed-loop equation*, can be solved to find the response characteristics of the system. For example, the new (effective) natural frequency is given by

$$\omega_n = \left(\frac{k + g_p}{m}\right)^{\frac{1}{2}} \tag{9.130}$$

and the new (effective) damping ratio by

$$\zeta = \frac{c + g_d}{2\sqrt{m(k + g_p)}} \tag{9.131}$$

The new (effective) time constant of the system, for $\zeta \leq 1$, is given by

$$\tau = \frac{2m}{c + g_d} \tag{9.132}$$

Thus the functioning of the active vibration control system can be described as follows: Given the values of m, c, and k, compute the control gains g_p and g_d to achieve the desired values of ω_n, ζ, or τ. In practice, the response of the system is continuously monitored, the computations are done, and the actuator is made to apply the control force f_c to the mass in real time so that the response of the system lies within the stated limits. Note that the gains g_p and g_d can be positive or negative depending on the measured and desired responses.

Vibration Control of a Precision Electronic System

EXAMPLE 9.13

It is proposed to control the vibration of a precision electronic system supported on an elastic pad (with no damping) by either a passive or an active method. The system has a mass of 15 kg and a natural frequency of 20 rad/s. It is estimated that the system requires a damping ratio of $\zeta = 0.85$ to control the vibration. Assume that the available dashpots can provide damping constants only in the range $0 \leq c \leq 400$ N-s/m.

Solution: First, we investigate the use of an available dashpot to control the vibration (passive control). From the known natural frequency of the system, we can find the stiffness of the elastic pad as

$$\omega_n = \sqrt{\frac{k}{m}} \quad \text{or} \quad k = m\omega_n^2 = 15(20)^2 = 6{,}000 \text{ N/m} \tag{E.1}$$

The required damping ratio of the system gives the necessary damping constant (c) as

$$\zeta = \frac{c}{2\sqrt{km}} = 0.85 \quad \text{or} \quad c = 2\zeta\sqrt{km} = 2(0.85)\sqrt{6{,}000(15)} = 510 \text{ N-s/m} \tag{E.2}$$

Since the available dashpots can provide damping constant values up to 400 N-s/m only, we cannot achieve the desired control using passive damping.

Thus we consider an active control system to create the required amount of damping into the system. Let the control force be of the form $f_c = -g_d\dot{x}$, so that the damping ratio, alternate form of Eq. (9.131), can be expressed (with $g_p = 0$):

$$2\zeta\omega_n = \frac{c + g_d}{m} \tag{E.3}$$

By adding the available dashpot, with a damping constant of 400 N-s/m, Eq. (E.3) can be rewritten as

$$400 + g_d = 2m\zeta\omega_n = 2(15)(0.85)(20) = 510 \text{ N-s/m}$$

or

$$g_d = 110 \text{ N-s/m}$$

This gives the value of the damping constant to be provided by the active control (also known as *derivative gain*) as $g_d = 110$ N-s/m.

∎

Active Control of a System with Rotating Unbalance

EXAMPLE 9.14

A single-degree-of-freedom system consists of a mass $(m) = 150$ kg, damping constant $(c) = 4{,}000$ N-s/m, and stiffness $(k) = 6 \times 10^6$ N/m. The mass is subjected to a rotating unbalanced force given by $f(t) = 100 \sin 60\pi t$ N. The following observations can be made from the given data:

(i) The natural frequency of the system, $\omega_n = \sqrt{\dfrac{k}{m}} = \sqrt{\dfrac{6(10^6)}{150}} = 200$ rad/s, is close to the frequency of the disturbance, $\omega = 60\pi = 188.4955$ rad/s.

(ii) The damping ratio of the system is small with a value of

$$\zeta = \frac{c}{2\sqrt{km}} = \frac{4{,}000}{2\sqrt{[6(10^6)(160)]}} = 0.06667$$

It is desired to change the natural frequency of the system to 100 rad/s and the damping ratio to 0.5. Because the values of k and c of the system cannot be altered, it is proposed to use an active control system. Determine the control gains required to achieve the desired values of ω_n and ζ. Also find the magnitude of the response and the actuator force of the system in the steady state.

Solution: When an active control system is used with control gains g_p and g_d, the natural frequency of the system can be expressed as

$$\omega_n = 100 = \sqrt{\frac{6(10^6) + g_p}{150}}$$

or

$$g_p = 150(10^4) - 6(10^6) = -4.5(10^6) \text{ N/m}$$

This implies that the stiffness of the system is to be reduced to 1.5×10^6 N/m. The new damping ratio of the system is given by

$$\zeta = 0.5 = \frac{c + g_d}{2\sqrt{km}} = \frac{4{,}000 + g_d}{2\sqrt{[1.5(10^6)](150)}}$$

or

$$g_d = 15{,}000 - 4{,}000 = 11{,}000 \text{ N-s/m}$$

This implies that the damping of the system is to be increased to 15,000 N-s/m.
The equation of motion of the actively controlled system can be written as

$$m\ddot{x} + c\dot{x} + kx = f(t) = f_0 \sin \omega t \tag{E.1}$$

which, in this case, takes the form

$$150\ddot{x} + 15{,}000\dot{x} + 1.5(10^6)x = f(t) = 100 \sin 60\pi t \tag{E.2}$$

From Eq. (E.1), the general transfer function of the system can be expressed as (see Section 3.12)

$$\frac{X(s)}{F(s)} = \frac{1}{ms^2 + cs + k} \tag{E.3}$$

The magnitude of the steady-state response of the system corresponding to Eq. (E.3) is given by (see Section 3.13)

$$X = \frac{f_0}{\left[(k - m\omega^2)^2 + (c\omega)^2\right]^{\frac{1}{2}}} \tag{E.4}$$

In the present case, $f_0 = 100$ N, $m = 150$ kg, $c = 15,000$ N-s/m, $k = 1.5 \times 10^6$ N-s/m, and $\omega = 188.4955$ rad/s. Thus Eq. (E.4) gives

$$X = \frac{150}{\left[\{1.5(10^6) - 150(188.4955)^2\}^2 + \{15,000(188.4955)\}^2 \right]^{\frac{1}{2}}}$$

$$= \frac{150}{4.7602(10^6)}$$

$$= 31.5113(10^{-6}) \text{ N}$$

The actuator (control) force, F_t, at steady state can be obtained from the relation

$$\frac{F_t(s)}{F(s)} = \frac{F_t(s)}{X(s)} \frac{X(s)}{F(s)} = \frac{k + cs}{ms^2 + cs + k} \qquad \text{(E.5)}$$

as

$$F_t(i\omega) = |4.5(10^6) - 11,000i\omega| X(i\omega)$$

$$= |4.5(10^6) - 11,000(188.4955)i| (31.5113(10^{-6}))$$

$$= \sqrt{\{4.5(10^6)\}^2 + \{11,000(188.4955)\}^2} \, (31.5113(10^{-6}))$$

$$= 156.1289 \text{ N}$$

∎

9.11   Vibration Absorbers

The *vibration absorber*, also called *dynamic vibration absorber*, is a mechanical device used to reduce or eliminate unwanted vibration. It consists of another mass and stiffness attached to the main (or original) mass that needs to be protected from vibration. Thus the main mass and the attached absorber mass constitute a two-degree-of-freedom system, hence the vibration absorber will have two natural frequencies. The vibration absorber is commonly used in machinery that operates at constant speed, because the vibration absorber is tuned to one particular frequency and is effective only over a narrow band of frequencies. Common applications of the vibration absorber include reciprocating tools, such as sanders, saws, and compactors, and large reciprocating internal combustion engines which run at constant speed (for minimum fuel consumption). In these systems, the vibration absorber helps balance the reciprocating forces. Without a vibration absorber, the unbalanced reciprocating forces might make the device impossible to

Transmission line

Vibration absorber

FIGURE 9.32

hold or control. Vibration absorbers are also used on high-voltage transmission lines. In this case, the dynamic vibration absorbers, in the form of dumbbell-shaped devices (Fig. 9.32), are hung from transmission lines to mitigate the fatigue effects of wind-induced vibration.

A machine or system may experience excessive vibration if it is acted upon by a force whose excitation frequency nearly coincides with a natural frequency of the machine or system. In such cases, the vibration of the machine or system can be reduced by using a *vibration neutralizer* or *dynamic vibration absorber*, which is simply another spring-mass system. The dynamic vibration absorber is designed such that the natural frequencies of the resulting system are away from the excitation frequency. We shall consider the analysis of a dynamic vibration absorber by idealizing the machine as a single-degree-of-freedom system.

9.11.1 Undamped Dynamic Vibration Absorber

When we attach an auxiliary mass m_2 to a machine of mass m_1 through a spring of stiffness k_2, the resulting two-degree-of-freedom system will look as shown in Fig. 9.33. The equations of motion of the masses m_1 and m_2 are

$$m_1\ddot{x}_1 + k_1 x_1 + k_2(x_1 - x_2) = F_0 \sin \omega t$$

$$m_2\ddot{x}_2 + k_2(x_2 - x_1) = 0 \tag{9.133}$$

By assuming harmonic solution,

$$x_j(t) = X_j \sin \omega t, \qquad j = 1, 2 \tag{9.134}$$

we can obtain the steady-state amplitudes of the masses m_1 and m_2 as

$$X_1 = \frac{(k_2 - m_2\omega^2)F_0}{(k_1 + k_2 - m_1\omega^2)(k_2 - m_2\omega^2) - k_2^2} \tag{9.135}$$

$$X_2 = \frac{k_2 F_0}{(k_1 + k_2 - m_1\omega^2)(k_2 - m_2\omega^2) - k_2^2} \tag{9.136}$$

$F_0 \sin \omega t$

Machine (m_1)

$x_1(t)$

k_2

Isolator
($k_1/2$)

Isolator
($k_1/2$)

$x_2(t)$

Rigid base

Dynamic vibration
absorber

FIGURE 9.33   Undamped dynamic vibration
absorber.

We are primarily interested in reducing the amplitude of the machine (X_1). In order to make the amplitude of m_1 zero, the numerator of Eq. (9.135) should be set equal to zero. This gives

$$\omega^2 = \frac{k_2}{m_2} \qquad (9.137)$$

if the machine, before the addition of the dynamic vibration absorber, operates near its resonance, $\omega^2 \simeq \omega_1^2 = k_1/m_1$. Thus if the absorber is designed such that

$$\omega^2 = \frac{k_2}{m_2} = \frac{k_1}{m_1} \qquad (9.138)$$

the amplitude of vibration of the machine, while operating at its original resonant frequency, will be zero. By defining

$$\delta_{st} = \frac{F_0}{k_1}, \qquad \omega_1 = \left(\frac{k_1}{m_1}\right)^{1/2}$$

as the natural frequency of the machine or main system, and

$$\omega_2 = \left(\frac{k_2}{m_2}\right)^{1/2} \qquad (9.139)$$

as the natural frequency of the absorber or auxiliary system, Eqs. (9.135) and (9.136) can be rewritten as

$$\frac{X_1}{\delta_{st}} = \frac{1 - \left(\dfrac{\omega}{\omega_2}\right)^2}{\left[1 + \dfrac{k_2}{k_1} - \left(\dfrac{\omega}{\omega_1}\right)^2\right]\left[1 - \left(\dfrac{\omega}{\omega_2}\right)^2\right] - \dfrac{k_2}{k_1}} \tag{9.140}$$

$$\frac{X_2}{\delta_{st}} = \frac{1}{\left[1 + \dfrac{k_2}{k_1} - \left(\dfrac{\omega}{\omega_1}\right)^2\right]\left[1 - \left(\dfrac{\omega}{\omega_2}\right)^2\right] - \dfrac{k_2}{k_1}} \tag{9.141}$$

Figure 9.34 shows the variation of the amplitude of vibration of the machine (X_1/δ_{st}) with the machine speed (ω/ω_1). The two peaks correspond to the two natural frequencies of the composite system. As seen before, $X_1 = 0$ at $\omega = \omega_1$. At this frequency, Eq. (9.141) gives

$$X_2 = -\frac{k_1}{k_2}\delta_{st} = -\frac{F_0}{k_2} \tag{9.142}$$

This shows that the force exerted by the auxiliary spring is opposite to the impressed force $(k_2 X_2 = -F_0)$ and neutralizes it, thus reducing X_1 to zero. The size of the dynamic vibration absorber can be found from Eqs. (9.138) and (9.142):

$$k_2 X_2 = m_2\omega^2 X_2 = -F_0 \tag{9.143}$$

Thus the values of k_2 and m_2 depend on the allowable value of X_2.

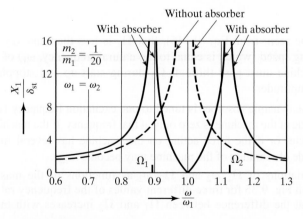

FIGURE 9.34   Effect of undamped vibration absorber on the response of machine.

It can be seen from Fig. 9.34 that the dynamic vibration absorber, while eliminating vibration at the known impressed frequency ω, introduces two resonant frequencies Ω_1 and Ω_2, at which the amplitude of the machine is infinite. In practice, the operating frequency ω must therefore be kept away from the frequencies Ω_1 and Ω_2. The values of Ω_1 and Ω_2 can be found by equating the denominator of Eq. (9.134) to zero. Noting that

$$\frac{k_2}{k_1} = \frac{k_2}{m_2}\frac{m_2}{m_1}\frac{m_1}{k_1} = \frac{m_2}{m_1}\left(\frac{\omega_2}{\omega_1}\right)^2 \tag{9.144}$$

and setting the denominator of Eq. (9.140) to zero leads to

$$\left(\frac{\omega}{\omega_2}\right)^4\left(\frac{\omega_2}{\omega_1}\right)^2 - \left(\frac{\omega}{\omega_2}\right)^2\left[1 + \left(1 + \frac{m_2}{m_1}\right)\left(\frac{\omega_2}{\omega_1}\right)^2\right] + 1 = 0 \tag{9.145}$$

The two roots of this equation are given by

$$\left.\begin{array}{c}\left(\dfrac{\Omega_1}{\omega_2}\right)^2 \\[2em] \left(\dfrac{\Omega_2}{\omega_2}\right)^2\end{array}\right\} = \frac{\left\{\left[1 + \left(1 + \dfrac{m_2}{m_1}\right)\left(\dfrac{\omega_2}{\omega_1}\right)^2\right] \pm \left\{\left[1 + \left(1 + \dfrac{m_2}{m_1}\right)\left(\dfrac{\omega_2}{\omega_1}\right)^2\right]^2 - 4\left(\dfrac{\omega_2}{\omega_1}\right)^2\right\}^{1/2}\right\}}{2\left(\dfrac{\omega_2}{\omega_1}\right)^2} \tag{9.146}$$

which can be seen to be functions of (m_2/m_1) and (ω_2/ω_1).

Notes

1. It can be seen, from Eq. (9.146), that Ω_1 is less than and Ω_2 is greater than the operating speed (which is equal to the natural frequency, ω_1) of the machine. Thus the machine must pass through Ω_1 during start-up and stopping. This results in large amplitudes.

2. Since the dynamic absorber is tuned to one excitation frequency (ω), the steady-state amplitude of the machine is zero only at that frequency. If the machine operates at other frequencies or if the force acting on the machine has several frequencies, then the amplitude of vibration of the machine may become large.

3. The variations of Ω_1/ω_2 and Ω_2/ω_2 as functions of the mass ratio m_2/m_1 are shown in Fig. 9.35 for three different values of the frequency ratio ω_2/ω_1. It can be seen that the difference between Ω_1 and Ω_2 increases with increasing values of m_2/m_1.

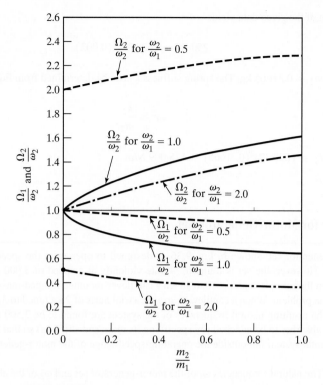

FIGURE 9.35 Variations of Ω_1 and Ω_2 given by Eq. (9.146).

EXAMPLE 9.15

Vibration Absorber for Diesel Engine

A diesel engine, weighing 3,000 N, is supported on a pedestal mount. It has been observed that the engine induces vibration into the surrounding area through its pedestal mount at an operating speed of 6,000 rpm. Determine the parameters of the vibration absorber that will reduce the vibration when mounted on the pedestal. The magnitude of the exciting force is 250 N, and the amplitude of motion of the auxiliary mass is to be limited to 2 mm.

Solution: The frequency of vibration of the machine is

$$f = \frac{6,000}{60} = 100 \text{ Hz} \quad \text{or} \quad \omega = 628.32 \text{ rad/s}$$

Since the motion of the pedestal is to be made equal to zero, the amplitude of motion of the auxiliary mass should be equal and opposite to that of the exciting force. Thus from Eq. (9.143), we obtain

$$|F_0| = m_2 \omega^2 X_2 \tag{E.1}$$

Substitution of the given data yields

$$250 = m_2 \,(628.32)^2(0.002)$$

Therefore, $m_2 = 0.31665$ kg. The spring stiffness k_2 can be determined from Eq. (9.138):

$$\omega^2 = \frac{k_2}{m_2}$$

Therefore, $k_2 = (628.32)^2(0.31665) = 125{,}009$ N/m.

∎

EXAMPLE 9.16

Absorber for Motor-Generator Set

A motor-generator set, shown in Fig. 9.36, is designed to operate in the speed range of 2,000 to 4,000 rpm. However, the set is found to vibrate violently at a speed of 3,000 rpm due to a slight unbalance in the rotor. It is proposed to attach a cantilever-mounted lumped-mass absorber system to eliminate the problem. When a cantilever carrying a trial mass of 2 kg tuned to 3,000 rpm is attached to the set, the resulting natural frequencies of the system are found to be 2,500 rpm and 3,500 rpm. Design the absorber to be attached (by specifying its mass and stiffness) so that the natural frequencies of the total system fall outside the operating-speed range of the motor-generator set.

Solution: The natural frequencies ω_1 of the motor-generator set and ω_2 of the absorber are given by

$$\omega_1 = \sqrt{\frac{k_1}{m_1}}, \qquad \omega_2 = \sqrt{\frac{k_2}{m_2}} \tag{E.1}$$

The resonant frequencies Ω_1 and Ω_2 of the combined system are given by Eq. (9.146). Since the absorber ($m = 2$ kg) is tuned, $\omega_1 = \omega_2 = 314.16$ rad/s (corresponding to 3,000 rpm). Using the notation

$$\mu = \frac{m_2}{m_1}, \qquad r_1 = \frac{\Omega_1}{\omega_2}, \qquad \text{and} \qquad r_2 = \frac{\Omega_2}{\omega_2}$$

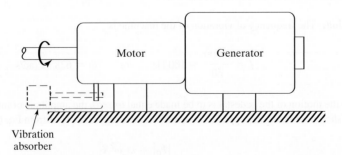

FIGURE 9.36   Motor-generator set.

Eq. (9.146) becomes

$$r_1^2, r_2^2 = \left(1 + \frac{\mu}{2}\right) \pm \sqrt{\left(1 + \frac{\mu}{2}\right)^2 - 1} \qquad \text{(E.2)}$$

Since Ω_1 and Ω_2 are known to be 261.80 rad/s (or 2,500 rpm) and 366.52 rad/s (or 3,500 rpm), respectively, we find that

$$r_1 = \frac{\Omega_1}{\omega_2} = \frac{261.80}{314.16} = 0.8333$$

$$r_2 = \frac{\Omega_2}{\omega_2} = \frac{366.52}{314.16} = 1.1667$$

Hence

$$r_1^2 = \left(1 + \frac{\mu}{2}\right) - \sqrt{\left(1 + \frac{\mu}{2}\right)^2 - 1}$$

or

$$\mu = \left(\frac{r_1^4 + 1}{r_1^2}\right) - 2 \qquad \text{(E.3)}$$

Since $r_1 = 0.8333$, Eq. (E.3) gives $\mu = m_2/m_1 = 0.1345$ and $m_1 = m_2/0.1345 = 14.8699$ kg. The specified lower limit of Ω_1 is 2,000 rpm or 209.44 rad/s, and so

$$r_1 = \frac{\Omega_1}{\omega_2} = \frac{209.44}{314.16} = 0.6667$$

With this value of r_1, Eq. (E.3) gives $\mu = m_2/m_1 = 0.6942$ and $m_2 = m_1(0.6942) = 10.3227$ kg. With these values, the second resonant frequency can be found from

$$r_2^2 = \left(1 + \frac{\mu}{2}\right) + \sqrt{\left(1 + \frac{\mu}{2}\right)^2 - 1} = 2.2497$$

which gives $\Omega_2 \simeq 4{,}499.4$ rpm, larger than the specified upper limit of 4,000 rpm. The spring stiffness of the absorber is given by

$$k_2 = \omega_2^2 m_2 = (314.16)^2(10.3227) = 1.0188 \times 10^6 \text{ N/m}$$

∎

The dynamic vibration absorber described in the previous section removes the original resonance peak in the response curve of the machine but introduces two new peaks. Thus the machine experiences large amplitudes as it passes through the first peak during start-up and stopping. The amplitude of the machine can be reduced by adding a damped vibration absorber, as shown in Fig. 9.37. The equations of motion of the two masses are given by

$$m_1\ddot{x}_1 + k_1 x_1 + k_2(x_1 - x_2) + c_2(\dot{x}_1 - \dot{x}_2) = F_0 \sin \omega t \qquad (9.147)$$

$$m_2\ddot{x}_2 + k_2(x_2 - x_1) + c_2(\dot{x}_2 - \dot{x}_1) = 0 \qquad (9.148)$$

By assuming the solution to be

$$x_j(t) = X_j e^{i\omega t}, \qquad j = 1, 2 \qquad (9.149)$$

the steady-state solution of Eqs. (9.147) and (9.148) can be obtained:

$$X_1 = \frac{F_0(k_2 - m_2\omega^2 + ic_2\omega)}{[(k_1 - m_1\omega^2)(k_2 - m_2\omega^2) - m_2 k_2\omega^2] + i\omega c_2(k_1 - m_1\omega^2 - m_2\omega^2)} \qquad (9.150)$$

$$X_2 = \frac{X_1(k_2 + i\omega c_2)}{(k_2 - m_2\omega^2 + i\omega c_2)} \qquad (9.151)$$

By defining

$$\mu = m_2/m_1 = \text{Mass ratio} = \text{Absorber mass/main mass}$$

$$\delta_{st} = F_0/k_1 = \text{Static deflection of the system}$$

$$\omega_a^2 = k_2/m_2 = \text{Square of natural frequency of the absorber}$$

$$\omega_n^2 = k_1/m_1 = \text{Square of natural frequency of the main mass}$$

$$f = \omega_a/\omega_n = \text{Ratio of natural frequencies}$$

FIGURE 9.37   Damped dynamic vibration absorber.

$$g = \omega/\omega_n = \text{Forced frequency ratio}$$

$$c_c = 2m_2\omega_n = \text{Critical damping constant}$$

$$\zeta = c_2/c_c = \text{Damping ratio}$$

the magnitudes, X_1 and X_2, can be expressed as

$$\frac{X_1}{\delta_{st}} = \left[\frac{(2\zeta g)^2 + (g^2 - f^2)^2}{(2\zeta g)^2(g^2 - 1 + \mu g^2)^2 + \{\mu f^2 g^2 - (g^2 - 1)(g^2 - f^2)\}^2} \right]^{1/2} \quad (9.152)$$

and

$$\frac{X_2}{\delta_{st}} = \left[\frac{(2\zeta g)^2 + f^4}{(2\zeta g)^2(g^2 - 1 + \mu g^2)^2 + \{\mu f^2 g^2 - (g^2 - 1)(g^2 - f^2)\}^2} \right]^{1/2} \quad (9.153)$$

Equation (9.152) shows that the amplitude of vibration of the main mass is a function of μ, f, g, and ζ. The graph of

$$\left| \frac{X_1}{\delta_{st}} \right|$$

against the forced frequency ratio $g = \omega/\omega_n$ is shown in Fig. 9.38 for $f = 1$ and $\mu = 1/20$ for a few different values of ζ.

   If damping is zero ($c_2 = \zeta = 0$), then resonance occurs at the two undamped resonant frequencies of the system, a result that is already indicated in Fig. 9.34. When the damping becomes infinite ($\zeta = \infty$), the two masses m_1 and m_2 are virtually clamped together, and the system behaves essentially as a single-degree-of-freedom system with a mass of $(m_1 + m_2) = (21/20)\,m$ and stiffness of k_1. In this case also, resonance occurs with $X_1 \to \infty$ at

$$g = \frac{\omega}{\omega_n} = \frac{1}{\sqrt{1 + \mu}} = 0.9759$$

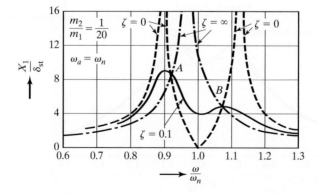

FIGURE 9.38   Effect of damped vibration absorber on the response of the machine.

Thus the peak of X_1 is infinite for $c_2 = 0$ as well as for $c_2 = \infty$. Somewhere in between these limits, the peak of X_1 will be a minimum.

Optimally-Tuned Vibration Absorber.   It can be seen from Fig. 9.38 that all the curves intersect at points A and B regardless of the value of damping. These points can be located by substituting the extreme cases of $\zeta = 0$ and $\zeta = \infty$ into Eq. (9.152) and equating the two. This yields

$$g^4 - 2g^2 \left(\frac{1 + f^2 + \mu f^2}{2 + \mu} \right) + \frac{2f^2}{2 + \mu} = 0 \qquad (9.154)$$

The two roots of Eq. (9.154) indicate the values of the frequency ratio, $g_A = \omega_A/\omega$ and $g_B = \omega_B/\omega$, corresponding to the points A and B. The ordinates of A and B can be found by substituting the values of g_A and g_B, respectively, into Eq. (9.146). It has been observed [9.35] that the most efficient vibration absorber is one for which the ordinates of the points A and B are equal. This condition requires that [9.35]

$$f = \frac{1}{1 + \mu} \qquad (9.155)$$

An absorber satisfying Eq. (9.155) can be correctly called the *tuned vibration absorber.* Although Eq. (9.155) indicates how to tune an absorber, it does not indicate the optimal value of the damping ratio ζ and the corresponding value of X_1/δ_{st}. The optimal value of ζ can be found by making the response curve X_1/δ_{st} as flat as possible at peaks A and B. This can be achieved by making the curve horizontal at either A or B, as shown in Fig. 9.39. For this, first Eq. (9.155) is substituted into Eq. (9.152) to make the resulting equation applicable to the case of optimum tuning. Then the modified Eq. (9.152) is differentiated with

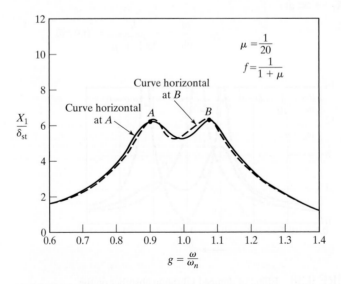

FIGURE 9.39   Tuned vibration absorber.

respect to g to find the slope of the curve of X_1/δ_{st}. By setting the slope equal to zero at points A and B, we obtain

$$\zeta^2 = \frac{\mu\left\{3 - \sqrt{\dfrac{\mu}{\mu + 2}}\right\}}{8(1 + \mu)^3} \qquad \text{for point } A \tag{9.156}$$

and

$$\zeta^2 = \frac{\mu\left\{3 + \sqrt{\dfrac{\mu}{\mu + 2}}\right\}}{8(1 + \mu)^3} \qquad \text{for point } B \tag{9.157}$$

A convenient average value of ζ^2 given by Eqs. (9.156) and (9.157) is used in design so that

$$\zeta^2_{\text{optimal}} = \frac{3\mu}{8(1 + \mu)^3} \tag{9.158}$$

The corresponding optimal value of $\left(\dfrac{X_1}{\delta_{st}}\right)$ becomes

$$\left(\frac{X_1}{\delta_{st}}\right)_{\text{optimal}} = \left(\frac{X_1}{\delta_{st}}\right)_{\text{max}} = \sqrt{1 + \frac{2}{\mu}} \tag{9.159}$$

Notes

1. It can be seen from Eq. (9.153) that the amplitude of the absorber mass (X_2) is always much greater than that of the main mass (X_1). Thus the design should be able to accommodate the large amplitudes of the absorber mass.

2. Since the amplitudes of m_2 are expected to be large, the absorber spring (k_2) needs to be designed from a fatigue point of view.

3. Most vibration absorbers used in practical applications are undamped. If damping is added, it defeats the purpose of the vibration absorber, which is to eliminate unwanted vibration. In a damped vibration absorber, the amplitude of vibration of the main mass will be nonzero. Damping is to be added only in situations in which the frequency band in which the absorber is effective is too narrow for operation.

4. Additional work relating to the optimum design of vibration absorbers can be found in [9.36–9.39].

9.12   Examples Using MATLAB

Plotting of Transmissibility

EXAMPLE 9.17

Using MATLAB, plot the variation of transmissibility of a single-degree-of-freedom system with the frequency ratio, given by Eq. (9.94), corresponding to $\zeta = 0.0, 0.1, 0.2, 0.3, 0.4,$ and 0.5.

Solution: The following MATLAB program plots the variation of transmissibility as a function of the frequency ratio using Eq. (9.94):

```
%Exam 9-17
for j = 1 : 5
    kesi = j * 0.1;
    for i = 1 : 1001
        w_wn(i) = 3 * (i - 1)/1000;
        T(i) = sqrt((1 + (2 * kesi * w_wn(i)) ^ 2)/((1 - w_wn(i) ^ 2)
        ^ 2 + ∠
2 * kesi * w_wn(i) ^2));
    end;
    plot(w_wn, T);
    hold on;
end;

xlabel ('w/w_n');
ylabel('Tr');
gtext('zeta = 0.1');
gtext('zeta = 0.2');
gtext('zeta = 0.3');
gtext('zeta = 0.4');
gtext('zeta = 0.5');
title('Ex9.2');
grid on;
```

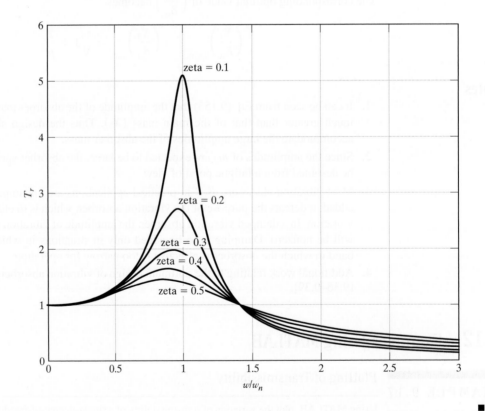

■

Vibration Amplitudes of Masses of Vibration Absorber

EXAMPLE 9.18

Using MATLAB, plot the variations of vibration amplitudes of the main and auxiliary masses of a vibration absorber, Eqs. (9.140) and (9.141), as functions of the frequency ratio.

Solution: Equations (9.140) and (9.141) are plotted for the following data: $f = \omega_a/\omega_n = 1$, $\zeta = 0.1$ and 0.5, $\mu = m_2/m_1 = 0.05$ and 0.1.

```
f = 1;
%------     zeta = 0.1, mu=0.05 -------------------------------------✓
-------
zeta = 0.1;
mu = 0.05;
g = 0.6 : 0.001 : 1.3;

tzg2 = (2.*zeta.*g).^2 ;%---   tzg2 = (2*zeta*g)^2
g2_f2_2 = (g.^2-f.^2).^2 ;% g2_f2_2 = (g^2-f^2)^2
g2_1mug2_2 = (g.^2-1+mu.*g.^2).^2;
muf2g2 = mu.*f.^2*g.^2 ;
g2_1 = g.^2-1 ;
g2_f2 = g.^2-f.^2 ;

x1r =sqrt((tzg2+g2_f2_2)./(tzg2.*g2_1mug2_2+(muf2g2-g2_1.*g2_f2).^2));
x2r =sqrt((tzg2+f.^4)./(tzg2.*g2_1mug2_2+(muf2g2-g2_1.*g2_f2).^2));
plot(g,x1r)
hold on
plot(g,x2r);
hold on
%------     zeta = 0.1, mu=0.01 -------------------------------------✓
------
zeta = 0.1;
mu = 0.1; 0.001:1.3;
g = 0.6:

tzg2 = (2.*zeta.*g).^2 ;%---- tzg2 = (2*zeta*g)^2
g2_f2_2 = (g.^2-f.^2).^2 ;% g2_f2_2 = (g^2-f^2)^2
g2_1mug2_2 = (g.^2-1+mu.*g.^2).^2;
muf2g2 = mu.*f.^2*g.^2 ;
g2_1 = g.^2-1 ;
g2_f2 = g.^2-f.^2 ;

x1r =sqrt((tzg2+g2_f2_2)./(tzg2.*g2_1mug2_2+(muf2g2-g2_1.*g2_f2).^2));
x2r =sqrt((tzg2+f.^4)./(tzg2.*g2_1mug2_2+(muf2g2-g2_1.*g2_f2).^2));
plot(g,x1r,'-.');
hold on
plot(g,x2r,'-.');
hold on
%------     zeta = 0.5, mu=0.05 -------------------------------------✓
------
zeta = 0.5;
mu = 0.05;
g = 0.6 : 0.001 : 1.3;

tzg2 = (2.*zeta.*g).^2 ;%---- tzg2 = (2*zeta*g)^2
g2_f2_2 = (g.^2-f.^2).^2) ;% g2_f2_2 = (g^2-f^2)^2
g2_1mug2_2 = (g.^2-1+mu.*g.^2).^2;
muf2g2 = mu.*f.^2*g.^2;
g2_1 = g.^2-1 ;
g2_f2 = g.^2-f.^2;

x1r =sqrt((tzg2+g2_f2_2)./(tzg2.*g2_1mug2_2+(muf2g2-g2_1.*g2_f2).^2));
x2r =sqrt((tzg2+f.^4)./(tzg2.*g2_1mug2_2+(muf2g2-g2_1.*g2_f2).^2));
plot(g,x1r,'--');
hold on
```

```
plot(g,x2r,'-');
hold on
%------    zeta = 0.5, mu=0.1 ----------------------------------------
------
zeta = 0.5;
mu = 0.1;
g = 0.6 : 0.001 : 1.3;

tzg2 = (2.*zeta.*g).^2 ;%---- tzg2 = (2*zeta*g)^2
g2_f2_2 = (g.^2-f.^2).^2 ;% g2_f2_2 = (g^2-f^2)^2
g2_1mug2_2 = (g.^2-1+mu.*g.^2).^2 ;
muf2g2 = mu.*f.^2*g.^2 ;
g2_1 = g.^2-1 ;
g2_f2 = g.^2-f.^2 ;


x1r =sqrt((tzg2+g2_f2_2)./(tzg2.*g2_1mug2_2+(muf2g2-g2_1.*g2_f2).^2));
x2r =sqrt((tzg2+f.^4)./(tzg2.*g2_1mug2_2+(muf2g2-g2_1.*g2_f2).^2));
plot(g,x1r,':');
hold on
plot(g,x2r,':');
xlabel('g')
ylabel('X1r and X2r')
axis ([0.6 1.3 0 16])
```

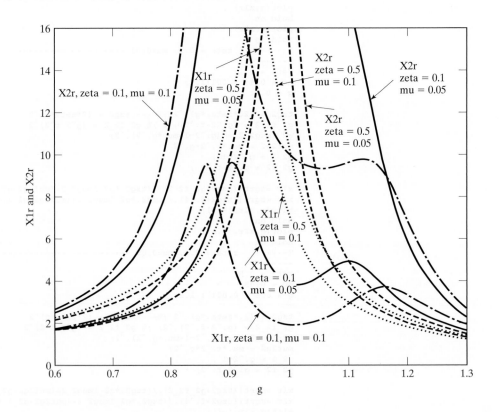

EXAMPLE 9.19

Resonant Frequencies of Vibration Absorber

Using MATLAB, plot the variations of the resonant frequency ratios given by Eq. (9.146) with the mass ratio, m_2/m_1.

Solution: The ratios Ω_1/ω_2 and Ω_2/ω_2, given by Eq. (9.146), are plotted for $\omega_2/\omega_1 = 0.5$, 1.0, and 2.0 over the range of $m_2/m_1 = 0$ to 1.

```
%------ omega2/omega1=0.5 ----------------------------------------
---
omega21=0.5
m21 = 0:0.001:1.0
X11 = sqrt(((1 + (1+m21)*omega21.^2) +((1+(1+m21).*omega21.^2).^2-
4.*omega21.^2).^0.5)...
      /(2.*omega21.^2))
plot(m21,X11,':')
axis([0 1.0 0.0 2.6])
hold on

X12 = sqrt(((1+(1+m21)*omega21.^2) - ((1 + (1+m21).*omega21.^2).^2-
4.*omega21.^2).^0.5)...
      /(2.*omega21.^2))
plot(m21,X12,':')
hold on

%------ omega2/omega1=1.0 ----------------------------------------
---

omega21=1.0
m21 = 0:0.001:1.0
X21 = sqrt(((1+(1+m21)*omega21.^2) +((1+(1+m21).*omega21.^2).^2-
4.*omega21.^2).^0.5)...
      /(2.*omega21.^2))
plot(m21,X21,'-')
axis([0 1.0 0.0 2.6])
hold on

X22 = sqrt(((1+(1+m21)*omega21.^2) -((1+(1+m21).*omega21.^2).^2-
4.*omega21.^2).^0.5)...
      /(2.*omega21.^2))
plot(m21,X22,'-')
hold on

%------ omega2/omega1=2.0 ----------------------------------------
---

omega21=2.0
m21 = 0 : 0.001 : 1.0
X31 = sqrt(((1+(1+m21)*omega21.^2) +((1+(1+m21).*omega21.^2).^2).
^2-4.*omega21.^2).^0.5)...
      / (2.*omega21.^2))
plot(m21,X31,'-.')
axis([0 1.0 0.0 2.6])
hold on

X32 = sqrt(((1+(1+m21)*omega21.^2) -((1+(1+m21).*omega21.^2).^2-4.
*omega21.^2) .^0.5)...
      /(2.*omega21.^2))
plot(m21,X32,'-.')
hold on
xlabel ('mr')
ylabel ('OM1 and OM2')
```

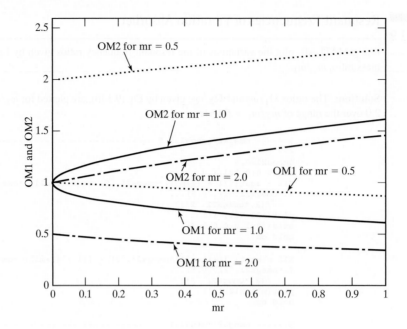

■

EXAMPLE 9.20

Two-Plane Balancing

Develop a general MATLAB program called **Program13.m** for the two-plane balancing of rotating machines. Use the program to solve Example 9.2.

Solution: Program13.m is developed to accept the vectors $\vec{V_A}$, $\vec{V_B}$, $\vec{V_A'}$, $\vec{V_B'}$, $\vec{V_A''}$, $\vec{V_B''}$, $\vec{W_L}$, and $\vec{W_R}$ as input in the form of two-dimensional arrays *VA*, *VB*, *VAP*, *VBP*, *VAPP*, *VBPP*, *WL*, *WR*, *BL*, and *BR*, respectively. The program gives the vectors B_L and B_R as output in the form of two-dimensional arrays *BL* and *BR* indicating the magnitude and position of the balancing weights in the left and right planes, respectively. The listing of the program and the output are given below.

```
%========================================================================
%
% Program13.m
% Two-plane balancing
%
%========================================================================
% Run "Program13" in MATLAB command window. Progrm13.m, balan.m,
vsub.m,
% vdiv.m and vmult.m should be in the same folder,and set the
MATLAB path
% to this folder.
% following 8 lines contain problem-dependent data
va=[8.5 60];
vap=[6 125];
wl=[10 270];
vb=[6.5 205];
```

```
vbp=[4.5 230];
vapp=[6 35];
vbpp=[10.5 160];
wr=[12 180];
% end of problem-dependent data
[bl,br]=balan(va,vb,vap,vbp,vapp,vbpp,wl,wr);
fprintf('                 Results of two-plane balancing \n\n');
fprintf('Left-plane balancing weight   Right-plane balancing weight');
fprintf('\n\n');
fprintf('Magnitude=%8.6f              Magnitude=%8.6f \n\n',bl(1),br(1));
fprintf('Angel=%8.6f                  Angle=%8.6f \n\n',bl(2),br(2));
%=====================================================================
%
%Function Balan.m
%
%=====================================================================
function [bl,br]=balan(va,vb,vap,vbp,vapp,vbpp,wl,wr);
pi=180/3.1415926;
va(2)=va(2)/pi;
p(1)=va(1);
p(2)=va(2);
va(1)=p(1)*cos(p(2));
va(2)=p(1)*sin(p(2));
vb(2)=vb(2)/pi;
p(1)=vb(1);
p(2)=vb(2);
vb(1)=p(1)*cos(p(2));
vb(2)=p(1)*sin(p(2));
vap(2)=vap(2)/pi;
p(1)=vap(1);
p(2)=vap(2);
vap(1)=p(1)*cos(p(2));
vap(2)=p(1)*sin(p(2));
vbp(2)=vbp(2)/pi;
p(1)=vbp(1);
p(2)=vbp(2);
vbp(1)=p(1)*cos(p(2));
vbp(2)=p(1)*sin(p(2));
vapp(2)=vapp(2)/pi;
p(1)=vapp(1);
p(2)=vapp(2);
vapp(1)=p(1)*cos(p(2));
vapp(2)=p(1)*sin(p(2));
vbpp(2)=vbpp(2)/pi;
p(1)=vbpp(1);
p(2)=vbpp(2);
vbpp(1)=p(1)*cos(p(2));
vbpp(2)=p(1)*sin(p(2));
wl(2)=wl(2)/pi;
p(1)=wl(1);
p(2)=wl(2);
wl(1)=p(1)*cos(p(2));
wl(2)=p(1)*sin(p(2));
wr(2)=wr(2)/pi;
p(1)=wr(1);
p(2)=wr(2);
wr(1)=p(1)*cos(p(2));
wr(2)=p(1)*sin(p(2));
[r]=vsub(vap,va);
[aal]=vdiv(r,wl);
[s]=vsub(vbp,vb);
[abl]=vdiv(s,wl);
[p]=vsub(vapp,va);
[aar]=vdiv(p,wr);
[q]=vsub(vbpp,vb);
[abr]=vdiv(q,wr);
```

```
[ar1]=sqrt(aar(1)^2+aar(2)^2);
[ar2]=atan(aar(2)/aar(1))*pi;
[al1]=sqrt(aal(1)^2+aal(2)^2);
[al2]=atan(aal(2)/aal(1))*pi;
[r]=vmult(abl,va);
[s]=vmult(aal,vb);
[vap]=vsub(r,s);
[r]=vmult(aar,abl);
[s]=vmult(aal,abr);
[vbp]=vsub(r,s);
[ur]=vdiv(vap,vbp);
[r]=vmult(abr,va);
[s]=vmult(aar,vb);
[vap]=vsub(r,s);
[r]=vmult(abr,aal);
[s]=vmult(aar,abl);
[vbp]=vsub(r,s);
[ul]=vdiv(vap,vbp);
bl(1)=sqrt(ul(1)^2+ul(2)^2);
a1=ul(2)/ul(1);
bl(2)=atan(ul(2)/ul(1));
br(1)=sqrt(ur(1)^2+ur(2)^2);
a2=ur(2)/ur(1);
br(2)=atan(ur (2)/ur (1));
bl(2)=bl(2)*pi;
br(2)=br(2)*pi;
bl(2)=bl(2)+180;
br(2)=br(2)+180;


%=====================================================================
%
%Function vdiv.m
%
%=====================================================================
function [c]=vdiv(a,b);
c(1)=(a(1)*b(1)+a(2)*b(2))/(b(1)^2+b(2)^2);
c(2)=(a(2)*b(1)-a(1)*b(2))/(b(1)^2+b(2)^2);
%=====================================================================
%
%Function vmult.m
%
%=====================================================================
function [c]=vmult(a,b);
c(1)=a(1)*b(1)-a(2)*b(2);
c(2)=a(2)*b(1)+a(1)*b(2);


%=====================================================================
%
%Function vsub.m
%
%=====================================================================
function [c]=vsub(a,b);
c(1)=a(1)-b(1);
c(2)=a(2)-b(2);
```

                     Results of two-plane balancing

Left-plane balancing weight   Right-plane balancing weight

Magnitude=10.056139   Magnitude=5.877362

Angle=145.554799   Angle=248.255931

■

CHAPTER SUMMARY

We discussed the use of vibration nomographs and vibration criteria to determine acceptable levels of vibration. We presented several methods, such as balancing of rotating and reciprocating machines, to eliminate or reduce vibration at the source. We outlined methods of changing mass and/or stiffness and dissipating energy by adding damping. We discussed methods of designing vibration isolators, vibration absorbers, and active vibration-control systems. We presented the solution of vibration-control problems using MATLAB.

Now that you have finished this chapter, you should be able to answer the review questions and solve the problems given below.

REFERENCES

9.1 J. E. Ruzicka, "Fundamental concepts of vibration control," *Sound and Vibration*, Vol. 5, July 1971, pp. 16–22.

9.2 J. A. Macinante, *Seismic Mountings for Vibration Isolation*, Wiley, New York, 1984.

9.3 International Organization for Standardization, *Mechanical Vibration of Machines with Operating Speeds from 10 to 200 rev/s—Basis for Specifying Evaluation Standards*, ISO 2372, 1974.

9.4 International Organization for Standardization, *Evaluation and Measurement of Vibration in Buildings*, Draft Proposal, ISO DP 4866, 1975.

9.5 R. J. Steffens, "Some aspects of structural vibration," in *Proceedings of the Symposium on Vibrations in Civil Engineering*, B. O. Skipp (ed.), Butterworths, London, 1966, pp. 1–30.

9.6 International Organization for Standardization, *Guide for the Evaluation of Human Exposure to Whole-Body Vibration*, ISO 2631, 1974.

9.7 C. Zenz, *Occupational Medicine: Principles and Practical Application* (2nd ed.), Year Book Medical Publishers, Chicago, 1988.

9.8 R. L. Fox, "Machinery vibration monitoring and analysis techniques," *Sound and Vibration*, Vol. 5, November 1971, pp. 35–40.

9.9 D. G. Stadelbauer, "Dynamic balancing with microprocessors," *Shock and Vibration Digest*, Vol. 14, December 1982, pp. 3–7.

9.10 J. Vaughan, *Static and Dynamic Balancing* (2nd ed.), Bruel and Kjaer Application Notes, Naerum, Denmark.

9.11 R. L. Baxter, "Dynamic balancing," *Sound and Vibration*, Vol. 6, April 1972, pp. 30–33.

9.12 J. H. Harter and W. D. Beitzel, *Mathematics Applied to Electronics*, Reston Publishing, Reston, VA, 1980.

9.13 R. G. Loewy and V. J. Piarulli, "Dynamics of rotating shafts," *Shock and Vibration Monograph SVM-4*, Shock and Vibration Information Center, Naval Research Laboratory, Washington, DC, 1969.

9.14 J. D. Irwin and E. R. Graf, *Industrial Noise and Vibration Control*, Prentice Hall, Englewood Cliffs, NJ, 1979.

9.15 T. Iwatsuba, "Vibration of rotors through critical speeds," *Shock and Vibration Digest*, Vol. 8, No. 2, February 1976, pp. 89–98.

9.16 R. J. Trivisonno, "Fortran IV computer program for calculating critical speeds of rotating shafts," NASA TN D-7385, 1973.

9.17 R. E. D. Bishop and G. M. L. Gladwell, "The vibration and balancing of an unbalanced flexible rotor," *Journal of Mechanical Engineering Science*, Vol. 1, 1959, pp. 66–77.

9.18 A. G. Parkinson, "The vibration and balancing of shafts rotating in asymmetric bearings," *Journal of Sound and Vibration*, Vol. 2, 1965, pp. 477–501.

9.19 C. E. Crede, *Vibration and Shock Isolation*, Wiley, New York, 1951.

9.20 W. E. Purcell, "Materials for noise and vibration control," *Sound and Vibration*, Vol. 16, July 1982, pp. 6–31.

9.21 B. C. Nakra, "Vibration control with viscoelastic materials," *Shock and Vibration Digest*, Vol. 8, No. 6, June 1976, pp. 3–12.

9.22 G. R. Tomlinson, "The use of constrained layer damping in vibration control," in *Modern Practice in Stress and Vibration Analysis*, J. E. Mottershead (ed.), Pergamon Press, Oxford, 1989, pp. 99–107.

9.23 D. E. Baxa and R. A. Dykstra, "Pneumatic isolation systems control forging hammer vibration," *Sound and Vibration*, Vol. 14, May 1980, pp. 22–25.

9.24 E. I. Rivin, "Vibration isolation of industrial machinery—Basic considerations," *Sound and Vibration*, Vol. 12, November 1978, pp. 14–19.

9.25 C. M. Salerno and R. M. Hochheiser, "How to select vibration isolators for use as machinery mounts," *Sound and Vibration*, Vol. 7, August 1973, pp. 22–28.

9.26 C. A. Mercer and P. L. Rees, "An optimum shock isolator," *Journal of Sound and Vibration*, Vol. 18, 1971, pp. 511–520.

9.27 M. L. Munjal, "A rational synthesis of vibration isolators," *Journal of Sound and Vibration*, Vol. 39, 1975, pp. 247–263.

9.28 C. Ng and P. F. Cunniff, "Optimization of mechanical vibration isolation systems with multi-degrees of freedom," *Journal of Sound and Vibration*, Vol. 36, 1974, pp. 105–117.

9.29 S. K. Hati and S. S. Rao, "Cooperative solution in the synthesis of multidegree of freedom shock isolation systems," *Journal of Vibration, Acoustics, Stress, and Reliability in Design*, Vol. 105, 1983, pp. 101–103.

9.30 S. S. Rao and S. K. Hati, "Optimum design of shock and vibration isolation systems using game theory," *Journal of Engineering Optimization*, Vol. 4, 1980, pp. 1–8.

9.31 J. E. Ruzicka, "Active vibration and shock isolation," Paper no. 680747, *SAE Transactions*, Vol. 77, 1969, pp. 2872–2886.

9.32 R. W. Horning and D. W. Schubert, "Air suspension and active vibration-isolation systems," in *Shock and Vibration Handbook* (3rd ed.), C. M. Harris (ed.), McGraw-Hill, New York, 1988.

9.33 O. Vilnay, "Active control of machinery foundation," *Journal of Engineering Mechanics*, ASCE, Vol. 110, 1984, pp. 273–281.

9.34 J. I. Soliman and M. G. Hallam, "Vibration isolation between non-rigid machines and non-rigid foundations," *Journal of Sound and Vibration*, Vol. 8, 1968, pp. 329–351.

9.35 J. Ormondroyd and J. P. Den Hartog, "The theory of the dynamic vibration absorber," *Transactions of ASME*, Vol. 50, 1928, p. APM-241.

9.36 H. Puksand, "Optimum conditions for dynamic vibration absorbers for variable speed systems with rotating and reciprocating unbalance," *International Journal of Mechanical Engineering Education*, Vol. 3, April 1975, pp. 145–152.

9.37 A. Soom and M.-S. Lee, "Optimal design of linear and nonlinear absorbers for damped systems," *Journal of Vibration, Acoustics, Stress, and Reliability in Design,* Vol. 105, 1983, pp. 112–119.

9.38 J. B. Hunt, *Dynamic Vibration Absorbers,* Mechanical Engineering Publications, London, 1979.

REVIEW QUESTIONS

9.1   Give brief answers to the following:

1. Name some sources of industrial vibration.
2. What are the various methods available for vibration control?
3. What is single-plane balancing?
4. Describe the two-plane balancing procedure.
5. What is whirling?
6. What is the difference between stationary damping and rotary damping?
7. How is the critical speed of a shaft determined?
8. What causes instability in a rotor system?
9. What considerations are to be taken into account for the balancing of a reciprocating engine?
10. What is the function of a vibration isolator?
11. What is a vibration absorber?
12. What is the difference between a vibration isolator and a vibration absorber?
13. Does spring mounting always reduce the vibration of the foundation of a machine?
14. Is it better to use a soft spring in the flexible mounting of a machine? Why?
15. Is the shaking force proportional to the square of the speed of a machine? Does the vibratory force transmitted to the foundation increase with the speed of the machine?
16. Why does dynamic balancing imply static balancing?
17. Explain why dynamic balancing can never be achieved by a static test alone.
18. Why does a rotating shaft always vibrate? What is the source of the shaking force?
19. Is it always advantageous to include a damper in the secondary system of a dynamic vibration absorber?
20. What is active vibration isolation?
21. Explain the difference between passive and active isolation.

9.2   Indicate whether each of the following statements is true or false:

1. Vibration can cause structural and mechanical failures.
2. The response of a system can be reduced by the use of isolators and absorbers.
3. Vibration control means the elimination or reduction of vibration.
4. The vibration caused by a rotating unbalanced disc can be eliminated by adding a suitable mass to the disc.
5. Any unbalanced mass can be replaced by two equivalent unbalanced masses in the end planes of the rotor.
6. The oil whip in the bearings can cause instability in a rotor system.
7. The natural frequency of a system can be changed by varying its damping.
8. The stiffness of a rotating shaft can be altered by changing the location of its bearings.
9. All practical systems have damping.

10. High loss factor of a material implies less damping.
11. Passive isolation systems require external power to function.
12. The transmissibility is also called the *transmission ratio*.
13. The force transmitted to the foundation of an isolator with rigid foundation can never be infinity.
14. Internal and external friction can cause instability in a rotating shaft at speeds above the first critical speed.

9.3   Fill in each of the following blanks with the appropriate word:

1. Even a small excitation force can cause an undesirably large response near _____.
2. The use of close tolerances and better surface finish for machine parts tends to make a machine _____ susceptible to vibration.
3. The presence of unbalanced mass in a rotating disc causes _____.
4. When the speed of rotation of a shaft equals one of the natural frequencies of the shaft, it is called _____ *speed*.
5. The moving elements of a reciprocating engine are the crank, the connecting rod, and the _____.
6. The vertical component of the inertia force of a reciprocating engine has primary and _____ parts.
7. Laminated structures have _____ damping.
8. Materials with a large value of the loss factor are subject to _____ stress.
9. Vibration isolation involves insertion of a resilient member between the vibrating mass and the _____ of vibration.
10. Cork is a _____ isolator.
11. An active isolator consists of a sensor, a signal processor, and an _____.
12. Vibration neutralizer is also known as *dynamic vibration* _____.
13. Although an undamped vibration absorber removes the original resonance peak of the response, it introduces _____ new peaks.
14. The single-plane balancing is also known as _____ *balancing*.
15. Phase marks are used in _____ plane balancing using a vibration analyzer.
16. Machine errors can cause _____ in rotating machines.
17. The combustion instabilities are a source of _____ in engines.
18. The deflection of a rotating shaft becomes very large at the _____ speed.
19. Oil whip in bearings can cause _____ in a flexible rotor system.

9.4   Select the most appropriate answer out of the multiple choices given:

1. An example of a source of vibration that cannot be altered is:
   a. atmospheric turbulance
   b. hammer blow
   c. tire stiffness of an automobile

2. The two-plane balancing is also known as:
   a. *static balancing*
   b. *dynamic balancing*
   c. *proper balancing*

3. The unbalanced force caused by an eccentric mass m rotating at an angular speed ω and located at a distance r from the axis of rotation is
   a. $mr^2\omega^2$ b. $mg\omega^2$ c. $mr\omega^2$

4. The following material has high internal damping:
   a. cast iron   **b.** copper   **c.** brass

5. Transmissibility is the ratio of
   a. force transmitted and exciting force
   b. force applied and the resulting displacement
   c. input displacement and output displacement

6. Mechanical impedance is the ratio of
   a. force transmitted and exciting force
   b. force applied and force transmitted
   c. applied force and displacement

7. Vibration can be eliminated on the basis of theoretical analysis
   a. sometimes   **b.** always   **c.** never

8. A long rotor can be balanced by adding weights in
   a. a single plane   **b.** any two planes   **c.** two specific planes

9. The damping caused by the internal friction of a shaft material is called
   a. *stationary damping*
   b. *external damping*
   c. *rotary damping*

10. The damping caused by the bearing support structure of a rotating shaft is called
    a. *stationary damping*
    b. *internal damping*
    c. *rotary damping*

11. An undamped vibration absorber removes the original resonance peak but introduces
    a. one new peak   **b.** two new peaks   **c.** several new peaks

9.5 Match the items in the two columns below.

1. Control natural frequency	**a.** Introduce damping
2. Avoid excessive response at resonance	**b.** Use vibration isolator
3. Reduce transmission of excitation force from one part to another	**c.** Add vibration absorber
4. Reduce response of the system	**d.** Avoid resonance

PROBLEMS

Section 9.2 Vibration Criteria

9.1 An automobile moving on a rough road, in the form of a sinusoidal surface, is modeled as a spring-mass system, as shown in Fig. 9.40. The sinusoidal surface has a wavelength of 5 m and an amplitude of $Y = 1$ mm. If the mass of the automobile, including the passengers, is 1,500 kg and the stiffness of the suspension system (k) is 400 kN/m, determine the range of speeds (v) of the automobile in which the passengers perceive the vibration. Suggest possible methods of improving the design for a more comfortable ride of the passengers.

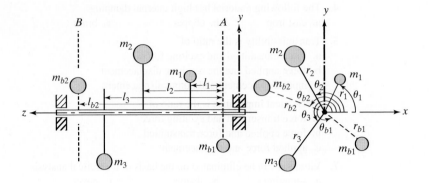

FIGURE 9.40

9.2 The root mean square value of a signal $x(t)$, x_{rms}, is defined as

$$x_{rms} = \left\{ \lim_{T \to \infty} \frac{1}{T} \int_0^T x^2(t)\, dt \right\}^{1/2}$$

Using this definition, find the root mean square values of the displacement (x_{rms}), velocity $(\dot{x}_{rms})$, and acceleration $(\ddot{x}_{rms})$ corresponding to $x(t) = X \cos \omega t$.

Section 9.4 Balancing of Rotating Machines

9.3 Two identical discs are connected by four bolts of different sizes and mounted on a shaft, as shown in Fig. 9.41. The masses and locations of three bolts are as follows:

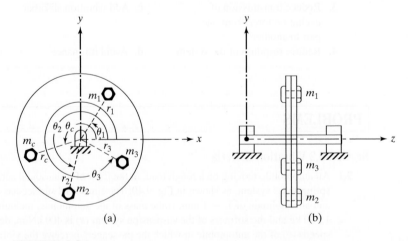

(a)   (b)

FIGURE 9.41

$m_1 = 35$ grams, $r_1 = 110$ mm, and $\theta_1 = 40°$; $m_2 = 15$ grams, $r_2 = 90$ mm, and $\theta_2 = 220°$; and $m_3 = 25$ grams, $r_3 = 130$ mm, and $\theta_3 = 290°$. Find the mass and location of the fourth bolt (m_c, r_c, and θ_c), which results in the static balance of the discs.

9.4 Four holes are drilled in a uniform circular disc at a radius of 0.1 m and angles of 0°, 60°, 120°, and 180°. The mass removed at holes 1 and 2 is 100 g each and the mass removed at holes 3 and 4 is 150 g each. If the disc is to be balanced statically by drilling a fifth hole at a radius of 0.12 m, find the mass to be removed and the angular location of the fifth hole.

9.5 Three masses of 0.5 kg, 0.7 kg, and 1.2 kg, are attached around the rim, of diameter 10 m, of a flywheel at the angular locations $\theta = 10°$, 100°, and 190°, respectively. Find the mass and the angular location of the fourth mass to be attached on the rim that leads to the dynamic balance of the flywheel.

9.6 The amplitude and phase angle due to original unbalance in a grinding wheel operating at 1,200 rpm are found to be 0.2 mm and 40° counterclockwise from the phase mark. When a trial mass $M = 0.2$ kg is added at 65° clockwise from the phase mark and at a radial distance 65 mm from the center of rotation, the amplitude and phase angle are observed to be 0.5 mm and 150° counterclockwise. Find the magnitude and angular position of the balancing mass if it is to be located 65 mm radially from the center of rotation.

9.7 An unbalanced flywheel shows an amplitude of 0.15 mm and a phase angle of 15° clockwise from the phase mark. When a trial mass of magnitude 50 g is added at an angular position 45° counterclockwise from the phase mark, the amplitude and the phase angle become 0.22 mm and 35° counterclockwise, respectively. Find the magnitude and angular position of the balancing mass required. Assume that the masses are added at the same radius.

9.8 In order to determine the unbalance in a grinding wheel rotating clockwise at 2,400 rpm, a vibration analyzer is used and an amplitude of 0.1 mm and a phase angle of 45° are observed with the original unbalance. When a trial mass $M = 0.1$ kg is added at 20° clockwise from the phase mark, the amplitude becomes 0.2 mm and the phase angle 145°. If the phase angles are measured counterclockwise from the right-hand horizontal, calculate the magnitude and location of the necessary balancing masses.

9.9 A turbine rotor is run at the natural frequency of the system. A stroboscope indicates that the maximum displacement of the rotor occurs at an angle 229° in the direction of rotation. At what angular position must mass be removed from the rotor in order to improve its balancing?

9.10 A rotor, having three eccentric masses in different planes, is shown in Fig. 9.42. The axial, radial, and angular locations of mass m_i are given by l_i, r_i, and θ_i, respectively, for $i = 1, 2, 3$. If the rotor is to be dynamically balanced by locating two masses m_{b1} and m_{b2} at radii r_{b1} and r_{b2} at the angular locations θ_{b1} and θ_{b2}, as shown in Fig. 9.42, derive expressions for $m_{b1}r_{b1}$, $m_{b2}r_{b2}$, θ_{b1}, and θ_{b2}.

9.11 The rotor shown in Fig. 9.43(a) is balanced temporarily in a balancing machine by adding the masses $M_1 = M_2 = 0.1$ kg in the plane A and $M_3 = M_4 = 0.1$ kg in the plane D at a radius of 0.08 m, as shown in Fig. 9.43(b). If the rotor is permanently balanced by drilling holes at a radius of 0.1 m in planes B and C, determine the position and amount of material to be removed from the rotor. Assume that the adjustable masses M_1 to M_4 will be removed from the planes A and D.

9.12 Masses of 1 kg, 2 kg, and 1.5 kg are located at radii 0.5 m, 0.8 m, and 0.3 m in the planes C, D, and E, respectively, on a shaft supported at the bearings B and F, as shown in Fig. 9.44. Find the masses and angular locations of the two balancing masses to be placed in the end planes A and G so that the dynamic load on the bearings will be zero.

FIGURE 9.42

(a)   (b)

FIGURE 9.43

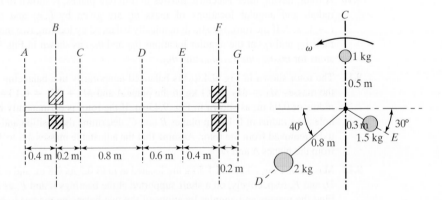

FIGURE 9.44

9.13 The data obtained in a two-plane balancing procedure are given in the table below. Determine the magnitude and angular position of the balancing masses, assuming that all angles are measured from an arbitrary phase mark and all masses are added at the same radius.

	Amplitude (mm)		Phase Angle	
Condition	Bearing A	Bearing B	Bearing A	Bearing B
Original unbalance	5	4	100°	180°
M_L = 2 kg added at 30° in the left plane	6.5	4.5	120°	140°
M_R = 2 kg added at 0° in the right plane	6	7	90°	60°

9.14 Figure 9.45 shows a rotating system in which the shaft is supported in bearings at A and B. The three masses m_1, m_2, and m_3 are connected to the shaft as indicated in the figure. (a) Find the bearing reactions at A and B if the speed of the shaft is 1,000 rpm. (b) Determine the locations and magnitudes of the balancing masses to be placed at a radius of 0.25 m in the planes L and R, which can be assumed to pass through the bearings A and B.

FIGURE 9.45

Section 9.5 Whirling of Rotating Shafts

9.15 A flywheel, with a mass of 45 kg and an eccentricity of 1 cm, is mounted at the center of a steel shaft of diameter 2 cm. If the length of the shaft between the bearings is 75 cm and the rotational speed of the flywheel is 1,200 rpm, find (a) the critical speed, (b) the vibration amplitude of the rotor, and (c) the force transmitted to the bearing supports.

9.16 Derive the expression for the stress induced in a shaft with an unbalanced concentrated mass located midway between two bearings.

9.17 A steel shaft of diameter 2.5 cm and length 1 m is supported at the two ends in bearings. It carries a turbine disc, of mass 20 kg and eccentricity 0.005 m, at the middle and operates at 6,000 rpm. The damping in the system is equivalent to viscous damping with ζ = 0.01. Determine the whirl amplitude of the disc at (a) operating speed, (b) critical speed, and (c) 1.5 times the critical speed.

9.18 Find the bearing reactions and the maximum bending stress induced in the shaft at (a) operating speed, (b) critical speed, and (c) 1.5 times the critical speed for the shaft-rotor system described in Problem 9.17.

9.19 Solve Problem 9.17 by assuming that the material of the shaft is aluminum rather than steel.

9.20 Solve Problem 9.18 by assuming that the material of the shaft is aluminum rather than steel.

9.21 A shaft, having a stiffness of 3.75 MN/m, rotates at 3,600 rpm. A rotor, having a mass of 60 kg and an eccentricity of 2,000 microns, is mounted on the shaft. Determine (a) the steady-state whirl amplitude of the rotor and (b) the maximum whirl amplitude of the rotor during start-up and stopping conditions. Assume the damping ratio of the system as 0.05.

Section 9.6 Balancing of Reciprocating Engines

9.22 The cylinders of a four-cylinder in-line engine are placed at intervals of 30 cm in the axial direction. The cranks have the same length, 10 cm, and their angular positions are given by 0°, 180°, 180°, and 0°. If the length of the connecting rod is 25 cm and the reciprocating mass is 1 kg for each cylinder, find the unbalanced forces and moments at a speed of 3,000 rpm, using the center line through cylinder 1 as the reference plane.

9.23 The reciprocating mass, crank radius, and connecting-rod length of each of the cylinders in a two-cylinder in-line engine are given by m, r, and l, respectively. The crank angles of the two cylinders are separated by 180°. Find the unbalanced forces and moments in the engine.

9.24 A four-cylinder in-line engine has a reciprocating mass of 1.5 kg, a stroke of 15 cm, and a connecting-rod length of 25 cm in each cylinder. The cranks are separated by 10 cm axially and 90° radially, as shown in Fig. 9.46. Find the unbalanced primary and secondary forces and moments with respect to the reference plane shown in Fig. 9.46 at an engine speed of 1,500 rpm.

FIGURE 9.46

9.25 The arrangement of cranks in a six-cylinder in-line engine is shown in Fig. 9.47. The cylinders are separated by a distance a in the axial direction, and the angular positions of the cranks are given by $\alpha_1 = \alpha_6 = 0°$, $\alpha_2 = \alpha_5 = 120°$, and $\alpha_3 = \alpha_4 = 240°$. If the crank length, connecting-rod length, and the reciprocating mass of each cylinder are r, l, and m, respectively, find the primary and secondary unbalanced forces and moments with respect to the reference plane indicated in Fig. 9.47.

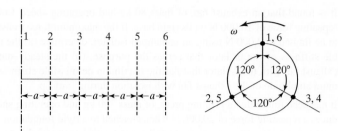

FIGURE 9.47

9.26 A single-cylinder engine has a total mass of 150 kg. Its reciprocating mass is 5 kg, and the rotating mass is 2.5 kg. The stroke ($2r$) is 15 cm, and the speed is 600 rpm. (a) If the engine is mounted floating on very weak springs, what is the amplitude of vertical vibration of the engine? (b) If the engine is mounted solidly on a rigid foundation, what is the alternating force amplitude transmitted? Assume the connecting rod to be of infinite length.

Section 9.10 Vibration Isolation

9.27 An electronic instrument is to be isolated from a panel that vibrates at frequencies ranging from 25 Hz to 35 Hz. It is estimated that at least 80 percent vibration isolation must be achieved to prevent damage to the instrument. If the instrument weighs 85 N, find the necessary static deflection of the isolator.

9.28* An exhaust fan, having a small unbalance, weighs 800 N and operates at a speed of 600 rpm. It is desired to limit the response to a transmissibility of 2.5 as the fan passes through resonance during start-up. In addition, an isolation of 90 percent is to be achieved at the operating speed of the fan. Design a suitable isolator for the fan.

9.29* An air compressor of mass 500 kg has an eccentricity of 50 kg-cm and operates at a speed of 300 rpm. The compressor is to be mounted on one of the following mountings: (a) an isolator consisting of a spring with negligible damping, and (b) a shock absorber having a damping ratio of 0.1 and negligible stiffness. Select a suitable mounting and specify the design details by considering the static deflection of the compressor, the transmission ratio, and the amplitude of vibration of the compressor.

9.30 The armature of a variable-speed electric motor, of mass 200 kg, has an unbalance due to manufacturing errors. The motor is mounted on an isolator having a stiffness of 10 kN/m and a dashpot having a damping ratio of 0.15. (a) Find the speed range over which the amplitude of the fluctuating force transmitted to the foundation will be larger than the exciting force. (b) Find the speed range over which the transmitted force amplitude will be less than 10 percent of the exciting force amplitude.

9.31 A dishwashing machine having a mass of 68 kg operates at 300 rpm. Find the minimum static deflection of an isolator that provides 60 percent isolation. Assume that the damping in the isolator is negligible.

9.32 A washing machine of mass 50 kg operates at 1,200 rpm. Find the maximum stiffness of an isolator that provides 75 percent isolation. Assume that the damping ratio of the isolator is 7 percent.

---

*The asterisk denotes a problem with no unique answer.

9.33 It is found that an exhaust fan, of mass 80 kg and operating speed 1,000 rpm, produces a repeating force of 10,000 N on its rigid base. If the maximum force transmitted to the base is to be limited to 2,000 N using an undamped isolator, determine (a) the maximum permissible stiffness of the isolator that serves the purpose; (b) the steady-state amplitude of the exhaust fan with the isolator that has the maximum permissible stiffness; and (c) the maximum amplitude of the exhaust fan with isolation during start-up.

9.34 It has been found that a printing press, of mass 300 kg and operating speed 3,000 rpm, produces a repeating force of 30,000 N when attached to a rigid foundation. Find a suitable viscously damped isolator to satisfy the following requirements: (a) the static deflection should be as small as possible; (b) the steady-state amplitude should be less than 2.5 mm; (c) the amplitude during start-up conditions should not exceed 20 mm; and (d) the force transmitted to the foundation should be less than 10,000 N.

9.35 A compressor of mass 120 kg has a rotating unbalance of 0.2 kg-m. If an isolator of stiffness 0.5 MN/m and damping ratio 0.06 is used, find the range of operating speeds of the compressor over which the force transmitted to the foundation will be less than 2,500 N.

9.36 An internal combustion engine has a rotating unbalance of 1.0 kg-m and operates between 800 and 2,000 rpm. When attached directly to the floor, it transmitted a force of 7,018 N at 800 rpm and 43,865 N at 2,000 rpm. Find the stiffness of the isolator that is necessary to reduce the force transmitted to the floor to 6,000 N over the operating-speed range of the engine. Assume that the damping ratio of the isolator is 0.08, and the mass of the engine is 200 kg.

9.37 A small machine tool of mass 100 kg operates at 600 rpm. Find the static deflection of an undamped isolator that provides 90 percent isolation.

9.38 A diesel engine of mass 300 kg and operating speed 1,800 rpm is found to have a rotating unbalance of 1 kg-m. It is to be installed on the floor of an industrial plant for the purpose of emergency power generation. The maximum permissible force that can be transmitted to the floor is 8,000 N and the only type of isolator available has a stiffness of 1 MN/m and a damping ratio of 5 percent. Investigate possible solutions to the problem.

9.39 The force transmitted by an internal combustion engine of mass 500 kg, when placed directly on a rigid floor, is given by

$$F_t(t) = (18,000 \cos 300t + 3,600 \cos 600\, t)\text{N}$$

Design an undamped isolator so that the maximum magnitude of the force transmitted to the floor does not exceed 12,000 N.

9.40 Design the suspension of an automobile such that the maximum vertical acceleration felt by the driver is less than $2g$ at all speeds between 40 and 80 mph while traveling on a road whose surface varies sinusoidally as $y(u) = 0.15 \sin 6.7u$ m, where u is the horizontal distance in meters. The mass of the automobile, with the driver, is 680 kg and the damping ratio of the suspension is to be 0.05. Use a single-degree-of-freedom model for the automobile.

9.41 Consider a single-degree-of-freedom system with Coulomb damping (which offers a constant friction force, F_c). Derive an expression for the force transmissibility when the mass is subjected to a harmonic force, $F(t) = F_0 \sin \omega t$.

9.42 Consider a single-degree-of-freedom system with Coulomb damping (which offers a constant friction force, F_c). Derive expressions for the absolute and relative displacement transmissibilities when the base is subjected to a harmonic displacement, $y(t) = Y \sin \omega t$.

9.43 When a washing machine, of mass 200 kg and an unbalance 0.02 kg-m, is mounted on an isolator, the isolator deflects by 5 mm under the static load. Find (a) the amplitude of the washing machine and (b) the force transmitted to the foundation at the operating speed of 1,200 rpm.

9.44 An electric motor, of mass 60 kg, rated speed 3,000 rpm, and an unbalance of 0.002 kg-m, is to be mounted on an isolator to achieve a force transmissibility of less than 0.25. Determine (a) the stiffness of the isolator, (b) the dynamic amplitude of the motor, and (c) the force transmitted to the foundation.

9.45 An engine is mounted on a rigid foundation through four springs. During operation, the engine produces an excitation force at a frequency of 3,000 rpm. If the weight of the engine causes the springs to deflect by 10 mm, determine the reduction in the force transmitted to the foundation.

9.46 A sensitive electronic system, of mass 30 kg, is supported by a spring-damper system on the floor of a building that is subject to a harmonic motion in the frequency range 10–75 Hz. If the damping ratio of the suspension is 0.25, determine the stiffness of the suspension if the amplitude of vibration transmitted to the system is to be less than 15 percent of the floor vibration over the given frequency range.

9.47 A machine having a mass of 1,200 kg is mounted on springs. A piston of mass $m = 30$ kg moves up and down in the machine at a speed of 600 rpm with a stroke of 38 cm. Considering the motion to be harmonic, determine the maximum force transmitted to the foundation if (a) $k = 1.8$ MN/m, and (b) $k = 4.4$ MN/m.

9.48 A printed circuit board of mass 1 kg is supported to the base through an undamped isolator. During shipping, the base is subjected to a harmonic disturbance (motion) of amplitude 2 mm and frequency 2 Hz. Design the isolator so that the displacement transmitted to the printed circuit board is to be no more than 5 percent of the base motion.

9.49 An electronic instrument of mass 10 kg is mounted on an isolation pad. If the base of the isolation pad is subjected to a shock in the form of a step velocity of 10 mm/s, find the stiffness of the isolation pad if the maximum permissible values of deflection and acceleration of the instrument are specified as 10 mm and $20g$, respectively.

9.50 A water tank of mass 10^5 kg is supported on a reinforced cement concrete column, as shown in Fig. 9.48(a). When a projectile hits the tank, it causes a shock, in the form of a step force, as shown in Fig. 9.48(b). Determine the stiffness of the column if the maximum deflection of the tank is to be limited to 0.5 m. The response spectrum of the shock load is shown in Fig. 9.48(c).

9.51 A viscously damped single-degree-of-freedom system has a body (mass) weighing 270 N with a spring constant of 700 N/cm. Its base is subjected to harmonic vibration. (a) When the base vibrates with an amplitude of 5 cm at resonance, the steady-state amplitude of the body is found to be 12.5 cm. Find the damping ratio of the system. (b) When the base vibrates at a frequency of 10 Hz, the steady-state amplitude of the body is found to be 4 cm. Find the magnitude of the force transmitted to the base.

9.52 A single-degree-of-freedom system is used to represent an automobile, of mass m, damping constant c, and stiffness k, which travels on a rough road that is in the form of a sinusoidal surface with an amplitude Y and wavelength l. If the automobile travels at a velocity v, derive an expression for the transmissibility of the vertical motion of the automobile mass (m).

9.53 A sensitive instrument of mass 100 kg is installed at a location that is subjected to harmonic motion with frequency 20 Hz and acceleration 0.5 m/s². If the instrument is supported on an isolator having a stiffness $k = 25 \times 10^4$ N/m and a damping ratio $\zeta = 0.05$, determine the maximum acceleration experienced by the instrument.

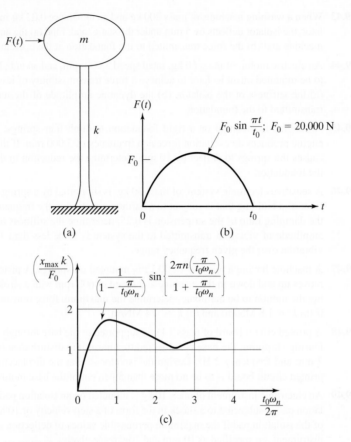

$F(t) \rightarrow$ m

k

(a)

$F(t)$

F_0

$F_0 \sin \dfrac{\pi t}{t_0}$; $F_0 = 20{,}000$ N

0 t_0 t

(b)

$\left(\dfrac{x_{max}\, k}{F_0}\right)$

$\dfrac{1}{\left(1 - \dfrac{\pi}{t_0 \omega_n}\right)} \sin\left[\dfrac{2\pi n\left(\dfrac{\pi}{t_0 \omega_n}\right)}{1 + \dfrac{\pi}{t_0 \omega_n}}\right]$

2

1

0 1 2 3 4 $\dfrac{t_0 \omega_n}{2\pi}$

(c)

FIGURE 9.48

9.54 An electronic instrument of mass 20 kg is to be isolated from engine vibrations with frequencies ranging from 1,000 rpm to 3,000 rpm. Find the stiffness of the undamped isolator to be used to achieve a 90% isolation.

9.55 A delicate instrument weighing 200 N is suspended by four identical springs, each with stiffness 50,000 N/m, in a rigid box as shown in Fig. 9.49. The box is transported by a truck. If the truck is subjected to a vertical harmonic motion given by $y(t) = 0.02 \sin 10t$ m, find the maximum displacement, velocity, and acceleration experienced by the instrument.

9.56 A damped torsional system is composed of a shaft and a rotor (disk). The torsional stiffness and the torsional damping constant of the shaft are given by $k_t = 6{,}000$ N-m/rad and $c_t = 100$ N-m-s/rad. The mass moment of inertia of the rotor is $J_0 = 5$ kg-m^2. The rotor is subjected to a harmonically varying torque of magnitude $M_t = 500$ N-m, which results in a steady-state angular displacement of $5°$. Find the frequency of the harmonically varying torque applied to the rotor and the maximum torque transmitted to the base or support of the system.

Rigid box

k k

$x(t)$

m

k k

$y(t) = 0.02 \sin 10t \ m$

Truck bed

FIGURE 9.49

9.57 The force transmissibility of a damped single-degree-of-freedom system with base motion is given by Eq. (9.106):

$$T_f = \frac{F_t}{kY} = r^2 \left\{ \frac{1 + (2\zeta r)^2}{(1 - r^2)^2 + (2\zeta r)^2} \right\}^{\frac{1}{2}}$$

where F_t is the magnitude of the force transmitted to the mass. Determine the frequency ratios (r) at which the force transmissibility attains maximum and minimum values. Discuss your results.

9.58 Derive an expression for the relative displacement transmissibility, $\dfrac{Z}{Y}$, where $Z = X - Y$, for a damped single-degree-of-freedom system subjected to the base motion, $y(t) = Y \sin \omega t$.

9.59 During operation, the compressor unit of a refrigerator, with mass 75 kg and rotational speed 900 rpm, experiences a dynamic force of 200 N. The compressor unit is supported on four identical springs, each with a stiffness of k and negligible damping. Find the value of k if only 15% of the dynamic force is to be transmitted to the support or base. Also find the clearance space to be provided to the compressor unit.

9.60 An electronic instrument, of mass 20 kg, is to be isolated to achieve a natural frequency of 15 rad/s and a damping ratio of 0.95. The available dashpots can produce a damping constant (c) in the range 10 N-s/m to 80 N-s/m. Determine whether the desired damping ratio can be achieved using a passive system. If a passive system cannot be used, design a suitable active control system to achieve the desired damping ratio.

9.61 A damped single-degree-of-freedom system has a mass (m) of 5 kg, stiffness (k) of 20 N/m, and a damping constant (c) of 5 N-s/m. Design an active controller to achieve a settling time of less than 15 s for the closed loop system.

Hint: The settling time is defined by Eqs. (4.68) and (4.69).

9.62 A damped single-degree-of-freedom system has an undamped natural frequency of 20 rad/s and a damping ratio of 0.20. Design an active control system which achieves an undamped natural frequency of 100 rad/s and a damping ratio of 0.8. Assume that the mass, stiffness, and damping constant of the original system remain in place.

Section 9.11 Vibration Absorbers

9.63 An air compressor of mass 200 kg, with an unbalance of 0.01 kg-m, is found to have a large amplitude of vibration while running at 1,200 rpm. Determine the mass and spring constant of the absorber to be added if the natural frequencies of the system are to be at least 20 percent from the impressed frequency.

9.64 An electric motor, having an unbalance of 2 kg-cm, is mounted at the end of a steel cantilever beam, as shown in Fig. 9.50. The beam is observed to vibrate with large amplitudes at the operating speed of 1,500 rpm of the motor. It is proposed to add a vibration absorber to reduce the vibration of the beam. Determine the ratio of the absorber mass to the mass of the motor needed in order to have the lower frequency of the resulting system equal to 75 percent of the operating speed of the motor. If the mass of the motor is 300 kg, determine the stiffness and mass of the absorber. Also find the amplitude of vibration of the absorber mass.

9.65* The pipe carrying feedwater to a boiler in a thermal power plant has been found to vibrate violently at a pump speed of 800 rpm. In order to reduce the vibrations, an absorber consisting of a spring of stiffness k_2 and a trial mass m_2' of 1 kg is attached to the pipe. This arrangement is found to give the natural frequencies of the system as 750 rpm and 1,000 rpm. It is desired to keep the natural frequencies of the system outside the operating speed range of the pump, which is 700 rpm to 1,040 rpm. Determine the values of k_2 and m_2 that satisfy this requirement.

FIGURE 9.50

9.66 A reciprocating engine is installed on the first floor of a building, which can be modeled as a rigid rectangular plate resting on four elastic columns. The equivalent mass of the engine and the floor is 900 kg. At the rated speed of the engine, which is 600 rpm, the operators experience large vibrations of the floor. It has been decided to reduce these vibrations by suspending a spring-mass system from the bottom surface of the floor. Assume that the spring stiffness is $k_2 = 900$ kN/m. (a) Find the mass to be attached to absorb the vibrations. (b) What will be the natural frequencies of the system after the absorber is added?

9.67* Find the values of k_2 and m_2 in Problem 9.54 in order to have the natural frequencies of the system at least 30 percent away from the forcing frequency.

9.68* A hollow steel shaft of outer diameter 5 cm, inner diameter 3.8 cm, and length 75 cm carries a solid disc of diameter 38 cm and mass 45 kg. Another hollow steel shaft of length 50 cm, carrying a solid disc of diameter 15 cm and mass 9 kg, is attached to the first disc, as shown in Fig. 9.51. Find the inner and outer diameters of the shaft such that the attached shaft-disc system acts as an absorber.

FIGURE 9.51

9.69* A rotor, having a mass moment of inertia $J_1 = 15$ kg-m², is mounted at the end of a steel shaft having a torsional stiffness of 0.6 MN-m/rad. The rotor is found to vibrate violently when subjected to a harmonic torque of 300 cos 200t N-m. A tuned absorber, consisting of a torsional spring and a mass moment of inertia (k_{t2} and J_2), is to be attached to the first rotor to absorb the vibrations. Find the values of k_{t2} and J_2 such that the natural frequencies of the system are away from the forcing frequency by at least 20 percent.

9.70 Plot the graphs of (Ω_1/ω_2) against (m_2/m_1) and (Ω_2/ω_2) against (m_2/m_1) as (m_2/m_1) varies from 0 to 1.0 when $\omega_2/\omega_1 = 0.1$ and 10.0.

9.71 Determine the operating range of the frequency ratio ω/ω_2 for an undamped vibration absorber to limit the value of $|X_1/\delta_{st}|$ to 0.5. Assume that $\omega_1 = \omega_2$ and $m_2 = 0.1m_1$.

9.72 When an undamped vibration absorber, having a mass 30 kg and a stiffness k, is added to a spring-mass system, of mass 40 kg and stiffness 0.1 MN/m, the main mass (40 kg) is found to have zero amplitude during its steady-state operation under a harmonic force of amplitude 300 N. Determine the steady-state amplitude of the absorber mass.

9.73 An electric motor, of mass 20 kg and operating speed 1,350 rpm, is placed on a fixed-fixed steel beam of width 15 cm and depth 12 cm, as shown in Fig. 9.52. The motor has a rotating unbalance of 0.1 kg-m. The amplitude of vibration of the beam under steady-state operation

FIGURE 9.52

of the motor is suppressed by attaching an undamped vibration absorber underneath the motor, as shown in Fig. 9.52. Determine the mass and stiffness of the absorber such that the amplitude of the absorber mass is less than 2 cm.

9.74 A bridge is found to vibrate violently when a vehicle, producing a harmonic load of magnitude 600 N, crosses it. By modeling the bridge as an undamped spring-mass system with a mass 15,000 kg and a stiffness 2 MN/m, design a suitably-tuned damped vibration absorber. Determine the improvement achieved in the amplitude of the bridge with the absorber.

9.75 A small motor, having a mass of 50 kg, is found to have a natural frequency of 100 rad/s. It is proposed that an undamped vibration absorber having a mass of 5 kg be used to suppress the vibrations when the motor operates at 80 rad/s. Determine the necessary stiffness of the absorber.

FIGURE 9.53

9.76 Consider the system shown in Fig. 9.53 in which a harmonic force acts on the mass m. Derive the condition under which the steady-state displacement of mass m will be zero.

Section 9.12 MATLAB Problems

9.77 Using MATLAB, plot Eq. (9.94) for $\zeta = 0, 0.25, 0.5, 0.75,$ and 1 over the range $0 \leq r \leq 3$.

9.78 Using MATLAB, plot Eqs. (9.140) and (9.141) for $f = 1, \zeta = 0.2, 0.3,$ and 0.4, and $\mu = 0.2$ and 0.5 over the range $0.6 \leq \omega/\omega_1$.

9.79 Using MATLAB, plot the ratios Ω_1/ω_2 and Ω_2/ω_2 given by Eq. (9.146) for $\omega_2/\omega_1 = 1.5,$ 3.0, and 4.5 and $m_2/m_1 = 0$ to 1.

9.80 Using **Program 13.m**, solve Problem 9.13.

9.81 Write a computer program to find the displacement of the main mass and the auxiliary mass of a damped dynamic vibration absorber. Use this program to generate the results of Fig. 9.38.

DESIGN PROJECT

9.82 Ground vibrations from a crane operation, a forging press, and an air compressor are transmitted to a nearby milling machine and are found to be detrimental to achieving specified accuracies during precision milling operations. The ground vibrations at the locations of the crane, forging press, and air compressor are given by $x_c(t) = A_c e^{-\omega_c \zeta_c t} \sin \omega_c t$, $x_f(t) = A_f \sin \omega_f t$, and $x_a(t) = A_a \sin \omega_a t$, respectively, where $A_c = 20\ \mu m$, $A_f = 30\ \mu m$, $A_a = 25\ \mu m$, $\omega_c = 10$ Hz, $\omega_f = 15$ Hz, $\omega_a = 20$ Hz, and $\zeta_c = 0.1$. The ground vibrations travel at the shear wave velocity of the soil, which is equal to 300 m/s, and the amplitudes attenuate according to the relation $A_r = A_0 e^{-0.016404 r}$, where A_0 is the amplitude at the source and A_r is the amplitude at a distance of r meters from the source. The crane, forging press, and air compressor are located at a distance of 18 m, 24 m, and 12 m, respectively, from the milling machine. The equivalent mass, stiffness, and damping ratio of the machine tool head in vertical vibration (at the location of the cutter) are experimentally determined to be 500 kg, 480 kN/m, and 0.15, respectively. The equivalent mass of the machine tool base is 1,000 kg. It is proposed that an isolator for the machine tool be used, as shown in Fig. 9.54, to improve the cutting accuracies [9.2]. Design a suitable vibration isolator, consisting of a mass, spring, and damper, as shown in Fig. 9.54(b), for the milling machine such that the maximum vertical displacement of the milling cutter, relative to the horizontal surface being machined, due to ground vibration from all the three sources does not exceed 5 μm peak-to-peak.

FIGURE 9.54

Gustav Robert Kirchhoff (1824–1887) was a German physicist and a friend of the famous chemist Robert Wilhelm Bunsen. Kirchhoff started teaching at Berlin University in 1848 and later moved to Heidelberg to occupy the chair of physics. There in 1859, he made his major contribution to physics, namely, the experimental discovery and theoretical analysis of a fundamental law of electromagnetic radiation. In addition, he made significant contributions to electrical circuits and the theory of elasticity. He published his important paper on the theory of plates in 1850 in which a satisfactory theory for the bending vibration of plates, along with the correct boundary conditions, was presented for the first time. In addition, he presented a paper on the vibration of bars of variable cross section. He moved to the University of Berlin in 1875 to occupy the chair of theoretical physics and published his famous book on mechanics in 1876. (Photo courtesy of *Applied Mechanics Reviews.*)

C H A P T E R   1 0

Vibration Measurement and Applications

Chapter Outline

In some practical situations, it might be difficult to develop a mathematical model of the system and predict its vibration characteristics through an analytical study. In such cases, we can use experimental methods to measure the vibration response of the system to a known input. This helps in identifying the system in terms of its mass, stiffness, and damping. This chapter presents the various aspects of vibration measurement and applications. The basic scheme of vibration measurement is outlined first. Descriptions are given of transducers—devices which transform physical variables into equivalent electrical signals—and of vibration pickups and frequency-measuring instruments used for vibration measurement. The working principles of mechanical and electrodynamic shakers or exciters, used to excite a machine or system to study its dynamic characteristics, are introduced. Signal analysis, which determines the response of a system under known excitation and presents it in a convenient form, is outlined along with descriptions of spectrum analyzer, bandpass filter, and bandwidth analyzers. The experimental modal analysis deals with the determination of natural frequencies, damping ratio, and mode shapes through vibration testing. The necessary equipment, digital signal processing, analysis of random signals, determination of modal data from observed peaks and Nyquist plot, and determination of mode shapes are described. Vibration severity criteria, machine maintenance techniques, machine-condition monitoring techniques, and instrumentation systems are presented for machine-condition monitoring and diagnosis. MATLAB programs are presented for plotting the Nyquist circle and the acceleration equation.

Learning Objectives

After you have finished studying this chapter, you should be able to do the following:

- Understand the various types of transducers, vibration pickups, and frequency-measuring instruments.
- Know the working principles of mechanical and electrodynamic shakers or exciters.
- Learn the process of signal analysis.
- Understand experimental modal analysis techniques to determine the natural frequencies, damping ratio, and mode shapes.
- Know the various aspects of machine-condition monitoring.
- Use MATLAB for plotting Nyquist circles and implementing methods of analysis discussed.

10.1   Introduction

In practice, the measurement of vibration becomes necessary for the following reasons:

1. The increasing demands of higher productivity and economical design lead to higher operating speeds of machinery[1] and efficient use of materials through lightweight structures. These trends make the occurrence of resonant conditions more frequent during the operation of machinery and reduce the reliability of the system. Hence the periodic measurement of vibration characteristics of machinery and structures becomes essential to ensure adequate safety margins. Any observed shift in the natural

[1]According to Eshleman, in [10.12], the average speed of rotating machines doubled—from 1,800 rpm to 3,600 rpm—during the period between 1940 and 1980.

frequencies or other vibration characteristics will indicate either a failure or a need for maintenance of the machine.

2. The measurement of the natural frequencies of a structure or machine is useful in selecting the operational speeds of nearby machinery to avoid resonant conditions.

3. The theoretically computed vibration characteristics of a machine or structure may be different from the actual values due to the assumptions made in the analysis.

4. The measurement of frequencies of vibration and the forces developed is necessary in the design and operation of active vibration-isolation systems.

5. In many applications, the survivability of a structure or machine in a specified vibration environment is to be determined. If the structure or machine can perform the expected task even after completion of testing under the specified vibration environment, it is expected to survive the specified conditions.

6. Continuous systems are often approximated as multidegree-of-freedom systems for simplicity. If the measured natural frequencies and mode shapes of a continuous system are comparable to the computed natural frequencies and mode shapes of the multidegree-of-freedom model, then the approximation will be proved to be valid.

7. The measurement of the input and the resulting output vibration characteristics of a system helps in identifying the system in terms of its mass, stiffness, and damping.

8. The information about ground vibrations due to earthquakes, fluctuating wind velocities on structures, random variation of ocean waves, and road surface roughness are important in the design of structures, machines, oil platforms, and vehicle suspension systems.

Vibration Measurement Scheme.   Figure 10.1 illustrates the basic features of a vibration measurement scheme. In this figure, the motion (or dynamic force) of the vibrating body is converted into an electrical signal by the vibration transducer or pickup. In general, a transducer is a device that transforms changes in mechanical quantities (such as displacement, velocity, acceleration, or force) into changes in electrical quantities (such as voltage or current). Since the output signal (voltage or current) of a transducer is too small to be recorded directly, a signal conversion instrument is used to amplify the signal to the required value. The output from the signal conversion instrument can be presented on a display unit for visual inspection, or recorded by a recording unit, or stored in a computer for later use. The data can then be analyzed to determine the desired vibration characteristics of the machine or structure.

Depending on the quantity measured, a vibration-measuring instrument is called a *vibrometer*, a *velocity meter*, an *accelerometer*, a *phase meter*, or a *frequency meter*. If the instrument is designed to record the measured quantity, then the suffix "meter" is to be replaced by "graph" [10.1]. In some applications, we need to vibrate a machine or structure to find its resonance characteristics. For this, electrodynamic vibrators, electrohydraulic vibrators, and signal generators (oscillators) are used.

FIGURE 10.1   Basic vibration measurement scheme.

The following considerations often dictate the type of vibration-measuring instruments to be used in a vibration test: (1) expected ranges of the frequencies and amplitudes, (2) sizes of the machine or structure involved, (3) conditions of operation of the machine, equipment or structure, and (4) types of data processing used (such as graphical display, graphical recording, or storing the record in digital form for computer processing).

10.2   Transducers

A transducer is a device that transforms values of physical variables into equivalent electrical signals. Several types of transducers are available; some of them are less useful than others due to their nonlinearity or slow response. Some of the transducers commonly used for vibration measurement are discussed below.

10.2.1 Variable-Resistance Transducers

In these transducers, a mechanical motion produces a change in electrical resistance (of a rheostat, a strain gage, or a semiconductor), which in turn causes a change in the output voltage or current. The schematic diagram of an electrical resistance strain gage is shown in Fig. 10.2. An electrical resistance strain gage consists of a fine wire whose resistance changes when it is subjected to mechanical deformation. When the strain gage is bonded to a structure, it experiences the same motion (strain) as the structure and hence its resistance change gives the strain applied to the structure. The wire is sandwiched between two sheets of thin paper. The strain gage is bonded to the surface where the strain is to be measured. The most common gage material is a copper-nickel alloy known as *Advance*. When the surface undergoes a normal strain (ϵ), the strain gage also undergoes the same strain and the resulting change in its resistance is given by [10.6]

$$K = \frac{\Delta R / R}{\Delta L / L} = 1 + 2v + \frac{\Delta r}{r} \frac{L}{\Delta L} \approx 1 + 2v \qquad (10.1)$$

where

$$K = \text{Gage factor for the wire}$$
$$R = \text{Initial resistance}$$

(a)   (b)

FIGURE 10.2   Electric resistance strain gage.

$$\Delta R = \text{Change in resistance}$$
$$L = \text{Initial length of the wire}$$
$$\Delta L = \text{Change in length of the wire}$$
$$v = \text{Poisson's ratio of the wire}$$
$$r = \text{Resistivity of the wire}$$
$$\Delta r = \text{Change in resistivity of the wire} \approx 0 \text{ for Advance}$$

The value of the gage factor K is given by the manufacturer of the strain gage, hence the value of ϵ can be determined, once ΔR and R are measured, as

$$\epsilon = \frac{\Delta L}{L} = \frac{\Delta R}{RK} \tag{10.2}$$

In a vibration pickup[2], the strain gage is mounted on an elastic element of a spring-mass system, as shown in Fig. 10.3. The strain at any point on the cantilever (elastic member) is proportional to the deflection of the mass, $x(t)$, to be measured. Hence the strain indicated by the strain gage can be used to find $x(t)$. The change in resistance of the wire ΔR can be measured using a Wheatstone bridge, potentiometer circuit, or voltage divider. A typical Wheatstone bridge, representing a circuit which is sensitive to small changes in the resistance, is shown in Fig. 10.4. A d.c. voltage V is applied across the points a and c. The resulting voltage across the points b and d is given by [10.6]:

$$E = \left[\frac{R_1 R_3 - R_2 R_4}{(R_1 + R_2)(R_3 + R_4)} \right] V \tag{10.3}$$

Initially, the resistances are balanced (adjusted) so that the output voltage E is zero. Thus, for initial balance, Eq. (10.3) gives

$$R_1 R_3 = R_2 R_4 \tag{10.4}$$

When the resistances (R_i) change by small amounts (ΔR_i), the change in the output voltage ΔE can be expressed as

$$\Delta E \approx V r_0 \left(\frac{\Delta R_1}{R_1} - \frac{\Delta R_2}{R_2} + \frac{\Delta R_3}{R_3} - \frac{\Delta R_4}{R_4} \right) \tag{10.5}$$

where

$$r_0 = \frac{R_1 R_2}{(R_1 + R_2)^2} = \frac{R_3 R_4}{(R_3 + R_4)^2} \tag{10.6}$$

---

[2]When a transducer is used in conjunction with other components that permit the processing and transmission of the signal, the device is called a *pickup*.

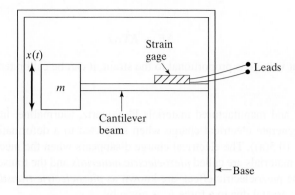

FIGURE 10.3   Strain gage as vibration pickup.

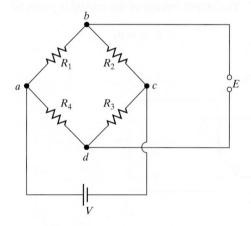

FIGURE 10.4   Wheatstone bridge.

If the strain gage leads are connected between the points a and b, $R_1 = R_g$, $\Delta R_1 = \Delta R_g$, and $\Delta R_2 = \Delta R_3 = \Delta R_4 = 0$, and Eq. (10.5) gives

$$\frac{\Delta R_g}{R_g} = \frac{\Delta E}{Vr_0} \qquad (10.7)$$

where R_g is the initial resistance of the gage. Equations (10.2) and (10.7) yield

$$\frac{\Delta R_g}{R_g} = \epsilon K = \frac{\Delta E}{Vr_0}$$

or

$$\Delta E = K V r_0 \epsilon \qquad (10.8)$$

Since the output voltage is proportional to the strain, it can be calibrated to read the strain directly.

10.2.2 Piezoelectric Transducers

Certain natural and manufactured materials like quartz, tourmaline, lithium sulfate, and Rochelle salt generate electrical charge when subjected to a deformation or mechanical stress (see Fig. 10.5(a)). The electrical charge disappears when the mechanical loading is removed. Such materials are called *piezoelectric materials* and the transducers, which take advantage of the piezoelectric effect, are known as *piezoelectric transducers*. The charge generated in the crystal due to a force F_x is given by

$$Q_x = k F_x = k A p_x \qquad (10.9)$$

where k is called the *piezoelectric constant*, A is the area on which the force F_x acts, and p_x is the pressure due to F_x. The output voltage of the crystal is given by

$$E = v t p_x \qquad (10.10)$$

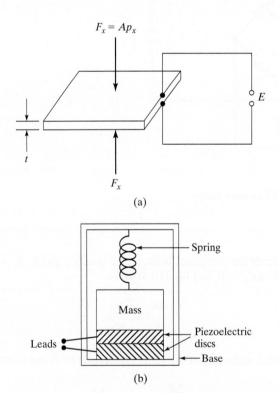

FIGURE 10.5   Piezoelectric accelerometer.

where v is called the voltage sensitivity and t is the thickness of the crystal. The values of the piezoelectric constant and voltage sensitivity for quartz are 2.25×10^{-12} C/N and 0.055 volt-meter/N, respectively [10.6]. These values are valid only when the perpendicular to the largest face is along the x-axis of the crystal. The electric charge developed and the voltage output will be different if the crystal slab is cut in a different direction.

A typical piezoelectric transducer (accelerometer) is shown in Fig. 10.5(b). In this figure, a small mass is spring loaded against a piezoelectric crystal. When the base vibrates, the load exerted by the mass on the crystal changes with acceleration, hence the output voltage generated by the crystal will be proportional to the acceleration. The main advantages of the piezoelectric accelerometer include compactness, ruggedness, high sensitivity, and high frequency range [10.5, 10.8].

EXAMPLE 10.1

Output Voltage of a Piezoelectric Transducer

A quartz crystal having a thickness of 2.5 mm is subjected to a pressure of 50 psi. Find the output voltage if the voltage sensitivity is 0.055 V-m/N.

Solution: With $t = 0.0025$ m, $p_x = 50$ psi $= 344.738$ N/m^2, and $v = 0.055$ V-m/N, Eq. (10.10) gives

$$E = (0.055)(0.0025)(344.738) = 47.4015 \text{ volts}$$

∎

**10.2.3
Electrodynamic
Transducers**

When an electrical conductor, in the form of a coil, moves in a magnetic field as shown in Fig. 10.6, a voltage E is generated in the conductor. The value of E in volts is given by

$$E = Dlv \qquad (10.11)$$

FIGURE 10.6   Basic idea behind electrodynamic transducer.

where D is the magnetic flux density (teslas), l is the length of the conductor (meters), and v is the velocity of the conductor relative to the magnetic field (meters/second). The magnetic field may be produced by either a permanent magnet or an electromagnet. Sometimes, the coil is kept stationary and the magnet is made to move. Since the voltage output of an electromagnetic transducer is proportional to the relative velocity of the coil, they are frequently used in "velocity pickups." Equation (10.11) can be rewritten as

$$Dl = \frac{E}{v} = \frac{F}{I} \tag{10.12}$$

where F denotes the force (Newtons) acting on the coil while carrying a current I (amperes). Equation (10.12) shows that the performance of an electrodynamic transducer can be reversed. In fact, Eq. (10.12) forms the basis for using an electrodynamic transducer as a "vibration exciter" (see Section 10.5.2).

10.2.4
Linear Variable
Differential
Transformer
Transducer

The schematic diagram of a *linear variable differential transformer* (*LVDT*) transducer is shown in Fig. 10.7. It consists of a primary coil at the center, two secondary coils at the ends, and a magnetic core that can move freely inside the coils in the axial direction. When an a.c. input voltage is applied to the primary coil, the output voltage will be equal to the difference of the voltages induced in the secondary coils. This output voltage depends on the magnetic coupling between the coils and the core, which in turn depends on the axial displacement of the core. The secondary coils are connected in phase opposition so that, when the magnetic core is in the exact middle position, the voltages in the two coils will be equal and 180° out of phase. This makes the output voltage of the LVDT as zero. When the core is moved to either side of the middle (zero) position, the magnetic coupling will be increased in one secondary coil and decreased in the other coil. The output polarity depends on the direction of the movement of the magnetic core.

The range of displacement for many LVDTs on the market is from 0.0002 cm to 40 cm. The advantages of an LVDT over other displacement transducers include insensitivity to

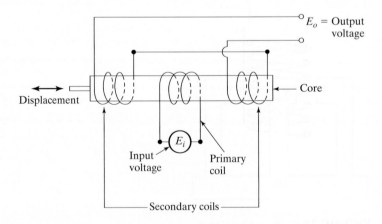

FIGURE 10.7   Schematic diagram of an LVDT transducer.

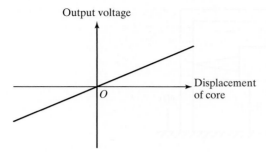

FIGURE 10.8   Linearity of voltage with displacement of core.

temperature and high output. The mass of the magnetic core restricts the use of the LVDT for high-frequency applications [10.4].

As long as the core is not moved very far from the center of the coil, the output voltage varies linearly with the displacement of the core, as shown in Fig. 10.8; hence the name *linear variable differential transformer*.

10.3   Vibration Pickups

When a transducer is used in conjunction with another device to measure vibrations, it is called a *vibration pickup*. The commonly used vibration pickups are known as *seismic instruments*. A seismic instrument consists of a mass-spring-damper system mounted on the vibrating body, as shown in Fig. 10.9. Then the vibratory motion is measured by finding the displacement of the mass relative to the base on which it is mounted.

The instrument consists of a mass m, a spring k, and a damper c inside a cage, which is fastened to the vibrating body. With this arrangement, the bottom ends of the spring and the dashpot will have the same motion as the cage (which is to be measured, y) and their vibration excites the suspended mass into motion. Then the displacement of the mass relative to the cage, $z = x - y$, where x denotes the vertical displacement of the suspended mass, can be measured if we attach a pointer to the mass and a scale to the cage, as shown in Fig. 10.9.[3]

The vibrating body is assumed to have a harmonic motion:

$$y(t) = Y \sin \omega t \tag{10.13}$$

The equation of motion of the mass m can be written as

$$m\ddot{x} + c(\dot{x} - \dot{y}) + k(x - y) = 0 \tag{10.14}$$

---

[3]The output of the instrument shown in Fig. 10.9 is the relative mechanical motion of the mass, as shown by the pointer and the graduated scale on the cage. For high-speed operation and convenience, the motion is often converted into an electrical signal by a transducer.

FIGURE 10.9 Seismic instrument.

By defining the relative displacement z as

$$z = x - y \tag{10.15}$$

Eq. (10.14) can be written as

$$m\ddot{z} + c\dot{z} + kz = -m\ddot{y} \tag{10.16}$$

Equations (10.13) and (10.16) lead to

$$m\ddot{z} + c\dot{z} + kz = m\omega^2 Y \sin \omega t \tag{10.17}$$

This equation is identical to Eq. (3.75); hence the steady-state solution is given by

$$z(t) = Z \sin(\omega t - \phi) \tag{10.18}$$

where Z and ϕ are given by (see Eqs. (3.76) and (3.77)):

$$Z = \frac{Y\omega^2}{[(k - m\omega^2)^2 + c^2\omega^2]^{1/2}} = \frac{r^2 Y}{[(1 - r^2)^2 + (2\zeta r)^2]^{1/2}} \tag{10.19}$$

$$\phi = \tan^{-1}\left(\frac{c\omega}{k - m\omega^2}\right) = \tan^{-1}\left(\frac{2\zeta r}{1 - r^2}\right) \tag{10.20}$$

$$r = \frac{\omega}{\omega_n} \tag{10.21}$$

and

$$\zeta = \frac{c}{2m\omega_n} \tag{10.22}$$

The variations of Z and ϕ with respect to r are shown in Figs. 10.10 and 10.11. As will be seen later, the type of instrument is determined by the useful range of the frequencies, indicated in Fig. 10.10.

10.3.1 Vibrometer

A *vibrometer* or a *seismometer* is an instrument that measures the displacement of a vibrating body. It can be observed from Fig. 10.10 that $Z/Y \approx 1$ when $\omega/\omega_n \geq 3$ (range II). Thus the relative displacement between the mass and the base (sensed by the transducer) is essentially the same as the displacement of the base. For an exact analysis, we consider Eq. (10.19). We note that

$$z(t) \simeq Y \sin(\omega t - \phi) \tag{10.23}$$

FIGURE 10.10   Response of a vibration-measuring instrument.

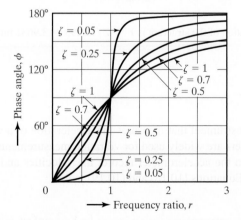

FIGURE 10.11   Variation of ϕ with r.

if

$$\frac{r^2}{[(1 - r^2)^2 + (2\zeta r)^2]^{1/2}} \approx 1 \qquad (10.24)$$

A comparison of Eq. (10.23) with $y(t) = Y \sin \omega t$ shows that $z(t)$ gives directly the motion $y(t)$ except for the phase lag ϕ. This phase lag can be seen to be equal to 180° for $\zeta = 0$. Thus the recorded displacement $z(t)$ lags behind the displacement being measured $y(t)$ by time $t' = \phi/\omega$. This time lag is not important if the base displacement $y(t)$ consists of a single harmonic component.

Since $r = \omega/\omega_n$ has to be large and the value of ω is fixed, the natural frequency $\omega_n = \sqrt{k/m}$ of the mass-spring-damper must be low. This means that the mass must be large and the spring must have a low stiffness. This results in a bulky instrument, which is not desirable in many applications. In practice, the vibrometer may not have a large value of r and hence the value of Z may not be equal to Y exactly. In such a case, the true value of Y can be computed by using Eq. (10.19), as indicated in the following example.

EXAMPLE 10.2

Amplitude by Vibrometer

A vibrometer having a natural frequency of 4 rad/s and $\zeta = 0.2$ is attached to a structure that performs a harmonic motion. If the difference between the maximum and the minimum recorded values is 8 mm, find the amplitude of motion of the vibrating structure when its frequency is 40 rad/s.

Solution: The amplitude of the recorded motion Z is 4 mm. For $\zeta = 0.2$, $\omega = 40.0$ rad/s, and $\omega_n = 4$ rad/s, $r = 10.0$, and Eq. (10.19) gives

$$Z = \frac{Y(10)^2}{[(1 - 10^2)^2 + \{2(0.2)(10)\}^2]^{1/2}} = 1.0093Y$$

Thus the amplitude of vibration of the structure is $Y = Z/1.0093 = 3.9631$ mm.

■

10.3.2 Accelerometer

An accelerometer is an instrument that measures the acceleration of a vibrating body (see Fig. 10.12). Accelerometers are widely used for vibration measurements [10.7] and also to record earthquakes. From the accelerometer record, the velocities and displacements are obtained by integration. Equations (10.18) and (10.19) yield

$$-z(t)\omega_n^2 = \frac{1}{[(1 - r^2)^2 + (2\zeta r)^2]^{1/2}}\{-Y\omega^2 \sin(\omega t - \phi)\} \qquad (10.25)$$

(a)

(b)

FIGURE 10.12   Accelerometers.
(Courtesy of Bruel and Kjaer Instruments, Inc.,
Marlborough, MA.)

This shows that if

$$\frac{1}{[(1 - r^2)^2 + (2\zeta r)^2]^{1/2}} \simeq 1 \tag{10.26}$$

Eq. (10.25) becomes

$$-z(t)\omega_n^2 \simeq -Y\omega^2 \sin(\omega t - \phi) \tag{10.27}$$

By comparing Eq. (10.27) with $\ddot{y}(t) = -Y\omega^2 \sin \omega t$, we find that the term $z(t)\omega_n^2$ gives the acceleration of the base $\ddot{y}$, except for the phase lag ϕ. Thus the instrument can be made to record (give) directly the value of $\ddot{y} = -z(t)\omega_n^2$. The time by which the record lags the acceleration is given by $t' = \phi/\omega$. If $\ddot{y}$ consists of a single harmonic component, the time lag will not be of importance.

The value of the expression on the left-hand side of Eq. (10.26) is shown plotted in Fig. 10.13. It can be seen that the left-hand side of Eq. (10.26) lies between 0.96 and 1.04 for $0 \le r \le 0.6$ if the value of ζ lies between 0.65 and 0.7. Since r is small, the natural frequency of the instrument has to be large compared to the frequency of vibration to be measured. From the relation $\omega_n = \sqrt{k/m}$, we find that the mass needs to be small and the spring needs to have a large value of k (that is, a short spring), so the instrument will be small in size.

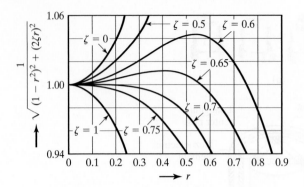

FIGURE 10.13   Variation of left-hand side of Eq. (10.26) with r.

Due to their small size and high sensitivity, accelerometers are preferred in vibration measurements. In practice, Eq. (10.26) may not be satisfied exactly; in such cases, the quantity

$$\frac{1}{[(1 - r^2)^2 + (2\zeta r)^2]^{1/2}}$$

can be used to find the correct value of the acceleration measured, as illustrated in the following example.

EXAMPLE 10.3

Design of an Accelerometer

An accelerometer has a suspended mass of 0.01 kg with a damped natural frequency of vibration of 150 Hz. When mounted on an engine undergoing an acceleration of 1g at an operating speed of 6,000 rpm, the acceleration is recorded as 9.5 m/s^2 by the instrument. Find the damping constant and the spring stiffness of the accelerometer.

Solution: The ratio of measured to true accelerations is given by

$$\frac{1}{[(1 - r^2)^2 + (2\zeta r)^2]^{1/2}} = \frac{\text{Measured value}}{\text{True value}} = \frac{9.5}{9.81} = 0.9684 \tag{E.1}$$

which can be written as

$$[(1 - r^2)^2 + (2\zeta r)^2] = (1/0.9684)^2 = 1.0663 \tag{E.2}$$

The operating speed of the engine gives

$$\omega = \frac{6,000(2\pi)}{60} = 628.32 \text{ rad/s}$$

The damped natural frequency of vibration of the accelerometer is

$$\omega_d = \sqrt{1 - \zeta^2}\omega_n = 150(2\pi) = 942.48 \text{ rad/s}$$

Thus

$$\frac{\omega}{\omega_d} = \frac{\omega}{\sqrt{1 - \zeta^2}\omega_n} = \frac{r}{\sqrt{1 - \zeta^2}} = \frac{628.32}{942.48} = 0.6667 \tag{E.3}$$

Equation (E.3) gives

$$r = 0.6667\sqrt{1 - \zeta^2} \quad \text{or} \quad r^2 = 0.4444(1 - \zeta^2) \tag{E.4}$$

Substitution of Eq. (E.4) into Eq. (E.2) leads to a quadratic equation in ζ^2 as

$$1.5801\zeta^4 - 2.2714\zeta^2 + 0.7576 = 0 \tag{E.5}$$

The solution of Eq. (E.5) gives

$$\zeta^2 = 0.5260, 0.9115$$

or

$$\zeta = 0.7253, 0.9547$$

By choosing $\zeta = 0.7253$ arbitrarily, the undamped natural frequency of the accelerometer can be found as

$$\omega_n = \frac{\omega_d}{\sqrt{1 - \zeta^2}} = \frac{942.48}{\sqrt{1 - 0.7253^2}} = 1,368.8889 \text{ rad/s}$$

Since $\omega_n = \sqrt{k/m}$, we have

$$k = m\omega_n^2 = (0.01)(1,368.8889)^2 = 18,738.5628 \text{ N/m}$$

The damping constant can be determined from

$$c = 2m\omega_n\zeta = 2(0.01)(1,368.8889)(0.7253) = 19.8571 \text{ N-s/m}$$

■

10.3.3
Velometer

A velometer measures the velocity of a vibrating body. Equation (10.13) gives the velocity of the vibrating body

$$\dot{y}(t) = \omega Y \cos \omega t \tag{10.28}$$

and Eq. (10.18) gives

$$\dot{z}(t) = \frac{r^2 \omega Y}{[(1 - r^2)^2 + (2\zeta r)^2]^{1/2}} \cos(\omega t - \phi) \tag{10.29}$$

If

$$\frac{r^2}{[(1 - r^2)^2 + (2\zeta r)^2]^{1/2}} \simeq 1 \tag{10.30}$$

then

$$\dot{z}(t) \simeq \omega Y \cos(\omega t - \phi) \tag{10.31}$$

A comparison of Eqs. (10.28) and (10.31) shows that, except for the phase difference ϕ, $\dot{z}(t)$ gives $\dot{y}(t)$ directly, provided that Eq. (10.30) holds true. In order to satisfy Eq. (10.30), r must be very large. In case Eq. (10.30) is not satisfied, then the velocity of the vibrating body can be computed using Eq. (10.29).

▬▬▬▬▬▬▬▬▬▬ **Design of a Velometer**

EXAMPLE 10.4 ──────────────────────────────

───────────────── Design a velometer if the maximum error is to be limited to 1 percent of the true velocity. The natural frequency of the velometer is to be 80 Hz and the suspended mass is to be 0.05 kg.

Solution: The ratio (R) of the recorded and the true velocities is given by Eq. (10.29):

$$R = \frac{r^2}{[(1 - r^2)^2 + (2\zeta r)^2]^{1/2}} = \frac{\text{Recorded velocity}}{\text{True velocity}} \tag{E.1}$$

The maximum of Eq. (E.1) occurs when (see Eq. (3.82))

$$r = r^* = \frac{1}{\sqrt{1 - 2\zeta^2}} \tag{E.2}$$

Substitution of Eq. (E.2) into Eq. (E.1) gives

$$\frac{\left(\dfrac{1}{1-2\zeta^2}\right)}{\sqrt{\left[1-\left(\dfrac{1}{1-2\zeta^2}\right)\right]^2+4\zeta^2\left(\dfrac{1}{1-2\zeta^2}\right)}}=R$$

which can be simplified as

$$\frac{1}{\sqrt{4\zeta^2-4\zeta^4}}=R \tag{E.3}$$

For an error of 1 percent, $R = 1.01$ or 0.99, and Eq. (E.3) leads to

$$\zeta^4-\zeta^2+0.245075=0 \tag{E.4}$$

and

$$\zeta^4-\zeta^2+0.255075=0 \tag{E.5}$$

Equation (E.5) gives imaginary roots and Eq. (E.4) gives

$$\zeta^2=0.570178,\ 0.429821$$

or

$$\zeta=0.755101,\ 0.655607$$

We choose the value $\zeta = 0.755101$ arbitrarily. The spring stiffness can be found as

$$k=m\omega_n^2=0.05(502.656)^2=12{,}633.1527\ \text{N/m}$$

since

$$\omega_n=80(2\pi)=502.656\ \text{rad/s}$$

The damping constant can be determined from

$$c=2\zeta\omega_n m=2(0.755101)(502.656)(0.05)=37.9556\ \text{N-s/m}$$

∎

10.3.4
Phase Distortion

As shown by Eq. (10.18), all vibration-measuring instruments exhibit phase lag. Thus the response or output of the instrument lags behind the motion or input it measures. The time lag is given by the phase angle divided by the frequency ω. The time lag is not important if we measure a single harmonic component. However, occasionally, the vibration to be recorded is not harmonic but consists of the sum of two or more harmonic components. In such a case, the recorded graph may not give an accurate picture of the vibration, because different harmonics may be amplified by different amounts and their phase shifts may also be different. The distortion in the waveform of the recorded signal is called the *phase distortion* or *phase-shift error*. To illustrate the nature of the phase-shift error, we consider a vibration signal of the form shown in Fig. 10.14(a) [10.10]:

$$y(t) = a_1 \sin \omega t + a_3 \sin 3\omega t \tag{10.32}$$

Let the phase shift be 90° for the first harmonic and 180° for the third harmonic of Eq. (10.32). The corresponding time lags are given by $t_1 = \theta_1/\omega = 90°/\omega$ and $t_2 = \theta_2/(3\omega) = 180°/(3\omega)$. The output signal is shown in Fig. 10.14(b). It can be seen that the output signal is quite different from the input signal due to phase distortion.

As a general case, let the complex wave being measured be given by the sum of several harmonics as

$$y(t) = a_1 \sin \omega t + a_2 \sin 2\omega t + \cdots \tag{10.33}$$

(a) Input signal

(b) Output signal

FIGURE 10.14   Phase-shift error.

If the displacement is measured using a vibrometer, its response to each component of the series is given by an equation similar to Eq. (10.18), so that the output of the vibrometer becomes

$$z(t) = a_1 \sin(\omega t - \phi_1) + a_2 \sin(2\omega t - \phi_2) + \cdots \qquad (10.34)$$

where

$$\tan \phi_j = \frac{2\zeta\left(j\dfrac{\omega}{\omega_n}\right)}{1 - \left(j\dfrac{\omega}{\omega_n}\right)^2}, \qquad j = 1, 2, \ldots \qquad (10.35)$$

Since ω/ω_n is large for this instrument, we can find from Fig. 10.11 that $\phi_j \simeq \pi$, $j = 1, 2, \ldots$, and Eq. (10.34) becomes

$$z(t) \simeq -[a_1 \sin \omega t + a_2 \sin 2\omega t + \cdots] \simeq -y(t) \qquad (10.36)$$

Thus the output record will be simply opposite to the motion being measured. This is unimportant and can easily be corrected.

By using a similar reasoning, we can show, in the case of a velometer, that

$$\dot{z}(t) \simeq -\dot{y}(t) \qquad (10.37)$$

for an input signal consisting of several harmonics. Next we consider the phase distortion for an accelerometer. Let the acceleration curve to be measured be expressed, using Eq. (10.33), as

$$\ddot{y}(t) = -a_1\omega^2 \sin \omega t - a_2(2\omega)^2 \sin 2\omega t - \cdots \qquad (10.38)$$

The response or output of the instrument to each component can be found as in Eq. (10.34), and so

$$\ddot{z}(t) = -a_1\omega^2 \sin(\omega t - \phi_1) - a_2(2\omega)^2 \sin(2\omega t - \phi_2) - \cdots \qquad (10.39)$$

where the phase lags ϕ_j are different for different components of the series in Eq. (10.39). Since the phase lag ϕ varies almost linearly from $0°$ at $r = 0$ to $90°$ at $r = 1$ for $\zeta = 0.7$ (see Fig. 10.11), we can express ϕ as

$$\phi \simeq \alpha r = \alpha \frac{\omega}{\omega_n} = \beta\omega \qquad (10.40)$$

where α and $\beta = \alpha/\omega_n$ are constants. The time lag is given by

$$t' = \frac{\phi}{\omega} = \frac{\beta\omega}{\omega} = \beta \tag{10.41}$$

This shows that the time lag of the accelerometer is independent of the frequency for any component, provided that the frequency lies in the range $0 \leq r \leq 1$. Since each component of the signal has the same time delay or phase lag, we have, from Eq. (10.39),

$$-\omega^2\ddot{z}(t) = -a_1\omega^2 \sin(\omega t - \omega\beta) - a_2(2\omega)^2 \sin(2\omega t - 2\omega\beta) - \cdots$$

$$= -a_1\omega^2 \sin \omega\tau - a_2(2\omega)^2 \sin 2\omega\tau - \cdots \tag{10.42}$$

where $\tau = t - \beta$. Note that Eq. (10.42) assumes that $0 \leq r \leq 1$—that is, even the highest frequency involved, $n\omega$, is less than ω_n. This may not be true in practice. Fortunately, no significant phase distortion occurs in the output signal, even when some of the higher-order frequencies are larger than ω_n. The reason is that, generally, only the first few components are important to approximate even a complex waveform; the amplitudes of the higher harmonics are small and contribute very little to the total waveform. Thus the output record of the accelerometer represents a reasonably true acceleration being measured [10.7, 10.11].

10.4 Frequency-Measuring Instruments

Most frequency-measuring instruments are of the mechanical type and are based on the principle of resonance. Two kinds are discussed in the following paragraphs: the Fullarton tachometer and the Frahm tachometer.

Single-Reed Instrument or Fullarton Tachometer. This instrument consists of a variable-length cantilever strip with a mass attached at one of its ends. The other end of the strip is clamped, and its free length can be changed by means of a screw mechanism (see Fig. 10.15(a)). Since each length of the strip corresponds to a different natural frequency, the reed is marked along its length in terms of its natural frequency. In practice, the clamped end of the strip is pressed against the vibrating body, and the screw mechanism is manipulated to alter its free length until the free end shows the largest amplitude of vibration. At that instant, the excitation frequency is equal to the natural frequency of the cantilever; it can be read directly from the strip.

Multireed-Instrument or Frahm Tachometer. This instrument consists of a number of cantilevered reeds carrying small masses at their free ends (see Fig. 10.15(b)). Each reed has a different natural frequency and is marked accordingly. Using a number of reeds makes it possible to cover a wide frequency range. When the instrument is mounted on a vibrating body, the reed whose natural frequency is nearest the unknown frequency of the

(a)   (b)

FIGURE 10.15   Frequency-measuring instruments.

body vibrates with the largest amplitude. The frequency of the vibrating body can be found from the known frequency of the vibrating reed.

Stroboscope.   A stroboscope is an instrument that produces light pulses intermittently. The frequency at which the light pulses are produced can be altered and read from the instrument. When a specific point on a rotating (vibrating) object is viewed with the stroboscope, it will appear to be stationary only when the frequency of the pulsating light is equal to the speed of the rotating (vibrating) object. The main advantage of the stroboscope is that it does not make contact with the rotating (vibrating) body. Due to the persistence of vision, the lowest frequency that can be measured with a stroboscope is approximately 15 Hz. A typical stroboscope is shown in Fig. 10.16.

FIGURE 10.16   A stroboscope. (Courtesy of Bruel and Kjaer Instruments, Inc., Marlborough, MA.)

10.5   Vibration Exciters

The vibration exciters or shakers can be used in several applications such as determination of the dynamic characteristics of machines and structures and fatigue testing of materials. The vibration exciters can be of mechanical, electromagnetic, electrodynamic, or hydraulic type. The working principles of mechanical and electromagnetic exciters are described in this section.

**10.5.1
Mechanical
Exciters**

As indicated in Section 1.10 (Fig. 1.46), a Scotch yoke mechanism can be used to produce harmonic vibrations. The crank of the mechanism can be driven either by a constant- or variable-speed motor. When a structure is to be vibrated, the harmonic force can be applied either as an inertia force, as shown in Fig. 10.17(a), or as an elastic spring force, as shown

(a)

(b)

FIGURE 10.17   Vibration of a structure through (a) an inertia force and (b) an elastic spring force.

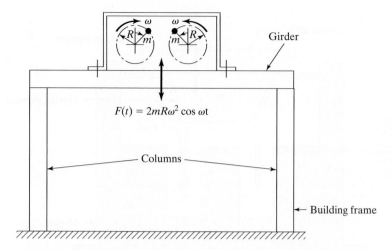

FIGURE 10.18 Vibration excitation due to unbalanced force.

in Fig. 10.17(b). These vibrators are generally used for frequencies less than 30 Hz and loads less than 700 N [10.1].

The unbalance created by two masses rotating at the same speed in opposite directions (see Fig. 10.18) can be used as a mechanical exciter. This type of shaker can be used to generate relatively large loads between 250 and 25,000 N. If the two masses, of magnitude m each, rotate at an angular velocity ω at a radius R, the vertical force $F(t)$ generated is given by

$$F(t) = 2mR\omega^2 \cos \omega t \qquad (10.43)$$

The horizontal components of the two masses cancel out, hence the resultant horizontal force will be zero. The force $F(t)$ will be applied to the structure to which the exciter is attached.

10.5.2 Electrodynamic Shaker

The schematic diagram of an *electrodynamic shaker*, also known as the *electromagnetic exciter*, is shown in Fig. 10.19(a). As stated in Section 10.2.3, the electrodynamic shaker can be considered as the reverse of an electrodynamic transducer. When current passes through a coil placed in a magnetic field, a force F (in Newtons) proportional to the current I (in amperes) and the magnetic flux intensity D (in teslas) is produced which accelerates the component placed on the shaker table:

$$F = DIl \qquad (10.44)$$

where l is the length of the coil (in meters). The magnetic field is produced by a permanent magnet in small shakers while an electromagnet is used in large shakers. The magnitude of acceleration of the table or component depends on the maximum current and the masses of

(a)

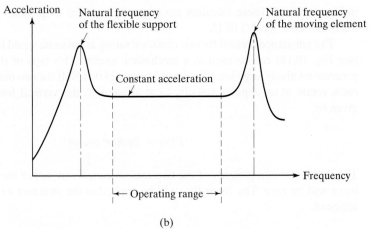

(b)

FIGURE 10.19   (a) Electrodynamic shaker. (b) Typical resonance characteristics of an electrodynamic exciter.

the component and the moving element of the shaker. If the current flowing through the coil varies harmonically with time (a.c. current), the force produced also varies harmonically. On the other hand, if direct current is used to energize the coil, a constant force is generated at the exciter table. The electrodynamic exciters can be used in conjunction with an inertia or a spring as in the case of Figs. 10.17(a) and (b) to vibrate a structure.

Since the coil and the moving element should have a linear motion, they are suspended from a flexible support (having a very small stiffness) as shown in Fig. 10.19(a). Thus the

FIGURE 10.20   An exciter with a general-purpose head. (Courtesy of Bruel and Kjaer Instruments, Inc., Marlborough, MA.)

electromagnetic exciter has two natural frequencies—one corresponding to the natural frequency of the flexible support and the other corresponding to the natural frequency of the moving element, which can be made very large. These two resonant frequencies are shown in Fig. 10.19(b). The operating-frequency range of the exciter lies between these two resonant frequencies, as indicated in Fig. 10.19(b) [10.7].

The electrodynamic exciters are used to generate forces up to 30,000 N, displacements up to 25 mm, and frequencies in the range of 5 to 20 kHz [10.1]. A practical electrodynamic exciter is shown in Fig. 10.20.

10.6 Signal Analysis

In signal analysis, we determine the response of a system under a known excitation and present it in a convenient form. Often, the time response of a system will not give much useful information. However, the frequency response will show one or more discrete frequencies around which the energy is concentrated. Since the dynamic characteristics of individual components of the system are usually known, we can relate the distinct frequency components (of the frequency response) to specific components [10.3].

For example, the acceleration-time history of a machine frame that is subjected to excessive vibration might appear as shown in Fig. 10.21(a). This figure cannot be used to identify the cause of vibration. If the acceleration-time history is transformed to the

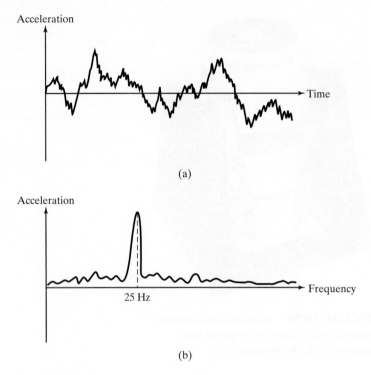

FIGURE 10.21   Acceleration history.

frequency domain, the resulting frequency spectrum might appear as shown in Fig. 10.21(b), where, for specificness, the energy is shown concentrated around 25 Hz. This frequency can easily be related, for example, to the rotational speed of a particular motor. Thus the acceleration spectrum shows strong evidence that the motor might be the cause of vibration. If the motor is causing the excessive vibrations, changing either the motor or its speed of operation might avoid resonance and hence the problem of excessive vibrations.

10.6.1
Spectrum
Analyzers

Spectrum or frequency analyzers can be used for signal analysis. These devices analyze a signal in the frequency domain by separating the energy of the signal into various frequency bands. The separation of signal energy into frequency bands is accomplished through a set of filters. The analyzers are usually classified according to the type of filter employed. For example, if an octave band filter is used, the spectrum analyzer is called an *octave band analyzer.*

In recent years, digital analyzers have become quite popular for real-time signal analysis. In a real-time frequency analysis, the signal is continuously analyzed over all the frequency bands. Thus the calculation process must not take more time than that taken to collect the signal data. Real-time analyzers are especially useful for machinery health monitoring, since a change in the noise or vibration spectrum can be observed at the same time that a change in the machine occurs. There are two types of real-time analysis procedures: the digital filtering method and the fast Fourier transform (FFT) method [10.13].

The digital filtering method is best suited for constant-percent bandwidth analysis; the FFT method for constant-bandwidth analysis. Before we consider the difference between those two approaches, we first discuss the basic component of a spectrum analyzer—namely, the bandpass filter.

10.6.2
Bandpass Filter

A bandpass filter is a circuit that permits the passage of frequency components of a signal over a frequency band and rejects all other frequency components of the signal. A filter can be built by using, for example, resistors, inductors, and capacitors. Figure 10.22 illustrates the response characteristics of a filter whose lower and upper cutoff frequencies are f_l and f_u, respectively. A practical filter will have a response characteristic deviating from the ideal rectangle, as shown by the full line in Fig. 10.22. For a good bandpass filter, the ripples within the band will be minimum and the slopes of the filter skirts will be steep to maintain the actual bandwidth close to the ideal value, $B = f_u - f_l$. For a practical filter, the frequencies f_l and f_u at which the response is 3 dB[4] below its mean bandpass response are called the *cutoff frequencies*.

There are two types of bandpass filters used in signal analysis: the constant-percent bandwidth filters and constant-bandwidth filters. For a constant-percent bandwidth filter, the ratio of the bandwidth to the center (tuned) frequency, $(f_u - f_l)/f_c$, is a constant. The octave,[5]

FIGURE 10.22   Response of a filter.

---

[4]A decibel (dB) of a quantity (such as power, P) is defined as

$$\text{Quantity in dB} = 10 \log_{10}\left(\frac{P}{P_{\text{ref}}}\right)$$

where P is the power and P_{ref} is a reference value of the power.

[5]An octave is the interval between any two frequencies $(f_2 - f_1)$, whose frequency ratio (f_2/f_1) is 2. Two frequencies f_1 and f_2 are said to be separated by a number of octaves N when

$$\frac{f_2}{f_1} = 2^N \qquad \text{or} \qquad N \text{ (in octaves)} = \log_2\left(\frac{f_2}{f_1}\right)$$

where N can be an integer or a fraction. If $N = 1$, we have an octave; if $N = 1/3$, we get a one-third octave, and so on.

TABLE 10.1

Lower cutoff limit (Hz)	5.63	11.2	22.4	44.7	89.2	178	355	709	1,410
Center frequency (Hz)	8.0	16.0	31.5	63.0	125	250	500	1,000	2,000
Upper cutoff limit (Hz)	11.2	22.4	44.7	89.2	178	355	709	1,410	2,820

one-half-octave, and one-third-octave band filters are examples of constant-percent bandwidth filters. Some of the cutoff limits and center frequencies of octave bands used in signal analysis are shown in Table 10.1. For a constant-bandwidth filter, the bandwidth, $f_u - f_l$, is independent of the center (tuned) frequency, f_c.

10.6.3 Constant-Percent Bandwidth and Constant-Bandwidth Analyzers

The primary difference between the constant-percent bandwidth and constant-bandwidth analyzers lies in the detail provided by the various bandwidths. The octave band filters, whose upper cutoff frequency is twice the lower cutoff frequency, give a less detailed (too coarse) analysis for practical vibration and noise encountered in machines. The one-half-octave band filter gives twice the information but requires twice the amount of time to obtain the data. A spectrum analyzer with a set of octave and one-third-octave filters can be used for noise (signal) analysis. Each filter is tuned to a different center frequency to cover the entire frequency range of interest. Since the lower cutoff frequency of a filter is equal to the upper cutoff frequency of the previous filter, the composite filter characteristic will appear as shown in Fig. 10.23. Figure 10.24 shows a real-time octave and fractional-octave digital frequency analyzer. A constant-bandwidth analyzer is used to obtain a more detailed analysis than in the case of a constant-percent bandwidth analyzer, especially in the high-frequency range of the signal. The constant-bandwidth filter, when used with a continuously varying center frequency, is called a *wave* or *heterodyne analyzer*. Heterodyne analyzers are available with constant filter bandwidths ranging from one to several hundred hertz. A practical heterodyne analyzer is shown in Fig. 10.25.

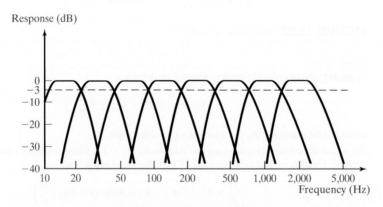

FIGURE 10.23 Response characteristic of a typical octave band filter set.

FIGURE 10.24   Octave and fractional-octave digital frequency analyzer. (Courtesy of Bruel and Kjaer Instruments, Inc., Marlborough, MA.)

FIGURE 10.25   Heterodyne analyzer. (Courtesy of Bruel and Kjaer Instruments, Inc., Marlborough, MA.)

10.7   Dynamic Testing of Machines and Structures

The dynamic testing of machines (structures) involves finding their deformation at a critical frequency. This can be done using the following two approaches [10.3].

10.7.1
Using Operational Deflection-Shape Measurements

In this method, the forced dynamic deflection shape is measured under the steady-state (operating) frequency of the system. For the measurement, an accelerometer is mounted at some point on the machine (structure) as a reference, and another moving accelerometer is placed at several other points, and in different directions, if necessary. Then the magnitudes and the phase differences between the moving and reference accelerometers at all the points under steady-state operation of the system are measured. By plotting these measurements, we can find how the various parts of the machine (structure) move relative to one another and also absolutely.

The deflection shape measured is valid only for the forces or frequencies associated with the operating conditions; as such, we cannot get information about deflections under other forces and/or frequencies. However, the measured deflection shape can be quite useful. For example, if a particular part or location is found to have excessive deflection, we can stiffen that part or location. This, in effect, increases the natural frequency beyond the operational frequency range of the system.

10.7.2
Using Modal Testing

Since any dynamic response of a machine (structure) can be obtained as a combination of its modes, a knowledge of the mode shapes, modal frequencies, and modal damping ratios constitutes a complete dynamic description of the machine (structure). The experimental modal analysis procedure is described in the following section.

10.8   Experimental Modal Analysis

10.8.1
The Basic Idea

Experimental modal analysis, also known as *modal analysis* or *modal testing*, deals with the determination of natural frequencies, damping ratios, and mode shapes through vibration testing. Two basic ideas are involved:

1. When a structure, machine, or any system is excited, its response exhibits a sharp peak at resonance when the forcing frequency is equal to its natural frequency when damping is not large.
2. The phase of the response changes by 180° as the forcing frequency crosses the natural frequency of the structure or machine, and the phase will be 90° at resonance.

10.8.2
The Necessary Equipment

The measurement of vibration requires the following hardware:

1. An exciter or source of vibration to apply a known input force to the structure or machine.
2. A transducer to convert the physical motion of the structure or machine into an electrical signal.

3. A signal conditioning amplifier to make the transducer characteristics compatible with the input electronics of the digital data acquisition system.

4. An analyzer to perform the tasks of signal processing and modal analysis using suitable software.

Exciter.   The exciter may be an electromagnetic shaker or an impact hammer. As explained in Section 10.5.2, the electromagnetic shaker can provide large input forces so that the response can be measured easily. Also, the output of the shaker can be controlled easily if it is of electromagnetic type. The excitation signal is usually of a swept sinusoidal or a random type signal. In the swept sinusoidal input, a harmonic force of magnitude F is applied at a number of discrete frequencies over a specific frequency range of interest. At each discrete frequency, the structure or machine is made to reach a steady state before the magnitude and phase of the response are measured. If the shaker is attached to the structure or machine being tested, the mass of the shaker will influence the measured response (known as the *mass loading effect*). As such, care is to be taken to minimize the effect of the mass of the shaker. Usually, the shaker is attached to the structure or machine through a short thin rod, called a *stringer*, to isolate the shaker, reduce the added mass, and apply the force to the structure or machine along the axial direction of the stringer. This permits the control of the direction of the force applied to the structure or machine.

The impact hammer is a hammer with a built-in force transducer in its head, as indicated in Examples 4.7 and 4.8. The impact hammer can be used to hit or impact the structure or machine being tested to excite a wide range of frequencies without causing the problem of mass loading. The impact force caused by the impact hammer, which is nearly proportional to the mass of the hammer head and the impact velocity, can be found from the force transducer embedded in the head of the hammer. As shown in Section 6.15, the response of the structure or machine to an impulse is composed of excitations at each of the natural frequencies of the structure or machine.

Although the impact hammer is simple, portable, inexpensive, and much faster to use than a shaker, it is often not capable of imparting sufficient energy to obtain adequate response signals in the frequency range of interest. It is also difficult to control the direction of the applied force with an impact hammer. A typical frequency response of a structure or machine obtained using an impact hammer is shown in Fig. 10.26. The shape of the frequency response is dependent on the mass and stiffness of both the hammer and the structure or machine. Usually, the useful range of frequency excitation is limited by a cutoff frequency, ω_c, which implies that the structure or machine did not receive sufficient energy to excite modes beyond ω_c. The value of ω_c is often taken as the frequency where the amplitude of the frequency response reduces by 10 to 20 dB from its maximum value.

Transducer.   Among the transducers, the piezoelectric transducers are most popular (see Section 10.2.2). A piezoelectric transducer can be designed to produce signals proportional to either force or acceleration. In an accelerometer, the piezoelectric material acts as a stiff spring that causes the transducer to have a resonant or natural frequency. Usually, the maximum measurable frequency of an accelerometer is a fraction of its natural

FIGURE 10.26 Frequency response of an impulse created by an impact hammer.

frequency. Strain gages can also be used to measure the vibration response of a structure or machine, as discussed in Section 10.2.1.

Signal Conditioner. Since the output impedance of transducers is not suitable for direct input into the signal analysis equipment, signal conditioners, in the form of charge or voltage amplifiers, are used to match and amplify the signals before signal analysis.

Analyzer. The response signal, after conditioning, is sent to an analyzer for signal processing. A type that is commonly used is the *fast Fourier transform (FFT) analyzer.* Such an analyzer receives analog voltage signals (representing displacement, velocity, acceleration, strain, or force) from a signal-conditioning amplifier, filter, and digitizer for computations. It computes the discrete frequency spectra of individual signals as well as cross-spectra between the input and the different output signals. The analyzed signals can be used to find the natural frequencies, damping ratios, and mode shapes in either numerical or graphical form.

The general arrangement for the experimental modal analysis of a structural or mechanical system is shown in Fig. 10.27. Note that all the equipment is to be calibrated before it is used. For example, an impact hammer is used more frequently in experimental stress analysis. The reason is that it is more convenient and faster to use than a shaker. An impact hammer consists of a force transducer or load cell built into the head (or tip) of the hammer. The built-in force transducer is to be calibrated dynamically whenever the head or tip is changed. Similarly, the transducers, along with the signal conditioners, are to be calibrated with respect to magnitude and phase over the frequency range of interest.

FIGURE 10.27   Experimental modal analysis.

**10.8.3
Digital Signal
Processing**

The analyzer converts the analog time-domain signals, $x(t)$, into digital frequency-domain data using Fourier series relations, given by Eqs. (1.97) to (1.99), to facilitate digital computation. Thus the analyzer accepts the analog output signals of accelerometers or force transducers, $x(t)$, and computes the spectral coefficients of these signals a_0, a_n, and b_n using Eqs. (1.97) to (1.99) in the frequency domain. The process of converting analog signals into digital data is indicated in Fig. 10.28 for two representative signals. In Fig. 10.28, $x(t)$ denotes the analog signal and $x_i = x(t_i)$ represents the corresponding digital record, with t_i indicating the ith discrete value of time. This process is performed by an analog-to-digital (A/D) converter, which is part of a digital analyzer. If N samples of $x(t)$ are collected at discrete values of time, t_i, the data $[x_1(t_i), x_2(t_i), \ldots, x_N(t_i)]$ can be used to obtain the discrete form of Fourier transform as

$$x_j = x(t_j) = \frac{a_0}{2} + \sum_{i=1}^{N/2}\left(a_i \cos\frac{2\pi i t_j}{T} + b_i \sin\frac{2\pi i t_j}{T}\right); \qquad j = 1, 2, \ldots, N \quad (10.45)$$

FIGURE 10.28   Representation of signals in different forms: (a) Signals in time domain. (b) Signals in frequency domain. (c) Digital records of $x(t)$.

where the digital spectral coefficients a_0, a_i, and b_i are given by (see Eqs. (1.97) to (1.99))

$$a_0 = \frac{1}{N} \sum_{j=1}^{N} x_j \tag{10.46}$$

$$a_i = \frac{1}{N} \sum_{j=1}^{N} x_j \cos \frac{2\pi i t_j}{N} \tag{10.47}$$

$$b_i = \frac{1}{N} \sum_{j=1}^{N} x_j \sin \frac{2\pi i t_j}{N} \tag{10.48}$$

with the number of samples N equal to some power of 2 (such as 256, 512, or 1,024) which is fixed for a given analyzer. Equations (10.46) to (10.48) denote N algebraic equations for each of the N samples. The equations can be expressed in matrix form as

$$\vec{X} = [A]\vec{d} \tag{10.49}$$

where $\vec{X} = \{x_1 \, x_2 \, \cdots \, x_N\}^T$ is the vector of samples, $\vec{d} = \{a_0 \, a_1 \, a_2 \cdots a_{N/2} \, b_1 \, b_2 \cdots b_{N/2}\}^T$ is the vector of spectral coefficients, and $[A]$ is the matrix composed of the coefficients $\cos \dfrac{2\pi i t_j}{T}$ and $\sin \dfrac{2\pi i t_j}{T}$ of Eqs. (10.46)–(10.48). The frequency content of the signal or response of the system can be determined from the solution

$$\vec{d} = [A]^{-1}\vec{X} \tag{10.50}$$

where $[A]^{-1}$ is computed efficiently using fast Fourier transform (FFT) by the analyzer.

10.8.4
Analysis of
Random Signals

The input and output data measured by the transducers usually contain some random component or noise that makes it difficult to analyze the data in a deterministic manner. Also, in some cases, random excitation is used in vibration testing. Thus random signal analysis becomes necessary in vibration testing. If $x(t)$ is a random signal, as shown in Fig. 10.29, its average or mean, denoted as $\bar{x}$, is defined as[6]

$$\bar{x}(t) = \lim_{T \to \infty} \frac{1}{N} \int_0^T x(t)\, dt \tag{10.51}$$

which, for a digital signal, can be expressed as

$$\bar{x} = \lim_{N \to \infty} \frac{1}{N} \sum_{j=1}^N x(t_j) \tag{10.52}$$

Corresponding to any random signal $y(t)$, we can always define a new variable $x(t)$ as $x(t) = y(t) - \bar{y}(t)$, so that the mean value of $x(t)$ is zero. Hence, without loss of generality, we can assume the signal $x(t)$ to have a zero mean and define the mean square value or variance of $x(t)$, denoted as $\bar{x}^2(t)$, as

$$\bar{x}^2(t) = \lim_{N \to \infty} \frac{1}{T} \int_0^T x^2(t)\, dt \tag{10.53}$$

which, for a digital signal with N samples of $x(t)$ at $t = t_1, t_2, \ldots, t_N$, can be expressed as

$$\bar{x}^2 = \lim_{n \to \infty} \frac{1}{N} \sum_{j=1}^N x^2(t_j) \tag{10.54}$$

The root mean square (RMS) value of $x(t)$ is given by

$$x_{\mathrm{RMS}} = \sqrt{\bar{x}^2} \tag{10.55}$$

$x(t)$

FIGURE 10.29   A random signal, $x(t)$.

[6]A detailed discussion of random signals (processes) and random vibration is given in Chapter 14.

The autocorrelation function of a random signal $x(t)$, denoted as $R(t)$, gives a measure of the speed with which the signal changes in the time domain and is defined as

$$R(t) = \overline{x^2} = \lim_{T \to \infty} \frac{1}{T} \int_0^T x(\tau)x(\tau + t) \, d\tau \tag{10.56}$$

which, for a digital signal, can be written as

$$R(n, \Delta t) = \frac{1}{N - n} \sum_{j=0}^{N-n} x_j x_{j+n} \tag{10.57}$$

where N is the number of samples, Δt is the sampling interval, and n is an adjustable parameter that can be used to control the number of points used in the computation. It can be seen that $R(0)$ denotes the mean square value, $\overline{x^2}$, of $x(t)$. The autocorrelation function can be used to identify the presence of periodic components present (buried) in a random signal. If $x(t)$ is purely random, then $R(t) \to 0$ as $T \to \infty$. However, if $x(t)$ is periodic or has a periodic component, then $R(t)$ will also be periodic.

The power spectral density (PSD) of a random signal $x(t)$, denoted as $S(\omega)$, gives a measure of the speed with which the signal changes in the frequency domain and is defined as the Fourier transform of $R(t)$:

$$S(\omega) = \frac{1}{2\pi} \int_{-\infty}^{\infty} R(\tau)e^{-i\omega\tau} \, d\tau \tag{10.58}$$

which, in digital form, can be expressed as

$$S(\Delta\omega) = \frac{|x(\omega)|^2}{N \, \Delta t} \tag{10.59}$$

where $|x(\omega)|^2$ represents the magnitude of the Fourier transform of the sampled data of $x(t)$. The definitions of autocorrelation and PSD functions can be extended for two different signals, such as a displacement signal $x(t)$ and an applied force signal $f(t)$. This leads to the cross-correlation function, $R_{xf}(t)$ and the cross-PSD $S_{xf}(\omega)$:

$$R_{xf}(t) = \lim_{T \to \infty} \frac{1}{T} \int_0^T x(\tau)f(\tau + t) \, d\tau \tag{10.60}$$

$$S_{xf}(\omega) = \frac{1}{2\pi} \int_{-\infty}^{\infty} R_{xf}(\tau)e^{-i\omega\tau} \, d\tau \tag{10.61}$$

Equations (10.60) and (10.61) permit the determination of the transfer functions of the structure or machine being tested. In Eq. (10.60), if $f(\tau + t)$ is replaced by $x(\tau + t)$, we

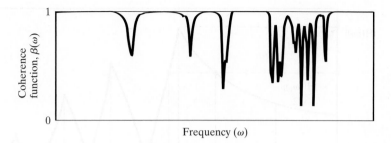

FIGURE 10.30   A typical coherence function.

obtain $R_{xx}(t)$, which when used in Eq. (10.61), leads to $S_{xx}(\omega)$. The frequency-response function, $H(i\omega)$, is related to the PSD functions as

$$S_{xx}(\omega) = |H(i\omega)|^2 S_{ff}(\omega) \tag{10.62}$$

$$S_{fx}(\omega) = H(i\omega) S_{ff}(\omega) \tag{10.63}$$

$$S_{xx}(\omega) = H(i\omega) S_{xf}(\omega) \tag{10.64}$$

with $f(t)$ and $x(t)$ denoting the random force input and the resulting output response, respectively. $S_{xx}(\omega)$, given by Eq. (10.62), contains information about the magnitude of the transfer function of the system (structure or machine), while $S_{xf}(\omega)$ and $S_{xx}(\omega)$, given by Eqs. (10.63) and (10.64), contain information about both magnitude and phase. In vibration testing, the spectrum analyzer first computes different spectral density functions from the transducer output, and then computes the frequency-response function $H(i\omega)$ of the system using Eqs. (10.63) and (10.64).

Coherence Function.   A function, known as *coherence function* (β), is defined as a measure of the noise present in the signals as

$$\beta(\omega) = \left(\frac{S_{fx}(\omega)}{S_{ff}(\omega)}\right)\left(\frac{S_{xf}(\omega)}{S_{xx}(\omega)}\right) = \frac{|S_{xf}(\omega)|^2}{S_{xx}(\omega) S_{ff}(\omega)} \tag{10.65}$$

Note that if the measurements of x and f are pure noises, then $\beta = 0$, and if the measurements of x and f are not contaminated at all with noise, then $\beta = 1$. The plot of a typical coherence function is shown in Fig. 10.30. Usually, $\beta \approx 1$ near the natural frequency of the system because the signals are large and are less influenced by the noise.

**10.8.5
Determination
of Modal Data
from Observed
Peaks**

The frequency-response function, $H(i\omega)$, computed from Eq. (10.63) or (10.64), can be used to find the natural frequencies, damping ratios, and mode shapes corresponding to all resonant peaks observed in the plot of $H(i\omega)$. Let the graph of the frequency-response function be as shown in Fig. 10.31, with its four peaks or resonances suggesting that the system being tested can be modeled as a four-degree-of-freedom system. Sometimes, it

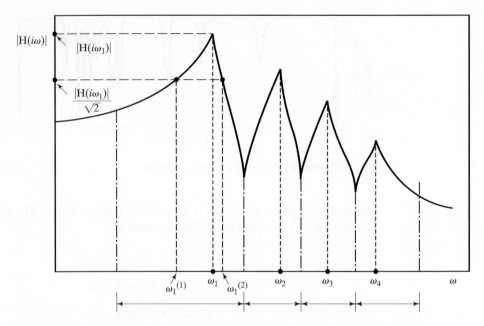

FIGURE 10.31   A typical graph of the frequency-response function of a structure or machine, obtained using Eq. (10.63) or (10.64).

becomes difficult to assign the number of degrees of freedom to the system, especially when the resonant peaks are closely spaced in the graph of $H(i\omega)$, which can be plotted by applying a harmonic force of adjustable frequency at a specific point of the structure or machine, measuring the response (for example, displacement) at another point, and finding the value of the frequency-response function using Eq. (10.63) or (10.64). The graph of $H(i\omega)$, similar to Fig. 10.31, can be plotted by finding the values of $H(i\omega)$ at a number of frequencies of the applied harmonic force.

A simple method of finding the modal data involves the use of a single-degree-of-freedom approach. In this method, the graph of $H(i\omega)$ is partitioned into several frequency ranges with each range bracketing one peak, as shown in Fig. 10.31. Each partitioned frequency range is then considered as the frequency-response function of a single-degree-of-freedom system. This implies that the frequency-response function in each frequency range is dominated by that specific single mode. As observed in Section 3.4, a peak denotes a resonance point corresponding to a phase angle of 90°. Thus the resonant frequencies can be identified as the peaks in the graph of $H(i\omega)$, which can be confirmed from an observation of the values of the phase angle to be 90° at each of the peaks. The damping ratio corresponding to peak j, with resonant frequency ω_j, in Fig. 10.31 denotes the modal damping ratio, ζ_j. This ratio can be found, using Eq. (3.45), as

$$\zeta_j = \frac{\omega_j^{(2)} - \omega_j^{(1)}}{2\,\omega_j} \tag{10.66}$$

where $\omega_j^{(1)}$ and $\omega_j^{(2)}$, known as *half-power points*, lie on either side of the resonant frequency ω_j and satisfy the relation

$$|H(i\omega_j^{(1)})| = |H(i\omega_j^{(2)})| = \frac{|H(i\omega_j)|}{\sqrt{2}} \tag{10.67}$$

Note that ω_j actually represents the damped natural frequency of the system being tested. However, when damping is small, ω_j can be considered approximately equal to the undamped natural frequency of the system. When the system being tested is approximated as a k-degree-of-freedom system ($k = 4$ for the system corresponding to Fig. 10.31), each peak observed in the graph of $H(i\omega)$ is assumed to be a single-degree-of-freedom system, and the k resonant frequencies (peaks) and the corresponding damping ratios are determined by repeating the above procedure (and using an application of Eq. (10.66)) k times.

EXAMPLE 10.5

Determination of Damping Ratio from Bode Diagram

The graphs showing the variations of the magnitude of the response and its phase angle with the frequency of a single-degree-of-freedom system, as indicated in Fig. 3.11, provide the frequency response of the system. Instead of dealing with the magnitude curves directly, if the logarithms of the magnitude ratios (in decibels) are used, the resulting plots are called *Bode diagrams*. Find the natural frequency and the damping ratio of a system whose Bode diagram is shown in Fig. 10.32.

Solution: The natural frequency, which corresponds approximately to the peak response of the system, can be seen to be 10 Hz and the peak response to be -35 dB. The half-power points

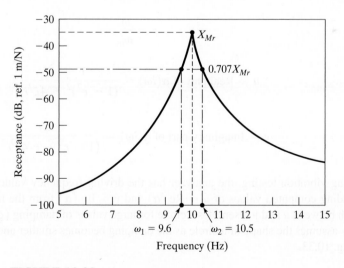

FIGURE 10.32 Bode diagram.

correspond to frequencies ω_1 and ω_2, where the response amplitude is equal to 0.707 times the peak response. From Fig. 10.32, the half-power points can be identified as $\omega_1 = 9.6$ Hz and $\omega_2 = 10.5$ Hz; thus the damping ratio can be determined by using Eq. (10.66) as

$$\zeta = \frac{\omega_2 - \omega_1}{2\omega_n} = \frac{10.5 - 9.6}{2(10.0)} = 0.045$$

■

The procedure described in this section for finding the modal parameters is basically a visual approach. A more systematic, computer-based approach that can be implemented by the analyzer in conjunction with suitable programming is presented in the next section.

10.8.6 Determination of Modal Data from Nyquist Plot

According to this method, a single mode is also assumed to dominate in the neighborhood of its natural frequency in the frequency-response function. When the real and imaginary parts of the frequency-response function of a single-degree-of-freedom system (given by Eq. (3.54)) are plotted along the horizontal and vertical axes of a graph for a range of frequencies, the resulting graph will be in the form of a circle, known as the *Nyquist circle* or *Nyquist plot*. The frequency-response function, given by Eq. (3.54), can be written as

$$\alpha(i\omega) = \frac{1}{1 - r^2 + i2\zeta r} = u + iv \tag{10.68}$$

where

$$r = \frac{\omega}{\omega_n} \tag{10.69}$$

$$u = \text{Real part of } \alpha(i\omega) = \frac{1 - r^2}{(1 - r^2)^2 + 4\zeta^2 r^2} \tag{10.70}$$

$$v = \text{Imaginary part of } \alpha(i\omega) = \frac{-2\zeta r}{(1 - r^2)^2 + 4\zeta^2 r^2} \tag{10.71}$$

During vibration testing, the analyzer has the driving frequency values ω and the corresponding computed values of $u = \text{Re}(\alpha)$ and $v = \text{Im}(\alpha)$ from the measured data. The graph between u and v resembles a circle for large values of damping (ζ), while it increasingly assumes the shape of a circle as the damping becomes smaller and smaller, as shown in Fig. 10.33.

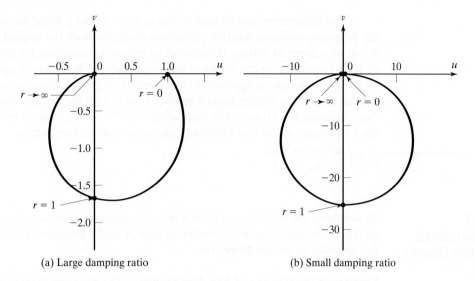

(a) Large damping ratio   (b) Small damping ratio

FIGURE 10.33   Nyquist circle.

Properties of Nyquist Circle.   To identify the properties of the Nyquist circle, we first observe that large values of u and v are attained in the vicinity of resonance, $r = 1$. In that region, we can replace $1 - r^2$ in Eqs. (10.70) and (10.71) as

$$1 - r^2 = (1 + r)(1 - r) \approx 2(1 - r) \qquad \text{and} \qquad 2\zeta r \approx 2\zeta$$

so that

$$u = \text{Re}(\alpha) \approx \frac{1 - r}{2[(1 - r)^2 + \zeta^2]} \tag{10.72}$$

$$v = \text{Im}(\alpha) \approx \frac{-\zeta}{2[(1 - r)^2 + \zeta^2]} \tag{10.73}$$

It can be easily verified that u and v, given by Eqs. (10.72) and (10.73), satisfy the relation

$$u^2 + \left(v + \frac{1}{4\zeta}\right)^2 = \left(\frac{1}{4\zeta}\right)^2 \tag{10.74}$$

which denotes the equation of a circle with its center at $\left(u = 0, v = -\frac{1}{4\zeta}\right)$ and radius $\frac{1}{4\zeta}$. The half-power points occur at $r = 1 \pm \zeta$, which correspond to $u = \pm\frac{1}{4\zeta}$ and $v = \frac{1}{4\zeta}$. These points are located at the two ends of the horizontal diameter of the circle, at which point u has its maximum magnitude.

These observations can be used to find ω_n $(r = 1)$ and ζ. Once the measured values of the frequency-response function $H(i\omega)$ are available (with the applied force magnitude fixed) for a range of driving frequencies ω, instead of searching for the peak in the plot of $H(i\omega)$ versus ω, we can construct the Nyquist plot of Re $(H(i\omega))$ against Im $(H(i\omega))$ by using a least-squares approach to fit a circle. This process also averages out the experimental errors. The intersection of the fitted circle with the negative imaginary axis will then correspond to $H(i\omega_n)$. The bandwidth, $(\omega^{(2)} - \omega^{(1)})$, is given by the difference of the frequencies at the two horizontal diametral points, from which ζ can be found as

$$\zeta = \left(\frac{\omega^{(2)} - \omega^{(1)}}{2\omega_n} \right).$$

10.8.7
Measurement
of Mode Shapes

To determine the mode shapes from vibration testing, we need to express the equations of motion of the multidegree-of-freedom system in modal coordinates [10.18]. For this, we first consider an undamped system.

Undamped Multidegree-of-Freedom System. The equations of motion of an undamped multidegree-of-freedom system in physical coordinates are given by

$$[m]\ddot{\vec{x}} + [k]\vec{x} = \vec{f} \tag{10.75}$$

For free harmonic vibration, Eq. (10.75) becomes

$$\left[[k] - \omega_i^2[m] \right]\vec{y}_i = \vec{0} \tag{10.76}$$

where ω_i is the ith natural frequency and $\vec{y}_i$ is the corresponding mode shape. The orthogonality relations for the mode shapes can be expressed as

$$[Y]^T[m][Y] = \mathrm{diag}[M] \equiv \left[\diagdown M_i \diagdown \right] \tag{10.77}$$

$$[Y]^T[k][Y] = \mathrm{diag}[K] \equiv \left[\diagdown K_i \diagdown \right] \tag{10.78}$$

where $[Y]$ is the modal matrix containing the modes $\vec{y}_1, \vec{y}_2, \ldots, \vec{y}_N$ as columns (N denotes the number of degrees of freedom of the system, also equal to the number of measured natural frequencies or peaks), M_i and K_i are the elements of diag $[M]$ and diag $[K]$, also called the *modal mass* and *modal stiffness*, respectively, corresponding to mode i, and

$$\omega_i^2 = \frac{K_i}{M_i} \tag{10.79}$$

When the forcing functions are harmonic, $\vec{f}(t) = \vec{F}e^{\tilde{i}\omega t}$, with $\tilde{i} = \sqrt{-1}$, Eq. (10.75) yields

$$\vec{x}(t) = \vec{X}e^{\tilde{i}\omega t} = \left[[k] - \omega^2[m] \right]^{-1}\vec{F}e^{\tilde{i}\omega t} \equiv [\alpha(\omega)]\vec{F}e^{\tilde{i}\omega t} \tag{10.80}$$

where $[\alpha(\omega)]$ is called the *frequency-response function* or *receptance matrix* of the system. Using the orthogonality relations of Eqs. (10.77) and (10.78), $[\alpha(\omega)]$ can be expressed as

$$[\alpha(\omega)] = [Y]\big[[K] - \omega^2[M]\big]^{-1}[Y]^T \tag{10.81}$$

An individual element of the matrix $[\alpha(\omega)]$ lying in row p and column q denotes the harmonic response of one coordinate, X_p, caused by a harmonic force applied at another coordinate, F_q (with no other forces), and can be written as

$$\alpha_{pq}(\omega) = [\alpha(\omega)]_{pq} = \frac{X_p}{F_q}\bigg|_{\text{with}\quad F_j=0;\, j=1,2,\ldots,N;\, j\neq q}$$

$$= \sum_{i=1}^{N} \frac{(\vec{y}_i)_p (\vec{y}_i)_q}{K_i - \omega^2 M_i} \tag{10.82}$$

where $(\vec{y}_i)_j$ denotes the jth component of mode $\vec{y}_i$. If the modal matrix $[Y]$ is further normalized (rescaled or mass-normalized) as

$$[\Phi] \equiv \big[\vec{\phi}_1\,\vec{\phi}_2\cdots\vec{\phi}_N\big] = [Y][M]^{-1/2} \tag{10.83}$$

the shape of the modes $\vec{\phi}_1, \vec{\phi}_2, \ldots, \vec{\phi}_N$ will not change, but Eq. (10.82) becomes

$$\alpha_{pq}(\omega) = \sum_{i=1}^{N} \frac{(\vec{\phi}_i)_p (\vec{\phi}_i)_q}{\omega_i^2 - \omega^2} \tag{10.84}$$

Damped Multidegree-of-Freedom System.   The equations of motion of a damped multi-degree-of-freedom system in physical coordinates are given by

$$[m]\ddot{\vec{x}} + [c]\dot{\vec{x}} + [k]\vec{x} = \vec{f} \tag{10.85}$$

For simplicity, we assume proportional damping, so that the damping matrix $[c]$ can be expressed as

$$[c] = a[k] + b[m] \tag{10.86}$$

where a and b are constants. Then the undamped mode shapes of the system, $\vec{y}_i$ and $\vec{\phi}_i$, diagonalize not only the mass and stiffness matrices, as indicated in Eqs. (10.77) and (10.78), but also the damping matrix:

$$[Y]^T[c][Y] = \text{diag}\,[C] = \big\lceil\,C_i\,\big\rfloor \tag{10.87}$$

Thus the mode shapes of the damped system will remain the same as those of the undamped system, but the natural frequencies will change and in general become complex. When the forcing vector $\vec{f}$ is assumed to be harmonic in Eq. (10.85), the frequency-response function or receptance can be derived as

$$\alpha_{pq}(\omega) = [\alpha(\omega)]_{pq} = \sum_{i=1}^{N} \frac{(\vec{y}_i)_p (\vec{y}_i)_q}{K_i - \omega^2 M_i + \widetilde{i}\omega C_i} \tag{10.88}$$

When mass-normalized mode shapes are used (see Eq. (10.83)), $\alpha_{pq}(\omega)$ becomes

$$\alpha_{pq}(\omega) = \sum_{i=1}^{N} \frac{(\vec{\phi}_i)_p (\vec{\phi}_i)_q}{\omega_i^2 - \omega^2 + 2\widetilde{i}\zeta_i\omega_i\omega} \tag{10.89}$$

where ζ_i is the damping ratio in mode i.

As indicated earlier, the element of the matrix $[\alpha(\omega)]$ in row p and column q, $\alpha_{pq}(\omega) = [\alpha(\omega)]_{pq}$, denotes the transfer function between the displacement or response at point $p(X_p)$ and the input force at point $q(F_q)$ of the system being tested (with all other forces equal to zero). Since this transfer function denotes the ratio $\dfrac{X_p}{F_q}$, it is given by $H_{pq}(\omega)$. Thus

$$\alpha_{pq}(\omega) = H_{pq}(\omega) \tag{10.90}$$

If the peaks or resonant (natural) frequencies of the system are well separated, then the term corresponding to the particular peak (ith peak) dominates all other terms in the summation of Eq. (10.88) or (10.89). By substituting $\omega = \omega_i$ in Eq. (10.89), we obtain

$$\alpha_{pq}(\omega_i) = H_{pq}(\omega_i) = \frac{(\vec{\phi}_i)_p (\vec{\phi}_i)_q}{\omega_i^2 - \omega_i^2 + \widetilde{i}2\zeta_i\omega_i^2}$$

or

$$|\alpha_{pq}(\omega_i)| = |H_{pq}(\omega_i)| = \frac{|(\vec{\phi}_i)_p (\vec{\phi}_i)_q|}{2\zeta_i\omega_i^2}$$

or

$$|(\vec{\phi}_i)_p (\vec{\phi}_i)_q| = 2\zeta_i\omega_i^2 |H_{pq}(\omega_i)| \tag{10.91}$$

It can be seen that Eq. (10.91) permits the computation of the absolute value of $(\vec{\phi}_i)_p (\vec{\phi}_i)_q$ using the measured values of the natural frequency (ω_i), damping ratio (ζ_i), and the transfer function $|H_{pq}(\omega_i)|$ at peak i. To determine the sign of the element $(\vec{\phi}_i)_p (\vec{\phi}_i)_q$, the phase plot of $H_{pq}(\omega_i)$ can be used. Since there are only N independent unknown components of $\vec{\phi}_i$ in the N^2 elements of the matrix $\left[(\vec{\phi}_i)_p (\vec{\phi}_i)_q\right] = \left[\vec{\phi}_i \, \vec{\phi}_i^{\mathrm{T}}\right]_{pq}$, N measurements of $|H_{pq}(\omega_i)|$ are required to determine the mode shape $\vec{\phi}_i$ corresponding to the modal frequency ω_i. This can be achieved by measuring the displacement or response of the system at point q with input at point 1 first, at point 2 next, ..., and at point N last.

10.9   Machine-Condition Monitoring and Diagnosis

Most machines produce low levels of vibration when designed properly. During operation, all machines are subjected to fatigue, wear, deformation, and foundation settlement. These effects cause an increase in the clearances between mating parts, misalignments in shafts, initiation of cracks in parts, and unbalances in rotors—all leading to an increase in the level of vibration, which causes additional dynamic loads on bearings. As time progresses, the vibration levels continue to increase, leading ultimately to the failure or breakdown of the machine. The common types of faults or operating conditions that lead to increased levels of vibration in machines include bent shafts, eccentric shafts, misaligned components, unbalanced components, faulty bearings, faulty gears, impellers with faulty blades, and loose mechanical parts.

10.9.1 Vibration Severity Criteria

The vibration severity charts, given by standards such as ISO 2372, can be used as a guide to determine the condition of a machine. In most cases, the root mean square (RMS) value of the vibratory velocity of the machine is compared against the criteria set by the standards. Although it is very simple to implement this procedure, the overall velocity signal used for comparison may not give sufficient warning of the imminent damage of the machine.

10.9.2 Machine Maintenance Techniques

The life of a machine follows the classic *bathtub curve* shown in Fig. 10.34. Since the failure of a machine is usually characterized by an increase in vibration and/or noise level, the vibration level also follows the shape of the same bathtub curve. The vibration level decreases during the initial running-in period, then increases very slowly during the normal operating period due to the normal wear, and finally increases rapidly due to excessive wear until failure or breakdown in the wearout period.

Three types of maintenance schemes can be used in practice:

1. *Breakdown maintenance.* The machine is allowed to fail, at which time the failed machine is replaced by a new one. This strategy can be used if the machine is inexpensive to replace and the breakdown does not cause any other damage. Otherwise, the cost of lost production, safety risks, and additional damage to other machines make this scheme unacceptable.

2. *Preventive maintenance.* Maintenance is performed at fixed intervals such as every 3,000 operating hours or once a year. The maintenance intervals are usually determined

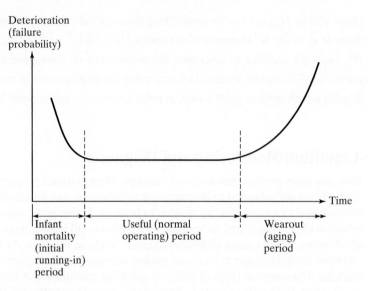

Deterioration
(failure
probability)

Time

Infant
mortality
(initial
running-in)
period

Useful (normal
operating) period

Wearout
(aging)
period

FIGURE 10.34   The bathtub curve for the life of a machine.

statistically from past experience. Although this method reduces the chance of unex-
pected breakdowns, it has been found to be uneconomical. The stoppage for maintenance involves not only lost production time but also a high risk of introducing
imperfections due to human error. In addition, the probability of failure of a machine
component cannot be reduced by replacing it with a new one during the wearout
period.

3. *Condition-based maintenance.* The fixed-interval overhauls are replaced by fixed-
interval measurements that permit the observation of changes in the running condition
of the machine regularly. Thus the onset of fault conditions can be detected and their
developments closely followed. The measured vibration levels can be extrapolated in
order to predict when the vibration levels reach unacceptable values and when the
machine must be serviced. Hence this scheme is also known as *predictive maintenance.*
In this method, the maintenance costs are greatly reduced due to fewer catastrophic
failures, better utilization of spare parts, and elimination of unnecessary preventive
maintenance. The vibration level (and hence the failure probability) of the machine due
to condition-based maintenance follows the shape indicated in Fig. 10.35.

**10.9.3
Machine-
Condition
Monitoring
Techniques**

Several methods can be used to monitor the condition of a machine, as indicated in
Fig. 10.36. Aural and visual methods are the basic forms of monitoring techniques in which
a skilled technician, having an intimate knowledge of machines, can identify a failure
simply by listening to the sounds and/or visually observing the large amplitudes of
vibration produced by a damaged machine. Sometimes a microphone or a stroboscope is
used to hear the machine noise. Similarly, devices ranging from magnifying glasses to

FIGURE 10.35   Condition-based maintenance.

FIGURE 10.36   Machine-condition monitoring techniques.

stroboscopes are used to visually monitor the condition of a machine. Current and voltage monitoring can be used for the condition monitoring of electrical drives such as large generators and motors.

In the operational-variables method of monitoring, also known as *performance* or *duty-cycle monitoring*, the performance of a machine is observed with regard to its intended duty. Any deviation from the intended performance denotes a malfunction of the machine. Temperature monitoring involves measuring the operational or surface temperature of a machine. This method can be considered as a kind of operational-variables method. A rapid increase in the temperature of a component, occurring mostly due to wear, is an indication of a malfunction such as inadequate lubricant in journal bearings. Temperature monitoring uses such devices as optical pyrometers, thermocouples, thermography, and resistance thermometers. In some cases, dye penetrants are used to identify cracks occurring

FIGURE 10.37   Machine-vibration monitoring techniques.

on the surface of a machine. This procedure requires the use of heat-sensitive paints, known as *thermographic paints*, to detect surface cracks on hot surfaces. In such cases, the most suitable paint matching the expected surface temperature is selected.

Wear debris is generated at relative moving surfaces of load-bearing machine elements. The wear particles that can be found in the lubricating oils or grease can be used to assess the extent of damage. As wear increases, the particles of the material used to construct machine components such as bearings and gears can be found in increasing concentration. Thus the severity of the wear can be assessed by observing the concentration (quantity), size, shape, and color of the particles. Note that the color of the particles indicates how hot they have been.

Vibration analysis is most commonly used for machine-condition monitoring. Vibration in machines is caused by cyclic excitation forces arising from imbalances, wear, or failure of parts. What type of changes occur in the vibration level, how these changes can be detected, and how the condition of the machine is interpreted has been the topic of several research studies in the past. The available vibration monitoring techniques can be classified as shown in Fig. 10.37. These techniques are described in the following section.

Time-Domain Analysis

10.9.4 Vibration Monitoring Techniques

Time Waveforms.   Time-domain analysis uses the time history of the signal (waveform). The signal is stored in an oscilloscope or a real-time analyzer and any nonsteady or transient impulses are noted. Discrete damages such as broken teeth in gears and cracks in inner or outer races of bearings can be identified easily from the waveform of the casing of a gearbox. As an example, Fig. 10.38 shows the acceleration signal of a single-stage gearbox. The pinion of the gear pair is coupled to a 5.6-kW, 2,865-rpm, a.c. electric motor. Since the pinion (shaft) speed is 2,865 rpm or 47.75 Hz, the period can be noted as 20.9 ms. The acceleration waveform indicates that pulses occur periodically with a period of 20 ms approximately. Noting that this period is the same as the period of the pinion, the origin of the pulses in the acceleration signal can be attributed to a broken gear tooth on the pinion.

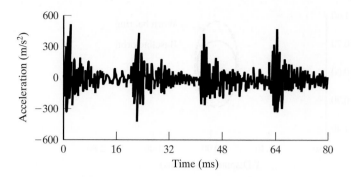

FIGURE 10.38 Time-domain waveform of a faulty gearbox [10.23].

Indices. In some cases, indices such as the peak level, the root mean square (RMS) level, and the crest factor are used to identify damage in machine-condition monitoring. Since the peak level occurs only once, it is not a statistical quantity and hence is not a reliable index to detect damage in continuously operating systems. Although the RMS value is a better index to detect damage in steady-state applications, it may not be useful if the signal contains information from more than one component, as in the case of vibration of a complete gearbox that consists of several gears, shafts, and bearings. The crest factor, defined as the ratio of the peak to RMS level, includes information from both the peak and the RMS levels. However, it may also not be able to identify failure in certain cases. For example, if the failure occurs progressively, the RMS level of the signal might be increasing gradually, although the crest factor might be showing a decreasing trend.

Orbits. Sometimes, certain patterns known as *Lissajous figures* can be obtained by displaying time waveforms obtained from two transducers whose outputs are shifted by 90° in phase. Any change in the pattern of these figures or orbits can be used to identify faults such as misalignment in shafts, unbalance in shafts, shaft rub, wear in journal bearings, and hydrodynamic instability in lubricated bearings. Figure 10.39 illustrates a change in orbit caused by a worn bearing. The enlarged orbit diameter in the vertical direction indicates that the bearing has become stiffer in the horizontal direction—that is, it has more bearing clearance in the vertical direction.

Statistical Methods

Probability Density Curve. Every vibration signal will have a characteristic shape for its probability density curve. The probability density of a signal can be defined as the probability of finding its instantaneous amplitude within a certain range, divided by the range. Usually, the waveform corresponding to good components will have a bell-shaped probability density curve similar to normal distribution. Thus any significant deviation from the bell shape can be associated with the failure of a component. Since the use of the

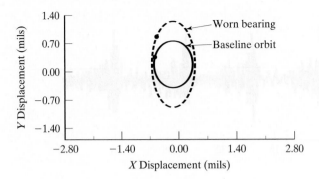

FIGURE 10.39   Change in orbit due to a bearing failure
[10.23].

probability density curve involves the comparison of variations in shape rather than variations in amplitudes, it is very useful in the diagnosis of faults in machines.

Moments.   In some cases, the moments of the probability density curve can be used for the machine-condition monitoring. The moments of the curve are similar to mechanical moments about the centroidal axis of the area. The first four moments of a probability density curve (with proper normalization) are known as the *mean, standard deviation, skewness,* and *kurtosis,* respectively. For practical signals, the odd moments are usually close to zero and the even moments denote the impulsiveness of the signal. The fourth-order moment, kurtosis, is commonly used in machine-condition monitoring. The kurtosis is defined as

$$k = \frac{1}{\sigma^4} \int_{-\infty}^{\infty} (x - \bar{x})^4 f(x)\, dx \tag{10.92}$$

where $f(x)$ is the probability density function of the instantaneous amplitude, $x(t)$, at time t, $\bar{x}$ is the mean value, and σ is the standard deviation of $x(t)$. Faults such as cracked races and spalling of rollers and balls in bearings cause relatively large pulses in the time-domain waveform of the signal, which in turn lead to large values of kurtosis. Thus an increase in the value of kurtosis can be attributed to the failure of a machine component.

Frequency-Domain Analysis

Frequency Spectrum.   The frequency-domain signal or frequency spectrum is a plot of the amplitude of vibration response versus the frequency and can be derived by using the digital fast Fourier analysis of the time waveform. The frequency spectrum provides valuable information about the condition of a machine. The vibration response of a machine is governed not only by its components but also by its assembly, mounting, and installation. Thus the vibration characteristics of any machine are somewhat unique to that particular machine; hence the vibration spectrum can be considered as the vibration signature of that machine. As long as the excitation forces are constant or vary by small amounts, the measured vibration level of the machine also remains constant or varies by small amounts. However, as the machine starts developing faults, its vibration level and hence the shape of the frequency spectrum change. By comparing the frequency spectrum of the machine in damaged condition with the reference frequency spectrum corresponding to the machine in good condition, the nature and location of the fault can be detected. Another important characteristic of a spectrum is that each rotating element in a machine generates identifiable frequency, as illustrated in Fig. 10.40; thus the changes in the spectrum at a given frequency can be attributed directly to the corresponding machine component. Since such changes can be detected more easily compared to changes in the overall vibration levels, this characteristic will be very valuable in practice.

Since the peaks in the spectrum relate to various machine components, it is necessary to be able to compute the fault frequencies. A number of formulas can be derived to find the fault frequencies of standard components like bearings, gearboxes, pumps, fans, and pulleys. Similarly, certain standard fault conditions can be described for standard faults such as unbalance, misalignment, looseness, oil whirl, and resonance.

(a)

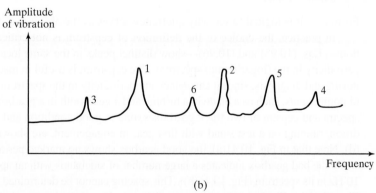

(b)

FIGURE 10.40   Relationship between machine components and the vibration spectrum.

Quefrency-Domain Analysis.   Quefrency serves as the abscissa (x-axis) for a parameter known as *cepstrum*, similar to frequency, that serves as the abscissa for the parameter of spectrum. Several definitions are available for the term *cepstrum* in the literature. Originally, cepstrum was defined as the power spectrum of the logarithm of the power spectrum. If $x(t)$ denotes a time signal, its power spectrum, $S_X(\omega)$, is given by

$$S_X(\omega) = |F\{x(t)\}|^2 \tag{10.93}$$

where $F\{\}$ denotes the Fourier transform of $\{\}$:

$$F\{x(t)\} = \frac{1}{T}\int_{-\frac{T}{2}}^{\frac{T}{2}} x(t)e^{i\omega t}\, dt \tag{10.94}$$

Thus the cepstrum, $c(\tau)$, is given by

$$c(\tau) = |F\{\log S_X(\omega)\}|^2 \tag{10.95}$$

Later, the cepstrum was defined as the inverse Fourier transform of the logarithm of the power spectrum, so that $c(\tau)$ becomes

$$c(\tau) = F^{-1}\{\log S_X(\omega)\} \tag{10.96}$$

The word *cepstrum* is derived by rearranging the letters in the word *spectrum*. The reason for this link is that the cepstrum is basically the spectrum of a spectrum. In fact, many of the terms used in spectrum analysis have been modified for use in cepstrum analysis. A few examples are given below:

> *Quefrency—Frequency*
>
> *Rahmonics—Harmonics*
>
> *Gamnitude—Magnitude*
>
> *Saphe—Phase*

From this, it is logical to see why quefrency serves as the abscissa of the cepstrum.

In practice, the choice of the definition of cepstrum is not critical, since both definitions—Eqs. (10.95) and (10.96)—show distinct peaks in the same location if there is strong periodicity in the (logarithmic) spectrum. The cepstrum is useful in machine-condition monitoring and diagnosis, since it can detect any periodicity in the spectrum caused by the failure of components, such as a blade in a turbine and a gear tooth in a gearbox. As an example, the spectra and cepstra of two truck gearboxes, one in good condition and the other in bad condition, running on a test stand with first gear in engagement, are shown in Figs. 10.41(a) to (d). Note that in Fig. 10.41(a), the good gearbox shows no marked periodicity in its spectrum while the bad gearbox indicates a large number of sidebands with an approximate spacing of 10 Hz in its spectrum (Fig. 10.41(b)). This spacing cannot be determined more accurately from Fig. 10.41(b). Similarly, the cepstrum of the good gearbox does not indicate any quefrencies prominently (Fig. 10.41(d)). However, the cepstrum of the bad gearbox (Fig. 10.41(c)) indicates three prominant quefrencies at 28.1 ms (35.6 Hz), 95.9 ms (10.4 Hz), and 191.0 ms (5.2 Hz). The first series of rahmonics corresponding to 35.6 Hz has been identified to cor-

FIGURE 10.41   Spectrum and cepstrum of a gearbox [10.24].

respond to the input speed of the gearbox. The theoretical output speed is 5.4 Hz. Thus the rahmonics corresponding to 10.4 Hz are not expected to be the same as the second harmonic of the output speed, which would be 10.8 Hz. A careful examination revealed that the rahmonics corresponding to the frequency 10.4 Hz are the same as the speed of the second gear. This indicates that the second gear was at fault although the first gear was in engagement.

**10.9.5
Instrumentation
Systems**

Based on their degree of sophistication, three types of instrumentation systems can be used for condition monitoring of machines—the basic system, the portable system, and the computer-based system. The first type, which can be labeled as the *basic system*, consists of a simple pocket-sized vibration meter, a stroboscope, and a headset. The vibration meter measures the overall vibration levels (RMS or peak values of acceleration or velocity) over suitable frequency ranges, the stroboscope indicates the speed of the machine, and the headset aids in hearing the machine vibration. The overall RMS velocity readings can be compared with published severity charts and any need for condition-based maintenance can be established. The overall vibration levels can also be plotted against time to find how rapidly the condition of the machine is changing. The vibration meter can also be used in conjunction with a pocket computer to collect and store the measurements. Sometimes, an experienced operator can hear the vibration (sound) of a machine over a period of time and assess its condition. In some cases, faults such as misalignment, unbalance, or looseness of parts can be observed visually.

The *portable condition-monitoring system* consists of a portable fast Fourier transform (FFT) vibration analyzer based on battery power. This vibration analyzer can be used for fault detection by recording and storing vibration spectra from each of the measurement points. Each newly recorded spectrum can be compared with a reference spectrum that was recorded at that particular measurement point when the machine was known to be in good condition. Any significant increase in the amplitudes in the new spectrum indicates a fault that needs further investigation. The vibration analyzer also has certain diagnostic capability to identify problems such as faulty belt drives and gearboxes and loose bearings. When the fault diagnosed requires a replacement of parts, it can be done by the operator. If a rotor requires balancing, the vibration analyzer can be used to compute the locations and magnitudes of the correction masses necessary to rebalance the rotor.

The *computer-based condition-monitoring system* is useful and economical when the number of machines, the number of monitoring points, and the complexity of fault detection increases. It consists of an FFT vibration analyzer coupled with a computer for maintaining a centralized database that can also provide diagnostic capabilities. The data are stored on a disk, allowing them to be used for spectrum comparison or for three-dimensional plots (see Fig. 10.42). Certain computer-based systems use tape recorders to record vibration signals from each machine at all the measurement points. These measurements can be played back into the computer for storage and postprocessing.

**10.9.6
Choice of
Monitoring
Parameter**

Piezoelectric accelerometers are commonly used for measuring the vibration of machines. They are preferred because of their smaller size, superior frequency and dynamic range, reliability over long periods, and robustness. When an accelerometer is used as the vibration pickup, the velocities and displacements can be obtained from the integrators built into the analyzer. Thus the user can choose between acceleration, velocity, and displacement as the monitoring parameter. Although any of these three spectra can be used for the condition monitoring of a machine, usually the velocity spectrum will be flattest (indicating that the range of velocity amplitudes is the smallest). Since a change in the amplitude of velocity can be observed easily in a flatter spectrum, velocity is commonly used as the parameter for monitoring the condition of machines.

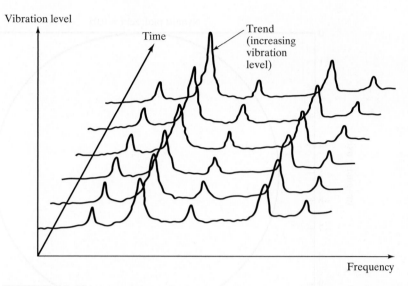

FIGURE 10.42   Three-dimensional plot of data.

10.10   Examples Using MATLAB

Plotting of Nyquist Circle

EXAMPLE 10.6

Using MATLAB, plot the Nyquist circle for the following data:

a. $\zeta = 0.75$

b. $\zeta = 0.05$

Solution: Equations (10.70) and (10.71) are plotted along the horizontal and vertical axes. The MATLAB program to plot the Nyquist circle is given below.

```
%Ex10_6.m
zeta = 0.05;
for i = 1: 10001
    r(i) = 50 * (i-1) / 10000;
    Re1(i) = ( 1-r(i)^2 )/( (1-r(i)^2)^2 + 4*zeta^2*r(i)^2 );
    Im1(i) = -2*zeta*r(i)/( (1-r(i)^2)^2 + 4*zeta^2*r(i)^2 );
end
zeta = 0.75;
for i = 1: 10001
    r(i) = 50 *(i-1) / 10000;
    Re2(i) = ( 1-r(i)^2 )/( (1-r(i)^2)^2 + 4*zeta^2*r(i)^2 );
    Im2(i) = -2*zeta*r(i)/( (1-r(i)^2)^2 + 4*zeta^2*r(i)^2 );
end
plot(Re1, Im1);
title('Nyquist plot: zeta = 0.05');
ylabel('Imaginary axis');
xlabel('Real axis');
pause;
plot(Re2, Im2);
title('Nyquist plot: zeta = 0.75');
ylabel('Imaginary axis');
xlabel('Real axis');
```

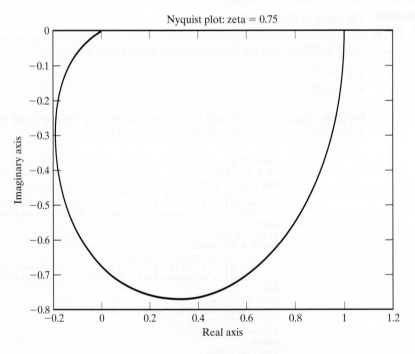

Plotting of Accelerometer Equation

EXAMPLE 10.7

Using MATLAB, plot the ratio of measured to true accelerations, given by

$$f(r) = \frac{1}{\{(1 - r^2)^2 + (2\zeta r)^2\}^{1/2}} \qquad (E.1)$$

for $\zeta = 0.0, 0.25, 0.5, 0.75$, and 1.0.

Solution: The MATLAB program to plot Eq. (E.1) in the range $0 \le r \le 1$ is given below.

```
%Ex10_7.m
zeta = 0.0;
for i = 1: 101
    r(i) = (i-1)/100;
    f1(i) = 1/sqrt((1-r(i)^2)^2 + (2*zeta*r(i))^2);
end
zeta = 0.25;
for i = 1: 101
    r(i) = (i-1)/100;
    f2(i) = 1/sqrt( (1-r(i)^2)^2 + (2*zeta*r(i))^2 );
end
zeta = 0.5;
for i = 1: 101
    r(i) = (i-1)/100;
    f3(i) = 1/sqrt( (1-r(i)^2)^2 + (2*zeta*r(i))^2 );
end
zeta = 0.75;
for i = 1: 101
    r(i) = (i-1)/100;
    f4(i) = 1/sqrt( (1-r(i)^2)^2 + (2*zeta*r(i))^2 );
end
zeta = 1.0;
for i = 1: 101
    r(i) = (i-1)/100;
    f-(i) = 1/sqrt( (1-r(i)^2)^2 + (2*zeta*r(i))^2 );
end
plot(r,f1);
axis([0 1 0 5]);
gtext('zeta = 0.00');
hold on;
plot(r,f2);
gtext('zeta = 0.25');
hold on;
plot(r,f3);
gtext('zeta = 0.50');
hold on;
plot(r,f4);
gtext('zeta = 0.75');
hold on;
plot(r,f5);
gtext('zeta = 1.00');
xlabel('r');
ylabel('f(r)');
```

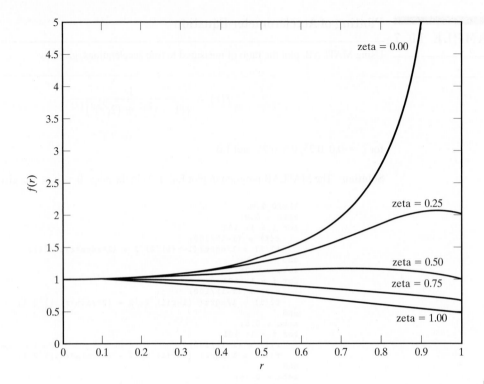

CHAPTER SUMMARY

In some practical applications, it might be difficult to develop a mathematical model, derive the governing equations, and conduct analysis to predict the vibration characteristics of the system. In such cases, we can measure the vibration characteristics of the system under known input conditions and develop a mathematical model of the system. We presented the various aspects of vibration measurement and applications. We discussed the various types of transducers, vibration pickups, frequency-measuring instruments, and shakers (exciters) available for vibration measurement. We described signal analysis, experimental modal analysis and determination of natural frequencies, damping ratio, and mode shapes. We presented machine-condition monitoring and diagnosis techniques. Finally, we presented MATLAB solutions for vibration-measurement-related analysis problems.

   Now that you have finished this chapter, you should be able to answer the review questions and solve the problems given below.

REFERENCES

10.1 G. Buzdugan, E. Mihailescu, and M. Rades, *Vibration Measurement*, Martinus Nijhoff, Dordrecht, The Netherlands, 1986.

10.2 *Vibration Testing*, Bruel & Kjaer, Naerum, Denmark, 1983.

10.3 O. Dossing, *Structural Testing. Part I. Mechanical Mobility Measurements*, Bruel & Kjaer, Naerum, Denmark, 1987.

10.4 D. N. Keast, *Measurements in Mechanical Dynamics*, McGraw-Hill, New York, 1967.

10.5 B. W. Mitchell (ed.), *Instrumentation and Measurement for Environmental Sciences* (2nd ed.), American Society of Agricultural Engineers, Saint Joseph, MI, 1983.

10.6 J. P. Holman, *Experimental Methods for Engineers* (4th ed.), McGraw-Hill, New York, 1984.

10.7 J. T. Broch, *Mechanical Vibration and Shock Measurements*, Bruel & Kjaer, Naerum, Denmark, 1976.

10.8 R. R. Bouche, *Calibration of Shock and Vibration Measuring Transducers*, Shock and Vibration Information Center, Washington, DC, SVM-11, 1979.

10.9 M. Rades, "Methods for the analysis of structural frequency-response measurement data," *Shock and Vibration Digest*, Vol. 8, No. 2, February 1976, pp. 73–88.

10.10 J. D. Irwin and E. R. Graf, *Industrial Noise and Vibration Control*, Prentice Hall, Englewood Cliffs, NJ, 1979.

10.11 R. K. Vierck, *Vibration Analysis*, Harper & Row, New York, 1979.

10.12 J. A. Macinante, *Seismic Mountings for Vibration Isolation*, Wiley, New York, 1984.

10.13 R. B. Randall and R. Upton, "Digital filters and FFT technique in real-time analysis,", in *Digital Signal Analysis Using Digital Filters and FFT Techniques*, Bruel & Kjaer, Naerum, Denmark, 1985, pp. 45–67.

10.14 G. Dovel, "A modal analysis—a dynamic tool for design and troubleshooting," *Mechanical Engineering*, Vol. 111, No. 3, March 1989, pp. 82–86.

10.15 C. W. deSilva and S. S. Palusamy, "Experimental modal analysis—a modeling and design tool," *Mechanical Engineering*, Vol. 106, No. 6, June 1984, pp. 56–65.

10.16 K. Zaveri, *Modal Analysis of Large Structures—Multiple Exciter Systems*, Bruel & Kjaer, Denmark, 1984.

10.17 O. Dossing, *Structural Testing—Part 2: Modal Analysis and Simulation*, Bruel & Kjaer, Naerum, Denmark, 1988.

10.18 D. J. Ewins, "Modal analysis as a tool for studying structural vibration," in *Mechanical Signature Analysis: Theory and Applications*, S. Braun (ed.), Academic Press, London, 1986, pp. 217–261.

10.19 B. A. Brinkman and D. J. Macioce, "Understanding modal parameters and mode shape scaling," *Sound and Vibration*, Vol. 19, No. 6, June 1985, pp. 28–30.

10.20 N. Tandon and B. C. Nakra, "Vibration and acoustic monitoring techniques for the detection of defects in rolling element bearings—a review," *Shock and Vibration Digest*, Vol. 24, No. 3, March 1992, pp. 3–11.

10.21 S. Braun, "Vibration monitoring," in *Mechanical Signature Analysis: Theory and Applications*, S. Braun (ed.), Academic Press, London, 1986, pp. 173–216.

10.22 A. El-Shafei, "Measuring vibration for machinery monitoring and diagnostics," *Shock and Vibration Digest*, Vol. 25, No. 1, January 1993, pp. 3–14.

10.23 J. Mathew, "Monitoring the vibrations of rotating machine elements—an overview," in *Diagnostics, Vehicle Dynamics and Special Topics*, T. S. Sankar (ed.), American Society of Mechanical Engineers, New York, 1989, pp. 15–22.

10.24 R. B. Randall, "Advances in the application of cepstrum analysis to gearbox diagnosis," in *Second International Conference Vibrations in Rotating Machinery (1980)*, Institution of Mechanical Engineers, London, 1980, pp. 169–174.

REVIEW QUESTIONS

10.1   Give brief answers to the following:

1. What is the importance of vibration measurement?
2. What is the difference between a vibrometer and a vibrograph?
3. What is a transducer?
4. Discuss the basic principle on which a strain gage works.
5. Define the gage factor of a strain gage.
6. What is the difference between a transducer and a pickup?
7. What is a piezoelectric material? Give two examples of such material.
8. What is the working principle of an electrodynamic transducer?
9. What is an LVDT? How does it work?
10. What is a seismic instrument?
11. What is the frequency range of a seismometer?
12. What is an accelerometer?
13. What is phase-shift error? When does it become important?
14. Give two examples of a mechanical vibration exciter.
15. What is an electromagnetic shaker?
16. Discuss the advantage of using operational deflection shape measurement.
17. What is the purpose of experimental modal analysis?
18. Describe the use of the frequency-response function in modal analysis.
19. Name two frequency-measuring instruments.
20. State three methods of representing the frequency-response data.
21. How are Bode plots used?
22. How is a Nyquist diagram constructed?
23. What is the principle of mode superposition? What is its use in modal analysis?
24. State the three types of maintenance schemes used for machinery.
25. How are orbits used in machine diagnosis?
26. Define the terms *kurtosis* and *cepstrum*.

10.2   Indicate whether each of the following statements is true or false:

1. A strain gage is a variable-resistance transducer.
2. The value of the gage factor of a strain gage is given by the manufacturer.
3. The voltage output of an electromagnetic transducer is proportional to the relative velocity of the coil.
4. The principle of the electrodynamic transducer can be used in vibration exciters.
5. A seismometer is also known as a *vibrometer*.
6. All vibration-measuring instruments exhibit phase lag.
7. The time lag is important when measuring harmonic motion of frequency ω.
8. The Scotch yoke mechanism can be used as a mechanical shaker.
9. The time response of a system gives better information on energy distribution than does the frequency response.
10. A spectrum analyzer is a device that analyzes a signal in the frequency domain.
11. The complete dynamic response of a machine can be determined through modal testing.
12. The damping ratio of a vibrating system can be found from the Bode diagram.
13. The spectrum analyzers are also known as *fast Fourier transform (FFT) analyzers*.

14. In breakdown maintenance, the machine is run until failure.

15. Time-domain waveforms can be used to detect discrete damages of machinery.

10.3 Fill in each of the following blanks with the appropriate word:

1. A device that transforms values of physical variables into equivalent electrical signals is called a _____.

2. Piezoelectric transducers generate electrical _____ when subjected to mechanical stress.

3. A seismic instrument consists of a _____ system mounted on the vibrating body.

4. The instrument that measures the acceleration of a vibrating body is called _____.

5. _____ can be used to record earthquakes.

6. The instrument that measures the velocity of a vibrating body is called a _____.

7. Most mechanical frequency-measuring instruments are based on the principle of _____.

8. The Frahm tachometer is a device consisting of several _____ carrying masses at free ends.

9. The main advantage of a stroboscope is that it can measure the speed without making _____ with the rotating body.

10. In real-time frequency analysis, the signal is continuously analyzed over all the _____ bands.

11. Real-time analyzers are useful for machinery _____ monitoring, since a change in the noise or vibration spectrum can be observed immediately.

12. An _____ is the interval between any two frequencies $(f_2 - f_1)$ whose frequency ratio $\left(\dfrac{f_2}{f_1}\right)$ is 2.

13. The dynamic testing of a machine involves finding the _____ of the machine at a critical frequency.

14. For vibration testing, the machine is supported to simulate a _____ condition of the system so that rigid body modes can also be observed.

15. The excitation force is measured by a _____ cell.

16. The response of a system is usually measured by _____.

17. The frequency response of a system can be measured using _____ analyzers.

18. The condition of a machine can be determined using _____ severity charts.

19. The life of a machine follows the classic _____ curve.

20. The _____ observed in Lissajous figures can be used to identify machinery faults.

21. Cepstrum can be defined as the power spectrum of the logarithm of the _____.

10.4 Select the most appropriate answer out of the choices given:

1. When a transducer is used in conjunction with another device to measure vibration, it is called a
  a. *vibration sensor*   **b.** *vibration pickup*   **c.** *vibration actuator*

2. The instrument that measures the displacement of a vibrating body is called a
  a. *seismometer*   **b.** *transducer*   **c.** *accelerometer*

3. The circuit that permits the passage of frequency components of a signal over a frequency band and rejects all other frequency components is called a
  a. *bandpass filter*   **b.** *frequency filter*   **c.** *spectral filter*

4. A decibel (dB) is a quantity, such as power (P), defined in terms of a reference value (P_{ref}), as
  a. $10 \log_{10}\left(\dfrac{P}{P_{ref}}\right)$ **b.** $\log_{10}\left(\dfrac{P}{P_{ref}}\right)$ **c.** $\dfrac{1}{P_{ref}} \log_{10}(P)$

5. The following function plays an important role in the experimental modal analysis:
  a. time-response function
  b. modal-response function
  c. frequency-response function

6. The method of subjecting a system to a known force as an initial condition and then releasing is known as
  a. *step relaxation*
  b. *excitation by electromagnetic shaker*
  c. *impactor*

7. The process of using an electrical signal, generalized by a spectrum analyzer, for applying a mechanical force on a system is known as
  a. *step relaxation*
  b. *excitation by electromagnetic shaker*
  c. *impactor*

8. The procedure of using a hammer with a built-in load cell to apply load at different points of a system is known as
  a. *step relaxation*
  b. *excitation by electromagnetic shaker*
  c. *impactor*

9. During the initial running-in period, usually the deterioration of a machine
  a. decreases   **b.** increases   **c.** remains constant

10. During the normal operating period, the deterioration of a machine usually
  a. decreases   **b.** increases   **c.** remains constant

11. During the aging or wearout period, the deterioration of a machine usually
  a. decreases   **b.** increases   **c.** remains constant

10.5 Match the items in the two columns below:

1. Piezoelectric accelerometer	**a.** produces light pulses intermittently
2. Electrodynamic transducer	**b.** has high output and is insensitive to temperature
3. LVDT transducer	**c.** frequently used in velocity pickups
4. Fullarton tachometer	**d.** has high sensitivity and frequency range
5. Stroboscope	**e.** variable-length cantilever with a mass at its free end

---

PROBLEMS

Section 10.2 Transducers

10.1 A Rochelle salt crystal, having a voltage sensitivity of 0.098 V-m/N and thickness 2 mm, produced an output voltage of 200 volts under pressure. Find the pressure applied to the crystal.

Section 10.3 Vibration Pickups

10.2 A spring-mass system with $m = 0.5$ kg and $k = 10,000$ N/m, with negligible damping, is used as a vibration pickup. When mounted on a structure vibrating with an amplitude of 4 mm, the total displacement of the mass of the pickup is observed to be 12 mm. Find the frequency of the vibrating structure.

10.3 The vertical motion of a machine is measured by using the arrangement shown in Fig. 10.43. The motion of the mass m relative to the machine body is recorded on a drum. If the damping constant c is equal to $c_{\text{cri}}/\sqrt{2}$, and the vertical vibration of the machine body is given by $y(t) = Y \sin \omega t$, find the amplitude of motion recorded on the drum.

FIGURE 10.43

10.4 It is proposed that the vibration of the foundation of an internal combustion engine be measured over the speed range 500 rpm to 1,500 rpm using a vibrometer. The vibration is composed of two harmonics, the first one caused by the primary inertia forces and the second one by the secondary inertia forces in the engine. Determine the maximum natural frequency of the vibrometer in order to have an amplitude distortion of less than 2 percent.

10.5 Determine the maximum percent error of a vibrometer in the frequency-ratio range $4 \leq r < \infty$ with a damping ratio of $\zeta = 0$.

10.6 Solve Problem 10.5 with a damping ratio of $\zeta = 0.67$.

10.7 A vibrometer is used to measure the vibration of an engine whose operating-speed range is from 500 to 2,000 rpm. The vibration consists of two harmonics. The amplitude distortion must be less than 3 percent. Find the natural frequency of the vibrometer if (a) the damping is negligible and (b) the damping ratio is 0.6.

10.8 A spring-mass system, having a static deflection of 10 mm and negligible damping, is used as a vibrometer. When mounted on a machine operating at 4,000 rpm, the relative amplitude is recorded as 1 mm. Find the maximum values of displacement, velocity, and acceleration of the machine.

10.9 A vibration pickup has a natural frequency of 5 Hz and a damping ratio of $\zeta = 0.5$. Find the lowest frequency that can be measured with a 1-percent error.

10.10 A vibration pickup has been designed for operation above a frequency level of 100 Hz without exceeding an error of 2 percent. When mounted on a structure vibrating at a frequency of 100 Hz, the relative amplitude of the mass is found to be 1 mm. Find the suspended mass of the pickup if the stiffness of the spring is 4,000 N/m and damping is negligible.

10.11 A vibrometer has an undamped natural frequency of 10 Hz and a damped natural frequency of 8 Hz. Find the lowest frequency in the range to infinity at which the amplitude can be directly read from the vibrometer with less than 2 percent error.

10.12 Determine the maximum percent error of an accelerometer in the frequency-ratio range $0 < r \leq 0.65$ with a damping ratio of $\zeta = 0$.

10.13 Solve Problem 10.12 with a damping ratio of 0.75.

10.14 Determine the necessary stiffness and the damping constant of an accelerometer if the maximum error is to be limited to 3 percent for measurements in the frequency range of 0 to 100 Hz. Assume that the suspended mass is 0.05 kg.

10.15 An accelerometer is constructed by suspending a mass of 0.1 kg from a spring of stiffness 10,000 N/m with negligible damping. When mounted on the foundation of an engine, the peak-to-peak travel of the mass of the accelerometer has been found to be 10 mm at an engine speed of 1,000 rpm. Determine the maximum displacement, maximum velocity, and maximum acceleration of the foundation.

10.16 A spring-mass-damper system, having an undamped natural frequency of 100 Hz and a damping constant of 20 N-s/m, is used as an accelerometer to measure the vibration of a machine operating at a speed of 3,000 rpm. If the actual acceleration is 10 m/s^2 and the recorded acceleration is 9 m/s^2, find the mass and the spring constant of the accelerometer.

10.17 A machine shop floor is subjected to the following vibration due to electric motors running at different speeds:

$$x(t) = 20 \sin 4\pi t + 10 \sin 8\pi t + 5 \sin 12\pi t \text{ mm}$$

If a vibrometer having an undamped natural frequency of 0.5 Hz and a damped natural frequency of 0.48 Hz is used to record the vibration of the machine shop floor, what will be the accuracy of the recorded vibration?

10.18 A machine is subjected to the vibration

$$x(t) = 20 \sin 50t + 5 \sin 150t \text{ mm} \qquad (t \text{ in seconds})$$

An accelerometer having a damped natural frequency of 80 rad/s and an undamped natural frequency of 100 rad/s is mounted on the machine to read the acceleration directly in mm/s^2. Discuss the accuracy of the recorded acceleration.

Section 10.4 Frequency-Measuring Instruments

10.19 A variable-length cantilever beam of rectangular cross section 1.6 mm $\times$ 25 mm, made of spring steel, is used to measure the frequency of vibration. The length of the cantilever can be varied between 5 cm and 25 cm. Find the range of frequencies that can be measured with this device.

Section 10.8 Experimental Modal Analysis

10.20 Show that the real component of the harmonic response of a viscously damped single-degree-of-freedom system (from X in Eq. (3.54)) attains a maximum at

$$R_1 = \frac{\omega_1}{\omega_n} = \sqrt{1 - 2\zeta}$$

and a minimum at

$$R_2 = \frac{\omega_2}{\omega_n} = \sqrt{1 + 2\zeta}$$

10.21 Find the value of the frequency at which the imaginary component of the harmonic response of a viscously damped single-degree-of-freedom system (from X in Eq. (3.54)) attains a minimum.

10.22 Construct the Nyquist diagram for a single-degree-of-freedom system with hysteretic damping.

10.23 The Bode plot of shaft vibration of a turbine obtained during coast-down is shown in Fig. 10.44. Determine the damping ratio of the system when the static deflection of the shaft is equal to 1.27 μm.

FIGURE 10.44

10.24 The vibratory response at the bearing of an internal combustion engine is shown in Fig. 10.45. Determine the equivalent viscous damping ratio of the system.

FIGURE 10.45   Response in time domain.

10.25 Suggest a method of using the Bode plot of phase angle versus frequency (Fig. 3.11(b)) to identify the natural frequency and the damping ratio of the system.

Section 10.9 Machine-Condition Monitoring and Diagnosis

10.26 Two ball bearings, each with 16 balls, are used to support the shaft of a fan that rotates at 750 rpm. Determine the frequencies, in hertz, corresponding to the following defects:* cage, inner race, outer race, and ball. Assume that $d = 15$ mm, $D = 100$ mm, and $\alpha = 30°$.

10.27 Determine the defect frequencies in hertz* corresponding to roller, inner race, outer race, and cage defects for a roller bearing with 18 rollers when installed in a machine that runs at a speed of 1,000 rpm. Assume $d = 2$ cm, $D = 15$ cm, and $\alpha = 20°$.

10.28 An angular contact thrust bearing consists of 18 balls, each of diameter 10 mm, and is mounted on a shaft that rotates at 1,500 rpm. If the contact angle of the bearing is 40° with a pitch diameter 80 mm, find the frequencies corresponding to cage, ball, inner race, and outer race faults.*

10.29 Find the value of kurtosis for a vibration signal that is uniformly distributed in the range 1–5 mm:

$$f(x) = \frac{1}{4}; \quad 1 \le x \le 5 \text{ mm}$$

10.30 Find the value of kurtosis for a vibration amplitude that can be approximated as a discrete random variable with the following probability mass function:

x (mm)	1	2	3	4	5	6	7
$f(x)$	$\dfrac{1}{32}$	$\dfrac{3}{32}$	$\dfrac{3}{16}$	$\dfrac{6}{16}$	$\dfrac{3}{16}$	$\dfrac{3}{32}$	$\dfrac{1}{32}$

Section 10.10 MATLAB Problems

10.31 Figure 10.46 shows the experimental transfer function of a structure. Determine the approximate values of ω_i and ζ_i.

---

*Each type of failure in ball and roller bearings generates frequency of vibration f (impact rate per minute) as follows. Inner race defect: $f = \frac{1}{2}nN(1 + c)$; outer race defect: $f = \frac{1}{2}nN(1 - c)$; ball or roller defect: $f = \frac{DN}{d}c(2 - c)$; cage defect: $f = \frac{1}{2}N(1 - c)$, where d = ball or roller diameter, D = pitch diameter, α = contact angle, n = number of balls or rollers, N = speed (rpm), and $c = \frac{d}{D}\cos\alpha$.

FIGURE 10.46

10.32 The experimental Nyquist circle of a structure is shown in Fig. 10.47. Estimate the modal damping ratio corresponding to this circle.

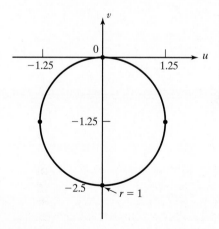

FIGURE 10.47

DESIGN PROJECTS

10.33 Design a vibration exciter to satisfy the following requirements:

    a. Maximum weight of the test specimen $= 10$ N
    b. Range of operating frequency $= 10$ to 50 Hz
    c. Maximum acceleration level $= 20g$
    d. Maximum vibration amplitude $= 0.5$ cm peak-to-peak.

10.34 Frahm tachometers are particularly useful to measure the speeds of engines whose rotating shafts are not easily accessible. When the tachometer is placed on the frame of a running engine, the vibration generated by the engine will cause one of the reeds to vibrate noticeably when the engine speed corresponds to the resonant frequency of a reed. Design a compact and lightweight Frahm tachometer with 12 reeds to measure engine speeds in the range 300–600 rpm.

10.35 A cantilever beam with an end mass m is fixed at the top of a multistory building to measure the acceleration induced at the top of the building during wind and earthquake loads (see Fig. 10.48). Design the beam (that is, determine the material, cross-sectional dimensions, and the length of the beam) such that the stress induced in the beam should not exceed the yield stress of the material under an acceleration of $0.2g$ at the top of the building. Assume that the end mass m is equal to one-half of the mass of the beam.

FIGURE 10.48

Nathan Newmark (1910–1981) was an American engineer and a professor of civil engineering at the University of Illinois at Urbana-Champaign. His research in earthquake-resistant structures and structural dynamics is widely known. The numerical method he presented in 1959 for the dynamic response computation of linear and nonlinear systems is known as the Newmark β-method.(Courtesy of University of Illinois Urbana-Champaign.)

CHAPTER 11

Numerical Integration Methods in Vibration Analysis

Chapter Outline

When the differential equation governing the free or forced vibration of a system cannot be integrated in closed form, a numerical approach is to be used for the vibration analysis. The finite difference method, which is based on the approximation of the derivatives appearing in the equation of motion and the boundary conditions, is presented. Specifically, both single- and multidegree-of-freedom systems using the central difference method are outlined. The free-vibration solution of continuous systems is also considered using the finite difference method in the context of the longitudinal vibration of bars and transverse vibration of beams with different boundary conditions. The fourth-order Runge-Kutta method is presented for the solution of differential equations governing the vibration of single- and multidegree-of-freedom systems. The Houbolt, Wilson, and Newmark methods are presented for the general solution of multidegree-of-freedom systems. Finally, MATLAB programs are given for the solution of multidegree-of-freedom systems with several numerical examples.

Learning Objectives

After you have finished studying this chapter, you should be able to do the following:

- Use the finite difference method for the solution of single- and multidegree-of-freedom vibration problems.
- Solve the vibration problems of continuous systems using the finite difference method.
- Solve the differential equations associated with discrete (multidegree-of-freedom) systems using the fourth-order Runge-Kutta, Houbolt, Wilson, and Newmark methods.
- Use MATLAB functions for solving discrete and continuous vibration problems.

11.1   Introduction

When the differential equation of motion of a vibrating system cannot be integrated in closed form, a numerical approach must be used. Several numerical methods are available for the solution of vibration problems [11.1–11.3].[1] Numerical integration methods have two fundamental characteristics. First, they are not intended to satisfy the governing differential equation(s) at all times t but only at discrete time intervals Δt apart. Second, a suitable type of variation of the displacement x, velocity $\dot{x}$, and acceleration $\ddot{x}$ is assumed within each time interval Δt. Different numerical integration methods can be obtained, depending on the type of variation assumed for the displacement, velocity, and acceleration, within each time interval Δt. We shall assume that the values of x and $\dot{x}$ are known to be x_0 and $\dot{x}_0$, respectively, at time $t = 0$ and that the solution of the problem is required from $t = 0$ to $t = T$. In the following, we subdivide the time duration T into n equal steps Δt so that $\Delta t = T/n$ and seek the solution at $t_0 = 0$, $t_1 = \Delta t$, $t_2 = 2\,\Delta t$, $\ldots$, $t_n = n\,\Delta t = T$. We shall derive formulas for finding the solution at $t_i = i\,\Delta t$ from the known solution at $t_{i-1} = (i-1)\Delta t$ according to five different numerical integration schemes: (1) the finite difference method, (2) the Runge-Kutta method, (3) the Houbolt

---

[1]A numerical procedure using different types of interpolation functions for approximating the forcing function $F(t)$ was presented in Section 4.9.

method, (4) the Wilson method, and (5) the Newmark method. In the finite difference and Runge-Kutta methods, the current displacement (solution) is expressed in terms of the previously determined values of displacement, velocity, and acceleration, and the resulting equations are solved to find the current displacement. These methods fall under the category of explicit integration methods. In the Houbolt, Wilson, and Newmark methods, the temporal difference equations are combined with the current equations of motion, and the resulting equations are solved to find the current displacement. These methods belong to the category of implicit integration methods.

11.2 Finite Difference Method

The main idea in the finite difference method is to use approximations to derivatives. Thus the governing differential equation of motion and the associated boundary conditions, if applicable, are replaced by the corresponding finite difference equations. Three types of formulas—forward, backward, and central difference formulas—can be used to derive the finite difference equations [11.4–11.6]. We shall consider only the central difference formulas in this chapter, since they are the most accurate.

In the finite difference method, we replace the solution domain (over which the solution of the given differential equation is required) with a finite number of points, referred to as *mesh* or *grid points*, and seek to determine the values of the desired solution at these points. The grid points are usually considered to be equally spaced along each of the independent coordinates (see Fig. 11.1). By using Taylor's series expansion, x_{i+1} and x_{i-1} can be expressed about the grid point i as

$$x_{i+1} = x_i + h\dot{x}_i + \frac{h^2}{2}\ddot{x}_i + \frac{h^3}{6}\dddot{x}_i + \cdots \tag{11.1}$$

FIGURE 11.1 Grid points.

$$x_{i-1} = x_i - h\dot{x}_i + \frac{h^2}{2}\ddot{x}_i - \frac{h^3}{6}\dddot{x}_i + \cdots \tag{11.2}$$

where $x_i = x(t = t_i)$ and $h = t_{i+1} - t_i = \Delta t$. By taking two terms only and subtracting Eq. (11.2) from Eq. (11.1), we obtain the central difference approximation to the first derivative of x at $t = t_i$:

$$\dot{x}_i = \frac{dx}{dt}\bigg|_{t_i} = \frac{1}{2h}(x_{i+1} - x_{i-1}) \tag{11.3}$$

By taking terms up to the second derivative and adding Eqs. (11.1) and (11.2), we obtain the central difference formula for the second derivative:

$$\ddot{x}_i = \frac{d^2x}{dt^2}\bigg|_{t_i} = \frac{1}{h^2}(x_{i+1} - 2x_i + x_{i-1}) \tag{11.4}$$

11.3   Central Difference Method for Single-Degree-of-Freedom Systems

The governing equation of a viscously damped single-degree-of-freedom system is

$$m\frac{d^2x}{dt^2} + c\frac{dx}{dt} + kx = F(t) \tag{11.5}$$

Let the duration over which the solution of Eq. (11.5) is required be divided into n equal parts of interval $h = \Delta t$ each. To obtain a satisfactory solution, we must select a time step Δt that is smaller than a critical time step Δt_{cri}.[2] Let the initial conditions be given by $x(t = 0) = x_0$ and $\dot{x}(t = 0) = \dot{x}_0$.

Replacing the derivatives by the central differences and writing Eq. (11.5) at grid point i gives

$$m\left\{\frac{x_{i+1} - 2x_i + x_{i-1}}{(\Delta t)^2}\right\} + c\left\{\frac{x_{i+1} - x_{i-1}}{2\,\Delta t}\right\} + kx_i = F_i \tag{11.6}$$

---

[2]Numerical methods that require the use of a time step (Δt) smaller than a critical time step (Δt_{cri}) are said to be *conditionally stable* [11.7]. If Δt is taken to be larger than Δt_{cri}, the method becomes unstable. This means that the truncation of higher-order terms in the derivation of Eqs. (11.3) and (11.4) (or rounding-off in the computer) causes errors that grow and make the response computations worthless in most cases. The critical time step is given by $\Delta t_{\text{cri}} = \tau_n/\pi$, where τ_n is the natural period of the system or the smallest such period in the case of a multidegree-of-freedom system [11.8]. Naturally, the accuracy of the solution always depends on the size of the time step. By using an unconditionally stable method, we can choose the time step with regard to accuracy only, not with regard to stability. This usually allows a much larger time step to be used for any given accuracy.

where $x_i = x(t_i)$ and $F_i = F(t_i)$. Solution of Eq. (11.6) for x_{i+1} yields

$$
x_{i+1} = \left\{ \cfrac{1}{\dfrac{m}{(\Delta t)^2} + \dfrac{c}{2\,\Delta t}} \right\} \left[\left\{ \frac{2m}{(\Delta t)^2} - k \right\} x_i \right.
$$

$$
\left. + \left\{ \frac{c}{2\,\Delta t} - \frac{m}{(\Delta t)^2} \right\} x_{i-1} + F_i \right] \tag{11.7}
$$

This is called the *recurrence formula*. It permits us to calculate the displacement of the mass (x_{i+1}) if we know the previous history of displacements at t_i and t_{i-1}, as well as the present external force F_i. Repeated application of Eq. (11.7) yields the complete time history of the behavior of the system. Note that the solution of x_{i+1} is based on the use of the equilibrium equation at time t_i—that is, Eq. (11.6). For this reason, this integration procedure is called an *explicit integration method*. Certain care has to be exercised in applying Eq. (11.7) for $i = 0$. Since both x_0 and x_{-1} are needed in finding x_1, and the initial conditions provide only the values of x_0 and $\dot{x}_0$, we need to find the value of x_{-1}. Thus the method is not self-starting. However, we can generate the value of x_{-1} by using Eqs. (11.3) and (11.4) as follows. By substituting the known values of x_0 and $\dot{x}_0$ into Eq. (11.5), $\ddot{x}_0$ can be found:

$$
\ddot{x}_0 = \frac{1}{m}[F(t = 0) - c\dot{x}_0 - kx_0] \tag{11.8}
$$

Application of Eqs. (11.3) and (11.4) at $i = 0$ yields the value of x_{-1}:

$$
x_{-1} = x_0 - \Delta t\,\dot{x}_0 + \frac{(\Delta t)^2}{2}\ddot{x}_0 \tag{11.9}
$$

EXAMPLE 11.1

Response of Single-Degree-of-Freedom System

Find the response of a viscously damped single-degree-of-freedom system subjected to a force

$$
F(t) = F_0\left(1 - \sin\frac{\pi t}{2t_0}\right)
$$

with the following data: $F_0 = 1$, $t_0 = \pi$, $m = 1$, $c = 0.2$, and $k = 1$. Assume the values of the displacement and velocity of the mass at $t = 0$ to be zero.

Solution: The governing differential equation is

$$m\ddot{x} + c\dot{x} + kx = F(t) = F_0\left(1 - \sin\frac{\pi t}{2t_0}\right) \tag{E.1}$$

The finite difference solution of Eq. (E.1) is given by Eq. (11.7). Since the initial conditions are $x_0 = \dot{x}_0 = 0$, Eq. (11.8) yields $\ddot{x}_0 = 1$; hence Eq. (11.9) gives $x_{-1} = (\Delta t)^2/2$. Thus the solution of Eq. (E.1) can be found from the recurrence relation

$$x_{i+1} = \frac{1}{\left[\frac{m}{(\Delta t)^2} + \frac{c}{2\,\Delta t}\right]}\left[\left\{\frac{2m}{(\Delta t)^2} - k\right\}x_i\right.$$

$$\left. + \left\{\frac{c}{2\,\Delta t} - \frac{m}{(\Delta t)^2}\right\}x_{i-1} + F_i\right], \qquad i = 0, 1, 2, \ldots \tag{E.2}$$

with $x_0 = 0$, $x_{-1} = (\Delta t)^2/2$, $x_i = x(t_i) = x(i\,\Delta t)$, and

$$F_i = F(t_i) = F_0\left(1 - \sin\frac{i\pi\,\Delta t}{2t_0}\right)$$

The undamped natural frequency and the natural period of the system are given by

$$\omega_n = \left(\frac{k}{m}\right)^{1/2} = 1 \tag{E.3}$$

and

$$\tau_n = \frac{2\pi}{\omega_n} = 2\pi \tag{E.4}$$

Thus the time step Δt must be less than $\tau_n/\pi = 2.0$. We shall find the solution of Eq. (E.1) by using the time steps $\Delta t = \tau_n/40, \tau_n/20$, and $\tau_n/2$. The time step $\Delta\tau = \tau_n/2 > \Delta t_{\text{cri}}$ is used to illustrate the unstable (diverging) behavior of the solution. The values of the response x_i obtained at different instants of time t_i are shown in Table 11.1.

This example can be seen to be identical to Example 4.17. The results obtained by idealization 4 (piecewise linear type interpolation) of Example 4.17 are shown in Table 11.1 up to time $t_i = \pi$ in the last column of the table. It can be observed that the finite difference method gives reasonably accurate results with time steps $\Delta t = \tau_n/40$ and $\tau_n/20$ (which are smaller than Δt_{cri}) but gives diverging results with $\Delta\tau = \tau_n/2$ (which is larger than Δt_{cri}).

∎

TABLE 11.1 Comparison of Solutions of Example 11.1

Time (t_i)	Values of $x_i = x(t_i)$ Obtained with			Value of x_i Given by Idealization 4 of Example 4.31
	$\Delta t = \dfrac{\tau_n}{40}$	$\Delta t = \dfrac{\tau_n}{20}$	$\Delta t = \dfrac{\tau_n}{2}$	
0	0.00000	0.00000	0.00000	0.00000
$\pi/10$	0.04638	0.04935	—	0.04541
$2\pi/10$	0.16569	0.17169	—	0.16377
$3\pi/10$	0.32767	0.33627	—	0.32499
$4\pi/10$	0.50056	0.51089	—	0.49746
$5\pi/10$	0.65456	0.66543	—	0.65151
$6\pi/10$	0.76485	0.77491	—	0.76238
$7\pi/10$	0.81395	0.82185	—	0.81255
$8\pi/10$	0.79314	0.79771	—	0.79323
$9\pi/10$	0.70297	0.70340	—	0.70482
π	0.55275	0.54869	4.9348	0.55647
2π	0.19208	0.19898	−29.551	—
3π	2.7750	2.7679	181.90	—
4π	0.83299	0.83852	−1,058.8	—
5π	−0.05926	−0.06431	6,253.1	—

11.4 Runge-Kutta Method for Single-Degree-of-Freedom Systems

In the Runge-Kutta method, the approximate formula used for obtaining x_{i+1} from x_i is made to coincide with the Taylor's series expansion of x at x_{i+1} up to terms of order $(\Delta t)^n$. The Taylor's series expansion of $x(t)$ at $t + \Delta t$ is given by

$$x(t + \Delta t) = x(t) + \dot{x}\,\Delta t + \ddot{x}\frac{(\Delta t)^2}{2!} + \dddot{x}\frac{(\Delta t)^3}{3!}$$

$$+ \ddddot{x}\frac{(\Delta t)^4}{4!} + \cdots \qquad (11.10)$$

In contrast to Eq. (11.10), which requires higher-order derivatives, the Runge-Kutta method does not require explicitly derivatives beyond the first [11.9–11.11]. For the solution of a second-order differential equation, we begin by reducing it to two first-order equations. For example, Eq. (11.5) can be rewritten as

$$\ddot{x} = \frac{1}{m}[F(t) - c\dot{x} - kx] = f(x, \dot{x}, t) \qquad (11.11)$$

By defining $x_1 = x$ and $x_2 = \dot{x}$, Eq. (11.11) can be written as two first-order equations:

$$\dot{x}_1 = x_2$$
$$\dot{x}_2 = f(x_1, x_2, t) \qquad (11.12)$$

By defining

$$\vec{X}(t) = \begin{Bmatrix} x_1(t) \\ x_2(t) \end{Bmatrix} \quad \text{and} \quad \vec{F}(t) = \begin{Bmatrix} x_2 \\ f(x_1, x_2, t) \end{Bmatrix}$$

the following recurrence formula is used to find the values of $\vec{X}(t)$ at different grid points t_i according to the fourth-order Runge-Kutta method

$$\vec{X}_{i+1} = \vec{X}_i + \tfrac{1}{6}[\vec{K}_1 + 2\vec{K}_2 + 2\vec{K}_3 + \vec{K}_4] \qquad (11.13)$$

where

$$\vec{K}_1 = h\vec{F}(\vec{X}_i, t_i) \qquad (11.14)$$

$$\vec{K}_2 = h\vec{F}(\vec{X}_i + \tfrac{1}{2}\vec{K}_1, t_i + \tfrac{1}{2}h) \qquad (11.15)$$

$$\vec{K}_3 = h\vec{F}(\vec{X}_i + \tfrac{1}{2}\vec{K}_2, t_i + \tfrac{1}{2}h) \qquad (11.16)$$

$$\vec{K}_4 = h\vec{F}(\vec{X}_i + \vec{K}_3, t_{i+1}) \qquad (11.17)$$

The method is stable and self-starting—that is, only the function values at a single previous point are required to find the function value at the current point.

EXAMPLE 11.2

Response of Single-Degree-of-Freedom System

Find the solution of Example 11.1 using the Runge-Kutta method.

Solution: We use a step size of $\Delta t = 0.3142$ and define

$$\vec{X}(t) = \begin{Bmatrix} x_1(t) \\ x_2(t) \end{Bmatrix} = \begin{Bmatrix} x(t) \\ \dot{x}(t) \end{Bmatrix}$$

and

$$\vec{F}(t) = \begin{Bmatrix} x_2 \\ f(x_1, x_2, t) \end{Bmatrix} = \begin{Bmatrix} \dot{x}(t) \\ \dfrac{1}{m}\left[F_0\left(1 - \sin\dfrac{\pi t}{2t_0}\right) - c\dot{x}(t) - kx(t) \right] \end{Bmatrix}$$

From the known initial conditions, we have

$$\vec{X}_0 = \begin{Bmatrix} 0 \\ 0 \end{Bmatrix}$$

The values of $\vec{X}_{i+1}, i = 0, 1, 2, \ldots$ obtained according to Eq. (11.13) are shown in Table 11.2.

TABLE 11.2

Step i	Time t_i	$x_1 = x$	$x_2 = \dot{x}$
1	0.3142	0.045406	0.275591
2	0.6283	0.163726	0.461502
3	0.9425	0.324850	0.547296
⋮			
19	5.9690	− 0.086558	0.765737
20	6.2832	0.189886	0.985565

■

11.5   Central Difference Method for Multidegree-of-Freedom Systems

The equation of motion of a viscously damped multidegree-of-freedom system (see Eq. (6.119)) can be expressed as

$$[m]\ddot{\vec{x}} + [c]\dot{\vec{x}} + [k]\vec{x} = \vec{F} \tag{11.18}$$

where $[m]$, $[c]$, and $[k]$ are the mass, damping, and stiffness matrices, $\vec{x}$ is the displacement vector, and $\vec{F}$ is the force vector. The procedure indicated for the case of a single-degree-of-freedom system can be directly extended to this case [11.12, 11.13]. The central difference formulas for the velocity and acceleration vectors at time $t_i = i \, \Delta t \, (\dot{\vec{x}}_i$ and $\ddot{\vec{x}}_i)$ are given by

$$\dot{\vec{x}}_i = \frac{1}{2 \, \Delta t}(\vec{x}_{i+1} - \vec{x}_{i-1}) \tag{11.19}$$

$$\ddot{\vec{x}}_i = \frac{1}{(\Delta t)^2}(\vec{x}_{i+1} - 2\vec{x}_i + \vec{x}_{i-1}) \tag{11.20}$$

which are similar to Eqs. (11.3) and (11.4). Thus the equation of motion, Eq. (11.18), at time t_i can be written as

$$[m]\frac{1}{(\Delta t)^2}(\vec{x}_{i+1} - 2\vec{x}_i + \vec{x}_{i-1}) + [c]\frac{1}{2 \, \Delta t}(\vec{x}_{i+1} - \vec{x}_{i-1}) + [k]\vec{x}_i = \vec{F}_i \tag{11.21}$$

where $\vec{x}_{i+1} = \vec{x}(t = t_{i+1})$, $\vec{x}_i = \vec{x}(t = t_i)$, $\vec{x}_{i-1} = \vec{x}(t = t_{i-1})$, $\vec{F}_i = \vec{F}(t = t_i)$, and $t_i = i\,\Delta t$. Equation (11.21) can be rearranged to obtain

$$\left(\frac{1}{(\Delta t)^2}[m] + \frac{1}{2\,\Delta t}[c]\right)\vec{x}_{i+1} + \left(-\frac{2}{(\Delta t)^2}[m] + [k]\right)\vec{x}_i$$

$$+ \left(\frac{1}{(\Delta t)^2}[m] - \frac{1}{2\,\Delta t}[c]\right)\vec{x}_{i-1} = \vec{F}_i$$

or

$$\left(\frac{1}{(\Delta t)^2}[m] + \frac{1}{2\,\Delta t}[c]\right)\vec{x}_{i+1} = \vec{F}_i - \left([k] - \frac{2}{(\Delta t)^2}[m]\right)\vec{x}_i$$

$$- \left(\frac{1}{(\Delta t)^2}[m] - \frac{1}{2\,\Delta t}[c]\right)\vec{x}_{i-1} \tag{11.22}$$

Thus Eq. (11.22) gives the solution vector $\vec{x}_{i+1}$ once $\vec{x}_i$ and $\vec{x}_{i-1}$ are known. Since Eq. (11.22) is to be used for $i = 1, 2, \ldots, n$, the evaluation of $\vec{x}_1$ requires $\vec{x}_0$ and $\vec{x}_{-1}$. Thus a special starting procedure is needed to find $\vec{x}_{-1} = \vec{x}(t = -\Delta t)$. For this, Eqs. (11.18) to (11.20) are evaluated at $i = 0$ to obtain

$$[m]\ddot{\vec{x}}_0 + [c]\dot{\vec{x}}_0 + [k]\vec{x}_0 = \vec{F}_0 = \vec{F}(t = 0) \tag{11.23}$$

$$\dot{\vec{x}}_0 = \frac{1}{2\,\Delta t}(\vec{x}_1 - \vec{x}_{-1}) \tag{11.24}$$

$$\ddot{\vec{x}}_0 = \frac{1}{(\Delta t)^2}(\vec{x}_1 - 2\vec{x}_0 + \vec{x}_{-1}) \tag{11.25}$$

Equation (11.23) gives the initial acceleration vector as

$$\ddot{\vec{x}}_0 = [m]^{-1}(\vec{F}_0 - [c]\dot{\vec{x}}_0 - [k]\vec{x}_0) \tag{11.26}$$

and Eq. (11.24) gives the displacement vector at t_1 as

$$\vec{x}_1 = \vec{x}_{-1} + 2\,\Delta t\dot{\vec{x}}_0 \tag{11.27}$$

Substituting Eq. (11.27) for $\vec{x}_1$, Eq. (11.25) yields

$$\ddot{\vec{x}}_0 = \frac{2}{(\Delta t)^2}[\Delta t \dot{\vec{x}}_0 - \vec{x}_0 + \vec{x}_{-1}]$$

or

$$\vec{x}_{-1} = \vec{x}_0 - \Delta t \dot{\vec{x}}_0 + \frac{(\Delta t)^2}{2}\ddot{\vec{x}}_0 \qquad (11.28)$$

where $\ddot{\vec{x}}_0$ is given by Eq. (11.26). Thus $\vec{x}_{-1}$ needed for applying Eq. (11.22) at $i = 1$ is given by Eq. (11.28). The computational procedure can be described by the following steps.

1. From the known initial conditions $\vec{x}(t = 0) = \vec{x}_0$ and $\dot{\vec{x}}(t = 0) = \dot{\vec{x}}_0$, compute $\ddot{\vec{x}}(t = 0) = \ddot{\vec{x}}_0$ using Eq. (11.26).
2. Select a time step Δt such that $\Delta t < \Delta t_{\text{cri}}$.
3. Compute $\vec{x}_{-1}$ using Eq. (11.28).
4. Find $\vec{x}_{i+1} = \vec{x}(t = t_{i+1})$, starting with $i = 0$, from Eq. (11.22), as

$$\vec{x}_{i+1} = \left[\frac{1}{(\Delta t)^2}[m] + \frac{1}{2\,\Delta t}[c]\right]^{-1}\left[\vec{F}_i - \left([k] - \frac{2}{(\Delta t)^2}[m]\right)\vec{x}_i\right.$$

$$\left. - \left(\frac{1}{(\Delta t)^2}[m] - \frac{1}{2\,\Delta t}[c]\right)\vec{x}_{i-1}\right] \qquad (11.29)$$

where

$$\vec{F}_i = (t = t_i) \qquad (11.30)$$

If required, evaluate accelerations and velocities at t_i:

$$\ddot{\vec{x}}_i = \frac{1}{(\Delta t)^2}[\vec{x}_{i+1} - 2\vec{x}_i + \vec{x}_{i-1}] \qquad (11.31)$$

and

$$\dot{\vec{x}}_i = \frac{1}{2\,\Delta t}[\vec{x}_{i+1} - \vec{x}_{i-1}] \qquad (11.32)$$

Repeat Step 4 until $\vec{x}_{n+1}$ (with $i = n$) is determined. The stability of the finite difference scheme for solving matrix equations is discussed in [11.14].

EXAMPLE 11.3

Central Difference Method for a Two-Degree-of-Freedom System

Find the response of the two-degree-of-freedom system shown in Fig. 11.2 when the forcing functions are given by $F_1(t) = 0$ and $F_2(t) = 10$. Assume the value of c as zero and the initial conditions as $\vec{x}(t = 0) = \dot{\vec{x}}(t = 0) = \vec{0}$.

Solution:

Approach: Use $\Delta t = \tau/10$, where τ is the smallest time period in the central difference method.

The equations of motion are given by

$$[m]\ddot{\vec{x}}(t) + [c]\dot{\vec{x}}(t) + [k]\vec{x}(t) = \vec{F}(t) \tag{E.1}$$

where

$$[m] = \begin{bmatrix} m_1 & 0 \\ 0 & m_2 \end{bmatrix} = \begin{bmatrix} 1 & 0 \\ 0 & 2 \end{bmatrix} \tag{E.2}$$

$$[c] = \begin{bmatrix} c & -c \\ -c & c \end{bmatrix} = \begin{bmatrix} 0 & 0 \\ 0 & 0 \end{bmatrix} \tag{E.3}$$

$$[k] = \begin{bmatrix} k_1 + k & -k \\ -k & k + k_2 \end{bmatrix} = \begin{bmatrix} 6 & -2 \\ -2 & 8 \end{bmatrix} \tag{E.4}$$

$$\vec{F}(t) = \begin{Bmatrix} F_1(t) \\ F_2(t) \end{Bmatrix} = \begin{Bmatrix} 0 \\ 10 \end{Bmatrix} \tag{E.5}$$

and

$$\vec{x}(t) = \begin{Bmatrix} x_1(t) \\ x_2(t) \end{Bmatrix} \tag{E.6}$$

The undamped natural frequencies and the mode shapes of the system can be found by solving the eigenvalue problem

FIGURE 11.2   Two-degree-of-freedom system.

$$\left[-\omega^2\begin{bmatrix}1 & 0\\0 & 2\end{bmatrix}+\begin{bmatrix}6 & -2\\-2 & 8\end{bmatrix}\right]\begin{Bmatrix}X_1\\X_2\end{Bmatrix}=\begin{Bmatrix}0\\0\end{Bmatrix} \tag{E.7}$$

The solution of Eq. (E.7) is given by

$$\omega_1=1.807747, \qquad \vec{X}^{(1)}=\begin{Bmatrix}1.0000\\1.3661\end{Bmatrix} \tag{E.8}$$

$$\omega_2=2.594620, \qquad \vec{X}^{(2)}=\begin{Bmatrix}1.0000\\-0.3661\end{Bmatrix} \tag{E.9}$$

Thus the natural periods of the system are

$$\tau_1=\frac{2\pi}{\omega_1}=3.4757 \quad \text{and} \quad \tau_2=\frac{2\pi}{\omega_2}=2.4216$$

We shall select the time step (Δt) as $\tau_2/10=0.24216$. The initial value of $\ddot{\vec{x}}$ can be found as follows:

$$\ddot{\vec{x}}_0=[m]^{-1}\{\vec{F}-[k]\vec{x}_0\}=\begin{bmatrix}1 & 0\\0 & 2\end{bmatrix}^{-1}\begin{Bmatrix}0\\10\end{Bmatrix}$$

$$=\frac{1}{2}\begin{bmatrix}2 & 0\\0 & 1\end{bmatrix}\begin{Bmatrix}0\\10\end{Bmatrix}=\begin{Bmatrix}0\\5\end{Bmatrix} \tag{E.10}$$

and the value of $\vec{x}_{-1}$ as follows:

$$\vec{x}_{-1}=\vec{x}_0-\Delta t\dot{\vec{x}}_0+\frac{(\Delta t)^2}{2}\ddot{\vec{x}}_0=\begin{Bmatrix}0\\0.1466\end{Bmatrix} \tag{E.11}$$

Now Eq. (11.29) can be applied recursively to obtain $\vec{x}_1$, $\vec{x}_2$, ... The results are shown in Table 11.3.

■

11.6 Finite Difference Method for Continuous Systems

11.6.1 Longitudinal Vibration of Bars

Equation of Motion. The equation of motion governing the free longitudinal vibration of a uniform bar (see Eqs. (8.20) and (8.49)) can be expressed as

$$\frac{d^2U}{dx^2}+\alpha^2U=0 \tag{11.33}$$

TABLE 11.3

Time ($t_i = i \, \Delta t$)	$\vec{x}_i = \vec{x}(t = t_i)$
t_1	$\begin{Bmatrix} 0 \\ 0.1466 \end{Bmatrix}$
t_2	$\begin{Bmatrix} 0.0172 \\ 0.5520 \end{Bmatrix}$
t_3	$\begin{Bmatrix} 0.0931 \\ 1.1222 \end{Bmatrix}$
t_4	$\begin{Bmatrix} 0.2678 \\ 1.7278 \end{Bmatrix}$
t_5	$\begin{Bmatrix} 0.5510 \\ 2.2370 \end{Bmatrix}$
t_6	$\begin{Bmatrix} 0.9027 \\ 2.5470 \end{Bmatrix}$
t_7	$\begin{Bmatrix} 1.2354 \\ 2.6057 \end{Bmatrix}$
t_8	$\begin{Bmatrix} 1.4391 \\ 2.4189 \end{Bmatrix}$
t_9	$\begin{Bmatrix} 1.4202 \\ 2.0422 \end{Bmatrix}$
t_{10}	$\begin{Bmatrix} 1.1410 \\ 1.5630 \end{Bmatrix}$
t_{11}	$\begin{Bmatrix} 0.6437 \\ 1.0773 \end{Bmatrix}$
t_{12}	$\begin{Bmatrix} 0.0463 \\ 0.6698 \end{Bmatrix}$

where

$$\alpha^2 = \frac{\omega^2}{c^2} = \frac{\rho \omega^2}{E} \tag{11.34}$$

To obtain the finite difference approximation of Eq. (11.33), we first divide the bar of length l into $n - 1$ equal parts each of length $h = l/(n - 1)$ and denote the mesh points as $1, 2, 3, \ldots, i, \ldots, n$, as shown in Fig. 11.3. Then, by denoting the value of U at mesh point i as U_i and using a formula for the second derivative similar to Eq. (11.4), Eq. (11.33) for mesh point i can be written as

$$\frac{1}{h^2}(U_{i+1} - 2U_i + U_{i-1}) + \alpha^2 U_i = 0$$

$$U_1 = U_2 = U_3 = \quad U_i = \qquad U_n =$$
$$U(x_1) \; U(x_2) \; U(x_3) \qquad U(x_i) \qquad U(x_n)$$

FIGURE 11.3 Division of a bar for finite difference approximation.

or

$$U_{i+1} - (2 - \lambda)U_i + U_{i-1} = 0 \qquad (11.35)$$

where $\lambda = h^2 \alpha^2$. The application of Eq. (11.35) at mesh points $i = 2, 3, \ldots, n - 1$ leads to the equations

$$
\begin{aligned}
U_3 - (2 - \lambda)U_2 + U_1 &= 0 \\
U_4 - (2 - \lambda)U_3 + U_2 &= 0 \\
&\vdots \\
U_n - (2 - \lambda)U_{n-1} + U_{n-2} &= 0
\end{aligned}
\qquad (11.36)
$$

which can be stated in matrix form as

$$
\begin{bmatrix}
-1 & (2 - \lambda) & -1 & 0 & 0 & \cdots & 0 & 0 & 0 \\
0 & -1 & (2 - \lambda) & -1 & 0 & \cdots & 0 & 0 & 0 \\
0 & 0 & -1 & (2 - \lambda) & -1 & \cdots & 0 & 0 & 0 \\
 & & & \cdot & & & & & \\
 & & & \cdot & & & & & \\
 & & & \cdot & & & & & \\
0 & 0 & 0 & 0 & 0 & \cdots & -1 & (2 - \lambda) & -1
\end{bmatrix}
$$

$$
\begin{Bmatrix}
U_1 \\
U_2 \\
U_3 \\
\cdot \\
\cdot \\
\cdot \\
U_n
\end{Bmatrix}
=
\begin{Bmatrix}
0 \\
0 \\
0 \\
\cdot \\
\cdot \\
\cdot \\
0
\end{Bmatrix}
\qquad (11.37)
$$

Boundary Conditions

Fixed End. The deflection is zero at a fixed end. Assuming that the bar is fixed at $x = 0$ and $x = l$, we set $U_1 = U_n = 0$ in Eq. (11.37) and obtain the equation

$$[[A] - \lambda[I]]\vec{U} = \vec{0} \tag{11.38}$$

where

$$[A] = \begin{bmatrix} 2 & -1 & 0 & 0 & \cdots & 0 & 0 & 0 \\ -1 & 2 & -1 & 0 & \cdots & 0 & 0 & 0 \\ 0 & -1 & 2 & -1 & \cdots & 0 & 0 & 0 \\ . & & & & & & & \\ . & & & & & & & \\ . & & & & & & & \\ 0 & 0 & 0 & 0 & \cdots & 0 & -1 & 2 \end{bmatrix} \tag{11.39}$$

$$\vec{U} = \begin{Bmatrix} U_2 \\ U_3 \\ . \\ . \\ . \\ U_{n-1} \end{Bmatrix} \tag{11.40}$$

and $[I]$ = identity matrix of order $n - 2$.

Note that the eigenvalue problem of Eq. (11.38) can be solved easily, since the matrix $[A]$ is a tridiagonal matrix [11.15–11.17].

Free End. The stress is zero at a free end, so $(dU)/(dx) = 0$. We can use a formula for the first derivative similar to Eq. (11.3). To illustrate the procedure, let the bar be free at $x = 0$ and fixed at $x = l$. The boundary conditions can then be stated as

$$\frac{dU}{dx}\bigg|_1 \simeq \frac{U_2 - U_{-1}}{2h} = 0 \quad \text{or} \quad U_{-1} = U_2 \tag{11.41}$$

$$U_n = 0 \tag{11.42}$$

In order to apply Eq. (11.41), we need to imagine the function $U(x)$ to be continuous beyond the length of the bar and create a fictitious mesh point -1 so that U_{-1} becomes the fictitious displacement of the point x_{-1}. The application of Eq. (11.35) at mesh point $i = 1$ yields

$$U_2 - (2 - \lambda)U_1 + U_{-1} = 0 \tag{11.43}$$

By incorporating the condition $U_{-1} = U_2$ (Eq. (11.41)), Eq. (11.43) can be written as

$$(2 - \lambda)U_1 - 2U_2 = 0 \tag{11.44}$$

By adding Eqs. (11.37) and (11.44), we obtain the final equations:

$$[[A] - \lambda[I]]\vec{U} = \vec{0} \tag{11.45}$$

where

$$[A] = \begin{bmatrix} 2 & -2 & 0 & 0 & \cdots & 0 & 0 & 0 \\ -1 & 2 & -1 & 0 & \cdots & 0 & 0 & 0 \\ 0 & -1 & 2 & -1 & \cdots & 0 & 0 & 0 \\ \cdot & & & & & & & \\ \cdot & & & & & & & \\ \cdot & & & & & & & \\ 0 & 0 & 0 & 0 & \cdots & -1 & 2 & -1 \\ 0 & 0 & 0 & 0 & \cdots & 0 & -1 & 2 \end{bmatrix} \tag{11.46}$$

and

$$\vec{U} = \begin{Bmatrix} U_1 \\ U_2 \\ \cdot \\ \cdot \\ \cdot \\ U_{n-1} \end{Bmatrix} \tag{11.47}$$

**11.6.2
Transverse
Vibration of
Beams**

Equation of Motion.   The governing differential equation for the transverse vibration of a uniform beam is given by Eq. (8.83):

$$\frac{d^4W}{dx^4} - \beta^4 W = 0 \tag{11.48}$$

where

$$\beta^4 = \frac{\rho A \omega^2}{EI} \tag{11.49}$$

By using the central difference formula for the fourth derivative,[3] Eq. (11.48) can be written at any mesh point i as

$$W_{i+2} - 4W_{i+1} + (6 - \lambda)W_i - 4W_{i-1} + W_{i-2} = 0 \qquad (11.50)$$

where

$$\lambda = h^4 \beta^4 \qquad (11.51)$$

Let the beam be divided into $n - 1$ equal parts with n mesh points and $h = l/(n - 1)$. The application of Eq. (11.50) at the mesh points $i = 3, 4, \ldots, n - 2$ leads to the equations

$$\begin{bmatrix} 1 & -4 & (6-\lambda) & -4 & 1 & 0 & 0 & \cdots & 0 & 0 & 0 & 0 & 0 \\ 0 & 1 & -4 & (6-\lambda) & -4 & 1 & 0 & \cdots & 0 & 0 & 0 & 0 & 0 \\ 0 & 0 & 1 & -4 & (6-\lambda) & -4 & 1 & \cdots & 0 & 0 & 0 & 0 & 0 \\ \vdots & & & & & & & & & & & & \\ 0 & 0 & 0 & 0 & 0 & 0 & 0 & \cdots & 1 & -4 & (6-\lambda) & -4 & 1 \end{bmatrix}$$

$$\begin{Bmatrix} W_1 \\ W_2 \\ W_3 \\ \cdot \\ \cdot \\ \cdot \\ W_n \end{Bmatrix} = \begin{Bmatrix} 0 \\ 0 \\ 0 \\ \cdot \\ \cdot \\ \cdot \\ 0 \end{Bmatrix} \qquad (11.52)$$

Boundary Conditions

Fixed End.   The deflection W and the slope $(dW)/(dx)$ are zero at a fixed end. If the end $x = 0$ is fixed, we introduce a fictitious node -1 on the left-hand side of the beam, as shown in Fig. 11.4, and state the boundary conditions, using the central difference formula for $(dW)/(dx)$, as

$$W_1 = 0$$

---

[3]The central difference formula for the fourth derivative (see Problem 11.3) is given by
$$\left.\frac{d^4f}{dx^4}\right|_i \simeq \frac{1}{h^4}(f_{i+2} - 4f_{i+1} + 6f_i - 4f_{i-1} + f_{i-2})$$

FIGURE 11.4   Beam with fixed ends.

$$\frac{dW}{dx}\bigg|_1 = \frac{1}{2h}(W_2 - W_{-1}) = 0 \qquad \text{or} \qquad W_{-1} = W_2 \tag{11.53}$$

where W_i denotes the value of W at node i. If the end $x = l$ is fixed, we introduce the fictitious node $n + 1$ on the right side of the beam, as shown in Fig. 11.4, and state the boundary conditions as

$$W_n = 0$$

$$\frac{dW}{dx}\bigg|_n = \frac{1}{2h}(W_{n+1} - W_{n-1}) = 0 \qquad \text{or} \qquad W_{n+1} = W_{n-1} \tag{11.54}$$

Simply Supported End.   If the end $x = 0$ is simply supported (see Fig. 11.5), we have

$$W_1 = 0$$

$$\frac{d^2W}{dx^2}\bigg|_1 = \frac{1}{h^2}(W_2 - 2W_1 + W_{-1}) = 0 \qquad \text{or} \qquad W_{-1} = -W_2 \tag{11.55}$$

Similar equations can be written if the end $x = l$ is simply supported.

FIGURE 11.5   Beam with simply supported ends.

FIGURE 11.6   Beam with free ends.

Free End.   Since bending moment and shear force are zero at a free end, we introduce two fictitious nodes outside the beam, as shown in Fig. 11.6, and use central difference formulas for approximating the second and the third derivatives of the deflection *W*. For example, if the end $x = 0$ is free, we have

$$\left.\frac{d^2 W}{dx^2}\right|_1 = \frac{1}{h^2}(W_2 - 2W_1 + W_{-1}) = 0$$

$$\left.\frac{d^3 W}{dx^3}\right|_1 = \frac{1}{2h^3}(W_3 - 2W_2 + 2W_{-1} - W_{-2}) = 0 \tag{11.56}$$

EXAMPLE 11.4

Pinned-Fixed Beam

Find the natural frequencies of the simply supported-fixed beam shown in Fig. 11.7. Assume that the cross section of the beam is constant along its length.

Solution: We shall divide the beam into four segments and express the governing equation

$$\frac{d^4 W}{dx^4} - \beta^4 W = 0 \tag{E.1}$$

in finite difference form at each of the interior mesh points. This yields the equations

$$W_0 - 4W_1 + (6 - \lambda)W_2 - 4W_3 + W_4 = 0 \tag{E.2}$$

$$W_1 - 4W_2 + (6 - \lambda)W_3 - 4W_4 + W_5 = 0 \tag{E.3}$$

$$W_2 - 4W_3 + (6 - \lambda)W_4 - 4W_5 + W_6 = 0 \tag{E.4}$$

where W_0 and W_6 denote the values of *W* at the fictitious nodes 0 and 6, respectively, and

$$\lambda = h^4 \beta^4 = \frac{h^4 \rho A \omega^2}{EI} \tag{E.5}$$

FIGURE 11.7 Simply supported-fixed beam.

The boundary conditions at the simply supported end (mesh point 1) are

$$W_1 = 0$$

$$W_0 = -W_2 \tag{E.6}$$

At the fixed end (mesh point 5), the boundary conditions are

$$W_5 = 0$$

$$W_6 = W_4 \tag{E.7}$$

With the help of Eqs. (E.6) and (E.7), Eqs. (E.2) to (E.4) can be reduced to

$$(5 - \lambda)W_2 - 4W_3 + W_4 = 0 \tag{E.8}$$

$$-4W_2 + (6 - \lambda)W_3 - 4W_4 = 0 \tag{E.9}$$

$$W_2 - 4W_3 + (7 - \lambda)W_4 = 0 \tag{E.10}$$

Equations (E.8) to (E.10) can be written in matrix form as

$$\begin{bmatrix} (5 - \lambda) & -4 & 1 \\ -4 & (6 - \lambda) & -4 \\ 1 & -4 & (7 - \lambda) \end{bmatrix} \begin{Bmatrix} W_2 \\ W_3 \\ W_4 \end{Bmatrix} = \begin{Bmatrix} 0 \\ 0 \\ 0 \end{Bmatrix} \tag{E.11}$$

The solution of the eigenvalue problem (Eq. (E.11)) gives the following results:

$$\lambda_1 = 0.7135, \quad \omega_1 = \frac{0.8447}{h^2}\sqrt{\frac{EI}{\rho A}}, \quad \begin{Bmatrix} W_2 \\ W_3 \\ W_4 \end{Bmatrix}^{(1)} = \begin{Bmatrix} 0.5880 \\ 0.7215 \\ 0.3656 \end{Bmatrix} \tag{E.12}$$

$$\lambda_2 = 5.0322, \quad \omega_2 = \frac{2.2433}{h^2}\sqrt{\frac{EI}{\rho A}}, \quad \begin{Bmatrix} W_2 \\ W_3 \\ W_4 \end{Bmatrix}^{(2)} = \begin{Bmatrix} 0.6723 \\ -0.1846 \\ -0.7169 \end{Bmatrix} \tag{E.13}$$

$$\lambda_3 = 12.2543, \quad \omega_3 = \frac{3.5006}{h^2}\sqrt{\frac{EI}{\rho A}}, \quad \begin{Bmatrix} W_2 \\ W_3 \\ W_4 \end{Bmatrix}^{(3)} = \begin{Bmatrix} 0.4498 \\ -0.6673 \\ 0.5936 \end{Bmatrix} \tag{E.14}$$

∎

11.7   Runge-Kutta Method for Multidegree-of-Freedom Systems

In the Runge-Kutta method, the matrix equations of motion, Eq. (11.18), are used to express the acceleration vector as

$$\ddot{\vec{x}}(t) = [m]^{-1}(\vec{F}(t) - [c]\dot{\vec{x}}(t) - [k]\vec{x}(t)) \tag{11.57}$$

By treating the displacements as well as velocities as unknowns, a new vector, $\vec{X}(t)$, is defined as $\vec{X}(t) = \begin{Bmatrix} \vec{x}(t) \\ \dot{\vec{x}}(t) \end{Bmatrix}$ so that

$$\dot{\vec{X}} = \begin{Bmatrix} \dot{\vec{x}} \\ \ddot{\vec{x}} \end{Bmatrix} = \begin{Bmatrix} \dot{\vec{x}} \\ [m]^{-1}(\vec{F} - [c]\dot{\vec{x}} - [k]\vec{x}) \end{Bmatrix} \tag{11.58}$$

Equation (11.58) can be rearranged to obtain

$$\dot{\vec{X}}(t) = \begin{bmatrix} [0] & [I] \\ -[m]^{-1}[k] & -[m]^{-1}[c] \end{bmatrix} \begin{Bmatrix} \vec{x}(t) \\ \dot{\vec{x}}(t) \end{Bmatrix} + \begin{Bmatrix} 0 \\ [m]^{-1}\vec{F}(t) \end{Bmatrix}$$

—that is,

$$\dot{\vec{X}}(t) = \vec{f}(\vec{X}, t) \tag{11.59}$$

where

$$\vec{f}(\vec{X}, t) = [A]\vec{X}(t) + \underset{\sim}{\vec{F}}(t) \tag{11.60}$$

$$[A] = \begin{bmatrix} [0] & [I] \\ -[m]^{-1}[k] & -[m]^{-1}[c] \end{bmatrix} \tag{11.61}$$

and

$$\underset{\sim}{\vec{F}}(t) = \begin{Bmatrix} \vec{0} \\ [m]^{-1}\vec{F}(t) \end{Bmatrix} \tag{11.62}$$

With this, the recurrence formula to evaluate $\vec{X}(t)$ at different grid points t_i according to the fourth-order Runge-Kutta method becomes [11.10]

$$\vec{X}_{i+1} = \vec{X}_i + \tfrac{1}{6}[\vec{K}_1 + 2\vec{K}_2 + 2\vec{K}_3 + \vec{K}_4] \tag{11.63}$$

where

$$\vec{K}_1 = h\vec{f}(\vec{X}_i, t_i) \tag{11.64}$$

$$\vec{K}_2 = h\vec{f}(\vec{X}_i + \tfrac{1}{2}\vec{K}_1, t_i + \tfrac{1}{2}h) \tag{11.65}$$

$$\vec{K}_3 = h\vec{f}(\vec{X}_i + \tfrac{1}{2}\vec{K}_2, t_i + \tfrac{1}{2}h) \tag{11.66}$$

$$\vec{K}_4 = h\vec{f}(\vec{X}_i + \vec{K}_3, t_{i+1}) \tag{11.67}$$

Runge-Kutta Method for a Two-Degree-of-Freedom System

EXAMPLE 11.5

Find the response of the two-degree-of-freedom system considered in Example 11.3 using the fourth-order Runge-Kutta method.

Solution:

Approach: Use the Runge-Kutta method with $\Delta t = 0.24216$.

Using the initial conditions $\vec{x}(t = 0) = \dot{\vec{x}}(t = 0) = \vec{0}$, Eq. (11.63) is sequentially applied with $\Delta t = 0.24216$ to obtain the results shown in Table 11.4.

TABLE 11.4

Time $t_i = i\,\Delta t$	$\vec{x}_i = \vec{x}(t = t_i)$
t_1	$\begin{Bmatrix} 0.0014 \\ 0.1437 \end{Bmatrix}$
t_2	$\begin{Bmatrix} 0.0215 \\ 0.5418 \end{Bmatrix}$
t_3	$\begin{Bmatrix} 0.0978 \\ 1.1041 \end{Bmatrix}$
t_4	$\begin{Bmatrix} 0.2668 \\ 1.7059 \end{Bmatrix}$
t_5	$\begin{Bmatrix} 0.5379 \\ 2.2187 \end{Bmatrix}$
t_6	$\begin{Bmatrix} 0.8756 \\ 2.5401 \end{Bmatrix}$
t_7	$\begin{Bmatrix} 1.2008 \\ 2.6153 \end{Bmatrix}$
t_8	$\begin{Bmatrix} 1.4109 \\ 2.4452 \end{Bmatrix}$
t_9	$\begin{Bmatrix} 1.4156 \\ 2.0805 \end{Bmatrix}$
t_{10}	$\begin{Bmatrix} 1.1727 \\ 1.6050 \end{Bmatrix}$
t_{11}	$\begin{Bmatrix} 0.7123 \\ 1.1141 \end{Bmatrix}$
t_{12}	$\begin{Bmatrix} 0.1365 \\ 0.6948 \end{Bmatrix}$

■

11.8   Houbolt Method

We shall consider the Houbolt method with reference to a multidegree-of-freedom system. In this method, the following finite difference expansions are employed:

$$\dot{\vec{x}}_{i+1} = \frac{1}{6\,\Delta t}(11\vec{x}_{i+1} - 18\vec{x}_i + 9\vec{x}_{i-1} - 2\vec{x}_{i-2}) \qquad (11.68)$$

$$\ddot{\vec{x}}_{i+1} = \frac{1}{(\Delta t)^2}(2\vec{x}_{i+1} - 5\vec{x}_i + 4\vec{x}_{i-1} - \vec{x}_{i-2}) \qquad (11.69)$$

To derive Eqs. (11.68) and (11.69), consider the function $x(t)$. Let the values of x at the equally spaced grid points $t_{i-2} = t_i - 2\,\Delta t$, $t_{i-1} = t_i - \Delta t$, t_i, and $t_{i+1} = t_i + \Delta t$ be given by x_{i-2}, x_{i-1}, x_i, and x_{i+1}, respectively, as shown in Fig. 11.8 [11.18]. The Taylor's series expansion, with backward step, gives several possibilities.

- With Step Size $= \Delta t$:

$$x(t) = x(t + \Delta t) - \Delta t\,\dot{x}(t + \Delta t) + \frac{(\Delta t)^2}{2!}\ddot{x}(t + \Delta t) - \frac{(\Delta t)^3}{3!}\dddot{x}(t + \Delta t)$$

or

$$x_i = x_{i+1} - \Delta t\,\dot{x}_{i+1} + \frac{(\Delta t)^2}{2}\ddot{x}_{i+1} - \frac{(\Delta t)^3}{6}\dddot{x}_{i+1} + \cdots \qquad (11.70)$$

- With Step Size $= 2\,\Delta t$:

$$x(t - \Delta t) = x(t + \Delta t) - (2\,\Delta t)\dot{x}(t + \Delta t)$$
$$+ \frac{(2\,\Delta t)^2}{2!}\ddot{x}(t + \Delta t) - \frac{(2\,\Delta t)^3}{3!}\dddot{x}(t + \Delta t) + \cdots$$

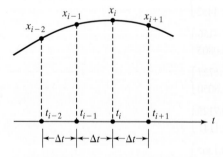

FIGURE 11.8   Equally spaced grid points.

or

$$x_{i-1} = x_{i+1} - 2\,\Delta t \dot{x}_{i+1} + 2(\Delta t)^2 \ddot{x}_{i+1} - \tfrac{4}{3}(\Delta t)^3 \dddot{x}_{i+1} + \cdots \quad (11.71)$$

- With Step Size $= 3\,\Delta t$:

$$x(t - 2\,\Delta t) = x(t + \Delta t) - (3\,\Delta t)\dot{x}(t + \Delta t)$$
$$+ \frac{(3\,\Delta t)^2}{2!}\ddot{x}(t + \Delta t) - \frac{(3\,\Delta t)^3}{3!}\dddot{x}(t + \Delta t) + \cdots$$

or

$$x_{i-2} = x_{i+1} - 3\,\Delta t\,\dot{x}_{i+1} + \tfrac{9}{2}(\Delta t)^2\ddot{x}_{i+1} - \tfrac{9}{2}(\Delta t)^3\dddot{x}_{i+1} + \cdots \quad (11.72)$$

By considering terms up to $(\Delta t)^3$ only, Eqs. (11.70) to (11.72) can be solved to express $\dot{x}_{i+1}$, $\ddot{x}_{i+1}$, and $\dddot{x}_{i+1}$ in terms of x_{i-2}, x_{i-1}, x_i, and x_{i+1}. This gives $\dot{x}_{i+1}$ and $\ddot{x}_{i+1}$ as in [11.18]:

$$\dot{x}_{i+1} = \frac{1}{6(\Delta t)}(11x_{i+1} - 18x_i + 9x_{i-1} - 2x_{i-2}) \quad (11.73)$$

$$\ddot{x}_{i+1} = \frac{1}{(\Delta t)^2}(2x_{i+1} - 5x_i + 4x_{i-1} - x_{i-2}) \quad (11.74)$$

Equations (11.68) and (11.69) represent the vector form of these equations.

To find the solution at step $i + 1(\vec{x}_{i+1})$, we consider Eq. (11.18) at t_{i+1}, so that

$$[m]\ddot{\vec{x}}_{i+1} + [c]\dot{\vec{x}}_{i+1} + [k]\vec{x}_{i+1} = \vec{F}_{i+1} \equiv \vec{F}(t = t_{i+1}) \quad (11.75)$$

By substituting Eqs. (11.68) and (11.69) into Eq. (11.75), we obtain

$$\left(\frac{2}{(\Delta t)^2}[m] + \frac{11}{6\,\Delta t}[c] + [k]\right)\vec{x}_{i+1}$$
$$= \vec{F}_{i+1} + \left(\frac{5}{(\Delta t)^2}[m] + \frac{3}{\Delta t}[c]\right)\vec{x}_i$$
$$- \left(\frac{4}{(\Delta t)^2}[m] + \frac{3[c]}{2\,\Delta t}\right)\vec{x}_{i-1} + \left(\frac{1}{(\Delta t)^2}[m] + \frac{[c]}{3\,\Delta t}\right)\vec{x}_{i-2} \quad (11.76)$$

Note that the equilibrium equation at time t_{i+1}, Eq. (11.75), is used in finding the solution $\vec{X}_{i+1}$ through Eq. (11.76). This is also true of the Wilson and Newmark methods. For this reason, these methods are called *implicit integration methods*.

It can be seen from Eq. (11.76) that a knowledge of $\vec{x}_i$, $\vec{x}_{i-1}$, and $\vec{x}_{i-2}$ is required to find the solution $\vec{x}_{i+1}$. Thus the values of $\vec{x}_{-1}$ and $\vec{x}_{-2}$ are to be found before attempting to find the vector $\vec{x}_1$ using Eq. (11.76). Since there is no direct method to find $\vec{x}_{-1}$ and $\vec{x}_{-2}$, we cannot use Eq. (11.76) to find $\vec{x}_1$ and $\vec{x}_2$. This makes the method non-self-starting. To start the method, we can use the central difference method described in Section 11.5 to find $\vec{x}_1$ and $\vec{x}_2$. Once $\vec{x}_0$ is known from the given initial conditions of the problem and $\vec{x}_1$ and $\vec{x}_2$ are known from the central difference method, the subsequent solutions $\vec{x}_3$, $\vec{x}_4$, ... can be found by using Eq. (11.76).

The step-by-step procedure to be used in the Houbolt method is as follows:

1. From the known initial conditions $\vec{x}(t = 0) = \vec{x}_0$ and $\dot{\vec{x}}(t = 0) = \dot{\vec{x}}_0$, find $\ddot{\vec{x}}_0 = \ddot{\vec{x}}(t = 0)$ using Eq. (11.26).
2. Select a suitable time step Δt.
3. Determine $\vec{x}_{-1}$ using Eq. (11.28).
4. Find $\vec{x}_1$ and $\vec{x}_2$ using the central difference equation (11.29).
5. Compute $\vec{x}_{i+1}$, starting with $i = 2$ and using Eq. (11.76):

$$\vec{x}_{i+1} = \left[\frac{2}{(\Delta t)^2}[m] + \frac{11}{6\,\Delta t}[c] + [k] \right]^{-1}$$

$$\times \left\{ \vec{F}_{i+1} + \left(\frac{5}{(\Delta t)^2}[m] + \frac{3}{\Delta t}[c] \right)\vec{x}_i \right.$$

$$- \left(\frac{4}{(\Delta t)^2}[m] + \frac{3}{2\,\Delta t}[c] \right)\vec{x}_{i-1}$$

$$\left. + \left(\frac{1}{(\Delta t)^2}[m] + \frac{1}{3\,\Delta t}[c] \right)\vec{x}_{i-2} \right\} \tag{11.77}$$

If required, evaluate the velocity and acceleration vectors $\dot{\vec{x}}_{i+1}$ and $\ddot{\vec{x}}_{i+1}$ using Eqs. (11.68) and (11.69).

EXAMPLE 11.6

Houbolt Method for a Two-Degree-of-Freedom System

Find the response of the two-degree-of-freedom system considered in Example 11.3 using the Houbolt method.

Solution

Approach: Use the Houbolt method with $\Delta t = 0.24216$.

The value of $\ddot{\vec{x}}_0$ can be found using Eq. (11.26):

$$\ddot{\vec{x}}_0 = \begin{Bmatrix} 0 \\ 5 \end{Bmatrix}$$

TABLE 11.5

Time $t_i = i\,\Delta t$	$\vec{x}_i = \vec{x}(t = t_i)$
t_1	$\begin{Bmatrix} 0.0000 \\ 0.1466 \end{Bmatrix}$
t_2	$\begin{Bmatrix} 0.0172 \\ 0.5520 \end{Bmatrix}$
t_3	$\begin{Bmatrix} 0.0917 \\ 1.1064 \end{Bmatrix}$
t_4	$\begin{Bmatrix} 0.2501 \\ 1.6909 \end{Bmatrix}$
t_5	$\begin{Bmatrix} 0.4924 \\ 2.1941 \end{Bmatrix}$
t_6	$\begin{Bmatrix} 0.7867 \\ 2.5297 \end{Bmatrix}$
t_7	$\begin{Bmatrix} 1.0734 \\ 2.6489 \end{Bmatrix}$
t_8	$\begin{Bmatrix} 1.2803 \\ 2.5454 \end{Bmatrix}$
t_9	$\begin{Bmatrix} 1.3432 \\ 2.2525 \end{Bmatrix}$
t_{10}	$\begin{Bmatrix} 1.2258 \\ 1.8325 \end{Bmatrix}$
t_{11}	$\begin{Bmatrix} 0.9340 \\ 1.3630 \end{Bmatrix}$
t_{12}	$\begin{Bmatrix} 0.5178 \\ 0.9224 \end{Bmatrix}$

By using a value of $\Delta t = 0.24216$, Eq. (11.29) can be used to find $\vec{x}_1$ and $\vec{x}_2$, and then Eq. (11.77) can be used recursively to obtain $\vec{x}_3$, $\vec{x}_4$, ..., as shown in Table 11.5.

■

11.9 Wilson Method

The Wilson method assumes that the acceleration of the system varies linearly between two instants of time. In particular, the two instants of time are taken as indicated in Fig. 11.9. Thus the acceleration is assumed to be linear from time $t_i = i\,\Delta t$ to time $t_{i+\theta} = t_i + \theta\,\Delta t$,

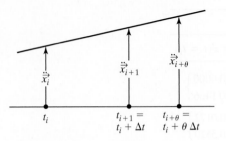

FIGURE 11.9   Linear acceleration
assumption of the Wilson method.

where $\theta \geq 1.0$ [11.19]. For this reason, this method is also called the *Wilson θ method*. If $\theta = 1.0$, this method reduces to the linear acceleration scheme [11.20].

A stability analysis of the Wilson method shows that it is unconditionally stable provided that $\theta \geq 1.37$. In this section, we shall consider the Wilson method for a multidegree-of-freedom system.

Since $\ddot{\vec{x}}(t)$ is assumed to vary linearly between t_i and $t_{i+\theta}$, we can predict the value of $\ddot{\vec{x}}$ at any time $t_i + \tau$, $0 \leq \tau \leq \theta \, \Delta t$:

$$\ddot{\vec{x}}(t_i + \tau) = \ddot{\vec{x}}_i + \frac{\tau}{\theta \, \Delta t}(\ddot{\vec{x}}_{i+\theta} - \ddot{\vec{x}}_i) \tag{11.78}$$

By integrating Eq. (11.78), we obtain[4]

$$\dot{\vec{x}}(t_i + \tau) = \dot{\vec{x}}_i + \ddot{\vec{x}}_i\tau + \frac{\tau^2}{2\theta \, \Delta t}(\ddot{\vec{x}}_{i+\theta} - \ddot{\vec{x}}_i) \tag{11.79}$$

and

$$\vec{x}(t_i + \tau) = \vec{x}_i + \dot{\vec{x}}_i\tau + \frac{1}{2}\ddot{\vec{x}}_i\tau^2 + \frac{\tau^3}{6\theta \, \Delta t}(\ddot{\vec{x}}_{i+\theta} - \ddot{\vec{x}}_i) \tag{11.80}$$

By substituting $\tau = \theta \, \Delta t$ into Eqs. (11.79) and (11.80), we obtain

$$\dot{\vec{x}}_{i+\theta} = \dot{\vec{x}}(t_i + \theta \, \Delta t) = \dot{\vec{x}}_i + \frac{\theta \, \Delta t}{2}(\ddot{\vec{x}}_{i+\theta} + \ddot{\vec{x}}_i) \tag{11.81}$$

$$\vec{x}_{i+\theta} = \vec{x}(t_i + \theta \, \Delta t) = \vec{x}_i + \theta \, \Delta t \, \dot{\vec{x}}_i + \frac{\theta^2(\Delta t)^2}{6}(\ddot{\vec{x}}_{i+\theta} + 2\ddot{\vec{x}}_i) \tag{11.82}$$

---

[4]$\dot{\vec{x}}_i$ and $\vec{x}_i$ have been substituted in place of the integration constants in Eqs. (11.79) and (11.80), respectively.

Equation (11.82) can be solved to obtain

$$\ddot{\vec{x}}_{i+\theta} = \frac{6}{\theta^2(\Delta t)^2}(\vec{x}_{i+\theta} - \vec{x}_i) - \frac{6}{\theta\,\Delta t}\dot{\vec{x}}_i - 2\ddot{\vec{x}}_i \tag{11.83}$$

By substituting Eq. (11.83) into Eq. (11.81), we obtain

$$\dot{\vec{x}}_{i+\theta} = \frac{3}{\theta\,\Delta t}(\vec{x}_{i+\theta} - \vec{x}_i) - 2\dot{\vec{x}}_i - \frac{\theta\,\Delta t}{2}\ddot{\vec{x}}_i \tag{11.84}$$

To obtain the value of $x_{i+\theta}$, we consider the equilibrium equation (11.18) at time $t_{i+\theta} = t_i + \theta\,\Delta t$ and write

$$[m]\ddot{\vec{x}}_{i+\theta} + [c]\dot{\vec{x}}_{i+\theta} + [k]\vec{x}_{i+\theta} = \underset{\sim}{F}_{i+\theta} \tag{11.85}$$

where the force vector $\underset{\sim}{F}_{i+\theta}$ is also obtained by using the linear assumption:

$$\underset{\sim}{F}_{i+\theta} = \vec{F}_i + \theta(\vec{F}_{i+1} - \vec{F}_i) \tag{11.86}$$

Substituting Eqs. (11.83), (11.84), and (11.86) for $\ddot{\vec{x}}_{i+\theta}$, $\dot{\vec{x}}_{i+\theta}$, and $\underset{\sim}{F}_{i+\theta}$, Eq. (11.85) gives

$$\left\{\frac{6}{\theta^2(\Delta t)^2}[m] + \frac{3}{\theta\,\Delta t}[c] + [k]\right\}\vec{x}_{i+1}$$

$$= \vec{F}_i + \theta(\vec{F}_{i+1} - \vec{F}_i) + \left\{\frac{6}{\theta^2(\Delta t)^2}[m] + \frac{3}{\theta\,\Delta t}[c]\right\}\vec{x}_i$$

$$+ \left\{\frac{6}{\theta\,\Delta t}[m] + 2[c]\right\}\dot{\vec{x}}_i + \left\{2[m] + \frac{\theta\,\Delta t}{2}[c]\right\}\ddot{\vec{x}}_i \tag{11.87}$$

which can be solved for $\vec{x}_{i+1}$.

The Wilson method can be described by the following steps:

1. From the known initial conditions $\vec{x}_0$ and $\dot{\vec{x}}_0$, find $\ddot{\vec{x}}_0$ using Eq. (11.26).
2. Select a suitable time step Δt and a suitable value of θ (θ is usually taken as 1.4).
3. Compute the effective load vector $\underset{\sim}{F}_{i+\theta}$, starting with $i = 0$:

$$\underset{\sim}{F}_{i+\theta} = \vec{F}_i + \theta(\vec{F}_{i+1} - \vec{F}_i) + [m]\left(\frac{6}{\theta^2(\Delta t)^2}\vec{x}_i + \frac{6}{\theta\,\Delta t}\dot{\vec{x}}_i + 2\ddot{\vec{x}}_i\right)$$

$$+ [c]\left(\frac{3}{\theta\,\Delta t}\vec{x}_i + 2\dot{\vec{x}}_i + \frac{\theta\,\Delta t}{2}\ddot{\vec{x}}_i\right) \tag{11.88}$$

4. Find the displacement vector at time $t_{i+\theta}$:

$$\vec{x}_{i+\theta} = \left[\frac{6}{\theta^2 (\Delta t)^2}[m] + \frac{3}{\theta \Delta t}[c] + [k] \right]^{-1} \underset{\approx}{F}_{i+\theta} \tag{11.89}$$

5. Calculate the acceleration, velocity, and displacement vectors at time t_{i+1}:

$$\ddot{\vec{x}}_{i+1} = \frac{6}{\theta^3 (\Delta t)^2}(\vec{x}_{i+\theta} - \vec{x}_i) - \frac{6}{\theta^2 \Delta t}\dot{\vec{x}}_i + \left(1 - \frac{3}{\theta}\right)\ddot{\vec{x}}_i \tag{11.90}$$

$$\dot{\vec{x}}_{i+1} = \dot{\vec{x}}_i + \frac{\Delta t}{2}(\ddot{\vec{x}}_{i+1} + \ddot{\vec{x}}_i) \tag{11.91}$$

$$\vec{x}_{i+1} = \vec{x}_i + \Delta t\, \dot{\vec{x}}_i + \frac{(\Delta t)^2}{6}(\ddot{\vec{x}}_{i+1} + 2\ddot{\vec{x}}_i) \tag{11.92}$$

EXAMPLE 11.7

Wilson Method for a Two-Degree-of-Freedom System

Find the response of the system considered in Example 11.3, using the Wilson θ method with $\theta = 1.4$.

Solution:
Approach: Use Wilson method with $\Delta t = 0.24216$.

The value of $\ddot{\vec{x}}_0$ can be obtained as in the case of Example 11.3:

$$\ddot{\vec{x}}_0 = \begin{Bmatrix} 0 \\ 5 \end{Bmatrix}$$

Then, by using Eqs. (11.90) to (11.92) with a time step of $\Delta t = 0.24216$, we obtain the results indicated in Table 11.6.

∎

11.10   Newmark Method

The Newmark integration method is also based on the assumption that the acceleration varies linearly between two instants of time. The resulting expressions for the velocity and displacement vectors $\dot{\vec{x}}_{i+1}$ and $\vec{x}_{i+1}$, for a multidegree-of-freedom system [11.21], are written as in Eqs. (11.79) and (11.80):

$$\dot{\vec{x}}_{i+1} = \dot{\vec{x}}_i + [(1 - \beta)\ddot{\vec{x}}_i + \beta\ddot{\vec{x}}_{i+1}] \Delta t \tag{11.93}$$

$$\vec{x}_{i+1} = \vec{x}_i + \Delta t\, \dot{\vec{x}}_i + [(\tfrac{1}{2} - \alpha)\ddot{\vec{x}}_i + \alpha\ddot{\vec{x}}_{i+1}](\Delta t)^2 \tag{11.94}$$

TABLE 11.6

Time $t_i = i\,\Delta t$	$\vec{x}_i = \vec{x}(t = t_i)$
t_1	$\begin{Bmatrix} 0.0033 \\ 0.1392 \end{Bmatrix}$
t_2	$\begin{Bmatrix} 0.0289 \\ 0.5201 \end{Bmatrix}$
t_3	$\begin{Bmatrix} 0.1072 \\ 1.0579 \end{Bmatrix}$
t_4	$\begin{Bmatrix} 0.2649 \\ 1.6408 \end{Bmatrix}$
t_5	$\begin{Bmatrix} 0.5076 \\ 2.1529 \end{Bmatrix}$
t_6	$\begin{Bmatrix} 0.8074 \\ 2.4981 \end{Bmatrix}$
t_7	$\begin{Bmatrix} 1.1035 \\ 2.6191 \end{Bmatrix}$
t_8	$\begin{Bmatrix} 1.3158 \\ 2.5056 \end{Bmatrix}$
t_9	$\begin{Bmatrix} 1.3688 \\ 2.1929 \end{Bmatrix}$
t_{10}	$\begin{Bmatrix} 1.2183 \\ 1.7503 \end{Bmatrix}$
t_{11}	$\begin{Bmatrix} 0.8710 \\ 1.2542 \end{Bmatrix}$
t_{12}	$\begin{Bmatrix} 0.3897 \\ 0.8208 \end{Bmatrix}$

where the parameters α and β indicate how much the acceleration at the end of the interval enters into the velocity and displacement equations at the end of the interval Δt. In fact, α and β can be chosen to obtain the desired accuracy and stability characteristics [11.22]. When $\beta = \frac{1}{2}$ and $\alpha = \frac{1}{6}$, Eqs. (11.93) and (11.94) correspond to the linear acceleration method (which can also be obtained using $\theta = 1$ in the Wilson method). When $\beta = \frac{1}{2}$ and $\alpha = \frac{1}{4}$, Eqs. (11.93) and (11.94) correspond to the assumption of constant acceleration between t_i and t_{i+1}. To find the value of $\vec{x}_{i+1}$, the equilibrium equation (11.18) is considered at $t = t_{i+1}$, so that

$$[m]\ddot{\vec{x}}_{i+1} + [c]\dot{\vec{x}}_{i+1} + [k]\vec{x}_{i+1} = \vec{F}_{i+1} \qquad (11.95)$$

Equation (11.94) can be used to express $\ddot{\vec{x}}_{i+1}$ in terms of $\vec{x}_{i+1}$, and the resulting expression can be substituted into Eq. (11.93) to express $\dot{\vec{x}}_{i+1}$ in terms of $\vec{x}_{i+1}$. By substituting these expressions for $\dot{\vec{x}}_{i+1}$ and $\ddot{\vec{x}}_{i+1}$ into Eq. (11.95), we can obtain a relation for finding $\vec{x}_{i+1}$:

$$\vec{x}_{i+1} = \left[\frac{1}{\alpha(\Delta t)^2}[m] + \frac{\beta}{\alpha\,\Delta t}[c] + [k]\right]^{-1}$$

$$\times \left\{\vec{F}_{i+1} + [m]\left(\frac{1}{\alpha(\Delta t)^2}\,\vec{x}_i + \frac{1}{\alpha\,\Delta t}\dot{\vec{x}}_i + \left(\frac{1}{2\alpha} - 1\right)\ddot{\vec{x}}_i\right)\right.$$

$$+ [c]\left(\frac{\beta}{\alpha\,\Delta t}\,\vec{x}_i + \left(\frac{\beta}{\alpha} - 1\right)\dot{\vec{x}}_i\right.$$

$$\left.\left. + \left(\frac{\beta}{\alpha} - 2\right)\frac{\Delta t}{2}\ddot{\vec{x}}_i\right)\right\} \tag{11.96}$$

The Newmark method can be summarized in the following steps:

1. From the known values of $\vec{x}_0$ and $\dot{\vec{x}}_0$, find $\ddot{\vec{x}}_0$ using Eq. (11.26).
2. Select suitable values of Δt, α, and β.
3. Calculate the displacement vector $\vec{x}_{i+1}$, starting with $i = 0$ and using Eq. (11.96).
4. Find the acceleration and velocity vectors at time t_{i+1}:

$$\ddot{\vec{x}}_{i+1} = \frac{1}{\alpha(\Delta t)^2}(\vec{x}_{i+1} - \vec{x}_i) - \frac{1}{\alpha\,\Delta t}\dot{\vec{x}}_i - \left(\frac{1}{2\alpha} - 1\right)\ddot{\vec{x}}_i \tag{11.97}$$

$$\dot{\vec{x}}_{i+1} = \dot{\vec{x}}_i + (1 - \beta)\,\Delta t\,\ddot{\vec{x}}_i + \beta\,\Delta t\,\ddot{\vec{x}}_{i+1} \tag{11.98}$$

It is important to note that unless β is taken as $\frac{1}{2}$, there is a spurious damping introduced, proportional to $(\beta - \frac{1}{2})$. If β is taken as zero, a negative damping results; this involves a self-excited vibration arising solely from the numerical procedure. Similarly, if β is greater than $\frac{1}{2}$, a positive damping is introduced. This reduces the magnitude of response even without real damping in the problem [11.21]. The method is unconditionally stable for $\alpha \geq \frac{1}{4}(\beta + \frac{1}{2})^2$ and $\beta \geq \frac{1}{2}$.

Newmark Method for a Two-Degree-of-Freedom System

EXAMPLE 11.8

Find the response of the system considered in Example 11.3, using the Newmark method with $\alpha = \frac{1}{6}$ and $\beta = \frac{1}{2}$.

Solution

Approach: Use the Newmark method with $\Delta t = 0.24216$.

TABLE 11.7

Time $t_i = i\,\Delta t$	$\vec{x}_i = \vec{x}(t = t_i)$
t_1	$\begin{Bmatrix} 0.0026 \\ 0.1411 \end{Bmatrix}$
t_2	$\begin{Bmatrix} 0.0246 \\ 0.5329 \end{Bmatrix}$
t_3	$\begin{Bmatrix} 0.1005 \\ 1.0884 \end{Bmatrix}$
t_4	$\begin{Bmatrix} 0.2644 \\ 1.6870 \end{Bmatrix}$
t_5	$\begin{Bmatrix} 0.5257 \\ 2.2027 \end{Bmatrix}$
t_6	$\begin{Bmatrix} 0.8530 \\ 2.5336 \end{Bmatrix}$
t_7	$\begin{Bmatrix} 1.1730 \\ 2.6229 \end{Bmatrix}$
t_8	$\begin{Bmatrix} 1.3892 \\ 2.4674 \end{Bmatrix}$
t_9	$\begin{Bmatrix} 1.4134 \\ 2.1137 \end{Bmatrix}$
t_{10}	$\begin{Bmatrix} 1.1998 \\ 1.6426 \end{Bmatrix}$
t_{11}	$\begin{Bmatrix} 0.7690 \\ 1.1485 \end{Bmatrix}$
t_{12}	$\begin{Bmatrix} 0.2111 \\ 0.7195 \end{Bmatrix}$

The value of $\ddot{\vec{x}}_0$ can be found using Eq. (11.26):

$$\ddot{\vec{x}}_0 = \begin{Bmatrix} 0 \\ 5 \end{Bmatrix}$$

With the values of $\alpha = \frac{1}{6}$, $\beta = 0.5$, and $\Delta t = 0.24216$, Eq. (11.96) gives the values of $\vec{x}_i = \vec{x}(t = t_i)$, as shown in Table 11.7.

■

11.11 Examples Using MATLAB

MATLAB Solution of a Single-Degree-of-Freedom System

EXAMPLE 11.9

Using the MATLAB function `ode23`, solve Example 11.1.

Solution: Defining $x_1 = x$ and $x_2 = \dot{x}$, Eq. (E.1) of Example 11.1 can be expressed as a set of two first-order differential equations:

$$\dot{x}_1 = x_2 \tag{E.1}$$

$$\dot{x}_2 = \frac{1}{m}\left[F_0\left(1 - \sin\frac{\pi t}{2t_0}\right) - cx_2 - kx_1\right] \tag{E.2}$$

with initial conditions $x_1(0) = x_2(0) = 0$. The MATLAB program to solve Eqs. (E.1) and (E.2) is given below.

```
% Ex11_9.m
tspan = [0: 0.1: 5*pi];
x0 = [0; 0];
[t,x] = ode23 ('dfunc11_9', tspan, x0);
plot (t,x(:,1));
xlabel ('t');
ylabel ('x(t) and xd(t)');
gtext ('x(t)');
hold on;
plot (t,x(:,2), '--');
gtext ('xd(t)')

%dfunc11_9.m
function f = dfunc11_9(t,x)
m = 1;
k = 1;
c = 0.2;
t0 = pi;
F0 = 1;
f = zeros (2,1);
f(1) = x(2);
f(2) = (F0* (1 - sin(pi*t/(2*t0))) - c*x(2) - k*x(1) )/m;
```

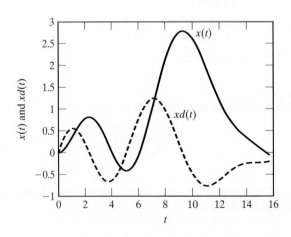

MATLAB Solution of Multidegree-of-Freedom System

EXAMPLE 11.10

Using the MATLAB function `ode23`, solve Example 11.3.

Solution: The equations of motion of the two-degree-of-freedom system in Eq. (E.1) of Example 11.3 can be expressed as a system of four first-order differential equations in terms of

$$y_1 = x_1, \qquad y_2 = \dot{x}_1, \qquad y_3 = x_2, \qquad y_4 = \dot{x}_2$$

as

$$\dot{y}_1 = y_2 \tag{E.1}$$

$$\dot{y}_2 = \frac{1}{m_1} \{F_1(t) - cy_2 + cy_4 - (k_1 + k)y_1 + ky_3\} = -6y_1 + y_3 \tag{E.2}$$

$$\dot{y}_3 = y_4 \tag{E.3}$$

$$\dot{y}_4 = \frac{1}{m_2} \{F_2(t) + cy_2 - cy_4 + ky_1 - (k + k_2)y_3\}$$

$$= \frac{1}{2} \{10 + 2\,y_1 - 8\,y_3\} = 5 + y_1 - 4\,y_3 \tag{E.4}$$

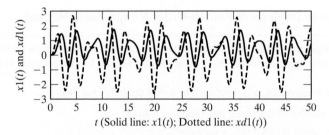

t (Solid line: $x1(t)$; Dotted line: $xd1(t)$)

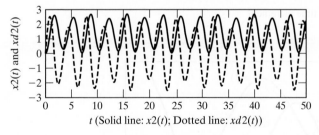

t (Solid line: $x2(t)$; Dotted line: $xd2(t)$)

with initial conditions $y_i(0) = 0$, $i = 1, 2, 3, 4$. The MATLAB program to solve Eqs. (E.1) to (E.4) is given below.

```
% Ex11_10.m
tspan = [0: 0.05: 50];
y0 = [0; 0; 0; 0];
[t,y] = ode23 ('dfunc11_10', tspan, y0);
```

```
subplot (211);
plot (t,y(:,1));
xlabel ('t ( Solid line: x1 (t) Dotted line: xd1 (t) ) ');
ylabel ('x1 (t) amd xd1 (t)');
hold on;
plot (t,y(:, 2), '--');
subplot (212);
plot (t,y(:, 3));
xlabel ('t ( Solid line: x2 (t) Dotted line: xd2 (t) )');
ylabel ('x2 (t) amd xd2 (t) ');
hold on;
plot (t,y (:,4), '--');
%dfunc11_10.m
function f = dfunc11_10 (t,y)
m1 = 1;
m2 = 2;
k1 = 4;
k2 = 6;
k = 2;
c = 0;
F1 = 0;
F2 = 10;
f = zeros (4,1);
f(1) = y(2);
f(2) = ( F1 2 c*y(2) + c*y(4) - (k1+k) *y(1) + k*y(3) )/m1;
f(3) = y(4);
f(4) = ( F2 + c*y(2) - c*y(4) + k*y(1) - (k + k2) *y(3) )/m2;
```

■

Program to Implement Fourth-Order Runge-Kutta Method

EXAMPLE 11.11

Develop a general MATLAB program called **Program14.m** for solving a set of first-order differential equations using the fourth-order Runge-Kutta method. Use the program to solve Example 11.2.

Solution: **Program14.m** is developed to accept the following input data:

n = number of first-order differential equations

xx = initial values $x_i(0)$, a vector of size n

dt = time increment

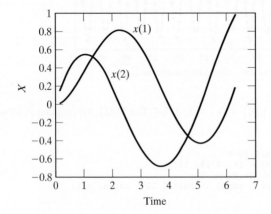

The program requires a subprogram to define the functions $f_i(\vec{x}, t), i = 1, 2, \ldots, n$. The program gives the values of $x_i(t), i = 1, 2, \ldots, n$ at different values of time t.

I	Time(I)	x(1)	x(2)
1	1.570800e−001	1.186315e−002	1.479138e−001
2	3.141600e−001	4.540642e−002	2.755911e−001
3	4.712400e−001	9.725706e−002	3.806748e−001
4	6.283200e−001	1.637262e−001	4.615022e−001
5	7.854000e−001	2.409198e−001	5.171225e−001
.			
.			
.			
36	5.654880e+000	−2.868460e−001	5.040887e−001
37	5.811960e+000	−1.969950e−001	6.388500e−001
38	5.969040e+000	−8.655813e−002	7.657373e−001
39	6.126120e+000	4.301693e−002	8.821039e−001
40	6.283200e+000	1.898865e−001	9.855658e−001

∎

EXAMPLE 11.12

Program for Central Difference Method

Using the central difference method, develop a general MATLAB program called **Program15.m** to find the dynamic response of a multidegree-of-freedom system. Use the program to find the solution of Example 11.3.

Solution: **Program15.m** is developed to accept the following input data:

n = degree of freedom of the system

m = mass matrix, of size $n \times n$

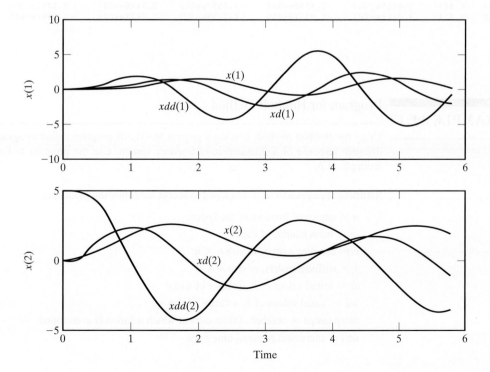

c = damping matrix, of size $n \times n$

k = stiffness matrix, of size $n \times n$

xi = initial values of x_i, a vector of size n

xdi = initial values of $\dot{x}^i$, a vector of size n

nstep (nstp) = number of time steps at which solution is to be found

delt = increment between time steps

The program requires a subprogram to define the forcing functions $f_i(t), i = 1, 2, \ldots, n$ at any time t. It gives the values of the response at different time steps i as $x_j(i), \dot{x}_j(i)$, and $\ddot{x}_j(i)$, $j = 1, 2, \ldots, n$.

```
Solution by central difference method
Given data:
n= 2 nstp= 24 delt=2.421627e-001
Solution:
```

step	time	x(i,1)	xd(i,1)	xdd(i,1)	x(i,2)	xd(i,2)	xdd(i,2)
1	0.0000	0.0000e+000	0.0000e+000	0.0000e+000	0.0000e+000	0.0000e+000	5.0000e+000
2	0.2422	0.0000e+000	0.0000e+000	0.0000e+000	1.4661e−001	0.0000e+000	5.0000e+000
3	0.4843	1.7195e−002	3.5503e−002	2.9321e−001	5.5204e−001	1.1398e+000	4.4136e+000
4	0.7265	9.3086e−002	1.9220e−001	1.0009e+000	1.1222e+000	2.0143e+000	2.8090e+000
5	0.9687	2.6784e−001	5.1752e−001	1.6859e+000	1.7278e+000	2.4276e+000	6.0429e−001
.							
.							
.							
21	4.8433	1.6034e+000	1.7764e+000	−4.0959e+000	2.2077e+000	1.6763e+000	−1.0350e+000
22	5.0854	1.6083e+000	6.5025e−001	−5.2053e+000	2.4526e+000	1.2813e+000	−2.2272e+000
23	5.3276	1.3349e+000	−5.5447e−001	−4.7444e+000	2.5098e+000	6.2384e−001	−3.2023e+000
24	5.5697	8.8618e−001	−1.4909e+000	−2.9897e+000	2.3498e+000	−2.1242e−001	−3.7043e+000
25	5.8119	4.0126e−001	−1.9277e+000	−6.1759e−001	1.9837e+000	−1.0863e+000	−3.5128e+000

∎

EXAMPLE 11.13

Program for Houbolt Method

Using the Houbolt method, develop a general MATLAB program called **Program16.m** to find the dynamic response of a multidegree-of-freedom system. Use the program to find the solution of Example 11.6.

Solution: **Program16.m** is developed to accept the following input data:

n = degree of freedom of the system

m = mass matrix, of size $n \times n$

c = damping matrix, of size $n \times n$

k = stiffness matrix, of size $n \times n$

xi = initial values of x_i, a vector of size n

xdi = initial values of $\dot{x}_i$, a vector of size n

nstep (nstp) = number of time steps at which solution is to be found

delt = increment between time steps

The program requires a subprogram to define the forcing functions $f_i(t), i = 1, 2, \ldots, n$ at any time t. It gives the values of the response at different time stations i as $x_j(i)$, $\dot{x}_j(i)$, and $\ddot{x}_j(i)$, $j = 1, 2, \ldots, n$.

```
Solution by Houbolt method
Given data:
n= 2 nstp= 24 delt=2.421627e-001
Solution:
step    time     x(i,1)      xd(i,1)      xdd(i,1)      x(i,2)      xd(i,2)      xdd(i,2)
   1   0.0000   0.0000e+000   0.0000e+000   0.0000e+000   0.0000e+000   0.0000e+000   5.0000e+000
   2   0.2422   0.0000e+000   0.0000e+000   0.0000e+000   1.4661e-001   0.0000e+000   5.0000e+000
   3   0.4843   1.7195e-002   3.5503e-002   2.9321e-001   5.5204e-001   1.1398e+000   4.4136e+000
   4   0.7265   9.1732e-002   4.8146e-001   1.6624e+000   1.1064e+000   2.4455e+000   6.6609e-001
   5   0.9687   2.5010e-001   8.6351e-001   1.8812e+000   1.6909e+000   2.3121e+000  -1.5134e+000
   .
   .
   .
  21   4.8433   8.7373e-001   1.7900e+000  -1.7158e+000   1.7633e+000   1.3850e+000  -1.1795e+000
  22   5.0854   1.2428e+000   1.1873e+000  -3.3403e+000   2.0584e+000   1.0125e+000  -1.9907e+000
  23   5.3276   1.4412e+000   3.6619e-001  -4.1553e+000   2.2460e+000   4.9549e-001  -2.5428e+000
  24   5.5697   1.4363e+000  -4.8458e-001  -4.0200e+000   2.2990e+000  -9.6748e-002  -2.7595e+000
  25   5.8119   1.2410e+000  -1.1822e+000  -3.0289e+000   2.2085e+000  -6.8133e-001  -2.5932e+000
```

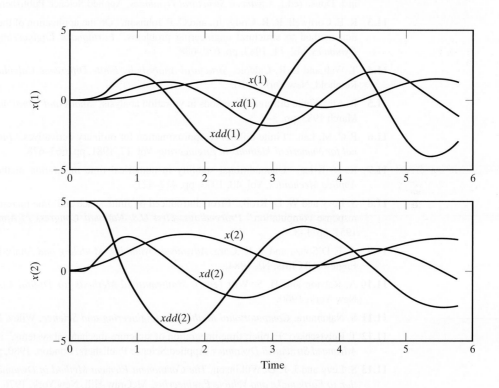

CHAPTER SUMMARY

Numerical methods are to be used in situations where the differential equations governing free and forced vibration cannot be solved to find closed-form solutions. We presented the finite difference method for the solution of the governing equations of discrete and continuous systems. We outlined the use of the fourth-order Runge-Kutta, Houbolt, Wilson, and Newmark methods for the solution of vibration problems related to multidegree-of-freedom systems. Finally, we presented the use of MATLAB for the numerical solution of vibration problems.

Now that you have finished this chapter, you should be able to answer the review questions and solve the problems given below.

REFERENCES

11.1 G. L. Goudreau and R. L. Taylor, "Evaluation of numerical integration methods in elastodynamics," *Computational Methods in Applied Mechanics and Engineering*, Vol. 2, 1973, pp. 69–97.

11.2 S. W. Key, "Transient response by time integration: Review of implicit and explicit operations," in J. Donéa (ed.), *Advanced Structural Dynamics*, Applied Science Publishers, London, 1980.

11.3 R. E. Cornwell, R. R. Craig, Jr., and C. P. Johnson, "On the application of the mode-acceleration method to structural engineering problems," *Earthquake Engineering and Structural Dynamics*, Vol. 11, 1983, pp. 679–688.

11.4 T. Wah and L. R. Calcote, *Structural Analysis by Finite Difference Calculus*, Van Nostrand Reinhold, New York, 1970.

11.5 R. Ali, "Finite difference methods in vibration analysis," *Shock and Vibration Digest*, Vol. 15, March 1983, pp. 3–7.

11.6 P. C. M. Lau, "Finite difference approximation for ordinary derivatives," *International Journal for Numerical Methods in Engineering*, Vol. 17, 1981, pp. 663–678.

11.7 R. D. Krieg, "Unconditional stability in numerical time integration methods," *Journal of Applied Mechanics*, Vol. 40, 1973, pp. 417–421.

11.8 S. Levy and W. D. Kroll, "Errors introduced by finite space and time increments in dynamic response computation," *Proceedings, First U.S. National Congress of Applied Mechanics*, 1951, pp. 1–8.

11.9 A. F. D'Souza and V. K. Garg, *Advanced Dynamics: Modeling and Analysis*, Prentice Hall, Englewood Cliffs, NJ, 1984.

11.10 A. Ralston and H. S. Wilf (eds.), *Mathematical Methods for Digital Computers*, Wiley, New York, 1960.

11.11 S. Nakamura, *Computational Methods in Engineering and Science*, Wiley, New York, 1977.

11.12 T. Belytschko, "Explicit time integration of structure-mechanical systems," in J. Donéa (ed.), *Advanced Structural Dynamics*, Applied Science Publishers, London, 1980, pp. 97–122.

11.13 S. Levy and J. P. D. Wilkinson, *The Component Element Method in Dynamics with Application to Earthquake and Vehicle Engineering*, McGraw-Hill, New York, 1976.

11.14 J. W. Leech, P. T. Hsu, and E. W. Mack, "Stability of a finite-difference method for solving matrix equations," *AIAA Journal*, Vol. 3, 1965, pp. 2172–2173.

11.15 S. D. Conte and C. W. DeBoor, *Elementary Numerical Analysis: An Algorithmic Approach* (2nd ed.), McGraw-Hill, New York, 1972.

11.16 C. F. Gerald and P. O. Wheatley, *Applied Numerical Analysis* (3rd ed.), Addison-Wesley, Reading, MA, 1984.

11.17 L. V. Atkinson and P. J. Harley, *Introduction to Numerical Methods with PASCAL*, Addison-Wesley, Reading, MA, 1984.

11.18 J. C. Houbolt, "A recurrence matrix solution for the dynamic response of elastic aircraft," *Journal of Aeronautical Sciences*, Vol. 17, 1950, pp. 540–550, 594.

11.19 E. L. Wilson, I. Farhoomand, and K. J. Bathe, "Nonlinear dynamic analysis of complex structures," *International Journal of Earthquake Engineering and Structural Dynamics*, Vol. 1, 1973, pp. 241–252.

11.20 S. P. Timoshenko, D. H. Young, and W. Weaver, Jr., *Vibration Problems in Engineering* (4th ed.), Wiley, New York, 1974.

11.21 N. M. Newmark, "A method of computation for structural dynamics," *ASCE Journal of Engineering Mechanics Division*, Vol. 85, 1959, pp. 67–94.

11.22 T. J. R. Hughes, "A note on the stability of Newmark's algorithm in nonlinear structural dynamics," *International Journal for Numerical Methods in Engineering*, Vol. 11, 1976, pp. 383–386.

REVIEW QUESTIONS

11.1 Give brief answers to the following:

1. Describe the procedure of the finite difference method.
2. Using Taylor's series expansion, derive the central difference formulas for the first and the second derivatives of a function,
3. What is a conditionally stable method?
4. What is the main difference between the central difference method and the Runge-Kutta method?
5. Why is it necessary to introduce fictitious mesh points in the finite difference method of solution?
6. Define a tridiagonal matrix.
7. What is the basic assumption of the Wilson method?
8. What is a linear acceleration method?
9. What is the difference between explicit and implicit integration methods?
10. Can we use the numerical integration methods discussed in this chapter to solve nonlinear vibration problems?

11.2 Indicate whether each of the following statements is true or false:

1. The grid points in the finite difference methods are required to be uniformly spaced.
2. The Runge-Kutta method is stable.
3. The Runge-Kutta method is self-starting.
4. The finite difference method is an implicit integration method.
5. The Newmark method is an implicit integration method.

6. For a beam with grid points $-1, 1, 2, 3, \ldots$, the central difference equivalence of the condition $\dfrac{dW}{dx}\Big|_1 = 0$ is $W_{-1} = W_2$.

7. For a beam with grid points $-1, 1, 2, 3, \ldots$, the central difference approximation of a simply supported end condition at grid point 1 is given by $W_{-1} = W_2$.

8. For a beam with grid points $-1, 1, 2, 3, \ldots$, the central difference approximation of $\dfrac{d^2W}{dx^2}\Big|_1 = 0$ yields $W_2 - 2W_1 + W_{-1} = 0$.

11.3 Fill in each of the following blanks with the appropriate word:

1. Numerical methods are to be used when the equations of motion cannot be solved in _____ form.

2. In finite difference methods, approximations are used for _____.

3. Finite difference equations can be derived using _____ different approaches.

4. In finite difference methods, the solution domain is to be replaced by _____ points.

5. The finite difference approximations are based on _____ series expansion.

6. Numerical methods that require the use of a time step (Δt) smaller than a critical value (Δt_{cri}) are said to be _____ stable.

7. In a conditionally stable method, the use of Δt larger than Δt_{cri} makes the method _____.

8. A _____ formula permits the computation of x_i from known values of x_{i-1}.

11.4 Select the most appropriate answer out of the choices given:

1. The central difference approximation of dx/dt at t_i is given by

   a. $\dfrac{1}{2h}(x_{i+1} - x_i)$

   b. $\dfrac{1}{2h}(x_i - x_{i-1})$

   c. $\dfrac{1}{2h}(x_{i+1} - x_{i-1})$

2. The central difference approximation of d^2x/dt^2 at t_i is given by

   a. $\dfrac{1}{h^2}(x_{i+1} - 2x_i + x_{i-1})$

   b. $\dfrac{1}{h^2}(x_{i+1} - x_{i-1})$

   c. $\dfrac{1}{h^2}(x_i - x_{i-1})$

3. An integration method in which the computation of x_{i+1} is based on the equilibrium equation at t_i is known as
   a. *explicit method*
   b. *implicit method*
   c. *regular method*

4. In a non-self-starting method, we need to generate the value of the following quantity using the finite difference approximations of $\dot{x}_i$ and $\ddot{x}_i$:
   a. $\dot{x}_{-1}$ **b.** $\ddot{x}_{-1}$ **c.** x_{-1}

5. Runge-Kutta methods find the approximate solutions of
   a. algebraic equations   **b.** differential equations   **c.** matrix equations

6. The finite difference approximation of $d^2U/dx^2 + \alpha^2 U = 0$ at x_i is given by

    a. $U_{i+1} - (2 - h^2\alpha^2)U_i + U_{i-1} = 0$

    b. $U_{i+1} - 2U_i + U_{i+1} = 0$

    c. $U_{i+1} - (2 - \alpha^2)U_i + U_{i-1} = 0$

7. The finite difference method requires the use of finite difference approximations in

    a. a governing differential equation only

    b. boundary conditions only

    c. a governing differential equation as well as boundary conditions

8. If a bar under longitudinal vibration is fixed at node 1, the forward difference formula gives

    a. $U_1 = 0$ **b.** $U_1 = U_2$ **c.** $U_1 = U_{-1}$

9. If a bar under longitudinal vibration is free at node 1, the forward difference formula gives

    a. $U_1 = 0$ **b.** $U_1 = U_2$ **c.** $U_1 = U_{-1}$

10. The central difference approximation of $d^4W/dx^4 - \beta^4 W = 0$ at grid point i with step size h is

    a. $W_{i+2} - 4W_{i+1} + (6 - h^4\beta^4)W_i - 4W_{i-1} + W_{i-2} = 0$

    b. $W_{i+2} - 6W_{i+1} + (6 - h^4\beta^4)W_i - 6W_{i-1} + W_{i-2} = 0$

    c. $W_{i+3} - 4W_{i+1} + (6 - h^4\beta^4)W_i - 4W_{i-1} + W_{i-3} = 0$

11.5 Match the items in the two columns below:

1. Houbolt method   **a.** Assumes that acceleration varies linearly between t_i and $t_i + \theta \Delta t$; $\theta \geq 1$

2. Wilson method   **b.** Assumes that acceleration varies linearly between t_i and t_{i+1}; can lead to negative damping

3. Newmark method   **c.** Based on the solution of equivalent system of first-order equations

4. Runge-Kutta method   **d.** Same as Wilson method with $\theta = 1$

5. Finite difference method   **e.** Uses finite difference expressions for $\dot{x}_{i+1}$ and $\ddot{x}_{i+1}$ in terms of x_{i-2}, x_{i-1}, x_i, and x_{i+1}

6. Linear acceleration method   **f.** Conditionally stable

PROBLEMS

Section 11.2 Finite Difference Approach

11.1 The forward difference formulas make use of the values of the function to the right of the base grid point. Thus the first derivative at point $i(t = t_i)$ is defined as

$$\frac{dx}{dt} = \frac{x(t + \Delta t) - x(t)}{\Delta t} = \frac{x_{i+1} - x_i}{\Delta t}$$

Derive the forward difference formulas for $(d^2x)/(dt^2)$, $(d^3x)/(dt^3)$, and $(d^4x)/(dt^4)$ at t_i.

11.2 The backward difference formulas make use of the values of the function to the left of the base grid point. Accordingly, the first derivative at point $i(t = t_i)$ is defined as

$$\frac{dx}{dt} = \frac{x(t) - x(t - \Delta t)}{\Delta t} = \frac{x_i - x_{i-1}}{\Delta t}$$

Derive the backward difference formulas for $(d^2x)/(dt^2)$, $(d^3x)/(dt^3)$, and $(d^4x)/(dt^4)$ at t_i.

11.3 Derive the formula for the fourth derivative, $(d^4x)/(dt^4)$, according to the central difference method.

Section 11.3 Central Difference Method for Single-Degree-Of-Freedom Systems

11.4 Find the free vibratory response of an undamped single-degree-of-freedom system with $m = 1$ and $k = 1$, using the central difference method. Assume $x_0 = 0$ and $\dot{x}_0 = 1$. Compare the results obtained with $\Delta t = 1$ and $\Delta t = 0.5$ with the exact solution $x(t) = \sin t$.

11.5 Integrate the differential equation

$$-\frac{d^2x}{dt^2} + 0.1x = 0 \qquad \text{for} \qquad 0 \le t \le 10$$

using the backward difference formula with $\Delta t = 1$. Assume the initial conditions as $x_0 = 1$ and $\dot{x}_0 = 0$.

11.6 Find the free-vibration response of a viscously damped single-degree-of-freedom system with $m = k = c = 1$, using the central difference method. Assume that $x_0 = 0$, $\dot{x}_0 = 1$, and $\Delta t = 0.5$.

11.7 Solve Problem 11.6 by changing c to 2.

11.8 Solve Problem 11.6 by taking the value of c as 4.

11.9 Find the solution of the equation $4\ddot{x} + 2\dot{x} + 3{,}000x = F(t)$, where $F(t)$ is as shown in Fig. 11.10 for the duration $0 \le t \le 1$. Assume that $x_0 = \dot{x}_0 = 0$ and $\Delta t = 0.05$.

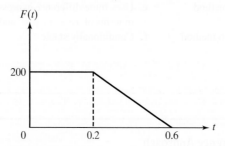

FIGURE 11.10

11.10 Find the solution of a spring-mass-damper system governed by the equation $m\ddot{x} + c\dot{x} + kx = F(t) = \delta F . t$ with $m = c = k = 1$ and $\delta F = 1$. Assume the initial values of x and $\dot{x}$ to be zero and $\Delta t = 0.5$. Compare the central difference solution with the exact solution given in Example 4.9.

Section 11.4 Runge-Kutta Method for Single-Degree-Of-Freedom Systems

11.11 Express the following nth-order differential equation as a system of n first-order differential equations:

$$a_n \frac{d^n x}{dt^n} + a_{n-1} \frac{d^{n-1} x}{dt^{n-1}} + \cdots + a_1 \frac{dx}{dt} = g(x, t)$$

11.12 Find the solution of the following equations by using the fourth-order Runge-Kutta method with $\Delta t = 0.1$:

(a) $\dot{x} = x - 1.5e^{-0.5t}$; $x_0 = 1$
(b) $\dot{x} = -tx^2$; $x_0 = 1$.

11.13 The second-order Runge-Kutta formula is given by

$$\vec{X}_{i+1} = \vec{X}_i + \tfrac{1}{2}(\vec{K}_1 + \vec{K}_2)$$

where

$$\vec{K}_1 = h\vec{F}(\vec{X}_i, t_i) \qquad \text{and} \qquad \vec{K}_2 = h\vec{F}(\vec{X}_i + \vec{K}_1, t_i + h)$$

Using this formula, solve the problem considered in Example 11.2.

11.14 The third-order Runge-Kutta formula is given by

$$\vec{X}_{i+1} = \vec{X}_i + \tfrac{1}{6}(\vec{K}_1 + 4\vec{K}_2 + \vec{K}_3)$$

where

$$\vec{K}_1 = h\vec{F}(\vec{X}_i, t_i)$$
$$\vec{K}_2 = h\vec{F}(\vec{X}_i + \tfrac{1}{2}\vec{K}_1, t_i + \tfrac{1}{2}h)$$

and

$$\vec{K}_3 = h\vec{F}(\vec{X}_i - \vec{K}_1 + 2\vec{K}_2, t_i + h)$$

Using this formula, solve the problem considered in Example 11.2.

11.15 Using the second-order Runge-Kutta method, solve the differential equation $\ddot{x} + 1{,}000x = 0$ with the initial conditions $x_0 = 5$ and $\dot{x}_0 = 0$. Use $\Delta t = 0.01$.

11.16 Using the third-order Runge-Kutta method, solve Problem 11.15.

11.17 Using the fourth-order Runge-Kutta method, solve Problem 11.15.

Section 11.5 Central Difference Method for Multidegree-Of-Freedom Systems

11.18 Using the central difference method, find the response of the two-degree-of-freedom system shown in Fig. 11.2 when $c = 2$, $F_1(t) = 0$, and $F_2(t) = 10$.

11.19 Using the central difference method, find the response of the system shown in Fig. 11.2 when $F_1(t) = 10 \sin 5t$ and $F_2(t) = 0$.

11.20 The equations of motion of a two-degree-of-freedom system are given by $2\ddot{x}_1 + 6x_1 - 2x_2 = 5$ and $\ddot{x}_2 - 2x_1 + 4x_2 = 20 \sin 5t$. Assuming the initial conditions as $x_1(0) = \dot{x}_1(0) = x_2(0) = \dot{x}_2(0) = 0$, find the response of the system, using the central difference method with $\Delta t = 0.25$.

Section 11.6 Central Difference Method for Continuous Systems

11.21 The ends of a beam are elastically restrained by linear and torsional springs, as shown in Fig. 11.11. Using the finite difference method, express the boundary conditions.

FIGURE 11.11

11.22 Using the fourth-order Runge-Kutta method, solve Problem 11.20.

11.23 Find the natural frequencies of a fixed-fixed bar undergoing longitudinal vibration, using three mesh points in the range $0 < x < l$.

11.24 Derive the finite difference equations governing the forced longitudinal vibration of a fixed-free uniform bar, using a total of n mesh points. Find the natural frequencies of the bar, using $n = 4$.

11.25 Derive the finite difference equations for the forced vibration of a fixed-fixed uniform shaft under torsion, using a total of n mesh points.

11.26 Find the first three natural frequencies of a uniform fixed-fixed beam.

11.27 Derive the finite difference equations for the forced vibration of a cantilever beam subjected to a transverse force $f(x, t) = f_0 \cos \omega t$ at the free end.

11.28 Derive the finite difference equations for the forced-vibration analysis of a rectangular membrane, using m and n mesh points in the x and y directions, respectively. Assume the membrane to be fixed along all the edges. Use the central difference formula.

Sections 11.7, 11.11 Runge-Kutta Method for Multidegree-Of-Freedom Systems and MATLAB Problems

11.29 Using `Program14.m` (fourth-order Runge-Kutta method), solve Problem 11.18 with $c = 1$.

11.30 Using `Program14.m` (fourth-order Runge-Kutta method), solve Problem 11.19.

11.31 Using `Program15.m` (central difference method), solve Problem 11.20.

Sections 11.8, 11.11 Houbolt Method

11.32 Using `Program15.m` (central difference method), solve Problem 11.18 with $c = 1$.

11.33 Using `Program16.m` (Houbolt method), solve Problem 11.19.

11.34 Using `Program16.m` (Houbolt method), solve Problem 11.20.

Section 11.9 Wilson Method

11.35 Using the Wilson method with $\theta = 1.4$, solve Problem 11.18.

11.36 Using the Wilson method with $\theta = 1.4$, solve Problem 11.19.

11.37 Using the Wilson method with $\theta = 1.4$, solve Problem 11.20.

Section 11.10 Newmark Method

11.38 Using the Newmark method with $\alpha = \frac{1}{6}$ and $\beta = \frac{1}{2}$, solve Problem 11.18.

11.39 Using the Newmark method with $\alpha = \frac{1}{6}$ and $\beta = \frac{1}{2}$, solve Problem 11.19.

11.40 Using the Newmark method with $\alpha = \frac{1}{6}$ and $\beta = \frac{1}{2}$, solve Problem 11.20.

Section 11.11 MATLAB Problems

11.41 Using MATLAB function `ode23`, solve the differential equation $5\ddot{x} + 4\dot{x} + 3x = 6 \sin t$ with $x(0) = \dot{x}(0) = 0$.

11.42 The equations of motion of a two-degree-of-freedom system are given by

$$\begin{bmatrix} 2 & 0 \\ 0 & 4 \end{bmatrix} \begin{Bmatrix} \ddot{x}_1 \\ \ddot{x}_2 \end{Bmatrix} + 5 \begin{bmatrix} 2 & -1 \\ -1 & 3 \end{bmatrix} \begin{Bmatrix} x_1 \\ x_2 \end{Bmatrix} = \begin{Bmatrix} F_1(t) \\ 0 \end{Bmatrix}$$

where $F_1(t)$ denotes a rectangular pulse of magnitude 5 acting over $0 \le t \le 2$. Find the solution of the equations using MATLAB.

11.43 Find the response of a simple pendulum numerically by solving the linearized equation:

$$\ddot{\theta} + \frac{g}{l}\theta = 0$$

with $\dfrac{g}{l} = 0.01$ and plot the response, $\theta(t)$, for $0 \le t \le 150$. Assume the initial conditions as $\theta(t = 0) = \theta_0 = 1$ rad and $\dot{\theta}(t = 0) = \dot{\theta}_0 = 1.5$ rad/s. Use the MATLAB function `ode23` for numerical solution.

11.44 Find the response of a simple pendulum numerically by solving the exact equation:

$$\ddot{\theta} + \frac{g}{l}\sin\theta = 0$$

with $\dfrac{g}{l} = 0.01$ and plot the response, $\theta(t)$, for $0 \le t \le 150$. Assume the initial conditions as $\theta(t = 0) = \theta_0 = 1$ rad and $\dot{\theta}(t = 0) = \dot{\theta}_0 = 1.5$ rad/s. Use the MATLAB function `ode23` for numerical solution.

11.45 Find the response of a simple pendulum numerically by solving the nonlinear equation:

$$\ddot{\theta} + \frac{g}{l}\left(\theta - \frac{\theta^3}{6}\right) = 0$$

with $\dfrac{g}{l} = 0.01$ and plot the response, $\theta(t)$, for $0 \leq t \leq 150$. Assume the initial conditions as $\theta(t=0) = \theta_0 = 1$ rad and $\dot{\theta}(t=0) = \dot{\theta}_0 = 1.5$ rad/s. Use the MATLAB function **ode23** for numerical solution.

11.46 Write a subroutine **WILSON** for implementing the Wilson method. Use this subroutine to find the solution of Example 11.7.

11.47 Write a subroutine **NUMARK** for implementing the Newmark method. Use this subroutine to find the solution of Example 11.8.

$\mathbf{A}$urel Boreslav Stodola (1859–1942) was a Swiss engineer who joined the Swiss Federal Institute of Technology in Zurich in 1892 to occupy the chair of thermal machinery. He worked in several areas including machine design, automatic controls, thermodynamics, rotor dynamics, and steam turbines. He published one of the most outstanding books, *Die Dampfturbin*, at the turn of the century. This book discussed not only the thermodynamic issues involved in turbine design but also the aspects of fluid flow, vibration, stress analysis of plates, shells, and rotating discs, thermal stresses, and stress concentrations at holes and fillets, and was translated into many languages. The approximate method he presented for the computation of natural frequencies of beams has become known as the Stodola method. (Photo courtesy of *Applied Mechanics Reviews*.)

C H A P T E R   1 2

Finite Element Method

Chapter Outline

The finite element method is a numerical method that can be used for an accurate (but approximate) solution of many complex vibration problems. The mass and stiffness matrices and force vectors needed for the finite element analysis are derived for the basic one-dimensional elements such as a bar in axial motion, a rod in torsional motion, and a beam

987

in bending motion. For the analysis of problems involving one-dimensional elements in two- and three-dimensional geometries (such as trusses and frames), the element matrices are to be transformed into the relevant higher-dimensional space. The details of transformation of element matrices and vectors and the use of the resulting matrices and vectors in formulating the finite element equations of motion for complex systems are explained. The incorporation of the boundary conditions to the assembled system matrices and equations is also discussed. Examples are presented to illustrate the finite element method in finding the natural frequencies of vibration of bars and beams and the stress analysis of a simple two-dimensional truss. The use of consistent- and lumped-mass matrices in the finite element analysis of vibration problems is outlined with illustrative examples. Finally, MATLAB programs are given to find the nodal displacements under specified axial loads, the natural frequencies of vibration of a stepped bar, and the eigenvalue analysis of a stepped beam.

Learning Objectives

After you have finished studying this chapter, you should be able to do the following:

- Identify the stiffness and mass matrices to be used for the solution of different types of vibration problems.
- Transform the element matrices from the local coordinate system to the global coordinate system.
- Assemble the element matrices and apply the boundary conditions.
- Conduct static analysis of problems involving bar, rod, and beam elements.
- Conduct dynamic analysis of problems involving bar, rod, and beam elements to find natural frequencies and mode shapes.
- Use consistent- and lumped-mass matrices in finite element vibration analysis.
- Use MATLAB for solving vibration problems using finite element analysis.

12.1   Introduction

The finite element method is a numerical method that can be used for the accurate solution of complex mechanical and structural vibration problems [12.1, 12.2]. In this method, the actual structure is replaced by several pieces or elements, each of which is assumed to behave as a continuous structural member called a *finite element*. The elements are assumed to be interconnected at certain points known as *joints* or *nodes*. Since it is very difficult to find the exact solution (such as the displacements) of the original structure under the specified loads, a convenient approximate solution is assumed in each finite element. The idea is that if the solutions of the various elements are selected properly, they can be made to converge to the exact solution of the total structure as the element size is reduced. During the solution process, the equilibrium of forces at the joints and the compatibility of displacements between the elements are satisfied so that the entire structure (assemblage of elements) is made to behave as a single entity.

The basic procedure of the finite element method, with application to simple vibration problems, is presented in this chapter. The element stiffness and mass matrices and force vectors are derived for a bar element, a torsion element, and a beam element. The transformation of element matrices and vectors from the local to the global coordinate system is presented. The equations of motion of the complete system of finite elements and the incorporation of the boundary conditions are discussed. The concepts of consistent- and lumped-mass matrices are presented along with a numerical example. Finally, a computer program for the eigenvalue analysis of stepped beams is presented. Although the techniques presented in this chapter can be applied to more complex problems involving two- and three-dimensional finite elements, only the use of one-dimensional elements is considered in the numerical treatment.

12.2   Equations of Motion of an Element

For illustration, the finite element model of a plano-milling machine structure (Fig. 12.1(a)) is shown in Fig. 12.1(b). In this model, the columns and the overarm are represented by triangular plate elements and the cross-slide and the tool holder are represented by beam elements [12.3]. The elements are assumed to be connected to each other only at the joints. The displacement within an element is expressed in terms of the displacements at the corners or joints of the element. In Fig. 12.1(b), the transverse displacement within a typical element e is assumed to be $w(x, y, t)$. The values of w, $(\partial w)/(\partial x)$, and $(\partial w)/(\partial y)$ at joints 1, 2, and 3—namely $w(x_1, y_1, t), (\partial w)/(\partial x)$ $(x_1, y_1, t), (\partial w)/(\partial y)(x_1, y_1, t), \ldots, (\partial w)/(\partial y)(x_3, y_3, t)$—are treated as unknowns and

(a) Plano-milling machine structure   (b) Finite element model

FIGURE 12.1   Finite element modeling.

are denoted as $w_1(t), w_2(t), w_3(t), \ldots, w_9(t)$. The displacement $w(x, y, t)$ can be expressed in terms of the unknown joint displacements $w_i(t)$ in the form

$$w(x, y, t) = \sum_{i=1}^{n} N_i(x, y) w_i(t) \tag{12.1}$$

where $N_i(x, y)$ is called the *shape function* corresponding to the joint displacement $w_i(t)$ and n is the number of unknown joint displacements ($n = 9$ in Fig. 12.1(b)). If a distributed load $f(x, y, t)$ acts on the element, it can be converted into equivalent joint forces $f_i(t)$ ($i = 1, 2, \ldots, 9$). If concentrated forces act at the joints, they can also be added to the appropriate joint force $f_i(t)$. We shall now derive the equations of motion for determining the joint displacements $w_i(t)$ under the prescribed joint forces $f_i(t)$. By using Eq. (12.1), the kinetic energy T and the strain energy V of the element can be expressed as

$$T = \frac{1}{2} \vec{\dot{W}}^T [m] \vec{\dot{W}} \tag{12.2}$$

$$V = \frac{1}{2} \vec{W}^T [k] \vec{W} \tag{12.3}$$

where

$$\vec{W} = \begin{Bmatrix} w_1(t) \\ w_2(t) \\ \cdot \\ \cdot \\ \cdot \\ w_n(t) \end{Bmatrix}, \qquad \vec{\dot{W}} = \begin{Bmatrix} \dot{w}_1(t) \\ \dot{w}_2(t) \\ \cdot \\ \cdot \\ \cdot \\ \dot{w}_n(t) \end{Bmatrix} = \begin{Bmatrix} dw_1/dt \\ dw_2/dt \\ \cdot \\ \cdot \\ \cdot \\ dw_n/dt \end{Bmatrix}$$

and $[m]$ and $[k]$ are the mass and stiffness matrices of the element. By substituting Eqs. (12.2) and (12.3) into Lagrange's equations, Eq. (6.44), the equations of motion of the finite element can be obtained as

$$[m]\vec{\ddot{W}} + [k]\vec{W} = \vec{f} \tag{12.4}$$

where $\vec{f}$ is the vector of joint forces and $\vec{\ddot{W}}$ is the vector of joint accelerations given by

$$\ddot{\vec{W}} = \left\{ \begin{array}{c} \ddot{w}_1 \\ \ddot{w}_2 \\ . \\ . \\ . \\ \ddot{w}_n \end{array} \right\} = \left\{ \begin{array}{c} d^2w_1/dt^2 \\ d^2w_2/dt^2 \\ . \\ . \\ . \\ d^2w_n/dt^2 \end{array} \right\}$$

Note that the shape of the finite elements and the number of unknown joint displacements may differ for different applications. Although the equations of motion of a single element, Eq. (12.4), are not useful directly (as our interest lies in the dynamic response of the assemblage of elements), the mass matrix $[m]$, the stiffness matrix $[k]$, and the joint force vector $\vec{f}$ of individual elements are necessary for the final solution. We shall derive the element mass and stiffness matrices and the joint force vectors for some simple one-dimensional elements in the next section.

12.3   Mass Matrix, Stiffness Matrix, and Force Vector

12.3.1
Bar Element

Consider the uniform bar element shown in Fig. 12.2. For this one-dimensional element, the two end points form the joints (nodes). When the element is subjected to axial loads $f_1(t)$ and $f_2(t)$, the axial displacement within the element is assumed to be linear in x as

$$u(x, t) = a(t) + b(t)x \tag{12.5}$$

When the joint displacements $u_1(t)$ and $u_2(t)$ are treated as unknowns, Eq. (12.5) should satisfy the conditions

$$u(0, t) = u_1(t), \qquad u(l, t) = u_2(t) \tag{12.6}$$

Equations (12.5) and (12.6) lead to

$$a(t) = u_1(t)$$

FIGURE 12.2   Uniform bar element.

and

$$a(t) + b(t)l = u_2(t) \qquad \text{or} \qquad b(t) = \frac{u_2(t) - u_1(t)}{l} \qquad (12.7)$$

Substitution for $a(t)$ and $b(t)$ from Eq. (12.7) into Eq. (12.5) gives

$$u(x, t) = \left(1 - \frac{x}{l}\right)u_1(t) + \frac{x}{l}u_2(t) \qquad (12.8)$$

or

$$u(x, t) = N_1(x)u_1(t) + N_2(x)u_2(t) \qquad (12.9)$$

where

$$N_1(x) = \left(1 - \frac{x}{l}\right), \qquad N_2(x) = \frac{x}{l} \qquad (12.10)$$

are the shape functions.

The kinetic energy of the bar element can be expressed as

$$
\begin{aligned}
T(t) &= \frac{1}{2}\int_0^l \rho A \left\{\frac{\partial u(x, t)}{\partial t}\right\}^2 dx \\
&= \frac{1}{2}\int_0^l \rho A \left\{\left(1 - \frac{x}{l}\right)\frac{du_1(t)}{dt} + \left(\frac{x}{l}\right)\frac{du_2(t)}{dt}\right\}^2 dx \\
&= \frac{1}{2}\frac{\rho Al}{3}(\dot{u}_1^2 + \dot{u}_1\dot{u}_2 + \dot{u}_2^2)
\end{aligned}
\qquad (12.11)
$$

where

$$\dot{u}_1 = \frac{du_1(t)}{dt}, \qquad \dot{u}_2 = \frac{du_2(t)}{dt}$$

ρ is the density of the material and A is the cross-sectional area of the element.

By expressing Eq. (12.11) in matrix form,

$$T(t) = \frac{1}{2}\vec{\dot{u}}(t)^T[m]\vec{\dot{u}}(t) \qquad (12.12)$$

where

$$\vec{\dot{u}}(t) = \left\{\begin{matrix} \dot{u}_1(t) \\ \dot{u}_2(t) \end{matrix}\right\}$$

and the superscript T indicates the transpose, the mass matrix $[m]$ can be identified as

$$[m] = \frac{\rho A l}{6} \begin{bmatrix} 2 & 1 \\ 1 & 2 \end{bmatrix} \tag{12.13}$$

The strain energy of the element can be written as

$$V(t) = \frac{1}{2} \int_0^l EA \left\{ \frac{\partial u(x,t)}{\partial x} \right\}^2 dx$$

$$= \frac{1}{2} \int_0^l EA \left\{ -\frac{1}{l} u_1(t) + \frac{1}{l} u_2(t) \right\}^2 dx$$

$$= \frac{1}{2} \frac{EA}{l} (u_1^2 - 2u_1 u_2 + u_2^2) \tag{12.14}$$

where $u_1 = u_1(t), u_2 = u_2(t)$, and E is Young's modulus. By expressing Eq. (12.14) in matrix form as

$$V(t) = \frac{1}{2} \vec{u}(t)^T [k] \vec{u}(t) \tag{12.15}$$

where

$$\vec{u}(t) = \begin{Bmatrix} u_1(t) \\ u_2(t) \end{Bmatrix} \qquad \text{and} \qquad \vec{u}(t)^T = \{u_1(t) u_2(t)\}$$

the stiffness matrix $[k]$ can be identified as

$$[k] = \frac{EA}{l} \begin{bmatrix} 1 & -1 \\ -1 & 1 \end{bmatrix} \tag{12.16}$$

The force vector

$$\vec{f} = \begin{Bmatrix} f_1(t) \\ f_2(t) \end{Bmatrix}$$

can be derived from the virtual work expression. If the bar is subjected to the distributed force $f(x, t)$, the virtual work δW can be expressed as

$$\delta W(t) = \int_0^l f(x,t) \, \delta u(x,t) \, dx$$

$$= \int_0^l f(x,t) \left\{ \left(1 - \frac{x}{l}\right) \delta u_1(t) + \left(\frac{x}{l}\right) \delta u_2(t) \right\} dx$$

$$= \left(\int_0^l f(x,t) \left(1 - \frac{x}{l} \right) dx \right) \delta u_1(t)$$

$$+ \left(\int_0^l f(x,t) \left(\frac{x}{l} \right) dx \right) \delta u_2(t) \tag{12.17}$$

By expressing Eq. (12.17) in matrix form as

$$\delta W(t) = \delta \vec{u}(t)^T \vec{f}(t) \equiv f_1(t)\, \delta u_1(t) + f_2(t)\, \delta u_2(t) \tag{12.18}$$

the equivalent joint forces can be identified as

$$\left. \begin{array}{l} f_1(t) = \displaystyle\int_0^l f(x,t) \left(1 - \frac{x}{l} \right) dx \\[2mm] f_2(t) = \displaystyle\int_0^l f(x,t) \left(\frac{x}{l} \right) dx \end{array} \right\} \tag{12.19}$$

12.3.2 Torsion Element

Consider a uniform torsion element with the x-axis taken along the centroidal axis, as shown in Fig. 12.3. Let I_p denote the polar moment of inertia about the centroidal axis and GJ represent the torsional stiffness ($J = I_p$ for a circular cross section). When the torsional displacement (rotation) within the element is assumed to be linear in x as

$$\theta(x,t) = a(t) + b(t)x \tag{12.20}$$

and the joint rotations $\theta_1(t)$ and $\theta_2(t)$ are treated as unknowns, Eq. (12.20) can be expressed, by proceeding as in the case of a bar element, as

$$\theta(x,t) = N_1(x)\theta_1(t) + N_2(x)\theta_2(t) \tag{12.21}$$

where $N_1(x)$ and $N_2(x)$ are the same as in Eq. (12.10). The kinetic energy, the strain energy, and the virtual work for pure torsion are given by

$$T(t) = \frac{1}{2} \int_0^l \rho I_p \left\{ \frac{\partial \theta(x,t)}{\partial t} \right\}^2 dx \tag{12.22}$$

$$V(t) = \frac{1}{2} \int_0^l GJ \left\{ \frac{\partial \theta(x,t)}{\partial x} \right\}^2 dx \tag{12.23}$$

$$\delta W(t) = \int_0^l f(x,t)\, \delta \theta(x,t)\, dx \tag{12.24}$$

FIGURE 12.3 Uniform torsion element.

where ρ is the mass density and $f(x, t)$ is the distributed torque per unit length. Using the procedures employed in Section 12.3.1, we can derive the element mass and stiffness matrices and the force vector:

$$[m] = \frac{\rho I_p l}{6}\begin{bmatrix} 2 & 1 \\ 1 & 2 \end{bmatrix} \tag{12.25}$$

$$[k] = \frac{GJ}{l}\begin{bmatrix} 1 & -1 \\ -1 & 1 \end{bmatrix} \tag{12.26}$$

$$\vec{f} = \begin{Bmatrix} f_1(t) \\ f_2(t) \end{Bmatrix} = \begin{Bmatrix} \int_0^l f(x, t)\left(1 - \frac{x}{l}\right)dx \\ \int_0^l f(x, t)\left(\frac{x}{l}\right)dx \end{Bmatrix} \tag{12.27}$$

12.3.3
Beam Element

We now consider a beam element according to the Euler-Bernoulli theory.[1] Figure 12.4 shows a uniform beam element subjected to the transverse force distribution $f(x, t)$. In this case, the joints undergo both translational and rotational displacements, so the unknown joint displacements are labeled as $w_1(t)$, $w_2(t)$, $w_3(t)$, and $w_4(t)$. Thus there will be linear joint forces $f_1(t)$ and $f_3(t)$ corresponding to the linear joint displacements $w_1(t)$ and $w_3(t)$ and rotational joint forces (bending moments) $f_2(t)$ and $f_4(t)$ corresponding to the rotational joint displacements $w_2(t)$ and $w_4(t)$, respectively. The transverse displacement within the element is assumed to be a cubic equation in x (as in the case of static deflection of a beam):

$$w(x, t) = a(t) + b(t)x + c(t)x^2 + d(t)x^3 \tag{12.28}$$

---

[1]The beam element, according to the Timoshenko theory, was considered in [12.4–12.7].

FIGURE 12.4 Uniform beam element.

The unknown joint displacements must satisfy the conditions

$$
\left.
\begin{aligned}
w(0, t) &= w_1(t), & \frac{\partial w}{\partial x}(0, t) &= w_2(t)\\
w(l, t) &= w_3(t), & \frac{\partial w}{\partial x}(l, t) &= w_4(t)
\end{aligned}
\right\}
\tag{12.29}
$$

Equations (12.28) and (12.29) yield

$$a(t) = w_1(t)$$

$$b(t) = w_2(t)$$

$$c(t) = \frac{1}{l^2}[-3w_1(t) - 2w_2(t)l + 3w_3(t) - w_4(t)l]$$

$$d(t) = \frac{1}{l^3}[2w_1(t) + w_2(t)l - 2w_3(t) + w_4(t)l] \tag{12.30}$$

By substituting Eqs. (12.30) into Eq. (12.28), we can express $w(x, t)$ as

$$w(x, t) = \left(1 - 3\frac{x^2}{l^2} + 2\frac{x^3}{l^3}\right)w_1(t) + \left(\frac{x}{l} - 2\frac{x^2}{l^2} + \frac{x^3}{l^3}\right)lw_2(t)$$

$$+ \left(3\frac{x^2}{l^2} - 2\frac{x^3}{l^3}\right)w_3(t) + \left(-\frac{x^2}{l^2} + \frac{x^3}{l^3}\right)lw_4(t) \tag{12.31}$$

This equation can be rewritten as

$$w(x, t) = \sum_{i=1}^{4} N_i(x) w_i(t) \tag{12.32}$$

where $N_i(x)$ are the shape functions given by

$$N_1(x) = 1 - 3\left(\frac{x}{l}\right)^2 + 2\left(\frac{x}{l}\right)^3 \tag{12.33}$$

$$N_2(x) = x - 2l\left(\frac{x}{l}\right)^2 + l\left(\frac{x}{l}\right)^3 \tag{12.34}$$

$$N_3(x) = 3\left(\frac{x}{l}\right)^2 - 2\left(\frac{x}{l}\right)^3 \tag{12.35}$$

$$N_4(x) = -l\left(\frac{x}{l}\right)^2 + l\left(\frac{x}{l}\right)^3 \tag{12.36}$$

The kinetic energy, bending strain energy, and virtual work of the element can be expressed as

$$T(t) = \frac{1}{2} \int_0^l \rho A \left\{ \frac{\partial w(x, t)}{\partial t} \right\}^2 dx \equiv \frac{1}{2} \dot{\vec{w}}(t)^T [m] \dot{\vec{w}}(t) \tag{12.37}$$

$$V(t) = \frac{1}{2} \int_0^l EI \left\{ \frac{\partial^2 w(x, t)}{\partial x^2} \right\}^2 dx \equiv \frac{1}{2} \vec{w}(t)^T [k] \vec{w}(t) \tag{12.38}$$

$$\delta W(t) = \int_0^l f(x, t) \, \delta w(x, t) \, dx \equiv \delta \vec{w}(t)^T \vec{f}(t) \tag{12.39}$$

where ρ is the density of the beam, E is Young's modulus, I is the moment of inertia of the cross section, A is the area of cross section, and

$$\vec{w}(t) = \begin{Bmatrix} w_1(t) \\ w_2(t) \\ w_3(t) \\ w_4(t) \end{Bmatrix}, \qquad \dot{\vec{w}}(t) = \begin{Bmatrix} dw_1/dt \\ dw_2/dt \\ dw_3/dt \\ dw_4/dt \end{Bmatrix}$$

$$\delta \vec{w}(t) = \begin{Bmatrix} \delta w_1(t) \\ \delta w_2(t) \\ \delta w_3(t) \\ \delta w_4(t) \end{Bmatrix}, \qquad \vec{f}(t) = \begin{Bmatrix} f_1(t) \\ f_2(t) \\ f_3(t) \\ f_4(t) \end{Bmatrix}$$

By substituting Eq. (12.31) into Eqs. (12.37) to (12.39) and carrying out the necessary integrations, we obtain

$$[m] = \frac{\rho A l}{420} \begin{bmatrix} 156 & 22l & 54 & -13l \\ 22l & 4l^2 & 13l & -3l^2 \\ 54 & 13l & 156 & -22l \\ -13l & -3l^2 & -22l & 4l^2 \end{bmatrix} \tag{12.40}$$

$$[k] = \frac{EI}{l^3} \begin{bmatrix} 12 & 6l & -12 & 6l \\ 6l & 4l^2 & -6l & 2l^2 \\ -12 & -6l & 12 & -6l \\ 6l & 2l^2 & -6l & 4l^2 \end{bmatrix} \tag{12.41}$$

$$f_i(t) = \int_0^l f(x,t) N_i(x)\, dx, \qquad i = 1, 2, 3, 4 \tag{12.42}$$

12.4   Transformation of Element Matrices and Vectors

As stated earlier, the finite element method considers the given dynamical system as an assemblage of elements. The joint displacements of an individual element are selected in a convenient direction, depending on the nature of the element. For example, for the bar element shown in Fig. 12.2, the joint displacements $u_1(t)$ and $u_2(t)$ are chosen along the axial direction of the element. However, other bar elements can have different orientations in an assemblage, as shown in Fig. 12.5. Here x denotes the axial direction of an individual element and is called a *local coordinate axis*. If we use $u_1(t)$ and $u_2(t)$ to denote the joint displacements of different bar elements, there will be one joint displacement at joint 1, three at joint 2, two at joint 3, and two at joint 4. However, the displacements of joints can be specified more conveniently using reference or global coordinate axes X and Y. Then the displacement components of joints parallel to the X- and Y-axes can be used as the joint displacements in the global coordinate system. These are shown as $U_i(t)$, $i = 1, 2, \ldots, 8$ in Fig. 12.5. The joint displacements in the local and the global coordinate system for a typical bar element e are shown in Fig. 12.6. The two sets of joint displacements are related as follows:

$$u_1(t) = U_{2i-1}(t) \cos \theta + U_{2i}(t) \sin \theta$$
$$u_2(t) = U_{2j-1}(t) \cos \theta + U_{2j}(t) \sin \theta \tag{12.43}$$

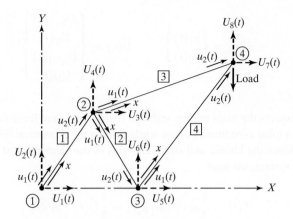

FIGURE 12.5   A dynamical system (truss) idealized as an assemblage of four bar elements.

These can be rewritten as

$$\vec{u}(t) = [\lambda]\vec{U}(t) \qquad (12.44)$$

where $[\lambda]$ is the coordinate transformation matrix given by

$$[\lambda] = \begin{bmatrix} \cos\theta & \sin\theta & 0 & 0 \\ 0 & 0 & \cos\theta & \sin\theta \end{bmatrix} \qquad (12.45)$$

and $\vec{u}(t)$ and $\vec{U}(t)$ are the vectors of joint displacements in the local and the global coordinate system, respectively, and are given by

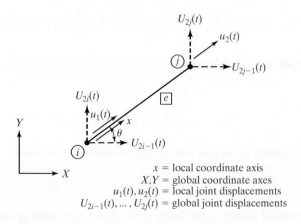

x = local coordinate axis
X,Y = global coordinate axes
$u_1(t), u_2(t)$ = local joint displacements
$U_{2i-1}(t), \dots, U_{2j}(t)$ = global joint displacements

FIGURE 12.6   Local and global joint displacements of element e.

$$\vec{u}(t) = \begin{Bmatrix} u_1(t) \\ u_2(t) \end{Bmatrix}, \qquad \vec{U}(t) = \begin{Bmatrix} U_{2i-1}(t) \\ U_{2i}(t) \\ U_{2j-1}(t) \\ U_{2j}(t) \end{Bmatrix}$$

It is useful to express the mass matrix, stiffness matrix, and joint force vector of an element in terms of the global coordinate system while finding the dynamical response of the complete system. Since the kinetic and strain energies of the element must be independent of the coordinate system, we have

$$T(t) = \frac{1}{2}\dot{\vec{u}}(t)^T[m]\dot{\vec{u}}(t) = \frac{1}{2}\dot{\vec{U}}(t)^T[\overline{m}]\dot{\vec{U}}(t) \tag{12.46}$$

$$V(t) = \frac{1}{2}\vec{u}(t)^T[k]\vec{u}(t) = \frac{1}{2}\vec{U}(t)^T[\overline{k}]\vec{U}(t) \tag{12.47}$$

where $[\overline{m}]$ and $[\overline{k}]$ denote the element mass and stiffness matrices, respectively, in the global coordinate system and $\dot{\vec{U}}(t)$ is the vector of joint velocities in the global coordinate system, related to $\dot{\vec{u}}(t)$ as in Eq. (12.44):

$$\dot{\vec{u}}(t) = [\lambda]\dot{\vec{U}}(t) \tag{12.48}$$

By inserting Eqs. (12.44) and (12.48) into Eqs. (12.46) and (12.47), we obtain

$$T(t) = \frac{1}{2}\dot{\vec{U}}(t)^T[\lambda]^T[m][\lambda]\dot{\vec{U}}(t) \equiv \frac{1}{2}\dot{\vec{U}}(t)^T[\overline{m}]\dot{\vec{U}}(t) \tag{12.49}$$

$$V(t) = \frac{1}{2}\vec{U}(t)^T[\lambda]^T[k][\lambda]\vec{U}(t) \equiv \frac{1}{2}\vec{U}(t)^T[\overline{k}]\vec{U}(t) \tag{12.50}$$

Equations (12.49) and (12.50) yield

$$[\overline{m}] = [\lambda]^T[m][\lambda] \tag{12.51}$$

$$[\overline{k}] = [\lambda]^T[k][\lambda] \tag{12.52}$$

Similarly, by equating the virtual work in the two coordinate systems,

$$\delta W(t) = \delta\vec{u}(t)^T \vec{f}(t) = \delta\vec{U}(t)^T \vec{\overline{f}}(t) \tag{12.53}$$

we find the vector of element joint forces in the global coordinate system $\vec{\overline{f}}(t)$:

$$\vec{\overline{f}}(t) = [\lambda]^T \vec{f}(t) \tag{12.54}$$

Equations (12.51), (12.52), and (12.54) can be used to obtain the equations of motion of a single finite element in the global coordinate system:

$$[\overline{m}]\ddot{\vec{U}}(t) + [\overline{k}]\vec{U}(t) = \vec{f}(t) \tag{12.55}$$

Although this equation is not of much use, since our interest lies in the equations of motion of an assemblage of elements, the matrices $[\overline{m}]$ and $[\overline{k}]$ and the vector $\vec{f}$ are useful in deriving the equations of motion of the complete system, as indicated in the following section.

12.5   Equations of Motion of the Complete System of Finite Elements

Since the complete structure is considered an assemblage of several finite elements, we shall now extend the equations of motion obtained for single finite elements in the global system to the complete structure. We shall denote the joint displacements of the complete structure in the global coordinate system as $U_1(t), U_2(t), \ldots, U_M(t)$ or, equivalently, as a column vector:

$$\underset{\sim}{\vec{U}}(t) = \begin{Bmatrix} U_1(t) \\ U_2(t) \\ \cdot \\ \cdot \\ \cdot \\ U_M(t) \end{Bmatrix}$$

For convenience, we shall denote the quantities pertaining to an element e in the assemblage by the superscript e. Since the joint displacements of any element e can be identified in the vector of joint displacements of the complete structure, the vectors $\vec{U}^{(e)}(t)$ and $\underset{\sim}{\vec{U}}(t)$ are related:

$$\vec{U}^{(e)}(t) = [A^{(e)}]\underset{\sim}{\vec{U}}(t) \tag{12.56}$$

where $[A^{(e)}]$ is a rectangular matrix composed of zeros and ones. For example, for element 1 in Fig. 12.5, Eq. (12.56) becomes

$$\vec{U}^{(1)}(t) \equiv \begin{Bmatrix} U_1(t) \\ U_2(t) \\ U_3(t) \\ U_4(t) \end{Bmatrix} = \begin{bmatrix} 1 & 0 & 0 & 0 & 0 & 0 & 0 & 0 \\ 0 & 1 & 0 & 0 & 0 & 0 & 0 & 0 \\ 0 & 0 & 1 & 0 & 0 & 0 & 0 & 0 \\ 0 & 0 & 0 & 1 & 0 & 0 & 0 & 0 \end{bmatrix} \begin{Bmatrix} U_1(t) \\ U_2(t) \\ \cdot \\ \cdot \\ \cdot \\ U_8(t) \end{Bmatrix} \tag{12.57}$$

The kinetic energy of the complete structure can be obtained by adding the kinetic energies of individual elements

$$T = \sum_{e=1}^{E} \frac{1}{2} \vec{U}^{(e)T} [\overline{m}] \vec{U}^{(e)} \tag{12.58}$$

where E denotes the number of finite elements in the assemblage. By differentiating Eq. (12.56), the relation between the velocity vectors can be derived:

$$\vec{U}^{(e)}(t) = [A^{(e)}] \vec{U}(t) \tag{12.59}$$

Substitution of Eq. (12.59) into Eq. (12.58) leads to

$$T = \frac{1}{2} \sum_{e=1}^{E} \vec{\underset{\sim}{U}}^{T} [A^{(e)}]^{T} [\overline{m}^{(e)}] [A^{(e)}] \vec{\underset{\sim}{U}} \tag{12.60}$$

The kinetic energy of the complete structure can also be expressed in terms of joint velocities of the complete structure $\vec{\underset{\sim}{U}}$:

$$T = \frac{1}{2} \vec{\underset{\sim}{U}}^{T} [M] \vec{\underset{\sim}{U}} \tag{12.61}$$

where $[\underset{\sim}{M}]$ is called the *mass matrix of the complete structure*. A comparison of Eqs. (12.60) and (12.61) gives the relation[2]

$$[\underset{\sim}{M}] = \sum_{e=1}^{E} [A^{(e)}]^{T} [\overline{m}^{(e)}][A^{(e)}] \tag{12.62}$$

Similarly, by considering strain energy, the stiffness matrix of the complete structure, $[\underset{\sim}{K}]$, can be expressed as

$$[\underset{\sim}{K}] = \sum_{e=1}^{E} [A^{(e)}]^{T} [\overline{k}^{(e)}] [A^{(e)}] \tag{12.63}$$

---

[2]An alternative procedure can be used for the assembly of element matrices. In this procedure, each of the rows and columns of the element (mass or stiffness) matrix is identified by the corresponding degree of freedom in the assembled structure. Then the various entries of the element matrix can be placed at their proper locations in the overall (mass or stiffness) matrix of the assembled system. For example, the entry belonging to the ith row (identified by the degree of freedom p) and the jth column (identified by the degree of freedom q) of the element matrix is to be placed in the pth row and qth column of the overall matrix. This procedure is illustrated in Example 12.3.

Finally, the consideration of virtual work yields the vector of joint forces of the complete structure, $\vec{F}$:

$$\vec{F} = \sum_{e=1}^{E} [A^{(e)}]^T \vec{f}^{(e)} \tag{12.64}$$

Once the mass and stiffness matrices and the force vector are known, Lagrange's equations of motion for the complete structure can be expressed as

$$[M]\ddot{\vec{U}} + [K]\vec{U} = \vec{F} \tag{12.65}$$

where $[K]$ is the stiffness matrix of the complete structure.

Note that the joint force vector $\vec{F}$ in Eq. (12.65) was generated by considering only the distributed loads acting on the various elements. If there is any concentrated load acting along the joint displacement $U_i(t)$, it must be added to the ith component of $\vec{F}$.

12.6   Incorporation of Boundary Conditions

In the preceding derivation, no joint was assumed to be fixed. Thus the complete structure is capable of undergoing rigid-body motion under the joint forces. This means that $[K]$ is a singular matrix (see Section 6.12). Usually the structure is supported such that the displacements are zero at a number of joints, to avoid rigid-body motion of the structure. A simple method of incorporating the zero displacement conditions is to eliminate the corresponding rows and columns from the matrices $[M]$ and $[K]$ and the vector $\vec{F}$. The final equations of motion of the restrained structure can be expressed as

$$\underset{N \times N}{[M]} \underset{N \times 1}{\ddot{\vec{U}}} + \underset{N \times N}{[K]} \underset{N \times 1}{\vec{U}} = \underset{N \times 1}{\vec{F}} \tag{12.66}$$

where N denotes the number of free joint displacements of the structure.

Note the following points concerning finite element analysis:

1. The approach used in the above presentation is called the *displacement method* of finite element analysis because it is the displacements of elements that are directly approximated. Other methods, such as the force method, the mixed method, and hybrid methods, are also available [12.8, 12.9].

2. The stiffness matrix, mass matrix, and force vector for other finite elements, including two-dimensional and three-dimensional elements, can be derived in a similar manner, provided the shape functions are known [12.1, 12.2].

3. In the Rayleigh-Ritz method discussed in Section 8.8, the displacement of the continuous system is approximated by a sum of assumed functions, where each function denotes a deflection shape of the entire structure. In the finite element method, an approximation using shape functions (similar to the assumed functions) is also used for a finite element

instead of the entire structure. Thus the finite element procedure can also be considered a Rayleigh-Ritz method.

4. Error analysis of the finite element method can also be conducted [12.10].

EXAMPLE 12.1

Analysis of a Bar

Consider a uniform bar, of length 0.5 m, area of cross section 5×10^{-4} m², Young's modulus 200 GPa, and density 7,850 kg/m³, which is fixed at the left end, as shown in Fig. 12.7.

a. Find the stress induced in the bar under an axial static load of 1,000 N applied at joint 2 along u_2.

b. Find the natural frequency of vibration of the bar.

Use a one-element idealization.

Solution:

a. Using the stiffness matrix of a bar element, Eq. (12.16), the equilibrium equations can be written as

$$\frac{AE}{l} \begin{bmatrix} 1 & -1 \\ -1 & 1 \end{bmatrix} \begin{Bmatrix} u_1 \\ u_2 \end{Bmatrix} = \begin{Bmatrix} f_1 \\ f_2 \end{Bmatrix} \qquad (E.1)$$

With $A = 5 \times 10^{-4}$, $E = 2 \times 10^{11}$, $l = 0.5$, $f_2 = 1,000$, Eq. (E.1) becomes

$$2 \times 10^8 \begin{bmatrix} 1 & -1 \\ -1 & 1 \end{bmatrix} \begin{Bmatrix} u_1 \\ u_2 \end{Bmatrix} = \begin{Bmatrix} f_1 \\ 1,000 \end{Bmatrix} \qquad (E.2)$$

where u_1 is the displacement and f_1 is the unknown reaction at joint 1. To incorporate the boundary condition $u_1 = 0$, we delete the first scalar equation (first row) and substitute $u_1 = 0$ in the resulting Eq. (E.2). This gives

$$2 \times 10^8 u_2 = 1,000 \qquad \text{or} \qquad u_2 = 500 \times 10^{-8} \text{ m} \qquad (E.3)$$

FIGURE 12.7   Uniform bar with two degrees of freedom.

The relation of stress (σ) versus strain (ε) gives

$$\sigma = E\varepsilon = E\frac{\Delta l}{l} = E\left(\frac{u_2 - u_1}{l}\right) \tag{E.4}$$

where $\Delta l = u_2 - u_1$ denotes the change in length of the element and $\dfrac{\Delta l}{l}$ indicates the strain. Equation (E.4) yields

$$\sigma = 2 \times 10^{11}\left(\frac{500 \times 10^{-8} - 0}{0.5}\right) = 2 \times 10^6 \text{ Pa} \tag{E.5}$$

b. Using the stiffness and mass matrices of the bar element, Eqs. (12.13) and (12.16), the eigen-value problem can be expressed as

$$\frac{AE}{l}\begin{bmatrix} 1 & -1 \\ -1 & 1 \end{bmatrix}\begin{Bmatrix} U_1 \\ U_2 \end{Bmatrix} = \omega^2 \frac{\rho\, Al}{6}\begin{bmatrix} 2 & 1 \\ 1 & 2 \end{bmatrix}\begin{Bmatrix} U_1 \\ U_2 \end{Bmatrix} \tag{E.6}$$

where ω is the natural frequency and U_1 and U_2 are the amplitudes of vibration of the bar at joints 1 and 2, respectively. To incorporate the boundary condition $U_1 = 0$, we delete the first row and first column in each of the matrices and vectors and write the resulting equation as

$$\frac{AE}{l}U_2 = \omega^2 \frac{\rho\, Al}{6}(2)U_2$$

or

$$\omega = \sqrt{\frac{3E}{\rho l^2}} = \sqrt{\frac{3(2 \times 10^{11})}{7{,}850\,(0.5)^2}} = 17{,}485.2076 \text{ rad/s} \tag{E.7}$$

∎

EXAMPLE 12.2

Natural Frequencies of a Simply Supported Beam

Find the natural frequencies of the simply supported beam shown in Fig. 12.8(a) using one finite element.

Solution: Since the beam is idealized using only one element, the element joint displacements are the same in both local and global systems, as indicated in Fig. 12.8(b). The stiffness and mass matrices of the beam are given by

$$[K] = [K^{(1)}] = \frac{EI}{l^3}\begin{bmatrix} 12 & 6l & -12 & 6l \\ 6l & 4l^2 & -6l & 2l^2 \\ -12 & -6l & 12 & -6l \\ 6l & 2l^2 & -6l & 4l^2 \end{bmatrix} \tag{E.1}$$

FIGURE 12.8   Simply supported beam.

$$[M] = [M^{(1)}] = \frac{\rho A l}{420} \begin{bmatrix} 156 & 22l & 54 & -13l \\ 22l & 4l^2 & 13l & -3l^2 \\ 54 & 13l & 156 & -22l \\ -13l & -3l^2 & -22l & 4l^2 \end{bmatrix} \tag{E.2}$$

and the vector of joint displacements by

$$\vec{W} = \begin{Bmatrix} W_1 \\ W_2 \\ W_3 \\ W_4 \end{Bmatrix} \equiv \begin{Bmatrix} w_1^{(1)} \\ w_2^{(1)} \\ w_3^{(1)} \\ w_4^{(1)} \end{Bmatrix} \tag{E.3}$$

The boundary conditions corresponding to the simply supported ends ($W_1 = 0$ and $W_3 = 0$) can be incorporated[3] by deleting the rows and columns corresponding to W_1 and W_3 in Eqs. (E.1) and (E.2). This leads to the overall matrices

$$[K] = \frac{2EI}{l} \begin{bmatrix} 2 & 1 \\ 1 & 2 \end{bmatrix} \tag{E.4}$$

$$[M] = \frac{\rho A l^3}{420} \begin{bmatrix} 4 & -3 \\ -3 & 4 \end{bmatrix} \tag{E.5}$$

and the eigenvalue problem can be written as

$$\left[\frac{2EI}{l} \begin{bmatrix} 2 & 1 \\ 1 & 2 \end{bmatrix} - \frac{\rho A l^3 \omega^2}{420} \begin{bmatrix} 4 & -3 \\ -3 & 4 \end{bmatrix} \right] \begin{Bmatrix} W_2 \\ W_4 \end{Bmatrix} = \begin{Bmatrix} 0 \\ 0 \end{Bmatrix} \tag{E.6}$$

---

[3]The bending moment cannot be set equal to zero at the simply supported ends explicitly, since there is no degree of freedom (joint displacement) involving the second derivative of the displacement w.

By multiplying throughout by $l/(2EI)$, Eq. (E.6) can be expressed as

$$\begin{bmatrix} 2 - 4\lambda & 1 + 3\lambda \\ 1 + 3\lambda & 2 - 4\lambda \end{bmatrix} \begin{Bmatrix} W_2 \\ W_4 \end{Bmatrix} = \begin{Bmatrix} 0 \\ 0 \end{Bmatrix} \tag{E.7}$$

where

$$\lambda = \frac{\rho A l^4 \omega^2}{840EI} \tag{E.8}$$

By setting the determinant of the coefficient matrix in Eq. (E.7) equal to zero, we obtain the frequency equation

$$\begin{vmatrix} 2 - 4\lambda & 1 + 3\lambda \\ 1 + 3\lambda & 2 - 4\lambda \end{vmatrix} = (2 - 4\lambda)^2 - (1 + 3\lambda)^2 = 0 \tag{E.9}$$

The roots of Eq. (E.9) give the natural frequencies of the beam as

$$\lambda_1 = \frac{1}{7} \quad \text{or} \quad \omega_1 = \left(\frac{120EI}{\rho A l^4} \right)^{1/2} \tag{E.10}$$

$$\lambda_2 = 3 \quad \text{or} \quad \omega_2 = \left(\frac{2,520EI}{\rho A l^4} \right)^{1/2} \tag{E.11}$$

These results can be compared with the exact values (see Fig. 8.15):

$$\omega_1 = \left(\frac{97.41EI}{\rho A l^4} \right)^{1/2}, \qquad \omega_2 = \left(\frac{1,558.56EI}{\rho A l^4} \right)^{1/2} \tag{E.12}$$

■

Stresses in a Two-Bar Truss

EXAMPLE 12.3

Find the stresses developed in the two members of the truss shown in Fig. 12.9(a), under a vertical load of 900 N at joint 3. The areas of cross section are 6.5 cm^2 for member 1 and 13 cm^2 for member 2, and the Young's modulus is 200×10^9 Pa.

Solution
Approach: Derive the static equilibrium equations and solve them to find the joint displacements. Use the elasticity relations to find the element stresses. Each member is to be treated as a bar element. From Fig. 12.9(a), the coordinates of the joints can be found as

$$(X_1, Y_1) = (0, 0.25) \text{ m}; \qquad (X_2, Y_2) = (0, 0) \text{ m}; \qquad (X_3, Y_3) = (0.25, 0.125) \text{ m}$$

FIGURE 12.9 Two bar truss.

The modeling of the truss as an assemblage of two bar elements and the displacement degrees of freedom of the joints are shown in Fig. 12.9(b). The lengths of the elements can be computed from the coordinates of the ends (joints) as

$$l^{(1)} = \{(X_3 - X_1)^2 + (Y_3 - Y_1)^2\}^{1/2} = \{(0.25 - 0)^2 + (0.125 - 0.25)^2\}^{1/2}$$
$$= 0.279508 \text{ m}$$

$$l^{(2)} = \{(X_3 - X_2)^2 + (Y_3 - Y_2)^2\}^{1/2} = \{(0.25 - 0)^2 + (0.125 - 0)^2\}^{1/2}$$

$$= 0.279508 \text{ m} \tag{E.1}$$

The element stiffness matrices in the local coordinate system can be obtained as

$$[k^{(1)}] = \frac{A^{(1)}E^{(1)}}{l^{(1)}} \begin{bmatrix} 1 & -1 \\ -1 & 1 \end{bmatrix} = \frac{(6.5 \times 10^{-4})(200 \times 10^9)}{0.279508} \begin{bmatrix} 1 & -1 \\ -1 & 1 \end{bmatrix}$$

$$= 4.65853 \times 10^8 \begin{bmatrix} 1 & -1 \\ -1 & 1 \end{bmatrix}$$

$$[k^{(2)}] = \frac{A^{(2)}E^{(2)}}{l^{(2)}} \begin{bmatrix} 1 & -1 \\ -1 & 1 \end{bmatrix} = \frac{(13 \times 10^{-4})(200 \times 10^9)}{0.279508} \begin{bmatrix} 1 & -1 \\ -1 & 1 \end{bmatrix}$$

$$= 9.30206 \times 10^8 \begin{bmatrix} 1 & -1 \\ -1 & 1 \end{bmatrix} \tag{E.2}$$

The angle between the local x-coordinate and the global X-coordinate is given by

$$\left. \begin{aligned} \cos\theta_1 &= \frac{X_3 - X_1}{l^{(1)}} = \frac{0.25 - 0}{0.279508} = 0.8944 \\ \sin\theta_1 &= \frac{Y_3 - Y_1}{l^{(1)}} = \frac{0.125 - 0.25}{0.279508} = -0.4472 \end{aligned} \right\} \text{ for element 1} \tag{E.3}$$

$$\left. \begin{aligned} \cos\theta_2 &= \frac{X_3 - X_2}{l^{(2)}} = \frac{0.25 - 0}{0.279508} = 0.8944 \\ \sin\theta_2 &= \frac{Y_3 - Y_2}{l^{(2)}} = \frac{0.125 - 0}{0.279508} = 0.4472 \end{aligned} \right\} \text{ for element 2} \tag{E.4}$$

The stiffness matrices of the elements in the global (X, Y) coordinate system can be derived as

$$[\bar{k}^{(1)}] = [\lambda^{(1)}]^T [k^{(1)}][\lambda^{(1)}]$$

$$= 4.65853 \times 10^8 \begin{matrix} 1 \quad2 \quad5 \quad6 \\ \begin{bmatrix} 0.8 & -0.4 & -0.8 & 0.4 \\ -0.4 & 0.2 & 0.4 & -0.2 \\ -0.8 & 0.4 & 0.8 & -0.4 \\ 0.4 & -0.2 & -0.4 & 0.2 \end{bmatrix} \begin{matrix} 1 \\ 2 \\ 5 \\ 6 \end{matrix} \end{matrix} \tag{E.5}$$

$$[\bar{k}^{(2)}] = [\lambda^{(2)}]^T [k^{(2)}][\lambda^{(2)}]$$

$$= 9.30206 \times 10^8 \begin{matrix} 3 \quad4 \quad5 \quad6 \\ \begin{bmatrix} 0.8 & 0.4 & -0.8 & -0.4 \\ 0.4 & 0.2 & -0.4 & -0.2 \\ -0.8 & -0.4 & 0.8 & 0.4 \\ -0.4 & -0.2 & 0.4 & 0.2 \end{bmatrix} \begin{matrix} 3 \\ 4 \\ 5 \\ 6 \end{matrix} \end{matrix} \tag{E.6}$$

where

$$[\lambda^{(1)}] = \begin{bmatrix} \cos\theta_1 & \sin\theta_1 & 0 & 0 \\ 0 & 0 & \cos\theta_1 & \sin\theta_1 \end{bmatrix}$$

$$= \begin{bmatrix} 0.8944 & -0.4472 & 0 & 0 \\ 0 & 0 & 0.8944 & -0.4472 \end{bmatrix} \tag{E.7}$$

$$[\lambda^{(2)}] = \begin{bmatrix} \cos\theta_2 & \sin\theta_2 & 0 & 0 \\ 0 & 0 & \cos\theta_2 & \sin\theta_2 \end{bmatrix}$$

$$= \begin{bmatrix} 0.8944 & 0.4472 & 0 & 0 \\ 0 & 0 & 0.8944 & 0.4472 \end{bmatrix} \tag{E.8}$$

Note that the top and right-hand sides of Eqs. (E.5) and (E.6) denote the global degrees of freedom corresponding to the rows and columns of the respective stiffness matrices. The assembled stiffness matrix of the system, $[\underset{\sim}{K}]$ can be obtained, by placing the elements of $[\bar{k}^{(1)}]$ and $[\bar{k}^{(2)}]$ at their proper places in $[\underset{\sim}{K}]$, as

$$[\underset{\sim}{K}] = 4.65853 \times 10^8 \begin{array}{c} \quad \\ \quad \end{array} \begin{bmatrix} \overset{1}{0.8} & \overset{2}{-0.4} & \overset{3}{} & \overset{4}{} & \overset{5}{-0.8} & \overset{6}{0.4} \\ -0.4 & 0.2 & & & 0.4 & -0.2 \\ & & 1.6 & 0.8 & -1.6 & -0.8 \\ & & 0.8 & 0.4 & -0.8 & -0.4 \\ -0.8 & 0.4 & -1.6 & -0.8 & (0.8 & (-0.4 \\ & & & & +1.6) & +0.8) \\ 0.4 & -0.2 & -0.8 & -0.4 & (-0.4 & (0.2 \\ & & & & +0.8) & +0.4) \end{bmatrix} \begin{array}{c} 1 \\ 2 \\ 3 \\ 4 \\ 5 \\ \\ 6 \\ \\ \end{array} \tag{E.9}$$

The assembled force vector can be written as

$$\vec{\underset{\sim}{F}} = \begin{Bmatrix} F_{X1} \\ F_{Y1} \\ F_{X2} \\ F_{Y2} \\ F_{X3} \\ F_{Y3} \end{Bmatrix} \tag{E.10}$$

where, in general, (F_{Xi}, F_{Yi}) denote the forces applied at joint i along (X, Y) directions. Specifically, (F_{X1}, F_{Y1}) and (F_{X2}, F_{Y2}) represent the reactions at joints 1 and 2, while $(F_{X3}, F_{Y3}) = (0, -900)$ N shows the external forces applied at joint 3. By applying the boundary conditions $U_1 = U_2 = U_3 = U_4 = 0$ (i.e., by deleting the rows and columns 1, 2, 3, and 4 in Eqs. (E.9) and (E.10)), we get the final assembled stiffness matrix and the force vector as

$$[K] = 4.65853 \times 10^8 \begin{array}{c} \quad 5 \quad\quad 6 \\ \begin{bmatrix} 2.4 & 0.4 \\ 0.4 & 0.6 \end{bmatrix} \begin{array}{c} 5 \\ 6 \end{array} \end{array} \tag{E.11}$$

$$\vec{F} = \begin{Bmatrix} 0 \\ -900 \end{Bmatrix} \begin{matrix} 5 \\ 6 \end{matrix} \tag{E.12}$$

The equilibrium equations of the system can be written as

$$[K]\vec{U} = \vec{F} \tag{E.13}$$

where $\vec{U} = \begin{Bmatrix} U_5 \\ U_6 \end{Bmatrix}$. The solution of Eq. (E.13) can be found as

$$U_5 = 603.731 \times 10^{-9}\,\text{m}, \qquad U_6 = -3.62239 \times 10^{-6}\,\text{m} \tag{E.14}$$

The axial displacements of elements 1 and 2 can be found as

$$\begin{Bmatrix} u_1 \\ u_2 \end{Bmatrix}^{(1)} = [\lambda^{(1)}] \begin{Bmatrix} U_1 \\ U_2 \\ U_5 \\ U_6 \end{Bmatrix}$$

$$= \begin{bmatrix} 0.8944 & -0.4472 & 0 & 0 \\ 0 & 0 & 0.8944 & -0.4472 \end{bmatrix} \begin{Bmatrix} 0 \\ 0 \\ 603.731 \times 10^{-9} \\ -3.62239 \times 10^{-6} \end{Bmatrix}$$

$$= \begin{Bmatrix} 0 \\ 2.15991 \times 10^{-6} \end{Bmatrix}\,\text{m} \tag{E.15}$$

$$\begin{Bmatrix} u_1 \\ u_2 \end{Bmatrix}^{(2)} = [\lambda^{(2)}] \begin{Bmatrix} U_3 \\ U_4 \\ U_5 \\ U_6 \end{Bmatrix}$$

$$= \begin{bmatrix} 0.8944 & 0.4472 & 0 & 0 \\ 0 & 0 & 0.8944 & 0.4472 \end{bmatrix} \begin{Bmatrix} 0 \\ 0 \\ 603.731 \times 10^{-9} \\ -3.62239 \times 10^{-6} \end{Bmatrix}$$

$$= \begin{Bmatrix} 0 \\ -1.079955 \times 10^{-6} \end{Bmatrix}\,\text{m} \tag{E.16}$$

The stresses in elements 1 and 2 can be determined as

$$\sigma^{(1)} = E^{(1)}\epsilon^{(1)} = E^{(1)}\frac{\Delta l^{(1)}}{l^{(1)}} = \frac{E^{(1)}(u_2 - u_1)^{(1)}}{l^{(1)}}$$

$$= \frac{(200 \times 10^9)(2.15991 \times 10^{-6})}{0.279508} = 1.545509\,\text{MPa} \tag{E.17}$$

$$\sigma^{(2)} = E^{(2)}\epsilon^{(2)} = \frac{E^{(2)}\Delta l^{(2)}}{l^{(2)}} = \frac{E^{(2)}(u_2 - u_1)^{(2)}}{l^{(2)}}$$

$$= \frac{(200 \times 10^9)(-1.079955 \times 10^{-6})}{0.279508} = -772.754 \text{ kPa} \qquad (E.18)$$

where $\sigma^{(i)}$ denotes the stress, $\varepsilon^{(i)}$ represents the strain, and $\Delta l^{(i)}$ indicates the change in length of element i $(i = 1, 2)$.

∎

12.7   Consistent- and Lumped-Mass Matrices

The mass matrices derived in Section 12.3 are called *consistent-mass matrices*. They are consistent because the same displacement model that is used for deriving the element stiffness matrix is used for the derivation of the mass matrix. It is of interest to note that several dynamic problems have been solved with simpler forms of mass matrices. The simplest form of the mass matrix, known as the *lumped-mass matrix*, can be obtained by placing point (concentrated) masses m_i at node points i in the directions of the assumed displacement degrees of freedom. The concentrated masses refer to translational and rotational inertia of the element and are calculated by assuming that the material within the mean locations on either side of the particular displacement behaves like a rigid body while the remainder of the element does not participate in the motion. Thus this assumption excludes the dynamic coupling that exists between the element displacements, hence the resulting element mass matrix is purely diagonal [12.11].

12.7.1 Lumped-Mass Matrix for a Bar Element

By dividing the total mass of the element equally between the two nodes, the lumped-mass matrix of a uniform bar element can be obtained as

$$[m] = \frac{\rho A l}{2}\begin{bmatrix} 1 & 0 \\ 0 & 1 \end{bmatrix} \qquad (12.67)$$

12.7.2 Lumped-Mass Matrix for a Beam Element

In Fig. 12.4, by lumping one-half of the total beam mass at each of the two nodes, along the translational degrees of freedom, we obtain the lumped-mass matrix of the beam element as

$$[m] = \frac{\rho A l}{2}\begin{bmatrix} 1 & 0 & 0 & 0 \\ 0 & 0 & 0 & 0 \\ 0 & 0 & 1 & 0 \\ 0 & 0 & 0 & 0 \end{bmatrix} \qquad (12.68)$$

Note that the inertia effect associated with the rotational degrees of freedom has been assumed to be zero in Eq. (12.68). If the inertia effect is to be included, we compute the mass moment of inertia of half of the beam segment about each end and include it at the diagonal locations corresponding to the rotational degrees of freedom. Thus, for a uniform beam, we have

$$I = \frac{1}{3}\left(\frac{\rho A l}{2}\right)\left(\frac{l}{2}\right)^2 = \frac{\rho A l^3}{24} \tag{12.69}$$

and hence the lumped mass matrix of the beam element becomes

$$[m] = \frac{\rho A l}{2} \begin{bmatrix} 1 & 0 & 0 & 0 \\ 0 & \left(\dfrac{l^2}{12}\right) & 0 & 0 \\ 0 & 0 & 1 & 0 \\ 0 & 0 & 0 & \left(\dfrac{l^2}{12}\right) \end{bmatrix} \tag{12.70}$$

12.7.3 Lumped-Mass Versus Consistent-Mass Matrices

It is not obvious whether the lumped-mass matrices or consistent-mass matrices yield more accurate results for a general dynamic response problem. The lumped-mass matrices are approximate in the sense that they do not consider the dynamic coupling present between the various displacement degrees of freedom of the element. However, since the lumped-mass matrices are diagonal, they require less storage space during computation. On the other hand, the consistent-mass matrices are not diagonal and hence require more storage space. They, too, are approximate in the sense that the shape functions, which are derived using static displacement patterns, are used even for the solution of dynamics problems. The following example illustrates the application of lumped- and consistent-mass matrices in a simple vibration problem.

EXAMPLE 12.4

Consistent- and Lumped-Mass Matrices of a Bar

Find the natural frequencies of the fixed-fixed uniform bar shown in Fig. 12.10 using consistent- and lumped-mass matrices. Use two bar elements for modeling.

Solution: The stiffness and mass matrices of a bar element are

$$[k] = \frac{AE}{l} \begin{bmatrix} 1 & -1 \\ -1 & 1 \end{bmatrix} \tag{E.1}$$

$$[m]_c = \frac{\rho A l}{6} \begin{bmatrix} 2 & 1 \\ 1 & 2 \end{bmatrix} \tag{E.2}$$

$$[m]_l = \frac{\rho A l}{2} \begin{bmatrix} 1 & 0 \\ 0 & 1 \end{bmatrix} \tag{E.3}$$

FIGURE 12.10   Fixed-fixed uniform bar.

where the subscripts c and l to the mass matrices denote the consistent and lumped matrices, respectively. Since the bar is modeled by two elements, the assembled stiffness and mass matrices are given by

$$[K] = \frac{AE}{l} \begin{array}{c} \\ \end{array} \begin{bmatrix} 1 & -1 & 0 \\ -1 & 1+1 & -1 \\ 0 & -1 & 1 \end{bmatrix} \begin{array}{c} 1 \\ 2 \\ 3 \end{array} = \frac{AE}{l} \begin{bmatrix} 1 & -1 & 0 \\ -1 & 2 & -1 \\ 0 & -1 & 1 \end{bmatrix} \tag{E.4}$$

$$[M]_c = \frac{\rho Al}{6} \begin{bmatrix} 2 & 1 & 0 \\ 1 & 2+2 & 1 \\ 0 & 1 & 2 \end{bmatrix} \begin{array}{c} 1 \\ 2 \\ 3 \end{array} = \frac{\rho Al}{6} \begin{bmatrix} 2 & 1 & 0 \\ 1 & 4 & 1 \\ 0 & 1 & 2 \end{bmatrix} \tag{E.5}$$

$$[M]_l = \frac{\rho Al}{2} \begin{bmatrix} 1 & 0 & 0 \\ 0 & 1+1 & 0 \\ 0 & 0 & 1 \end{bmatrix} \begin{array}{c} 1 \\ 2 \\ 3 \end{array} = \frac{\rho Al}{2} \begin{bmatrix} 1 & 0 & 0 \\ 0 & 2 & 0 \\ 0 & 0 & 1 \end{bmatrix} \tag{E.6}$$

The dashed boxes in Eqs. (E.4) through (E.6) enclose the contributions of elements 1 and 2. The degrees of freedom corresponding to the columns and rows of the matrices are indicated at the top and the right-hand side of the matrices. The eigenvalue problem, after applying the boundary conditions $U_1 = U_3 = 0$, becomes

$$[[K] - \omega^2[M]] \{U_2\} = \{0\} \tag{E.7}$$

The eigenvalue ω^2 can be determined by solving the equation

$$|[K] - \omega^2[M]| = 0 \tag{E.8}$$

which, for the present case, becomes

$$\left| \frac{AE}{l}[2] - \omega^2 \frac{\rho Al}{6}[4] \right| = 0 \qquad \text{with consistent-mass matrices} \tag{E.9}$$

and

$$\left| \frac{AE}{l}[2] - \omega^2 \frac{\rho Al}{2}[2] \right| = 0 \qquad \text{with lumped-mass matrices} \tag{E.10}$$

Equations (E.9) and (E.10) can be solved to obtain

$$\omega_c = \sqrt{\frac{3E}{\rho l^2}} = 3.4641 \sqrt{\frac{E}{\rho L^2}} \tag{E.11}$$

$$\omega_l = \sqrt{\frac{2E}{\rho l^2}} = 2.8284 \sqrt{\frac{E}{\rho L^2}} \tag{E.12}$$

These values can be compared with the exact value (see Fig. 8.7)

$$\omega_1 = \pi \sqrt{\frac{E}{\rho L^2}} \tag{E.13}$$

■

12.8   Examples Using MATLAB

Finite Element Analysis of a Stepped Bar

EXAMPLE 12.5

Consider the stepped bar shown in Fig. 12.11 with the following data: $A_1 = 16 \times 10^{-4}\,\text{m}^2$, $A_2 = 9 \times 10^{-4}\,\text{m}^2$, $A_3 = 4 \times 10^{-4}\,\text{m}^2$, $E_i = 20 \times 10^{10}\,\text{Pa}$, $i = 1, 2, 3$, $\rho_i = 7.8 \times 10^3\,\text{kg/m}^3$, $i = 1, 2, 3$, $l_1 = 1\,\text{m}$, $l_2 = 0.5\,\text{m}$, $l_3 = 0.25\,\text{m}$.
Write a MATLAB program to determine the following:

a.   Displacements u_1, u_2, and u_3 under load $p_3 = 1,000\,\text{N}$

b.   Natural frequencies and mode shapes of bar

Solution: The assembled stiffness and mass matrices of the stepped bar are given by

$$[\underset{\sim}{K}] = \begin{bmatrix} \dfrac{A_1 E_1}{l_1} & \dfrac{-A_1 E_1}{l_1} & 0 & 0 \\[2mm] \dfrac{-A_1 E_1}{l_1} & \dfrac{A_1 E_1}{l_1} + \dfrac{A_2 E_2}{l_2} & \dfrac{-A_2 E_2}{l_2} & 0 \\[2mm] 0 & \dfrac{-A_2 E_2}{l_2} & \dfrac{A_2 E_2}{l_2} + \dfrac{A_3 E_3}{l_3} & \dfrac{-A_3 E_3}{l_3} \\[2mm] 0 & 0 & \dfrac{-A_3 E_3}{l_3} & \dfrac{A_3 E_3}{l_3} \end{bmatrix} \tag{E.1}$$

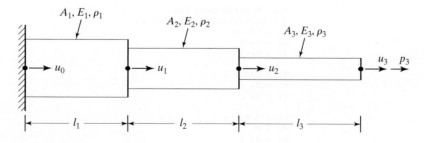

FIGURE 12.11   Stepped bar.

$$[\underset{\sim}{M}] = \frac{1}{6}\begin{bmatrix} 2\rho_1 A_1 l_1 & \rho_1 A_1 l_1 & 0 & 0 \\ \rho_1 A_1 l_1 & 2\rho_1 A_1 l_1 + 2\rho_2 A_2 l_2 & \rho_2 A_2 l_2 & 0 \\ 0 & \rho_2 A_2 l_2 & 2\rho_2 A_2 l_2 + 2\rho_3 A_3 l_3 & \rho_3 A_3 l_3 \\ 0 & 0 & \rho_3 A_3 l_3 & 2\rho_3 A_3 l_3 \end{bmatrix} \tag{E.2}$$

The system matrices $[K]$ and $[M]$ can be obtained by incorporating the boundary condition $u_0 = 0$ —that is, by deleting the first row and first column in Eqs. (E.1) and (E.2).

a. The equilibrium equations under the load $p_3 = 1,000$ N are given by

$$[K]\vec{U} = \vec{P} \tag{E.3}$$

where

$$[K] = \begin{bmatrix} \dfrac{A_1 E_1}{l_1} + \dfrac{A_2 E_2}{l_2} & \dfrac{-A_2 E_2}{l_2} & 0 \\ \dfrac{-A_2 E_2}{l_2} & \dfrac{A_2 E_2}{l_2} + \dfrac{A_3 E_3}{l_3} & \dfrac{-A_3 E_3}{l_3} \\ 0 & \dfrac{-A_3 E_3}{l_3} & \dfrac{A_3 E_3}{l_3} \end{bmatrix} \tag{E.4}$$

$$\vec{U} = \begin{Bmatrix} u_1 \\ u_2 \\ u_3 \end{Bmatrix}, \qquad \vec{P} = \begin{Bmatrix} 0 \\ 0 \\ 1,000 \end{Bmatrix}$$

b. The eigenvalue problem can be expressed as

$$\left[[K] - \omega^2 [M] \right] \vec{U} = \vec{0} \tag{E.5}$$

where $[K]$ is given by Eq. (E.4) and $[M]$ by

$$[M] = \frac{1}{6}\begin{bmatrix} 2\rho_1 A_1 l_1 + 2\rho_2 A_2 l_2 & \rho_2 A_2 l_2 & 0 \\ \rho_2 A_2 l_2 & 2\rho_2 A_2 l_2 + 2\rho_3 A_3 l_3 & \rho_3 A_3 l_3 \\ 0 & \rho_3 A_3 l_3 & 2\rho_3 A_3 l_3 \end{bmatrix} \tag{E.6}$$

The MATLAB solution of Eqs. (E.3) and (E.5) is given below.

```
%--- Program Ex12_5.m
%---Initialization of values-------------
A1 = 16e-4 ;
A2 = 9e-4 ;
A3 = 4e-4 ;

E1 = 20e10 ;
E2 = E1 ;
E3 = E1 ;

R1 = 7.8e3 ;
R2 = R1 ;
R3 = R1 ;
```

```
L1 = 1 ;
L2 = 0.5 ;
L3 = 0.25 ;

%---Definition of [K]-----------------

K11 = A1*E1/L1+A2*E2/L2 ;
K12 = -A2*E2/L2 ;
K13 = 0 ;

K21 = K12 ;
K22 = A2*E2/L2+A3*E3/L3 ;
K23 = -A3*E3/L3 ;

K31 = K13 ;
K32 = K23 ;
K33 = A3*E3/L3 ;

K = [ K11 K12 K13; K21 K22 K23; K31 K32 K33 ]

%---- Calculation of matrix

P = [ 0 0 1000]'

U = inv(K)*P

%---- Definition of [M] -------------

M11 = (2*R1*A1*L1+2*R2*A2*L2) / 6;
M12 = (R2*A2*L2) / 6;
M13 = 0;

M21 = M12;
M22 = (2*R2*A2*L2+2*R3*A3*L3) / 6;
M23 = R3*A3*L3;

M31 = M13;
M32 = M23;
M33 = 2*M23;

M= [M11 M12 M13; M21 M22 M23; M31 M32 M33 ]

MI = inv (M)

KM = MI*K

%-------Calculation of eigenvector and eigenvalue-------

[L, V] = eig (KM)

>> Ex12_5
K =

   680000000   -360000000            0
  -360000000    680000000   -320000000
           0   -320000000    320000000
P =
           0
           0
        1000

U =
   1.0e-005 *
     0.3125
     0.5903
     0.9028

M =
```

```
              5.3300         0.5850              0
              0.5850         1.4300         0.7800
                   0         0.7800         1.5600
    MI =

              0.2000        -0.1125         0.0562
             -0.1125         1.0248        -0.5124
              0.0562        -0.5124         0.8972

    KM =

              1.0e+008*
              1.7647        -1.6647         0.5399
             -4.4542         9.0133        -4.9191
              2.2271        -6.5579         4.5108

     L =

             -0.1384         0.6016         0.3946
              0.7858        -0.1561         0.5929
             -0.6028        -0.7834         0.7020

     V =
              1.0e+009*
              1.3571              0              0
                   0         0.1494              0
                   0              0         0.0224
    >>
```

■

Program for Eigenvalue Analysis of a Stepped Beam

EXAMPLE 12.6

Develop a MATLAB program called **Program17.m** for the eigenvalue analysis of a fixed-fixed stepped beam of the type shown in Fig. 12.12.

Solution: **Program17.m** is developed to accept the following input data:

$xl(i)$ = length of element (step) i

$xi(i)$ = moment of inertia of element i

$a(i)$ = area of cross section of element i

$bj(i, j)$ = global degree-of-freedom number corresponding to the local jth degree of freedom of element i

e = Young's modulus

rho = mass density

The program gives the natural frequencies and mode shapes of the beam as output.

```
      Natural frequencies of the stepped beams
      4.0052e+002  1.5449e+003  5.6344e+003  1.7831e+004
      Mode shapes
      1    1.2656e-003       2.3166e-003   1.7346e-003   5.4518e-004
      2   -4.6124e-004       2.4859e-003   5.7696e-004   3.1782e-003
      3    2.0595e-005      -2.2252e-003   1.6620e-004   2.5424e-003
      4    2.2442e-005       7.4391e-004   4.5870e-005   2.0069e-003
```

■

$$E = 200 \times 10^9 \text{ Pa}, \rho = 7,800 \text{ kg/m}^3$$

FIGURE 12.12 Stepped beam.

CHAPTER SUMMARY

The finite element method is a popular numerical procedure for finding accurate solutions of complex practical systems. We presented an introduction to the method as applied to vibration problems. We outlined the method of deriving stiffness and mass matrices of simple structural elements such as bars, rods, and beams, the transformation of the matrices to a global coordinate system, assembly of element matrices, and solution of finite element equations. We presented the application of the method through several static and dynamic (vibration) examples. The solution of finite-element-based vibration problems using MATLAB was also considered.

Now that you have finished this chapter, you should be able to answer the review questions and solve the problems given below.

REFERENCES

12.1 O. C. Zienkiewicz, *The Finite Element Method* (4th ed.), McGraw-Hill, London, 1987.

12.2 S. S. Rao, *The Finite Element Method in Engineering* (3rd ed.), Butterworth-Heinemann, Boston, 1999.

12.3 G. V. Ramana and S. S. Rao, "Optimum design of plano-milling machine structure using finite element analysis," *Computers and Structures*, Vol. 18, 1984, pp. 247–253.

12.4 R. Davis, R. D. Henshell, and G. B. Warburton, "A Timoshenko beam element," *Journal of Sound and Vibration*, Vol. 22, 1972, pp. 475–487.

12.5 D. L. Thomas, J. M. Wilson, and R. R. Wilson, "Timoshenko beam finite elements," *Journal of Sound and Vibration*, Vol. 31, 1973, pp. 315–330.

12.6 J. Thomas and B. A. H. Abbas, "Finite element model for dynamic analysis of Timoshenko beams," *Journal of Sound and Vibration*, Vol. 41, 1975, pp. 291–299.

12.7 R. S. Gupta and S. S. Rao, "Finite element eigenvalue analysis of tapered and twisted Timoshenko beams," *Journal of Sound and Vibration*, Vol. 56, 1978, pp. 187–200.

12.8 T. H. H. Pian, "Derivation of element stiffness matrices by assumed stress distribution," *AIAA Journal*, Vol. 2, 1964, pp. 1333–1336.

12.9 H. Alaylioglu and R. Ali, "Analysis of an automotive structure using hybrid stress finite elements," *Computers and Structures*, Vol. 8, 1978, pp. 237–242.

12.10 I. Fried, "Accuracy of finite element eigenproblems," *Journal of Sound and Vibration*, Vol. 18, 1971, pp. 289–295.

12.11 P. Tong, T. H. H. Pian, and L. L. Bucciarelli, "Mode shapes and frequencies by the finite element method using consistent and lumped matrices," *Computers and Structures*, Vol. 1, 1971, pp. 623–638.

REVIEW QUESTIONS

12.1 Give brief answers to the following:

1. What is the basic idea behind the finite element method?
2. What is a shape function?
3. What is the role of transformation matrices in the finite element method?
4. What is the basis for the derivation of transformation matrices?
5. How are fixed boundary conditions incorporated in the finite element equations?
6. How do you solve a finite element problem having symmetry in geometry and loading by modeling only half of the problem?
7. Why is the finite element approach presented in this chapter called the *displacement method*?
8. What is a consistent-mass matrix?
9. What is a lumped-mass matrix?
10. What is the difference between the finite element method and the Rayleigh-Ritz method?
11. How is the distributed load converted into an equivalent joint force vector in the finite element method?

12.2 Indicate whether each of the following statements is true or false:

1. For a bar element of length l with two nodes, the shape function corresponding to node 2 is given by x/l.
2. The element stiffness matrices are always singular.
3. The element mass matrices are always singular.
4. The system stiffness matrix is always singular unless the boundary conditions are incorporated.
5. The system mass matrix is always singular unless the boundary conditions are incorporated.
6. The lumped-mass matrices are always diagonal.
7. The coordinate transformation of element matrices is required for all systems.
8. The element stiffness matrix in the global coordinate system, $[\bar{k}]$, can be expressed in terms of the local matrix $[k]$ and the coordinate transformation matrix $[\lambda]$ as $[\lambda]^T[k][\lambda]$.
9. The derivation of system matrices involves the assembly of element matrices.
10. Boundary conditions are to be imposed to avoid rigid-body motion of the system.

12.3 Fill in each of the following blanks with the appropriate word:

1. In the finite element method, the solution domain is replaced by several _____.
2. In the finite element method, the elements are assumed to be interconnected at certain points known as _____.
3. In the finite element method, an _____ solution is assumed within each element.
4. The displacement within a finite element is expressed in terms of _____ functions.
5. For a thin beam element, _____ degrees of freedom are considered at each node.
6. For a thin beam element, the shape functions are assumed to be polynomials of degree _____.
7. In the displacement method, the _____ of elements is directly approximated.
8. If the displacement model used in the derivation of the element stiffness matrices is also used to derive the element mass matrices, the resulting mass matrix is called _____ *mass matrix*.
9. If the mass matrix is derived by assuming point masses at node points, the resulting mass matrix is called _____ *mass*.
10. The lumped-mass matrices do not consider the _____ coupling between the various displacement degrees of freedom of the element.
11. Different orientations of finite elements require _____ of element matrices.

12.4 Select the most appropriate answer out of the choices given:

1. For a bar element of length l with two nodes, the shape function corresponding to node 1 is given by

    a. $\left(1 - \dfrac{x}{l}\right)$ **b.** $\dfrac{x}{l}$ **c.** $\left(1 + \dfrac{x}{l}\right)$

2. The simplest form of mass matrix is known as
    a. *lumped-mass matrix*
    b. *consistent-mass matrix*
    c. *global mass matrix*

3. The finite element method is
    a. an approximate analytical method
    b. a numerical method
    c. an exact analytical method

4. The stiffness matrix of a bar element is given by

    a. $\dfrac{EA}{l}\begin{bmatrix} 1 & 1 \\ 1 & 1 \end{bmatrix}$ **b.** $\dfrac{EA}{l}\begin{bmatrix} 1 & -1 \\ -1 & 1 \end{bmatrix}$ **c.** $\dfrac{EA}{l}\begin{bmatrix} 1 & 0 \\ 0 & 1 \end{bmatrix}$

5. The consistent mass matrix of a bar element is given by

    a. $\dfrac{\rho Al}{6}\begin{bmatrix} 2 & 1 \\ 1 & 2 \end{bmatrix}$ **b.** $\dfrac{\rho Al}{6}\begin{bmatrix} 2 & -1 \\ -1 & 2 \end{bmatrix}$ **c.** $\dfrac{\rho Al}{6}\begin{bmatrix} 1 & 0 \\ 0 & 1 \end{bmatrix}$

6. The finite element method is similar to
    a. Rayleigh's method
    b. the Rayleigh-Ritz method
    c. the Lagrange method

7. The lumped-mass matrix of a bar element is given by

 a. $\rho Al \begin{bmatrix} 1 & 0 \\ 0 & 1 \end{bmatrix}$ **b.** $\dfrac{\rho Al}{6} \begin{bmatrix} 2 & 1 \\ 1 & 2 \end{bmatrix}$ **c.** $\dfrac{\rho Al}{2} \begin{bmatrix} 1 & 0 \\ 0 & 1 \end{bmatrix}$

8. The element mass matrix in the global coordinate system, $[\overline{m}]$, can be expressed in terms of the element mass matrix in the local coordinate system $[m]$ and the coordinate transformation matrix $[\lambda]$ as

 a. $[\overline{m}] = [\lambda]^T[m]$ **b.** $[\overline{m}] = [m][\lambda]$ **c.** $[\overline{m}] = [\lambda]^T[m][\lambda]$

12.5   Match the items in the two columns below. Assume a fixed-fixed bar with one middle node:

Element matrices: $[k] = \dfrac{AE}{l} \begin{bmatrix} 1 & -1 \\ -1 & 1 \end{bmatrix}$, $[m]_c = \dfrac{\rho Al}{6} \begin{bmatrix} 2 & 1 \\ 1 & 2 \end{bmatrix}$, $[m]_l = \dfrac{\rho Al}{2} \begin{bmatrix} 1 & 0 \\ 0 & 1 \end{bmatrix}$

Steel bar: $E = 200 \times 10^9$ Pa, $\rho = 7{,}800$ kg/m^3, $L = 0.3$ m

Aluminum bar: $E = 72 \times 10^9$ Pa, $\rho = 2{,}800$ kg/m^3, $L = 0.3$ m

1. Natural frequency of steel bar given by lumped-mass matrices   **a.** 58,470.5346 rad/s

2. Natural frequency of aluminum bar given by consistent-mass matrices   **b.** 47,809.1444 rad/s

3. Natural frequency of steel bar given by consistent-mass matrices   **c.** 56,554.0044 rad/s

4. Natural frequency of aluminum bar given by lumped-mass matrices   **d.** 47,740.9916 rad/s

PROBLEMS

Section 12.3 Derivation of Element Matrices and Vectors

12.1   Derive the stiffness matrix of the tapered bar element (which deforms in the axial direction) shown in Fig. 12.13. The diameter of the bar decreases from D to d over its length.

FIGURE 12.13

12.2 Derive the stiffness matrix of the bar element in longitudinal vibration whose cross-sectional area varies as $A(x) = A_0 e^{-(x/l)}$, where A_0 is the area at the root (see Fig. 12.14).

FIGURE 12.14

12.3 The tapered cantilever beam shown in Fig. 12.15 is used as a spring to carry a load P. Derive the stiffness matrix of the beam using a one-element idealization. Assume $B = 25$ cm, $b = 10$ cm, $t = 2.5$ cm, $l = 2$ m, $E = 2.07 \times 10^{11}$ N/m^2, and $P = 1,000$ N.

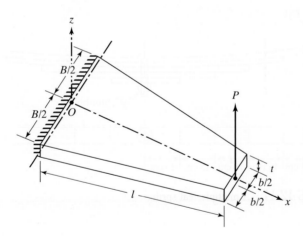

FIGURE 12.15

12.4 Derive the stiffness and mass matrices of the planar frame element (general beam element) shown in Fig. 12.16 in the global XY-coordinate system.

12.5 A multiple-leaf spring used in automobiles is shown in Fig. 12.17. It consists of five leaves, each of thickness $t = 0.65$ cm and width $w = 3.8$ cm. Find the deflection of the leaves under a load of $P = 8,900$ N. Model only a half of the spring for the finite element analysis. The Young's modulus of the material is 200×10^9 N/m^2.

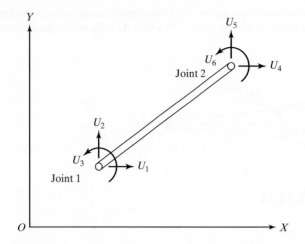

FIGURE 12.16   A frame element in global system.

FIGURE 12.17   A multiple-leaf spring.

12.6 A seven-member planar truss (with pin joints) is shown in Fig. 12.18. Each of the seven members has an area of cross section of 4 cm^2 and a Young's modulus of 207 GPa.

    (a) Label the complete set of local and global nodal displacement degrees of freedom of the truss. Assume the X- and Y- coordinates shown in Fig. 12.18 as global coordinates.

    (b) Find the coordinate transformation matrix of each member.

    (c) Find the local and global stiffness matrices of each member.

12.7 Find the stiffness and mass matrices of the beam supported on springs as shown in Fig. 12.19. Model the beam using one finite element. Assume the material of the beam as steel with a Young's modulus of 207 GPa and weight density of 7,650 N/m^3. Neglect the weights of the springs.

FIGURE 12.18

FIGURE 12.19

12.8 For the beam shown in Fig. 12.20, one end (point A) is fixed and a spring-mass system is attached to the other end (point B). Assume the cross section of the beam to be circular with radius 2 cm and the material of the beam to be steel with Young's modulus of 207 GPa and weight density of 7,650 N/m^3. Using two beam elements of equal length, derive the element stiffness and mass matrices of both elements.

Section 12.4 Transformation Matrix

12.9 Find the global stiffness matrix of each of the four bar elements of the truss shown in Fig. 12.5 using the following data:

Nodal coordinates: $(X_1, Y_1) = (0, 0)$ cm, $(X_2, Y_2) = (125, 250)$ cm, $(X_3, Y_3) = (250, 0)$ cm, $(X_4, Y_4) = (500, 375)$ cm.

Cross-sectional areas: $A_1 = A_2 = A_3 = A_4 = 13$ cm^2.

Young's modulus of all members: 200×10^9 Pa.

12.10 For the seven-member planar truss considered in Problem 12.6 (Fig. 12.18), determine the assembled stiffness matrix of the system before applying the boundary conditions.

FIGURE 12.20

Section 12.5 Assembly of Matrices and Vectors

12.11 Using the result of Problem 12.9, find the assembled stiffness matrix of the truss and formulate the equilibrium equations if the vertical downward load applied at node 4 is 5,000 N.

12.12 For the beam considered in Problem 12.8 (Fig. 12.20), derive the assembled stiffness and mass matrices of the system.

Section 12.6 Application of Boundary Conditions and Solution of Problem

12.13 For the tapered beam considered in Problem 12.3 (Fig. 12.15), find the stress induced in the beam using a one-element idealization.

12.14 For the multiple-leaf spring described in Problem 12.5 (Fig. 12.17), derive the assembled stiffness and mass matrices. Consider only one-half of the spring for modeling using five beam elements of equal length.

12.15 Find the nodal displacements of the crane shown in Fig. 12.21 when a vertically downward load of 4,000 N is applied at node 4. The Young's modulus is 200×10^9 Pa and the cross-sectional area is 13 cm^2 for elements 1 and 2 and 6.5 cm^2 for elements 3 and 4.

FIGURE 12.21

12.16 Find the tip deflection of the cantilever beam shown in Fig. 12.22 when a vertical load of $P = 500$ N is applied at point Q using (a) a one-element approximation and (b) a two-element approximation. Assume $l = 0.25$ m, $h = 25$ mm, $b = 50$ mm, $E = 2.07 \times 10^{11}$ Pa, and $k = 10^5$ N/m.

FIGURE 12.22

12.17 Find the stresses in the stepped beam shown in Fig. 12.23 when a moment of 1,000 N-m is applied at node 2 using a two-element idealization. The beam has a square cross section 50×50 mm between nodes 1 and 2 and 25×25 mm between nodes 2 and 3. Assume the Young's modulus as 2.1×10^{11} Pa.

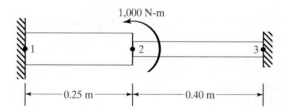

FIGURE 12.23

12.18 Find the transverse deflection and slope of node 2 of the beam shown in Fig. 12.24 using a two-element idealization. Compare the solution with that of the simple beam theory.

FIGURE 12.24

12.19 Find the displacement of node 3 and the stresses in the two members of the truss shown in Fig. 12.25. Assume that the Young's modulus and the cross-sectional areas of the two members are the same with $E = 200$ GPa and $A = 0.5 \times 10^{-3}$ m^2.

FIGURE 12.25

12.20 A simplified model of a radial drilling machine is shown in Fig. 12.26. If a vertical force of 5,000 N along the z direction and a bending moment of 500 N-m in the xz-plane are developed at point A during a metal-cutting operation, find the stresses developed in the machine. Use two beam elements for the column and one beam element for the arm. Assume the material of the machine as steel.

FIGURE 12.26   A radial drilling-machine structure.

12.21 The crank in the slider-crank mechanism shown in Fig. 12.27 rotates at a constant clockwise angular speed of 1,000 rpm. Find the stresses in the connecting rod and the crank when the pressure acting on the piston is 1 MPa and $\theta = 30°$. The diameter of the piston is 0.3 m and the material of the mechanism is steel. Model the connecting rod and the crank by one beam element each. The lengths of the crank and connecting rod are 0.3 m and 1.2 m, respectively.

FIGURE 12.27   A slider-crank mechanism.

12.22 A water tank of weight M is supported by a hollow circular steel column of inner diameter d, wall thickness t, and height l. The wind pressure acting on the column can be assumed to vary linearly from 0 to P_{max} as shown in Fig. 12.28. Find the bending stress induced in the column under the loads using a one-beam element idealization. Data: $M = 5,000$ kg, $l = 12$ m, $d = 0.6$ m, $t = 0.02$ m, and $P_{max} = 700$ kPa.

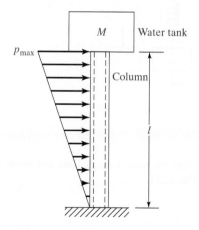

FIGURE 12.28

12.23 For the seven-member planar truss considered in Problem 12.6 (Fig. 12.18), determine the following:

(a) The system stiffness matrix after applying the boundary conditions.

(b) The nodal displacements of the truss under the loads indicated in Fig. 12.18.

12.24 Using one beam element, find the natural frequencies of the uniform pinned-free beam shown in Fig. 12.29.

FIGURE 12.29

12.25 Using one beam element and one spring element, find the natural frequencies of the uniform spring-supported cantilever beam shown in Fig. 12.22.

12.26 Using one beam element and one spring element, find the natural frequencies of the system shown in Fig. 12.30.

FIGURE 12.30

12.27 Using two beam elements, find the natural frequencies and mode shapes of the uniform fixed-fixed beam shown in Fig. 12.31.

FIGURE 12.31

12.28* An electric motor, of mass $m = 100$ kg and operating speed $= 1{,}800$ rpm, is fixed at the middle of a clamped-clamped steel beam of rectangular cross section, as shown in Fig. 12.32. Design the beam such that the natural frequency of the system exceeds the operating speed of the motor.

FIGURE 12.32

12.29 Find the natural frequencies of the beam shown in Fig. 12.33, using three finite elements of length l each.

FIGURE 12.33

12.30 Find the natural frequencies of the cantilever beam carrying an end mass M shown in Fig. 12.34, using a one-beam element idealization.

FIGURE 12.34

---

*An asterisk denotes a problem with no unique answer.

12.31 Find the natural frequencies of vibration of the beam shown in Fig. 12.35, using two beam elements. Also find the load vector if a uniformly distributed transverse load p is applied to element 1.

FIGURE 12.35

12.32 Find the natural frequencies of a beam of length l, which is pin-connected at $x = 0$ and fixed at $x = l$, using one beam element.

12.33 Find the natural frequencies of torsional vibration of the stepped shaft shown in Fig. 12.36. Assume that $\rho_1 = \rho_2 = \rho, G_1 = G_2 = G, I_{p1} = 2I_{p2} = 2I_p, J_1 = 2J_2 = 2J$, and $l_1 = l_2 = l$.

FIGURE 12.36

12.34 Find the dynamic response of the stepped bar shown in Fig. 12.37(a) when its free end is subjected to the load given in Fig. 12.37(b).

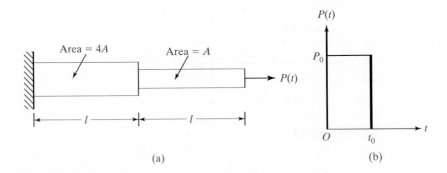

(a)   (b)

FIGURE 12.37

12.35 Find the natural frequencies of a cantilever beam of length l, cross-sectional area A, moment of inertia I, Young's modulus E, and density ρ, using one finite element.

12.36 Find the natural frequencies of vibration of the radial drilling machine considered in Problem 12.20 (Fig. 12.26).

12.37 Find the natural frequencies of the water tank considered in Problem 12.22 (Fig. 12.28) using a one-beam element idealization.

12.38 Find the natural frequencies of vibration of the beam considered in Problem 12.7 using one finite element (Fig. 12.19).

Section 12.7 Consistent- and Lumped-Mass Matrices

12.39 Derive the consistent- and lumped-mass matrices of the tapered bar element (which deforms in the axial direction) shown in Fig. 12.13. The diameter of the bar decreases from D to d over its length.

12.40 Find the natural frequencies of the stepped bar shown in Fig. 12.38 with the following data using consistent- and lumped-mass matrices: $A_1 = 0.001$ m^2, $A_2 = 0.0006$ m^2, $E = 200 \times 10^9$ Pa, $\rho_w = 7{,}750$ kg/m^3, and $l_1 = l_2 = 1$ m.

FIGURE 12.38

12.41 Find the undamped natural frequencies of longitudinal vibration of the stepped bar shown in Fig. 12.39 with the following data using consistent- and lumped-mass matrices: $l_1 = l_2 = l_3 = 0.2$ m, $A_1 = 2A_2 = 4A_3 = 0.4 \times 10^{-3}$ m^2, $E = 2.1 \times 10^{11}$ N/m^2, and $\rho = 7.8 \times 10^3$ kg/m^3.

FIGURE 12.39

Section 12.8 MATLAB Problems

12.42 Consider the stepped bar shown in Fig. 12.11 with the following data: $A_1 = 25 \times 10^{-4}$ m^2, $A_2 = 16 \times 10^{-4}$ m^2, $A_3 = 9 \times 10^{-4}$ m^2, $E_i = 2 \times 10^{11}$ Pa, $i = 1, 2, 3, \rho_i = 7.8 \times 10^3$ kg/m^3, $i = 1, 2, 3, l_1 = 3$ m, $l_2 = 2$ m, $l_3 = 1$ m. Using MATLAB, find the axial displacements u_1, u_2, and u_3 under the load $p_3 = 500$ N.

12.43 Using MATLAB, find the natural frequencies and mode shapes of the stepped bar described in Problem 12.42.

12.44 Use `Program17.m` to find the natural frequencies of a fixed-fixed stepped beam, similar to the one shown in Fig. 12.12, with the following data:

Cross sections of elements: 1, 2, 3: 0.1 m × 0.1m, 0.08m × 0.08 m, 0.05m × 0.05m

Lengths of elements: 1, 2, 3: 0.8 m, 0.5m, 0.2m

Young's modulus of all elements: 70 GPa

Weight density of all elements: 3,000 kg/m^3

12.45 Write a computer program for finding the assembled stiffness matrix of a general planar truss.

DESIGN PROJECTS

12.46 Derive the stiffness and mass matrices of a uniform beam element in transverse vibration rotating at an angular velocity of Ω rad/s about a vertical axis as shown in Fig. 12.40(a). Using these matrices, find the natural frequencies of transverse vibration of the rotor blade of a helicopter (see Fig. 12.40(b)) rotating at a speed of 300 rpm. Assume a uniform rectangular cross section 0.02 m × 0.3 m and a length 1.2 m for the blade. The material of the blade is aluminum.

Beam element. ρ, A, I, E

(a)

Rotor blade

(b)

FIGURE 12.40

12.47   An electric motor having a mass of 500 kg operates on the first floor of a building frame that can be modeled by a steel girder supported by two reinforced concrete columns, as shown in Fig. 12.41. If the operating speed of the motor is 1,500 rpm, design the girder and the columns such that the fundamental frequency of vibration of the building frame is greater than the operating speed of the motor. Use two beam and two bar elements for the idealization. Assume the following data:

$$\text{Girder: } E = 200 \text{ GPa}, \qquad \rho = 250 \text{ kg/m}^3, \qquad h/b = 2$$

$$\text{Columns: } e = 30 \text{ GPa}, \qquad \rho = 75 \text{ kg/m}^3$$

FIGURE 12.41

APPENDIX A

Mathematical Relations and Material Properties

Some of the relationships from trigonometry, algebra, and calculus that are frequently used in vibration analysis are given below.

$$\sin(\alpha \pm \beta) = \sin\alpha\cos\beta \pm \cos\alpha\sin\beta$$

$$\cos(\alpha \pm \beta) = \cos\alpha\cos\beta \pm \sin\alpha\sin\beta$$

$$\sin(\alpha + \beta)\sin(\alpha - \beta) = \sin^2\alpha - \sin^2\beta = \cos^2\beta - \cos^2\alpha$$

$$\cos(\alpha + \beta)\cos(\alpha - \beta) = \cos^2\alpha - \sin^2\beta = \cos^2\beta - \sin^2\alpha$$

$$\sin\alpha\sin\beta = \frac{1}{2}[\cos(\alpha - \beta) - \cos(\alpha + \beta)]$$

$$\cos\alpha\cos\beta = \frac{1}{2}[\cos(\alpha - \beta) + \cos(\alpha + \beta)]$$

$$\sin\alpha\cos\beta = \frac{1}{2}[\sin(\alpha + \beta) + \sin(\alpha - \beta)]$$

$$\sin\alpha + \sin\beta = 2\sin\left(\frac{\alpha + \beta}{2}\right)\cos\left(\frac{\alpha - \beta}{2}\right)$$

$$\sin \alpha - \sin \beta = 2 \cos\left(\frac{\alpha + \beta}{2}\right) \sin\left(\frac{\alpha - \beta}{2}\right)$$

$$\cos \alpha + \cos \beta = 2 \cos\left(\frac{\alpha + \beta}{2}\right) \cos\left(\frac{\alpha - \beta}{2}\right)$$

$$\cos \alpha - \cos \beta = -2 \sin\left(\frac{\alpha + \beta}{2}\right) \sin\left(\frac{\alpha - \beta}{2}\right)$$

$$A \sin \alpha + B \cos \alpha = \sqrt{A^2 + B^2} \cos(\alpha - \phi_1)$$

$$= \sqrt{A^2 + B^2} \sin(\alpha + \phi_2)$$

$$\text{where} \quad \phi_1 = \tan^{-1}\frac{A}{B}, \quad \phi_2 = \tan^{-1}\frac{B}{A}$$

$$\sin^2\alpha + \cos^2\alpha = 1$$

$$\cos 2\alpha = 1 - 2 \sin^2 \alpha = 2 \cos^2 \alpha - 1 = \cos^2 \alpha - \sin^2 \alpha$$

Law of cosines for triangles: $\quad c^2 = a^2 + b^2 - 2ab \cos C$

$$\pi = 3.14159265 \text{ rad}, \quad 1 \text{ rad} = 57.29577951°, \quad 1° = 0.017453292 \text{ rad}$$

$$e = 2.71828183$$

$$\log a^b = b \log a, \quad \log_{10} x = 0.4343 \log_e x, \quad \log_e x = 2.3026 \log_{10} x$$

$$e^{ix} = \cos x + i \sin x$$

$$\sin x = \frac{e^{ix} - e^{-ix}}{2i}, \quad \cos x = \frac{e^{ix} + e^{-ix}}{2}$$

$$\sinh x = \frac{1}{2}(e^x - e^{-x}), \quad \cosh x = \frac{1}{2}(e^x + e^{-x})$$

$$\cosh^2 x - \sinh^2 x = 1$$

$$\frac{d}{dx}(uv) = u\frac{dv}{dx} + v\frac{du}{dx}$$

$$\frac{d}{dx}\left(\frac{u}{v}\right) = \frac{1}{v}\frac{du}{dx} - \frac{u}{v^2}\frac{dv}{dx} = \frac{v\dfrac{du}{dx} - u\dfrac{dv}{dx}}{v^2}$$

$$\int e^{ax} \, dx = \frac{1}{a}e^{ax}$$

$$\int uv \, dx = u \int v \, dx - \int \left(\frac{du}{dx} \int v \, dx\right) dx$$

Complex algebra:

$$z = x + iy \equiv Ae^{i\theta} \quad \text{with} \quad A = \sqrt{x^2 + y^2} \quad \text{and} \quad \theta = \tan^{-1}\left(\frac{y}{x}\right)$$

If $\quad z_1 = x_1 + iy_1 \quad$ and $\quad z_2 = x_2 + iy_2,$

$$z_1 \pm z_2 = (x_1 \pm x_2) + i(y_1 \pm y_2)$$

$$z_1 z_2 = (x_1 x_2 - y_1 y_2) + i(x_1 y_1 + x_2 y_2)$$

$$\frac{z_1}{z_2} = \frac{(x_1 x_2 + y_1 y_2) + i(x_2 y_1 - x_1 y_2)}{\sqrt{x_2^2 + y_2^2}}$$

If $\quad z_1 = A_1 e^{i\theta_1} \quad$ and $\quad z_2 = A_2 e^{i\theta_2},$

$$z_1 + z_2 = Ae^{i\theta}$$

with $\quad A = \left[A_1^2 + A_2^2 - 2A_1 A_2 \cos(\theta_1 - \theta_2)\right]^{\frac{1}{2}}$

and $\quad \theta = \tan^{-1}\left[\dfrac{A_1 \sin\theta_1 + A_2 \sin\theta_2}{A_1 \cos\theta_1 + A_2 \cos\theta_2}\right]$

$$z_1 z_2 = A_1 A_2 e^{i(\theta_1 + \theta_2)}$$

$$\frac{z_1}{z_2} = \frac{A_1}{A_2} e^{i(\theta_1 - \theta_2)}$$

Material Properties

Material	Young's modulus (E)	Modulus of rigidity (G)	Poisson's ratio (ν)	Specific weight (ρ_w)
Steel (Carbon steel)	207 GPa	79.3 GPa	0.292	76.5 kN/m^3
Aluminum (Aluminum alloys)	71 GPa	26.2 GPa	0.334	26.6 kN/m^3
Brass	106 GPa	40 GPa	0.324	83.8 kN/m^3
Copper	119 GPa	44.7 GPa	0.326	87.3 kN/m^3

Carl Gustov Jacob Jacobi (1804–1851), a German mathematician, was educated at the University of Berlin and became a full professor at the University of Konigsberg in 1832. The method he developed for finding the eigen solution of real symmetric matrices has become known as the Jacobi method. He made significant contributions to the fields of elliptic functions, number theory, differential equations, and mechanics and introduced the definition of Jacobian in the theory of determinants. (Courtesy of Dirk J. Struik, *A Concise History of Mathematics* (2nd ed.), Dover Publications, New York, 1948.)

APPENDIX B

Deflection of Beams and Plates

Cantilever Beam

$$y(x) = \begin{cases} \dfrac{Px^2}{6EI}(3a - x); & 0 \le x \le a \\[2mm] \dfrac{Pa^2}{6EI}(3x - a); & a \le x \le l \end{cases}$$

Simply Supported Beam

$$y(x) = \begin{cases} \dfrac{Pbx}{6EIl}(l^2 - x^2 - b^2); & 0 \le x \le a \\[2mm] \dfrac{Pa(l - x)}{6EIl}(2lx - x^2 - a^2); & a \le x \le l \end{cases}$$

Fixed-fixed Beam

$$y(x) = \begin{cases} \dfrac{Pb^2x^2}{6EIl^3}[3al - x(3a + b)]; & 0 \le x \le a \\[2mm] \dfrac{Pa^2(l - x)^2}{6EIl^3}[3bl - (l - x)(3b + a)]; & a \le x \le l \end{cases}$$

Simply Supported Beam with Overhang

$$y(x) = \begin{cases} \text{Same as in the case of a simply supported beam} \\ \text{for } 0 \leq x \leq a \text{ and } a \leq x \leq l \\ \dfrac{Pa}{6EIl}(l^2 - a^2)(x - l); \quad l \leq x \leq l + c \end{cases}$$

Simply Supported Beam with Overhanging Load

$$y(x) = \begin{cases} \dfrac{Pax}{6EIl}(x^2 - l^2); \quad 0 \leq x \leq l \\ \dfrac{P(x - l)}{6EIl}[a(3x - l) - (x - l)^2]; \quad l \leq x \leq l + a \end{cases}$$

Fixed-fixed Beam with End Displacement

$$y(x) = \frac{P}{12EI}(3lx^2 - 2x^3)$$

Simply Supported Circular Plate

$$y_{\text{center}} = \frac{Pr^2(3 + v)}{16\pi D(1 + v)}$$

where $D = \dfrac{Et^3}{12(1 - v^2)}$, t = plate thickness,

and v = Poisson's ratio

Fixed Circular Plate

$$y_{\text{center}} = \frac{Pr^2}{16\pi D}$$

Square Plate Simply Supported on All Sides

$$y_{\text{center}} = \frac{\alpha Pa^2}{Et^3} \text{ with } \alpha = 0.1267 \text{ for } v = 0.3$$

Square Plate Fixed on All Sides

$$y_{\text{center}} = \frac{\alpha Pa^2}{Et^3} \text{ with } \alpha = 0.0611 \text{ for } v = 0.3$$

A P P E N D I X   C

Matrices

C.1 Definitions

Matrix. A matrix is a rectangular array of numbers. An array having m rows and n columns enclosed in brackets is called an *m-by-n matrix*. If $[A]$ is an $m \times n$ matrix, it is denoted as

$$[A] = [a_{ij}] = \begin{bmatrix} a_{11} & a_{12} & \cdot & \cdot & \cdot & a_{1n} \\ a_{21} & a_{22} & \cdot & \cdot & \cdot & a_{2n} \\ \cdot & & & & & \\ \cdot & & & & & \\ \cdot & & & & & \\ a_{m1} & a_{m2} & \cdot & \cdot & \cdot & a_{mn} \end{bmatrix} \tag{C.1}$$

where the numbers a_{ij} are called the *elements* of the matrix. The first subscript i denotes the row and the second subscript j specifies the column in which the element a_{ij} appears.

Square Matrix.   When the number of rows (m) is equal to the number of columns (n), the matrix is called a *square matrix of order n*.

Column Matrix.   A matrix consisting of only one column—that is, an $m \times 1$ matrix—is called a *column matrix* or more commonly a *column vector*. Thus if $\vec{a}$ is a column vector having m elements, it can be represented as

$$\vec{a} = \begin{Bmatrix} a_1 \\ a_2 \\ \cdot \\ \cdot \\ \cdot \\ a_m \end{Bmatrix} \tag{C.2}$$

Row Matrix.   A matrix consisting of only one row—that is, a $1 \times n$ matrix—is called a *row matrix* or a *row vector*. If $\lfloor b \rfloor$ is a row vector, it can be denoted as

$$\lfloor b \rfloor = [b_1 \ b_2 \ \cdots \ b_n] \tag{C.3}$$

Diagonal Matrix.   A square matrix in which all the elements are zero except those on the principal diagonal is called a *diagonal matrix*. For example, if $[A]$ is a diagonal matrix of order n, it is given by

$$[A] = \begin{bmatrix} a_{11} & 0 & 0 & \cdot & \cdot & \cdot & 0 \\ 0 & a_{22} & 0 & \cdot & \cdot & \cdot & 0 \\ 0 & 0 & a_{33} & \cdot & \cdot & \cdot & 0 \\ \cdot & & & & & & \\ \cdot & & & & & & \\ \cdot & & & & & & \\ 0 & 0 & 0 & \cdot & \cdot & \cdot & a_{nn} \end{bmatrix} \tag{C.4}$$

Identity Matrix.   If all the elements of a diagonal matrix have a value 1, then the matrix is called an *identity matrix* or a *unit matrix* and is usually denoted as $[I]$.

Zero Matrix.   If all the elements of a matrix are zero, it is called a *zero* or *null matrix* and is denoted as $[0]$. If $[0]$ is of order 2×4, it is given by

$$[0] = \begin{bmatrix} 0 & 0 & 0 & 0 \\ 0 & 0 & 0 & 0 \end{bmatrix} \tag{C.5}$$

Symmetric Matrix. If the element in the ith row and jth column is the same as the one in the jth row and ith column in a square matrix, it is called a *symmetric matrix*. This means that if $[A]$ is a symmetric matrix, we have $a_{ji} = a_{ij}$. For example,

$$[A] = \begin{bmatrix} 4 & -1 & -3 \\ -1 & 0 & 7 \\ -3 & 7 & 5 \end{bmatrix} \tag{C.6}$$

is a symmetric matrix of order 3.

Transpose of a Matrix. The transpose of an $m \times n$ matrix $[A]$ is the $n \times m$ matrix obtained by interchanging the rows and columns of $[A]$, and is denoted as $[A]^T$. Thus if

$$[A] = \begin{bmatrix} 2 & 4 & 5 \\ 3 & 1 & 8 \end{bmatrix} \tag{C.7}$$

then $[A]^T$ is given by

$$[A]^T = \begin{bmatrix} 2 & 3 \\ 4 & 1 \\ 5 & 8 \end{bmatrix} \tag{C.8}$$

Note that the transpose of a column matrix (vector) is a row matrix (vector), and vice versa.

Trace. The sum of the main diagonal elements of a square matrix $[A] = [a_{ij}]$ is called the *trace* of $[A]$ and is given by

$$\text{Trace}[A] = a_{11} + a_{22} + \cdots + a_{nn} \tag{C.9}$$

Determinant. If $[A]$ denotes a square matrix of order n, then the determinant of $[A]$ is denoted as $|[A]|$. Thus

$$|[A]| = \begin{vmatrix} a_{11} & a_{12} & \cdot & \cdot & \cdot & a_{1n} \\ a_{21} & a_{22} & \cdot & \cdot & \cdot & a_{2n} \\ \cdot & & & & & \\ \cdot & & & & & \\ \cdot & & & & & \\ a_{n1} & a_{n2} & \cdot & \cdot & \cdot & a_{nn} \end{vmatrix} \tag{C.10}$$

The value of a determinant can be found by obtaining the minors and cofactors of the determinant.

The *minor* of the element a_{ij} of the determinant $|[A]|$ of order n is a determinant of order $(n-1)$ obtained by deleting the row i and the column j of the original determinant. The minor of a_{ij} is denoted as M_{ij}.

The *cofactor* of the element a_{ij} of the determinant $|[A]|$ of order n is the minor of the element a_{ij}, with either a plus or a minus sign attached; it is defined as

$$\text{Cofactor of } a_{ij} = \beta_{ij} = (-1)^{i+j}M_{ij} \tag{C.11}$$

where M_{ij} is the minor of a_{ij}. For example, the cofactor of the element a_{32} of

$$\det[A] = \begin{vmatrix} a_{11} & a_{12} & a_{13} \\ a_{21} & a_{22} & a_{23} \\ a_{31} & a_{32} & a_{33} \end{vmatrix} \tag{C.12}$$

is given by

$$\beta_{32} = (-1)^5 M_{32} = -\begin{vmatrix} a_{11} & a_{13} \\ a_{21} & a_{23} \end{vmatrix} \tag{C.13}$$

The value of a second-order determinant $|[A]|$ is defined as

$$\det[A] = \begin{vmatrix} a_{11} & a_{12} \\ a_{21} & a_{22} \end{vmatrix} = a_{11}a_{22} - a_{12}a_{21} \tag{C.14}$$

The value of an nth-order determinant $|[A]|$ is defined as

$$\det[A] = \sum_{j=1}^{n} a_{ij}\beta_{ij} \text{ for any specific row } i$$

or

$$\det[A] = \sum_{i=1}^{n} a_{ij}\beta_{ij} \text{ for any specific column } j \tag{C.15}$$

For example, if

$$\det[A] = |[A]| = \begin{vmatrix} 2 & 2 & 3 \\ 4 & 5 & 6 \\ 7 & 8 & 9 \end{vmatrix} \tag{C.16}$$

then, by selecting the first column for expansion, we obtain

$$\det[A] = 2\begin{vmatrix} 5 & 6 \\ 8 & 9 \end{vmatrix} - 4\begin{vmatrix} 2 & 3 \\ 8 & 9 \end{vmatrix} + 7\begin{vmatrix} 2 & 3 \\ 5 & 6 \end{vmatrix}$$

$$= 2(45 - 48) - 4(18 - 24) + 7(12 - 15) = -3 \qquad \text{(C.17)}$$

Properties of Determinants

1. The value of a determinant is not affected if rows (or columns) are written as columns (or rows) in the same order.
2. If all the elements of a row (or a column) are zero, the value of the determinant is zero.
3. If any two rows (or two columns) are interchanged, the value of the determinant is multiplied by -1.
4. If all the elements of one row (or one column) are multiplied by the same constant a, the value of the new determinant is a times the value of the original determinant.
5. If the corresponding elements of two rows (or two columns) of a determinant are proportional, the value of the determinant is zero. For example,

$$\det[A] = \begin{vmatrix} 4 & 7 & -8 \\ 2 & 5 & -4 \\ -1 & 3 & 2 \end{vmatrix} = 0 \qquad \text{(C.18)}$$

Adjoint Matrix. The adjoint matrix of a square matrix $[A] = [a_{ij}]$ is defined as the matrix obtained by replacing each element a_{ij} by its cofactor β_{ij} and then transposing. Thus

$$\text{Adjoint}[A] = \begin{bmatrix} \beta_{11} & \beta_{12} & \cdots & \beta_{1n} \\ \beta_{21} & \beta_{22} & \cdots & \beta_{2n} \\ \vdots & & & \\ \beta_{n1} & \beta_{n2} & \cdots & \beta_{nn} \end{bmatrix}^T = \begin{bmatrix} \beta_{11} & \beta_{21} & \cdots & \beta_{n1} \\ \beta_{12} & \beta_{22} & \cdots & \beta_{n2} \\ \vdots & & & \\ \beta_{1n} & \beta_{2n} & \cdots & \beta_{nn} \end{bmatrix} \qquad \text{(C.19)}$$

Inverse Matrix. The inverse of a square matrix $[A]$ is written as $[A]^{-1}$ and is defined by the following relationship:

$$[A]^{-1}[A] = [A][A]^{-1} = [I] \qquad \text{(C.20)}$$

where $[A]^{-1}[A]$, for example, denotes the product of the matrix $[A]^{-1}$ and $[A]$. The inverse matrix of $[A]$ can be determined (see [A.1]):

$$[A]^{-1} = \frac{\text{adjoint}[A]}{\det[A]} \qquad \text{(C.21)}$$

when det $[A]$ is not equal to zero. For example, if

$$[A] = \begin{bmatrix} 2 & 2 & 3 \\ 4 & 5 & 6 \\ 7 & 8 & 9 \end{bmatrix} \qquad\qquad (C.22)$$

its determinant has a value det$[A] = -3$. The cofactor of a_{11} is

$$\beta_{11} = (-1)^2 \begin{vmatrix} 5 & 6 \\ 8 & 9 \end{vmatrix} = -3 \qquad\qquad (C.23)$$

In a similar manner, we can find the other cofactors and determine

$$[A]^{-1} = \frac{\text{adjoint}[A]}{\det[A]} = \frac{1}{-3}\begin{bmatrix} -3 & 6 & -3 \\ 6 & -3 & 0 \\ -3 & -2 & 2 \end{bmatrix} = \begin{bmatrix} 1 & -2 & 1 \\ -2 & 1 & 0 \\ 1 & 2/3 & -2/3 \end{bmatrix} \quad (C.24)$$

Singular Matrix.   A square matrix is said to be singular if its determinant is zero.

C.2   Basic Matrix Operations

Equality of Matrices.   Two matrices $[A]$ and $[B]$, having the same order, are equal if and only if $a_{ij} = b_{ij}$ for every i and j.

Addition and Subtraction of Matrices.   The sum of the two matrices $[A]$ and $[B]$, having the same order, is given by the sum of the corresponding elements. Thus if $[C] = [A] + [B] = [B] + [A]$, we have $c_{ij} = a_{ij} + b_{ij}$ for every i and j. Similarly, the difference between two matrices $[A]$ and $[B]$ of the same order, $[D]$, is given by $[D] = [A] - [B]$ with $d_{ij} = a_{ij} - b_{ij}$ for every i and j.

Multiplication of Matrices.   The product of two matrices $[A]$ and $[B]$ is defined only if they are conformable—that is, if the number of columns of $[A]$ is equal to the number of rows of $[B]$. If $[A]$ is of order $m \times n$ and $[B]$ is of order $n \times p$, then the product $[C] = [A][B]$ is of order $m \times p$ and is defined by $[C] = [c_{ij}]$, with

$$c_{ij} = \sum_{k=1}^{n} a_{ik}b_{kj} \qquad\qquad (C.25)$$

This means that c_{ij} is the quantity obtained by multiplying the ith row of $[A]$ and the jth column of $[B]$ and summing these products. For example, if

$$[A] = \begin{bmatrix} 2 & 3 & 4 \\ 1 & -5 & 6 \end{bmatrix} \quad \text{and} \quad [B] = \begin{bmatrix} 8 & 0 \\ 2 & 7 \\ -1 & 4 \end{bmatrix} \tag{C.26}$$

then

$$[C] = [A][B] = \begin{bmatrix} 2 & 3 & 4 \\ 1 & -5 & 6 \end{bmatrix} \begin{bmatrix} 8 & 0 \\ 2 & 7 \\ -1 & 4 \end{bmatrix}$$

$$= \begin{bmatrix} 2 \times 8 + 3 \times 2 + 4 \times (-1) & 2 \times 0 + 3 \times 7 + 4 \times 4 \\ 1 \times 8 + (-5) \times 2 + 6 \times (-1) & 1 \times 0 + (-5) \times 7 + 6 \times 4 \end{bmatrix}$$

$$= \begin{bmatrix} 18 & 37 \\ -8 & -11 \end{bmatrix} \tag{C.27}$$

If the matrices are conformable, the matrix multiplication process is associative

$$([A][B])[C] = [A]([B][C]) \tag{C.28}$$

and is distributive

$$([A] + [B])[C] = [A][C] + [B][C] \tag{C.29}$$

The product $[A][B]$ denotes the premultiplication of $[B]$ by $[A]$ or the postmultiplication of $[A]$ by $[B]$. It is to be noted that the product $[A][B]$ is not necessarily equal to $[B][A]$.

The transpose of a matrix product can be found to be the product of the transposes of the separate matrices in reverse order. Thus, if $[C] = [A][B]$,

$$[C]^T = ([A][B])^T = [B]^T[A]^T \tag{C.30}$$

The inverse of a matrix product can be determined from the product of the inverse of the separate matrices in reverse order. Thus if $[C] = [A][B]$,

$$[C]^{-1} = ([A][B])^{-1} = [B]^{-1}[A]^{-1} \tag{C.31}$$

REFERENCE

C.1 Barnett, *Matrix Methods for Engineers and Scientists*, McGraw-Hill, New York, 1982.

Pierre Simon Laplace (1749–1827) was a French mathematician remembered for his fundamental contributions to probability theory, mathematical physics, and celestial mechanics; the name Laplace occurs in both mechanical and electrical engineering. Much use is made of Laplace transforms in vibrations and applied mechanics, and the Laplace equation is applied extensively in the study of electric and magnetic fields. (Courtesy of Dirk J. Struik, *A Concise History of Mathematics* (2nd ed.), Dover Publications, New York, 1948.)

APPENDIX D

Laplace Transform

The Laplace transform method is a powerful method for solving linear ordinary differential equations with constant coefficients, particularly when the forcing function is in the form of discontinuous functions (which cannot be solved easily by other methods). This appendix gives a brief outline of the Laplace transform method. The basic idea is to transform a differential equation into a polynomial type of equation and then use the inverse Laplace transform to find the solution of the original problem.

D.1 Definition

If a function $f(t)$ is defined for all positive values of time, $t \geq 0$, the Laplace transform of $f(t)$, denoted, $\mathscr{L}[f(t)]$, is defined as

$$\mathscr{L}[f(t)] = F(s) = \int_0^\infty e^{-st} f(t)\, dt \tag{D.1}$$

where e^{-st} is known as the *kernel* of the transform and s is a subsidiary variable, called the *Laplace variable*, that is in general a complex quantity.

Notation: The original functions depend on t and their Laplace transforms depend on s. The original functions are denoted by lower-case letters while their Laplace transforms are denoted by the same letters in capitals. For example, the Laplace transforms of $f(t)$ and $y(t)$ are denoted as $F(s)$ and $Y(s)$, respectively. The Laplace transform of $\alpha f(t)$ where α is a constant is given by $\alpha F(s)$. Similarly, the Laplace transform of a linear sum of two functions $f_1(t)$ and $f_2(t)$, $\alpha_1 f_1(t) + \alpha_2 f_2(t)$, is given by

$$\mathcal{L}[\alpha_1 f_1(t) + \alpha_2 f_2(t)] = \alpha_1 F_1(s) + \alpha_2 F_2(s) \tag{D.2}$$

Inverse Laplace Transform:

To find the original function $f(t)$ from the transformed function $F(s)$, we need to use the inverse Laplace transform defined as

$$\mathcal{L}^{-1}[F(s)] = \frac{1}{2\pi i} \int_{s=\sigma-i\infty}^{\sigma+i\infty} F(s)e^{st} \ ds = f(t)u(t) \tag{D.3}$$

where $u(t)$ is the unit step function defined as

$$u(t) = \begin{cases} 1 & \text{for} \quad t > 0 \\ 0 & \text{for} \quad t < 0 \end{cases} \tag{D.4}$$

and σ is a value selected to the right of all the singularities of $F(s)$ in the s-plane. In practice, Eq. (D.3) is rarely used. Instead, Laplace transforms of complex expressions are broken down into simpler ones for which the inverse Laplace transforms can be found from Laplace transform tables.

D.2 Transform of Derivatives

While applying Laplace transforms to solve differential equations, we need to use the Laplace transforms of various order derivatives of a function. The Laplace transform of the first derivative of $f(t)$ is defined as

$$\mathcal{L}\left[\frac{df(t)}{dt}\right] = \int_0^\infty e^{-st}\frac{df(t)}{dt}dt \tag{D.5}$$

By using integration by parts, Eq. (D.5) can be expressed as

$$\mathcal{L}\left[\frac{df(t)}{dt}\right] = e^{-st}f(t)|_0^\infty - \int_0^\infty (-se^{-st})f(t) \ dt = -f(0) + sF(s) \tag{D.6}$$

where $f(0)$ is the initial value of $f(t)$—namely, the value of $f(t = 0)$. Using a similar approach, the Laplace transform of the second derivative of $f(t)$ is defined as

$$\mathscr{L}\left[\frac{d^2 f(t)}{dt^2}\right] = \int_0^\infty e^{-st} \frac{d^2 f(t)}{dt^2} dt \tag{D.7}$$

Equation (D.7) can be simplified to obtain

$$\mathscr{L}\left[\frac{d^2 f(t)}{dt^2}\right] = -\dot{f}(0) - sf(0) + s^2 F(s) \tag{D.8}$$

where $\dot{f}(0)$ is the value of $\dfrac{df}{dt}$ at $t = 0$. By using a similar approach, the Laplace transform of the nth derivative of $f(t)$ can be found as

$$\mathscr{L}\left[\frac{d^n f(t)}{dt^n}\right] \equiv \mathscr{L}[f^{(n)}] = s^n F(s) - s^{n-1} f(0) - s^{n-2} f^{(1)}(0) - \ldots - sf^{(n-1)}(0) \tag{D.9}$$

where $f^{(n)}$ is used to denote the nth derivative of f, $\dfrac{d^n f}{dt^n}$.

D.3　Shifting Theorems

In some applications, the function $f(t)$ appears together with the term e^{at} as $f(t)\, e^{at}$, where a is a real or complex number. The Laplace transform of this product, $F_1(s)$, is given by

$$F_1(s) = \mathscr{L}[f(t)e^{at}] = \int_0^\infty \{f(t)e^{at}\}e^{-st}\, dt = \int_0^\infty f(t)e^{-(s-a)}\, dt \equiv F(s-a) \tag{D.10}$$

Thus we have

$$\mathscr{L}[f(t)e^{at}] = F(s-a) \tag{D.11}$$

This shows that the effect of multiplying the function $f(t)$ by e^{at} is to shift the Laplace transform of $f(t)$ by the amount a in the s-domain. The result indicated by Eq. (D.11) is known as the *shifting theorem*.

D.4　Method of Partial Fractions

In some problems, the function $F(s)$ is of the form

$$F(s) = \frac{B(s)}{A(s)} \tag{D.12}$$

where $B(s)$ and $A(s)$ are polynomials in s and the degree of $A(s)$ is usually higher than that of $B(s)$. The inverse transform of $F(s)$ can be simplified if the right-hand side of Eq. (D.12) is expanded into partial fractions. In order to apply the partial fraction expansion technique, we need to know the roots of the denominator polynomial, $A(s)$, first.

When all the roots of $A(s)$ are distinct: Let $-p_1, -p_2, \ldots, -p_n$ be the roots, also called the *poles*, of $A(s)$. Then Eq. (D.12) can be expressed as

$$F(s) = \frac{B(s)}{A(s)} = \frac{a_1}{s + p_1} + \frac{a_2}{s + p_2} + \cdots + \frac{a_n}{s + p_n} \tag{D.13}$$

where a_k is an unknown constant, called the *residue*, at the pole $s = -p_k, k = 1, 2, \ldots, n$. The value of a_k can be determined by multiplying both sides of Eq. (D.13) by $(s + p_k)$ and letting $s = -p_k$ so that

$$a_k = \left\{ (s + p_k)\frac{B(s)}{A(s)} \right\}_{s=-p_k} \tag{D.14}$$

Note that $f(t)$ is a real function of time t; if the roots p_1 and p_2 of $A(s)$ are complex conjugates, then the corresponding residues or constants a_1 and a_2 will also be complex conjugates. Once the partial fraction expansion of $F(s)$, Eq. (D.13), is known, the inverse Laplace transform of $F(s)$ can be determined using relations of the type

$$\mathcal{L}^{-1}\left[\frac{a_k}{s + p_k}\right] = a_k e^{-p_k t} \tag{D.15}$$

Thus $f(t)$ can be found as

$$f(t) = \mathcal{L}^{-1}[F(s)] = a_1 e^{-p_1 t} + a_2 e^{-p_2 t} + \cdots + a_n e^{-p_n t} \tag{D.16}$$

When $A(s)$ involves multiple roots: Let the polynomial $A(s)$ have a multiple root of order k at $s = -p_1$ which implies that $F(s)$ has a pole of order k at $-p_1$ in addition to the simple roots (or poles) at $-p_2, -p_3, \ldots, -p_n$ considered above. Thus $A(s)$ can be expressed as

$$A(s) = (s + p_1)^k(s + p_2)(s + p_3)\ldots(s + p_n) \tag{D.17}$$

The partial fraction expansion of $F(s)$ is written in the form

$$F(s) = \frac{B(s)}{A(s)} = \frac{a_{11}}{(s + p_1)^k} + \frac{a_{12}}{(s + p_1)^{k-1}} + \cdots$$
$$+ \frac{a_{1k}}{(s + p_1)} + \frac{a_2}{s + p_2} + \frac{a_3}{s + p_3} + \cdots + \frac{a_n}{s + p_n} \tag{D.18}$$

It can be verified that the constants $a_{11}, a_{12}, \ldots, a_{1k}$ can be determined as

$$a_{1r} = \frac{1}{(1 - r)!} \frac{d^{r-1}}{ds^{r-1}} [(s + p_1)^k F(s)]_{s=-p_1}; \qquad r = 1, 2, \ldots, k \qquad (D.19)$$

Noting that

$$\mathcal{L}[t^{r-1}] = \frac{(r - 1)!}{s^r} \qquad (D.20)$$

the inverse Laplace transform of the terms due to the higher-order pole can be obtained, using the shifting theorem, as

$$\mathcal{L}^{-1}\left[\frac{1}{(s + p_1)^r}\right] = \frac{t^{r-1}}{(r - 1)!} e^{-p_1 t} \qquad (D.21)$$

Thus the inverse Laplace transform of $F(s)$ given by Eq. (D.18) becomes

$$f(t) = \mathcal{L}^{-1}[F(s)] = \left[a_{11} \frac{t^{k-1}}{(k - 1)!} + a_{12} \frac{t^{k-2}}{(k - 2)!} + \cdots + a_{1k} \right] e^{-p_1 t}$$
$$+ a_2 e^{-p_2 t} + \cdots + a_n e^{-p_n t} \qquad (D.22)$$

D.5   Convolution Integral

Let two functions $f_1(t)$ and $f_2(t)$, defined for $t > 0$, possess the Laplace transforms $F_1(s)$ and $F_2(s)$, respectively. Then consider the function $f(t)$ defined in several ways as

$$f(t) = f_1(t) * f_2(t) = \int_0^t f_1(\tau) f_2(t - \tau) \, d\tau = \int_0^\infty f_1(\tau) f_2(t - \tau) \, d\tau \qquad (D.23)$$

This function $f(t)$ is called the *convolution* of the functions $f_1(t)$ and $f_2(t)$ over the interval $0 < t < \infty$. Note that the upper limits of the integrals in Eq. (D.23) are interchangeable because $f_2(t - \tau) = 0$ for $\tau > t$, which is the same as $t - \tau < 0$. The Laplace transform of Eq. (D.23) can be expressed as

$$F(s) = \mathcal{L}[f_1(t) * f_2(t)] = F_1(s) F_2(s) \qquad (D.24)$$

where

$$F_1(s) = \int_0^\infty e^{-s\tau} f_1(\tau)\, d\tau, \qquad F_2(s) = \int_0^\infty e^{-s\sigma} f_2(\sigma)\, d\sigma \qquad \text{(D.25)}$$

The inverse Laplace transform of Eq. (D.24) is given by

$$f(t) = \mathcal{L}^{-1}[F(s)] = \mathcal{L}^{-1}[F_1(s)F_2(s)] \qquad \text{(D.26)}$$

$$= \int_0^t f_1(\tau) f_2(t-\tau)\, d\tau = \int_0^t f_1(t-\tau) f_2(\tau)\, d\tau$$

EXAMPLE   D.1

Expand the function $F(s) = \dfrac{4s + 7}{(s + 3)(s + 4)^2}$ using partial fractions.

Solution The partial fraction expansion of the function $F(s)$ can be expressed as

$$F(s) = \frac{4s + 7}{(s + 3)(s + 4)^2} = \frac{C_1}{(s + 3)} + \frac{C_2}{(s + 4)} + \frac{C_3}{(s + 4)^2} \qquad \text{(E.1)}$$

where the constants C_i, $i = 1, 2, 3$, can be evaluated as

$$C_1 = (s + 3)F(s)\Big|_{s=-3} = \frac{4s + 7}{(s + 4)^2}\Big|_{s=-3} = -5 \qquad \text{(E.2)}$$

$$C_2 = \frac{d}{dx}[(s + 4)^2 F(s)]\Big|_{s=-4} = \frac{d}{dx}\left[\frac{4s + 7}{s + 3}\right]\Big|_{s=-4} = \frac{(s + 3)4 - (4s + 7)1}{(s + 3)}\Big|_{s=-4} = 5 \quad \text{(E.3)}$$

$$C_3 = [(s + 4)^2 F(s)]\Big|_{s=-4} = \left[\frac{4s + 7}{s + 3}\right]\Big|_{s=-4} = 9 \qquad \text{(E.4)}$$

Thus the partial fraction expansion of $F(s)$ is given by

$$F(s) = \frac{4s + 7}{(s + 3)(s + 4)^4} = -\frac{5}{(s + 3)} + \frac{5}{(s + 4)} - \frac{9}{(s + 4)^2} \qquad \text{(E.5)}$$

The partial fraction expansion can also be carried out using MATLAB's residue function as follows:

```
num = [4  7];              % coefficients of polynomial in the numerator
den = conv([1 3], [1 8 8]);  % coefficients of the two polynomials in the denominator
[r, p, k] = residue(num, den)  % compute residue and print the result
R = [-5 5 9]', p = [-3 - 4 - 4], and k = [].
```

This result can be seen to be the same as the one computed above.

A short table of Laplace transform pairs is given on the next page.

∎

Laplace Transform Pairs

Laplace Domain	**Time Domain** $f(t)$
$$F(s) = \int_0^\infty f(t)e^{-st}\,dt$$	

---

1. $c_1 F(s) + c_2 G(s)$	$c_1 f(t) + c_2 g(t)$
2. $F\!\left(\dfrac{s}{a}\right)$	$f(a \cdot t)a$
3. $F(s)G(s)$	$\displaystyle\int_0^t f(t-\tau)g(\tau)\,d\tau = \int_0^t f(\tau)g(t-\tau)\,d\tau$
4. $s^n F(s) - \displaystyle\sum_{j=1}^{n} s^{n-j}\dfrac{d^{j-1}f}{dt^{j-1}}(0)$	$\dfrac{d^n f}{dt^n}(t)$
5. $\dfrac{1}{s^n}F(s)$	$\underbrace{\displaystyle\int_0^t \cdots \int_0^t f(\tau)\,d\tau \cdots d\tau}_{n}$
6. $F(s+a)$	$e^{-at}f(t)$
7. $\dfrac{1}{s^{n+1}}$	$t^n;\ n = 1, 2, \ldots, t$
8. $\dfrac{1}{s+a}$	e^{-at}
9. $\dfrac{1}{(s+a)^2}$	te^{-at}
10. $\dfrac{a}{s(s+a)}$	$1 - e^{-at}$
11. $\dfrac{s+a}{s^2}$	$1 + at$
12. $\dfrac{a^2}{s^2(s+a)}$	$at - (1 - e^{-at})$
13. $\dfrac{s+b}{s(s+a)}$	$\dfrac{b}{a}\left\{1 - \left(1 - \dfrac{a}{b}\right)e^{-at}\right\}$
14. $\dfrac{a}{s^2 + a^2}$	$\sin at$
15. $\dfrac{s}{s^2 + a^2}$	$\cos at$
16. $\dfrac{a^2}{s(s^2 + a^2)}$	$1 - \cos at$
17. $\dfrac{s}{s^2 - a^2}$	$\cosh at$
18. $\dfrac{a}{s^2 - a^2}$	$\sinh at$
19. $\dfrac{a(s^2 - a^2)}{(s^2 + a^2)^2}$	$at \cos at$

Laplace Transform Pairs

Laplace Domain $$F(s) = \int_0^\infty f(t)e^{-st}\,dt$$	Time Domain $f(t)$
20. $\dfrac{2sa^2}{(s^2 + a^2)^2}$	$at \sin at$
21. $\dfrac{s + a}{(s + a)^2 + b^2}$	$e^{-at} \cos bt$
22. $\dfrac{b}{(s + a)^2 + b^2}$	$e^{-at} \sin bt$
23. $\dfrac{1}{(s + a)(s + b)}$	$\dfrac{e^{-at} - e^{-bt}}{(b - a)}$
24. $\dfrac{s + w}{(s + a)(s + b)}$	$\dfrac{\{(w - a)e^{-at} - (w - b)e^{-bt}\}}{(b - a)}$
25. $\dfrac{1}{(s + a)(s + b)(s + c)}$	$\dfrac{e^{-at}}{(b - a)(c - a)} + \dfrac{e^{-bt}}{(a - b)(c - b)} + \dfrac{e^{-ct}}{(a - c)(b - c)}$
26. $\dfrac{s + w}{(s + a)(s + b)(s + c)}$	$\dfrac{(w - a)e^{-at}}{(b - a)(c - a)} + \dfrac{(w - b)e^{-bt}}{(a - b)(c - b)} + \dfrac{(w - c)e^{-ct}}{(a - c)(b - c)}$
27.* $\dfrac{1}{s^2 + 2\zeta\omega_n s + \omega_n^2}$	$\dfrac{1}{\omega_d} e^{-\zeta\omega_n t} \sin \omega_d t$
28. $\dfrac{s}{s^2 + 2\zeta\omega_n s + \omega_n^2}$	$-\dfrac{\omega_n}{\omega_d} e^{-\zeta\omega_n t} \sin(\omega_d t - \phi_1)$
29.* $\dfrac{s + 2\zeta\omega_n s}{s^2 + 2\zeta\omega_n s + \omega_n^2}$	$\dfrac{\omega_n}{\omega_d} e^{-\zeta\omega_n t} \sin(\omega_d t + \phi_1)$
30.* $\dfrac{\omega_n^2}{s(s^2 + 2\zeta\omega_n s + \omega_n^2)}$	$1 - \dfrac{\omega_n}{\omega_d} e^{-\zeta\omega_n t} \sin(\omega_d t + \phi_1)$
31.* $\dfrac{s + \zeta\omega_n}{s(s^2 + 2\zeta\omega_n s + \omega_n^2)}$	$e^{-\zeta\omega_n t} \sin(\omega_d t + \phi_1)$
32. 1	Unit impulse at $t = 0$
33. $\dfrac{e^{-as}}{s}$	Unit step function at $t = a$

*$\omega_d = \omega_n\sqrt{1 - \zeta^2}$; $\zeta < 1$

$\phi_1 = \cos^{-1}\zeta$; $\zeta < 1$

Heinrich Rudolf Hertz (1857–1894), a German physicist and a professor of physics at the Polytechnic Institute in Karlsruhe and later at the University of Bonn, gained fame through his experiments on radio waves. His investigations in the field of elasticity formed a relatively small part of his achievements but are of vital importance to engineers. His work on the analysis of elastic bodies in contact is referred to as "Hertzian stresses," and is very important in the design of ball and roller bearings. The unit of frequency of periodic phenomena, measured in cycles per second, is named hertz in SI units. (Photo courtesy of *Applied Mechanics Reviews*.)

APPENDIX E

Units

The English system of units is now being replaced by the International System of units (SI). The SI system is the modernized version of the metric system of units. Its name in French is *Système International*; hence the abbreviation *SI*. The SI system has seven basic units. All other units can be derived from these seven [E.1–E.2]. The three basic units of concern in the study of vibrations are meter for length, kilogram for mass, and second for time.

The common prefixes for multiples and submultiples of SI units are given in Table E.1. In the SI system, the combined units must be abbreviated with care. For example, a torque of 4 N × 2 m must be stated as 8 N m or 8 N · m with either a space or a dot between N and m. It should not be written as Nm. Another example is 8 m × 5 s = 40 m s or 40 m · s or 40 meter-seconds. If it is written as 40 ms, it means 40 milliseconds.

Conversion of Units

To convert the units of any given quantity from one system to another, we use the equivalence of units given in Table E.2. The following examples illustrate the procedure.

TABLE E.1 Prefixes for multiples and submultiples of SI units

Multiple	Prefix	Symbol	Submultiple	Prefix	Symbol
10	deka	da	10^{-1}	deci	d
10^2	hecto	h	10^{-2}	centi	c
10^3	kilo	k	10^{-3}	milli	m
10^6	mega	M	10^{-6}	micro	μ
10^9	giga	G	10^{-9}	nano	n
10^{12}	tera	T	10^{-12}	pico	p

TABLE E.2 Conversion of units

Quantity	SI Equivalence	English Equivalence
Mass	$1\ \text{lb}_\text{f}\text{-sec}^2/\text{ft (slug)} = 14.5939\ \text{kg}$	$1\ \text{kg} = 2.204623\ \text{lb}_\text{m}$
	$= 32.174\ \text{lb}_\text{m}$	
	$1\ \text{lb}_\text{m} = 0.45359237\ \text{kg}$	$= 0.06852178\ \text{slug}$
		$(\text{lb}_\text{f}\text{-sec}^2/\text{ft})$
Length	$1\ \text{in.} = 0.0254\ \text{m}$	$1\ \text{m} = 39.37008\ \text{in.}$
	$1\ \text{ft} = 0.3048\ \text{m}$	$= 3.28084\ \text{ft}$
	$1\ \text{mile} = 5,280\ \text{ft} = 1.609344\ \text{km}$	$1\ \text{km} = 3,280.84\ \text{ft} = 0.621371\ \text{mile}$
Area	$1\ \text{in.}^2 = 0.00064516\ \text{m}^2$	$1\ \text{m}^2 = 1,550.0031\ \text{in.}^2$
	$1\ \text{ft}^2 = 0.0929030\ \text{m}^2$	$= 10.76391\ \text{ft}^2$
Volume	$1\ \text{in.}^3 = 16.3871 \times 10^{-6}\ \text{m}^3$	$1\ \text{m}^3 = 61.0237 \times 10^3\ \text{in.}^3$
	$1\ \text{ft}^3 = 28.3168 \times 10^{-3}\ \text{m}^3$	$= 35.3147\ \text{ft}^3$
	$1\ \text{US gallon} = 3.7853\ \text{litres}$	$= 10^3\ \text{litres} = 0.26418\ \text{US gallon}$
	$= 3.7853 \times 10^{-3}\ \text{m}^3$	
Force or weight	$1\ \text{lb}_\text{f} = 4.448222\ \text{N}$	$1\ \text{N} = 0.2248089\ \text{lb}_\text{f}$
Torque or moment	$1\ \text{lb}_\text{f}\text{-in.} = 0.1129848\ \text{N}\cdot\text{m}$	$1\ \text{N}\cdot\text{m} = 8.850744\ \text{lb}_\text{f}\text{-in.}$
	$1\ \text{lb}_\text{f}\text{-ft} = 1.355818\ \text{N}\cdot\text{m}$	$= 0.737562\ \text{lb}_\text{f}\text{-ft}$
Stress, pressure, or	$1\ \text{lb}_\text{f}/\text{in.}^2\ (\text{psi}) = 6,894.757\ \text{Pa}$	$1\ \text{Pa} = 1.450377 \times 10^{-4}\ \text{lb}_\text{f}/\text{in.}^2\ (\text{psi})$
elastic modulus	$1\ \text{lb}_\text{f}/\text{ft}^2 = 47.88026\ \text{Pa}$	$= 208.8543 \times 10^{-4}\ \text{lb}_\text{f}/\text{ft}^2$
Mass density	$1\ \text{lb}_\text{m}/\text{in.}^3 = 27.6799 \times 10^3\ \text{kg/m}^3$	$1\ \text{kg/m}^3 = 36.127 \times 10^{-6}\ \text{lb}_\text{m}/\text{in.}^3$
	$1\ \text{lb}_\text{m}/\text{ft}^3 = 16.0185\ \text{kg/m}^3$	$= 62.428 \times 10^{-3}\ \text{lb}_\text{m}/\text{ft}^3$

EXAMPLE E.1

Mass moment of inertia:

$$\begin{pmatrix} \text{Mass moment of} \\ \text{inertia in SI units} \end{pmatrix} = \begin{pmatrix} \text{Mass moment of inertia} \\ \text{in English units} \end{pmatrix} \times \begin{pmatrix} \text{Multiplication} \\ \text{factor} \end{pmatrix}$$

$$(\text{kg} \cdot \text{m}^2) \equiv (\text{N} \cdot \text{m} \cdot \text{s}^2) = \left(\frac{\text{N}}{\text{lb}_f} \cdot \text{lb}_f \right) \left(\frac{\text{m}}{\text{in.}} \cdot \text{in.} \right) (\text{sec}^2)$$

$$= (\text{N per 1 lb}_f)(\text{m per 1 in.})(\text{lb}_f\text{-in.-sec}^2)$$

$$= (4.448222)(0.0254)(\text{lb}_f\text{-in.-sec}^2)$$

$$= 0.1129848 \ (\text{lb}_f\text{-in.-sec}^2)$$

■

EXAMPLE E.2

Stress:

$$(\text{Stress in SI units}) = (\text{Stress in English units}) \times \begin{pmatrix} \text{Multiplication} \\ \text{factor} \end{pmatrix}$$

$$(\text{Pa}) \equiv (\text{N/m}^2) = \left(\frac{\text{N}}{\text{lb}_f} \cdot \text{lb}_f \right) \frac{1}{\left(\dfrac{\text{m}}{\text{in.}} \cdot \text{in.} \right)^2} = \frac{\text{N}}{\text{lb}_f} \frac{1}{\left(\dfrac{\text{m}}{\text{in.}} \right)^2} (\text{lb}_f/\text{in.}^2)$$

$$= \frac{(\text{N per 1 lb}_f)}{(\text{m per 1 in.})^2} (\text{lb}_f/\text{in.}^2)$$

$$= \frac{(4.448222)}{(0.0254)^2} (\text{lb}_f/\text{in.}^2)$$

$$= 6,894.757 \ (\text{lb}_f/\text{in.}^2)$$

■

REFERENCES

E.1   E. A. Mechtly, "The International System of Units" (2nd rev. ed.), NASA SP-7012, 1973.

E.2   C. Wandmacher, *Metric Units in Engineering—Going SI*, Industrial Press, New York, 1978.

Thomas Young (1773–1829) was a British physicist and physician who introduced Young's modulus and principle of interference of light. He studied medicine and received his M.D. in 1796. He was appointed as a professor of natural philosophy at the Royal Institution in 1801, but he resigned in 1803 as his lectures were disappointing to popular audiences. He joined St. George's hospital in London as a physician in 1811 and continued there until his death. Young made many contributions to mechanics. He was the first to use the terms "energy" and "labor expended" (i.e., work done) for the quantities mv^2 and Fx, respectively, where m is the mass of the body, v is its velocity, F is a force, and x is the distance by which F is moved, and to state that the two terms are proportional to one another. He defined the term "*modulus*" (which has become known as "*Young's modulus*") as the weight which would double the length of a rod of unit cross section.

APPENDIX F

Introduction to MATLAB

MATLAB, derived from *MATrix LABoratory*, is a software package that can be used for the solution of a variety of scientific and engineering problems including linear algebraic equations, nonlinear equations, numerical differentiation and integration, curve fitting, ordinary and partial differential equations, optimization, and graphics. It uses matrix notation extensively; in fact, the only data type in MATLAB is a complex-valued matrix. Thus it handles scalars, vectors, and real- and integer-valued matrices as special cases of complex matrices. The software can be used to execute a single statement or a list of statements, called a *script file*. MATLAB provides excellent graphing and programming capabilities. It can also be used to solve many types of problems symbolically. Simple computations can be done by entering an instruction, similar to what we do on a calculator, at the prompt. The symbols to be used for the basic arithmetic operations of addition, subtraction, multiplication, division, and exponentiation are $+$, $-$, $*$, $/$, and $\char94$, respectively. In any expression, the computations are performed from left to right, with exponentiation having the highest priority, followed by multiplication and division (with equal priority) and then addition and subtraction (with equal priority). It uses the symbol *log* to denote the natural logarithm (ln). MATLAB uses double precision during computations but prints results on the screen using a shorter format. This default can be changed by using the *format* command.

F.1   Variables

Variable names in MATLAB should start with a letter and can have a length of up to 31 characters in any combination of letters, digits, and underscores. Uppercase and lowercase letters are treated separately. As stated earlier, MATLAB treats all variables as matrices, although scalar quantities need not be given as arrays.

F.2   Arrays and Matrices

The name of a matrix must start with a letter and may be followed by any combination of letters or digits. The letters may be uppercase or lowercase. Before performing arithmetic operations such as addition, subtraction, multiplication, and division on matrices, the matrices must be created using statements such as the following:

Row vector

$$\gg A = \begin{bmatrix} 1 & 2 & 3 \end{bmatrix};$$

A row vector is treated as a 1-by-n matrix; its elements are enclosed in brackets and are separated by either spaces or commas. Note that the command-line prompt in the professional version of MATLAB is $\gg$ while it is EDU $\gg$ in the student edition of MATLAB. If a semicolon is not put at the end of the line, MATLAB displays the results of the line on the screen.

Column vector

$$\gg A = \begin{bmatrix} 1 \\ 2 \\ 3 \end{bmatrix} \quad \text{or} \quad A = \begin{bmatrix} 1; & 2; & 3 \end{bmatrix} \quad \text{or} \quad A = \begin{bmatrix} 1 & 2 & 3 \end{bmatrix}';$$

A column vector is treated as an n-by-1 matrix. Its elements can be entered in different lines, in a single line using a semicolon to separate them, or in a single line using a row vector with a prime on the right bracket (to denote the transpose).

Matrix

To define the matrix

$$[A] = \begin{bmatrix} 1 & 2 & 3 \\ 4 & 5 & 6 \\ 7 & 8 & 6 \end{bmatrix}$$

the following specification can be used:

$$\gg A = \begin{bmatrix} 1 & 2 & 3 \\ 4 & 5 & 6 \\ 7 & 6 & 9 \end{bmatrix} \quad \text{or} \quad A = \begin{bmatrix} 1 & 2 & 3; & 4 & 5 & 6; & 7 & 8 & 9 \end{bmatrix};$$

F.3   Arrays with Special Structure

In some cases, the special structure of an array is used to specify the array in a simpler manner. For example, $A = 1:10$ denotes a row vector

$$A = \begin{bmatrix} 1 & 2 & 3 & 4 & 5 & 6 & 7 & 8 & 9 & 10 \end{bmatrix}$$

and $A = 2:0.5:4$ represents the row vector

$$A = \begin{bmatrix} 2.5 & 3.0 & 3.5 & 4.0 \end{bmatrix}$$

F.4   Special Matrices

Some of the special matrices are identified as follows:
$\gg A =$ eye (3); implies an identity matrix of order 3,

$$A = \begin{bmatrix} 1 & 0 & 0 \\ 0 & 1 & 0 \\ 0 & 0 & 1 \end{bmatrix}$$

$\gg A =$ ones (3); implies a square matrix of order 3 with all elements equal to one,

$$A = \begin{bmatrix} 1 & 1 & 1 \\ 1 & 1 & 1 \\ 1 & 1 & 1 \end{bmatrix}$$

$\gg A =$ zeros (2, 3); implies a 2 × 3 matrix with all elements equal to zero,

$$A = \begin{bmatrix} 0 & 0 & 0 \\ 0 & 0 & 0 \end{bmatrix}$$

F.5   Matrix Operations

To add the matrices $[A]$ and $[B]$ to get $[C]$, we use the statement

$$\gg C = A + B;$$

To solve a system of linear equations $[A]\vec{X} = \vec{B}$, we define the matrix A and the vector B and use the following statement:

$$\gg X = A \backslash B$$

F.6   Functions in MATLAB

MATLAB has a large number of built-in functions, such as the following:

Square root of x: sqrt(x)

Sine of x: sin(x)

Logarithm of x to base 10: log10(x)

Gamma function of x: gamma(x)

To generate a new vector y having 11 values given by the function $y = e^{-2x} \cos x$ with $x = 0.0, 0.1, \ldots, 1.0$, we type the following:

$\gg x = [0{:}0.1{:}1];$

$\gg y = \exp(-2{*}x){.}{*}\cos(x){.};$

F.7   Complex Numbers

MATLAB considers complex-number algebra automatically. The symbol i or j can be used to represent the imaginary part with no need of an asterisk between i or j and a number. For example, $a = 1 - 3i$ is a complex number with real and imaginary parts equal to 1 and -3, respectively. The magnitude and angle of a complex number can be found using the statements

```
>> a = 1 - 3i;
>> abs (a)
ans =
      ...
>> angle (a)
ans =
      ... (in radians)
```

F.8   M-files

MATLAB can be used in an interactive mode by typing each command from the keyboard. In this mode, MATLAB performs the operations much like an extended calculator. However, there are situations in which this mode of operation is inefficient. For example, if the same set of commands is to be repeated a number of times with different values of the input parameters, developing a MATLAB program will be quicker and more efficient.

A MATLAB program consists of a sequence of MATLAB instructions written outside MATLAB and then executed in MATLAB as a single block of commands. Such a program is called a *script file* or *M-file*. It is necessary to give a name to the script file. The name should end with **.m** (a dot (.) followed by the letter **m**). A typical *M*-file (called **fibo.m**) is given below:

```
file "fibo.m"
% m-file to compute Fibonacci numbers
f=[1 1];
```

```
i=1;
while f(i)+f(i+1)<1000
    f(i+2)=f(i)+f(i+1);
    i=i+1;
end
```

An *M*-file can also be used to write function subroutines. For example, the solution of a quadratic equation

$$Ax^2 + Bx + C = 0$$

can be determined using the following program:

```
% roots_quadra.m (Note: Line starting with % denotes a
comment line)
function [x1, x2] = roots_quadra(A, B, C)
% det = determinant
det = ^2 - 4 * A * C;
if (det < 0.0);
    x1 = (-B + j * sqrt(-det))/(2 * A);
    x2 = (-B - j * sqrt(-det))/(2 * A);
    disp('Roots are complex conjugates');
elseif (abs(det) < 1e-8); % det = 0.0
    x1 = -B / (2 * A);
    x2 = -B / (2 * A);
    disp('Roots are identical');
else (det > 0);
    x1 = (-B + sqrt(det))/(2 * A);
    x2 = (-B - sqrt(det))/(2 * A);
    disp('Roots are real and distinct');
end
```

The program **roots_quadra.m** can be used to find the roots of a quadratic with $A = 2$, $B = 2$, and $C = 1$—for example, as follows:

```
>> [x1,x2]=roots_quadra(2, 2, 1)
Roots are complex conjugates

x1 =
   -0.5000 + 0.5000i

x2 =
   -0.5000 - 0.5000i
```

F.9 Plotting of Graphs

To plot a graph in MATLAB, we define a vector of values of the independent variable x (array x) and a vector of values of the dependent variable y corresponding to the values of x (array y). Then the x-y graph can be plotted using the command:

```
plot (x,y)
```

As an example, the following commands can be used to plot the function $y = x^2 + 1$ in the range $0 \le x \le 3$:

```
x = 0 : 0.2 : 3;
y = x² + 1;
plot (x,y);
hold on
x1 = [0 3];
y1 = [0 0];
plot (x1,y1);
grid on
hold off
```

Note that the first two lines are used to generate the arrays x and y (using increments of 0.2 for x); the third line plots the graph (using straight lines between the indicated points); the next six lines permit the plotting of x- and y-axes along with the setting up of the grid (using the **grid on** command).

F.10   Roots of Nonlinear Equations

To find the roots of a nonlinear equation, the MATLAB function **fzero(y,x1)** can be used. Here **y** defines the nonlinear function and **x1** denotes the initial estimate (starting value) of the root. The roots of polynomials can be determined using the function **roots(p)**, where **p** is a row vector of coefficients of the polynomial in descending order of the power of the variable.

```
>> f='tan (x)-tanh (x)'

f =

tan (x)-tanh (x)

>> root=fzero(f,1.0)

root =

    1.5708

>> roots([1 0 0 0 0 0 -2])

ans =

   -1.1225
   -0.5612 +0.9721i
   -0.5612 -0.9721i
    0.5612 +0.9721i
    0.5612 -0.9721i
    1.1225

>>
```

F.11   Solution of Linear Algebraic Equations

A set of simultaneous linear algebraic equations $[A]\vec{x} = \vec{b}$ can be solved using MATLAB in two different ways: Find $\vec{x}$ as $[A]^{-1}\vec{b}$ or find $\vec{x}$ directly as indicated by the following examples:

```
>> A=[4 -3 2; 2 3 1; 5 4 7]
A =
         4     -3      2
         2      3      1
         5      4      7

>> b=[16; -1; 18]

b =

        16
        -1
        18
>> C=inv(A)

C =

    0.2099     0.3580    -0.1111
   -0.1111     0.2222     0.0000
   -0.0864    -0.3827     0.2222

>> x=C*b

x =

1.0000
-2.0000
3.0000

>> x=A\b

x =

1.0000
-2.0000
3.0000

>>
```

F.12   Solution of Eigenvalue Problem

An algebraic eigenvalue problem is defined by $[A]\vec{X} = \lambda\vec{X}$, where $[A]$ is a square matrix of size $n \times n$, $\vec{X}$ is a column vector of size n, and λ is a scalar. For any given matrix $[A]$, the solution can be found using two types of commands. The use of the command **b = eig(A)** gives the eigenvalues of the matrix $[A]$ as elements of the vector d. The use of the command **[V,D] = eig(A)** gives the eigenvalues as diagonal elements of the matrix $[D]$

and the eigenvectors as corresponding columns of the matrix $[V]$. The following example illustrates the procedure:

```
>> A=[2 1 3 4; 1 -3 1 5; 3 1 6 -2; 4 5 -2 -1]

A =

          2     1     3     4
          1    -3     1     5
          3     1     6    -2
          4     5    -2    -1

>> b=eig(A)

b=

7.9329
5.6689
-1.5732
-8.0286

>> [V, d] = eig(A)

V =

0.5601     0.3787     0.6880     0.2635
0.2116     0.3624    -0.6241     0.6590
0.7767    -0.5379    -0.2598    -0.1996
0.1954     0.6602    -0.2638    -0.6756

d =

7.9329          0          0          0
     0     5.6689          0          0
     0          0    -1.5732          0
     0          0          0    -8.0286

>>
```

F.13   Solution of Differential Equations

MATLAB has several functions or solvers, based on the use of Runge-Kutta methods, that can be used for the solution of a system of first-order ordinary differential equations. Note that an nth-order ordinary differential equation is to be converted into a system of n first-order ordinary differential equations before using MATLAB functions. The MATLAB function `ode23` implements a combination of second- and third-order Runge-Kutta methods, while the function `ode45` is based on a combination of fourth- and fifth-order Runge-Kutta methods. To solve a system of first-order differential equations $\dot{y} = f(t, y)$ using the MATLAB function `ode23`, the following command can be used:

```
>>[t,y] = ode('dfunc',tspan,y0)
```

where `'dfunc'` is the name of the function **m-file** whose input must be **t** and **y** and whose output must be a column vector denoting dy/dt—that is, **f(t,y)**. The number of rows in the column vector must be equal to the number of first-order equations. The vector **tspan** should contain the initial and final values of the independent variable **t** and, optionally, any intermediate values of **t** at which the solution is desired. The vector **y0** should contain the initial values of **y(t)**. Note that the function **m-file** should have two input arguments **t** and **y** even if the function **f(t,y)** does not involve **t**. A similar procedure can be used with the MATLAB function **ode45**.

As an example, consider the solution of the differential equation with $c = 0.1$ and $k = 10.0$:

$$\frac{d^2y}{dt^2} + c\,\frac{dy}{dt} + ky = 0; \qquad y(0) = 0, \qquad \frac{dy}{dt}(0) = 0$$

This equation can be written as a set of two first-order differential equations by introducing

$$y_1 = y$$

and

$$y_2 = \frac{dy}{dt} = \frac{dy_1}{dt}$$

as

$$\frac{d\vec{y}}{dt} = \vec{f} = \begin{Bmatrix} f_1(t,\vec{y}) \\ f_2(t,\vec{y}) \end{Bmatrix} = \begin{Bmatrix} y_2 \\ -cy_2 - ky_1 \end{Bmatrix}$$

with

$$\vec{y}(0) = \begin{Bmatrix} 1 \\ 0 \end{Bmatrix}$$

The following MATLAB program finds the solution of the above differential equations:

```
% ProbappendixF.m
tspan = [0: 0.05: 3];
y0 = [1; 0];
[t,y] = ode23 ('dfunc', tspan, y0);
[t y]
plot (t, y(:,1));
xlabel ('t');
ylabel ('y(1) and y(2)')
gtext ('y(1)');
hold on
plot (t,y (:,2));
gtext ('y(2)');

%dfunc.m
function f = dfunc(t,y)
f = zeros (2,1);

f(1) = y(2);
f(2) = -0.1 * y(2) - 10.0 * y(1);
```

```
>> ProbappendixF
ans =

        0    1.0000         0
   0.0500    0.9875   -0.4967
   0.1000    0.9505   -0.9785
   0.1500    0.8901   -1.4335
   0.2000    0.8077   -1.8505
   0.2500    0.7056   -2.2191
   0.3000    0.5866   -2.5308
   0.3500    0.4534   -2.7775
   0.4000    0.3098   -2.9540
   0.4500    0.1592   -3.0561
   0.5000    0.0054   -3.0818
   0.5500   -0.1477   -3.0308
      .
      .
      .
   2.7500   -0.6380   -1.8279
   2.8000   -0.7207   -1.4788
   2.8500   -0.7851   -1.0949
   2.9000   -0.8296   -0.6858
   2.9500   -0.8533   -0.2617
   3.0000   -0.8556    0.1667
```

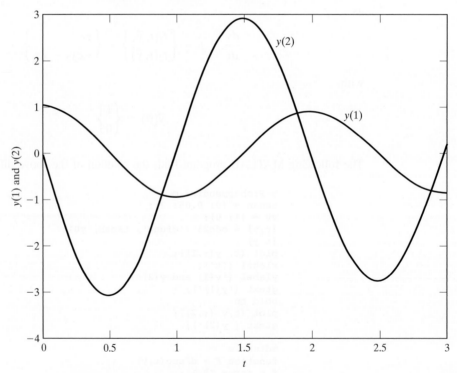

Answers to Selected Problems

Chapter 1

1.7 $k_{eq} = \dfrac{k_2 k_3 k_4 k_5 + 2 k_1 k_3 k_4 k_5 + k_1 k_2 k_4 k_5 + 2 k_1 k_2 k_3 k_5}{(k_2 k_3 k_4 + k_2 k_3 k_5 + 2 k_1 k_3 k_4 + 2 k_1 k_3 k_5 + k_1 k_2 k_4 + k_1 k_2 k_5 + 2 k_1 k_2 k_3)}$

1.11 (a) $k = 37.08 \times 10^7$ N/m, (b) $k = 12.36 \times 10^7$ N/m, (c) $k = 4.12 \times 10^7$ N/m

1.15 $k_{eq} = 456.21$ N/cm   **1.17** $k_{eq} = 3 k \cos^2 \alpha$ **1.21** $k_{eq} = 2\gamma A$

1.24 $l_{eq} = \dfrac{4t\,(d + t)}{Dd}$ **1.29** $k = \dfrac{p\gamma A^2}{v}$ **1.32** $k = 77.4414$ N/m   **1.36** $F(x) = (32{,}000x - 80)$ N

1.39 $k_{eq} = \dfrac{1}{l}(E_s A_s + E_a A_a)$ **1.43**   (a) $k_{t_{eq}} = 5.54811 \times 10^6$ N-m/rad, (b) $k_{t_{eq}} = 5.59597 \times 10^6$ N-m/rad

1.45 (a) $k_{eq} = 15{,}356.63$ N/m, (b) $k_{eq} = 509.23$ N/m   **1.47** $k_{axial} = 2{,}789{,}736$ N/m; $k_{torsion} = 15.8524$ N-m/rad

1.49 $m_{eq} = m_1 \left(\dfrac{a}{b}\right)^2 + m_2 + J_0\left(\dfrac{1}{b^2}\right)$ **1.52** $m_{eq} = m_h + \dfrac{J_b}{l_3^2} + J_c\left[\dfrac{l_2}{l_3 \, r_c}\right]^2$

1.55 (a) $c_{eq} = c_1 + c_2 + c_3$,   (c) $c_{eq} = c_1 + c_2\left(\dfrac{l_2}{l_1}\right)^2 + c_3\left(\dfrac{l_3}{l_1}\right)^2$,

(b) $\dfrac{1}{c_{eq}} = \dfrac{1}{c_1} + \dfrac{1}{c_2} + \dfrac{1}{c_3}$,   (d) $c_{t eq} = c_{t1} + c_{t2}\left(\dfrac{n_1}{n_2}\right)^2 + c_{t_3}\left(\dfrac{n_1}{n_3}\right)^2$

1.59 $c_t = \dfrac{\pi\mu D^2(l - h)}{2d} + \dfrac{\pi\mu D^3}{32h}$ **1.64** $c = 3{,}225.8$ N-s/m   **1.71** $c = 4{,}205.64$ N-s/m

1.76 $A = 4.4721$, $\theta = -26.5651°$ **1.78** $z = 11.1803\, e^{0.1798\,i}$ **1.81** $X = 9.8082$ mm, $Y = 9.4918$ mm, $\phi = 39.2072°$

1.85 $x_2(t) = 6.1966 \sin(\omega t + 83.7938°)$ **1.87**   Not harmonic

1.90 $X = 2.5$ mm, $\omega = 5.9092$ rad/s, $\omega + \delta\omega = 6.6572$ rad/s   **1.92** $A = 0.5522$ mm, $\dot{x}_{max} = 52.04$ mm/s

1.104 $x_{rms} = X/\sqrt{2}$ **1.108** $x(t) = \dfrac{A}{\pi} + \dfrac{A}{2}\sin \omega t - \dfrac{2A}{\pi}\displaystyle\sum_{n=2,4,6,\ldots}^{\infty} \dfrac{\cos n\omega t}{(n^2 - 1)}$ **1.110** $x(t) = \dfrac{8A}{\pi^2}\displaystyle\sum_{n=1,3,5,\ldots}^{\infty} (-1)^{\frac{n-1}{2}}\dfrac{\sin n\omega t}{n^2}$

1.114 $p(t) = \dfrac{a_0}{2} + \displaystyle\sum_{m=1}^{\infty} [a_m \cos m\omega t + b_m \sin m\omega t]$ kPa where $a_0 = 350$, $a_1 = 222.817$, $a_2 = 0$, $a_3 = -74.272$,

$b_1 = 222.817$, $b_2 = 222.817$, $b_3 = 74.272$

1.117 $a_0 = 19.92$, $a_1 = -20.16$, $a_2 = 3.31$, $a_3 = 3.77$; $b_1 = 23.52$, $b_2 = 12.26$, $b_3 = -0.41$

Chapter 2

2.2 (a) 0.1715 s,   (b) 0.2970 s   **2.4**   0.0993 s

2.6 (a) $A = 0.03183$ m,   (c) $\ddot{x}_{max} = 0.31415$ m/s^2,

(b) $\dot{x}_0 = 0.07779$ m/s,   (d) $\phi_0 = 51.0724°$

2.8 $\omega_n = 22.1472$ rad/s   **2.10** $\omega_n = 3.9597$ rad/s   **2.13** $\omega_n = [k/(4m)]^{1/2}$

2.15 (a) $\omega_n = \sqrt{\dfrac{4k}{M}}$,   (b) $\omega_n = \sqrt{\dfrac{4k}{m+M}}$ **2.17** $\omega_n = \sqrt{\dfrac{g}{W}\left(\dfrac{3E_1I_1}{l_1^3} + \dfrac{48E_2I_2}{l_2^3}\right)}$

2.19 $k = 52.6381$ N/m, $m = 1/3$ kg   **2.21** (a) $\omega_n = \sqrt{\dfrac{kg\,\mathrm{cosec}^2\,\theta}{W}}$,   (b) $\omega_n = \sqrt{\dfrac{kg}{W}}$

2.23 (a) $\omega_n = \sqrt{\dfrac{k}{2m}}$,   (b) $\omega_n = \sqrt{\dfrac{8m}{b^2}\left[l^2 - \dfrac{b^2}{4}\right]}$ **2.26** (a) $m\ddot{x} + \left(\dfrac{1}{a} + \dfrac{1}{b}\right)Tx = 0$,   (b) $\omega_n = \sqrt{\dfrac{T(a+b)}{mab}}$

2.28 $T = 7{,}346.9388$ N   **2.30** (a) $N = 81.914$ rpm,   (b) $\omega_n = 37.5851$ rad/s   **2.32** $\omega_n = \sqrt{\dfrac{2g}{L}}$

2.34 $A = 0.9536 \times 10^{-4}$ m^2 **2.37** Torsion about z-axis   **2.39** $\omega_n = 2{,}066.8429$ rad/s

2.42 $\mu = \sqrt{\left(\dfrac{\omega^2 Wc - 2kgc}{Wg + Wa\omega^2 - 2kga}\right)}$ **2.44** $m\ddot{x} + (k_1 + k_2)x = 0$ **2.47** $\left(m + \dfrac{J_0}{r^2}\right)\ddot{x} + 16kx = 0$

2.49 $\omega_n = 359.6872$ rad/s   **2.51** $x(t) = 0.1\cos 15.8114t + 0.3162 \sin 15.8114t$ m

2.53 $x_0 = 0.007864$ m; $\dot{x}_0 = -0.013933$ m/s   **2.55** $\dot{x}_0 = 4$ m/s   **2.57** $d = 0.00324203$ m, $N = 26.569397$

2.64 $\omega_n = 2$ rad/s, $l = 2.4525$ m   **2.66** $\tau_n = 1.4185$ s   **2.68** $\omega_n = 22.044$ rad/s

2.70 $\tau_n = 0.04693$ s   **2.72** $\omega_n = 17.7902$ rad/s   **2.74** $\omega_n = \left\{\dfrac{(k_1 + k_2)(R+a)^2}{1.5mR^2}\right\}^{1/2}$

2.76 $\frac{1}{3}ml^2\ddot{\theta} + (k_t + k_1a^2 + k_2l^2)\theta = 0$ **2.86** $m_{eff} = \dfrac{17}{35}m$ **2.88** $\omega_n = \sqrt{\dfrac{k}{4m}}$

2.91 45.1547 rad/s   **2.93** $\omega_n = \sqrt{\dfrac{\rho_0 g}{\rho_w h}}$ **2.95** $\omega_n = \sqrt{\dfrac{16kr^2}{mr^2 + J_0}}$ **2.98** (a) $14{,}265.362$, (b) 3.8296

2.100 $x_{\max} = \left(x_0 + \dfrac{\dot{x}_0}{\omega_n}\right)e^{-(\dot{x}_0/(\dot{x}_0 + \omega_n x_0))}$ **2.103** (a) $c_c = 1{,}000$ N-s/m,   (b) $\omega_d = 8.6603$ rad/s,   (c) $\delta = 3.6276$

2.105 $\theta = 0.09541°$ **2.107** $\zeta = 0.0109801$ **2.109** $m = 500$ kg, $k = 27{,}066.403$ N/m

2.112 $\omega_n = \sqrt{\dfrac{2k}{3m}}$ **2.116** $\frac{3}{2}m\ddot{x} + c\dot{x} + 2kx = 0$ **2.118** $\rho_0 = 2{,}682.8816$ kg/m^3

2.120 (a) $J_0 = 1.9436 \times 10^{-4}$ N-m-s^2,   (c) $c_t = 5.3887 \times 10^{-4}$ N-m-s/rad,
(b) $\tau_n = 1.8297$ s,   (d) $k_t = 2.2917 \times 10^{-3}$ N-m/rad

2.121 (a) $\zeta = 0.75$, $\omega_d = 6.6144$ rad/s,   (b) $\zeta = 1.0$, $\omega_d = 0$,   (c) $\zeta = 1.25$

2.123 (a) 60.8368 J,   (b) 124.6784 J   **2.139** Coulomb, 5N, 14.1421 rad/s   **2.141** 5.8 mm

2.143 (a) 5 (b) 0.7025 sec   (c) 1.9620 cm   **2.145** $c_{eq} = \dfrac{4\mu N}{\pi\omega X}$ **2.148** 1.40497 s

2.150 1.7022 sec, 0.004 m   **2.152** $\beta = 0.03032$, $c_{eq} = 0.04288$ N-s/m, $\Delta W = 19.05 \times 10^{-6}$ N-m   **2.154** $h = 0.583327$ N/m

Chapter 3

3.2 5 s

3.4 (a) $x(t) = 0.1\cos 20t + t\sin 20t$,   (b) $x(t) = (0.5 + t)\sin 20t$,   (c) $x(t) = 0.1\cos 20t + (0.5 + t)\sin 20t$

3.6 (a) $x(t) = 0.18\cos 20t - 0.08\cos 30t$,   (b) $x(t) = 0.08\cos 20t + 0.5\sin 20t - 0.08\cos 30t$,
(c) $x(t) = 0.18\cos 20t + 0.5\sin 20t - 0.08\cos 30t$

3.8 9.1189 kg   **3.16** $X = \left|\dfrac{mrl^3 N^2}{22.7973\,Eba^3 - 0.2357\,\rho abl^4 N^2}\right|$ **3.18** $\omega = 731.815$ Hz   **3.22** 0.676 s

3.24 $\theta_p(t) = \Theta \sin \omega t$ with $\Theta = -8.5718 \times 10^{-4}$ rad and $\omega = 104.72$ rad/s

3.26 $x_p(t) = 0.06610 \cos(10t - 0.1325)$ m

$x_{\text{total}}(t) = 0.0345e^{-2t} \cos(19.8997t + 0.0267) + 0.0661 \cos(10t - 0.1325)$ m

3.28 $x_p(t) = 0.25 \cos\left(20t - \dfrac{\pi}{2}\right)$ m

$x_{\text{total}}(t) = 0.2611e^{-2t} \cos(19.8997t + 1.1778) + 0.25 \cos\left(20t - \dfrac{\pi}{2}\right)$ m

3.30 $k = 12.0006 \times 10^6$ N/m, $c = 636.412$ N-s/m   **3.32** $r = \sqrt{1 - 2\zeta^2}$, $X_{\max} = \dfrac{\delta_{\text{st}}}{2\zeta\sqrt{1 - \zeta^2}}$ **3.34** $\zeta = 0.1180$

3.41 (a) 64.16 rad/s,   (b) 967.2 N-m

3.43 (a) $\zeta = 0.25$,   (b) $\omega_1 = 22.2145$ rad/s, $\omega_2 = 38.4766$ rad/s   **3.45** 169.5294×10^{-6} m

3.47 $k = 1.0070 \times 10^5$ N/m, $c = 633.4038$ N-s/m   **3.53** $0.3339 \sin 25t$ mm   **3.55** $X = 0.106$ m, $s = 246.73$ km/h

3.57 $c = (k - m\omega^2)/\omega$ **3.59** $\theta(t) = 0.01311 \sin(10t - 0.5779)$ rad   **3.63** $x_p(t) = 110.9960 \times 10^{-6} \sin(314.16t + 0.07072)$ m

3.66 0.4145×10^{-3} m, 1.0400×10^{-3} m   **3.68** 1.4195 N-m   **3.70** $\zeta = 0.1333$ **3.74** Maximum force = 175.88 N

3.82 $\mu = 0.1$ **3.85** (a) 0.8162 J,   (b) 3.2646 J   **3.88** $\dfrac{1}{\left\{\dfrac{4\mu N}{\pi X k} + \dfrac{3}{4k}c\omega^3 X^2\right\}}$

3.91 (a) 1.0623 Hz,   (b) 1.2646 m/s,   (c) 5.557×10^{-4} m

Chapter 4

4.2 $x(t) = \dfrac{F_0}{2k} - \dfrac{4F_0}{\pi^2 k} \displaystyle\sum_{n=1,3,\dots}^{\infty} \dfrac{1}{n^2} \dfrac{1}{\sqrt{(1 - r^2 n^2)^2 + (2\zeta nr)^2}} \cos(n\omega t - \phi_n)$

with $r = \omega/\omega_n$ and $\phi_n = \tan^{-1}\left(\dfrac{2\zeta nr}{1 - n^2 r^2}\right)$

4.6 $\theta(t) = 0.0023873 + \displaystyle\sum_{n=1}^{\infty}\left\{\dfrac{318.3091 \sin 5.8905n \cos n\omega t + 318.3091(1 - \cos 5.8905n)\sin n\omega t}{n(392{,}700.0 - 1{,}096.6278n^2)}\right\}$ rad

4.13 $x_p(t) = 6.6389 \times 10^{-4} - 13.7821 \times 10^{-4}\cos(10.472t - 0.0172)$

$+ 15.7965 \times 10^{-4}\sin(10.472t - 0.0172) + \dots$ m

4.16 $x(t) = \dfrac{F_0}{k}\left\{1 + \dfrac{\sin \omega_n(t - t_0) - \sin \omega_n t}{\omega_n t_0}\right\}$;   for $t \geq t_0$

4.19 $x(t) = \dfrac{F_0}{2k\left(1 - \dfrac{\omega^2}{\omega_n^2}\right)}\left[2 - \dfrac{\omega^2}{\omega_n^2}\left(1 - \cos\dfrac{\omega_n\pi}{\omega}\right)\right] + \dfrac{F_0}{k}\left[1 - \cos \omega_n\left(t - \dfrac{\pi}{\omega}\right)\right]$ for $t > \pi/\omega$.

4.25 $x(t) = 1.7689 \sin 6.2832(t - 0.018) - 0.8845 \sin 6.2832t$

$- 0.8845 \sin 6.2832(t - 0.036)$ m;   for $t > 0.036$ s

4.29 $x_p(t) = 0.002667$ m   **4.32** $\theta(t) = 0.3094e^{-t} + 0.05717 \sin 5.4127t - 0.3094 \cos 5.4127t$ rad

4.35 $x(t) = 0.04048e^{-t} + 0.01266 \sin 3.198t - 0.04048 \cos 3.198t$ m   **4.37** $x(t) = 0.5164e^{-t} \sin 3.8729t$ m

4.50 $x_m = \dfrac{F_0}{k\omega_n t_0}[(1 - \cos \omega_n t_0)^2 + (\omega_n t_0 - \sin \omega_n t_0)^2]^{1/2}$; for $t > t_0$ **4.53** $d = 20$ mm   **4.56** $k = 2{,}270{,}019.63$ N/m

4.60 $x(t) = \begin{cases} \dfrac{F_0}{m\omega_n^2}(1 - \cos \omega_n t); & 0 \leq t \leq t_0 \\[3mm] \dfrac{F_0}{m\omega_n^2}[\cos \omega_n(t - t_0) - \cos \omega_n t]; & t \geq t_0 \end{cases}$

Chapter 5

5.5 $\omega_1 = 3.6603$ rad/s, $\omega_2 = 13.6603$ rad/s   **5.7** $\omega_1 = \sqrt{\dfrac{k}{m}}$, $\omega_2 = \sqrt{\dfrac{2k}{m}}$ **5.9** 0.7×10^3 m^2

5.10 $\omega_1 = 7.3892$ rad/s, $\omega_2 = 58.2701$ rad/s   **5.11** $\omega_{1,2}^2 = \dfrac{48}{7}\dfrac{EI}{m_1 m_2}\left[(m_1 + 8m_2) \pm \sqrt{(m_1 - 8m_2)^2 + 25m_1 m_2}\right]$

5.13 $\omega_1 = 0.7654\sqrt{\dfrac{g}{l}}$, $\omega_2 = 1.8478\sqrt{\dfrac{g}{l}}$ **5.16** $\omega_1 = 12.8817$ rad/s, $\omega_2 = 30.5624$ rad/s

5.19 $x_1(t) = 0.1046 \sin 40.4225t + 0.2719 \sin 58.0175t$,
$x_2(t) = 0.1429 \sin 40.4225t - 0.09952 \sin 58.0175t$

5.21 $\omega_1 = 3.7495\sqrt{\dfrac{EI}{mh^3}}$, $\omega_2 = 9.0524\sqrt{\dfrac{EI}{mh^3}}$ **5.23** $\vec{X}^{(1)} = \begin{Bmatrix} 1.0 \\ 2.3029 \end{Bmatrix}$, $\vec{X}^{(2)} = \begin{Bmatrix} 1.0 \\ -1.3028 \end{Bmatrix}$

5.25 $x_2(0) = r_1 x_1(0) = \dfrac{x_1(0)}{\sqrt{3} - 1}$, $\dot{x}_2(0) = r_1 \dot{x}_1(0) = \dfrac{\dot{x}_1(0)}{\sqrt{3} - 1}$

5.29 $x_1(t) = 0.5 \cos 2t + 0.5 \cos \sqrt{12}t$; $x_2(t) = 0.5 \cos 2t - 0.5 \cos \sqrt{12}t$

5.36 $\omega_1 = 0.5176\sqrt{k_t/J_0}$, $\omega_2 = 1.9319\sqrt{k_t/J_0}$ **5.39** $\omega_1 = 0.38197\sqrt{k_t/J_0}$, $\omega_2 = 2.61803\sqrt{k_t/J_0}$

5.41 Frequency equation:

$$\omega^4 (m_1 m_2 \, l_1^2 \, l_2^2) - \omega^2 \left\{ m_2 l_2^2 (W_1 l_1 + k l_1^2) + m_1 l_1^2 (W_2 l_2 + k \, l_2^2) \right\}$$
$$+ (W_1 l_1 W_2 l_2 + W_2 l_2 k \, l_1^2 + W_1 l_1 k \, l_2^2) = 0$$

5.43 $\omega_{1,2}^2 = \left\{ \dfrac{(J_0 k + m k_t) \pm \sqrt{(J_0 k + m k_t)^2 - 4(J_0 - m e^2) m k k_t}}{2m(J_0 - m e^2)} \right\}$

5.46 $1{,}000\ddot{x} + 40{,}000x + 15{,}000\,\theta = 900 \sin 8.7267t + 1{,}100 \sin(8.7267t - 1.5708)$
$810\,\ddot{\theta} + 15{,}000x + 67{,}500\,\theta = 1{,}650 \sin(8.7267t - 1.5708) - 900 \sin 8.7267t$

5.49 (a) $\begin{bmatrix} m & 0 \\ 0 & J_0 \end{bmatrix}\begin{Bmatrix} \ddot{x} \\ \ddot{\theta} \end{Bmatrix} + \begin{bmatrix} 3k & kl/6 \\ kl/6 & 17kl^2/36 \end{bmatrix}\begin{Bmatrix} x \\ \theta \end{Bmatrix} = \begin{Bmatrix} F(t) \\ l\,F(t)/3 \end{Bmatrix}$ where $J_0 = ml^2/12$ and $F(t) = F_0 \sin \omega t$,
(b) Static coupling

5.53 (a) $\omega_1 = 12.2474$ rad/s, $\omega_2 = 38.7298$ rad/s

5.56 $x_j(t) = X_j e^{i\omega t}$
with $X_1 = (-8.0230 - i\,5.8964 \times 10^{-6}) \times 10^{-5}$ m,
$X_2 = (2.2145 + i\,0.5536) \times 10^{-6}$ m

5.57 $k_2 = m_2 \omega^2$ **5.58** $x_2(t) = \left\{ \dfrac{k_2 F_0}{(-m_1 \omega^2 + k_1 + k_2)(-m_2 \omega^2 + k_2) - k_2^2} \right\} \sin \omega t$

5.60 $x_1(t) = (17.2915\, F_0 \cos \omega t + 6.9444\, F_0 \sin \omega t)10^{-4}$
$x_2(t) = (17.3165\, F_0 \cos \omega t + 6.9684\, F_0 \sin \omega t)10^{-4}$

5.62 $x_1(t) = 0.009773 \sin 4\pi t$ m, $x_2(t) = 0.016148 \sin 4\pi t$ m   **5.64** $x_2(t) = (\frac{1}{60} - \frac{1}{40}\cos 10t + \frac{1}{120}\cos 10\sqrt{3}t)u(t)$

5.66 $\omega_1 = 0$, $\omega_2 = \sqrt{\dfrac{4k}{3m}}$ **5.67** $b_1 c_2 - c_1 b_2 = 0$

5.69 $\ddot{\alpha} + \left(\dfrac{k_t}{J_1} + \dfrac{k_t}{J_2}\right)\alpha = 0$ where $\alpha = \theta_1 - \theta_2$

5.71 $\omega_1 = 0$, $\omega_2 = \sqrt{\dfrac{6k\,(m + M)}{mM}}$

5.77 $k \geq \dfrac{mg}{2l}$

apter 6

6.1
$$\begin{bmatrix} m_1 & 0 & 0 \\ 0 & m_2 & 0 \\ 0 & 0 & m_3 \end{bmatrix} \begin{Bmatrix} \ddot{x}_1 \\ \ddot{x}_2 \\ \ddot{x}_3 \end{Bmatrix} + k \begin{bmatrix} 7 & -1 & -5 \\ -1 & 2 & -1 \\ -5 & -1 & 7 \end{bmatrix} \begin{Bmatrix} x_1 \\ x_2 \\ x_3 \end{Bmatrix} = \begin{Bmatrix} F_1(t) \\ F_2(t) \\ F_3(t) \end{Bmatrix}$$

6.3
$$\frac{m}{3}\begin{bmatrix} 1 & 0 & 2 \\ 2 & 0 & 1 \\ 0 & 15 & 0 \end{bmatrix} \begin{Bmatrix} \ddot{x}_1 \\ \ddot{x}_2 \\ \ddot{x}_3 \end{Bmatrix} + \frac{c}{25} \begin{bmatrix} 6 & -10 & 4 \\ 9 & -15 & 6 \\ -15 & 25 & -10 \end{bmatrix} \begin{Bmatrix} \dot{x}_1 \\ \dot{x}_2 \\ \dot{x}_3 \end{Bmatrix}$$
$$+ \frac{k}{25} \begin{bmatrix} 6 & -10 & 29 \\ 34 & -15 & 6 \\ -15 & 25 & -10 \end{bmatrix} \begin{Bmatrix} x_1 \\ x_2 \\ x_3 \end{Bmatrix} = \begin{Bmatrix} F_3(t) \\ F_1(t) \\ F_2(t) \end{Bmatrix}$$

6.5 $\quad I_1 \ddot{\theta}_1 + k_{t_1}(\theta_1 - \theta_2) = M_1 \cos \omega t$

$$\left(I_2 + I_3 \frac{n_2^2}{n_3^2}\right)\ddot{\theta}_2 + k_{t_1}(\theta_2 - \theta_1) + k_{t_2}\frac{n_2}{n_3}\left(\theta_2 \frac{n_2}{n_3} - \theta_3\right) = 0$$

$$\left(I_4 + I_5 \frac{n_4^2}{n_5^2}\right)\ddot{\theta}_3 + k_{t_2}\left(\theta_3 - \theta_2\frac{n_2}{n_3}\right) + k_{t_3}\frac{n_4}{n_5}\left(\theta_3\frac{n_4}{n_5} - \theta_4\right) = 0$$

$$I_6 \ddot{\theta}_4 + k_{t_3}\left(\theta_4 - \theta_3\frac{n_4}{n_5}\right) = 0$$

6.12 $\quad k \begin{bmatrix} 7 & -1 & -5 \\ -1 & 2 & -1 \\ -5 & -1 & 7 \end{bmatrix}$

6.14 $\quad \dfrac{k}{25} \begin{bmatrix} 34 & -15 & 6 \\ -15 & 25 & -10 \\ 6 & -10 & 29 \end{bmatrix}$

6.16
$$\begin{bmatrix} k_{t_1} & -k_{t_1} & 0 & 0 \\ -k_{t_1} & k_{t_1} + k_{t_2}\left(\dfrac{n_2}{n_3}\right)^2 & -k_{t_2}\left(\dfrac{n_2}{n_3}\right) & 0 \\ 0 & -k_{t_2}\left(\dfrac{n_2}{n_3}\right) & k_{t_2} + k_{t_3}\left(\dfrac{n_4}{n_5}\right)^2 & -k_{t_3}\left(\dfrac{n_4}{n_5}\right) \\ 0 & 0 & -k_{t_3}\left(\dfrac{n_4}{n_5}\right) & k_{t_3} \end{bmatrix}$$

6.18 $\quad \begin{bmatrix} \dfrac{k_1 + k_2}{k_1 k_2} & \dfrac{1}{k_1 r} \\ \dfrac{1}{k_1 r} & \dfrac{1}{k_1 r^2} \end{bmatrix}$

6.20 $\quad \begin{bmatrix} \dfrac{2}{3k} & -\dfrac{1}{3kl} \\ -\dfrac{1}{3kl} & \dfrac{2}{3kl^2} \end{bmatrix}$

6.22 $\quad \begin{bmatrix} m & 0 \\ 0 & 4ml^2 \end{bmatrix}$

6.24 $\quad [k] = \begin{bmatrix} (k_1 + k_2) & -k_2 & 0 \\ -k_2 & (k_2 + k_3) & -k_3 \\ 0 & -k_3 & (k_3 + k_4) \end{bmatrix}$

6.26 $\quad [a] = \dfrac{l^3}{EI}\begin{bmatrix} 9/64 & 1/6 & 13/192 \\ 1/6 & 1/3 & 1/6 \\ 13/192 & 1/6 & 9/64 \end{bmatrix}$

6.30 $\quad 2k$

6.32 $\quad \begin{bmatrix} m_1 & 0 & 0 \\ 0 & m_2 & 0 \\ 0 & 0 & m_3 \end{bmatrix}$

6.34 $\quad \dfrac{m}{3}\begin{bmatrix} 2 & 0 & 1 \\ 0 & 15 & 0 \\ 1 & 0 & 2 \end{bmatrix}$

6.36 $\quad \begin{bmatrix} I_1 & 0 & 0 & 0 \\ 0 & I_2 + I_3\left(\dfrac{n_2}{n_3}\right)^2 & 0 & 0 \\ 0 & 0 & I_4 + I_5\left(\dfrac{n_4}{n_5}\right)^2 & 0 \\ 0 & 0 & 0 & I_6 \end{bmatrix}$

6.39 $\quad 2m\ddot{x} + kx = 0,\ l\ddot{\theta} + g\theta = 0$

6.41 $\quad m_1\ddot{x}_1 + (k_1 + k_2)x_1 - k_2 x_2 = 0$
$\qquad m_2\ddot{x}_2 - k_2 x_1 + (k_2 + k_3)x_2 - k_3 x_3 = 0$
$\qquad m_3\ddot{x}_3 - k_3 x_2 + (k_3 + k_4)x_3 = 0$

6.44 $\quad m_1\ddot{x}_1 + 7kx_1 - kx_2 - 5kx_3 = F_1(t)$
$\qquad m_2\ddot{x}_2 - kx_1 + 2kx_2 - kx_3 = F_2(t)$
$\qquad m_3\ddot{x}_3 - 5kx_1 - kx_2 + 7kx_3 = F_3(t)$

6.47 $\quad \left(M + \dfrac{J_0}{9r^2}\right)\ddot{x}_1 - \dfrac{J_0}{9r^2}\ddot{x}_2 + \dfrac{41}{9}kx_1 - \dfrac{8}{9}kx_2 - \dfrac{8}{3}kx_3 = F_1(t)$

$$-\frac{J_0}{9r^2}\ddot{x}_1 + \left(3m + \frac{J_0}{9r^2}\right)\ddot{x}_2 - \frac{8}{9}kx_1 + \frac{2}{9}kx_2 + \frac{2}{3}kx_3 = F_2(t)$$

$$m\ddot{x}_3 - \frac{8}{3}kx_1 + \frac{2}{3}kx_2 + 5kx_3 = F_3(t)$$

6.49 $\omega_1 = 0.44504\sqrt{k/m}$, $\omega_2 = 1.2471\sqrt{k/m}$, $\omega_3 = 1.8025\sqrt{k/m}$

6.52 $\omega_1 = 0.533399\sqrt{k/m}$, $\omega_2 = 1.122733\sqrt{k/m}$, $\omega_3 = 1.669817\sqrt{k/m}$

6.55 $\lambda_1 = 2.21398$, $\lambda_2 = 4.16929$, $\lambda_3 = 10.6168$ **6.58** $\omega_1 = 0.644798\sqrt{g/l}$, $\omega_2 = 1.514698\sqrt{g/l}$, $\omega_3 = 2.507977\sqrt{g/l}$

6.61 $\omega_1 = 0.562587\sqrt{\dfrac{P}{ml}}$, $\omega_2 = 0.915797\sqrt{\dfrac{P}{ml}}$, $\omega_3 = 1.584767\sqrt{\dfrac{P}{ml}}$ **6.64** $[X] = \dfrac{1}{2}\begin{Bmatrix} 1 & 1 & 0 \\ -1 & 1 & \sqrt{2/3} \\ 1 & 1 & \sqrt{8/3} \end{Bmatrix}$

6.67 $\omega_1 = 0.7653\sqrt{\dfrac{k}{m}}$, $\omega_2 = 1.8478\sqrt{\dfrac{k}{m}}$, $\omega_3 = 3.4641\sqrt{\dfrac{k}{m}}$

6.75 $\omega_1 = 0$, $\omega_2 = 0.752158\sqrt{k/m}$, $\omega_3 = 1.329508\sqrt{k/m}$

6.77 $x_3(t) = x_{10}\left\{0.5\cos 0.4821\sqrt{\dfrac{k}{m}}t - 0.3838\cos\sqrt{\dfrac{k}{m}}t + 0.8838\cos 1.1976\sqrt{\dfrac{k}{m}}t\right\}$

6.79 $x_3(t) = x_{20}\left\{0.1987\cos 0.5626\sqrt{\dfrac{P}{lm}}t - 0.06157\cos 0.9158\sqrt{\dfrac{P}{lm}}t - 0.1372\cos 1.5848\sqrt{\dfrac{P}{lm}}t\right\}$

6.82 $x_1(t) = \dot{x}_0\left\{\dfrac{t}{3} + \sqrt{\dfrac{m}{4k}}\sin\sqrt{\dfrac{k}{m}}t + \sqrt{\dfrac{m}{108k}}\sin\sqrt{\dfrac{3k}{m}}t\right\}$

6.84 $x_1(t) = \dfrac{1}{2}\left[\cos 2t + \dfrac{1}{2}\sin 2t + \cos\sqrt{12}t - \dfrac{1}{\sqrt{12}}\sin\sqrt{12}t\right]$; $x_2(t) = \dfrac{1}{2}\left[\cos 2t + \dfrac{1}{2}\sin 2t - \cos\sqrt{12}t + \dfrac{1}{\sqrt{12}}\sin\sqrt{12}t\right]$

6.90 **(a)** $\omega_1 = 0.44497\sqrt{k_t/J_0}$, $\omega_2 = 1.24700\sqrt{k_t/J_0}$, $\omega_3 = 1.80194\sqrt{k_t/J_0}$ **(b)** $\vec{\theta}(t) = \begin{Bmatrix} -0.0000025 \\ 0.0005190 \\ -0.0505115 \end{Bmatrix}\cos 100t$ radians

6.92 $\vec{x}(t) = \begin{Bmatrix} 5.93225 \\ 10.28431 \\ 12.58863 \end{Bmatrix}\dfrac{F_0}{k}\cos\omega t$ **6.95** $\vec{x}(t) = \begin{Bmatrix} 0.03944\,(1 - \cos 18.3013t) + 0.01057\,(1 - \cos 68.3015t) \\ 0.05387\,(1 - \cos 18.3013t) - 0.00387\,(1 - \cos 68.3015t) \end{Bmatrix}$

6.99 $x_3(t) = 0.0256357\cos(\omega t + 0.5874°)$ m

Chapter 7

7.1 **(a)** $\omega_1 \simeq 2.6917\sqrt{\dfrac{EI}{ml^3}}$,   **(b)** $\omega_1 \simeq 2.7994\sqrt{\dfrac{EI}{ml^3}}$ **7.3** $3.5987\sqrt{\dfrac{EI}{ml^3}}$ **7.5** $0.3015\sqrt{k/m}$

7.10 $0.4082\sqrt{k/m}$ **7.12** $1.0954\sqrt{\dfrac{T}{lm}}$ **7.19** $\omega_1 = 0$, $\omega_2 \simeq 6.2220$ rad/s, $\omega_3 \simeq 25.7156$ rad/s   **7.22** $\omega_1 = \sqrt{k/m}$

7.27 $\omega_1 = 0.3104$, $\omega_2 = 0.4472$, $\omega_3 = 0.6869$ where $\omega_i = 1/\sqrt{\lambda_i}$

7.30 $\widetilde{\omega}_1 = 0.765366$, $\widetilde{\omega}_2 = 1.414213$, $\widetilde{\omega}_3 = 1.847759$ with $\omega_i = \widetilde{\omega}_i\sqrt{\dfrac{GJ}{lJ_0}}$

7.37 $\omega_1 = 0.2583$, $\omega_2 = 3.0$, $\omega_3 = 7.7417$ **7.41** $[U]^{-1} = \begin{bmatrix} 0.44721359 & 0.083045475 & -0.12379687 \\ 0 & 0.41522738 & 1.1760702 \\ 0 & 0 & 1.7950547 \end{bmatrix}$

Chapter 8

8.1 28.2843 m/s   **8.3** $\omega_3 = 9,000$ Hz, both increased by 9.54%

8.6 **(a)** 0.1248×10^6 N,   **(b)** 3.12×10^6 N   **8.8** $w(x,t) = \dfrac{8al}{\pi^3 c}\sum_{n=1,3,5,\dots}(-1)^{\frac{n-1}{2}}\dfrac{1}{n^3}\sin\dfrac{n\pi x}{l}\sin\dfrac{n\pi ct}{l}$

8.11 $w\left(x, \dfrac{l}{c}\right) = -\dfrac{\sqrt{3}\,9h}{2\pi^2}\sin\dfrac{\pi x}{l} + \dfrac{\sqrt{3}\,9h}{8\pi^2}\sin\dfrac{2\pi x}{l} - \dfrac{\sqrt{3}\,9h}{32\pi^2}\sin\dfrac{4\pi x}{l} + \dfrac{\sqrt{3}\,9h}{50\pi^2}\sin\dfrac{5\pi x}{l}$

17 $\tan\dfrac{\omega l}{c} = \dfrac{AE\omega c(k - M\omega^2)}{A^2 E^2 \omega^2 - M\omega^2 kc^2}$ **8.20** $\tan\dfrac{\omega l_1}{c_1}\tan\dfrac{\omega l_2}{c_2} = \dfrac{A_1 E_1 c_2}{A_2 E_2 c_1}$ **8.23** $\omega_n = \dfrac{n\pi}{l}\sqrt{\dfrac{G}{\rho}}; n = 1, 2, 3, \dots$

25 $\omega_n = \dfrac{(2n + 1)\pi}{2}\sqrt{\dfrac{G}{\rho l^2}}; n = 0, 1, 2, \dots$ **8.28** 5,030.59 rad/sec   **8.31** $\cos\beta l\cosh\beta l = -1$

34 $\tan\beta l - \tanh\beta l = 0$ **8.36** 20.2328 N-m   **8.39** $\cos\beta l\cosh\beta l = 1$, and $\tan\beta l - \tanh\beta l = 0$

41 $\omega \approx \sqrt{120}\left(\dfrac{EI_0}{\rho A_0\, l^4}\right)^{1/2}$ **8.46** $w(x, t) = \dfrac{F_0}{2\rho Ac^2}\left\{\cos\beta x + \cosh\beta x + \tan\dfrac{\beta l}{2}\sin\beta x - \tanh\dfrac{\beta l}{2}\sinh\beta x - 2\right\}\sin\omega t$

49 $w(x, t) = \displaystyle\sum_{n=1}^{\infty} W_n(x)\, q_n(t)$ where $q_n(t) = \dfrac{M_0}{\rho Al\omega_n^2}\dfrac{dW_n}{dx}\Big|_{x=l}(1 - \cos\omega_n t)$

60 $w(x, y, t) = \dfrac{\dot{w}_0}{\omega_{12}}\sin\dfrac{\pi x}{a}\sin\dfrac{2\pi y}{b}\sin\omega_{12}t$ **8.62** $\omega_{mn}^2 = \dfrac{\gamma_n P}{\rho}$, where $J_m(\gamma_n R) = 0; m = 0, 1, 2, \dots; n = 1, 2, \dots$

63 $22.4499\sqrt{\dfrac{EI}{\rho Al^4}}$ **8.65** $\omega = 15.4510\sqrt{\dfrac{EI}{\rho Al^4}}$ **8.67** $7.7460\sqrt{\dfrac{EI_0}{\rho A_0 l^4}}$ **8.70** $2.4146\sqrt{\dfrac{EA_0}{m_0\, l^2}}$ **8.72** $\omega \approx 13,867.3328$ rad/s

74 (a) $1.73205\sqrt{\dfrac{E}{\rho l^2}}$,   (b) $1.57669\sqrt{\dfrac{E}{\rho l^2}}$, $5.67280\sqrt{\dfrac{E}{\rho l^2}}$ **8.77** $\omega_1 = 3.142\sqrt{\dfrac{P}{\rho l^2}}$, $\omega_2 = 10.12\sqrt{\dfrac{P}{\rho l^2}}$

apter 9

9.1 Around 46.78 km/h   **9.3** $m_c r_c = 3,354.6361$ g-mm, $\theta_c = -25.5525°$ **9.5** $m_4 = 0.98996$ oz, $\theta_4 = -35.0045°$

9.8 0.0419 kg, $\alpha = 75.6440°$ CW

9.11 Remove 0.07124 kg at 10.8178° CCW at radius 0.1 m in plane B and 0.11 kg at 1.3931° CCW in radius 0.1 m in plane C.

9.14 (a) $\vec{R}_A = -28.4021\vec{j} - 3.5436\vec{k}$, $\vec{R}_B = 13.7552\,\vec{j} + 4.7749\,\vec{k}$,   (b) $m_L = 10.44$ g, $\theta_L = 7.1141°$

9.17 (a) 0.005124 m,   (b) 0.06074 m,   (c) 0.008457 m

9.20 (a) 0.5497×10^8 N/m^2,   (b) 6.4698×10^8 N/m^2,   (c) 0.9012×10^8 N/m^2

9.22 $F_{xp} = 0$, $F_{xs} = 15,791.37$ N, $M_{zp} = M_{zs} = 0$ **9.25** The engine is completely force and moment balanced.

9.27 0.2385 mm   **9.30** (a) $\omega < 95.4927$ rpm,   (b) $\omega > 276.7803$ rpm   **9.32** $k = 152,243.1865$ N/m

9.35 79.7808 rad/s $-$ 1,419.8481 rad/s   **9.37** $\delta_{st} = 0.02733$ m   **9.40** $k = 18.9530$ kN/m

9.43 (a) $X = 11.4188 \times 10^{-3}$ m,   (b) $F_T = 44.8069$ N   **9.45** 98.996%   **9.47** (a) 13,110.73 N,   (b) 293.429 kN

9.49 49,752.86 N/m   **9.64** $\mu = 0.3403$; $m_2 = 102.09$ kg, $k_2 = 2.519$ MN/m; $X_2 = -0.1959$ mm

9.66 (a) 2,236.40 N,   (b) $\Omega_1 = 467.7191$ rpm, $\Omega_2 = 769.6938$ rpm   **9.68** For $D/d = 4/3$, $d = 0.015656$ m, $D = 0.0208743$ m

9.71 $0.9764 \le \dfrac{\omega}{\omega_2} \le 1.05125$ **9.73** $m_2 = 10$ kg, $k_2 = 0.19986$ MN/m   **9.75** 32 kN/m

apter 10

10.2 18.3777 Hz   **10.4** 3.6935 Hz   **10.6** 0.53%   **10.9** 35.2635 Hz   **10.12** 73.16%

10.14 $k = 33,623.85$ N/m, $c = 50.55$ N-s/m   **10.16** $m = 19.41$ g, $k = 7,622.8$ N/m   **10.19** 131.573 rad/s $-$ 3,289.33 rad/s

10.21 $r \approx 1$ **10.23** $\zeta = 0.1111$ **10.26** Cage (51.93 Hz), Inner race (1,078.97 Hz), Outer race (830.88 Hz), Ball (193.31 Hz)

10.29 1.8   **10.30** 2.9630   **10.32** $\zeta = 0.2$

apter 11

11.2 $\dfrac{d^4 x}{dt^4}\Big|_i = \dfrac{x_i - 4x_{i-1} + 6x_{i-2} - 4x_{i-3} + x_{i-4}}{(\Delta t)^4}$ **11.4** $x(t = 5) = -1$ with $\Delta t = 1$ and -0.9733 with $\Delta t = 0.5$

11.6 $x_{10} = -0.0843078$, $x_{15} = 0.00849639$ **11.9** $x(t = 0.1) = 0.131173$, $x(t = 0.4) = -0.0215287$, $x(t = 0.8) = -0.0676142$

11.14 With $\Delta t = 0.07854$, $x_1 = x$ and $x_2 = \dot{x}$, $x_1(t = 0.2356) = 0.100111$, $x_2(t = 0.2356) = 0.401132$,
$x_1(t = 1.5708) = 1.040726$, $x_2 = (t = 1.5708) = -0.378066$

11.20

t	x_1	x_2
0.25	0.07813	1.1860
1.25	2.3360	−0.2832
3.25	−0.6363	2.3370

11.23 $\omega_1 = 3.06147\sqrt{\dfrac{E}{\rho l^2}}, \omega_2 = 5.65685\sqrt{\dfrac{E}{\rho l^2}}, \omega_3 = 7.39103\sqrt{\dfrac{E}{\rho l^2}}$

11.26 $\omega_1 = 17.9274\sqrt{\dfrac{EI}{\rho A l^4}}, \omega_2 = 39.1918\sqrt{\dfrac{EI}{\rho A l^4}}, \omega_3 = 57.1193\sqrt{\dfrac{EI}{\rho A l^4}}$

11.38 With $\Delta t = 0.24216267$,

t	x_1	x_2
0.2422	0.01776	0.1335
2.4216	0.7330	1.8020
4.1168	0.1059	0.8573

Chapter 12

12.2 $[k] = \dfrac{EA_0}{l}(0.6321)\begin{bmatrix} 1 & -1 \\ -1 & 1 \end{bmatrix}$

12.3 $3.3392 \times 10^7 \text{ N/m}^2$

12.15 0.00133645 m under load

12.18 Deflection $= 0.002197\dfrac{Pl^3}{EI}$, slope $= 0.008789\dfrac{Pl^3}{EI}$

12.19 $\sigma^{(1)} = -25.04$ kPa, $\sigma^{(2)} = 26.9266$ kPa

12.21 Maximum bending stresses: -7.6463×10^7 Pa (in both connecting rod and crank), maximum axial stresses: -16.67 MPa (in connecting rod), -20 MPa (in crank)

12.26 $\omega_1 = 0.8587\sqrt{\dfrac{EI}{\rho A l^4}}, \omega_2 = 4.0965\sqrt{\dfrac{EI}{\rho A l^4}}, \omega_3 = 34.9210\sqrt{\dfrac{EI}{\rho A l^4}}$

12.29 $\omega_1 = 15.1357\sqrt{\dfrac{EI}{\rho A l^4}}, \omega_2 = 28.9828\sqrt{\dfrac{EI}{\rho A l^4}}$

12.32 $\omega_1 = 20.4939\sqrt{\dfrac{EI}{\rho A l^4}}$

12.40 $\omega_1 = 4.4728 \times 10^3$ rad/s, $\omega_2 = 9.1224 \times 10^3$ rad/s